Lecture Notes in Computer Science 8866

Commenced Publication in 1973
Founding and Former Series Editors:
Gerhard Goos, Juris Hartmanis, and Jan van Leeuwen

More information about this series at http://www.springer.com/series/7407

Zhigang Zeng · Yangmin Li
Irwin King (Eds.)

Advances in Neural Networks – ISNN 2014

11th International Symposium
on Neural Networks, ISNN 2014
Hong Kong and Macao, China
November 28 – December 1, 2014
Proceedings

 Springer

Editors
Zhigang Zeng
Huazhong University of Science
    and Technology
Wuhan
China

Irwin King
The Chinese University of Hong Kong
Hong Kong
China

Yangmin Li
University of Macau
Macao
China

ISSN 0302-9743   ISSN 1611-3349 (electronic)
ISBN 978-3-319-12435-3   ISBN 978-3-319-12436-0 (eBook)
DOI 10.1007/978-3-319-12436-0

Library of Congress Control Number: 2014953830

LNCS Sublibrary: SL1 – Theoretical Computer Science and General Issues

Springer Cham Heidelberg New York Dordrecht London

Printed on acid-free paper

Springer is part of Springer Science+Business Media (www.springer.com)

Preface

This volume of Lecture Notes in Computer Science, vol. 8866, constitutes the Proceedings of the 11th International Symposium on Neural Networks (ISNN 2014) held during November 28–December 1, 2014 in Hong Kong and Macao, the twin city, following the successes of previous events. Known as the Special Administrative Regions of China, Hong Kong and Macao are two modern metropolies situated on the southern coast of China by the Pearl River Delta. ISNN is a prestigious annual symposium on neural networks with past events held in Dalian (2004), Chongqing (2005), Chengdu (2006), Nanjing (2007), Beijing (2008), Wuhan (2009), Shanghai (2010), Guilin (2011), Shenyang (2012) and Dalian (2013). Over the past few years, ISNN has matured into a well-established series of international conference on neural networks and their applications to other fields. Following this tradition, ISNN 2014 provided an academic forum for the participants to disseminate their new research findings and discuss emerging areas of research. Also, it created a stimulating environment for the participants to interact and exchange information on future research challenges and opportunities of neural networks and their applications.

ISNN 2014 received submissions from about 218 authors in 14 countries and regions (Belgium, Canada, China, Czech Republic, Hong Kong, India, Japan, Macao, Pakistan, Poland, Republic of Korea, Tunisia, UK and USA). Based on the rigorous peer reviews by the Program Committee members and the reviewers, 71 high-quality papers were selected for publications in the LNCS proceedings. These papers cover all major topics of the theoretical research, empirical study, and applications of neural networks research.

ISNN 2014 would not have achieved its success without the support and contributions of many volunteers and organizations. We would like to express our sincere thanks to The Chinese University of Hong Kong, and University of Macau, to European Neural Network Society, International Neural Network Society, IEEE Computational Intelligence Society, and Asia Pacific Neural Network Assembly for their technical co-sponsorship.

We would also like to sincerely thank the General Chair and General Co-chairs for their overall organization of the symposium, members of the Advisory Committee and Steering Committee for their guidance for every aspects of the entire conference, and the members of the Organizing Committee, Special Sessions Committee, Publication Committee, Publicity Committee, Finance Committee, Registration Committee, and Local Arrangements Committee for all their great efforts and time in organizing such an event. We would also like to take this opportunity to express our deepest gratitude to the members of the International Program Committee and all reviewers for their professional review of the papers; their expertise guaranteed the high qualify of the technical program of the ISNN 2014!

Furthermore, we would also like to thank Springer for publishing the proceedings in the prestigious series of Lecture Notes in Computer Science. Meanwhile, we also would like to express our heartfelt appreciations to the plenary and panel speakers for their vision and discussions on the latest research development in the field as well as critical future research directions, opportunities, and challenges.

Finally, we would also like to thank all the speakers, authors, and participants for their great contribution and support that made ISNN 2014 a great success.

November 2014   Zhigang Zeng
                                                                  Yangmin Li
                                                                  Irwin King

Organization

General Chair

Jun Wang   The Chinese University of Hong Kong,
Hong Kong

Advisory Chairs

Philip Chen   University of Macau, Macao
Gary G. Yen   Oklahoma State University, USA

Steering Chair

Derong Liu   Institute of Automation, Chinese Academy
of Sciences, China

Program Committee Chairs

Irwin King   The Chinese University of Hong Kong,
Hong Kong
Yangmin Li   University of Macau, Macao
Zhigang Zeng   Huazhong University of Science and
Technology, China

Special Sessions Chairs

Zeng-Guang Hou   Institute of Automation, Chinese Academy
of Sciences, China
Tieshan Li   Dalian Maritime University, China
Yunong Zhang   Sun Yat-sen University, China

Publications Chairs

Xinzhe Wang   Dalian University of Technology, China
Tao Xiang   Chongqing University, China
Wei Yao   South-Central University for Nationalities,
China

Publicity Chairs

Tingwen Huang   Texas A&M University at Qatar, Qatar
Yi Shen   Huazhong University of Science and
                                Technology, China
Changyin Sun   Southeast University, China

Registration Chairs

Shenshen Gu   Shanghai University, China
Qingshan Liu   Southeast University, China
Li Zhang   The Chinese University of Hong Kong,
                                Hong Kong

Local Arrangements Chair

Feng Wan   University of Macau, Macao

Program Committee Members

Gang Bao	Jin Hu
Salim Bouzerdoum	Jinglu Hu
Jonathan Chan	Guangbin Huang
Guici Chen	Jigui Jian
Huanhuan Chen	Danchi Jiang
Zheru Chi	Haijun Jiang
Mingcong Deng	Sungshin Kim
Ji-Xiang Du	Bin Li
Andries Engelbrecht	Bo Li
Peter Erdi	Chuandong Li
Mauro Forti	Kang Li
Wai-Keung Fung	Jie Lian
Marcus Gallagher	Hualou Liang
John Gan	Jing Liu
Yang Gao	Ju Liu
Weimin Ge	Meiqin Liu
Erol Gelenbe	Xiaoming Liu
Shenshen Gu	Jianquan Lu
Chengan Guo	Wenlian Lu
Fei Han	Jinwen Ma
Haibo He	Vanessa Ng
Hanlin He	Yew-Soon Ong
Shan He	Qiankun Song
Daniel Ho	Ponnuthurai Suganthan

Qingquan Sun
Kay Chen Tan
Jinshan Tang
Ke Tang
Peter Tino
Christos Tjortjis
Michel Verleysen
Feng Wan
Dianhui Wang
Ailong Wu

Shane Xie
Bjingji Xu
Chenguang Yang
Yingjie Yang
Yongqing Yang
Wen Yu
Xiaoqin Zeng
Jie Zhang
Hongyong Zhao
Song Zhu

Contents

Analysis

A Less Conservative Guaranteed Cost Stabilization of Time-Varying
Delayed CNNs 3
   Mei Jiang, Hanlin He, and Lu Yan

Existence and Uniqueness of Almost Automorphic Solutions
to Cohen-Grossberg Neural Networks with Delays 12
   Xianyun Xu, Tian Liang, Fei Wang, and Yongqing Yang

A Proof of a Key Formula in the Error-Backpropagation Learning
Algorithm for Multiple Spiking Neural Networks 19
   Wenyu Yang, Dakun Yang, and Yetian Fan

Anti-Synchronization Control for Memristor-Based Recurrent Neural
Networks 27
   Ning Li, Jinde Cao, and Mengzhe Zhou

Adaptive Pinning Synchronization of Coupled Inertial Delayed
Neural Networks 35
   Jianqiang Hu, Jinde Cao, and Quanxin Cheng

The Relationship between Neural Avalanches and Neural Oscillations 43
   Yan Liu, Jiawei Chen, and Liujun Chen

Reinforcement-Learning-Based Controller Design for Nonaffine
Nonlinear Systems 51
   Xiong Yang, Derong Liu, and Qinglai Wei

Multistability and Multiperiodicity Analysis of Complex-Valued
Neural Networks 59
   Jin Hu and Jun Wang

Exponential Synchronization of Coupled Stochastic and Switched
Neural Networks with Impulsive Effects 69
   Yangling Wang and Jinde Cao

The Research of Building Indoor Temperature Prediction Based
on Support Vector Machine 80
   Wenbiao Wang, Qi Cai, and Siyuan Wang

Towards the Computation of a Nash Equilibrium 90
   Yu Lu and Ying He

Sensorless PMSM Speed Control Based on NN Adaptive Observer   100
Longhu Quan, Zhanshan Wang, Xiuchong Liu, and Mingguo Zheng

Neural-Network-Based Adaptive Fault Estimation for a Class
of Interconnected Nonlinear System with Triangular Forms.   110
Lei Liu, Zhanshan Wang, Jinhai Liu, and Zhenwei Liu

Big Data Matrix Singular Value Decomposition Based on Low-Rank
Tensor Train Decomposition   121
Namgil Lee and Andrzej Cichocki

CSP-Based EEG Analysis on Dissociated Brain Organization
for Single-Digit Addition and Multiplication   131
Lihan Wang, John Qiang Gan, and Haixian Wang

Lazy Fully Probabilistic Design of Decision Strategies   140
Miroslav Kárný, Karel Macek, and Tatiana V. Guy

Memristive Radial Basis Function Neural Network for Parameters
Adjustment of PID Controller   150
Xiaojuan Li, Shukai Duan, Lidan Wang, Tingwen Huang, and Yiran Chen

Finite-Time Stability of Switched Static Neural Networks   159
Yuanyuan Wu and Jinde Cao

Fixed-Priority Scheduling Policies and Their Non-utilization Bounds   167
Guangyi Chen and Wenfang Xie

Modeling Dynamic Hysteresis of Smart Actuators with Fuzzy Tree   175
Yongxin Guo, Zhen Zhang, Jianqin Mao, Haishan Ding, and Yanhua Ma

Neurodynamics-Based Model Predictive Control for Trajectory
Tracking of Autonomous Underwater Vehicles.   184
Xinzhe Wang and Jun Wang

Feedback-Dependence of Correlated Firing in Globally Coupled Networks . . .   192
Jinli Xie, Zhijie Wang, and Jianyu Zhao

Wilcoxon-Norm-Based Robust Extreme Learning Machine   200
Xiao-Liang Xie, Gui-Bin Bian, Zeng-Guang Hou,
Zhen-Qiu Feng, and Jian-Long Hao

Modeling

An Artificial Synaptic Plasticity Mechanism for Classical
Conditioning with Neural Networks   213
Caroline Rizzi Raymundo and Colin Graeme Johnson

An Application of Dynamic Regression Model and Residual Auto-Regressive
Model in Time Series   222
   Yu-zhen Lu, Ming-hui Qu, and Min Zhang

Reweighted l_2-Regularized Dual Averaging Approach for Highly Sparse
Stochastic Learning.   232
   Vilen Jumutc and Johan A.K. Suykens

Perceptual Learning Model on Recognizing Chinese Characters   243
   Jiawei Chen, Yan Liu, Xiaomeng Li, and Liujun Chen

Hierarchical Solving Method for Large Scale TSP Problems   252
   Jingqing Jiang, Jingying Gao, Gaoyang Li, Chunguo Wu, and Zhili Pei

Gaussian Process Learning: A Divide-and-Conquer Approach   262
   Wenye Li

A Kernel ELM Classifier for High-Resolution Remotely Sensed Imagery
Based on Multiple Features   270
   Wei Yao, Zhigang Zeng, Cheng Lian, and Huiming Tang

Early-Stopping Regularized Least-Squares Classification   278
   Wenye Li

Z-Type Model for Real-Time Solution of Complex ZLE   286
   Long Jin, Hongzhou Tan, Ziyi Luo, Zhan Li, and Yunong Zhang

A Dynamic Generation Approach for Ensemble of Extreme Learning
Machines   294
   Hualong Yu, Yulong Yuan, Xibei Yang, and Yuanyuan Dan

Saliency Detection: A Divisive Normalization Approach   303
   Ying Yu, Jie Lin, and Jian Yang

A Novel Neural Network Based Adaptive Control for a Class
of Uncertain Strict-Feedback Nonlinear Systems   312
   Baobin Miao and Tieshan Li

Rough Sets Theory Approch to Extenics Risk Assessment Model
in Social Risk.   321
   Guangyao Gu, Yaodong Le, Fajie Wei, and Chen Li

Kernel Parameter Optimization for KFDA Based on the Maximum
Margin Criterion.   330
   Yue Zhao and Jinwen Ma

A Polynomial Time Solvable Algorithm to Binary Quadratic Programming
Problems with Q Being a Seven-Diagonal Matrix and Its
Neural Network Implementation.................................. 338
  Shenshen Gu, Jiao Peng, and Rui Cui

A New Nonlinear Neural Network for Solving QP Problems............ 347
  Yinhui Yan

Untrained Method for Ensemble Pruning and Weighted Combination...... 358
  Bartosz Krawczyk and Michał Woźniak

An Improved Learning Algorithm with Tunable Kernels for Complex-Valued
Radial Basis Function Neural Networks.......................... 366
  Xia Mo, He Huang, and Tingwen Huang

Combined Methodology Based on Kernel Regression and Kernel Density
Estimation for Sign Language Machine Translation 374
  Mehrez Boulares and Mohamed Jemni

Adaptive Intelligent Control for Continuous Stirred Tank Reactor
with Output Constraint 385
  Dong-Juan Li and Yan-Jun Liu

Applications

Modeling and Application of Principal Component Analysis in Industrial
Boiler 395
  Wenbiao Wang, Lan Chen, Xinjie Han, Zhanyuan Ge, and Siyuan Wang

An Orientation Column-Inspired Contour Representation and Its
Application in Shape-Based Recognition.......................... 405
  Hui Wei and Wentao Ge

Region Based Image Preprocessor for Feed-Forward Perceptron Based
Systems.. 414
  Keith A. Greenhow and Colin G. Johnson

Image Denoising with Signal Dependent Noise Using Block
Matching and 3D Filtering.................................... 423
  Guangyi Chen, Wenfang Xie, and Shuling Dai

Different-Level Simultaneous Minimization with Aid of Ma Equivalence
for Robotic Redundancy Resolution 431
  Binbin Qiu, Dongsheng Guo, Hongzhou Tan, Zhi Yang, and Yunong Zhang

Content-Adaptive Rain and Snow Removal Algorithms for Single Image ... 439
  Shujian Yu, Yixiao Zhao, Yi Mou, Jinghui Wu, Lu Han,
  Xiaopeng Yang, and Baojun Zhao

Data-Driven Bridge Detection in Compressed Domain from Panchromatic
Satellite Imagery... 449
    Yixiao Zhao, Shujian Yu, Jinghui Wu, Lu Han, Zijing Chen,
    Xiaopeng Yang, and Baojun Zhao

Fast Nonnegative Tensor Factorization by Using Accelerated
Proximal Gradient.. 459
    Guoxu Zhou, Qibin Zhao, Yu Zhang, and Andrzej Cichocki

Multiclassifier System with Fuzzy Inference Method Applied to the
Recognition of Biosignals in the Control of Bioprosthetic Hand.......... 469
    Marek Kurzynski and Andrzej Wolczowski

Target Detection Using Radar in Heavy Sea Clutter by Polarimetric
Analysis and Neural Network..................................... 479
    Ji Eun Kim, Sang Min Lee, Seung-Phil Lee, SooBum Kim,
    Young-Soo Kim, and Chan Hong Kim

Human Action Recognition Based on Difference Silhouette and Static
Reservoir... 489
    Danchen Zheng and Min Han

Metric-Based Multi-Task Grouping Neural Network for Traffic
Flow Forecasting 499
    Haikun Hong, Wenhao Huang, Guojie Song, and Kunqing Xie

Object Tracking with a Novel Method Based on FS-CBWH
within Mean-Shift Framework 508
    Dejun Wang, Yongtao Shi, Weiping Sun, and Shengsheng Yu

Bayesian Covariance Tracking with Adaptive Feature Selection.......... 516
    Dejun Wang, Lin Li, Wei Liu, Weiping Sun, and Shengsheng Yu

Single-Trial Detection of Error-Related Potential by One-Unit SOBI-R
in SSVEP-Based BCI 524
    Janir Nuno da Cruz, Ze Wang, Chi Man Wong, and Feng Wan

Generalized Regression Neural Networks with K-Fold Cross-Validation
for Displacement of Landslide Forecasting 533
    Ping Jiang, Zhigang Zeng, Jiejie Chen, and Tingwen Huang

One-Class Classification Ensemble with Dynamic Classifier Selection 542
    Bartosz Krawczyk and Michał Woźniak

A Massive Sensor Data Streams Multi-dimensional Analysis
Strategy Using Progressive Logarithmic Tilted Time Frame
for Cloud-Based Monitoring Application........................... 550
    Xin Song, Cuirong Wang, Yanjun Chen, and Jing Gao

Evolutionary Clustering Detection of Similarity in Neuronal Spike
Patterns   558
    Hu Lu, Zhe Liu, and Yuqing Song

Fast and Effective Image Segmentation via Superpixels and Adaptive
Thresholding   568
    Yunsheng Jiang and Jinwen Ma

Diagonal Log-Normal Generalized RBF Neural Network for Stock
Price Prediction   576
    Wenli Zheng and Jinwen Ma

Design of a Greedy V2G Coordinator Achieving Microgrid-Level Load
Shift   584
    Junghoon Lee and Gyung-Leen Park

PolSAR Image Segmentation Based on the Modified Non-negative Matrix
Factorization and Support Vector Machine   594
    Jianchao Fan, Jun Wang, and Dongzhi Zhao

Solving Path Planning of UAV Based on Modified Multi-Population
Differential Evolution Algorithm   602
    Zhengxue Li, Jie Jia, Mingsong Cheng, and Zhiwei Cui

Consensus for Higher-Order Multi-agent Networks with External
Disturbances.   611
    Deqiang Ouyang, Haijun Jiang, Cheng Hu, and Yingying Liu

The Research of Document Clustering Topical Concept Based
on Neural Networks   621
    Xian Fu and Yi Ding

The Research of Reducing Misregistration Based on Image Mosaicing   629
    Yi Ding and Xian Fu

A Parallel Image Segmentation Method Based on SOM and GPU with
Application to MRI Image Processing.   637
    Ailing De, Yuan Zhang, and Chengan Guo

Author Index   647

Analysis

A Less Conservative Guaranteed Cost Stabilization of Time-Varying Delayed CNNs

Mei Jiang[✉], Hanlin He, and Lu Yan

College of Science, Naval University of Engineering, Wuhan 430033, China
jjmm_73@163.com, hanlinhe62@sina.com, y30117891@126.com

Abstract. This paper deals with the guaranteed cost stabilization problem of time-varying delayed cellular neural networks. By introducing the saturation degree function and applying the convex hull theory to handle the activation function, the main contribution of the paper lies in its proposal of a new controller for time-varying delayed CNNs with guaranteed cost according to Lyapunov-Krasovskii theorem, which extends the earlier results and gets less conservative guaranteed cost stabilization. Then we make use of Schur complement to convert the QMI (quadratic matrix inequality) to an LMI (linear matrix inequality) and thus it can be easily used as controller synthesis. The minimization of the guaranteed cost is further studied, and the corresponding LMI criterion to get the controller is given. Finally, numerical examples are given to show the effectiveness of the proposed controller and its corresponding minimization problem.

Keywords: Time-Delayed Cellular Neural Networks (DCNNs) · Stabilization · Convex hull · Guaranteed cost

1 Introduction

As signal transmissions along cells in cellular neural networks (CNNs) usually cause time delays, the oscillation phenomenon or network instability may occurs in the interaction between the neurons. The delayed-type CNNs (DCNNs) were proposed [1] and stability of DCNNs has attracted many researchers' attention recently. Many delay-independent and delay-dependent stability criteria for DCNNs have been proposed over the past years, mainly based on Razumikhin techniques, the Lyapunov–Krasovskii functionals and linear matrix inequalities (LMIs) formulation [2-4]. However, the stability of many practical neural networks cannot always be guaranteed by these techniques. Zhou investigated global exponential stability with multi-proportional delays and present a new delay-dependent sufficient conditions to ensure global exponential stability [5]. Noticing the saturation property of the activation function in CNNs, many results have been proposed to the processing of saturated nonlinear terms over the past years. In [6-8] the saturated control is formulated into a convex combination of a feedback matrix and an auxiliary feedback matrix, but

National Natural Science Foundation of China (Grant No. 61374003).

Z. Zeng et al. (Eds.): ISNN 2014, LNCS 8866, pp. 3–11, 2014.
DOI: 10.1007/978-3-319-12436-0_1

these papers do not give the effective method to search for the optimal auxiliary feedback matrix. Without free-weighting matrices, a new delay-dependent sufficient condition for the exponential synchronization with memoryless hybrid feedback control are first established in terms of LMIs [9]. In [10-11], the authors presented the zoned discussion and maximax synthesis method (ZDMS) and considered guaranteed cost synchronization of DCNNs with ZDMS.

In this paper, we mainly focus on the design of a new controller with guaranteed cost for time-varying delayed cellular neural networks. First, the activation functions are handled together with the convex hull theory. Then a less conservative guaranteed cost controller is derived by Lyapunov-Krasovskii functional method. With the help of Schur complements, the LMI criterion is obtained from transformation of QMI condition. Further, the minimization of the guaranteed cost and its' corresponding LMI condition are derived. Analysis of the control cost and numerical examples are given to show the effectiveness of proposed method.

2   Problem Statement

The dynamic behavior of a continuous time-varying delayed CNNs can be described by the following state equations:

$$\dot{x} = Ax + B\mathrm{sat}(x) + C\mathrm{sat}(x(t-\tau)) + u \tag{1}$$

where $x(t-\tau) \in \Omega$, $x(t)$ (Ω is a compact subset of $\mathbb{R}^n$) are the state vectors with and without time delays, respectively. $u(t) \in \mathbb{R}^n$ is the control input (simply written as $\bar{x}, x, u$). $A = \mathrm{diag}(a_1, a_2, \cdots, a_n)$ represents the rate with which the i th neuron will reset its potential to the resting state in isolation when disconnected from the network and external inputs, where $a_i < 0 (i = 1, 2, \cdots, n)$. $B, C \in \mathbb{R}^{n \times n}$ are the interconnection matrices representing the weighting coefficients of the neurons, the conjunction matrices of the saturated parts. Here $0 < \tau(t) < +\infty$, $\dot{\tau}(t) \le \beta < 1$. Saturated function $\mathrm{sat}(x): \mathbb{R}^n \to \mathbb{R}^n$ is the activation function of the neurons defined as follows:

$$\mathrm{sat}(x_i) = \mathrm{sign}(x_i)\min\{x_i, |\rho_i|\}, \ i = 1, 2, \cdots, n, \ \rho_i \in \mathbb{R}^n$$

As saturation exists, the system has strong nonlinear characteristics. So we should firstly process the saturated terms. Define saturation degree function $\alpha(x) = (\alpha_1(x), \alpha_2(x), \cdots, \alpha_n(x))^{\mathrm{T}}$, where

$$\alpha_i(x) = \begin{cases} \dfrac{\rho_i}{x_i}; & \rho_i < x_i \\ 1; & -\rho_i < x_i < \rho_i \\ \dfrac{-\rho_i}{x_i}; & x_i < -\rho_i \end{cases}, \ i = 1, \cdots n \tag{2}$$

$\alpha(x)$, $\alpha(\overline{x})$ are simply written as $\alpha, \overline{\alpha}$. Define $\hat{\alpha}_i = \inf\limits_{x \in \Omega}\{\alpha_i(x)\}$, $\hat{\overline{\alpha}}_i = \inf\limits_{\overline{x} \in \Omega}\{\overline{\alpha}_i(\overline{x})\}$.

$(i = 1, \cdots, n)$. Clearly we have $0 < \hat{\alpha}_i \le \alpha_i \le 1$ and $\hat{\alpha} = \hat{\overline{\alpha}}$. $D(\alpha) = \mathrm{diag}(\alpha)$ denotes a diagonal matrix obtained from vector α, we get $\mathrm{sat}(x) = D(\alpha)x$, $\mathrm{sat}(\overline{x}) = D(\overline{\alpha})\overline{x}$, so the system (1) can be rewritten as $\dot{x} = (A + BD(\alpha))x + CD(\overline{\alpha})\overline{x} + u$.

Considering the set $\Lambda = \{v \in \mathbb{R}^n : v_i = 1 \text{ or } \hat{\alpha}_i \; ; i = 1, \cdots, n\}$, then there are 2^n elements in it. Denote the element in Λ as $r_i, i = 1, \cdots, 2^n$. According to the convex hull theory, we have

$$\mathrm{sat}(x) \in \mathrm{Co}\{D(r_i)x \; ; \; i = 1, \ldots, 2^n\}, \quad \mathrm{sat}(\overline{x}) \in \mathrm{Co}\{D(r_i)\overline{x} \; ; \; i = 1, \ldots, 2^n\},$$

the system (1) can be changed into

$$\dot{x} = \sum_{i=1}^{2^n}\sum_{j=1}^{2^n}\lambda_i\mu_j\left[(A + BD(r_i))x + CD(r_j)\overline{x} + u\right] \tag{3}$$

where $\lambda_i \ge 0, \sum\limits_{i=1}^{2^n}\lambda_i = 1, \mu_j \ge 0, \sum\limits_{j=1}^{2^n}\mu_j = 1$.

Remark 1: If we can find the vector $\hat{\alpha}$, then the convex combination of the right hand in Eq. (3) could be the smallest, then the designed controller based on the convex combination could be less conservative.

3   Design of the Guaranteed Cost Controller

For the system (1), we select a commonly used quadratic cost function

$$J = \int_0^{\infty}[x^{\mathrm{T}}(t)Qx(t) + u^{\mathrm{T}}(t)Ru(t)]\mathrm{d}t \tag{4}$$

where $Q > 0, R > 0$. Then we have the following theorem.

Theorem 1: For the nonlinear system (1), if there exist a positive definite matrix $P \in \mathbb{R}^{n \times n}$ and a matrix $K \in \mathbb{R}^{n \times n}$ such that

$$\begin{pmatrix} P(A + BD(r_i) - K) + (A + BD(r_i) - K)^{\mathrm{T}}P + P + Q + K^{\mathrm{T}}RK & PCD(r_j) \\ (PCD(r_j))^{\mathrm{T}} & -(1 - \beta)P \end{pmatrix} < 0 \tag{5}$$

for all $r_i, r_j \in \Lambda$, then $u(t) = -Kx(t)$ is a guaranteed cost controller of the system (1), making the system globally asymptotically stable. And the upper bound of guaranteed cost J^* satisfies

$$J^* = x(0)^{\mathrm{T}}Px(0) + \int_{-\tau_0}^0 x^{\mathrm{T}}(\theta)Px(\theta)\mathrm{d}\theta \tag{6}$$

where $x_0 = x(s), -\tau(0) \le s \le 0$, $\tau_0 = \tau(0)$.

Proof: Define a Lyapunov–Krasovskii functional $V(x_t) = V_1 + V_2$, where

$$V_1 = x(t)^T P x(t), \quad V_2 = \int_{t-\tau(t)}^{t} x^T(\theta) P x(\theta) \, d\theta.$$

It can be derived that

$$\dot{V}_1 = \sum_{i=1}^{2^n} \sum_{j=1}^{2^n} \lambda_i \mu_j \left[x^T \left(P(A + BD(r_i)) + (A + BD(r_i))^T P \right) x + 2x^T PCD(r_j)\overline{x} + 2x^T Pu \right]$$

$$\dot{V}_2 = x^T P x - (1 - \dot{\tau}(t))\overline{x}^T P\overline{x} \le x^T P x - (1 - \beta)\overline{x}^T P\overline{x}$$

Define $\xi^T = [x^T, \overline{x}^T]$, we get

$$\dot{V} \le \sum_{i=1}^{2^n} \sum_{j=1}^{2^n} \lambda_i \mu_j \left[\xi^T \begin{pmatrix} P(A + BD(r_i) - K) + (A + BD(r_i) - K)^T P + P & PCD(r_j) \\ (PCD(r_j))^T & -(1-\beta)P \end{pmatrix} \xi \right]$$

$$\le -x^T(Q + K^T RK)x$$

If the inequality (5) holds, then when $\xi \ne 0$, we have $\dot{V} < 0$. Because $V(x_t)$ is radially unbounded, by the Lyapunov–Krasovskii theorem, the system (1) is globally asymptotically stable. Meanwhile, we can further get from the upper inequality

$$J = \int_0^\infty x^T(Q + K^T RK)x dt \le -\int_0^\infty \dot{V}(x) dt = V(x_0) = J^*.$$

J^* is the upper bound of J. This completes the proof of Theorem 1.

Remark 2: The controller is robust to some extent. Suppose that there exists an uncertain signal Δ in system (1), thus the system should be rewritten as

$$\dot{x} = Ax + B\mathrm{sat}(x) + C\mathrm{sat}(\overline{x}) + u + \Delta \tag{7}$$

According to the proof of Theorem 1, $\dot{V} \le -x^T(Q + K^T RK)\, x + 2x^T P\Delta$. Suppose that $\|\Delta\| \le l\|x\|$, then we get $\dot{V} \le -(\lambda_{\min}(Q + K^T RK) - 2l\lambda_{\max}(P))\|x\|^2$. Thus, if $l \le \lambda_{\min}(Q + K^T RK)/(2\lambda_{\max}(P))$, the system (7) is globally asymptotically stable.

4   LMI Formulation

Corollary 1: There exist a positive definite matrix P_1 and a matrix $N_1 \in \mathbb{R}^{n \times n}$ such that

$$\begin{pmatrix} \begin{matrix} P_1(A + BD(r_i) + \frac{1}{2}I)^T + \\ (A + BD(r_i) + \frac{1}{2}I)P_1 - N_1 - N_1^T \end{matrix} & P_1 & N_1 & CD(r_j)P_1 \\ P_1 & -Q^{-1} & 0 & 0 \\ N_1^T & 0 & -R^{-1} & 0 \\ P_1(CD(r_j))^T & 0 & 0 & -(1-\beta)P_1 \end{pmatrix} < 0 \tag{8}$$

if and only if there exist a positive definite matrix P and a matrix $K \in \mathbb{R}^{n \times n}$ such that (5) is satisfied with $P_1 = P^{-1}$, $N_1 = P_1 K^{\mathrm{T}}$.

Proof:

Sufficiency: If (5) holds, according to Schur complement, we have

$$\begin{pmatrix} P(A+BD(r_i)-K+\frac{1}{2}I)+ & & & \\ (A+BD(r_i)-K+\frac{1}{2}I)^{\mathrm{T}}P & I & K^{\mathrm{T}} & PCD(r_j) \\ I & -Q^{-1} & 0 & 0 \\ K & 0 & -R^{-1} & 0 \\ (PCD(r_j))^{\mathrm{T}} & 0 & 0 & -(1-\beta)P \end{pmatrix} < 0 \qquad (9)$$

Multiplying both sides of the matrix (9) by $\mathrm{diag}(P^{-1},I,I,P^{-1})$, then it follows that

$$\begin{pmatrix} P^{-1}(A+BD(r_i)+\frac{1}{2}I)^{\mathrm{T}}+ & & & \\ (A+BD(r_i)+\frac{1}{2}I)P^{-1}-P^{-1}K^{\mathrm{T}}-KP^{-1} & P^{-1} & P^{-1}K^{\mathrm{T}} & CD(r_j)P^{-1} \\ P^{-1} & -Q^{-1} & 0 & 0 \\ KP^{-1} & 0 & -R^{-1} & 0 \\ P^{-1}(CD(r_j))^{\mathrm{T}} & 0 & 0 & -(1-\beta)P^{-1} \end{pmatrix} < 0 \quad (10)$$

By substituting $P_1 = P^{-1}$, $N_1 = P_1 K^{\mathrm{T}}$ into (10), we have (8).

It is also easy to prove the necessity by using Schur complement lemma. So it completes the proof of Corollary 1.

Remark 3: The following problem is the minimization of the guaranteed cost stabilization control. Since only P is undetermined in (8), the minimization of J^* is equivalent to the minimization of $x_0^{\mathrm{T}} P x_0$. Notice that if x_0 be a random variable with $E\{x_0 x_0^{\mathrm{T}}\} = I$, then it can de derived that $E\{x_0^{\mathrm{T}} P x_0\} = \mathrm{Trace}(P)$. Hence, the essence of the minimization of J^* is the minimization of $\mathrm{Trace}(P)$, which can be expressed as: $\min\limits_{P>0,K} \mathrm{Trace}(P)$, s.t. (8). for all $r_i, r_j \in \Lambda$.

The above problem can be transformed into following LMI problem:

$$\min\limits_{P_1>0,X>0,N_1} \mathrm{Trace}(X), \text{ s.t. } \begin{pmatrix} X & I \\ I & P_1 \end{pmatrix} > 0 \text{ and (8) for all } r_i, r_j \in \Lambda. \qquad (11)$$

5   Analysis of the Control Cost

If processing $2x^{\mathrm{T}} PB\mathrm{sat}(x)$ by the convex hull theory, where $B = (b_1, b_2, \cdots, b_n)$, we have $2x^{\mathrm{T}} PB\mathrm{sat}(x) = \sum\limits_{i=1}^{n} 2x^{\mathrm{T}} Pb_i \alpha_i x_i$. And it is known that $0 < \alpha_i \leq 1$, therefore if $x^T P b_i x_i < 0$, then $2x^T P b_i \alpha_i x_i \le 2x^T P b_i \hat{\alpha}_i x_i \le 0 = \max\{2x^T P b_i x_i, 0\}$;

if $x^T P b_i x_i \ge 0$, then $2x^T P b_i \alpha_i x_i \le 2x^T P b_i x_i = \max\{2x^T P b_i x_i, 0\}$.

Combining the two cases, we get $2x^T P b_i \alpha_i x_i \le \max\{2x^T P b_i x_i, 0\}$.

So the control cost of the controller designing with the convex hull theory is relatively lower than the method of zoned discussion and maximum synthesis (ZDMS). Because the design of feedback control $u = -Kx$ via Lyapunov stability analysis is essentially counteracting the positive term emerging in the amplification process, and the convex hull theory doesn't involve amplification, offering less conservative gain K of the controller.

6   Simulations of Guaranteed Cost Stabilization

6.1   Example of Guaranteed Cost Stabilization

Consider DCNNs (1) with the following coefficient matrices:

$$A = \begin{pmatrix} -1 & 0 \\ 0 & -1 \end{pmatrix}, \quad B = \begin{pmatrix} 1 + \pi/4 & 20 \\ 0.1 & 1 + \pi/4 \end{pmatrix}, \quad C = \begin{pmatrix} -1.3\sqrt{2}\,\pi/4 & 0.1 \\ 0.1 & -1.3\sqrt{2}\,\pi/4 \end{pmatrix}$$

and time delay $\tau(t) = 1 + 0.2\sin t$. so we have $\beta = 0.2$. With the initial states chosen as $x(s) = [0.1.0.6]^T$, $s \in [-1, 0]$. When $u = 0$, without loss of generality, we suppose the compact subset

$$\Omega = \{(x_1, x_2) | x_1 \in [-17.6903, 18.1368], x_2 \in [-1.1727, 1.4081]\},$$

hence $\hat{\alpha} = (1/18.1368, 1/1.4081)$. Now we use theorem 1 and corollary 1 to design the controller with guaranteed cost for the DCNN. We choose $Q = R = I$. Since there are 4 elements in Λ. Hence, we get 16 blocks in the form of LMI (8). It is convenient to solve by the help of LMI toolbox of MATLAB:

$$P_1 = \begin{pmatrix} 0.7481 & -0.1406 \\ -0.1406 & 0.0758 \end{pmatrix}, \quad N_1 = \begin{pmatrix} 0.9408 & 0.0167 \\ 0.0185 & 0.9678 \end{pmatrix}, \quad K = \begin{pmatrix} 1.9257 & 3.8124 \\ 3.7123 & 19.6394 \end{pmatrix},$$

thus $P = \begin{pmatrix} 2.0507 & 3.8004 \\ 3.8004 & 20.2272 \end{pmatrix}$ and $\mathrm{Trace}(P) = 22.2779$.

Then the states curves are shown in Fig. 1.

We use J_1 and J_1^* to denote J and J^* in this case. According to the data of the states, we obtain $J_1 \approx 5.6748$, and $J_1^* = V(x_0) = 15.5167$. It's obvious that $J_1 < J_1^*$.

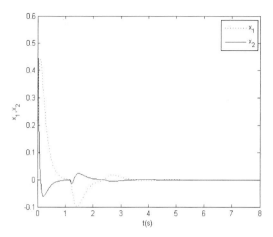

Fig. 1. Stabilization under the controller with guaranteed cost

6.2 Minimization of the Guaranteed Cost

To minimize the guaranteed cost's upper bound J^*, it has been elaborated in Remark 3 that we can find the minimum of Trace(P). According to (11), we should find the solution of 17 LMIs. The solution obtained is as follows

$$P_1 = \begin{pmatrix} 0.5870 & -0.1312 \\ -0.1312 & 0.0969 \end{pmatrix}, \quad N_1 = \begin{pmatrix} 0.0093 & 0.3278 \\ 0.3278 & 0.8113 \end{pmatrix}, \quad K = \begin{pmatrix} 1.1067 & 4.8810 \\ 3.4833 & 13.0872 \end{pmatrix},$$

thus $P = \begin{pmatrix} 2.4426 & 3.3065 \\ 3.3065 & 14.7951 \end{pmatrix}$ and Trace(P) = 17.2377.

The states curves are shown in Fig. 2.

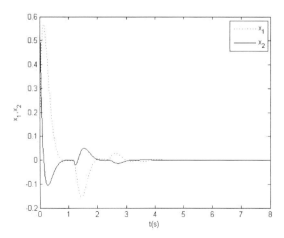

Fig. 2. Stabilization under minimal controller with guaranteed cost

In this section, we use J_2 and J_2^* to denote J and J^* in this case. According to the data of the states, we obtain $J_2 \approx 3.0692$, and $J_2^* = V(x_0) = 11.4949$, so $J_2 < J_2^*$.

6.3   Results Analysis

By the zoned discussion and maximum synthesis (ZDMS) method, we calculate the guaranteed cost J_3 and its' upper bound J_3^*, and the minimization of guaranteed cost J_4 and its' upper bound J_4^*. The results are listed as follows:

$$J_3 = 7.7850, J_3^* = 19.7611, \quad J_4 = 6.1699, J_4^* = 14.4578.$$

As expected, the values obtained by our method are smaller than the corresponding ones obtained by ZDMS method.

7   Conclusion

In this paper, we mainly designed a new controller with guaranteed cost for time–varying delayed cellular neural network. Firstly, we make use of the saturation degree function combining with the convex hull theory to handle saturated terms. Then we put forward the sufficient condition for the controller according to the Lyapunov–Krasovskii theorem, and proved that the designed controller has certain robustness. Because the condition is a QMI, so we applied Schur complement lemma to convert it to an LMI to be easily solved by computer. We also discussed the minimization problem of guaranteed cost, trying to optimize the controller. Further, we proved the control cost of the designed controller is lower than the zoned discussion and maximum synthesis method. Finally, simulations demonstrate the effectiveness of the method.

References

1. Roska, T., Chua, L.O.: Cellular neural networks with nonlinear and delay-type template elements and nonuniform grids. International Journal of Circuit Theory and Applications **20**, 469–481 (1992)
2. Zheng, C.D., Zhang, H.G.: Novel delay-dependent criteria for global robust exponential stability of delayed cellular neural networks with norm-bounded uncertainties. Neurocomputing **72**, 1744–1754 (2009)
3. Tan, M.C.: Global asymptotic stability of fuzzy cellular neural networks with unbounded distributed delays. Neural Processing Letters **31**, 147–157 (2010)
4. Akhmet, M.U., Arugaslan, D., Yilmaz, E.: Stability in cellular neural networks with a piecewise constant argument. Journal of Computational and Applied Mathematics **233**, 2365–2373 (2010)
5. Zhou, L.: Delay-Dependent Exponential Stability of Cellular Neural Networks with Multi-Proportional Delays. Neural Processing Letters **38**, 347–359 (2013)

6. Hu, T., Lin, Z.: Control systems with Actuator Saturation. Analysis and Design, Boston (2001)
7. Hu, T., Lin, Z., Chen, B.: An analysis and design method for linear systems subject to actuator saturation and disturbance. Automatica **38**, 351–359 (2002)
8. Hu, T., Lin, Z., Chen, B.: Analysis and design for discrete-time systems subject to actuator saturation. System & Control Letters **45**, 87–112 (2002)
9. Botmart, T., Weera, W.: Guaranteed Cost Control for Exponential Synchronization of Cellular Neural Networks with Mixed Time-Varying Delays via Hybrid Feedback Control. Abstract and Applied Analysis (2013)
10. He, H., Yan, L., Tu, J.: Guaranteed cost stabilization of time–varying delay cellular neural networks via Riccati inequality approach. Neural Processing Letters **35**, 151–158 (2012)
11. He, H., Yan, L., Tu, J.: Guaranteed cost stabilization of cellular neural networks with time-varying delay. Asian Journal of Control **15**, 1224–1227 (2013)

Existence and Uniqueness of Almost Automorphic Solutions to Cohen-Grossberg Neural Networks with Delays

Xianyun Xu, Tian Liang, Fei Wang, and Yongqing Yang[✉]

School of Science, Jiangnan University, 214122,
Wuxi, People's Republic of China
yyq640613@gmail.com

Abstract. The almost automorphic solution is a generalization of the almost periodic solution. In this paper, the almost automorphic solutions of Cohen-Grossberg neural networks with delays are considered. Using the semi-discretization method and the contraction mapping principle, some sufficient conditions are obtained to ensure the existence and the uniqueness of almost automorphic solutions to Cohen-Grossberg neural networks with delays.

Keywords: Cohen-Grossberg neural network · Almost automorphic solution · Contraction mapping principle

1 Introduction

In 1983, Cohen and Grossberg constructed an important kind of simplified neural networks model which is now called Cohen-Grossberg neural networks (CGNNS) [1]. This kind of neural networks is very general and includes Hopfield neural networks, cellular neural networks and BAM neural networks as its special cases. It has received increasing interest due to its applications in many fields such as pattern recognition, parallel computing, associative memory and combinatorial optimization. In recent years, the Cohen-Grossberg neural networks have been widely studied and many useful and interesting results have been obtained (see [6] and its references).

The concept of almost automorphy was introduced by Bochner [2] in 1964. It is a natural generalization of the classical almost periodicity. According to the properties of periodic functions, we know that periodic functions are all uniformly continuous. However, there exist some functions that have the similar properties to periodic functions, and meanwhile they are not uniformly continuous, such as $f(k) = sign(cos2\pi k\theta)$. This kind of function is almost automorphic.

This work was jointly supported by the National Natural Science Foundation of China under Grant 11226116, the Fundamental Research Funds for the Central Universities JUSRP51317B.

ⓒ Springer International Publishing Switzerland 2014
Z. Zeng et al. (Eds.): ISNN 2014, LNCS 8866, pp. 12–18, 2014.
DOI: 10.1007/978-3-319-12436-0_2

At present, the almost automorphic functions have been used in many different kind of fields [3–5], [10, 15, 16], such as ordinary differential equation, partial
differential equation, integral equation and dynamic system and so on. In [5], the
authors studied a kind of partial differential equation based on biology, and the
natural function classes of the solutions about this kind differential equation are
almost automorphic functions. However, there is no paper discussed the almost
automorphic solution to Cohen-Grossberg neural networks, so it is meaningful
to discuss it and ours is the first one.

Generally speaking, the Cohen-Grossberg neural networks with delays can
be described as following:

$$\dot{x}_i(t) = -a_i(x_i(t))\big[b_i(x_i(t)) - \sum_{j=1}^{n} d_{ij}(t)g_j(x_j(t-\tau_j)) - I_i(t)\big] \qquad (1)$$

where $i = 1, 2, \cdots, n$, $x_i(t)$ is the state variable associated with the i_{th} neuron,
$a_i(\cdot)$ is an amplification function and $b_i(\cdot)$ represents a behaved function, $d_{ij}(t)$
presents the strength of connectivity between cells i and j at time t, the activation function $g_i(\cdot)$ tells how the i_{th} neuron reacts to the input, τ_i corresponds to
the time delay. The initial condition of (1) is $x_i(t) = \varphi_i(t)$, $t \in [-\tau_i, 0]$.

In reality for the applications of neural networks to some practical problems,
such as experiment, image processing, computational purposes and so on, it is essential to formulate a discrete-time system which is a version of the continuous-time
system. The discrete-time system is desired to preserve the dynamical characteristics of the continuous-time system. There are many numerical schemes such as
Euler scheme and Runge-Kutta scheme that can be utilized to obtain the discrete-
time version of the continuous-time system. In this paper, we will use the semi-
discretization scheme to obtain the discrete-time analogues of the continuous-time
(1). The semi-discretization idea was originally used in the partial differential equations and then introduced to the ordinary differential equations. It has been proved
that such kind of method can preserve the dynamical characteristics of the
continuous-time systems to some extent, we can find examples in [7]-[9].

Using the semi-discretization method, the model (1) can be written as:

$$\dot{x}_i(t) = -a_i(x_i(t))\big[b_i(x_i(t)) - \sum_{j=1}^{n} d_{ij}(t)g_j(x_j([\tfrac{t}{h}]h - [\tfrac{\tau_j}{h}]h)) - I_i(t)\big] \qquad (2)$$

$t \in [nh, (n+1)h)$, $[\tfrac{t}{h}] = n$, h is the discretization step-size, it is a fixed positive
real number.

In this paper, we consider the existence and the uniqueness of almost automorphic solutions of (2).

The remainder of this paper is organized as following: some definitions and
assumptions are given in Section 2, and in Section 3, some sufficient conditions
are given to ensure the existence of the almost automorphic solutions of (2). In
the last section, Section 4, some conclusions about this paper are presented.

2   Preliminaries

For the readers' convenience, we first give some definitions (for details, see
[11]-[14]).

Definition 1. *A continuous function $f : R \times X \to R$ is called almost automorphic for x in compact subsets of X, if for every compact subset K of x and every real sequence s_n, there exists a subsequence s_{n_k}, such that*

$$\lim_{n \to +\infty} f(t + s_{n_k}, x) = g(t, x) \quad and \quad \lim_{n \to +\infty} g(t - s_{n_k}, x) = f(t, x), \quad t \in R, \ x \in K$$

Definition 2. *A continuous $f : Z \times X \to X$ is a called almost automorphic sequence for $x \in X$ if for every sequence of integer $\{n\}$, there exists a subsequence $\{n_l\}_{l \in N}$, such that*

$$f(n + hn_l, x) \to g(n, x) \quad and \quad g(n - hn_l, x) \to f(n, x), \quad n \in Z \quad and \quad x \in X$$

The set of all such functions are denoted by $AAS(Z \times X \to X)$, AAS for short.

The following are some assumptions which will be used later.

$\mathbf{A_1}$: $a_i(\cdot)$, $d_{ij}(\cdot)$, and $I_i(\cdot)$ are almost automorphic to the variable t, and $0 < \underline{a_i} \le a_i(\cdot) \le \overline{a_i}$.

$\mathbf{A_2}$: for any $x, y \in R$, there exist some constants A_i, L_j, and G_j, such that $|a_i(x) - a_i(y)| < A_i|x - y|, \ |g_j(x) - g_j(y)| \le L_j|x - y|, \ |g_j(x)| \le G_j$.

$\mathbf{A_3}$: There exist positive almost automorphic functions $\underline{\beta}_i(t)$, $\overline{\beta}_i(t)$, such that $\forall x_i, y_i \in R$, $i = 1, 2, \cdots, n$, the following inequality holds:

$$0 < \underline{\beta}_i(t) \le \frac{a_i(x_i(t))b_i(x_i(t)) - a_i(y_i(t))b_i(y_i(t))}{x_i(t) - y_i(t)} \le \overline{\beta}_i(t)$$

and $\underline{\beta}_i = \inf_{t \ge 0} |\underline{\beta}_i(t)|$, $\overline{\beta}_i = \sup_{t \ge 0} |\overline{\beta}_i(t)|$, $b_i(0) \equiv 0$.

3   Main Results

According to A_3, the model (2) can be written as following:

$$\dot{x}_i(t) = -r_i(t)x_i(t) + a_i(x_i(t))\left[\sum_{j=1}^{n} d_{ij}(t)g_j(x_j([\tfrac{t}{h}]h - [\tfrac{\tau_j}{h}]h)) + I_i(t)\right] \quad (3)$$

From (3) we can obtain:

$$
\begin{aligned}
x_i^h(n + 1) = {}& x_i^h(n)e^{-\int_{nh}^{(n+1)h} r_i(u)du} + a_i(x_i^h(n)) \\
& \times \left\{ \int_{nh}^{(n+1)h} \left[\sum_{j=1}^{n} d_{ij}(s)g_j(x_j(n - \tau_j^*)) + I_i(s)\right]e^{-\int_s^{(n+1)h} r_i(u)du}ds \right\}
\end{aligned}
$$
$$(4)$$

where $x_i^h(n) = x_i(nh)$, and $\tau_j^* = [\tfrac{\tau_j}{h}]$.

Let

$$
\begin{aligned}
R_i(n) &= e^{-\int_{nh}^{(n+1)h} r_i(u)du}, \\
D_{ij}(n) &= \int_{nh}^{(n+1)h} d_{ij}(s)e^{-\int_s^{(n+1)h} r_i(u)du}ds, \\
E_i(n) &= \int_{nh}^{(n+1)h} I_i(s)e^{-\int_s^{(n+1)h} r_i(u)du}ds
\end{aligned}
$$

then (4) is reformulated as:

$$x_i^h(n + 1) = R_i(n)x_i^h(n) + \sum_{j=1}^{n} a_i(x_i^h(n))D_{ij}(n)g_j(x_j(n - \tau_j^*)) + a_i(x_i^h(n))E_i(n)$$
$$(5)$$

Denote $\overline{R}_i = \sup_{n \in Z}\{R_i(n)\}$, $\overline{D}_{ij} = \sup_{n \in Z}\{D_{ij}(n)\}$, $\overline{E}_i = \sup_{n \in Z}\{E_i(n)\}$.

Theorem 1. *Suppose that the assumptions $A_1 - A_3$ hold, then there exists a unique almost automorphic solution of (5) if*

$$\max_{1 \le i,j \le n} \{\overline{R}_i + \sum_{j=1}^{n} \overline{a}_i \overline{D}_{ij} L_j + \sum_{j=1}^{n} \overline{A}_i \overline{D}_{ij} G_j + \overline{A}_i \overline{I}_i\} < 1.$$

Proof. There are three steps to complete the proof.

Step1. To start the proof, we show that $R_i(n)$, $D_{ij}(n)$, $E_i(n)$ are almost automorphic for $i, j = 1, 2, \cdots, n$, firstly.

For $r_i(t)$ is almost automorphic, then for any sequence t_n, there exists a subsequence t_{n_l} such that $r_i(t + t_{n_l}) \to \overline{r}_i(t)$ and $\overline{r}_i(t - t_{n_l}) \to r_i(t)$ for $n_l \to \infty$. so

$$|R_i(n + t_{n_l}) - \overline{R}_i(n)| = |e^{-\int_{(n+t_{n_l})h}^{(n+1+t_{n_l})h} r_i(u)du} - e^{-\int_{nh}^{(n+1)h} \overline{r}_i(u)du}|$$
$$= |e^{-\int_{nh}^{(n+1)h} r_i(u+t_{n_l})du} - e^{-\int_{nh}^{(n+1)h} \overline{r}_i(u)du}| \to 0$$

Thus $R_i(n + t_{n_l}) \to \overline{R}_i(n)$. Likewise, $\overline{R}_i(n - t_{n_l}) \to R_i(n)$.

Under assumption A_1, $d_{ij}(t)$ is almost automorphic and $d_{ij}(t + t_{n_l}) \to \overline{d}_{ij}(t)$. Let $\overline{D}_{ij}(n) = \int_{nh}^{(n+1)h} \overline{d}_{ij}(s) e^{-\int_{s}^{(n+1)h} \overline{r}_i(u)du} ds$, then for $\{t_{n_l}\} \in Z$,

$$|D_{ij}(n + t_{n_l}) - \overline{D}_{ij}(n)|$$
$$= |\int_{(n+t_{n_l})h}^{(n+1+t_{n_l})h} d_{ij}(s) e^{-\int_{s}^{(n+1+t_{n_l})h} r_i(u)du} ds - \int_{nh}^{(n+1)h} \overline{d}_{ij}(s) e^{-\int_{s}^{(n+1)h} \overline{r}_i(u)du} ds|$$
$$\le |\int_{nh}^{(n+1)h} d_{ij}(s + t_{n_l})[e^{-\int_{s}^{(n+1)h} r_i(u+t_{n_l})du} - e^{-\int_{s}^{(n+1)h} r_i(u)du}] ds|$$
$$+ |\int_{nh}^{(n+1)h} [d_{ij}(s + t_{n_l}) - \overline{d}_{ij}(s)] e^{-\int_{s}^{(n+1)h} \overline{r}_i(u)du} ds|$$
$$\to 0.$$

Likewise, $\overline{D}_{ij}(n - t_{n_l}) \to D_{ij}(n)$. Then by the similar analysis, $E_i(n + t_{n_l}) \to \overline{E}_i(n)$, $\overline{E}_i(n - t_{n_l}) \to E_i(n)$. That is to say, $A_i(n)$, $D_{ij}(n)$, $E_i(n) \in AAS$.

Step2. Consider the following equation:

$$x_i^h(n + 1) = R_i(n)x_i^h(n) + a_i(x_i^h(n))E_i(n) \tag{6}$$

Next, we will show that (6) has a unique almost automorphic sequence solution. Using the method of induction, according to (6), we can obtain

$$x_i^h(n + 1) = \prod_{l=0}^{n} R_i(l)x_i^h(0) + \sum_{q=0}^{n} a_i(x_i^h(n - q)) \int_{(n-q)h}^{(n+1-q)h} I_i(s) e^{-\int_{s}^{(n+1)h} r_i(u)du} ds$$

Let

$$\tilde{x}_i^h(n) = \sum_{q=0}^{n-1} a_i(\tilde{x}_i^h(n - 1 - q)) \int_{(n-1-q)h}^{(n-q)h} I_i(s) e^{-\int_{s}^{nh} r_i(u)du} ds,$$

then

$$|\tilde{x}_i^h(n)| \le |\sum_{q=0}^{n-1} \overline{a}_i \frac{\overline{I}_i}{\underline{\beta}_i} (e^{-qh\underline{\beta}_i} - e^{-(q+1)h\underline{\beta}_i})| < |\frac{\overline{a}_i \overline{I}_i}{\underline{\beta}_i} (1 - e^{-nh\underline{\beta}_i})| < |\frac{\overline{a}_i \overline{I}_i}{\underline{\beta}_i}|.$$

We can easily verify that
$$\widetilde{x}_i^h(n+1) = R_i(n)\widetilde{x}_i^h(n) + a_i(\widetilde{x}_i^h(n))E_i(n).$$
Let $\widetilde{x}_{i_*}^h(n) = \sum_{q=0}^{n-1}\overline{a}_i(\widetilde{x}_{i_*}^h(n-1-q))\int_{(n-1-q)h}^{(n-q)h}\overline{I}_i(s)e^{-\int_s^{(n-q)h}\overline{r}_i(u)du}ds$, where $a_i(\widetilde{x}_i^h(n+t_{n_l})) \to \overline{a}_i(\widetilde{x}_{i_*}^h(n))$, and $\overline{a}_i(\widetilde{x}_{i_*}^h(n-t_{n_l})) \to a_i(\widetilde{x}_i^h(n))$.

Then for any given sequence $\{t_{n_l}\} \in Z$,

$$|\widetilde{x}_i^h(n+t_{n_l}) - \widetilde{x}_{i_*}^h(n)|$$
$$\leq \sum_{q=0}^{n-1}\Big|[a_i(\widetilde{x}_i^h(n-1-q)) - \overline{a}_i(\widetilde{x}_{i_*}^h(n-1-q))]$$
$$\int_{(n-1-q)h}^{(n-q)h}I_i(s+t_{n_l})e^{-\int_s^{(n-q)h}r_i(u+t_{n_l})du}ds\Big|$$
$$+\sum_{q=0}^{n-1}\Big|\overline{a}_i(\widetilde{x}_{i_*}^h(n-1-q))\int_{(n-1-q)h}^{(n-q)h}[I_i(s+t_{n_l}) - \overline{I}_i(s)]e^{-\int_s^{(n-q)h}r_i(u+t_{n_l})du}ds\Big|$$
$$+\sum_{q=0}^{n-1}\Big|\overline{a}_i(\widetilde{x}_{i_*}^h(n-1-q))\int_{(n-1-q)h}^{(n-q)h}\overline{I}_i(s)$$
$$[e^{-\int_s^{(n-q)h}r_i(u+t_{n_l})du} - e^{-\int_s^{(n-q)h}r_i(u)du}]ds\Big| \to 0$$

So $\widetilde{x}_i^h(n+t_{n_l}) \to \widetilde{x}_{i_*}^h(n)$. Likewise, $\widetilde{x}_{i_*}^h(n-t_{n_l}) \to \widetilde{x}_i^h(n)$. Thus, $\widetilde{x}_i^h(n)$ is almost automorphic. In addition, $\widetilde{x}_i^h(n+1) = R_i(n)\widetilde{x}_i^h(n) + a_i(\widetilde{x}_i^h(n))E_i(n)$, then $\widetilde{x}_i^h(n)$ is the almost automorphic solution of (6).

Step3. Assume that

$$\theta = \max_{1\leq i\leq n}\frac{\overline{a}_i\overline{I}_i}{\underline{\beta}_i}, \quad \omega = \max_{1\leq i\leq n}\{\overline{R}_i + \overline{A}_i\overline{E}_i\}, \quad \gamma = \max_{1\leq i\leq n}\{\overline{R}_i + \sum_{j=1}^n\overline{a}_i\overline{D}_{ij}L_j\}$$

Define a mapping F: $AAS \to AAS$, $x \to Fx$, $Fx = ((Fx)_2,\cdots,(Fx)_n)^T$,

$$(Fx)_i(n+1) = R_i(n)x_i^h(n) + \sum_{j=1}^n a_i(x_i^h(n))D_{ij}(n)g_j(x_j(n-\tau_j^*)) + a_i(x_i^h(n))E_i(n).$$

Denote $\|x\| = \sup_{n\in Z}\max_{1\leq i\leq n}|x_i(n)|$, let $\Omega = \{x : x$ is almost automorphic, $\|x - \widetilde{x}\| \leq \frac{\omega+\gamma}{1-\gamma}\theta\}$, then $\|x\| \leq \|x - \widetilde{x}\| + \|\widetilde{x}\| = \frac{\omega+1}{1-\gamma}\theta$.
$\forall x, y \in \Omega$, we have:

$$\|Fx - \widetilde{x}\| = \sup_{n\in Z}\max_{1\leq i\leq n}|R_i(n)(x_i^h(n) - \widetilde{x}_i^h(n))$$
$$+\sum_{j=1}^n a_i(x_i^h(n))D_{ij}(n)g_j(x_j(n-\tau_j^*)) + [a_i(x_i^h(n)) - a_i(\widetilde{x}_i^h(n))]E_i(n)|$$
$$\leq \overline{R}_i\|x\| + \overline{R}_i\|\widetilde{x}\| + \sum_{j=1}^n\overline{a}_i\overline{D}_{ij}L_j\|x\| + \overline{A}_i\overline{E}_i\|\widetilde{x}\|$$
$$\leq (\overline{R}_i + \sum_{j=1}^n\overline{a}_i\overline{D}_{ij}L_j)\|x\| + \omega\|\widetilde{x}\|$$
$$\leq \frac{\omega+\gamma}{1-\gamma}\theta$$

$$\|Fx - Fy\| = \sup_{n \in Z} \max_{1 \le i \le n} \left\{ |R_i(n)(x_i^h(n) - y_i^h(n)) \right.$$

$$+ \sum_{j=1}^{n} [a_i(x_i^h(n))g_j(x_j(n - \tau_j^*)) - \sum_{j=1}^{n} a_i(y_i^h(n))g_j(y_j(n - \tau_j^*))]D_{ij}(n)$$

$$+ [a_i(x_i^h(n)) - a_i(y_i^h(n))]I_i(n)| \right\}$$

$$\le \max_{1 \le i \le n} \left\{ (\overline{R}_i + \sum_{j=1}^{n} \overline{a}_i \overline{D}_{ij} L_j + \sum_{j=1}^{n} \overline{A}_i \overline{D}_{ij} G_j + \overline{A}_i \overline{I}_i) \|x - y\| \right\} < \|x - y\|$$

Then F is a construction mapping, thus (5) has a unique almost automorphic solution which satisfies that $\|x - \widetilde{x}\| < \frac{\omega + \gamma}{1 - \gamma} \theta$. This completes the proof.

4 Conclusions

In this paper, the almost automorphic solutions of delayed Cohen-Grossberg neural networks are investigated. The almost automorphic solution is a generalization of the almost periodic solution, and it has been used in ordinary differential equation, partial differential equation, integral equation and dynamic system and so on. Our paper is the first one to discuss such solutions on Cohen-Grossberg neural networks. By the contraction mapping principle, the existence and the uniqueness of almost automorphic solutions are discussed, and some new results are obtained.

References

1. Cohen, M., Grossberg, S.: Absolute Stability of Global Pattern Formation and Parallel Memory Storage by Competitive Neural Networks. IEEE Trans. Sys. Man Cyber. **3**, 815–826 (1983)
2. Bochner, S.: Continuous Mappings of Almost Automorphic and Almost Periodic Functions. PNAS **52**, 907–910 (1964)
3. N'Guérékata, G.M.: Almost Automorphic Functiolls and Almost Periodic Functions in Abstract Spaces. Kluwer Academic/Plnum Publishers, New York (2001)
4. N'Guérékata, G.M.: Topics Almost Automorphy. Springer, New York (2005)
5. Hetzer, G., Shen, W.: Uniform Persistence, Coexistence, and Extinction in Almost Periodic/Nonautonomous Competition Diffusion Systems. SIAM J. Mathe. Anal. **34**(1), 204–227 (2002)
6. Balasubramaniam, P., Ali, M.: Stability Analysis of Takagi-Sugeno Fuzzy Cohen-Grossberg BAM Neural Networks with Discrete and Distributed Time-varying Delays. Mathe. Compu. Model. **53**, 151–160 (2011)
7. Mohamad, S., Gopalsamy, K.: Exponential Stability of Continuous-time and Discrete-time Cellular Neural Networks with Delays. Appl. Math. Comput. **135**, 17–38 (2003)
8. Insperger, T., Stepan, G.: Semi-discretization Method for Delayed Systems. Int. J. Numer. Meth. Eng. **55**, 503–518 (2002)
9. Liang, J., Cao, J.: Exponential Stability of Continuous-time and Discrete-time Bidirectional Associative Memory Networks with Delays. Chaos Soli. Fract. **22**, 773–785 (2004)

10. Bugajewski, D., N'Guérékata, G.M.: On the Topological Structure of Amost Auto-
    morphic and Asymptotically Almost Automorphhic Solutios of Differerential and
    Integral Equations in Abstract Spaces. Nonl. Anal. **59**, 1333–1345 (2004)
11. Chang, Y., Zhao, Z., N'Guérékata, G.M.: A New Composition Theorem for Square-
    mean Almost Automorphic Functions and Applications to Stochastic Differential
    Equations. Nonl. Anal. **74**, 2210–2219 (2011)
12. Chen, Z., Lin, W.: Square-mean Pseudo Almost Automorphic Process and Its
    Application to Stochastic Evolution Equations. J. Func. Anal. **261**, 69–89 (2011)
13. Fu, M.: Almost Automorphic Solutions for Nonautonomous Stochastic Differential
    Equations. J. Mathe. Anal. Appl. **393**, 231–238 (2012)
14. Fu, M., Chen, F.: Almost Automorphic Solutions for Some Stochastic Differential
    Equations. Nonl. Anal. **80**, 66–75 (2013)
15. Diagana, T.: Existence of Globally Attracting Almost Automorphic Solutions to
    Some Nonautonomous Higher-order Difference Equations. Appl. Mathe. Comp.
    219, 6510–6519 (2013)
16. Abbas, S., Xia, Y.: Existence and Global Attractivity of K-almost Automorphic
    Sequence Solution of a Model of Cellular Neural Networks with Delays. Acta
    Mathe. Sci. **33B**(1), 290–302 (2013)

A Proof of a Key Formula in the Error-Backpropagation Learning Algorithm for Multiple Spiking Neural Networks

Wenyu Yang[1], Dakun Yang[2(✉)], and Yetian Fan[3]

[1] College of Science, Huazhong Agricultural University, Wuhan, China
[2] School of Information Science and Technology, Sun Yat-sen University,
Guangzhou, China
ydk1026@gmail.com
[3] School of Mathematical Sciences, Dalian University of Technology,
Dalian, China

Abstract. In the error-backpropagation learning algorithm for spiking neural networks, solving the differentiation of the firing time t^α with respect to the weight w is essential. Bohte et al. see the firing time t^α as a functional of the state variable x(t). But the differentiation of the firing time t^α with respect to the state variable x(t) is impossible to perform directly. To overcome this problem, Bohte et al. assume that the state variable x(t) is a linear function of the time t around $t = t^\alpha$. Then, it seems that the solution of Bohte et al. is used by all related Literatures. In particular, Ghosh-Dastidar and Adeli offer another explanation. In this paper, we consider the firing time t^α as a function of the time t and the weight w and prove that the key formula for multiple spiking neural networks is in fact mathematically correct through the implicit function theorem.

Keywords: Spiking neuron · Error-backpropagation · Differentiation of the firing time · Implicit function theorem

1 Introduction

In recent years, spiking neural networks (SNNs) are more biologically plausible and often referred to as the third generation of neural networks [1]. SNNs have been the subject of significant research and have been applied extensively and successfully for practical applications [2–8]. The aim of this paper is to prove the following equality of differentiations

$$\frac{\partial t^\alpha}{\partial w} = -(\frac{\partial x}{\partial t}(t))^{-1}\frac{\partial x(t)}{\partial w}, \quad \text{at } t = t^\alpha \tag{1}$$

Research funded by the Fundamental Research Funds for the Central Universities (2662013BQ049, 2662014QC011).

© Springer International Publishing Switzerland 2014
Z. Zeng et al. (Eds.): ISNN 2014, LNCS 8866, pp. 19–26, 2014.
DOI: 10.1007/978-3-319-12436-0_3

which is a key formula in the error-backpropagation learning algorithm for spiking neural networks and multiple spiking neural networks. To solve the differentiation $\partial t^\alpha / \partial w$, Bohte et al. use the chain rule

$$\frac{\partial t^\alpha}{\partial w} = \frac{\partial t^\alpha}{\partial x(t)} \frac{\partial x(t)}{\partial w} \tag{2}$$

by considering the firing time t^α as a functional of the state variable x(t) which is a funtion of the weight w. The only and the most difficult problem for Bohte et al. is solving $\partial t^\alpha / \partial x(t)$ since t^α cannot be formulated in a standard form as a functional of x(t). To overcome the problem, Bohte et al. [9] assume that the state x(t) is a linear function of the time t around $t = t^\alpha$. Therefore, the Frechet derivative

$$\begin{aligned}
\frac{\partial t^a}{\partial x(t)} &= \lim_{\triangle x \to \infty} \frac{t^a(x + \triangle x) - t^a(x)}{\triangle x} \\
&= \frac{-\triangle t}{\triangle x} \\
&= -\left(\frac{\partial x(t)}{\partial t}\right)^{-1}, \quad \text{at } t = t^\alpha
\end{aligned} \tag{3}$$

where $\triangle x$ and $\triangle t$ represent an infinitesimal change in x and t, respectively. It seems that the solution of Bohte et al. is used by all related literatures [4,10–14], whether for spiking neural networks or for multiple spiking neural networks. In particular, Ghosh-Dastidar [4] et al. give another explanation. They consider the relationship between the threshold value θ and the time t and assume that there is a linear relationship between them around $t = t^\alpha$. As a result of the consideration and the assumption, there is a relationship as follows

$$\frac{\partial x(t)}{\partial t} = \frac{\partial \theta}{\partial t}, \quad \text{at } t = t^\alpha \tag{4}$$

Moreover, they give the following equality

$$\frac{\partial t^\alpha}{\partial x(t)} = -\frac{\partial t^\alpha}{\partial \theta}, \quad \text{at } t = t^\alpha \tag{5}$$

to model the opposite relationship between the state variable and threshold, with respect to the firing time. From Eqs. (4) and (5), the Frechet derivative becomes

$$\frac{\partial t^\alpha}{\partial x(t)} = -\left(\frac{\partial x(t)}{\partial t}\right)^{-1}, \quad \text{at } t = t^\alpha \tag{6}$$

Our contribution in this paper is to prove that the key equality (4) is in fact mathematically correct for multiple spiking neural networks, without the help of the linearity assumption. Spiking neural networks can be considered a special case of multiple spiking neural networks. Hence, with the same method of this paper, the key equality (4) is in fact mathematically correct for spiking neural networks.

The rest of this paper is organized as follows. In Section 2 we introduce the spiking neuron and the multi-spiking neuron. Then, in Section 3 we describe the errorback-propagation algorithm. Finally in Section 4, we devote to the proof of the key equality.

2 Spiking Neuron and Multi-Spiking Neuron

In the spike response model (SRM), the output of a neuron is described by the firing time of the spike it produced. The firing time t^α of the postsynaptic neuron is defined as the time when the state variable $x(t)$ exceeds a given threshold from low to high.

2.1 Spiking Neuron

In the spike response model (SRM) of spiking neuron, the state variable of the postsynaptic neuron j is influenced by the spike times t_i of its presynaptic neurons i as follows:

$$x_j(t) = \sum_{i=1}^{I} \sum_{k=1}^{K} w_{ij}^k \epsilon(t - t_i - d_{ij}^k) \tag{7}$$

where w_{ij}^k and d_{ij}^k are the kth synaptic weight and the kth synaptic delay, between the presynaptic neuron i and the postsynaptic neuron j, respectively. The response function $\epsilon(\cdot)$ is chosen as

$$\epsilon(t) = \begin{cases} \dfrac{t}{\tau} e^{1 - \frac{t}{\tau}} & t > 0 \\ \\ 0 & t \leq 0 \end{cases} \tag{8}$$

where τ is the time decay constant. The firing time t_j^α is defined as

$$t_j^\alpha = \min\{t \mid x_j(t) = \theta \wedge x_j'(t) > 0\} \tag{9}$$

2.2 Multi-Spiking Neuron

In the spike response model (SRM) of multi-spiking neuron, the internal state of the postsynaptic neuron j at time t is modeled as

$$x_j(t) = \sum_{i=1}^{n} \sum_{k=1}^{K} \sum_{g=1}^{G_i} w_{ij}^k \epsilon(t - t_i^{(g)} - d_{ij}^k) + \rho(t - t_j^{(f)}) \tag{10}$$

where $\{t_i^{(1)}, t_i^{(2)}, \ldots, t_i^{(G_i)}\}$ are spikes produced by the presynaptic neuron i and $t_j^{(f)}$ is the timing of the most recent, the fth, output spike from neuron j prior to time t. The refractoriness function ρ is chosen as

$$\rho(t) = \begin{cases} -2\theta e^{-\frac{t}{\tau_s}} & t > 0 \\ 0 & t \leq 0 \end{cases} \tag{11}$$

where θ is the neuron threshold value and τ_s is the time decay constant. The firing time t_j^α is expressed as

$$t_j^\alpha = \{t \mid x_j(t) = \theta \wedge x_j'(t) > 0\} \tag{12}$$

From the description above, we can see that the spiking neural networks is a special case of the multi-spiking neural networks. Therefore, we prove the key equality (4) for multiple spiking neural networks.

3   Error-Backpropagation Learning Algorithm for Multiple Spiking Neural Networks

For the sake of brevity, let us consider one layer multiple spiking neural networks composed of I input neurons and J output neurons. The error function is defined as

$$E(W) = \frac{1}{2} \sum_{j=1}^{J} \sum_{h=1}^{H_j} (t_j^{d(h)} - t_j^{\alpha(h)})^2 \tag{13}$$

where

$$W = \begin{pmatrix} w_{11}^1 \cdots w_{11}^K \cdots w_{I1}^1 \cdots w_{I1}^K \\ w_{12}^1 \cdots w_{12}^K \cdots w_{I2}^1 \cdots w_{I2}^K \\ \cdots\cdots\cdots\cdots \\ w_{1J}^1 \cdots w_{1J}^K \cdots w_{IJ}^1 \cdots w_{IJ}^K \end{pmatrix} , \quad t_j^{d(h)} \text{ represents the } h\text{th desired fir-}$$

ing time, and $t_j^{\alpha(f)}$ is the hth actual firing time, namely an element of the set $\{t \mid x_j(t) = \theta \wedge x_j'(t) > 0\}$. A simple and widely used supervised learning rule for the weight W is the gradient descent method: Update each component w_{ij}^k of the present weight W iteratively by adding the following increment

$$\triangle w_{ij}^k = -\eta \frac{\partial E}{\partial w_{ij}^k} = -\eta \sum_{h=1}^{H_j} \frac{\partial E}{\partial t_j^{\alpha(h)}} \frac{\partial t_j^{\alpha(h)}}{\partial w_{ij}^k} \tag{14}$$

where η is the learning rate.

4   Main Result

For the sake of simple representation, let the number of synaptic delays be zero, and let the number of output neurons be one. Considering that the state variable depends on not only the time t but also the weight W, we rewrite the state variable

$$x(W,t) = \sum_{i=1}^{n} \sum_{g=1}^{G_i} w_i \epsilon(t - t_i^{(g)}) + \rho(t - t^{(f)}) \tag{15}$$

and the firing time

$$t^\alpha = \{t \mid x(W,t) = \theta \wedge x'_t(W,t) > 0\} \tag{16}$$

where $W = (w_1, w_2, \ldots, w_n)^T \in \mathbf{R}^n$, $t \in \mathbf{R}$. Let $D \subset \mathbf{R}^{n+1}$ be an open set, we define a function

$$F(W,t) = x(W,t) - \theta \tag{17}$$

on D and rewrite the firing time

$$t^\alpha = \{t \mid F(W,t) = 0 \wedge F'_t(W,t) > 0\} \tag{18}$$

If a spiking neuron can fire, then there is a point $(W^0, t^\alpha) \in \mathbf{R}^{n+1}$ such that

$$F(W^0, t^\alpha) = x(W^0, t^\alpha) - \theta = 0 \tag{19}$$

$$\frac{\partial F(W^0, t^\alpha)}{\partial t} = \frac{\partial x(W^0, t^\alpha)}{\partial t} > 0 \tag{20}$$

Furthermore, if the condition $F \in C^{(1)}(D)$ holds, by the implicit function theorem, then there is a neighbourhood $A \times B$ of the point (W^0, t^α), where B is an open interval, such that:

(a) for all $W \in A$, the equality $F(W,t) = 0$ has a unique solution in B, denoted by $f(W)$;
(b) $t^\alpha = f(W^0)$;
(c) $f \in C^{(1)}(A)$;
(d) when $W \in A$, $\partial f(W)/\partial w_i = -\frac{\partial F(W,t)}{\partial w_i} / \frac{\partial F(W,t)}{\partial t}$ where $t = f(W) \in B$.

Hence, we have

$$\frac{\partial t}{\partial w_i} = -(\frac{\partial x(W,t)}{\partial t})^{-1} \frac{\partial x(W,t)}{\partial w_i}, \quad t \in B \tag{21}$$

In particular, we have

$$\frac{\partial t^\alpha}{\partial w_i} = -(\frac{\partial x(W^0, t^\alpha)}{\partial t})^{-1} \frac{\partial x(W^0, t^\alpha)}{\partial w_i} \tag{22}$$

which is what we want to prove in this section. Therefore, the task needed to complete is to prove the condition $F \in C^{(1)}(D)$.

Theorem 1. *Let $D \subset \mathbf{R}^{n+1}$ be an open set. Let the component $w_i (i = 1, 2, \ldots, n)$ of the W be uniformly bounded and $t \in (0, T)$, where T is a constant. Then the function $F(W,t) \in C^{(1)}(D)$.*

Proof. To prove the function $F \in C^{(1)}(D)$, namely to prove that each partial derivative of the function F is continuous on D.

(1) We first prove that

$$\frac{\partial F(W,t)}{\partial w_i} = \sum_{g=1}^{G_i} \epsilon(t - t_i^{(g)}) \tag{23}$$

is continuous on D. By the function $\epsilon(t - t_i) = (t - t_i)/\tau \cdot e^{(1-(t-t_i)/\tau)}$ is continuous on the point $t^1 \in (0,T)$, we have that its finite sum function $\sum_{g=1}^{G_i} \epsilon(t - t_i^{(g)})$ is continuous on the point $t^1 \in (0,T)$. Therefore, for every $\xi > 0$ there is a $\delta_0 > 0$, such that

$$| \sum_{g=1}^{G_i} (\epsilon(t - t_i^{(g)}) - \epsilon(t^1 - t_i^{(g)})) | < \xi \tag{24}$$

for all points t for which$| t - t^1 | < \delta_0$. Let $V = (W, t)^T$, by $| t - t^1 | < \|V - V^1\|$ and (24), we have that for the $\xi > 0$ there is a $\delta = \delta_0 > 0$ such that

$$| \frac{\partial F(W,t)}{\partial w_i} - \frac{\partial F(W^1, t^1)}{\partial w_i} | \\ = | \sum_{g=1}^{G_i} (\epsilon(t - t_i^{(g)}) - \epsilon(t^1 - t_i^{(g)})) | < \xi \tag{25}$$

for all points V for which $\|V - V^1\| < \delta$. Since the point (W^1, t^1) is arbitrary, $\partial F(W,t)/\partial w_i$ is continuous on D.

(2) Now we prove that

$$\frac{\partial F(W,t)}{\partial t} = \sum_{i=1}^{n} \sum_{g=1}^{G_i} w_i \epsilon'(t - t_i^{(g)}) + \rho'(t - t^{(f)}) \tag{26}$$

is continuous on D. By $\epsilon'(t - t_i) = 1/\tau \cdot (1 - (t - t_i)/\tau) \cdot e^{(1-(t-t_i)/\tau)}$ is continuous on the interval $(0,T)$, we have $M_i = \sup\limits_{t \in (o,T), 1 \leq g \leq G_i} \epsilon'(t - t_i^{(g)})$. Let $M = \max\limits_i M_i$ and $G = \max\limits_i G_i$. Let $\epsilon_l'(t) = (\epsilon'(t - t_1^{(l)}), \ldots, \epsilon'(t - t_n^{(l)}))^T$, where $1 \leq l \leq G$ and if $l > G_i$ let $\epsilon'(t - t_i^{(l)}) = 0$. Then, for an arbitrary point (W^2, t^2) in D, we have

$$| \frac{\partial F(W,t)}{\partial t} - \frac{\partial F(W^2, t)}{\partial t} | = | \sum_{l=1}^{G} (W - W^2) \cdot \epsilon_l'(t) | \\ \leq \sum_{l=1}^{G} \|W - W^2\| \|\epsilon_l'(t)\| \tag{27}$$

Therefore, for every $\xi_1 > 0$ there is a $\delta_1 = \xi_1/(\sqrt{n}MG) > 0$, such that

$$| \frac{\partial F(W,t)}{\partial t} - \frac{\partial F(W^2, t)}{\partial t} | < \delta_1 \sqrt{n}MG = \xi_1 \tag{28}$$

for all points W for which $\|W - W^2\| < \delta_1$. The w_i is uniformly bounded on D, namely existing a constant N such that $| w_i | < N$ $(i = 1, 2, \ldots, n)$. By the function $\epsilon'(t - t_i)$ is continuous on the point $t^2 \in (0, T)$, we have that its finite sum function $\sum_{g=1}^{G_i} \epsilon'(t - t_i^{(g)})$ is continuous on the point $t^2 \in (0, T)$. Thus for the $\xi_2 = \xi_1/(nN + 1) > 0$, there is a $\delta_2 > 0$, such that

$$| \sum_{g=1}^{G_i} (\epsilon'(t - t_i^{(g)}) - \epsilon'(t^2 - t_i^{(g)})) | < \xi_2 \qquad (29)$$

for all points t for which $| t - t^2 | < \delta_2$. By $\rho'(t) = 2\theta/\tau_s e^{-t/\tau_s}$ is continuous on the point t^2, we have that for the $\xi_2 = \xi_1/(nN + 1) > 0$, there is a $\delta_3 > 0$, such that

$$| \rho'(t - t^{(f)}) - \rho'(t^2 - t^{(f)}) | < \xi_2 \qquad (30)$$

for all points t for which $| t - t^2 | < \delta_3$. Therefore, for the $\xi_1 > 0$, there is a $\delta_4 = min\{\delta_2, \delta_3\}$, such that

$$| \frac{\partial F(W^2, t)}{\partial t} - \frac{\partial F(W^2, t^2)}{\partial t} |$$

$$= | \sum_{i=1}^{n} \sum_{g=1}^{G_i} w_i^2 (\epsilon'(t - t_i^{(g)}) - \epsilon'(t^2 - t_i^{(g)}))$$

$$+ \rho'(t - t^{(f)}) - \rho'(t^2 - t^{(f)}) | \qquad (31)$$

$$\leq \sum_{i=1}^{n} | w_i^2 | | \sum_{g=1}^{G_i} (\epsilon'(t - t_i^{(g)}) - \epsilon'(t^2 - t_i^{(g)})) |$$

$$+ | \rho'(t - t^{(f)}) - \rho'(t^2 - t^{(f)}) |$$

$$< nN\xi_2 + \xi_2 = \xi_1$$

for all points t for which $| t - t^2 | < \delta_4$. Let $V = (W, t)^T$ and $\xi' = 2\xi_1$. By $| t - t^2 | < \|V - V^2\|$, $\|W - W^2\| < \|V - V^2\|$, (28) and (31), we have that for the $\xi' > 0$, there is a $\delta' = min\{\delta_1, \delta_4\}$, such that

$$| \frac{\partial F(W, t)}{\partial t} - \frac{\partial F(W^2, t^2)}{\partial t} |$$

$$\leq | \frac{\partial F(W, t)}{\partial t} - \frac{\partial F(W^2, t)}{\partial t} | + | \frac{\partial F(W^2, t)}{\partial t} - \frac{\partial F(W^2, t^2)}{\partial t} | \qquad (32)$$

$$< \xi_1 + \xi_1 = \xi'$$

for all points V for which $\|V - V^2\| < \delta'$. Since the point (W^2, t^2) is arbitrary, $\partial F(W, t)/\partial t$ is continuous on D.

Therefore, the function $F \in C^{(1)}(D)$.

References

1. Maass, W.: Networks of spiking neurons: the third generation of neural network. Neural Networks **10**, 1659–1671 (1997)
2. Voutsas, K., Adamy, J.: A biologically inspired spiking neural network for sound source lateralization. IEEE Trans. Neural Netw. **18**(6), 1785–1799 (2007)

3. Wysoski, S.G., Benuscova, L., Kasabov, N.: Fast and adaptive network of spiking neurons for multi-view visual pattern recongnition. Neurocomputing **71**, 2563–2575 (2008)
4. Ghosh-Dastidar, S., Adeli, H.: A new supervised learning algorithm for multiple spiking neural networks with application in epilepsy and seizure detection. Neural Networks **22**, 1419–1431 (2009)
5. Liu, J., Perez-Gonzalez, D., Rees, A., Erwin, H., Wermter, S.: A biologically inspired spiking neural network model of the auditory midbrain for sound source localisation. Neurocomputing **74**(1–3), 129–139 (2010)
6. Wysoski, S.G., Benuskova, L., Kasabov, N.: Evolving spiking neural networks for audiovisual information processing. Neural Networks **23**, 819–835 (2010)
7. Wade, J., McDaid, L., Santos, J., Sayers, H.: SWAT: A spiking neural network training algorithm for classification problems. IEEE Trans. Neural Netw. **21**(11), 1817–1830 (2010)
8. Wall, J.A., McDaid, L.J., Maguire, L.P., McGinnity, T.M.: Spiking neural network model of sound localization using the interaural intensity difference. IEEE Transactions on Neural Network and Learning System **23**(4), 574–586 (2012)
9. Bohte, S.M., Kok, J.N., La Poutre, J.A.: Error-backpropagation in temporally encoded networks of spiking neurons. Neurocomputing **48**(1–4), 17–37 (2002)
10. Ghosh-Dastidar, S., Adeli, H.: Improved spiking neural networks for EEG classification and epilepsy and seizure detection. Integrated Computer-Aided Engineering **14**, 187–212 (2007)
11. Booij, O., Nguyen, H.T.: A gradient descent rule for spiking neurons emitting multiple spikes. Inf. Process. Lett. **95**(6), 552–558 (2005)
12. McKennoch, S., Liu, D., Bushnell, L.G.: Fast modifications of the SpikeProp algorithm. In: Proceedings of the International Joint Conference on Neural Networks, pp. 3970–3977 (2006)
13. Wu, Q.X., McGinnity, T.M., Maguire, L.P., Glackin, B., Belatreche, A.: Learning under weight constraints in networks of temporal encoding spiking neurons. Neurocomputing **69**, 1912–1922 (2006)
14. Xu, Y., Zeng, X.Q., Han, L.X., Yang, J.: A supervised multi-spike learning algorithm based on gradient descent for spiking neural networks. Neural Networks **43**, 99–113 (2013)

Anti-Synchronization Control
for Memristor-Based Recurrent Neural Networks

Ning Li[1]([envelope]), Jinde Cao[1,2], and Mengzhe Zhou[1]

[1] Department of Mathematics, Research Center for Complex Systems and Network
Sciences, Southeast University, Nanjing 210096, China
ningning12345@126.com
[2] Department of Mathematics, Faculty of Science, King Abdulaziz University,
Jeddah 21589, Saudi Arabia

Abstract. In this paper, we consider the simplified memristor-based
neural networks with time-varying delay, under the framework of Filippov's solution and differential inclusion theory, by structuring novel
Lyapunov functional and employing feedback control technique, adopt-
ing feedback controller, anti-synchronization criteria for memristor-based
neural networks with time-varying delay are derived, which depend
on the jumps parameter T_i, hence the proposed criteria are more general than existing reference. Finally, an example is provided to show the
effectiveness of theoretical result.

Keywords: Memristor-based neural networks · Time delay · Feedback
control · Anti-Synchronization

1 Introduction

Memristor [1,2], which is known as the fourth circuit element, has been applied to
neural networks [3,4] to imitate human's brain due to its memory and forgetting
ability. Recently, the dynamical behaviors of memritor-based neural networks
[5,6] have been extensively studied, especially, synchronization [7–9] or anti-
synchronization control problem. Different from traditional neural networks, the
model of memritor-based neural networks is differential equations with discontinuous right-hand side, this make research of this kinds of neural networks
more difficult. Anti-synchronization control [12,13] of memritor-based neural
networks [10,11] is very useful in practical application, and it has been applied
to image processing, secure communication, and so on. In this paper, we give a
new anti-synchronization criteria for memristor-based neural networks by using
new analysis method.

The rest of this paper is organized as follows: In section 2, the model formulation and some preliminaries are presented. In section 3, anti-synchronization
criteria for memristor-based neural network are derived by feedback control. An
example is given to demonstrate the validity of the proposed results in Section
4. Some conclusions are drawn in Section 5.

© Springer International Publishing Switzerland 2014
Z. Zeng et al. (Eds.): ISNN 2014, LNCS 8866, pp. 27–34, 2014.
DOI: 10.1007/978-3-319-12436-0_4

Notations. Throughout this paper, $\mathbb{R}$ denotes the set of real numbers, $\mathbb{R}^n$ denotes the n-dimensional Euclidean space. For $\tau > 0$, $C([-\tau, 0]; \mathbb{R}^n)$ denotes the family of continuous function φ from $[-\tau, 0]$ to $\mathbb{R}^n$ with the norm $\|\varphi\| = \sup_{-\tau \le s \le 0} \max_{1 \le i \le n} |\varphi_i(s)|$. Solution of memristor networks is considered in Filippov's sense, $[\cdot, \cdot]$ represents the interval. $co(Q)$ denotes the closure of the convex hull of Q, and function $sign(\cdot)$ denotes sign function.

2   Model Description and Preliminaries

According to the current-voltage characteristic of memristor, we consider simplified memristor-based recurrent neural networks as follows:

$$\dot{x}_i(t) = -c_i(x_i(t)) + \sum_{j=1}^{n} a_{ij}(x_i(t))f_j(x_j(t)) + \sum_{j=1}^{n} b_{ij}(x_i(t))g_j(x_j(t - \tau(t)))$$

$$+ I_i, \qquad t \ge 0, \quad i = 1, 2, \cdots, n, \tag{1}$$

where $f_j(\cdot)$ and $g_j(\cdot)$ are activation functions, $\tau(t)$ is time-varying transmission delay, it satisfies $0 < \tau(t) \le \tau$, I_i is the external input, and the connection weight coefficients satisfy:

$$a_{ij}(x_i(t)) = \begin{cases} \acute{a}_{ij}, & |x_i(t)| \le T_i, \\ \grave{a}_{ij}, & |x_i(t)| > T_i, \end{cases} \qquad b_{ij}(x_i(t)) = \begin{cases} \acute{b}_{ij}, & |x_i(t)| \le T_i, \\ \grave{b}_{ij}, & |x_i(t)| > T_i. \end{cases} \tag{2}$$

in which switching jumps $T_i > 0$, $\acute{a}_{ij}$, $\grave{a}_{ij}$, $\acute{b}_{ij}$, $\grave{b}_{ij}$, $i, j = 1, 2, \cdots, n$, are constants;

Remark 1 Actually, the memristive neural networks (1) with different nonlinearity of memductance functions evolve into different forms: a state-dependent switched system or a state-dependent continuous system, in this paper, we consider the state-dependent switched case.

Throughout this paper, we make the following assumptions:

$(\mathcal{H}_1)$: Behaved function $c_i(x)$ is odd function and satisfies $\dot{c}_i(x) \ge \beta_i > 0$, $i = 1, 2, \ldots, n$.

$(\mathcal{H}_2)$: The function f_j, g_j, are odd functions, and satisfy Lipschitz conditions:

$$|f_j(x) - f_j(y)| \le h_j|x - y|, \quad |g_j(x) - g_j(y)| \le k_j|x - y|, \qquad \forall x, y \in \mathbb{R}$$

where $h_j, k_j (j = 1, 2, \ldots, n)$ are positive constants.

$(\mathcal{H}_3)$: For any $x \in \mathbb{R}$, there exist positive constants M_j, N_j, such that

$$|f_j(x)| \le M_j, \quad |g_j(x)| \le N_j.$$

The model of memristor-based recurrent networks (1) are discontinuous, hence, its solutions are considered in Filippov's sense.

Definition 1. ([14]) For differential system $\frac{dx}{dt} = f(t, x)$, where $f(t, x)$ is discontinuous in x. The set-valued map of $f(t, x)$ is defined as

$$F(t, x) = \bigcap_{\delta > 0} \bigcap_{\mu(N) = 0} co[f(B(x, \delta) \backslash N)],$$

where $B(x, \delta) = \{y : \|y-x\| \leq \delta\}$ is the ball of center x with radius δ; intersection is taken over all sets N of measure zero and over all $\delta > 0$; $\mu(N)$ is Lebesgue measure of set N.

For convenience, we define some notations: $\underline{a}_{ij} = \min\{\acute{a}_{ij}, \grave{a}_{ij}\}$, $\overline{a}_{ij} = \max\{\acute{a}_{ij}, \grave{a}_{ij}\}$, $\underline{b}_{ij} = \min\{\acute{b}_{ij}, \grave{b}_{ij}\}$, $\overline{b}_{ij} = \max\{\acute{b}_{ij}, \grave{b}_{ij}\}$, $a_{ij}^+ = \max\{|\underline{a}_{ij}|, |\overline{a}_{ij}|\}$, $b_{ij}^+ = \max\{|\underline{b}_{ij}|, |\overline{b}_{ij}|\}$.

By applying the above theory of differential inclusion, the memristor-based networks (1) can be written as the following: there exist $\gamma_{ij}(t) \in co(\underline{a}_{ij}, \overline{a}_{ij})$, $\delta_{ij}(t) \in co(\underline{b}_{ij}, \overline{b}_{ij})$, such that

$$\dot{x}_i(t) = -c_i(x_i(t)) + \sum_{j=1}^{n} \gamma_{ij}(t) f_j(x_j(t)) + \sum_{j=1}^{n} \delta_{ij}(t)$$
$$g_j(x_j(t - \tau(t))) + I_i, \quad t \geq 0. \tag{3}$$

The initial value associated with system (3) is $\phi(s) = (\phi_1(s), \phi_2(s), \cdots, \phi_n(s))^T \in \mathcal{C}([-\tau, 0]; \mathbb{R}^n)$. Similar to (3), the response system can be rewritten as: there exist $\bar{\gamma}_{ij}(t) \in co(\underline{a}_{ij}, \overline{a}_{ij})$, $\bar{\delta}_{ij}(t) \in co(\underline{b}_{ij}, \overline{b}_{ij})$, such that

$$\dot{y}_i(t) = -c_i(y_i(t)) + \sum_{j=1}^{n} \bar{\gamma}_{ij}(t) f_j(y_j(t)) + \sum_{j=1}^{n} \bar{\delta}_{ij}(t)$$
$$g_j(y_j(t - \tau(t))) + I_i + u_i(t), \quad t \geq 0. \tag{4}$$

The initial value associated with system (4) is $\varphi(s) = (\varphi_1(s), \varphi_2(s), \cdots, \varphi_n(s))^T \in \mathcal{C}([-\tau, 0]; \mathbb{R}^n)$.

Define the anti-synchronization error as $e_i(t) = y_i(t) + x_i(t)$, from (3) and (4), we have

$$\dot{e}_i(t) = -\tilde{c}_i(e_i(t)) + \sum_{j=1}^{n} \gamma_{ij}(t) \tilde{f}_j(e_j(t)) + \sum_{j=1}^{n} \delta_{ij}(t) \tilde{g}_j(e_j(t - \tau(t))) +$$
$$\sum_{j=1}^{n} (\bar{\gamma}_{ij}(t) - \gamma_{ij}(t)) f_j(y_j(t)) + \sum_{j=1}^{n} (\bar{\delta}_{ij}(t) - \delta_{ij}(t)) g_j(y_j(t - \tau(t))) + \tag{5}$$
$$2I_i + u_i(t),$$

where $\tilde{c}_i(e_i(t)) = c_i(e_i(t) - x_i(t)) + c_i(x_i(t))$; $\tilde{f}_i(e_i(t)) = f_i(e_i(t) - x_i(t)) + f_i(x_i(t))$; $\tilde{g}_i(e_i(t - \tau(t))) = g_i(e_i(t - \tau(t)) - x_i(t - \tau(t))) + g_i(x_i(t - \tau(t)))$. The feedback controller $u_i(t) = -p_i e_i(t) - \eta_i \mathrm{sign} e_i(t) - 2I_i$, and $p_i > 0$, $\eta_i > 0$ are control gains to be determined.

Definition 2. *Systems (3) and (4) are said to be exponentially anti-synchronized if there exist positive scalars $\alpha > 0$ and $\beta > 0$ such that*

$$|e_i(t)| \leq \beta \|\phi + \varphi\|^{-\alpha t}, \quad t \geq 0.$$

Lemma 1. ([15]) For $-\infty < a < b \leq +\infty$, let $\psi_i(t) \in \mathcal{C}([a,b];\mathbb{R})$, $(i = 1, 2, \cdots, n)$ satisfy the following integral delay inequality:

$$
\begin{cases}
\psi_i(t) \leq e^{-\check{\chi}_i(t-a)}\psi_i(a) + \int_a^t e^{-\check{\chi}_i(t-s)}[\sum_{j=1}^n \tilde{\chi}_{ij}\psi_j(s) + \sum_{j=1}^n \hat{\chi}_{ij}\psi_j(s - \tau(s))]ds, \\
\quad t \in [a,b), \\
\psi_i(a+s) = \sigma_i(s) \quad s \in [-\tau, 0],
\end{cases}
\tag{6}
$$

where $\check{\chi}_i, \tilde{\chi}_{ij}$ and $\hat{\chi}_{ij}$, $(i,j = 1,2,\cdots,n)$ are positive constants. Assume that

$$
-\check{\chi}_i + \sum_{j=1}^n (\tilde{\chi}_{ij} + \hat{\chi}_{ij}) < 0, and \quad \psi_i(t) \leq M\|\sigma\|^2, \quad t \in [a - \tau, a], \quad i = 1, 2, \cdots, n
$$

then $\psi_i(t) \leq M\|\sigma\|^2$, $t \in (a, b)$, where $M > 0$ is a positive constant.

3   Feedback Control for Memristor-Based Neural Network

Theorem 1. Under Assumptions $(\mathcal{H}_1) - (\mathcal{H}_3)$, if there exists a constant $\alpha > 0$, such that the following inequalities hold:

$$
a) \quad \eta_i > \sum_{j=1}^n |\acute{a}_{ij} - \grave{a}_{ij}|h_j T_j + \sum_{j=1}^n |\acute{b}_{ij} - \grave{b}_{ij}|N_j;
$$

$$
b) \quad -(p_i + \beta_i - \alpha) + \sum_{j=1}^n (|\acute{a}_{ij}|h_j + |\acute{b}_{ij}|k_j e^{\alpha\tau}) < 0;
$$

$$
c) \quad -(p_i + \beta_i - \alpha) + \sum_{j=1}^n (|\grave{a}_{ij}|h_j + |\grave{b}_{ij}|k_j e^{\alpha\tau}) < 0;
$$

then the response system (4) will globally exponentially anti-synchronize with the drive system (3) under the feedback controller.

Proof: Define a Lyapunov functional by

$$
V_i = e^{\alpha t}|e_i(t)|, \quad i = 1, 2, \cdots, n.
$$

According the special character of memristive neural networks, we will divide the proof into four steps.

Step 1. If $|x_i(t)| \leq T_i$, $|y_i(t)| \leq T_i$ at time t, then from system (3) and (4), the error system can be written as

$$
\dot{e}_i(t) = -\tilde{c}_i(e_i(t)) + \sum_{j=1}^n \acute{a}_{ij}\tilde{f}_j(e_j(t)) + \sum_{j=1}^n \acute{b}_{ij}\tilde{g}_j(e_j(t - \tau(t))) -
$$
$$
p_i e_i(t) - \eta_i sign e_i(t), \quad t \geq 0, \quad i = 1, 2, \cdots, n.
\tag{7}
$$

Calculating the time derivative of $V(t)$, we can get

$$\dot{V}_i(t) \le \alpha e^{\alpha t}|e_i(t)| + e^{\alpha t}(-\beta_i - p_i)|e_i(t)| + \sum_{j=1}^{n}|\acute{a}_{ij}|h_j e^{\alpha t}|e_j(t)| +$$

$$\sum_{j=1}^{n}|\acute{b}_{ij}|k_j e^{\alpha \tau}e^{\alpha(t-\tau(t))}|e_j(t-\tau(t))|$$

$$\le (\alpha - \beta_i - p_i)V_i(t) + \sum_{j=1}^{n}|\acute{a}_{ij}|h_j V_j(t) + \sum_{j=1}^{n}|\acute{b}_{ij}|k_j e^{\alpha \tau}V_j(t-\tau(t)).$$

Let $-\check{\chi}_i = \alpha - \beta_i - p_i$, $\tilde{\chi}_{ij} = |\acute{a}_{ij}|h_j$, $\hat{\chi}_{ij} = |\acute{b}_{ij}|k_j e^{\alpha \tau}$, we have

$$\dot{V}_i(t) \le -\check{\chi}_i V_i(t) + \sum_{j=1}^{n}\tilde{\chi}_{ij}V_j(t) + \sum_{j=1}^{n}\hat{\chi}_{ij}V_j(t-\tau(t))$$

Since

$$|e_i(t)| \le \|\varphi + \phi\| \le \|\varphi + \phi\|e^{-\alpha t}, \quad t \le 0, \quad i = 1, 2, \cdots, n.$$

One has

$$V_i(t) \le \|\varphi + \phi\|, \quad t \le 0.$$

From condition (b) and Lemma 1, it follows that $|e_i(t)| \le \|\varphi + \phi\|e^{-\alpha t}$, $t \ge 0$.

Step 2. If $|x_i(t)| > T_i$, $|y_i(t)| > T_i$ at time t, then the error system can be written as

$$\dot{e}_i(t) = -\tilde{c}_i(e_i(t)) + \sum_{j=1}^{n}\grave{a}_{ij}\tilde{f}_j(e_j(t)) + \sum_{j=1}^{n}\grave{b}_{ij}\tilde{g}_j(e_j(t-\tau(t))) - \tag{8}$$

$$p_i e_i(t) - \eta_i sign e_i(t), \quad t \ge 0, \quad i = 1, 2, \cdots, n.$$

Similar to the proof of step 1, let $-\check{\chi}_i = \alpha - \beta_i - p_i, \tilde{\chi}_{ij} = |\grave{a}_{ij}|h_j$, $\hat{\chi}_{ij} = |\grave{b}_{ij}|k_j e^{\alpha \tau}$, then we have

$$\dot{V}_i(t) \le -\check{\chi}_i V_i(t) + \sum_{j=1}^{n}\tilde{\chi}_{ij}V_j(t) + \sum_{j=1}^{n}\hat{\chi}_{ij}V_j(t-\tau(t)).$$

From condition (c) and Lemma 1, it follows that $|e_i(t)| \le \|\varphi + \phi\|e^{-\alpha t}$, $t \ge 0$.

Step 3. If $|x_i(t)| \le T_i$, $|y_i(t)| > T_i$ at time t, the error system changes as

$$\dot{e}_i(t) = -\tilde{c}_i(e_i(t)) + \sum_{j=1}^{n}\grave{a}_{ij}\tilde{f}_j(e_j(t)) + \sum_{j=1}^{n}\grave{b}_{ij}\tilde{g}_j(e_j(t-\tau(t))) +$$

$$\sum_{j=1}^{n}(\grave{a}_{ij} - \acute{a}_{ij})f_j(x_j(t)) + \sum_{j=1}^{n}(\grave{b}_{ij} - \acute{b}_{ij})g_j(x_j(t-\tau(t))) \tag{9}$$

$$- p_i e_i(t) - \eta_i sign e_i(t), \quad t \ge 0, \quad i = 1, 2, \cdots, n.$$

Calculating the time derivative of $V(t)$, one obtains:

$$\dot{V}_i(t) \le \alpha e^{\alpha t}|e_i(t)| + e^{\alpha t}[-\beta_i|e_i(t)| + \sum_{j=1}^{n} |\grave{a}_{ij}|h_j|e_j(t)|$$

$$+ \sum_{j=1}^{n} |\grave{b}_{ij}|k_j|e_j(t - \tau(t))| + \sum_{j=1}^{n} |\grave{a}_{ij} - \acute{a}_{ij}|h_j|x_j(t)|$$

$$+ \sum_{j=1}^{n} |\grave{b}_{ij} - \acute{b}_{ij}|N_j - p_i|e_i(t)| - \eta_i].$$

Note that $|x_i(t)| \le T_i$ and condition (a), one has

$$\dot{V}_i(t) \le (\alpha - \beta_i - p_i)V_i(t) + \sum_{j=1}^{n} |\grave{a}_{ij}|h_j V_j(t) + \sum_{j=1}^{n} |\grave{b}_{ij}|k_j e^{\alpha \tau} V_j(t - \tau(t)).$$

Let $-\check{\chi}_i = \alpha - \beta_i - p_i$, $\tilde{\chi}_{ij} = |\grave{a}_{ij}|h_j$, $\hat{\chi}_{ij} = |\grave{b}_{ij}|k_j e^{\alpha \tau}$, we can get

$$\dot{V}_i(t) \le -\check{\chi}_i V_i(t) + \sum_{j=1}^{n} \tilde{\chi}_{ij} V_j(t) + \sum_{j=1}^{n} \hat{\chi}_{ij} V_j(t - \tau(t)).$$

In view of Lemma 1, it follows that $|e_i(t)| \le \|\varphi + \phi\|e^{-\alpha t}$, $t \ge 0$.

Step 4. If $|x_i(t)| > T_i$, $|y_i(t)| \le T_i$ at time t, correspondingly, the error system can be written as

$$\dot{e}_i(t) = -\tilde{c}_i(e_i(t)) + \sum_{j=1}^{n} \grave{a}_{ij}\tilde{f}_j(e_j(t)) + \sum_{j=1}^{n} \grave{b}_{ij}\tilde{g}_j(e_j(t - \tau(t))) +$$

$$\sum_{j=1}^{n}(\acute{a}_{ij} - \grave{a}_{ij})f_j(y_j(t)) + \sum_{j=1}^{n}(\acute{b}_{ij} - \grave{b}_{ij})g_j(y_j(t - \tau(t))) \quad (10)$$

$$- p_i e_i(t) - \eta_i sign e_i(t), \quad t \ge 0, \quad i = 1, 2, \cdots, n.$$

Note that $|y_i(t)| \le T_i$, similar to step 3, let $-\check{\chi}_i = \alpha - \beta_i - p_i$, $\tilde{\chi}_{ij} = |\grave{a}_{ij}|h_j$, $\hat{\chi}_{ij} = |\grave{b}_{ij}|k_j e^{\alpha \tau}$, one can obtain $|e_i(t)| \le \|\varphi + \phi\|e^{-\alpha t}$, $t \ge 0$.

Based on the above analysis, one always has $|e_i(t)| \le \|\varphi + \phi\|e^{-\alpha t}$, for $t \ge 0$, if the conditions (a)-(c) in Theorem 1 hold, from the Definition 2, the proof is completed.

4   An Illustrative Example

Example 1. onsider the following two-order mermristor-based recurrent neural networks with time delay:

$$\begin{cases} \dot{x}_1(t) = -c_1(x_1(t)) + a_{11}(x_1(t))f_1(x_1(t)) + a_{12}(x_1(t))f_2(x_2(t)) \\ \quad + b_{11}(x_1(t))g_1(x_1(t - \tau(t))) + b_{12}(x_1(t))g_2(x_2(t - \tau(t))) + I_1, \\ \dot{x}_2(t) = -c_2(x_2(t)) + a_{21}(x_2(t))f_1(x_1(t)) + a_{22}(x_2(t))f_2(x_2(t)) \\ \quad + b_{21}(x_2(t))g_1(x_1(t - \tau(t))) + b_{22}(x_2(t))g_2(x_2(t - \tau(t))) + I_2, \end{cases}$$

$$(11)$$

where

$$a_{11}(x_1(t)) = \begin{cases} 1.78, & |x_1(t)| \leq 1, \\ 1.68, & |x_1(t)| > 1, \end{cases} \qquad a_{12}(x_1(t)) = \begin{cases} 20, & |x_1(t)| \leq 1, \\ 19.5, & |x_1(t)| > 1, \end{cases}$$

$$a_{21}(x_2(t)) = \begin{cases} 0.09, & |x_2(t)| \leq 1, \\ 0.1, & |x_2(t)| > 1, \end{cases} \qquad a_{22}(x_2(t)) = \begin{cases} 1.68, & |x_2(t)| \leq 1, \\ 1.78, & |x_2(t)| > 1, \end{cases}$$

$$b_{11}(x_1(t)) = \begin{cases} -1.5, & |x_1(t)| \leq 1, \\ -1.4, & |x_1(t)| > 1, \end{cases} \qquad b_{12}(x_1(t)) = \begin{cases} 0.1, & |x_1(t)| \leq 1, \\ 0.09, & |x_1(t)| > 1, \end{cases}$$

$$b_{21}(x_2(t)) = \begin{cases} 0.09 & |x_2(t)| \leq 1, \\ 0.1, & |x_2(t)| > 1, \end{cases} \qquad b_{22}(x_2(t)) = \begin{cases} -1.5, & |x_2(t)| \leq 1, \\ -1.4, & |x_2(t)| > 1, \end{cases}$$

and behaved functions $c_i(x_i(t)) = x_i(t)$, activation functions $f_i(x_i) = g_i(x_i) = \sin(x_i)$, $i = 1, 2$, $\tau(t) = 1$, $I_1 = I_2 = 0$, we choose the control gains $p_1 = 23$, $p_2 = 25$, $\eta_1 = \eta_2 = 1$, it is obvious that the conditions (a)-(c) in Theorem 1 hold. For numerical simulation, we take initial conditions as $(\phi_1(t), \phi_2(t)) = (0.1 * \sin(3t), 0.1 * \cos(2t))$, $(\varphi_1(t), \varphi_2(t)) = (0.5 * \sin(3t), 0.5 * \cos(2t))$. Fig. 1 shows the anti-synchronization error states converge to zero, this verifies the effectiveness of Theorem 1.

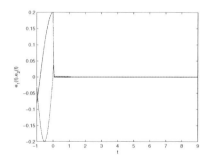

Fig. 1. The error state trajectories of variables $e_1(t)$, $e_2(t)$

5   Conclusion

In this paper, anti-synchronization control problem of memristor-based neural networks with time-varying delay has been studied, the analysis method is different with references [10, 11], four different cases have been discussed, the proposed results depend on jumps parameter T_i, which are more general than existing literature. Furthermore, an example is given to verify the proposed anti-synchronization criteria, the analysis method in this paper may open up a new view for memristor-based neural networks.

Acknowledgments. This work was jointly supported by the National Natural Science Foundation of China (NSFC) under Grants Nos. 61272530 and 11072059, and the Natural Science Foundation of Jiangsu Province of China under Grants No. BK2012741, and supported by the Fundamental Research Funds for the Central Universities, the JSPS Innovation Program under Grant CXLX13_075, and the Scientific Research Foundation of Graduate School of Southeast University YBJJ1407.

References

1. Chua, L.O.: Memristor-the missing circuit element. IEEE Transactions on Circuit Theory **18**(5), 507–519 (1971)
2. Tour, J.M., He, T.: The fourth element. Nature **453**(7191), 42–43 (2008)
3. Cao, J., Wang, J.: Global asymptotic stability of a general class of recurrent neural networks with time-varying delays. IEEE Transactions on Circuits and Systems I: Fundamental Theory and Applications **50**(1), 34–44 (2003)
4. Nara, S., Davis, P., Kawachi, M.: Chaotic memory dynamics in a recurrent neural network with cycle memories embedded by pseudo-inverse method. International Journal of Bifurcation and Chaos **5**(04), 1205–1212 (1995)
5. Pershin, Y.V., Ventra, D.M.: Experimental demonstration of associative memory with memristive neural networks. Neural Networks **23**(7), 881–886 (2010)
6. Wen, S., Zeng, Z., Huang, T.: Exponential stability analysis of memristor-based recurrent neural networks with time-varying delays. Neurocomputing **97**, 233–240 (2012)
7. Wu, A., Zeng, Z., Zhu, X.: Exponential synchronization of memristor-based recurrent neural networks with time delays. Neurocomputing **74**(17), 3043–3050 (2011)
8. Yang, X., Cao, J., Yu, W.: Exponential synchronization of memristive Cohen-Grossberg neural networks with mixed delays. Cognitive Neurodynamics **8**(3), 239–249 (2014)
9. Wang, W., Li, L., Peng, H.: Synchronization control of memristor-based recurrent neural networks with perturbations. Neural Networks **53**, 8–14 (2014)
10. Zhang, G., Shen, Y., Wang, L.: Global anti-synchronization of a class of chaotic memristive neural networks with time-varying delays. Neural Networks **46**, 1–8 (2013)
11. Wu, A., Zeng, Z.: Anti-synchronization control of a class of memristive recurrent neural networks. Communications in Nonlinear Science and Numerical Simulation **18**(2), 373–385 (2013)
12. Pan, Z., Basar, T.: Adaptive controller design for tracking and disturbance attenuation in parametric strict-feedback nonlinear systems. IEEE Transactions on Automatic Control **43**(8), 1066–1083 (1998)
13. Yang, T., Chua, L.O.: Impulsive stabilization for control and synchronization of chaotic systems: theory and application to secure communication. IEEE Transactions on Circuits and Systems I: Fundamental Theory and Applications **44**(10), 976–988 (1997)
14. Filippov, A.F.: Differential equations with discontinuous right-hand side. Matematicheskii Sbornik **93**(1), 99–128 (1960)
15. Zhao, H.Y., Zhang, Q.: Global impulsive exponential anti-synchronization of delayed chaotic neural networks. Neurocomputing **74**(4), 563–567 (2011)

Adaptive Pinning Synchronization of Coupled Inertial Delayed Neural Networks

Jianqiang Hu[1](✉), Jinde Cao[2], and Quanxin Cheng[2]

[1] School of Automation, Southeast University, Nanjing 210096, China
jqhuseu@seu.edu.cn
[2] Department of Mathematics, Research Center for Complex Systems
and Network Sciences, Southeast University, Nanjing 210096, China
jdcao@seu.edu.cn

Abstract. In this paper, adaptive pinning synchronization (i.e., leader-following synchronization) is considered for an array of linearly coupled inertial delayed neural network. By applying feedback control on a small fraction of network nodes with the dynamical feedback gains turning adaptively and combining the Lyapunov function method, an easy-to-verify sufficient condition is derived for globally asymptotically synchronization for the coupled network. Meanwhile, the coupling configuration matrix is not necessary to be symmetric or irreducible. Finally, an illustrative example is given to show the effectiveness of the obtained theoretical results.

Keywords: Inertial delayed neural networks · Asymptotical synchronization · Adaptive pinning control

1 Introduction

Synchronization of complex networks has received notable attentions in the past decade due to its potential applications in various fields, see [1–3]. As a special class of complex networks, neural networks have also been intensively investigated [4,5], where the network nodes are neurons and the network coupling is the connection weight matrix. Synchronization of coupled neural networks means multiple neural networks can achieve a common trajectory, such as a common equilibrium, limit cycle or chaotic trajectory. Based on Lyapunov functional methods, global synchronization was investigated in [6,7] for linearly and diffusively coupled identical delayed neural networks.

Inertial electronic neural networks with one or two neurons were considered in [8], where it was found that when the neuron couplings were of an inertial nature, the dynamics could be more complex compared with the simpler behavior displayed in the standard resistor-capacitor variety. The dynamical behaviors of a single delayed neuron model with inertial terms were investigated in [9]; bifurcation problems were investigated in [10,11] for low-order neural networks. While most of the published investigations in the literature concerning inertial

© Springer International Publishing Switzerland 2014
Z. Zeng et al. (Eds.): ISNN 2014, LNCS 8866, pp. 35–42, 2014.
DOI: 10.1007/978-3-319-12436-0_5

neural networks are always focusing on small-scale neural networks with only one or two neurons, the general network coupled by multiple neural networks are rarely seen in the literature.

Recently, the authors in [12] considered the stability and existence of periodic solutions for the general inertial BAM neural networks with time delays. Furthermore, stability analysis was carried out in [13] for the general inertial Cohen–Grossberg-type neural networks with time delays. While in [14], the stability of an inertial delayed neural network was investigated by matrix measure strategies and drive-response synchronization was considered as an application at the end of the paper. On the other hand, pinning synchronization of coupled neural networks has been investigated, such as the synchronization of a general weighted neural network with coupling delay was investigated in [15, 16] by adaptive pinning control. More studies concerning pinning synchronization of neural networks can be found in [17,18] and references cited therein. Inspired by the above discussions, this paper investigates the pinning synchronization of coupled inertial delayed neural networks.

2   Model Description and Preliminaries

Consider an array of linearly coupled inertial delayed neural networks consisted of N identical nodes with dynamics of the ith node described by the following equation:

$$\frac{\mathrm{d}^2 x_i(t)}{\mathrm{d}t^2} = -D\frac{\mathrm{d}x_i(t)}{\mathrm{d}t} - Cx_i(t) + Af(x_i(t)) + Bf(x_i(t - \tau(t))) + I(t)$$

$$+ c\sum_{j=1}^{N} G_{ij}\Gamma\left(\frac{\mathrm{d}x_j(t)}{\mathrm{d}t} + x_j(t)\right) + u_i(t), \quad i = 1, \ldots, N, \tag{1}$$

where $x_i(t) = (x_{i1}(t), \ldots, x_{in}(t))^T \in \mathbb{R}^n$ is the state vector of the ith neural network, and $u_i(t)$ is the control input imposed on the ith node; $D = \mathrm{diag}\{d_1, \ldots, d_n\}$, $C = \mathrm{diag}\{c_1, \ldots, c_n\}$ are constant positive definite matrices. $A = (a_{ij})_{n \times n}$ and $B = (b_{ij})_{n \times n}$ denote the connection weight matrix and the delayed connection weight matrix, respectively. The nonlinear function $f(x_i) = (f_1(x_{i1}), \ldots, f_n(x_{in}))^T$ is the activation function for the inertial neural network; and $I(t) = (I_1(t), \ldots, I_n(t))^T$ is the external input vector. The second derivative of $x_i(t)$ is called an inertial term of system (1). The positive constant c is the network coupling strength and Γ is the inner coupling matrix. $G = (G_{ij})_{N \times N}$ is the constant coupling configuration matrix defined to be diffusive: $G_{ij} \geq 0 (i \neq j)$ and $G_{ii} = -\sum_{j=1, j \neq i}^{N} G_{ij}$. The coupling matrix G is not required to be symmetric or irreducible.

The initial conditions associated with system (1) are given as $x_i(\omega) = \phi_i(\omega) \in \mathcal{C}^{(1)}([-\tau, 0], \mathbb{R}^n)$, $i = 1, \ldots, N$, where $\mathcal{C}^{(1)}([-\tau, 0], \mathbb{R}^n)$ denotes the set of all n-dimensional continuous differentiable functions defined on the interval $[-\tau, 0]$ with $\tau = \sup_{t \geq 0}\{\tau(t)\}$.

The isolated node of network (1) is given by the following inertial delayed neural network:

$$\frac{d^2 s(t)}{dt^2} = -D\frac{ds(t)}{dt} - Cs(t) + Af(s(t)) + Bf(s(t - \tau(t))) + I(t), \qquad (2)$$

where $s(t) = (s_1(t), \ldots, s_n(t)) \in \mathbb{R}^n$. The initial condition for system (2) is given as $s(\omega) = \varphi(\omega) \in \mathcal{C}^{(1)}([-\tau, 0], \mathbb{R}^n)$.

To proceed, the following assumptions and definition are given.

Assumption 1. *The activation functions $f_i(\cdot) : \mathbb{R} \to \mathbb{R}, 1 \leq i \leq n$ are bounded and satisfy Lipschitz condition, i.e., there exist constants F_i and M_i such that $|f_i(x) - f_i(y)| \leq F_i|x - y|$ and $|f_i(x)| \leq M_i$ for all $x, y \in \mathbb{R}$.*

Assumption 2. *The time delay $\tau(t) \geq 0$ in systems (1) and (2) is a bounded and differentiable function of time t satisfying $\dot{\tau}(t) \leq \rho < 1$ for all $t \geq 0$, where $\rho > 0$.*

Definition 1. *The coupled inertial neural network (1) is said to be globally asymptotically synchronizable to the goal trajectory $s(t)$ if the discriminant relations $\lim_{t \to \infty} \|x_i(t) - s(t)\| = 0$, $i = 1, 2 \ldots, N$ hold for all initial functions.*

3   Main Results

In this section, we will investigate the global synchronization of the coupled inertial neural network by adaptive pinning control. The feedback injections are only placed on a small fraction of the total network nodes and the feedback gains are turned adaptively.

By letting the synchronization error $e_i(t) = x_i(t) - s(t)$, one can derive the following error system:

$$\frac{d^2 e_i(t)}{dt^2} = -D\frac{de_i(t)}{dt} - Ce_i(t) + Ag(e_i(t)) + Bg(e_i(t - \tau(t)))$$

$$+ c\sum_{j=1}^{N} G_{ij}\Gamma\left(\frac{de_j(t)}{dt} + e_j(t)\right) + u_i(t) \quad i = 1, \ldots, N, \qquad (3)$$

where $g(e_i) = (f_1(e_{i1} + s_1) - f_1(s_1), \ldots, f_n(e_{in} + s_n) - f_n(s_n))^T$.

Next, by introducing the following variable transformation:

$$r_i(t) = \frac{de_i(t)}{dt} + e_i(t), \qquad i = 1, \ldots, n,$$

the error system (3) can be written as

$$\begin{cases} \dfrac{de_i(t)}{dt} = -e_i(t) + r_i(t), \\[2mm] \dfrac{dr_i(t)}{dt} = -Ce_i(t) - Dr_i(t) + Ag(e_i(t)) + Bg(e_i(t - \tau(t))) \\[2mm] \qquad + c\displaystyle\sum_{j=1}^{N} G_{ij}\Gamma r_j(t) + u_i(t), \end{cases} \qquad (4)$$

for $i = 1, \ldots, N$, where $\boldsymbol{C} \triangleq C + I_n - D$ and $\boldsymbol{D} \triangleq D - I_n$.

The pinning controller is designed as follows:

$$u_i(t) = -\sigma_i(t)\Gamma r_i(t) \quad i = 1, \ldots, N, \tag{5}$$

where $\sigma_i(t)$ is the time-varying feedback control gain designed as

$$\dot{\sigma}_i(t) = \begin{cases} \sigma_i r_i^T(t)\Gamma r_i(t), \ \sigma_i(0) > 0, \text{ for } i \in \mathcal{V}_{pin}, \\ 0, \qquad\qquad \sigma_i(0) = 0, \text{ for } i \notin \mathcal{V}_{pin}, \end{cases}$$

where $\sigma_i > 0$ is a constant and $\mathcal{V}_{pin}$ is the set of the pinning nodes.

Thus, under the control input (5), the error system (4) turns out to be the following one

$$\begin{cases} \dfrac{de_i(t)}{dt} = -e_i(t) + r_i(t), \\ \dfrac{dr_i(t)}{dt} = -\boldsymbol{C}e_i(t) - \boldsymbol{D}r_i(t) + Ag(e_i(t)) + Bg(e_i(t - \tau(t))) \\ \qquad\qquad + c\displaystyle\sum_{j=1}^{N} G_{ij}\Gamma r_j(t) - \sigma_i(t)\Gamma r_i(t). \end{cases} \tag{6}$$

The coupled inertial neural network (1) can be synchronized if the above error system (6) is globally asymptotically stable. The following theorem gives the synchronization criterion.

Theorem 1. *Under Assumptions 1 and 2, the coupled inertial neural network (1) is globally asymptotically synchronized if there exists a positive definite matrix P such that*

$$\Phi = \begin{bmatrix} I_N \otimes [-P + (\frac{1}{2}F^2 + \eta)I_n] & \frac{1}{2}I_N \otimes (P - \boldsymbol{C}) \\ * & Q \end{bmatrix} < 0, \tag{7}$$

where $F = \max\limits_{1 \leq i \leq N}\{F_i\}$, $\eta > \max\limits_{1 \leq i \leq N}\left\{\frac{F_i^2}{2(1-\rho)}\right\}$ *is a positive constant,* $\boldsymbol{C} \triangleq C + I_n - D$, $\boldsymbol{D} \triangleq D - I_n$ *and* $Q = I_N \otimes (-\boldsymbol{D} + \frac{AA^T + BB^T}{2} + c\frac{G + G^T}{2} \otimes \Gamma - M \otimes \Gamma)$ *with* $M = \mathrm{diag}\{\sigma_1^*, \ldots, \sigma_N^*\} \geq 0$, *in which* $\sigma_i^* = 0$ *for* $i \in \mathcal{V}_{pin}$ *and* $\sigma_i^* > 0$ *when* $i \notin \mathcal{V}_{pin}$.

Proof. To prove the result, one just need to show that the error system (6) is globally asymptotically stable. Consider the following Lyapunov-Krasovskii functional candidate:

$$V(t) = \frac{1}{2}\sum_{i=1}^{N} e_i^T(t)Pe_i(t) + \eta\sum_{i=1}^{N} \int_{t-\tau(t)}^{t} e_i^T(s)e_i(s)ds$$

$$+ \frac{1}{2}\sum_{i=1}^{N} r_i^T(t)r_i(t) + \sum_{i=1}^{N} \frac{(\sigma_i(t) - \sigma_i^*)^2}{2\sigma_i}. \tag{8}$$

Calculating the time derivative of $V(t)$ along the trajectories of system (6), one can obtain

$$\dot{V}(t) \leq \sum_{i=1}^{N} e_i^T(t)P\big(-e_i(t) + r_i(t)\big) + \eta \sum_{i=1}^{N} e_i^T(t)e_i(t)$$

$$- \eta(1-\rho)\sum_{i=1}^{N} e_i^T(t-\tau(t))e_i(t-\tau(t))$$

$$- \sum_{i=1}^{N} r_i^T(t)Ce_i(t) - \sum_{i=1}^{N} r_i^T(t)Dr_i(t) + \sum_{i=1}^{N} r_i^T(t)Ag(e_i(t))$$

$$+ \sum_{i=1}^{N} r_i^T(t)Bg(e_i(t-\tau(t))) + c\sum_{i=1}^{N}\sum_{j=1}^{N} r_i^T(t)G_{ij}\Gamma r_j(t)$$

$$- \sum_{i=1}^{N} \sigma_i(t)r_i^T(t)\Gamma r_i(t) + \sum_{i\in \mathcal{V}_{pin}} \sigma_i(t)r_i^T(t)\Gamma r_i(t)$$

$$- \sum_{i\notin \mathcal{V}_{pin}} \sigma_i^* r_i^T(t)\Gamma r_i(t).$$

It follows from Assumption 1 that

$$\sum_{i=1}^{N} r_i^T(t)Ag_i(e_i(t)) \leq \sum_{i=1}^{N} \left(\frac{1}{2}r_i^T(t)AA^T r_i(t) + \frac{1}{2}F_i^2 e_i^T(t)e_i(t)\right) \qquad (9)$$

and

$$\sum_{i=1}^{N} r_i^T(t)Bg_i(e_i(t-\tau(t))) \leq \sum_{i=1}^{N} \Big(\frac{1}{2}r_i^T(t)BB^T r_i(t)$$

$$+ \frac{1}{2}F_i^2 e_i^T(t-\tau(t))e_i(t-\tau(t))\Big). \qquad (10)$$

Combining inequalities (9) and (10), we have

$$\dot{V}(t) \leq \sum_{i=1}^{N} e_i^T(t)\big(-P + (\eta + \frac{1}{2}F_i^2)I_n\big)e_i(t) + \sum_{i=1}^{N} e_i^T(t)(P - C)r_i(t)$$

$$+ \sum_{i=1}^{N} r_i^T(t)\big(-D + \frac{1}{2}(AA^T + BB^T)\big)r_i(t)$$

$$+ c\sum_{i=1}^{N}\sum_{j=1}^{N} r_i^T(t)G_{ij}\Gamma r_j(t) - \sum_{i=1}^{N} \sigma_i^* r_i^T(t)\Gamma r_i(t)$$

$$= \psi^T(t)\Phi\psi(t),$$

where $\psi(t) = [e^T(t), r^T(t)]^T$. Thus, by LMI (7) we have $\dot{V}(t) < 0$ for $\psi(t) \neq 0$, which .indicates that $\lim_{t\to\infty} e(t) = \mathbf{0}$ and $\lim_{t\to\infty} r(t) = \mathbf{0}$. Therefore, the pinning controlled network (6) can be globally asymptotically synchronized to the objective trajectory.

4   Illustrative Example

In this section, one illustrative example is presented to demonstrate the effectiveness of the obtained theoretical results.

Example 1. Consider the following coupling inertial delayed neural networks with 12 nodes:

$$\frac{\mathrm{d}^2 x_i(t)}{\mathrm{d}t^2} = -D\frac{\mathrm{d}x_i(t)}{\mathrm{d}t} - Cx_i(t) + Af(x_i(t)) + Bf(x_i(t - \tau(t))) + I_i(t)$$

$$+ c\sum_{j=1}^{12} G_{ij}\Gamma\left(\frac{\mathrm{d}x_j(t)}{\mathrm{d}t} + x_j(t)\right) + u_i(t), \quad i = 1, \ldots, 12, \tag{11}$$

where $x_i(t) = (x_{i1}(t), x_{i2}(t))^T$, $f(x_i(t)) = (\tanh(x_{i1}(t)), \tanh(x_{i2}(t)))^T$, $I(t) = (2, 4)^T, 1 \leq i \leq 12$ and the time delay $\tau(t) = 0.15e^t/(1 + e^t)$. So, it is easy to get $F_i = 1$, $\tau = 0.15$ and $\rho = 0.0375$. The coefficient matrices and inner coupling matrix of (11) are given as

$$D = \begin{bmatrix} 2.6 & 0 \\ 0 & 2.4 \end{bmatrix}, \ C = \begin{bmatrix} 4.6 & 0 \\ 0 & 3.8 \end{bmatrix}, \ A = \begin{bmatrix} 0.2 & -0.2 \\ -0.4 & 0.3 \end{bmatrix}, \ B = \begin{bmatrix} -4 & -5 \\ -2 & -5 \end{bmatrix}, \ \Gamma = \begin{bmatrix} 6 & 1 \\ 1 & 4 \end{bmatrix}.$$

The coupling matrix G is determined by the directed topology given in Fig. 1 with $G_{ij} = 0, 1(i \neq j)$.

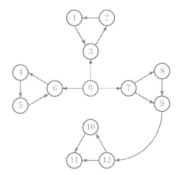

Fig. 1. Communication topology $\mathcal{G}$ and node 0 is the isolated objective node

Let the initial state of the objective system be $\tilde{\phi} = [3, -3]^T$ on the interval $[-0.15, 0]$ and initial functions for system (11) are chosen randomly. We use the quantity $E(t) = \sqrt{(1/12)\sum_{i=1}^{12} e_i^T(t)e_i(t)}$ to measure the quality of the synchronization process. Setting the pinning node set $\mathcal{V}_{pin} = \{3, 6, 7\}$ (see Fig. 1),

$\eta = 0.5205$ and the coupling strength $c = 30$, it is easy to check that the LMI (7) has a positive definite solution. Theorem 1 ensures that the whole coupled neural network system (11) can be synchronized to the given goal trajectory asymptotically.

The objective trajectory of the pinning controlled system (1) is shown in Fig. 2; and the state trajectories of (11) are given in Fig. 3. The synchronization error and the pinning feedback gains $\sigma_3(t)$, $\sigma_6(t)$ and $\sigma_7(t)$ are illustrated, respectively, in Fig. 4 and Fig. 5.

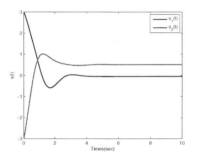

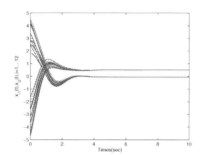

Fig. 2. State trajectory $s(t)$ in system (2)

Fig. 3. State trajectories $x_i(t)$ in system (11)

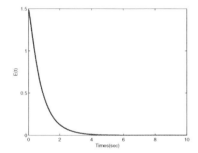

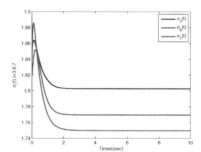

Fig. 4. Time evolution of synchronization error $E(t)$

Fig. 5. Variations of pinning feedback gains

5   Conclusions

In this paper, the synchronization control problem of coupled inertial neural network systems is formulated based on adaptive pinning control strategy. By Lyapunov stability theory and LMI technique, some sufficient criteria have been established for the global asymptotically synchronization of the coupled system. A numerical example has been given to illustrate the usefulness of the obtained results.

Acknowledgments. This work was jointly supported by the National Natural Science Foundation of China under Grant 61272530 and 11072059, the Natural Science Foundation of Jiangsu Province of China under Grant BK2012741, and the Specialized Research Fund for the Doctoral Program of Higher Education under Grant 20110092110017 and 20130092110017.

References

1. Arenas, A., Diaz-Guilera, A., Kurths, J., Moreno, Y., Zhou, C.: Synchronization in complex networks. Phys. Rep. **469**(3), 93–153 (2008)
2. Hu, J., Liang, J., Cao, J.: Cluster syhchronization of delayed complex networks with nonidential community structures. In: 5th International Conference on Advanced Computational Intelligence, pp. 48–53. IEEE, Nanjing (2012)
3. Hu, J., Liang, J., Cao, J.: Synchronization of hybrid-coupled heterogeneous networks: pinning control and impulsive control schemes. J. Franklin Inst. **361**(5), 2600–2622 (2014)
4. Arik, S.: Stability analysis of delayed neural networks. IEEE Trans. Circuits Syst. I **47**(7), 1089–1092 (2000)
5. Hu, J., Liang, J., Karimi, H.R., Cao, J.: Sliding intermittent control for BAM neural networks with delays. Abstr. Appl. Anal. Article ID:615947, 15 pages (2013)
6. Chen, G., Zhou, J., Liu, Z.: Global synchronization of coupled delayed neural networks and applications to chaotic CNN models. Int. J. Bifurcation and Chaos **14**(07), 2229–2240 (2004)
7. Lu, W., Chen, T.: Synchronization of coupled connected neural networks with delays. IEEE Trans. Circuits Syst. I **51**(12), 2491–2503 (2004)
8. Babcock, K.L., Westervelt, R.M.: Dynamics of simple electronic neural networks. Physica D **28**(3), 305–316 (1987)
9. Li, C., Chen, G., Liao, X., Yu, J.: Hopf bifurcation and chaos in a single inertial neuron model with time delay. Eur. Phys. J. B **41**(3), 337–343 (2004)
10. Liu, Q., Liao, X., Guo, S., Wu, Y.: Stability of bifurcating periodic solutions for a single delayed inertial neuron model under periodic excitation. Nonlinear Anal.: RWA **10**(4), 2384–2395 (2009)
11. He, X., Li, C., Shu, Y.: Bogdanov-Takens bifurcation in a single inertial neuron model with delay. Neurocomputing **89**, 193–201 (2012)
12. Ke, Y., Miao, C.: Stability and existence of periodic solutions in inertial BAM neural networks with time delay. Neural Comp. Appl. **23**(3–4), 1089–1099 (2013)
13. Ke, Y., Miao, C.: Stability analysis of inertial Cohen-Grossberg-type neural networks with time delays. Neurocomputing **117**, 196–205 (2013)
14. Cao, J., Wan, Y.: Matrix measure strategies for stability and synchronization of inertial BAM neural network with time delays. Neural Netw. **53**, 165–172 (2014)
15. Zhou, J., Wu, X., Yu, W., Small, M., Lu, J.: Pinning synchronization of delayed neural networks. Chaos **18**(4), 043111 (2008)
16. Song, Q., Cao, J., Liu, F.: Pinning synchronization of linearly coupled delayed neural networks. Math. Compu. in Simulation **86**, 39–51 (2012)
17. Zhou, J., Yu, W., Li, X., Small, M., Lu, J.: Identifying the topology of a coupled Fitzhugh-Nagumo neurobiological network via a pinning mechanism. IEEE Trans. Neural Netw. **20**(10), 1679–1684 (2009)
18. Lu, J., Ho, D.W.C., Wang, Z.: Pinning stabilization of linearly coupled stochastic neural networks via minimum number of controllers. IEEE Trans. Neural Netw. **20**(10), 1617–1629 (2009)

The Relationship between Neural Avalanches and Neural Oscillations

Yan Liu, Jiawei Chen, and Liujun Chen[✉]

School of Systems Science, Beijing Normal University,
Beijing 100875, People's Republic of China
chenlj@bnu.edu.cn

Abstract. Recent experiments found that the spontaneous cortical activity showed a dynamic firing pattern called neuronal avalanches, in which the distribution of firing activity followed a power law and the firing pattern was accompanied by nested θ-and β-/γ-oscillations. These results provided important evidence that neural oscillations could be formed during neuronal avalanches. The relationship between neuronal avalanches and oscillations has not been discussed in previous models. In this paper, we analyzed the relationship between neuronal avalanches and nested oscillations. Our results showed that the excitation-inhibition balance was a crucial mechanism for the formation of oscillation, but it was not enough for neuronal avalanches. The excitation-inhibition balance and synaptic plasticity were both necessary for a neural network to access the critical state and form neuronal avalanches. With the dynamic excitatory and inhibitory synaptic transmission processes and STDP rule, neuronal avalanches and nested oscillations could emerge simultaneously in a neural network.

Keywords: Neuronal avalanches · Neural oscillations · Nested oscillations

1 Introduction

Recently, a set of experiments found a convincing evidence that the spontaneous cortical activity had a dynamic firing pattern called 'neuronal avalanches' [1–4]. The experiment results showed that the spatio-temporal form of spontaneous activities distributed according to a power law with exponent lies between -1 and -2, which supported the conjecture that the brain might operate at criticality [1]. At the same time, the firing pattern was accompanied by nested θ-and β-/γ- oscillations [5], which provided an important evidence that neural oscillations might be formed during individual neurons triggering action potential firing in subsequent neurons [6–8]. So neuronal avalanches and nested oscillations give us a new way to analyse mechanisms of neural oscillations [9,10].

The paper was supported by MOE (Ministry of Education in China) Youth Fund Project of Humanities and Social Sciences (Project No.11YJC840006) and Fundamental Research Funds for the Central Universities (Fund number: 2013YB76).

© Springer International Publishing Switzerland 2014
Z. Zeng et al. (Eds.): ISNN 2014, LNCS 8866, pp. 43–50, 2014.
DOI: 10.1007/978-3-319-12436-0_6

The dynamic mechanism of neuronal avalanches is still in discussion. Previous models were mainly based on the mechanism of the sandpile model, seldom considering the neural dynamics and the dynamic synaptic transmission process [11–13]. Although neuronal avalanches showed a similar power law distribution as other physical systems (such as a sandpile system), their mechanisms might be different, because neurosystem has its typical properties which could not be explained through the models of other physical systems [14,15]. What's more, these models had not discussed the relationship between the nested oscillations and neuronal avalanches. In fact, there are many properties should be included in the model according to the neuronal avalanches experiments: (a)the neuronal avalanches predominantly depend on the GABA$_A$ and glutamatergic NMDA receptor [5]; (b)the neural system access the critical state only under a typical excitatory and inhibitory ratio [1]; (c)the spatio-temporal form of neuronal avalanches is accompanied by nested θ and β-/γ-oscillations [5].

Therefore, in this paper, we analysed the mechanisms of neuronal avalanches and discussed the relationship between neuronal avalanches and nested oscillations. The research was based on two neural network models, both of which are important models for discussing oscillations and include dynamic neurons and synaptic transmissions properties. Our results showed that the excitation-inhibition balance and synaptic plasticity were both necessary for a neural network to form neuronal avalanches. With the dynamic excitatory and inhibitory synaptic transmission processes and STDP rule, neuronal avalanches and nested oscillations could emerge simultaneously in a neural network.

2   Leaky Integrate-and-Fire Neurons Model

2.1   The Structure of the Model

The neural network is composed of $N_E = 400$ pyramidal cells and $N_I = 100$ interneurons, randomly connected with probability 20%. Both interneurons and pyramidal cells are described as leaky integrate-and-fire (LIF) neuron model [16],

$$C_m \frac{dV_m}{dt} = -g_L(V_m - V_{rest}) + C_m \Delta V \sum_i \delta(t - t_i) + I_{syn}(t) \qquad (1)$$

If $V_m > V_{th}$, then a spike is discharged and V_m is reset to V_{reset}. C_m is the capacitance, g_L is the leak conductance, so the time constant $\tau_m = C_m/g_L$. For pyramidal cells, $C_m = 0.5nF$ and $g_L = 0.025\mu S$, so that $\tau_m = 20ms$. For interneurons, $\tau_m = 10ms$. $\sum_i \delta(t - t_i)$ represents a random Poisson spike train, $\triangle V = 1mV$. The spike threshold $V_{th} = -52mV$, the leak (resting) membrane potential $V_{rest} = -70mV$, and the reset potential $V_{reset} = -59mV$. The absolute refractory period is 2ms (pyramidal cells) and 1ms (interneurons).

There are three types of synaptic currents in the model, including fast excitatory synaptic current (AMPA-type), slow excitatory synaptic current

(NMDA-type), and inhibitory synaptic current (GABAergic). All the three synaptic currents are described as the following equation,

$$I_{syn}(t) = g_{syn}(V - V_{syn})s(t) \tag{2}$$

where g_{syn} is the synaptic conductance, V_{syn} is the corresponding reversal potential, and $s(t)$ describes the time course of synaptic currents. After a presynaptic spike, $s(t)$ follows the following function with a latency of τ_l,

$$s(t) = \frac{\tau_m}{\tau_d - \tau_r}[exp(-\frac{t - \tau_l}{\tau_d}) - exp(-\frac{t - \tau_l}{\tau_r})] \tag{3}$$

where τ_l is latency, τ_r is rise time and τ_d is decay time. See Table 1 for the details on the parameters we used in the simulation.

Table 1. Synaptic parameters

	g_{syn} (pyramids) nS	g_{syn} (interneurons) nS	V_{syn} mV	τ_l mS	τ_r mS	τ_d mS
AMPA	0.19	0.3	0	1	0.5	2
NMDA	0.06	0.1	0	1	2	100
GABA	2.5	4	-70	1	0.5	5

2.2 Results

Rhythms of Oscillations. The population firing activity is calculated by the number of firing neurons in each millisecond. As shown in Fig.1, the activity of neurons in the network shows a γ-oscillation with rhythm of ca. 50Hz.

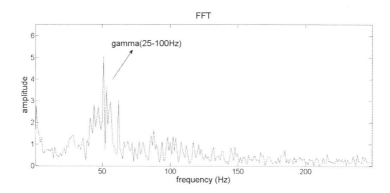

Fig. 1. The rhythm of firing activity calculated by Fast Fourier Transformation(FFT). The result shows a γ-oscillation with rhythm of ca. 50Hz.

Measurements of Neuronal Avalanches. In the experiments, neuronal avalanches are characterized based on two terms, a frame and an avalanche [1]. Data are binned at width $\delta t = 4ms$. The firing activity during one time bin δt is called a frame. A sequence of consecutively active frames that is preceded by a bland frame and ended by a blank frame is called an avalanche. The number of firing neurons in an avalanche is called the avalanche size. If the distribution of avalanches size follows a power law, then the network forms neuronal avalanches.

In our simulation, the firing activity of the network might not form neuronal avalanches, so here we call the number of firing neurons in consecutively active frames as 'number of consecutively active neurons'(NCANs) instead of 'avalanche size'. If the distribution follows a power law, then we call 'avalanche size' as in the experiments.

Under this definition, the simulation results of this model show that the cumulative probability distribution of NCANs is exponential (as shown in Fig.2), which indicates that many neurons fire synchronously in most frames and the firing state is in a super-critical state.

Therefore, a network with γ-oscillation firing pattern is not necessarily access the critical state. Although the network can form a γ-oscillation with the excitation-inhibition balance, it is not enough to form neuronal avalanches.

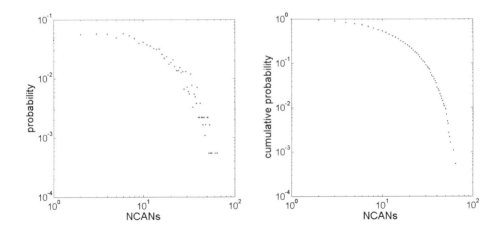

Fig. 2. The distribution of the number of consecutively active neurons(NCANs). The left figure shows that the probability distribution is exponential. The right figure shows that the cumulative probability distribution is exponential, which indicates that many neurons fire synchronously in most frames and the firing state is in a super-critical state.

3   Simple Spiking Model with STDP

3.1   The Structure of the Model

We model a neural network of $N = 500$ coupled dynamic neurons. The coupled structure is based on the neural model created by Izhikevich [17]. N neurons in the network include the excitatory and inhibitory neurons with proportion of 4:1. Each neuron has M synapses connecting to other neurons. Each synapse of each excitatory synapse randomly connects to a neuron, no matter whether the postsynaptic neuron is excitatory or inhibitory. Each neuron is described by the spiking model [8]:

$$\begin{cases} \frac{dv}{dt} = 0.04v^2 + 5v + 140 - u + I \\ \frac{du}{dt} = a(bv - u) \end{cases} \tag{4}$$

Here v denotes the membrane voltage of the neuron; u represents a membrane recovery variable, which accounts for the activity of Na^+ and K^+. If $v = 30mV$, then $v = c$, $u = u + d$. For all the neurons, $(b, c) = (0.2, -65)$. For excitatory neurons, $(a, d) = (0.02, 8)$ corresponding to cortical pyramidal neurons with the regular spiking pattern. For inhibitory ones, $(a, d) = (0.1, 2)$ corresponding to cortical inter-neurons exhibiting fast spiking firing patterns.

The input signal variable I in the model is composed by two parts. One part is the random input from the outside that a random chosen neuron will receive a random input current at each time step. The other part is the spiking input from the other neurons calculated by the synaptic weight. The delay time of the inhibitory synapse is fixed to $1ms$ and the delay time of the excitatory synapse is set between $1ms$ and t_{max}, in which t_{max} is tuned between $10ms$ and $30ms$.

3.2   Spike-Timing-Dependent Plasticity (STDP) Rule

The synaptic connection in the network is modified according to the spike-timing-dependent plasticity (STDP) rule [18]. If a spike from an excitatory pre-synaptic neuron arrives at a postsynaptic neuron before the postsynaptic neuron fired, then this synapse is potentiated (strengthened). On the contrary, if the spike arrives after the postsynaptic neuron fired, the synapse is depressed.

The magnitude of potentiation or depression relies on the time interval between the spikes. When a neuron fires, the variable $STDP$ is reset to 0.1. Every millisecond (one time step is one millisecond), $STDP$ decreases by $0.95 * STDP$, so that it decays to zero as $0.1e^{-t/20}$.

3.3   Results

Rhythms of Oscillations. The population firing activity is calculated by the number of fired neurons in each millisecond. As shown in Fig.3, the activity of neurons in the network shows a nested θ-and γ-oscillation, which is consistent with the experiment results [5].

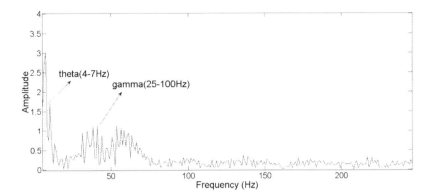

Fig. 3. The rhythms of firing are calculated by FFT. The result shows a nested θ-and γ-oscillation.

Neuronal Avalanches. In the simulation, data are binned at width $\delta t = 4ms$. Here, we have also tried the other time bins from $1ms$ to $10ms$, and the results show that the width of the time bin will not impact the distribution.

The probability distribution of NCANs displays a power law with characteristic slope of -1.5 and the cumulative probability distribution is almost a line (as shown in Fig. 4), which means the firing activity of the network form neuronal avalanches. So here we can call 'avalanches size' instead of NCANs in Fig. 4. This distribution result is consistent with the experiment result [1].

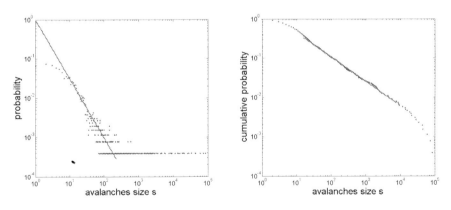

Fig. 4. The distribution of neuronal avalanches size. The left figure shows that the exponent of the probability distribution is -1.5. The right figure shows the cumulative probability distribution.

4   Conclusion

In this paper, the dynamic mechanism of neuronal avalanches and the relationship between neuronal avalanches and oscillations were discussed based on two neural network models.

In the leaky integrate-and-fire neurons model, we simulated a neural network composed of pyramidal cells and interneurons with fast excitatory synaptic current, slow excitatory synaptic current and inhibitory synaptic current. Our result showed that the activity of neurons in this network formed a γ-oscillation. The distribution of NCANs was exponential, which meant that many neurons fired synchronously and the firing state was in a super-critical state.

In the spiking neural network model under the STDP rule, the simulation result showed that the activity of neurons in this network formed a nested θ-and γ-oscillation. At the same time, the size distribution of the neuronal avalanches displayed a power law, which meant that the firing pattern was in the critical state.

These results indicated that the excitation-inhibition balance was a crucial mechanism for the formation of oscillation, but it was not enough for neuronal avalanches. The excitation-inhibition balance and synaptic plasticity were both necessary for a neural network to access the critical state and form neuronal avalanches. With the dynamic excitatory and inhibitory synaptic transmission processes and STDP rule, neuronal avalanches and nested oscillations could emerge simultaneously in a neural network.

References

1. Beggs, J.M., Plenz, D.: Neuronal Avalanches in Neocortical Circuits. J. Neurosci. **23**, 11167–11177 (2003)
2. Petermann, T., Thiagarajan, T.C., Lebedev, M.A., Nicolelis, M., Chialvo, D.R., Plenz, D.: Spontaneous Cortical Activity in Awake Monkeys Composed of Neuronal Avalanches. A-4658-2009. Proc. Natl. Acad. Sci. USA **106**(37), 15921–15926 (2009)
3. Yang, H.D., Shew, W.L., Roy, R., Plenz, D.: Maximal Variability of Phase Synchrony in Cortical Networks with Neuronal Avalanches. J. Neurosci. **32**, 1061–1072 (2012)
4. Pasquale, V., Massobrio, P., Bologna, L.L., Chiappalonea, M., Martinoia, S.: Self-Organization and Neuronal Avalanches in Networks of Dissociated Cortical Neurons. Neuroscience **153**, 1354–1369 (2008)
5. Girees, E.D., Plenz, D.: Neuronal Avalanches Organize as Nested Theta- and Beta/Gamma-Oscillations during Development of Cortical Layer 2/3. Proc. Natl. Acad. Sci. USA **105**, 7576–7581 (2008)
6. Shew, W.L., Plenz, D.: The Functional Benefits of Criticality in the Cortex. Neuroscient. **19**(1), 88–100 (2013)
7. Plenz, D., Thiagarajan, T.C.: The Organizing Principles of Neuronal Avalanches: Cell Assemblies in The Cortex? Tren. Neurosci. **30**(3), 101–110 (2007)
8. Chialvo, D.R.: Emergent Complex Neural Dynamics. Nature Physics **6**(10), 744–750 (2010)

9. Buzsaki, G., Wang, X.-J.: Mechanisms of Gamma Oscillations. Annu. Rev. Neurosci. **35**, 203–225 (2012)

10. Ainsworth, M., Lee, S., Cunningham, M.O., Traub, R.D., Kopell, N.J., Whittington, M.A.: Rates and Rhythms: A Synergistic View of Frequency and Temporal Coding in Neuronal Networks. Neuron **75**, 572–583 (2012)

11. Klaus, A., Yu, S., Plenz, D.: Statistical Analyses Support Power Law Distributions Found in Neuronal Avalanches. PLoS One **6**, e19779 (2011)

12. Abbott L.F., Rohrkemper R.: A Simple Growth Model Constructs Critical Avalanche Networks. Comput. Neurosci.: Theo. Insi. Brain Func. **165**, 13–19 (2007)

13. Sato, Y.D.: An Optimal Power-Law for Synchrony and Lognormally Synaptic Weighted Hub Networks. Chin. Phys. Lett. **30**, 098701 (2013)

14. Abbott, L.F., Regehr, W.G.: Synaptic Computation. Nature **431**(7010), 796–803 (2004)

15. Wu, J.J.S., Chang, W.P., Shih, H.C., Yen, C.T., Shyu, B.C.: Cingulate Seizure-Like Activity Reveals Neuronal Avalanche Regulated by Network Excitability and Thalamic Inputs. BMC Neurosci. **15**, 3 (2014)

16. Brunel, N., Wang, X.-J.: What Determines the Frequency of Fast Network Oscillations with Irregular Neural Discharges? I. Synaptic Dynamics and Excitation-Inhibition Balance. J. Neurophysiol. **90**, 415–430 (2003)

17. Izhikevich, E.M.: Polychronization: Computation with Spikes. Neural Comput. **18**, 245–282 (2006)

18. Bi, G.Q., Poo, M.M.: Synaptic Modification by Correlated Activity: Hebb's Postulate Revisited. Ann. Rev. Neurosci. **24**, 139–166 (2001)

Reinforcement-Learning-Based Controller Design for Nonaffine Nonlinear Systems

Xiong Yang, Derong Liu$^{(\boxtimes)}$, and Qinglai Wei

The State Key Laboratory of Management and Control for Complex Systems
Institute of Automation, Chinese Academy of Sciences, Beijing 100190, China
{xiong.yang,derong.liu,qinglai.wei}@ia.ac.cn

Abstract. In this paper, we develop an online learning control for a class of unknown nonaffine nonlinear discrete-time systems with unknown bounded disturbances. Under the framework of reinforcement learning, we employ two neural networks (NNs): an action NN is used to generate the control signal, and a critic NN is utilized to estimate the prescribed cost function. By using Lyapunov's direct method, we prove the stability of the closed-loop system. Moreover, based on the developed adaptive scheme, we show that all signals involved are uniformly ultimately bounded. Finally, we provide an example to demonstrate the effectiveness and applicability of the present approach.

Keywords: Neural network · Nonaffine system · Reinforcement learning · Online control

1 Introduction

Reinforcement learning (RL) is a type of machine learning in which an agent revises its actions based on responses from dynamic environment [1,2]. The typical structure of implementing RL algorithm is the actor-critic architecture, where the actor performs actions by interacting with its surroundings, and the critic evaluates actions and offers feedback information to the actor, leading to the improvement in performance of the subsequent actor [3,4].

Recently, RL is extensively used to derive the feedback control for nonlinear systems [5–11]. Most of the existing literature develop the adaptive actor–critic neural-network (NN)-based control to optimize the prescribed cost function for affine nonlinear systems. Nevertheless, few of them propose adaptive actor–critic NN-based control laws for nonaffine nonlinear systems, especially unknown nonaffine nonlinear discrete-time (DT) systems. Though there exist some researches about nonaffine nonlinear DT systems, they either focus on the feedback control problem of nonlinear autoregressive moving average with exogenous inputs (NARMAX) systems [12] or transfer the systems to NARMAX systems [13]. In comparison to state-form of nonaffine nonlinear systems, NARMAX systems do not take the consideration with the inner structure of systems. Therefore, this type of systems is less convenient than the state-form of nonaffine nonlinear

© Springer International Publishing Switzerland 2014
Z. Zeng et al. (Eds.): ISNN 2014, LNCS 8866, pp. 51–58, 2014.
DOI: 10.1007/978-3-319-12436-0_7

systems for purposes of adaptive control using NNs. On the other hand, in real engineering applications, control approaches of affine nonlinear systems do not always hold for nonaffine nonlinear systems. Hence, control methods for nonaffine nonlinear systems are necessary.

The objective of this paper is to develop an online learning control for a class of nonaffine nonlinear DT systems with unknown bounded disturbances. An action NN and a critic NN are employed to derive the control and the cost function, respectively. The stability of the closed-loop system is verified by using Lyapunov's direct method. Furthermore, the overall adaptive scheme guarantees all signals involved are uniformly ultimately bounded (UUB).

The paper is organized as follows. Preliminaries are provided in Section 2. A generalized feedback-linearization-method for nonaffine nonlinear DT systems is developed in Section 3. RL-based controller is designed in Section 4. Stability analysis is presented in Section 5. Simulation results are given in Section 6 to show the effectiveness of the developed control scheme. Finally, several conclusions are provided in Section 7.

2   Preliminaries

Consider the nonaffine nonlinear DT system described by the form

$$x_1(k+1) = x_2(k)$$

$$\vdots$$

$$x_{n-1}(k+1) = x_n(k)$$

$$x_n(k+1) = h\big(x(k), u(x(k))\big) + d(k) \tag{1}$$

where the state $x(k) = [x_1(k), \ldots, x_n(k)]^{\mathsf{T}} \in \Omega \subset \mathbb{R}^n$, the control $u(x(k)) \in \mathbb{R}$ is a continuous function with respect to $x(k)$. For convenience, we denote $\upsilon(k) = u(x(k))$. $d(k) \in \mathbb{R}$ is an unknown disturbance but bounded by a constant $d_M > 0$, i.e., $\|d(k)\| \le d_M$. $h\big(x(k), \upsilon(k)\big) \in \mathbb{R}$ is an unknown nonaffine function with $h(0,0) = 0$.

Assumption 1. *The state $x(k)$ is available from measurement at the time k.*

Assumption 2. *Let the desired system trajectory be $x_d(k) = [x_{1d}(k), \ldots, x_{nd}(k)]^{\mathsf{T}}$. $x_{id}(k)$ is arbitrarily selected and satisfies that $x_{id}(k+1) = x_{(i+1)d}(k), i = 1, 2, \ldots, n$. The desired trajectory $x_d(k)$ is bounded over the compact region Ω.*

Assumption 3. *$\partial h(x(k), \upsilon(k))/\partial \upsilon(k) \neq 0$ for $\forall \big(x(k), \upsilon(k)\big) \in \Omega \times \mathbb{R}$ with the compact region Ω.*

3   Feedback-Linearization-Method for Nonaffine Systems

From system (1), we have

$$
\begin{aligned}
x_n(k+1) &= h\big(x(k), \upsilon(k)\big) + d(k) \\
&= \alpha\upsilon(k) + \mathcal{F}\big(x(k), \upsilon(k)\big) + d(k) \tag{2}
\end{aligned}
$$

where $\mathcal{F}(x(k), v(k)) = h\big(x(k), v(k)\big) - \alpha v(k)$, and $\alpha > 0$ is a design constant (Note: α is actually an arbitrary positive constant. In order to get better performance, we often choose α by experience). Define the control input $v(k)$ as

$$v(k) = \big(v_s(k) - v_a(k)\big)/\alpha \tag{3}$$

where $v_s(k)$ is a feedback controller to stabilize linearized error dynamics, $v_a(k)$ is an adaptive controller designed to approximate $\mathcal{F}(x(k), v(k))$ by using a two-layer NN. From (2) and (3), we obtain

$$x_n(k+1) = \mathcal{F}\big(x(k), v(k)\big) - v_a(k) + v_s(k) + d(k). \tag{4}$$

Due to the objective of controllers $v_s(k)$ and $v_a(k)$, we define

$$\begin{cases} v_a(k) = \hat{\mathcal{F}}(x(k), v(k)) \\ v_s(k) = x_{nd}(k+1) + \lambda_1 e_{n-1}(k) + \cdots + \lambda_n e_0(k) \end{cases} \tag{5}$$

where $\hat{\mathcal{F}}(x(k), v(k))$ is an estimates of $\mathcal{F}(x(k), v(k))$, $e_{n-1}(k), \ldots, e_0(k)$ are the delayed values of the error $e_n(k)$, $\lambda_1, \ldots, \lambda_n$ are constant values selected such that $|z^n + \lambda_1 z^{n-1} + \cdots + \lambda_n|$ is stable.

The approximation error of $\mathcal{F}(x(k), v(k))$ is defined as

$$\tilde{\mathcal{F}}(x(k), v(k)) = \hat{\mathcal{F}}(x(k), v(k)) - \mathcal{F}(x(k), v(k)). \tag{6}$$

Meanwhile, the tracking error of the system state is given as

$$e_i(k) = x_{(i+1)d}(k) - x_{i+1}(k). \tag{7}$$

Theorem 1. *Let $v_s(k)$ be proposed as (5), and the tracking error $e_i(k)$ be given as (7). Then, the error dynamics is derived as*

$$e(k+1) = Ae(k) + B\big[\tilde{\mathcal{F}}(x(k), v(k)) - d(k)\big] \tag{8}$$

where

$$e(k) = \big[e_0(k), \ldots, e_{n-1}(k)\big]^{\mathsf{T}}, \; A = \begin{bmatrix} 0 & 1 & \cdots & 0 \\ \vdots & \vdots & & \vdots \\ -\lambda_n & -\lambda_{n-1} & \cdots & -\lambda_1 \end{bmatrix}, \; B = \begin{bmatrix} 0 \\ \vdots \\ 1 \end{bmatrix}. \tag{9}$$

Proof. By using Assumption 2 and (7), we obtain $e_i(k) = x_{(i+1)d}(k) - x_{i+1}(k) = x_{id}(k+1) - x_i(k+1) = e_{i-1}(k+1)$, $i = 1, \ldots, n-1$. Meanwhile, from (4) to (6), we can get $e_n(k) = -\lambda^{\mathsf{T}} e(k) + \tilde{\mathcal{F}}(x(k), v(k)) - d(k)$, where $\lambda = [\lambda_n, \ldots, \lambda_1]^{\mathsf{T}}$. Noticing $e_n(k) = e_{n-1}(k+1)$, we have $e_{n-1}(k+1) = -\lambda^{\mathsf{T}} e(k) + \tilde{\mathcal{F}}(x(k), v(k)) - d(k)$. By rewriting the tracking error in the vector form defined as in (9), we can derive (8). This completes the proof.

Remark 1: By the definition of the matrix A given in (9), one can conclude that A is the Hurwitz matrix .

Define $\mathcal{G}\big(x(k), v_a(k), v_s(k)\big) = \mathcal{F}\big(x(k), (v_s(k) - v_a(k))/\alpha\big) - v_a(k)$. Now, we show the existence of the solution of $\mathcal{G}\big(x(k), v_a(k), v_s(k)\big) = 0$.

Theorem 2. *Let Assumption 3 hold. Then, there exists $v_a(k)$ defined on Ω such that $\mathcal{G}\big(x(k), v_a(k), v_s(k)\big) = 0$.*

Proof. By using Assumption 3, we can obtain $\partial \mathcal{G}\big(x(k), v_a(k), v_s(k)\big)/\partial v_a(k) = \big(\partial[h(x(k), v(k)) - \alpha v(k)]/\partial v(k)\big) \cdot \big(\partial v(k)/\partial v_a(k)\big) - 1 = \big(\partial h(x(k), v(k))/\partial v(k)\big) \times (-1/\alpha) \neq 0$. By using Implicit Function Theorem, we can derive the conclusion.

4   Controller Design via Reinforcement Learning

4.1   Critic NN Design

A utility function depending on the tracking error $e(k)$ is defined as

$$r(k) = \begin{cases} 0, & \text{if } \tilde{e}(k) \leq \epsilon \\ 1, & \text{if } \tilde{e}(k) > \epsilon \end{cases}$$

where $\tilde{e}(k) = \lambda^{\mathsf{T}} e(k)$, and $\epsilon > 0$ is a prescribed threshold. The utility function $r(k)$ is considered to be the performance index: $r(k) = 0$ and $r(k) = 1$ represent the good and poor tracking performances, respectively. The cost function $J(k) \in \mathbb{R}$ [6] is defined as $J(k) = \tau^N r(k+1) + \tau^{N-1} r(k+2) + \cdots + \tau^{k+1} r(N)$, where $0 < \tau \leq 1$ is a design parameter, and N is the horizon. Then, we can get $J(k) = \tau J(k-1) - \tau^{N+1} r(k)$. The prediction error is $e_c(k) = \hat{J}(k) - \tau \hat{J}(k-1) + \tau^{N+1} r(k)$, where $\hat{J}(k) \in \mathbb{R}$ is an approximation of $J(k)$.

The output of the critic NN is given as

$$\hat{J}(k) = \hat{w}_c^{\mathsf{T}}(k) \sigma(\vartheta_c^{\mathsf{T}} x(k)) = \hat{w}_c^{\mathsf{T}}(k) \sigma_c(x(k)) \tag{10}$$

where $\vartheta_c \in \mathbb{R}^{n \times s_1}$ and $\hat{w}_c(k) \in \mathbb{R}^{s_1}$ are the weights of the critic NN, s_1 is the number of the nodes in the hidden layer. Since $\sigma(v_c^{\mathsf{T}} x(k))$ is initialized randomly and kept constant, it is written as $\sigma_c(x(k))$.

By using the gradient-based adaptation to minimize the objective function $E_c(k) = e_c^{\mathsf{T}}(k) e_c(k)/2$, we derive the weight update law for the critic NN as

$$\begin{aligned} \hat{w}_c(k+1) &= \hat{w}_c(k) + l_c \left[-\frac{\partial E_c(k)}{\partial \hat{w}_c(k)} \right] \\ &= \hat{w}_c(k) - l_c \sigma_c(x(k)) \Big[\hat{w}_c^{\mathsf{T}}(k) \sigma_c(x(k)) + \tau^{N+1} r(k) \\ &\quad - \tau \hat{w}_c^{\mathsf{T}}(k-1) \sigma_c\big(x(k-1)\big) \Big]^{\mathsf{T}}. \end{aligned} \tag{11}$$

where $0 < l_c < 1$ is the learning rate of the critic NN.

4.2   Action NN Design

The prediction error for the action NN is defined as $e_a(k) = \hat{J}(k) - J_d(k) + \tilde{\mathcal{F}}(x(k), x_d(k))$. Because $J_d(k)$ is often considered to be zero, the prediction error becomes $e_a(k) = \hat{J}(k) + \tilde{\mathcal{F}}(x(k), x_d(k))$.

The output of the action NN is given as

$$\hat{\mathcal{F}}(k) = \hat{w}_a^{\mathsf{T}}(k)\sigma\big(\vartheta_a^{\mathsf{T}} z(k)\big) = \hat{w}_a^{\mathsf{T}}(k)\sigma_a(z(k)) \tag{12}$$

where $\hat{\mathcal{F}}(k)$ stands for $\hat{\mathcal{F}}(x(k), x_d(k))$, $\vartheta_a \in \mathbb{R}^{(n+1)\times s_2}$, $\hat{w}_a \in \mathbb{R}^{s_2}$ and s_2 are defined in the same way as in the critic NN, $z(k) = [x^{\mathsf{T}}(k), x_{nd}(k) + e^{\mathsf{T}}(k)\lambda]^{\mathsf{T}} \in \mathbb{R}^{n+1}$. Suppose $\mathcal{F}(k) = w_a^{\mathsf{T}}\sigma_a(z(k)) + \varepsilon(k)$, where $\mathcal{F}(k) = \mathcal{F}(x(k), x_d(k))$, w_a is the ideal weight, $\varepsilon(k)$ is the action NN approximation error. Define $\tilde{w}_a(k) = \hat{w}_a(k) - w_a$. Then the function approximation error can be derived as

$$\tilde{\mathcal{F}}(k) = \hat{\mathcal{F}}(k) - \mathcal{F}(k) = \tilde{w}_a^{\mathsf{T}}(k)\sigma_a(z(k)) - \varepsilon(k). \tag{13}$$

Remark 2: By using Theorem 2 and (5), we can conclude that $\mathcal{F}(x(k), \upsilon(k))$ is actually a function with respect to $x(k)$ and $x_d(k)$. Hence, in the rest of the paper, we denote $\mathcal{F}(x(k), x_d(k)) = \mathcal{F}(x(k), \upsilon(k))$, $\tilde{\mathcal{F}}(x(k), x_d(k)) = \tilde{\mathcal{F}}(x(k), \upsilon(k))$.

Combining (6) and (13) and using Theorem 1, we derive $\tilde{w}_a^{\mathsf{T}}(k)\sigma_a(z(k)) - \varepsilon(k) = B^{\mathsf{T}}\big(e(k+1) - Ae(k)\big) + d(k)$. By using the gradient-based adaptation to minimize $E_a(k) = e_a^{\mathsf{T}}(k)e_a(k)/2$ and noticing that $d(k)$ is often chosen to be zero, i.e., $d(k) = 0$, we obtain the weight update rule for the action NN as

$$\hat{w}_a(k+1) = \hat{w}_a(k) + l_a\left[-\frac{\partial E_a(k)}{\partial \hat{w}_a(k)}\right]$$

$$= \hat{w}_a(k) - l_a\sigma_a(z(k))\Big[\hat{w}_c^{\mathsf{T}}(k)\sigma_c(k) + B^{\mathsf{T}}\big(e(k+1) - Ae(k)\big)\Big]^{\mathsf{T}}. \tag{14}$$

where $0 < l_a < 1$ is the learning rate of the action NN.

5   Stability Analysis

Assumption 4. *Let the ideal out layer weights w_a and w_c be bounded as $\|w_a\| \le w_{aM}$, $\|w_c\| \le w_{cM}$, where $w_{aM} > 0$ and $w_{cM} > 0$. Let $\varepsilon(k)$ be bounded by a constant $\varepsilon_M > 0$ over the compact set Ω, i.e., $\|\varepsilon(k)\| \le \varepsilon_M$.*

Assumption 5. *The activation functions are bounded over the compact set Ω as $\|\sigma_a(z(k))\| \le \sigma_{aM}$, $\|\sigma_c(x(k))\| \le \sigma_{cM}$, where $\sigma_{aM} > 0$ and $\sigma_{cM} > 0$.*

Fact 1. *Since the matrix A is Hurwitz, there exists $P \in \mathbb{R}^{n\times n}$ satisfying $A^{\mathsf{T}}PA - P = -\mu I_n$, where $\mu > 0$ is a constant. There exist two known positive constants ϱ and ρ ($\varrho < \rho$) such that $\varrho \le B^{\mathsf{T}}PB \le \rho$.*

Theorem 3. *Consider the nonaffine nonlinear system described by (1). Let Assumptions 1–5 hold. Take the control input for system (1) as (3) with (5) and the critic NN (10), as well as the action NN (12). Moreover, let the weight update law for the critic NN and the action NN be (11) and (14), respectively. Then, the tracking error $e(k)$, the weight of the estimation error for the action NN $\tilde{w}_a(k)$, and the weight of the estimation error for the critic NN $\tilde{w}_c(k)$ are UUB as long as one of the following conditions hold:*

$$(a)\ 0 < l_c\|\sigma_c(x(k))\|^2 < 1, \quad (b)\ \mu > 1,$$

$$(c)\ 0 < l_a\|\sigma_a(z(k))\|^2 < 1, \quad (d)\ 0 < \tau < \sqrt{2}/2. \tag{15}$$

Proof. We provide an outline of the proof due to the space limit. Consider the Lyapunov function candidate

$$L(k) = \beta_1 e^{\mathsf{T}}(k) P e(k) + \frac{\beta_2}{l_c} \mathrm{tr}\big(\tilde{w}_c^{\mathsf{T}}(k) \tilde{w}_c(k)\big)$$

$$+ \frac{\beta_3}{l_a} \mathrm{tr}\big(\tilde{w}_a^{\mathsf{T}}(k) \tilde{w}_a(k)\big) + \beta_4 \|\xi_c(k-1)\|^2 \tag{16}$$

where $\beta_j > 0$, $j = 1, 2, 3, 4$, and $\xi_c(k-1) = \tilde{w}_c^{\mathsf{T}}(k-1) \sigma_c\big(x(k-1)\big)$.

By computing the first difference of $L(k)$ defined as in (16), we derive

$$\Delta L(k) \leq - \beta_1(\mu - 1)\|e(k)\|^2 + \mathcal{D}^2(k) - (\beta_2 - 2\beta_3 - \beta_4)\|\xi_c(k)\|^2$$

$$- (\beta_4 - 2\tau^2 \beta_2)\|\xi_c(k-1)\|^2 - (\beta_3 - 2\beta_1(\rho + \eta))\|\xi_a(k)\|^2$$

$$- \beta_2\big(1 - l_c\|\sigma_c(x(k))\|^2\big)\|\xi_c(k) + M(k)\|^2$$

$$- \beta_3\big(1 - l_a\|\sigma_a(z(k))\|^2\big)\|N(k)\|^2 \tag{17}$$

where $\xi_c(k) = \tilde{w}_c^{\mathsf{T}}(k)\sigma_c(x(k))$, $\xi_a(k) = \tilde{w}_a^{\mathsf{T}}(k)\sigma_a(z(k))$, $M(k) = w_c^{\mathsf{T}}\sigma_c(x(k)) + \tau^{N+1} r(k) - \tau \hat{w}_c^{\mathsf{T}}(k-1)\sigma_c(x(k-1))$, $N(k) = \hat{w}_c^{\mathsf{T}}(k)\sigma_c(x(k)) + \xi_a(k) - \varepsilon(k) - d(k)$, $\eta = \|A^{\mathsf{T}} P B\|^2$, and

$$\mathcal{D}^2(k) = 2\beta_2 \|P(k)\|^2 + 2\beta_3 \|w_c^{\mathsf{T}}\sigma_c(k) - \varepsilon(k) - d(k)\|^2$$

$$+ 2\beta_1(\rho + \eta)\|\varepsilon(k) + d(k)\|^2,$$

and $P(k) = w_c^{\mathsf{T}}\sigma_c(z(k)) + \tau^{N+1} r(k) - \tau w_c^{\mathsf{T}}\sigma_c(x(k-1))$.

Using Assumptions 1–5 and Fact 1, we obtain

$$\mathcal{D}^2(k) \leq (12\beta_2 + 6\beta_3) w_{cM}^2 \sigma_{cM}^2 + 6\beta_2 r_M^2$$

$$+ \big(6\beta_3 + 4\beta_1(\rho + \eta)\big)(\varepsilon_M^2 + d_M^2) \triangleq \mathcal{D}_M^2 \tag{18}$$

where $r_M > 0$ is the bound of $|r(k)|$, i.e., $|r(k)| \leq r_M$.

The parameters β_i ($i = 1, 2, 3$) are selected to satisfy $\beta_1 < \beta_3/2(\rho + \eta)$, $\beta_2 = \beta_4/2\tau^2$, $\beta_2 > 2\beta_3/(1 - 2\tau^2)$. Then, by using (15), we can derive that (17) and (18) implies $\Delta L(k) < 0$ as long as one of the following conditions holds: $\|e(k)\| > |\mathcal{D}_M|/\sqrt{\beta_1(\mu - 1)}$, or $\|\tilde{w}_a(k)\| > \frac{|\mathcal{D}_M|}{\sigma_{aM}}/\sqrt{\beta_3 - 2\beta_1(\rho + \eta)}$, or $\|\tilde{w}_c(k)\| > \frac{|\mathcal{D}_M|}{\sigma_{cM}}/\sqrt{\beta_2 - 2\tau^2\beta_2 - 2\beta_3}$. By using standard Lyapunov extension theorem [14], we can obtain that the tracking error $e(k)$, the weight of the estimation error for the action NN $\tilde{w}_a(k)$, and the weight of the estimation error for the critic NN $\tilde{w}_c(k)$ are UUB.

6   Simulation Results

Consider the nonaffine nonlinear DT system given by

$$x_1(k+1) = x_2(k)$$

$$x_2(k+1) = x_1^2(k) + 0.15u^3(k) + 0.08u(k) + \sin(0.08u(k))$$

$$+ 0.01\big(x_2(k) - x_1(k)\big)^2 u(k) + 2x_2(k) - x_1(k) + 0.5\sin(3\pi k). \tag{19}$$

The desired system trajectory is $x_d(k) = [y(k), y(k+1)]^{\mathsf{T}}$, where $y(k) = 0.8\sin(k\pi/200) + 0.8\cos(k\pi/400)$. From (19), we have $\partial h(x(k), u(k))/\partial u(k) \neq 0$. The initial state is $x_0 = [1, 0.31]^{\mathsf{T}}$. Let $\lambda_1 = 1$, $\lambda_2 = 0.25$ (i.e., $s^2 + \lambda_1 s + \lambda_2$ is stable). We select $\alpha = 2$, $\mu = 2$, $\tau = 0.7$, $\epsilon = 0.008$, $l_a = 0.1$ and $l_c = 0.001$. The initial weight for the inner layer is selected randomly within an interval of $[0, 1]$ and held constant. Meanwhile, the initial weights for the output layer is chosen randomly within an interval of $[-0.2, 0.2]$. The structure of the action NN is 3–15–1. And the structure of the critic NN is 2–10–1. The computer simulation results are shown by Figs. 1(a)–(d). Fig. 1(a) and Fig. 1(b) indicate the 2-norm of the weights of the action NN and the critic NN, respectively. Fig. 1(c) presents the control input. Fig. 1(d) shows the trajectories of $x_1(k)$ and $x_{1d}(k)$ (Since the trajectories of $x_2(k)$ and $x_{2d}(k)$ is same with Fig. 1(d), we omit it here). From simulation results, it is observed that the state $x(k)$ tracks the desired trajectory $x_d(k)$ very well, and all signals involved are bounded.

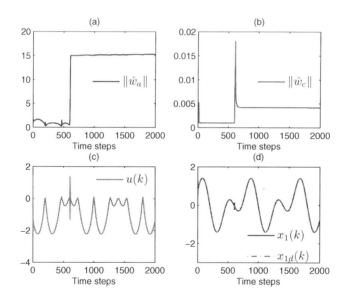

Fig. 1. (a) 2-norm of the action NN weight $\|\hat{W}_a\|$ (b) 2-norm of the critic NN weight $\|\hat{W}_c\|$ (c) Control input $u(k)$ (d) Trajectories of $x_1(k)$ and $x_{1d}(k)$

7   Conclusion

We have developed an RL-based online learning control for unknown nonaffine nonlinear DT systems with unknown bounded disturbances. By using Lyapunov's method, the stability of the closed-loop system is verified, and all signals involved are UUB. The computer simulation results show that the online controller can perform successfully control and attain the desired performance. In our future work, we focus on deriving online control for nonaffine nonlinear CT systems.

Acknowledgments. This work was supported in part by the National Natural Science Foundation of China under Grants 61034002, 61233001, 61273140, 61304086, and 61374105, and in part by Beijing Natural Science Foundation under Grant 4132078.

References

1. Sutton, R.S., Barto, A.G.: Reinforcement Learning-An Introduction. MIT Press, Cambridge (1998)
2. Du, K., Swamy, M.N.S.: Neural Networks and Statistical Learning. Springer, London (2014)
3. Werbos, P.J.: Neural networks and the experience and cultivation of mind. Neural Networks **32**, 86–95 (2012)
4. Lewis, F.L., Vrabie, D., Vamvoudakis, K.G.: Reinforcement learning and feedback control: Using natural decision methods to design optimal adaptive controllers. IEEE Control Systems **32**(6), 76–105 (2012)
5. Lewis, F.L., Liu, D.: Reinforcement Learning and Approximate Dynamic Programming for Feedback Control. Wiley, New Jersey (2012)
6. He, P., Jagannathan, S.: Reinforcement learning neural-network-based controller for nonlinear discrete-time systems with input constraints. IEEE Transactions on Systems, Man, and Cybernetics, Part B: Cybernetics **37**(2), 425–436 (2007)
7. Liu, D., Yang, X., Li, H.: Adaptive optimal control for a class of continuous-time affine nonlinear systems with unknown internal dynamics. Neural Computing and Applications **23**(7–8), 1843–1850 (2013)
8. Liu, D., Wei, Q.: Finite-approximation-error-based optimal control approach for discrete-time nonlinear systems. IEEE Transactions on Cybernetics **43**(2), 779–789 (2013)
9. Yang, Q., Jagannathan, S.: Reinforcement learning controller design for affine nonlinear discrete-time systems using online approximators. IEEE Transactions on Systems, Man, and Cybernetics, Part B: Cybernetics **42**(2), 377–390 (2012)
10. Yang, X., Liu, D., Wang, D.: Reinforcement learning for adaptive optimal control of unknown continuous-time nonlinear systems with input constraints. International Journal of Control **87**(3), 553–566 (2014)
11. Yang, X., Liu, D., Wang, D.: Observer-based adaptive output feedback control for nonaffine nonlinear discrete-time systems using reinforcement learning. In: 20th International Conference on Neural Information Processing, Daegu, Korea, pp. 631–638 (2013)
12. Yang, Q., Vance, J.B., Jagannathan, S.: Control of nonaffine nonlinear discrete-time systems using reiforcement-learning-based linearly parameterized neural networks. IEEE Transactions on Systems, Man, Cybernetics, Part B, Cybernetics **38**(4), 994–1001 (2008)
13. Xu, B., Yang, C., Shi, Z.: Reinforcement learning output feedback NN control using deterministic learning technique. IEEE Transactions on Neural Networks and Learning Systems **25**(3), 635–641 (2014)
14. Lewis, F.L., Jagannathan, S., Yesildirek, A.: Neural Network Control of Robot Manipulators and Nonlinear Systems. Taylor & Francis, London (1999)

Multistability and Multiperiodicity Analysis of Complex-Valued Neural Networks

Jin Hu[1] and Jun Wang[2]([✉])

[1] School of Science, Chongqing Jiaotong University, Chongqing 400074, China
[2] Department of Mechanical and Automation Engineering, The Chinese University of Hong Kong, Shatin, N.T., Hong Kong
jwang@mae.cuhk.edu.hk

Abstract. Multistability and multiperiodicity of neual networks are usually considered in the application of associative memory. In this paper, we study the multistability and multiperiodicity of complex-valued neural networks (CVNNs for short) with one step piecewise linear activation functions. By separating the CVNN into its real and imaginary parts and using state decomposition, we can easily increase the storage capacity by using less neurons. Simulation results are given to illustrative the effectiveness of the theoretical results.

Keywords: Complex-valued neural networks · Multistability · Multiperiodicity

1 Introduction

Recently, complex-valued neural networks (CVNNs for short), are proposed to handle various applications involving complex-valued numbers, see [2,7,8,16,17] and the references herein. Most of these literatures studied the mono-stability and mono-periodicity of CVNNs. However, in some neurodynamic systems, there exist multiple stable equilibrium points or periodic orbits. For example, when a neural network is employed as an associative memory storage for pattern recognition, the existence of multiple equilibrium points is a necessary feature. The studies of multistability and multiperiodicity of recurrent neural networks have attracted considerable research interests, see [1,3–6,13–15].

It is shown that the n-neuron recurrent neural networks with one step piecewise linear activation function can have 2^n locally exponentially stable equilibrium points located in saturation regions (see [3,14,15]). In order to increase storage capacity, in [13] the authors employed a stair-style activation function with k steps and obtained the existence condition of $(4k-1)^n$ equilibrium points. However, stair-style activation functions with k steps are quite more complicated

The work described in this paper was supported by the National Natural Science Foundation of China under Grants 61273021 and 61403051 and grants from the Research Grants Council of the Hong Kong Special Administrative Region, China (Project nos. CUHK417209E and CUHK416811E).

Z. Zeng et al. (Eds.): ISNN 2014, LNCS 8866, pp. 59–68, 2014.
DOI: 10.1007/978-3-319-12436-0_8

and difficult to implement than one step piecewise linear activation functions. Since CVNNs can be separated into their real and imaginary parts, if we use CVNNs in the design of multistate associative memory models, it is easy to increase the storage capacity of the models. In recent years, some CVNN models have been applied in the design and synthesis of associative memory, see [9,11,12]. In those papers, the multistate associative memory models are discussed by employing a class of amplitude-phase-type activation functions wich are suitable for processing information meaningful in rotation around the origin of coordinate. In this paper, we investigate the multistability and multiperiodicity of CVNNs with real-imaginary-type one step piecewise linear complex-valued activation functions which are suitable for dealing with complex information that must have symmetry concerning. The existence conditions of multiple stable equilibrium points and periodic orbits are obtained which improves a lot in the design and synthesis in associative memory. The rest of this paper is organized as follows: In Section 2 we give the description of the CVNN model as well as some definitions and lemmas. In Section 3, we present the sufficient condition for multistabilty and multiperiodicity of the CVNN models and extend the results to complex-valued cellular neural networks. In Section 4, we provide a numerical simulation to illustrative the effectiveness of the results. Section 5 concludes the paper.

2   Model Description and Preliminaries

In this paper, we investigate the following CVNN model

$$\dot{z}(t) = -Dz(t) + Af(z(t)) + Bf(z(t - \tau)) + u(t) \tag{1}$$

where $z = (z_1, z_2, \cdots, z_n)^{\mathrm{T}} \in C^n$ is the state vector, $D = \mathrm{diag}(d_1, d_2, \cdots, d_n) \in R^{n \times n}$ with $d_j > 0$ $(j = 1, 2, \cdots, n)$ is the self-feedback connection weight matrix, $A = (a_{jk})_{n \times n} \in C^{n \times n}$ and $B = (b_{jk})_{n \times n} \in C^{n \times n}$ are, respectively, the connection weight matrix without and with time delays, $f(z(t)) = (f_1(z_1(t)), f_2(z_2(t)), \cdots, f_n(z_n(t)))^{\mathrm{T}} : C^n \to C^n$ is the vector-valued activation function whose elements consist of complex-valued nonlinear functions, τ_j $(j = 1, 2, \cdots, n)$ are constant time delays, $u(t) = (u_1(t), u_2(t), \cdots, u_n(t))^{\mathrm{T}} \in C^n$ is the external input vector-valued function with period ω.

We consider the following one step piecewise linear activation functions $f_j(z)$ that can be separated into its real and imaginary part as follows::

$$f_j(z) = f_j^R(x) + \mathrm{i} f_j^I(y)$$

where

$$f_j^R(x) = \begin{cases} l_j^R, & -\infty < x < q_j^R \\ h(x), & q_j^R \le x \le p_j^R \\ m_j^R, & p_j^R < x < \infty \end{cases}, f_j^I(y) = \begin{cases} l_j^I, & -\infty < y < q_j^I \\ w(y), & q_j^I \le y \le p_j^I \\ m_j^I, & p_j^I < y < \infty \end{cases}, \tag{2}$$

where

$$h(x) = l_j^R + \frac{m_j^R - l_j^R}{p_j^R - q_j^R}(x - q_j^R), w(y) = l_j^I + \frac{m_j^I - l_j^I}{p_j^I - q_j^I}(y - q_j^I), l_j^R < m_j^R, l_j^I < m_j^I.$$

Let $C([t_0 - \tau, t_0], \Omega)$ be the Banach space of continuous functions $\phi : [t_0 - \tau, t_0] \to \Omega \subset R^n$ (C^n) with norm defined by $||\phi||_{t_0} = \sup_{-\tau \leq \theta \leq 0} ||\phi(t_0 + \theta)||$.

Definition 1. *A set Ω is said to be an invariant set of CVNN (1), if the solution $z(t; t_0, \phi)$ of (1) with any initial condition $\phi(s) \in C([t_0 - \tau, t_0], \Omega)$, satisfies $z(t; t_0, \phi) \in \Omega$ for $t > t_0$.*

Definition 2. *A periodic orbit $\hat{z}(t)$ is said to be a limit cycle of CVNN (1) if $\hat{z}$ is an isolated periodic orbit; that is, there exists $\omega > 0$ such that $\forall t \geq t_0, \hat{z}(t + \omega) = \hat{z}(t)$, and there exists $\delta > 0$ such that $\forall z(t) \in \{z(t) \mid 0 < ||z(t) - \hat{z}|| < \delta, t \geq t_0\}$, $z(t)$ is not a point on any periodic orbit of neural network (11).*

Definition 3. *A periodic orbit $\hat{z}(t)$ of CVNN (1) is said to be locally exponentially periodic in region Ω if there exist constants $\alpha > 0$ and $\beta > 0$ such that $\forall t \geq t_0$,*

$$||z(t; t_0, \phi) - \hat{z}(t)|| \leq \beta ||\phi - \hat{\phi}||_{t_0} \exp\{-\alpha(t - t_0)\}$$

where $z(t; t_0, \phi)$ is the state of (1) with any initial condition $\phi(s) \in C([t_0 - \tau, t_0], \Omega)$. When $\Omega = C^n$, $\hat{z}$ is said to be globally exponentially periodic.

Lemma 1. *[10] Let H be a mapping on complete metric space $(C([t_0 - \tau, t_0], \Omega), ||\cdot||_{t_0})$. If $H(C([t_0 - \tau, t_0], \Omega)) \subset C([t_0 - \tau, t_0], \Omega)$, and there exists a constant $\gamma < 1$ such that $\forall \varphi, \phi \in C([t_0 - \tau, t_0], \Omega)$, $||H(\varphi) - H(\phi)|| \leq \gamma ||\varphi - \phi||_{t_0}$, then there exists one unique $\bar{\varphi} \in C([t_0 - \tau, t_0], \Omega)$ such that $H(\bar{\varphi}) = \bar{\varphi}$.*

3   Main Results

Assume that $N_1 \cup N_2 \cup N_3 \cup N_4 = \{1, 2, 3, \cdots, n\}$, N_1, N_2, N_3 and N_4 are mutually exclusive sets, that is, $N_j \cap N_k = \varnothing$ $(j, k = 1, 2, 3, 4, j \neq k)$. Let $w = (x^T, y^T)^T$, and

$$\Omega_1 = \{w \in R^{2n} | -\infty < x_j < q_j^R, -\infty < y_j < q_j^I, j \in N_1; -\infty < x_j < q_j^R, p_j^I < y_j < \infty,$$
$$j \in N_2; p_j^R < x < \infty, -\infty < y_j < q_j^I, j \in N_3; p_j^R < x < \infty, p_j^I < y_j < \infty, j \in N_4\}.$$
$$(3)$$

Theorem 1. *CVNN (1) has 4^n locally exponentially stable limit cycles located in Ω_1 if $\forall t \geq t_0$ the following conditions hold*

$$|u_j^R(t)| < -d_j \max\{|q_j^R|, |p_j^R|\} + \alpha_j^R - \sum_{k=1, k\neq j}^{n} [|a_{jk}^R| \max\{|l_k^R|, |m_k^R|\} + |a_{jk}^I| \max\{|l_k^I|, |m_k^I|\}]$$

$$- \sum_{k=1}^{n} [|b_{jk}^R| \max\{|l_k^R|, |m_k^R|\} + |b_{jk}^I| \max\{|l_k^I|, |m_k^I|\}],$$

$$|u_j^I(t)| < -d_j \max\{|q_j^I|, |p_j^I|\} + \alpha_j^I - \sum_{k=1, k\neq j}^{n} [|a_{jk}^I| \max\{|l_k^R|, |m_k^R|\} + |a_{jk}^R| \max\{|l_k^I|, |m_k^I|\}]$$

$$- \sum_{k=1}^{n} [|b_{jk}^I| \max\{|l_k^R|, |m_k^R|\} + |b_{jk}^R| \max\{|l_k^I|, |m_k^I|\}],$$

$$(4)$$

where

$$\alpha_j^R = \min\{-a_{jj}^R l_j^R + a_{jj}^I l_j^I, -a_{jj}^R l_j^R + a_{jj}^I m_j^I, a_{jj}^R m_j^R - a_{jj}^I l_j^I, a_{jj}^R m_j^R - a_{jj}^I m_j^I\} > 0,$$
$$\alpha_j^I = \min\{-a_{jj}^I l_j^R - a_{jj}^R l_j^I, a_{jj}^I l_j^R + a_{jj}^R m_j^I, -a_{jj}^I m_j^R - a_{jj}^R l_j^I, a_{jj}^I m_j^R + a_{jj}^R m_j^I\} > 0.$$

Proof. By separating the state vector, connection weight matrix, vector-valued activation function and the external input vector of CVNN (1) into its real and imaginary part, we have that $\forall t \in [t_0 - \tau, t_0]$, $w(t) \in \Omega_1$,

$$\begin{aligned}
\dot{x}_j = {}& -d_j x_j + \sum_{k \in N_1 \cup N_2} (a_{jk}^R + b_{jk}^R) l_k^R + \sum_{k \in N_3 \cup N_4} (a_{jk}^R + b_{jk}^R) m_k^R \\
& - \sum_{k \in N_1 \cup N_3} (a_{jk}^I + b_{jk}^I) l_k^I - \sum_{k \in N_2 \cup N_4} (a_{jk}^I + b_{jk}^I) m_k^I + u_j^R(t), \\
\dot{y}_j = {}& -d_j y_j + \sum_{k \in N_1 \cup N_2} (a_{jk}^I + b_{jk}^I) l_k^R + \sum_{k \in N_3 \cup N_4} (a_{jk}^I + b_{jk}^I) m_k^R \\
& + \sum_{k \in N_1 \cup N_3} (a_{jk}^R + b_{jk}^R) l_k^I + \sum_{k \in N_2 \cup N_4} (a_{jk}^R + b_{jk}^R) m_k^I + u_j^I(t).
\end{aligned} \tag{5}$$

If $j \in N_1$ and $x_j(t_0) = q_j^R$, $y_j(t_0) = q_j^I$, according to (4), we have

$$\begin{aligned}
\dot{x}_j(t_0) = {}& -d_j q_j^R + a_{jj}^R l_j^R - a_{jj}^I l_j^I + \sum_{k \in N_1 \cup N_2, k \neq j} a_{jk}^R l_k^R + \sum_{k \in N_1 \cup N_2} b_{jk}^R l_k^R + \sum_{k \in N_3 \cup N_4} (a_{jk}^R \\
& + b_{jk}^R) m_k^R - \sum_{k \in N_1 \cup N_3} b_{jk}^I l_k^I - \sum_{k \in N_1 \cup N_3, k \neq j} a_{jk}^I l_k^I - \sum_{k \in N_2 \cup N_4} (a_{jk}^I + b_{jk}^I) m_k^I + u_j^R(t) < 0, \\
\dot{y}_j(t_0) = {}& -d_j q_j^I + a_{jj}^I l_j^R + a_{jj}^R l_j^I + \sum_{k \in N_1 \cup N_2, k \neq j} a_{jk}^I l_k^R + \sum_{k \in N_1 \cup N_2} b_{jk}^I l_k^R + \sum_{k \in N_3 \cup N_4} (a_{jk}^I \\
& + b_{jk}^I) m_k^R + \sum_{k \in N_1 \cup N_3} b_{jk}^R l_k^I + \sum_{k \in N_1 \cup N_3, k \neq j} a_{jk}^R l_k^I + \sum_{k \in N_2 \cup N_4} (a_{jk}^R + b_{jk}^R) m_k^I + u_j^I(t) < 0.
\end{aligned} \tag{6}$$

If $j \in N_2$ and $x_j(t_0) = q_j^R$, $y_j(t_0) = p_j^I$, according to (4), we have

$$\begin{aligned}
\dot{x}_j(t_0) = {}& -d_j q_j^R + a_{jj}^R l_j^R - a_{jj}^I m_j^I + \sum_{k \in N_1 \cup N_2, k \neq j} a_{jk}^R l_k^R + \sum_{k \in N_1 \cup N_2} b_{jk}^R l_k^R + \sum_{k \in N_3 \cup N_4} (a_{jk}^R \\
& + b_{jk}^R) m_k^R - \sum_{k \in N_2 \cup N_4} b_{jk}^I m_k^I - \sum_{k \in N_2 \cup N_4, k \neq j} a_{jk}^I m_k^I - \sum_{k \in N_1 \cup N_3} (a_{jk}^I + b_{jk}^I) l_k^I + u_j^R(t) < 0, \\
\dot{y}_j(t_0) = {}& -d_j p_j^I + a_{jj}^I l_j^R + a_{jj}^R m_j^I + \sum_{k \in N_1 \cup N_2, k \neq j} a_{jk}^I l_k^R + \sum_{k \in N_1 \cup N_2} b_{jk}^I l_k^R + \sum_{k \in N_3 \cup N_4} (a_{jk}^I \\
& + b_{jk}^I) m_k^R + \sum_{k \in N_2 \cup N_4} b_{jk}^R m_k^I + \sum_{k \in N_2 \cup N_4, k \neq j} a_{jk}^R m_k^I + \sum_{k \in N_1 \cup N_3} (a_{jk}^R + b_{jk}^R) l_k^I + u_j^I(t) > 0.
\end{aligned} \tag{7}$$

If $j \in N_3$ and $x_j(t_0) = p_j^R$, $y_j(t_0) = q_j^I$, according to (4), we have

$$\begin{aligned}
\dot{x}_j(t_0) = {}& -d_j p_j^R + a_{jj}^R m_j^R - a_{jj}^I l_j^I + \sum_{k \in N_1 \cup N_2} (a_{jk}^R + b_{jk}^R) l_k^R - \sum_{k \in N_1 \cup N_3, k \neq j} (a_{jk}^I + b_{jk}^I) l_k^I \\
& - \sum_{k \in N_2 \cup N_4} (a_{jk}^I + b_{jk}^I) m_k^I + \sum_{k \in N_3 \cup N_4, k \neq j} a_{jk}^R m_k^R + \sum_{k \in N_3 \cup N_4} b_{jk}^R m_k^R + u_j^R(t) > 0,
\end{aligned}$$

$$\dot{y}_j(t_0) = -d_j q_j^I + a_{jj}^I m_j^R + a_{jj}^R l_j^I + \sum_{k \in N_1 \cup N_2} (a_{jk}^I + b_{jk}^I) l_k^R + \sum_{k \in N_2 \cup N_4} (a_{jk}^R + b_{jk}^R) m_k^I$$

$$+ \sum_{k \in N_3 \cup N_4, k \neq j} (a_{jk}^I + b_{jk}^I) m_k^R + \sum_{k \in N_1 \cup N_3} b_{jk}^R l_k^I + \sum_{k \in N_1 \cup N_3, k \neq j} a_{jk}^R l_k^I + u_j^I(t) < 0.$$

$$(8)$$

If $j \in N_4$ and $x_j(t_0) = p_j^R$, $y_j(t_0) = p_j^I$, according to (4), we have

$$\dot{x}_j(t_0) = -d_j p_j^R + a_{jj}^R m_j^R - a_{jj}^I m_j^I + \sum_{k \in N_1 \cup N_2} (a_{jk}^R + b_{jk}^R) l_k^R + \sum_{k \in N_3 \cup N_4, k \neq j} a_{jk}^R m_k^R$$

$$+ \sum_{k \in N_3 \cup N_4} b_{jk}^R m_k^R - \sum_{k \in N_1 \cup N_3} (a_{jk}^I + b_{jk}^I) l_k^I - \sum_{k \in N_2 \cup N_4, k \neq j} a_{jk}^I m_k^I$$

$$- \sum_{k \in N_2 \cup N_4, k \neq j} b_{jk}^I m_k^I + u_j^R(t) > 0,$$

$$\dot{y}_j(t_0) = -d_j p_j^I + a_{jj}^I m_j^R + a_{jj}^R m_j^I + \sum_{k \in N_1 \cup N_2} (a_{jk}^I + b_{jk}^I) l_k^R + \sum_{k \in N_3 \cup N_4, k \neq j} a_{jk}^I m_k^R$$

$$+ \sum_{k \in N_3 \cup N_4} b_{jk}^I m_k^R + \sum_{k \in N_1 \cup N_3} (a_{jk}^R + b_{jk}^R) l_k^I + \sum_{k \in N_2 \cup N_4, k \neq j} a_{jk}^R m_k^I$$

$$+ \sum_{k \in N_2 \cup N_4} b_{jk}^R m_k^I + u_j^I(t) > 0.$$

$$(9)$$

From Eq. (6)-(9), we can conclude that Ω_1 is an invariant set, that is, $\forall t \geq t_0 - \tau$, $w(t) \in \Omega_1$.

Let $w(t) = w(t; t_0, \phi)$ and $\overline{w}(t) = w(t; t_0, \psi)$ be two states of (5) with initial conditions ϕ and ψ, respectively. Since $d_j > 0$, we can choose a small ε such that $0 < \varepsilon < d_j$, $1 \leq j \leq n$. Let

$$V(t) = e^{\varepsilon t} \sum_{j=1}^{2n} |w_j(t) - \overline{w}_j(t)|.$$

By calculating the upper right Dini derivative of $V(t)$ along (5), we can obtain that

$$\dot{V}(t) = e^{\varepsilon t} \left[\varepsilon \sum_{j=1}^{2n} |w_j(t) - \overline{w}_j(t)| + \sum_{j=1}^{2n} (\dot{w}_j(t) - \dot{\overline{w}}_j(t)) \mathrm{sgn}(w_j(t) - \overline{w}_j(t)) \right]$$

$$= e^{\varepsilon t} \sum_{j=1}^{2n} (\varepsilon - d_j) |w_j(t) - \overline{w}_j(t)| < 0$$

and this means $V(t) \leq V(t_0)$, thus we have

$$||w(t) - \overline{w}(t))|| \leq ||\phi - \phi_0||_{t_0} \exp\{-\varepsilon(t - t_0)\}. \tag{10}$$

Let $z_t(\phi) = z(t, \phi)$ be a solution of (1) with $z_0 = \phi \in \Omega_1$. Since Ω_1 is an invariant set of CVNN (1), we have $z_t(\phi) \in \Omega_1$. Define a Poincaré mapping $H : \Omega_1 \to \Omega_1$ by $H(\phi) = z_\omega(\phi)$.Then $H(\Omega_1) \subset \Omega_1$. We can choose a positive integer N such that $\exp\{-\varepsilon(N\omega - t_0)\} \leq \gamma < 1$, thus from (10) we have

$$||H(\phi) - H(\phi_0)|| \leq ||\phi - \phi_0||_{t_0} \exp\{-\varepsilon(N\omega - t_0)\} \leq \gamma ||\phi - \phi_0||_{t_0}$$

By Lemma 1, we conclude that there exists a unique fixed point $\bar{\phi} \in \Omega_1$ such that $H^N(\bar{\phi}) = \bar{\phi}$. In addition,

$$H^N(H(\bar{\phi})) = H(H^N(\bar{\phi})) = H(\bar{\phi}).$$

that is, $H(\bar{\phi})$ is also a fixed point of H^N. By the uniqueness of the fixed point of the mapping H^N, $H(\bar{\phi}) = \bar{\phi}$, i.e., $z_\omega \bar{\phi} = \bar{\phi}$. Let $\hat{z}(t) = \hat{x}(t) + i\hat{y}(t)$ be a state of (1) through $z_0 = \bar{\phi}$, then $((\hat{x}(t))^{\mathrm{T}}, (\hat{y}(t))^{\mathrm{T}})^{\mathrm{T}}$ is a state of (5), thus, $\forall j = 1, 2, \cdots, n$, $t + \omega \geq t_0$, we have

$$\dot{\hat{x}}_j(t+\omega) = -d_j x_j(t) + \sum_{k \in N_1 \cup N_2} (a_{jk}^R + b_{jk}^R) l_k^R + \sum_{k \in N_3 \cup N_4} (a_{jk}^R + b_{jk}^R) m_k^R$$
$$- \sum_{k \in N_1 \cup N_3} (a_{jk}^I + b_{jk}^I) l_k^I - \sum_{k \in N_2 \cup N_4} (a_{jk}^I + b_{jk}^I) m_k^I + u_j^R(t),$$

$$\dot{\hat{y}}_j(t+\omega) = -d_j \hat{y}_j(t+\omega) + \sum_{k \in N_1 \cup N_2} (a_{jk}^I + b_{jk}^I) l_k^R + \sum_{k \in N_3 \cup N_4} (a_{jk}^I + b_{jk}^I) m_k^R$$
$$+ \sum_{k \in N_1 \cup N_3} (a_{jk}^R + b_{jk}^R) l_k^I + \sum_{k \in N_2 \cup N_4} (a_{jk}^R + b_{jk}^R) m_k^I + u_j^I(t).$$

Hence $\bar{z}(t)$ is an isolated periodic orbit of CVNN (1) with period ω. Since Ω_1 has 4^n elements, there are 4^n limit cycles in Ω_1, and from (10), we know that such 4^n limit cycles are locally exponentially stable. This completes the proof. □

Remark 1. In the design and applications of associative memory, to determine the one step piecewise linear activation functions defined in Eq. (2), we shall first find the location of the equilibrium points or periodic orbits. And then, we can carefully choose parameters $p_j^R, p_j^I, q_j^R, q_j^I, (j = 1, 2, \cdots, n)$ such that the decomposed state space defined in (3) just match the location of the equilibrium points or periodic orbits.

In CVNN (1) if the external input is a constant vector, that is, $u(t)$ is replaced by a constant vector $u = (u_1, u_2, \cdots, u_n)^{\mathrm{T}}$, then we obtain the following CVNN:

$$\dot{z}(t) = -Dz(t) + Af(z(t)) + Bf(z(t - \tau)) + u. \tag{11}$$

Since the multistability of CVNN (11) is a special case of the multiperiodicity of CVNN (1) with arbitrary period or zero magnitude, according to Theorem 1, we can obtain the following corollary for the multistability of CVNN (11):

Corollary 1. *CVNN (11) has 4^n locally exponentially stable equilibrium points located in Ω_1 if $\forall t \geq t_0$ conditions (4) hold.*

Let $q_j^R = q_j^I = -1$, $p_j^R = p_j^I = 1$, $l_j^R = l_j^I = -1$, $m_j^R = m_j^I = 1$ $(j = 1, 2, \cdots, n)$, then the real and imaginary parts of activation functions $f_j(z)$ $(j = 1, 2, \cdots, n)$ of CVNN (1) are degenerated to

$$f_j^R(x) = \frac{1}{2}[|x+1| - |x-1|], \quad f_j^I(y) = \frac{1}{2}[|y+1| - |y-1|].$$

Then we obtain the following complex-valued cellular neural networks:

$$\dot{z}(t) = -Dz(t) + Af(z(t)) + Bf(z(t - \tau)) + u \tag{12}$$

and

$$\dot{z}(t) = -Dz(t) + Af(z(t)) + Bf(z(t - \tau)) + u(t) \tag{13}$$

and we consider the following region

$$\Omega_2 = \{w \in R^{2n} | -\infty < x_j < -1, -\infty < y_j < -1, j \in N_1; -\infty < x_j < -1, 1 < y_j < \infty,$$
$$\times j \in N_2; 1 < x < \infty, -\infty < y_j < -1, j \in N_3; 1 < x < \infty, 1 < y_j < \infty, j \in N_4\}$$

According to Theorem 1 and Corollary 1 we can obtain the multistability and multiperiodicity conditions for CVNNs (12) and (13):

Corollary 2. *CVNN (12) has 4^n locally exponentially stable equilibrium points located in Ω_2 if $\forall t \geq t_0$ the following conditions hold*

$$|u_j^R| < -d_j + \alpha_j - \sum_{k-1, k \neq j}^n (|a_{jk}^R| + |a_{jk}^I|) - \sum_{k=1}^n (|b_{jk}^R| + |b_{jk}^I|),$$

$$|u_j^I| < -d_j + \alpha_j - \sum_{k=1, k \neq j}^n (|a_{jk}^R| + |a_{jk}^I|) - \sum_{k=1}^n (|b_{jk}^R| + |b_{jk}^I|),$$

(14)

where $\alpha_j = \min\{a_{jj}^R + a_{jj}^I, a_{jj}^R - a_{jj}^I\} > 0$.
    CVNN (13) has 4^n locally exponentially limit cycles located in Ω_2 if $\forall t \geq t_0$ the above conditions hold.

Remark 2. The conditions in Corollary 2 include criteria for multistability and multiperiodicity of real-valued cellular neural networks as special cases when $a_{jk}^I = b_{jk}^I = 0$, $u_j^I = u_j^I(t) = 0$, $j, k = 1, 2, \cdots, n$, e.g., Theorem 1 and Corollary 1 in [14].

4   Illustrative Examples

In this section, we give a numerical example to demonstrate the above result.

Example 1. Consider a two-neuron CVNN described as follows:

$$\dot{z}(t) = -Dz(t) + Af(z(t)) + Bf(z(t-\tau)) + u$$ (15)

where

$$D = \begin{pmatrix} 2 & 0 \\ 0 & 1.5 \end{pmatrix}, A = \begin{pmatrix} 4.8 + 1.5i & 0.2 - 0.1i \\ -0.1 + 0.15i & 4 + 2i \end{pmatrix}, B = \begin{pmatrix} 0.25 + 0.1i & -0.15 - 0.2i \\ -0.2 + 0.2i & -0.15 + 0.1i \end{pmatrix},$$

$$u = (-2 + 0.8i, 0.4 - 0.6i)^{\mathrm{T}},$$

$$f_1^R(x) = \begin{cases} -3, & x < -1.5 \\ 1.5x - 0.75, & -1.5 \leq x \leq 2.5 \\ 3, & x > 2.5 \end{cases}, f_2^R(x) = \begin{cases} -3.5, & x < -0.8 \\ 3.6x - 9.1, & -0.8 \leq x \leq 1 \\ 3, & x > 1 \end{cases},$$

$$f_1^I(y) = \begin{cases} -2.5, & x < -0.2 \\ 6.4y - 2.2, & -0.2 \leq x \leq 0.5 \\ 2, & x > 0.5 \end{cases}, f_2^I(y) = \begin{cases} -2.8, & x < -0.6 \\ 3.6y - 0.6, & -0.6 \leq x \leq 1.2 \\ 3.7, & x > 1.2 \end{cases}.$$

It can be checked that the conditions in Corollary 1 hold. Thus, CVNN (15) has 16 locally exponentially stable equilibrium points. Simulation results with 100 random initial states are depicted in Fig. 1 and Fig. 2.

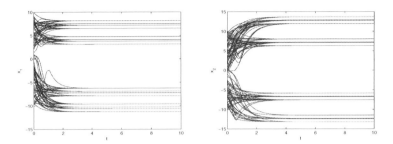

Fig. 1. Transient state of the real part of CVNN (15) in Example 1

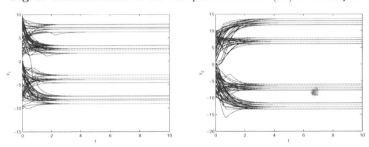

Fig. 2. Transient state of the imaginary part of CVNN (15) in Example 1

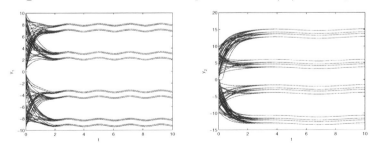

Fig. 3. Transient state of the real part of CVNN (16) in Example 1

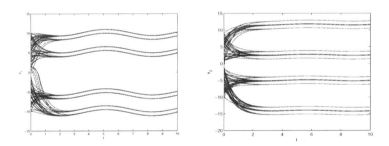

Fig. 4. Transient state of the imaginary part of CVNN (16) in Example 1

Consider another two-neuron CVNN described as follows:

$$\dot{z}(t) = -Dz(t) + Af(z(t)) + Bf(z(t - \tau)) + u(t) \tag{16}$$

where $u(t) = (-1.8\sin t + 0.6\mathrm{i}\sin(3t), -0.3\cos(2 - t) - 0.45\mathrm{i}\cos(t))^{\mathrm{T}}$, D, A, B are the same as in (15). It can be checked that the conditions in Theorem 1 hold. Thus, CVNN (16) has 16 locally exponentially periodic orbits. Simulation results with 100 random initial states are depicted in Fig. 3 and Fig. 4.

5   Conclusion and Future Works

In this paper, we investigate the multistability and multiperiodicity of CVNNs with one step piecewise linear activation functions. Since a complex-valued neuron contains more information than the real-valued counterpart, we can easily increase the storage capacity by using less number of neurons. This is quite helpful to the design and application of associative memory and other applications. Future works include the multistability and multiperiodicity of CVNNs with other types of activation functions and the applications in associative memory.

References

1. Bao, G., Zeng, Z.: Analysis and design of associative memories based on recurrent neural network with discontinuous activation functions. Neurocomputing **77**(1), 101–107 (2012)
2. Bohner, M., Rao, V.S.H., Sanyal, S.: Global stability of complex-valued neural networks on time scales. Differential Equations and Dynamical Systems **19**(1–2), 3–11 (2011)
3. Cao, J., Feng, G., Wang, Y.: Multistability and multiperiodicity of delayed Cohen-Grossberg neural networks with a general class of activation functions. Physica D **237**(13), 1734–1749 (2008)
4. Cheng, C., Lin, K., Shih, C.: Multistability in recurrent neural networks. SIAM Journal of Applied Mathematics **66**(4), 1301–1320 (2006)
5. Di Marco, M., Forti, M., Grazzini, M., Pancioni, L.: Limit set dichotomy and multistability for a class of cooperative neural networks with delays. IEEE Transactions on Neural Networks and Learning Systems **23**(9), 1473–1485 (2012)
6. Di Marco, M., Forti, M., Grazzini, M., Pancioni, L.: Convergent dynamics of nonreciprocal differential variational inequalities modeling neural networks. IEEE Transactions on Circuits and Systems I: Regular Papers **60**(12), 3227–3238 (2013)
7. Duan, C., Song, Q.: Boundedness and stability for discrete-time delayed neural network with complex-valued linear threshold neurons. Discrete Dynamics in Nature and Society **2010**, 1–19 (2010)
8. Hu, J., Wang, J.: Global stability of complex-valued recurrent neural networks with time-delays. IEEE Transcations on Neural Networks and Learning Systems **23**(6), 853–865 (2012)
9. Jankowski, S., Lozowski, A., Zurada, J.: Complex-valued multistate neural associative memory. IEEE Transactions on Neural Networks **7**(6), 1491–1496 (1996)
10. Kosaku, Y.: Functional Analysis. Springer, Berlin (1978)

11. Muezzinoglu, M.K., Guzelis, C., Zurada, J.M.: A new design method for the complex-valued multistate hopfield associative memory. IEEE Transactions on Neural Networks **14**(4), 891–899 (2003)
12. Tanaka, G., Aihara, K.: Complex-valued multistate associative memory with non-linear multilevel functions for gray-level image reconstruction. IEEE Transactions on Neural Networks **20**(9), 1463–1473 (2009)
13. Zeng, Z., Huang, T., Zheng, W.X.: Multistability of recurrent neural networks with time-varying delays and the piecewise linear activation function. IEEE Transcations on Neural Networks **21**(8), 1371–1377 (2010)
14. Zeng, Z., Wang, J.: Multiperiodicity and exponential attractivity evoked by periodic external inputs in delayed cellular neural networks. Neural Computation **18**(4), 848–870 (2006)
15. Zeng, Z., Wang, J., Liao, X.: Stability analysis of delayed cellular neural networks described using cloning templates. IEEE Transcations on Circuits and Systems-I:Regular Papers **51**(11), 2313–2324 (2004)
16. Zhou, B., Song, Q.: Boundedness and complete stability of complex-valued neural networks with time delay. IEEE Transactions on Neural Networks and Learning Systems **24**(8), 1227–1238 (2013)
17. Zhou, W., Zurada, J.M.: Discrete-time recurrent neural networks with complex-valued linear threshold neurons. IEEE Transactions on Circuits and Systems II: Express Briefs 56(8), 669–673 (2009)

Exponential Synchronization of Coupled Stochastic and Switched Neural Networks with Impulsive Effects

Yangling Wang[1,2](✉) and Jinde Cao[1]

[1] Department of Mathematics, Research Center for Complex Systems and Network Sciences, Southeast University, Nanjing 210096, People's Republic of China
wyangling@126.com
[2] School of Mathematics and Information Technology,
Nanjing Xiaozhuang University, Nanjing 211171, People's Republic of China
jdcao@seu.edu.cn

Abstract. In this paper, we investigate the exponential synchronization of coupled stochastic and switched neural networks (CSSNNs) with mixed time-varying delays. By exerting impulsive controller to the considered dynamical systems in each switching interval, and combining the multiple Lyapunov theory, we obtain a class of sufficient exponential synchronization criteria in terms of nonlinear equations and LMIs, which are easy to check. A simple example is presented to show the application of the criteria obtained in this paper.

Keywords: Coupled stochastic and switched neural networks · Exponential synchronization · Mixed time-varying delays · Impulsive effects

1 Introduction

Synchronization is an important and interesting collective behavior in coupled networks, and the study on synchronization of coupled neural networks can help us understand brain science and design coupled neural networks for real world applications. So synchronization of coupled neural networks has become a hot topic and been extensively investigated in recent years [1-8]. It is well known that time delays are unavoidable in the information processing of neurons due to various reasons, so most of the above-mentioned literatures are on synchronization of delayed neural networks.

It should be noted that because of link failures and the creation of new links in the information processing of neurons, the communication topology may change

This work was supported by the National Natural Science Foundation of China under grant 61272530, the Natural Science Foundation of Jiangsu Province of China under grant BK2012741, the Specialized Research Fund for the Doctoral Program of Higher Education under Grant 20130092110017, the JSPS Innovation Program under Grant CXZZ13_00, and the Natural Science Foundation of the Jiangsu Higher Education Institutions under Grant 14KJB110019.

ⓒ Springer International Publishing Switzerland 2014
Z. Zeng et al. (Eds.): ISNN 2014, LNCS 8866, pp. 69–79, 2014.
DOI: 10.1007/978-3-319-12436-0_9

in time, thus it is more natural and important to design switching signal when modeling real-world networks. There have been some works on the synchronization of switched neural networks, see for example, [9-11]. On the other hand, impulsive control has been proved to be an important and economical control method, because it acts only at the discrete times and synchronize the coupled systems quickly. Recently, hybrid impulsive switched systems have received increasing attentions due to their wide applications in various fields, one can refer to [12,13]. Zhang et al. have investigated in [13] the exponential synchronization of coupled impulsive switched neural networks by using average dwell time approach and comparison principle, but the coupling is linear and coupling delay was not taken into account in the associated networks. As discussed in [14], sometimes state variables $x_i(t)$ may be unobservable, but $g(x_i(t))$ can be observed easily, so nonlinear coupling is more practical. Additionally, Haykin pointed out in [15] that synaptic transmission is a noisy process brought on by random fluctuations from the release of neurotransmitters and other probabilistic causes. Practically, the stochastic phenomenon often appears in the electrical circuit design of neural networks. Hence, stochastic disturbances should also be considered in the dynamical behaviors of neural networks. However, the authors of [13] did't consider stochastic perturbation either.

Motivated by above discussions, this paper aims to analyze the exponential synchronization of nonlinearly coupled impulsive switched neural networks with stochastic perturbation and mixed time-varying delays. The rest of this paper is organized as follows: in Section 2, we first give the problem statement, and then present some definitions, lemmas and assumptions required throughout this paper; in Section 3, we will give a novel criterion to ensure the exponential synchronization for the considered neural networks in terms of LMIS and nonlinear equations; in Section 4, a simple example is provided to show the application of the theoretical results obtained in this paper.

2 Preliminaries

In this paper, we consider the following switched coupled neural networks with stochastic perturbations and impulsive effects:

$$\begin{cases} dx_i(t) = [-Cx_i(t) + B\tilde{f}(t, x_i(t)) + D\tilde{f}(t, x_i(t - \tau(t)))]dt \\ \quad + \tilde{g}(t, x_i(t), x_i(t - \rho(t)))dw(t) + \sum_{j=1}^{N} a_{ij,\sigma(t)}\tilde{h}(x_j(t))dt, \quad t \neq t_{k,l_k} \\ x_i(t_{k,l_k}) = (1 + \mu_{l_k})x_i(t_{k,l_k}^-), \qquad\qquad t = t_{k,l_k} \end{cases} \quad (1)$$

where $t \in [t_k, t_{k+1}), i = 1, \cdots, N$, $x_i(t) = [x_{i1}(t), \cdots, x_{in}(t)]^T \in \mathbb{R}^n$ is the state of the ith node at time t; $\tau(t), \rho(t)$ are the time-varying connected delay of neurons and coupled delay of nodes, respectively, and satisfying $0 < \tau(t) < \tau, 0 < \rho(t) < \rho$ with τ, ρ are positive constants; $\sigma(t) : [0, +\infty) \to \mathfrak{M} = \{1, 2, \cdots, m\}$ is a piecewise right-continuous function representing the switching signal. The switching time instants t_k satisfy $0 = t_0 < t_1 < \cdots < t_k < t_{k+1} < \cdots$, $\lim_{k \to +\infty} t_k = +\infty$

and $\inf\limits_{0\leq k<\infty}\{t_{k+1}-t_k\}\geq\aleph$ where $\aleph=\max\{\tau,\rho\};\{t_{k,l_k},l_k\in\mathbb{N}^+\}\subset[t_k,t_{k+1})$
are impulsive instances satisfying $t_k<t_{k,1}<t_{k,2}<\cdots<t_{k,l_k}<\cdots<t_{k+1}$,
and $x_i(t_{k,l_k}^+)=x_i(t_{k,l_k});\mu_{l_k}$ is the impulsive strength satisfying $(1+\mu_{l_k})^2\leq\mu<$
1; $C=diag\{c_1,\cdots,c_n\},c_l>0(l=1,\cdots,n)$ denotes the rate with which the l-th
neuron $x_{il}(t)$ reset their potential to the resting state when disconnected from the
networks and inputs, $B,D\in\mathbb{R}^{n\times n}$ denote the connection weight matrices of the
neurons, $\tilde{f}(t,x_i(t))=(\tilde{f}_1(t,x_i(t)),\cdots,\tilde{f}_n(t,x_i(t)))^T\in\mathbb{R}^n$ is the activation function of the neurons; $\tilde{g}(t,x_i(t),x_i(t-\rho(t)))\in\mathbb{R}^{n\times m}$ is the noise intensity function
matrix; $w(t)=(w_1(t),w_2(t),\cdots,w_m(t))^T\in\mathbb{R}^m$ is a Brownian motion defined on
a complete probability space $(\Omega,\mathcal{F},P)$ with a nature filtration $\{\mathcal{F}_t\}_{t\geq0}$ satisfying
$E(w_j(t))=0,E(w_j^2(t))=1,E(w_j(t)w_k(t))=0\ (j\neq k)$. The configuration cou-
pling matrices $A_{\sigma(t)}=(a_{ij,\sigma(t)})_{N\times N}$ are defined as follows: if there is a directed
edge from node j to node i, then $a_{ij,\sigma(t)}>0$, otherwise, $a_{ij,\sigma(t)}=0$, and $a_{ii,\sigma(t)}=$
$-\sum\limits_{j=1,j\neq i}^N a_{ij,\sigma(t)}$ for $i=1,\cdots,N$; $\tilde{h}(x_j(t))=(\tilde{h}_1(x_j(t)),\cdots,\tilde{h}_n(x_j(t)))^T\in\mathbb{R}^n$ is
the inner coupling vector function between two connected nodes i and j.

The initial condition of system (1) is given by $x_i(t)=\varphi_i(t)\in C([-\aleph,0],\mathbb{R}^n)$,
where $C([-\aleph,0],\mathbb{R}^n)$ is the set of continuous functions from $[-\aleph,0]$ to $\mathbb{R}^n$. Let $s(t)$
be a solution of the following stochastic delayed dynamical system of an isolate
neural network:

$$ds(t)=[-Cs(t)+B\tilde{f}(t,s(t))+D\tilde{f}(t,s(t-\tau(t)))]dt+\tilde{g}(t,s(t),s(t-\rho(t)))dw(t),(2)$$

which is the same as other neural networks. $s(t)$ can be any desired state: equilibrium point, a nontrivial periodic orbit, or even a chaotic orbit. In this paper,
we hope to force the network (1) to globally exponentially synchronize with $s(t)$.
The initial condition (2) is given by $s(t)=\phi(t)\in C([-\aleph,0],\mathbb{R}^n)$.

Let $e_i(t)=x_i(t)-s(t),f(t,e_i(t))=\tilde{f}(t,e_i(t)+s(t))-\tilde{f}(t,s(t)),g(t,e_i(t),e_i(t-\tau(t)))=\tilde{g}(t,e_i(t)+s(t),e_i(t-\tau(t)+s(t-\tau(t)))-\tilde{g}(t,s(t),s(t-\tau(t))),h(e_j(t-\rho(t)))=\tilde{h}(e_j(t-\rho(t))+s(t-\rho(t)))-\tilde{h}(s(t-\rho(t)));e(t)=(e_1^T(t),\cdots,e_N^T(t))^T,C^N=I_N\otimes C,B^N=I_N\otimes B,D^N=I_N\otimes D,\mathbf{A}=A\otimes I_n,F(t,e(t))=(f^T(t,e_1(t)),\cdots,f^T(t,e_N(t)))^T,H(e(t-\rho(t)))=(h^T(e_1(t-\rho(t))),\cdots,h^T(e_N(t-\rho(t))))^T,G(t,e(t),e(t-\tau(t)))=diag\{g(t,e_1(t),e_1(t-\tau(t))),\cdots,g(t,e_N(t),e_N(t-\tau(t)))\},dW(t)=1_N\otimes dw(t)$, where $1_N=(1,1,\cdots,1)^T,\sigma(t)=r_k\in\mathfrak{M},t\in[t_k,t_{k+1})$. Then we can write the error system of the coupled neural
networks (1) in the following compact form when $t\in[t_k,t_{k+1})$:

$$\begin{cases} de(t)=[-C^Ne(t)+B^NF(t,e(t))+D^NF(t,e(t-\tau(t)))dt \\ \qquad+G(t,e(t),e(t-\tau(t)))dW(t)+\mathbf{A}_{r_k}H(e(t-\rho(t)))]dt, \quad t\neq t_{k,l_k} \\ e(t_{k,l_k})=(1+\mu_{l_k})e(t_{k,l_k}^-). \qquad\qquad\qquad\qquad t=t_{k,l_k} \end{cases} \quad(3)$$

Definition 1. *The dynamical neural networks (1) is said to be globally exponentially synchronized to $s(t)$ in mean square if for any initial condition $x_i(t_0)$,
there exist constants $\lambda>0$ and $M>1$ such that the following inequality is*

satisfied for $t \geq t_0$:

$$E\left(\sum_{i=1}^{N} \|x_i(t) - s(t)\|^2\right) \leq M \sup_{t_0 - \aleph \leq \iota \leq t_0} E\left(\sum_{i=1}^{N} \|x_i(\iota) - s(\iota)\|^2\right) e^{-\lambda(t - t_0)}.$$

Definition 2. [10]: *An impulsive sequence $\varsigma = \{t_1, t_2, \cdots\}$ is said to have average impulsive interval T_a if there exist positive integer δ and positive constant T_a such that*

$$\frac{T - t}{T_a} - \delta \leq N_\delta(T, t) \leq \frac{T - t}{T_a} + \delta, \quad \forall T \geq t \geq 0,$$

where $N_\delta(T, t)$ denotes the number of impulsive times of the impulsive sequence $\{t_1, t_2, \cdots\}$ on the interval (t, T).

Assumption 1: Assume that there exist diagonal matrices L_1 and L_2 such that for $\forall x, y \in \mathbb{R}^n$, the function $\tilde{f}(t, \cdot)$ and $\tilde{h}(\cdot)$ satisfy the following Lipschitz conditions:

$$\|\tilde{f}(t, x) - \tilde{f}(t, y)\| \leq \|L_1(x - y)\|; \quad \|\tilde{h}(x) - \tilde{h}(y)\| \leq \|L_2(x - y)\|.$$

Assumption 2: Assume that there exist positive constants η_1, η_2 such that

$$\text{trace}\left\{[\tilde{g}(t, x_1, y_1) - \tilde{g}(t, x_2, y_2)]^T \cdot [\tilde{g}(t, x_1, y_1) - \tilde{g}(t, x_2, y_2)]\right\}$$
$$\leq \eta_1 \|x_1 - y_1\|^2 + \eta_2 \|x_2 - y_2\|^2, \ \forall x_1, y_1, x_2, y_2 \in \mathbb{R}^n, \ t \in \mathbb{R}^+.$$

Lemma 1. [13]: *Let $0 \leq \tau_i(t) \leq \tau, F(t, u, \bar{u}_1, \cdots, \bar{u}_m) : \mathbb{R}^+ \times \underbrace{\mathbb{R} \times \cdots \times \mathbb{R}}_{m+1}$ be nondecreasing in $\bar{u}_i$ for each fixed $(t, u, \bar{u}_1, \cdots, \bar{u}_{i-1}, \bar{u}_{i+1}, \cdots, \bar{u}_m)$, $i = 1, \cdots, m$, and $I_k(u) : \mathbb{R} \to \mathbb{R}$ be nondecreasing in u. Suppose that*

$$\begin{cases} D^+ u(t) \leq F(t, u(t), u(t - \tau_1(t)), \cdots, u(t - \tau_m(t))) \\ u(t_k^+) \leq I_k(u(t_k^-)), \ k \in \mathbb{N}_+ \end{cases}$$

and

$$\begin{cases} D^+ v(t) > F(t, v(t), v(t - \tau_1(t)), \cdots, v(t - \tau_m(t))) \\ v(t_k^+) \geq I_k(v(t_k^-)), \ k \in \mathbb{N}_+ \end{cases}$$

where the upper-right Dini derivative is defined as $D^+ y(t) = \overline{\lim}_{h \to 0^+} \frac{y(t+h) - y(t)}{h}$. Then $u(t) \leq v(t)$ for $-\tau \leq t \leq 0$ implies that $u(t) \leq v(t)$ for $t \geq 0$.

Lemma 2. [17]: *For any real matrices X, Y and any positive matrix U, the following inequality holds:*

$$2X^T Y \leq X^T U X + Y^T U^{-1} Y.$$

3   Exponential Synchronization Analysis

Theorem: Under Assumptions 1-2, the coupled neural networks (1) can be globally exponentially synchronized to $s(t)$, if there exist positive constants $\varepsilon_{1,r_k}, \varepsilon_{2,r_k}, \varepsilon_{3,r_k}$, positive matrices $P_{r_k} \in \mathbb{R}^{nN \times nN}$ satisfying $P_{r_k} \le \theta_{r_k} I_{nN}$ with θ_{r_k} are positive constants, such that the following conditions are satisfied:

$$(\mathbf{H_1}) \quad \Phi_{r_k} = \begin{pmatrix} \Phi_{11,r_k} & P_{r_k}B^N & P_{r_k}D^N & P_{r_k}\mathbf{A}_{r_k} & 0 & 0 \\ (B^N)^T P_{r_k} & -\varepsilon_{1,r_k}I_{nN} & 0 & 0 & 0 & 0 \\ (D^N)^T P_{r_k} & 0 & -\varepsilon_{2,r_k}I_{nN} & 0 & 0 & 0 \\ \mathbf{A}_{r_k}^T P_{r_k} & 0 & 0 & -\varepsilon_{3,r_k}I_{nN} & 0 & 0 \\ 0 & 0 & 0 & 0 & \Phi_{55,r_k} & 0 \\ 0 & 0 & 0 & 0 & 0 & \Phi_{66,r_k} \end{pmatrix} < 0,$$

where $r_k \in \mathfrak{M}$, $\Phi_{11,r_k} = -2P_{r_k}C^N + \varepsilon_{1,r_k}(L_1^T L_1)^N + \eta_1\theta_{r_k}I_{nN} + \alpha_{r_k}P_{r_k}$ and $\Phi_{55,r_k} = \varepsilon_{2,r_k}(L_1^T L_1)^N + \eta_2\theta_{r_k}I_{nN} - \beta_{r_k}P_{r_k}$, $\Phi_{66,r_k} = \varepsilon_{3,r_k}(L_2^T L_2)^N - \gamma_{r_k}P_{r_k}$.

$$(\mathbf{H_2}) \qquad -\alpha + \frac{ln\mu}{T_a} + \mu^{-\delta}(\beta + \gamma) < 0,$$

where $\alpha = \min_{r_k \in \mathfrak{M}}\{\alpha_{r_k}\}, \beta = \max_{r_k \in \mathfrak{M}}\{\beta_{r_k}\}, \gamma = \max_{r_k \in \mathfrak{M}}\{\gamma_{r_k}\}$.

$$(\mathbf{H_3}) \qquad \lambda - \frac{ln\Upsilon}{T_a} > 0,$$

where λ is the sole positive solution of the equation $-\alpha + \frac{ln\mu}{T_a} + \lambda + \mu^{-\delta}(\beta e^{\lambda\tau} + \gamma e^{\lambda\rho}) = 0$, $\Upsilon = \mu^{-\delta}\max\{\frac{\bar{p}}{\underline{p}}, e^{\lambda\aleph}\}, \bar{p} = \max_{r_k \in \mathfrak{M}}\{\lambda_{max}(P_{r_k})\}, \underline{p} = \min_{r_k \in \mathfrak{M}}\{\lambda_{min}(P_{r_k})\}$.

Proof: Define the following Lyapunov functions for system (3):

$$V(t) = e^T(t)P_{r_k}e(t), t \in [t_k, t_{k+1}), k \in \mathbb{N}^+.$$

Differentiating $V(t)$ along the trajectories of system (3) for $t \in [t_k, t_{k+1})$, we can obtain

$$dV(t) = \mathcal{L}V(t)dt + 2e^T(t)P_{r_k}G(t, e(t), e(t-\tau(t)))dW(t). \tag{4}$$

By applying the Itô's formula to $\bar{V}(t)$ we can obtain

$$\begin{aligned}\mathcal{L}V(t) = {} & 2e^T(t)P_{r_k}[-C^N e(t) + B^N F(t, e(t)) + D^N F(t, e(t-\tau(t)))] \\ & + trace[G^T(t, e(t), e(t-\tau(t)))P_{r_k}G(t, e(t), e(t-\tau(t)))] \\ & + 2e^T(t)P_{r_k}\mathbf{A}_{r_k}H(e(t-\rho(t))).\end{aligned}$$

By using Lemma 2 and Assumption 1 we get

$$\begin{aligned}& 2e^T(t)P_{r_k}B^N F(t, e(t)) \\ \le {} & \frac{1}{\varepsilon_{1,r_k}}e^T(t)P_{r_k}B^N(P_{r_k}B^N)^T e(t) + \varepsilon_{1,r_k}F^T(t, e(t))F(t, e(t)) \\ \le {} & \frac{1}{\varepsilon_{1,r_k}}e^T(t)P_{r_k}B^N(P_{r_k}B^N)^T e(t) + \varepsilon_{1,r_k}e^T(t)(L_1^T L_1)^N e(t). \tag{5}\end{aligned}$$

Similar to (5), we can obtain the following inequalities:

$$2e^T(t)P_{r_k}D^N F(t, e(t - \tau(t)))$$
$$\leq \frac{1}{\varepsilon_{2,r_k}} e^T(t)P_{r_k}D^N (P_{r_k}D^N)^T e(t) + \varepsilon_{2,r_k} e^T(t - \tau(t))(L_1^T L_1)^N e(t - \tau(t)), \quad (6)$$

$$2e^T(t)P_{r_k}\mathbf{A}_{r_k}H(e(t - \rho(t)))$$
$$\leq \frac{1}{\varepsilon_{3,r_k}} e^T(t)P_{r_k}\mathbf{A}_{r_k}\mathbf{A}_{r_k}^T P_{r_k}^T e(t) + \varepsilon_{3,r_k} e^T(t - \rho(t))(L_2^T L_2)^N e(t - \rho(t)). \quad (7)$$

According to Assumption 2 we have

$$trace[G^T(t, e(t), e(t - \tau(t)))P_{r_k}G(t, e(t), e(t - \tau(t)))]$$
$$\leq \theta_{r_k} \sum_{i=1}^{N} \left(\eta_1 \|e_i(t)\|^2 + \eta_2 \|e_i(t - \tau(t))\|^2 \right)$$
$$= \theta_{r_k} \left(\eta_1 \ e^T(t)e(t) + \eta_2 \ e^T(t - \tau(t))e(t - \tau(t)) \right). \quad (8)$$

It follows from (5)-(8) that for $t \in [t_k, t_{k+1})$,

$$\mathcal{L}V(t) \leq e^T(t)\Big\{ -P_{r_k}C^N - (P_{r_k}C^N)^T + \frac{1}{\varepsilon_{1,r_k}}P_{r_k}B^N(P_{r_k}B^N)^T$$
$$+\varepsilon_1(L_1^T L_1)^N + \frac{1}{\varepsilon_{2,r_k}}P_{r_k}D^N(P_{r_k}D^N)^T + \frac{1}{\varepsilon_{3,r_k}}P_{r_k}\mathbf{A}_{r_k}\mathbf{A}_{r_k}^T P_{r_k} + \theta_{r_k}\eta_1 I_{nN}$$
$$+\alpha_{r_k}P_{r_k}\Big\}e(t) - \alpha_{r_k}e^T(t)P_{r_k}e(t) + \beta_{r_k}e^T(t - \tau(t))P_{r_k}e(t - \tau(t))$$
$$+e^T(t - \tau(t))\Big[\varepsilon_{2,r_k}(L_1^T L_1)^N + \theta_{r_k}\eta_2 I_{nN} - \beta_{r_k}P_{r_k}\Big]e(t - \tau(t))$$
$$+e^T(t - \rho(t))\Big[\varepsilon_{3,r_k}(L_2^T L_2)^N - \gamma_{r_k}P_{r_k}\Big]e(t - \rho(t))$$
$$+\gamma_{r_k}e^T(t - \rho(t))P_{r_k}e(t - \rho(t))$$
$$\leq -\alpha_{r_k}V(t) + \beta_{r_k}V(t - \tau(t)) + \gamma_{r_k}V(t - \rho(t)). \quad (9)$$

Integrating on both sides of (9) from t to $t + \triangle t$ for any $\triangle t > 0$ and taking mathematical expectation. Let $m(t) = EV(t)$, associating with the properties of the Itô's integral and Dini derivation, we can derive from (9) that

$$D^+ m(t) \leq -\alpha_{r_k}m(t) + \beta_{r_k}m(t - \tau(t)) + \gamma_{r_k}m(t - \rho(t)), \quad t \in [t_k, t_{k+1}).$$

When $t = t_{k,l_k}$, we can easily derive that

$$m(t_{k,l_k}) = (1 + \mu_{l_k})^2 E[e^T(t_{k,l_k}^-)P_{r_k}e(t_{k,l_k}^-)] \leq \mu m(t_{k,l_k}^-).$$

For any $\varepsilon > 0$, let $y(t)$ be a unique solution of the following delay system:

$$\begin{cases} \dot{y}(t) = -\alpha y(t) + \beta y(t - \tau(t)) + \gamma y(t - \rho(t)) + \varepsilon, \ t \neq t_{k,l_k} \\ y(t_{k,l_k}) = \mu y(t_{k,l_k}^-), \quad t = t_{k,l_k} \\ y(t) = m(t), \ t_k - \aleph \leq t \leq t_k. \end{cases} \quad (10)$$

By the formula for the variation of parameters, it follows from (10) that for $t \in [t_k, t_{k+1})$,

$$y(t) = P(t, t_k)y(t_k) + \int_{t_k}^t P(t, s)\big[\beta y(s - \tau(s)) + \gamma y(s - \rho(s)) + \varepsilon\big]ds, \qquad (11)$$

where $P(t, s)$, $t, s > t_k$ is the Cauchy matrix of the linear system

$$\begin{cases} \dot{y}(t) = -\alpha y(t), \ t \neq t_{k,l_k} \\ y(t_{k,l_k}) = \mu y(t_{k,l_k}^-), \quad t = t_{k,l_k}. \end{cases} \qquad (12)$$

According to the representation of Cauchy matrix, one can get the following estimation:

$$P(t, s) = e^{-\alpha(t-s)}\mu^{N_\delta(s,t)} \leq \mu^{-\delta}e^{-\alpha^*(t-s)},$$

where $\alpha^* = \alpha - \frac{\ln\mu}{T_a}$. Define $s(\varsigma) = \varsigma - \alpha^* + \mu^{-\delta}(\beta e^{\varsigma\tau} + \gamma e^{\varsigma\rho})$. From ($\mathbf{H_2}$) we know $s(0) = -\alpha^* + \mu^{-\delta}(\beta + \gamma) < 0$. Since $\dot{s}(\varsigma) > 0$ and $\lim\limits_{\varsigma \to +\infty} s(\varsigma) = +\infty$, there exists a unique $\lambda > 0$ such that $s(\lambda) = 0$, i. e., $\lambda - \alpha^* + \mu^{-\delta}(\beta e^{\lambda\tau} + \gamma e^{\lambda\rho}) = 0$. Let $\xi = \mu^{-\delta}\|y(t_k)\|_\aleph = \mu^{-\delta} \sup\limits_{t_k-\aleph \leq t \leq t_k} \|y(t)\|$. In the following, we shall prove the following inequality is satisfied:

$$y(t) < \xi e^{-\lambda(t-t_k)} + \frac{\varepsilon}{\alpha^*\mu^\delta - \beta - \gamma}, \quad t_k - \aleph \leq t \leq t_{k+1}. \qquad (13)$$

It is obvious that $y(t) \leq \mu^\delta\xi < \xi < \xi e^{-\lambda(t-t_k)} + \frac{\varepsilon}{\alpha^*\mu^\delta-\beta-\gamma}$ for $t_k - \aleph \leq t \leq t_k$. When $t_k < t < t_{k+1}$, we will prove the inequality (13) is still satisfied by the way of contradiction. If there exists a $t^* \in (t_k, t_{k+1})$ such that

$$y(t^*) \geq \xi e^{-\lambda(t^*-t_k)} + \frac{\varepsilon}{\alpha^*\mu^\delta - \beta - \gamma}, \qquad (14)$$

and

$$y(t) < \xi e^{-\lambda(t-t_k)} + \frac{\varepsilon}{\alpha^*\mu^\delta - \beta - \gamma}, \quad t \in (t_k - \aleph, t^*) \qquad (15)$$

Note that $\tau(t) \leq \tau, \rho(t) \leq \rho$ and $e^{\lambda\tau}\beta + e^{\lambda\rho}\gamma = \mu^\delta(\alpha^* - \lambda)$, then by some simple computation, we can derive from (11) and (15) that

$$y(t^*)$$
$$< \xi e^{-\alpha^*(t^*-t_k)} + \int_{t_k}^{t^*} \mu^{-\delta}e^{-\alpha^*(t^*-s)}\Big[\xi(e^{\lambda\tau}\beta + e^{\lambda\rho}\gamma)e^{-\lambda(s-t_k)} + \frac{\alpha^*\mu^\delta\varepsilon}{\alpha^*\mu^\delta - \beta - \gamma}\Big]ds$$
$$= \xi e^{-\lambda(t^*-t_k)} + \frac{\varepsilon}{\alpha^*\mu^\delta - \beta - \gamma} - \frac{\varepsilon}{\alpha^*\mu^\delta - \beta - \gamma}e^{-\alpha^*(t^*-t_k)}$$
$$< \xi e^{-\lambda(t^*-t_k)} + \frac{\varepsilon}{\alpha^*\mu^\delta - \beta - \gamma},$$

which contradicts with (14). Thus, (13) is always satisfied for $t_k - \aleph \leq t < t_{k+1}$. Let $\varepsilon \to 0$, one can obtain $y(t) \leq \xi e^{-\lambda(t-t_k)}$. Then it follows from Lemma 1 that

$m(t) \leq y(t) \leq \xi e^{-\lambda(t-t_k)} = \mu^{-\delta}\|m(t_k)\|_{\aleph}e^{-\lambda(t-t_k)}$ for $t_k \leq t < t_{k+1}$. We will show by induction that

$$m(t) \leq \mu^{-\delta}\Upsilon^k\|m(t_0)\|_{\aleph}e^{-\lambda(t-t_0)}, \qquad t_k \leq t < t_{k+1}, \tag{16}$$

where $\Upsilon = \mu^{-\delta}\max\{\frac{\overline{p}}{\underline{p}}, e^{\lambda\aleph}\}$ and $\overline{p} = \max\limits_{r_k \in \mathfrak{M}}\{\lambda_{max}(P_{r_k})\}, \underline{p} = \min\limits_{r_k \in \mathfrak{M}}\{\lambda_{min}(P_{r_k})\}$. When $t \in [t_0, t_1), m(t) \leq \mu^{-\delta}\|m(t_0)\|_{\aleph}e^{-\lambda(t-t_0)}$. Assume (16) holds for $1 \leq k \leq j, j \in \mathbb{N}^+$, then we will show that (16) holds for $k = j + 1$. Since

$$m(t) \leq \mu^{-\delta}\Upsilon^j\|m(t_0)\|_{\aleph}e^{-\lambda(t_{j+1}-\aleph-t_0)} = \mu^{-\delta}\Upsilon^j e^{\lambda\aleph}\|m(t_0)\|_{\aleph}e^{-\lambda(t_{j+1}-t_0)}$$

for $t_{j+1} - \aleph \leq t < t_{j+1}$, and note that $t_{j+1} < t_{j+1,1}$, which follows that

$$m(t_{j+1}) = E(e^T(t_{j+1})P_{r_{j+1}}e(t_{j+1})) = E(e^T(t_{j+1}^-)P_{r_{j+1}}e(t_{j+1}^-))$$
$$\leq \frac{\overline{p}}{\underline{p}}m(t_{j+1}^-) \leq \frac{\overline{p}}{\underline{p}}\mu^{-\delta}\Upsilon^j\|m(t_0)\|_{\aleph}e^{-\lambda(t_{j+1}-t_0)},$$

then it follows that

$$\|m(t_{j+1})\|_{\aleph} \leq \Upsilon\Upsilon^j\|m(t_0)\|_{\aleph}e^{-\lambda(t_{j+1}-t_0)} = \Upsilon^{j+1}\|m(t_0)\|_{\aleph}e^{-\lambda(t_{j+1}-t_0)}.$$

Thus we can get

$$m(t) \leq \mu^{-\delta}\|m(t_{j+1})\|_{\aleph}e^{-\lambda(t-t_{j+1})} \leq \mu^{-\delta}\Upsilon^{j+1}\|m(t_0)\|_{\aleph}e^{-\lambda(t_{j+1}-t_0)}e^{-\lambda(t-t_{j+1})}$$
$$= \mu^{-\delta}\Upsilon^{j+1}\|m(t_0)\|_{\aleph}e^{-\lambda(t-t_0)}, \quad t_{j+1} \leq t < t_{j+2}$$

Thus, (16) can be derived for $\forall t \in [t_k, t_{k+1})$ and $\forall k \in \mathbb{N}^+$ by the induction principle. For an arbitrarily given $t > t_0, \exists k \in \mathbb{N}^+$, such that $t \in [t_k, t_{k+1})$. Note that $k \leq N_\delta(t, t_0)$, then it follows from (16) that

$$m(t) \leq \mu^{-\delta}\Upsilon^k\|m(t_0)\|_{\aleph}e^{-\lambda(t-t_0)} \leq \mu^{-\delta}\Upsilon^{N_\delta(t,t_0)}\|m(t_0)\|_{\aleph}e^{-\lambda(t-t_0)}$$
$$\leq \mu^{-\delta}\Upsilon^{\delta+\frac{t-t_0}{T_a}}\|m(t_0)\|_{\aleph}e^{-\lambda(t-t_0)} = \left(\frac{\Upsilon}{\mu}\right)^\delta\|m(t_0)\|_{\aleph}e^{-\lambda^*(t-t_0)},$$

where $\lambda^* = \lambda - \frac{ln\Upsilon}{T_a} > 0$. Then we have

$$\underline{p}E(\|e(t)\|^2) \leq m(t) \leq \left(\frac{\Upsilon}{\mu}\right)^\delta\|m(t_0)\|_{\aleph}e^{-\lambda^*(t-t_0)} \leq \overline{p}\left(\frac{\Upsilon}{\mu}\right)^\delta E(\|e(t_0)\|_{\aleph}^2)e^{-\lambda^*(t-t_0)},$$

which follows that

$$E\left(\sum_{i=1}^N\|x_i(t) - s(t)\|^2\right) \leq M \sup_{t_0-\aleph\leq\iota\leq t_0} E\left(\sum_{i=1}^N\|x_i(\iota) - s(\iota)\|^2\right)e^{-\lambda^*(t-t_0)},$$

where $M = \frac{\overline{p}}{\underline{p}}(\frac{\Upsilon}{\mu})^\delta > 1$. This shows that the dynamical network (1) is globally exponentially synchronized to $s(t)$ in mean square. This completes the proof.

4   Numerical Simulation

Example 1. In system (1), we select $\tilde{f}(t, x_i(t)) = (\tilde{f}_1(t, x_i(t)), \tilde{f}_2(t, x_i(t)))^T$ and $\tilde{f}_1(t, x_i(t)) = \frac{\sqrt{2}}{8} x_{i1}(t) + \frac{\sqrt{3}}{8}(|x_{i2}(t)+1| - |x_{i2}(t)-1|), \tilde{f}_2(t, x_i(t)) = \frac{\sqrt{6}}{8}(|x_{i2}(t)+1|$ $- |x_{i2}(t) - 1|)$, which follows that $L_1 = diag\{0.25, 0.75\}, L_2 = diag\{0.5, 0.25\}$. Let $\mu_{l_k} = -0.1$ for $\forall l_k \in \mathbb{N}^+$, and $C_1 = 4.5I_2, C_2 = 3.8I_2$,

$$\tilde{g}(t, x_i(t), x_i(t - \tau(t))) = 0.1 = diag\{x_i(t), x_i(t - \tau(t))\}, \tau(t) = 0.3sint,$$

$$\rho(t) = 0.2cost, B_1 = \begin{pmatrix} 0.8 & 0.9 \\ -0.6 & 0.8 \end{pmatrix}, B_2 = \begin{pmatrix} 1 & 0.5 \\ -0.9 & 1 \end{pmatrix}, D_1 = \begin{pmatrix} 0.5 & 0.4 \\ 0.8 & 0.5 \end{pmatrix},$$

$$D_2 = \begin{pmatrix} 0.9 & 0.5 \\ 0.6 & 0.8 \end{pmatrix}, A_1 = \begin{pmatrix} -0.9 & 0.5 & 0.4 \\ 0.8 & -1 & 0.2 \\ 0.5 & 0.5 & -1 \end{pmatrix}, A_2 = \begin{pmatrix} -1 & 0.4 & 0.6 \\ 0.6 & -1.1 & 0.5 \\ 0.4 & 0.6 & -1 \end{pmatrix}.$$

Assuming that the coupled neural networks switches in a random order between two networks, i. e., $\mathfrak{M} = \{1, 2\}$. The switching scheme is shown in Fig. 1. Select $\alpha_1 = 4.5, \beta_1 = \gamma_1 = 0.42, \alpha_2 = 4.25, \beta_2 = \gamma_2 = 0.78$, then by using Matlab LMI tool we can obtain $\varepsilon_{1,1} = 1.3931, \varepsilon_{1,2} = 1.2767, \varepsilon_{2,1} = 0.5423, \varepsilon_{2,2} = 0.9266, \varepsilon_{3,1} = 1.4451, \varepsilon_{3,2} = 3.1661, \theta_1 = 2.1717, \theta_2 = 1.6614, p = 0.6353, \bar{p} = 1.3182$. The impulsive sequence is constructed by taking $T_a = 4.6$ and $\delta = 4$, then by solving the nonlinear equation $-\alpha + \frac{ln\mu}{T_a} + \lambda + \mu^{-\delta}(\beta e^{\lambda\tau} + \gamma e^{\lambda\rho}) = 0$, we can get $\lambda = 0.3431$. So by virtue of the Theorem in this paper, it can be concluded that the considered network can be exponentially synchronized onto the objective trajectory. The following Figure shows that the errors between the networks' states and converge to zero under the given conditions.

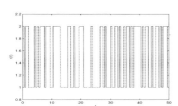

Fig.1. The switching scheme

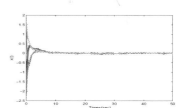

Fig.2. The state variables $x_{ir}(t)$

5   Conclusion

The exponential synchronization of switched coupled neural networks with mixed time-varying delays and stochastic disturbances is investigated in this paper. The main contribution of this paper contains three aspects. Firstly, as discussed in the section of Introduction, the network model considered in this paper is more practical in real world. Secondly, different from the average dwell time approach used in many existing literatures, there is no upper bound for switching interval, which is only assumed to be greater than the maximum of delays. As for the impulsive scheme, the named average impulsive interval is utilized to get less conservative synchronization criterion. Thirdly, by using multiple Lyapunov function, we have shown that the exponential synchronization can be achieved by solve some LMIs and nonlinear equations, which are easy to check.

References

1. Chen, G.R., Zhou, J., Liu, Z.R.: Global synchronization of coupled delayed neural networks and applications to chaotic CNN models. Int. J. Bifurcat. Chaos **14**, 2229–2240 (2004)
2. Chang, C.L., Fan, K.W., Chung, I.F., Lin, C.H.: A recurrent fuzzy coupled cellular neural network system with automatic structure and template learning. IEEE Trans. Circuits Syst. Express Briefs **53**, 602–606 (2006)
3. Liang, J.L., Wang, Z.D., Liu, Y.Y., Liu, X.H.: Robust synchronization of an array of coupled stochastic discrete-time delayed neural networks. IEEE Trans. Neural Netw. **19**, 1910–1921 (2008)
4. Wu, W., Chen, T.P.: Global synchronization criteria of linearly coupled neural network systems with time-varying coupling. IEEE Trans. Neural Netw. **19**, 319–332 (2008)
5. Cao, J.D., Chen, G.R., Li, P.: Global synchronization in an array of delayed neural networks with hybrid coupling. IEEE Trans. Syst. Man Cybern B **38**(2), 488–498 (2008)
6. Yang, X.S., Cao, J.D., Long, Y., Rui, W.G.: Adaptive lag synchronization for competitive neural networks with mixed delays and uncertain hybrid perturbations. IEEE Trans. Neural Netw. **21**, 1656–1667 (2010)
7. Wu, Z.G., Shi, P., Su, H., Chu, J.: Exponential synchronization of neural networks with discrete and distributed delays under time-varying sampling. IEEE Trans. Neural Netw. Learn Syst. **23**, 1368–1376 (2012)
8. Wang, G., Shen, Y.: Exponential synchronization of coupled memristive neural networks with time delays. Neural Comput. Appl. **24**, 1421–1430 (2014)
9. Tang, Y., Fang, J.A., Miao, Q.Y.: Synchronization of stochastic delayed neural networks with Marokovian switching and its application. Int. J. Neural Syst. **19**, 43–56 (2009)
10. Lu, J.Q., Ho, D.W.C., Cao, J.D., Kurths, J.: Exponential synchronization of linearly coupled neural networks with switching topology. IEEE Trans. Neural Netw. **22**, 169–175 (2011)
11. Shi, G.D., Ma, Q.: Synchronization of stochastic Markovian jump neural networks with reaction-diffusion terms. Neurocomputing **77**, 275–280 (2012)
12. Guan, Z.H., Hill, D.J., Shen, X.: On hybrid impulsive and switching systems and application to nonlinear control. IEEE Trans. Autom. Control **50**, 1058–1062 (2005)
13. Li, C.D., Feng, G., Huang, T.: On hybrid impulsive and switching neural networks. IEEE Trans. Syst. Man Cybern B Cybern 38, 1549–1560 (2008)
14. Zhang, W.B., Tang, Y., Miao, Q.Y., Du, W.: Exponential synchronization of coupled switched neural networks with mode-dependent impulsive effects. IEEE Trans. Neural Netw. Learn Syst. **24**, 1368–1376 (2013)
15. Wang, J.Y., Feng, J.W., Chen, X., Zhao, Y.: Cluster synchronization of nonlinearly-coupled complex networks with nonidentical nodes and asymmetrical coupling matrix. Nonlinear Dyn. **67**, 1635–1646 (2012)

16. Haykin, S.: Neural Networks. Prentice-Hall, Englewood Cliffs (1994)
17. Cao, J.D., Liang, J.L., Lam, J.: Exponential stability of high-order bidirectional associative memory neural networks with time delays. Physica D **199**, 425–436 (2004)

The Research of Building Indoor Temperature Prediction Based on Support Vector Machine

Wenbiao Wang, Qi Cai[✉], and Siyuan Wang

School of Information Science and Technology, Dalian Maritime University, Dalian, China
wwb201@163.com, 1213743426@qq.com, dl_wsy@sina.com

Abstract. Aiming at the problems for predicting the building indoor temperature so as to set up a reasonable indoor environment, four building indoor
temperature prediction models were established in this paper. The theory of
Support Vector Machine (SVM) and the LibSVM toolbox were used to predict
the indoor temperature. The experimental results shown that the prediction effect of the model which the input are the outdoor temperature, the solar radiation, the wind speed and the time series, the output is the indoor temperature is
the best. It's really effective to use the support vector regression (SVR) model
to predict the building indoor temperature. This predicting method based on
SVM can be promoted and applied in the field of prediction.

Keywords: Building indoor temperature · Time series prediction · SVM ·
Kernel function · LibSVM toolbox

1   Introduction

In recent years, with the increase of the urban buildings and the modernization buildings, the research of the building indoor temperature has been widely carried out
throughout the world [1]. It's very important to predict the building indoor temperature so as to set up a reasonable indoor environment and save the energy and resource,
the most importantly, it's really convenient to create an appropriate living environment by predicting the indoor temperature [2].

The current predicting methods mainly include the linear regression method, the
exponential smoothing method, the analytic hierarchy process method, and the neural
network method and so on [3]. Among these predicting methods, a new method of
machine learning, support vector machine (SVM) based on the statistical theory, was
put forward by Vapnik and other researchers. SVM has been successfully applied in
many areas such as the building energy consumption simulation, the power grid load
and the gas load predicting because of its excellent learning ability [4]. The building
indoor temperature prediction can be also regarded as a complex nonlinear approximation problem between the building indoor temperature characteristic and its influence factors.

Dalian Maritime University, Dalian 116026, China.

Z. Zeng et al. (Eds.): ISNN 2014, LNCS 8866, pp. 80–89, 2014.
DOI: 10.1007/978-3-319-12436-0_10

Therefore, the theory of SVM was attempt to applied in the building indoor temperature prediction in this paper. On the basis of the establishment of the building indoor temperature prediction models, the theory of SVM was applied in an office building indoor temperature prediction so as to test the feasibility and validity of this predicting method based on SVM.

2   The Theory of SVM

2.1   SVM for Regression Estimation

SVM is a method of machine learning, which built its approximation function based on underlying concepts that rise from the statistical learning theory as well as on a set of input/output examples that come up from process under analysis. SVM has been widely used in practice for the classification and regression problems. SVM was originally used for the classification purposes but its principle can be easily extended to the task of the regression and the time series prediction [5]. This method has been proven to be very effective for addressing the general purpose classification and regression problems. Because of it's out of the scope of this paper to explain the theory of SVM completely, and the theme of this paper is the building indoor temperature prediction which is a regression estimation problem, so this paper focuses on the theory of support vector regression (SVR) which is the theory of SVM for regression estimation.

The basic idea of SVR is to introduce the kernel function, and map the input space into a high dimensional feature space by a nonlinear mapping, in final, the original nonlinear regression problem performs a linear regression problem in this feature space [6].

Suppose that all the input parameters of the SVR model compose a vector X_i^* (i represents one input sample), and all the output parameters of the SVR model compose a vector Y_i^* (i represents one output sample).

In order to improve the calculation efficiency, and prevent individual data from overflowing during the calculation, input and output parameters should be normalized as follows:

$$X_i = \frac{X_i^* - X_{min}^*}{X_{max}^* - X_{min}^*} \tag{1}$$

$$Y_i = \frac{Y_i^* - Y_{min}^*}{Y_{max}^* - Y_{min}^*} \tag{2}$$

When the total number of the sample is N , the sample set is defined as $\{(X_i, Y_i)\}_{i=1}^{N}$. Therefore, the SVR model approximates the relationship between the input and output parameters using the following form:

$$Y = f(X) = w \times \phi(X) + b \tag{3}$$

Where $\phi(X)$ represents the high-dimensional feature space which is non-linearly mapped from the input space X, the coefficients w and b are estimated by minimizing the regularized risk function, as shown in Eq. (4):

$$Minimize : \frac{1}{2}\|w\|^2 + C\frac{1}{2}\sum_{i=1}^{N} L_\varepsilon(Y_i, f(X_i)) \tag{4}$$

$$L_\varepsilon(Y_i, f(X_i)) = \begin{cases} 0 & |Y_i - f(X_i)| \le \varepsilon \\ |Y_i - f(X_i)| - \varepsilon & |Y_i - f(X_i)| > \varepsilon \end{cases} \tag{5}$$

In Eq. (4), the first term $\|w\|^2$ is called regularized term. Minimizing $\|w\|^2$ will make a function as flat as possible. The second term is the empirical error measured by the ε-insensitive loss function, which is defined as Eq. (5). This defines a ε tube (Fig. 1) so that if predicted value is within the tube, the loss is zero, while if predicted point is outside the tube, the loss is the magnitude of the difference between the predicted value and the radius e of the tube. C is penalty parameter, which is a regularized constant to determine the trade-off between training error and model flatness. To get the estimations of w and b, Eq. (5) is transformed to the primal objective function (6) by introducing positive slack variables ξ_i and ξ_i^*.

$$\begin{aligned} \underset{\xi_i, \xi_i^*, w, b}{Minimize} &: \frac{1}{2}\|w\|^2 + C\frac{1}{N}\sum_{i=1}^{N}(\xi_i + \xi_i^*) \\ Subject\ to &: \quad Y_i - w \times \phi(X_i) - b \le \varepsilon + \xi_i \\ & w \times \phi(X_i) + b - Y_i \le \varepsilon + \xi_i^*, i = 1, \cdots\cdots, N \\ & \xi_i \ge 0 \quad \xi_i^* \ge 0 \end{aligned} \right\} \tag{6}$$

By introducing kernel function $K(X_i, X)$, the dual form of Eq. (6) is obtained as:

$$\begin{aligned} \underset{\{\alpha_i\}, \{\alpha_i^*\}}{Minimize} &: \quad -\frac{1}{2}\sum_{i=1}^{N}\sum_{j=1}^{N}(\alpha_i - \alpha_i^*)(\alpha_j - \alpha_j^*) \times K(X_i, X_j) \\ & -\varepsilon\sum_{i=1}^{N}(\alpha_i - \alpha_i^*) + \sum_{i=1}^{N}Y_i(\alpha_i - \alpha_i^*) \\ Subjct\ to &: \quad \sum_{i=1}^{N}(\alpha_i - \alpha_i^*) = 0 \quad \alpha_i, \alpha_i^* \in [0, C] \end{aligned} \right\} \tag{7}$$

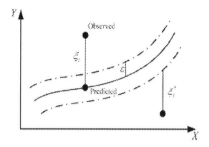

Fig. 1. The parameters for the support vector regression

where α_i, α_i^* are Lagrange multipliers, i and j represent different samples respectively. Thus, the function (3) becomes the following explicit form:

$$Y = f(X) = \sum_{i=1}^{N} (\alpha_i - \alpha_i^*) K(X_i, X) + b \qquad (8)$$

Through selecting the appropriate kernel function, the non-linear relation between the building cooling load and its correlative influence parameters based on SVM is established. After the prediction output Y from the SVR model is gotten, it should be transformed into the actual prediction value by Eq. (9):

$$\tilde{Y} = Y_{min}^* + Y \times (Y_{max}^* - Y_{min}^*) \qquad (9)$$

where $\tilde{Y}$ is the predicted value of the actual value Y^*.

2.2   The Network Structure of SVR

Here K is the Kernel function, α_i and α_i^* are the parameters of the model, s is the total number of the sample characteristic, X_i data vector is for network learning, X is an independent vector. The parameters of the SVR model are determined with maximizing the objective of function. The structure of model is shown in Fig. 2.

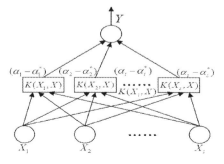

Fig. 2. The network structure of the SVR model

2.3   Choosing Kernel Function and Evaluation Indices

Any function satisfying Mercer's condition can be used as the kernel function, and the typical kernel functions include linear function, polynomial function, Gaussian function and Sigmoid function [7]. Among these functions, the Gaussian function can map the sample set from the input space into a high dimensional feature space effectively, which is good for representing the complex non-linear relationship between the output and input samples. Because of the above advantages, in this paper, Gaussian function is also selected as the kernel function, whose expression is shown as follows:

$$K(X_i, X) = \exp(-\gamma \|X_i - X\|^2) \tag{10}$$

where γ is the width parameter of the Gaussian kernel function.

The evaluation indices of the SVR model adopted throughout this paper are the mean square error (MSE) and mean relative error (MRE), which are defined as follows:

$$MSE = \frac{1}{N} \sum_{i=1}^{N} (\tilde{Y}_i - Y_i^*)^2 \tag{11}$$

$$MRE = \frac{1}{N} \sum_{i=1}^{N} \left| \frac{\tilde{Y}_i - Y_i^*}{Y_i^*} \right| \times 100\% \tag{12}$$

where $\tilde{Y}_i$ is the predicted value of the actual value Y_i^*, N is the sample number.

3   The Application of the SVR Model

3.1   The Establishment of the Building Indoor Temperature Prediction Model

Generally speaking, the main influence factors of the building indoor temperature are the outdoor temperature, the solar radiation and the wind speed. The structure of the building indoor temperature prediction model is shown in Fig. 3 below.

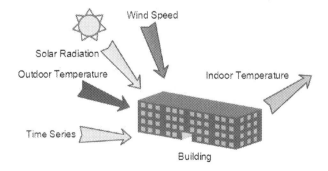

Fig. 3. The structure of the building indoor temperature prediction model

As is shown in Fig. 3 above, apparently, the primary influence factor of the building indoor temperature is the outdoor temperature, the secondary influence factors are the solar radiation and the wind speed and so on. Because of the the building indoor temperature is changing with the time, so the time series should be considered into the building indoor temperature prediction model.

Based on the analysis above, the sample data of the prediction model can be established. The input data of the sample data includes the outdoor temperature, the solar radiation, the wind speed and the time series. The output data of the sample data includes the indoor temperature. So the four corresponding building indoor temperature prediction model can be established as is shown in Tab. 1 below.

Table 1. The four building indoor temperature prediction models

Prediction Model (Output)		Prediction Model (Input)
Model 1	M	T, L, W, t
Model 2	M	T, L, t
Model 3	M	T, W, t
Model 4	M	T, t

As is shown in Tab. 1 above, the indoor temperature is defined as M, it's the output of the prediction model. The outdoor temperature is defined as T, the solar radiation is defined as L, the wind speed is defined as W and the time series is defined as t, they are the input of the prediction model.

3.2   Choosing the Sample Data

After the four building indoor temperature prediction model was established, so the next step is to choose the sample.

The form of the sample is divided into the input sample and output sample. Apparently, the input sample includes the outdoor temperature T, the solar radiation L, the wind speed W and the time series t, the output sample only includes the indoor temperature M.

The variety of the sample is divided into the training and predicting sample. The training sample would be trained in the SVR model, and then, after the predicting sample has been predicted in the SVR model, the output of the SVR model would be the predicted vlaue. Afer that, the predicted vlaue can be compared with the actual value.

The training sample presents the sample data which is range from May 1 to May 5 during this five days, the training sample data was acquired per ten minutes, so the total number of the training sample points is 720. Apparently, the input training sample includes the outdoor temperature T_{train}, the solar radiation L_{train}, the wind speed

W_{train} and the time series t_{train}, the output training sample only includes the indoor temperature M_{train}.

The predicting sample presents the sample data which is range from May 6 to May 10 during this five days, the predicting sample data was also acquired per ten minutes, so the total number of the predicting sample points is also 720. Apparently, the input predicting sample includes the outdoor temperature $T_{predict}$, the solar radiation $L_{predict}$, the wind speed $W_{predict}$ and the time series $t_{predict}$, the output predicting sample only includes the indoor temperature $M_{predict}$.

Because of the total number of all the sample points is 1440, is too large, so the part of the training and predicting sample is displayed in this paper as is shown in Tab. 2 and Tab. 3 below.

Table 2. The part of the training sample data

No	Input				Output
	T_{train} /°C	L_{train} /lx	W_{train} /m/s	t_{train} /min	M_{train} /°C
1	11.1	9	0.9	0	14.2
2	11.2	9	1.3	10	14.4
……	……	……	……	……	……
720	13.3	14	1.4	1430	15.5

Table 3. The part of the predicting sample data

No	Input				Output
	$T_{predict}$ /°C	$L_{predict}$ /lx	$W_{predict}$ /m/s	$t_{predict}$ /min	$M_{predict}$ /°C
1	15.5	14	2.5	0	16.3
2	15.6	14	2.8	10	16.4
……	……	……	……	……	……
720	14.4	9	1.4	1430	15.3

3.3   The Application of the SVR Model Based on LibSVM Toolbox

LibSVM toolbox is a simple software toolbox which is easy to use for the classification and regression problems. LibSVM toolbox was designed by Chih-Jen Lin who is a professor of the Nation Taiwan University. The algorithm of LibSVM is the sequential minimal optimization (SMO) algorithm. The program scale of LibSVM toolbox is small, it's convenient to do the secondary research and develop. The number of the parameters which are need to adjust is very less, the function of the cross validation is provided by LibSVM, so it's easy to be extended. Therefore, LibSVM toolbox is well known around the world for the uses of the classification and regression.

Therefore, LibSVM toolbox was used for predicting the building indoor temperature in this paper. LibSVM and MATLAB were combined to apply in the building indoor temperature prediction. The corresponding SVR predicting steps based on LibSVM toolbox are as follow:

- Step 1: Scaling the sample data. The scope of all the sample data should be scaled to the range $[0,1]$.
- Step 2: Choosing the kernel function. The Gaussian kernel function was selected in LibSVM, and choose any three parameters (C, γ, ε) to train the scaled sample.
- Step 3: Determing the best parameters. The three parameters (C, γ, ε) would be adjusted by the SMO algorithm so as to find the best three parameters.
- Step 4: Establishing the final model. The final SVR model would be established by the determined best three parameters (C, γ, ε).
- Step 5: Predicting the target value. The scaled sample data would be put into the final SVR model, then, the output of the final SVR model is the predicted value.
- Step 6: Reverse scaling the predicted value. The scope of the predicted value of the range $[0,1]$ should be reverse scaled to the actual range, like $[15.1, 23.8]$.

The final predicted value would be acquired after finished the six steps. The corresponding contrast curves between the predicted value and actual value of the four prediction models are shown in Fig. 4 below.

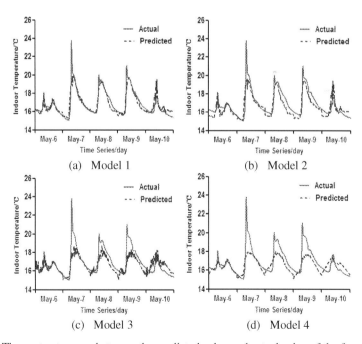

Fig. 4. The contrast curves between the predicted value and actual value of the four models

4   The Analysis of the Experimental Results

According to the mentioned above in this paper, the best three parameters (C, γ, ε) is $(2, 0.5, 0.01)$. As is shown in Fig. 4 above, the contrast time scope between the predicted value and actual value is range from May 6 to May 10 during this five days. The overall predicting effect of the four models is well. Apparently, the predicting effects of the model 1 and model 2 are the best. In order to compare the differences between the predicting effects of the four models better, the corresponding predicted relative error curves of the four prediction models are shown in Fig. 5 below.

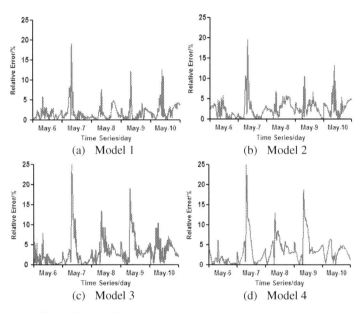

Fig. 5. The predicted relative error curves of the four models

As is shown in Fig. 5 above, the scope the overall predicted relative error of the four models is about 5%. Apparently, the predicted relative error of the model 3 and model 4 changed with the time more acuter than the model 1 and model 2. In order to make a quantitative comparison between the predicting effects of the four models, the corresponding predicted evaluation indices of the four prediction models are shown in Tab. 4 below.

Table 4. The predicted evaluation indices of the four models

Evaluation Indices	*MSE*	*MRE* /%
Model 1	0.2859	1.8695
Model 2	0.4326	2.8536
Model 3	0.8116	3.2693
Model 4	0.8854	3.4677

As is shown in Tab. 4 above, the predicted MSE and MRE of the model 1 are the minimum. The minimum $MSE = 0.2859$, the minimum $MRE = 1.8695\%$. Apparently, the more are the MSE and MRE closer to 0, the better is the predicted performance. Therefore, the predicted performance of the building indoor temperature prediction model 1 is the best.

5   Conclusion

The theory of SVM was used to apply in the building indoor temperature prediction in this paper. First of all, the four building indoor temperature prediction models were established in this paper. And then, based on LibSVM toolbox and MATLAB, the sample data was trained in the four prediction models, the output of the four prediction models is the target predicted value. In final, the predicted value and actual value were compared by a series evaluation indices so as to evaluate the predicted performance of the four models. The experimental results shown that the predicted performance of the building indoor temperature prediction model 1 is the best. It's really effective to use the SVR model to predict the building indoor temperature. The successful application of SVM on the building indoor temperature prediction presented the SVM's feasibility and validity.

Acknowledgements. The authors wish to thank "The Fundamental Research Funds for The Central Universities" for its support.

References

1. Li, J.H., Luo, X., Huang, C., Song, Y.: Block Model and Numerical Simulation for Predicting Indoor Temperature Distributions. Journal of Engineering Thermophysics **28**(2), 124–126 (2007)
2. Ji, X.L., Li, G.Z., Dai, Z.Z.: Influencing Factors and The Research Progress on Forecasting and Evaluating Indoor Thermal Comfort. Journal of Hygiene Research **32**(3), 295–299 (2003)
3. Sun, B., Yao, H.T.: The Short-term Wind Speed Forecast Analysis Based on The PSO-LSSVM Predict Model. Journal of Power System Protection and Control **40**(5), 85–89 (2012)
4. Yang, J.F., Cheng, H.Z.: Application of SVM to Power System Short-term Load Forecast. Journal of Electric Power Automation Equipment **24**(2), 30–32 (2004)
5. Wang, H.Q., Lei, G.: A Method for Forecasting Wind Speed by LIBSVM. Journal of Science Technology and Engineering **11**(22), 5440–5442 (2011)
6. Tang, W.H., Li, W.F.: Application of Support Vector Machines Based on Time Sequence in Logistics Forecasting. Journal of Logistics Science Technology **28**(113), 8–11 (2004)
7. Zhang, Q., Yang, Y.Q.: Research on The Kernel Function of SVM. Journal of Electric Power Science and Engineering **28**(5), 42–45 (2012)

Towards the Computation
of a Nash Equilibrium

Yu Lu$^{(\boxtimes)}$ and Ying He

School of Computing Science, University of Glasgow, Glasgow G12 8RZ, UK
y.lu.3@research.gla.ac.uk, yingh@dcs.gla.ac.uk

Abstract. Game theory has played a progressively more noticeable and important role in computer science topics, such as artificial intelligence, computer networking, and distributed computing, in recent years. In this paper, we provide a preliminary review of where efforts on this topic have been focused over the past several decades and find that currently, the most remarkable interface between algorithmic game theory and theoretical computer science is the computational complexity of computing a Nash equilibrium.

Keywords: Game theory · Multi-agent systems · Nash equilibrium · Computational complexity

1 Introduction

There is a long-lasting and close relationship between game theory and computer science. This type of relationship has become closer and more essential over the last decade due to the advent of the Internet. Strategic behaviour has become relevant to the design of computer systems, and considerable economic activity now takes place on computational platforms. We focus our attention on the concept of Nash equilibrium, explained below, which is one of the most important and arguably one of the most influential solution concepts in game theory.

Game theory is used to mathematically describe reasoning in strategic circumstances. In game theory, Nash equilibrium is a formal rule for predicting a game involving two or more players, where each player is supposed to understand others' strategies to create a stable situation and no player can benefit from changing only his or her own strategy unilaterally. If each player has chosen a strategy and no player can benefit by changing his or her strategy while the other players keep theirs unchanged, then the current set of strategy choices and the corresponding payoffs constitute a Nash equilibrium. Nash equilibrium has been universally considered and accepted as one of the most important concepts in the research of many subjects involved with game theory.

The rapid development of the Internet and distributed systems and the research of their cooperative environments have made the theory of Nash equilibrium one of the essential constituents of computer science. In the past few years, algorithmic issues of related problems have become the main focus of the

© Springer International Publishing Switzerland 2014
Z. Zeng et al. (Eds.): ISNN 2014, LNCS 8866, pp. 90–99, 2014.
DOI: 10.1007/978-3-319-12436-0_11

intersection between the disciplines of theoretical computer science and algorithmic game theory. Naturally, the study of computing the algorithmic complexity of Nash equilibrium and related topics has become one of theoreticians' favourite topics. This topic has attracted many researchers, led to the derivation of many novel methodologies, and demanded extensive mental efforts. New and advanced methods have been created and applied to solve various forms of such problems.

The remainder of this paper is organised as follows. We provide some background information in Section 2. Then, we review the related literature in Section 3. We demonstrate the recent convergence of Nash equilibrium and theoretical computer science in Section 4. In Section 5, we provide an in-depth discussion of the computational issues for finding the Nash equilibrium. Finally, we conclude in Section 6 with unresolved questions and directions for future research.

2   Background

2.1   Game Theory and Nash Equilibrium

Game theory is used to mathematically describe reasoning in strategic circumstances, in which the outcome of a participant's achievement in making decisions relies on the decisions of someone else. It is a branch of applied mathematics that is often used in economics as well as other fields, such as international relations, political science, and computer science. As presented in [1], the purpose of games is to help us understand economic behaviour by predicting how players will act in each particular game.

To realise games, we must employ a natural and ultimately mathematical language, which lead the games to be viewed as merely a collection of mathematical formulas to the general population. Fortunately, game theory is concerned with everyday economic life, and thus, it cannot live in vain as immortal. In fact, this theory borrows terminology from problems based on competition, confrontation and decision making, such as chess, poker, and war. Although game theory may initially sound purely theoretical, it has significant practical significance. Specialists consider economic and social problems, such as playing chess, and the game itself often contains profound truth.

Game theory is mathematical in nature and constitutes a branch of operations research. It studies games, thought experiments modelling various situations of conflict [1]. One commonly studied model aims to capture two players interacting in a single round. To further discuss Nash's significant contribution, the problem of non-cooperative games must first be considered. At present, almost all textbooks on game theory refer to the classical example of the "prisoner's dilemma"; the examples provided in books are typically highly similar, exhibiting only slight differences. The prisoner's dilemma demonstrates why two suspects might not cooperate even though it is in their best interest to do so.

In its classical form, the prisoner's dilemma is presented as follows. Two suspects are under arrest, and the police have inadequate evidence for a final judgment of guilty. Having separated both prisoners, the police visit each of them to offer the same deal. If one testifies (defects from the other) for the prosecution against the other and the other remains silent (cooperates with the other), the defector is let go and the silent accomplice receives the full 10-year sentence. If both cooperate, both prisoners are sentenced to only six months in jail for a minor charge. If each defects the other, each receives a five-year sentence. Each prisoner must choose to either defect the other or cooperate. Each is assured that the other would not know about the defect before the end of the investigation. How should the prisoners act? Table 1 can summarise the prisoner's dilemma:

Table 1. The Prisoner's Dilemma

	Prisoner B cooperate	Prisoner B defect
Prisoner A cooperate	*Each serves 6 months*	*A: 10 years & B: goes free*
Prisoner A defect	*A: goes free & B: 10 years*	*Each serves 5 years*

There are two prisoners, one choosing a row and one choosing a column; the choices of a prisoner are his actions, and the prisoners receive the corresponding payoffs shown in the table: the first and second numbers denote the payoff of the row and column, respectively. The predictions made by game theorists regarding prisoner behaviour are called equilibrium. One such prediction is the pure Nash equilibrium, in which each prisoner chooses an action that is the best response to the other prisoner's choice. For example, the best response is the highest payoff for the prisoner in the row or column chosen by the other prisoner. In our example, the only pure Nash equilibrium is when both prisoners choose "defect". The payoff matrix of the prisoner's dilemma is shown in Table 2.

Table 2. Payoff Matrix

	Cooperate	Defect
Cooperate	*(0, 0)*	*(-5, 1)*
Defect	*(1, -5)*	*(-4, -4)*

2.2   Computational Complexity

The theory of computational complexity is a part of the theory of computation that classifies computational problems based on their intrinsic level of difficulty. From this perspective, a computational problem is typically not intractable for computers. Informally, "a computational problem can be viewed as an infinite collection of instances together with a solution for every problem instance" [1].

After Nash's theorem was published in 1951, numerous researchers have sought algorithms for computing the Nash equilibrium. In the case of a zero-sum game, such as the rock-paper-scissors game, the Nash equilibrium results from the contribution of John von Neumann in the 1920s that it can be proposed with respect to linear programming. Although linear programming was only fully implemented in the 1970s, linear programs can be settled effectively. In the case of a non-zero-sum game, various algorithms have been formulated during the last 50 years. Unfortunately, "all of them are either of unknown complexity, or known to have to need exponential time in the worst case" [1].

One can assume that finding the mixed Nash equilibrium is an NP-complete problem, as there no algorithm for computing the mixed Nash equilibrium has been proposed to date. However, the situation is not that simple. The mixed Nash equilibrium is unlike any NP-complete problem because according to Nash's theorem, this problem always has a solution [2]. In contrast, NP-complete problems, such as SAT, "draw their intractability from the possibility that a solution might not exist, and this possibility is used heavily in the NP-completeness proof" [1].

Because NP-completeness is not an option, one must reconsider the path leading to NP-completeness to understand the complexity of mixed Nash equilibrium. Namely, we must define a class of problems that contains some other well-known hard problems along with the mixed Nash equilibrium and then prove that the mixed Nash equilibrium is complete for that class. Indeed, in this essay, we introduce a proven significant finding that the mixed Nash equilibrium is PPAD-complete, where PPAD is a subclass of NP that contains several important problems that are suspected to be hard. "All problems in PPAD share the same style of proof that every instance has a solution" [1].

3   Literature Review

Over the past few years, computational complexity in game theory has become increasingly important for solving natural questions. This focus began at the start of the 1970s, and game theorists have devoted significant effort to investigating the complexity of playing particular highly structured games [3,4]. These types of games are often zero-sum games containing a large state space; however, such games can be succinctly expressed due to the simple rules managing the state interaction. Therefore, the solutions of general categories of games obtained via research are often concerned with complicated languages. Particular highly structured games can be expressed accurately using such languages. "Algorithms for analysing this more general class of games strategically are a necessary component of sophisticated agents that are to play such games" [5].

Non-cooperative game theory can represent relatively enormous classes of strategic environments and provide refined conceptions of underlying understanding to resolve these games. These games are notion activities for realising and predicting the reaction of rational strategic individual players. The predicted outcome of the game is the equilibrium, and the well-known solution concept is the refined concept of Nash equilibrium. Determining its complexity been recognised as a most fundamental computational problem whose complexity is wide open and the most important concrete open question on the boundary of P today [6]. Nash proved that every game has a Nash equilibrium [2]. This is, the individuals' strategies have distributions in terms of each individual's best strategy is considering the other individuals' strategies in the game. This significant generality theorem constitutes the fundamental equilibrium concept of Nash equilibrium game theory, and all succeeding developments and improvements originated from this generality theorem.

Generality is an attractive characteristic for computing the Nash equilibrium. Such a model must also be normal and reliable for calculating the actions of a set of participants, e.g., the pure Nash equilibrium is more suitable than the mixed Nash equilibrium because a pure Nash equilibrium does exist. However, there is one more computational requirement for computing the Nash equilibrium: the Nash equilibrium should be computationally efficient when used to calculate how a set of participants tends to act. This requirement results from the notion that "if computing a particular kind of equilibrium is an intractable problem, of the kind that take lifetimes of the universe to solve on the world's fastest computers, it is ludicrous to expect that it can be arrived at in real life" [1]. This notion leads to the following significant question: Is there an efficient algorithm for computing the mixed Nash equilibrium?

Because this question remains unresolved, significant specific progress has been made in calculating the complexity of relative questions. For example, 2-person zero-sum games can be solved using linear programming in polynomial time [7]. As another example, determining the existence of a joint strategy where each player obtains an expected payoff of at least k is NP-complete in a concisely representable extensive form game where both players receive the same utility [8]. Similarly, in 2-player general-sum normal-form games, determining the existence of a Nash equilibrium with certain properties is NP-hard [9]. Finally, the complexity of best-responding, of guaranteeing payoffs, and of finding an equilibrium in repeated and sequential games has been studied in [10–14].

The computational problem of finding the Nash equilibrium was recently determined to be PPAD-complete [1], and thus presumably intractable, for the case of 4 players; this result was subsequently extended to three players [15,16] and two players [17,18]. In particular, the combined results of [1,17,19] establish that the general Nash equilibrium problem for normal-form games, which is the standard and most explicit representation, and for graphical games, which are an important succinct representation, can be reduced to 2-player games. Two-player games can in turn be solved by several techniques, such as the Lemke-Howson algorithm [20,21], a simplex-like technique that is empirically known to behave well, even though exponential counterexamples do exist [22]. The authors in [23] extended these results to all known classes of succinct representations of games and to more sophisticated concepts of equilibrium.

Several decades earlier, it was rather predictable to understand the complexity of computing equilibrium algorithmic game theory "given our field's obsessions, but it also entails an important contribution to the other side, as algorithmic issues have influenced and shaped the debate on equilibrium concepts"[24]. We now recognise that computing Nash equilibrium is PPAD-complete. Therefore, the most significant question in this field is now whether there is a polynomial-time approximation scheme (PTAS) for computing the Nash equilibrium. A negative answer has been obtained for relative multiplicative approximation when negative payoffs are allowed [25]. A quasi-polynomial time approximation scheme for this problem has been known for some time [26]; in fact the algorithm in [26] is of a special type called oblivious, as it examines possible solutions without looking at the game except to check the quality of the approximation. It can be shown [27] that the algorithm in [26] is nearly optimal among oblivious algorithms.

4 Computing a Nash Equilibrium

4.1 Who Cares About It?

There are many famous practical uses of computing the Nash equilibrium in the field of economics and game theory, such as forecasting the result and evaluating appropriate constraint standards of a model, contrasting empirical outcomes with model predictions, testing the design of a mechanism, and automatically generating conjectures and counterexamples.

Computer scientists and game theorists also focus on finding the Nash equilibrium. The initial incentive was the implementation of automatic game-playing programs, such as for chess; a sample of modern applications is reviewed by [28]. For example, "automated agents - anything from a physical robot to an automated bidding strategy in an Internet auction - are often programmed to optimise some payoff function while participating in a competitive or cooperative environment" [29]. Accurate and computationally efficient algorithms for computing the Nash equilibrium are essential applications for controlling and operating such agents to reach decisions automatically.

Finally, complexity theorists have been attracted by problems related to computing the Nash equilibrium because they include several of the most natural problems believed to be medium hard, i.e., between easy and difficult. Their studies have resulted in many advancements in the intersection of algorithmic game theory and complexity theory.

4.2 Why Studying the Complexity of It?

A majority of games are more complex than the prisoner's dilemma, and their Nash equilibriums are more difficult to compute. However, John Nash was the first scientist to verify that every game must have a Nash equilibrium. Most explanations of the concept of equilibrium entail individuals deciding equilibrium. For example, the thought that markets implicitly compute a solution to a significant computational problem goes back to Adam Smith's notion of the invisible hand, if not earlier. If every participant is reasonable in a limited manner, then equilibrium cannot be explained as a reliable prediction unless it can be calculated with a rational attempt. Rigorous intractability results thus cast doubt on the predictive power of equilibrium concepts [30].

From a virtual implementation perspective, a method, which is essentially designed, was improved to be computationally and easily addressed or controlled as well as easy to run and play. Thus, on this basis, participants should not need to perform hard and complex calculations. Many researchers believe that although the Nash equilibrium for a specific market may be difficult to obtain, it will precisely explain all activities and strategies in the market as soon as it is found.

4.3 How Difficult is It?

The authors in [1] proved that the Nash equilibrium is rather difficult to compute for several games. In some cases, the Nash equilibrium cannot be computed in a finite amount of time even when using all of the computers and computing resources around the world. Considering such situations, human beings playing the game probably have not found it either.

In the context of real life, business competitors in commercial activities or drivers on a high road actually do not commonly compute the Nash equilibrium for their specific games and subsequently take their corresponding counter-plan. However, when one player changes his or her counter-plan, the other players will also change their counter-plans accordingly, which will force the first player to change his or her counter-plan one more time, and so on. This nature of responses will eventually converge to the Nash equilibrium.

Many complexity theorists have demonstrated that the Nash equilibrium is classified in a collection of problems that has been systematically researched in algorithmic computer science: those whose answers might be difficult to calculate but whose correctness is still reasonably easy to prove. The simplest accepted form of an example of these problems in mathematics is the factoring of a large number. The answer looks to require testing a large number of various distinct probabilities while proving that a given solution only requires multiplying several numbers at the same time.

4.4 What is the Complexity of It?

Generally, the procedures are rather unexpectedly more complex than the above example of computing prime numbers in the situation of finding Nash equilibrium. Any person who is interested in computer science will acknowledge the collection of problems whose answers can be proven efficiently: it is the class of complexity that complexity theorists and computer researchers call NP. Daskalakis demonstrated that the Nash equilibrium is classified to a sub-collection of NP and that it contains difficult problems such that an answer to one problem can be modified to solve any other problem.

The algorithm specialist will imply that it belongs to a complexity class called NP-complete. However, the fact that there exists a Nash equilibrium at all times makes it unsuitable to be considered as NP-complete. Actually, it is classified to another complexity class named PPAD-complete. That result is one of the biggest yet in the approximately 10-year-old field of algorithmic game theory [31]. It formalises the suspicion that the Nash equilibrium is not likely to be an accurate predictor of rational behaviour in all strategic environments.

In terms of the unpredictability of the Nash equilibrium, there are three routes that one can go. One is to say, we know that there exist games that are hard, but maybe most of them are not hard. In that situation, we can seek to identify classes of games that are easy, that are tractable [32]. The second route is to "find mathematical models other than Nash equilibrium to characterise markets - models that describe transition states on the way to equilibrium, for example, or other types of equilibrium that are not so hard to calculate" [32]. Finally, "it may be that where the Nash equilibrium is hard to calculate, some approximation of it - where the players' strategies are almost the best responses to their opponents' strategies - might not be" [32]. In those cases, the approximate equilibrium could describe the behaviour of real-world systems.

5. Discussion

Nash equilibrium is the main universally and frequently used equilibrium concept in game theory. The insight into Nash equilibrium is that it describes a likely stable situation of game playing. It is a fixed point where each player holds correct beliefs about what other players are doing, and plays a best response to those beliefs. The Nash equilibrium is unique because a Nash equilibrium occurs in game playing whenever each player has only finitely many possible deterministic strategies and we allow for mixed strategies.

In many games, thoughts with respect to the Nash equilibrium require a deep understanding of an individual's activity. However, as many scientists have noted, the Nash equilibrium experiences several problems. For instance, there exists a flaw in the repeated prisoner's dilemma in terms of the Nash equilibrium in game playing. It is rather difficult to present a situation in which reasonable participants are supposed to play the Nash equilibrium in such a game but unreasonable participants who collaborate for a short time obtain a better outcome. In addition, if a game is played only once, then why should the solution converge to the Nash equilibrium when there are numerous Nash equilibriums? In fact, participants of a game do not have a means of distinguishing which one of multiple Nash equilibriums tends to occur.

As one might expect, extensive work has been performed in the field for evolving progressive solutions of game models. Numerous alternatives to and improvements of Nash equilibrium have been proposed, including rationalisability, sequential equilibrium, trembling hand perfect equilibrium, proper equilibrium, and iterated deletion of weakly dominated strategies. Although a number of alternatives achieve success to some extent, none of these solutions address the following three problems, all originating from concerns of computer science perspectives on Nash equilibrium.

First, game theory focuses on the participant itself strategically. Reasonable participants choose strategies that are best responses to strategies chosen by other player, and the focus in distributed computing has been on problems such as fault tolerance and asynchrony. It cannot solve problems with defective or unpredicted activities, nor can it solve problems with individuals conspiring with each other. However, we look for both situations in games, such as large game playing. Second, computational efficiency is also a consideration when discussing it. Solution concepts must be developed and improved to resolve resource limited agents. For example, this issue is critical for cryptography. Finally, it assumes that participants have a universal understanding of the complex factors of the game and of every possible decision that can be taken in each state and every participant in the game. However, this assumption is not rational in many situations, such as in large auctions performed through the Internet at all times.

6 Concluding Remarks

The Internet and distributed systems owe much of their complexity to a large number of individuals who manage them and make them work. These individuals have distinct and often contradictory focuses and concerns. Thus, these individuals are rational and their communications are naturally strategic. Thus, concepts from computer science and game theory are essential to understand these communications. Nash equilibrium provides individuals with an accurate approach to making a prediction about the activities of strategic individuals in circumstances of contradiction. However, the reliability of it as a structure for action calculation relies on whether it is computationally efficient. Why should we presume that a set of reasonable individuals act in a way that requires exponential computation time? Motivated by this question, we study the computational complexity of the Nash equilibrium in this paper.

References

1. Daskalakis, C., Goldberg, P.W., Papadimitriou, C.H.: The Complexity of Computing a Nash Equilibrium. Communications of the ACM **52**(2), 89–97 (2009)
2. Nash, J.: Non-Cooperative Games. Annals of Mathematics **2**(2), 286–295 (1951)
3. Lichtenstein, D., Sipser, M.: GO Is Polynomial-Space Hard. Journal of the ACM **27**(2), 393–401 (1980)
4. Stockmeyer, L.J., Meyer, A.R.: Word problems requiring exponential time (Preliminary Report). In: 5th Annual ACM Symposium on Theory of Computing, pp. 1–9. ACM (1973)
5. Conitzer, V., Sandholm, T.: New complexity results about Nash equilibria. Games and Economic Behavior **63**(2), 621–641 (2008)
6. Papadimitriou, C.: Algorithms, games, and the internet. In: 33rd Annual ACM Symposium on Theory of Computing, pp. 749–753 (2001)
7. Luce, R.D., Raiffa, H.: Games and Decisions: Introduction and Critical Survey. Dover Publications (1989)
8. Chu, F., Halpern, J.: On the NP-completeness of finding an optimal strategy in games with common payoffs. International J. of Game Theory **30**(1), 99–106 (2001)
9. Gilboa, I., Zemel, E.: Nash and correlated equilibria: Some complexity considerations. Games and Economic Behavior **1**(1), 80–93 (1989)

10. Ben-porath, E.: The complexity of computing a best response automaton in repeated games with mixed strategies. Games and Eco. Behavior **2**(1), 1–12 (1990)
11. Koller, D., Megiddo, N.: The complexity of two-person zero-sum games in extensive form. Games and Economic Behavior **4**(4), 528–552 (1992)
12. Littman, M.L., Stone, P.: A polynomial-time nash equilibrium algorithm for repeated games. Decision Support Systems **39**(1), 55–66 (2005)
13. Papadimitriou, C.H., Yannakakis, M.: On complexity as bounded rationality. In: 26th Annual ACM Symposium on Theory of Computing, pp. 726–733 (1994)
14. Nachbar, J.H., Zame, W.R.: Non-computable strategies and discounted repeated games. Economic Theory **8**(1), 103–122 (1996)
15. Chen, X., Deng, X.: 3-Nash is PPAD-complete. In: Electronic Colloquium on Computational Complexity (2005)
16. Daskalakis, C., Papadimitriou, C.H.: Three-Player Games Are Hard. In: Electronic Colloquium on Computational Complexity (2005)
17. Chen, X., Deng, X.: Settling the Complexity of Two-Player Nash Equilibrium. In: Electronic Colloquium on Computational Complexity (2006)
18. Chen, X., Deng, X., Teng, S.H.: Settling the complexity of computing two-player Nash equilibria. Journal of the ACM **56**(3) (2009)
19. Goldberg, P.W., Papadimitriou, C.H.: Reducibility Among Equilibrium Problems. In: 38th Annual ACM Symposium on Theory of Computing, pp. 61–70 (2006)
20. Lemke, C.E., Howson, J.T.: Equilibrium Points of Bimatrix Games. Journal of the Society for Industrial and Applied Mathematics **12**(2), 413–423 (1964)
21. Stengel, B.V.: 45. In: Computing Equilibria for Two-Person Games, vol. 3, pp. 1723–1759. Springer (2002)
22. Savani, R., von Stengel, B.: Exponentially Many Steps for Finding a Nash Equilibrium in a Bimatrix Game. In: Proceedings of the 45th Annual IEEE Symposium on Foundations of Computer Science, pp. 258–267 (2004)
23. Daskalakis, C., Fabrikant, A., Papadimitriou, C.H.: The Game World Is Flat: The Complexity of Nash Equilibria in Succinct Games. In: Bugliesi, M., Preneel, B., Sassone, V., Wegener, I. (eds.) ICALP 2006. LNCS, vol. 4051, pp. 513–524. Springer, Heidelberg (2006)
24. Papadimitriou, C.H.: Algorithmic Game Theory: A Snapshot. In: Albers, S., Marchetti-Spaccamela, A., Matias, Y., Nikoletseas, S., Thomas, W. (eds.) ICALP 2009, Part I. LNCS, vol. 5555, pp. 3–11. Springer, Heidelberg (2009)
25. Daskalakis, C.: On the Complexity of Approximating a Nash Equilibrium. ACM Transactions on Algorithms **9**(3) (2013)
26. Lipton, R.J., Markakis, E., Mehta, A.: Playing large games using simple strategies. In: 4th ACM Conference on Electronic Commerce, pp. 36–41 (2003)
27. Daskalakis, C., Papadimitriou, C.H.: On oblivious PTAS's for Nash equilibrium. In: 41st Annual ACM Symposium on Theory of Computing, pp. 75–84. ACM (2009)
28. Koller, D., Pfeffer, A.: Representations and solutions for game-theoretic problems. Artificial Intelligence **94**(1–2), 167–215 (1997)
29. Shoham, Y., Leyton-Brown, K.: Multiagent Systems: Algorithmic, Game-Theoretic, and Logical Foundations. Cambridge University Press (2008)
30. Roughgarden, T.: Computing equilibria: a computational complexity perspective. Economic Theory **42**(1), 193–236 (2010)
31. Papadimitriou, C.H.: Algorithmic Game Theory. Cambridge Uni. Press (2007)
32. Daskalakis, C.: The Complexity of Nash Equilibria. PhD thesis, University of California, Berkeley (2004)

Sensorless PMSM Speed Control
Based on NN Adaptive Observer

Longhu Quan[1], Zhanshan Wang[1(✉)], Xiuchong Liu[1], and Mingguo Zheng[2]

[1] College of Information Science and Engineering, Northeastern University, Shenyang, China
jonryongho911@163.com, wangzhanshan@ise.neu.edu.cn
[2] College of Information Science and Technique, Kimchaek University,
Pyongyang, Democratic People's Republic of Korea
jongmyongguk@163.com

Abstract. The AC motor control by neural networks includes the reconstruction errors in a certain degree, which can cause the non-convergence in the control results. To learn the complete system dynamics of the sensorless PMSM, a neural network adaptive speed control strategy is proposed to eliminate the NN reconstruction errors. A robust modification term, which is a function of estimation error and an additional tunable parameter, is introduced to guarantee the asymptotic stability of the speed estimation. A rotor-flux-oriented vector control is employed as the basic control strategy for the sensorless PMSM drive system. The simulation results demonstrated the validity and feasibility of the proposed control strategy.

Keywords: Neural network adaptive speed control · Permanent magnet synchronous motor (PMSM) · NN reconstruction error · Robust modification term · Speed estimation

1 Introduction

In recent years, permanent magnet synchronous motors (PMSM) are widely used in high-performance applications such as industrial robots and machine tools, because PMSMs have the advantages of compact size, high-power density, high air-gap flux density, high-torque/inertia ratio, high torque capability, high efficiency and free maintenance. In the applications of PMSM, the use of the high resolution speed and position sensors increase the size of the machine, raise the cost of control system and require additional cabling. Especially, the measurements are highly sensitive to inaccuracies of faults which could affect sensor. The desire to eliminate the use of sensors from PMSM applications has resulted in the several techniques for sensorless motor operation [1-4].

This work was supported by the National Natural Science Foundation of China (Grant Nos. 61074073 and 61034005), the Fundamental Research Funds for the Central Universities (Grant Nos. N130504002 and N130104001), and SAPI Fundamental Research Funds (Grant No. 2013ZCX01).

© Springer International Publishing Switzerland 2014
Z. Zeng et al. (Eds.): ISNN 2014, LNCS 8866, pp. 100–109, 2014.
DOI: 10.1007/978-3-319-12436-0_12

The various techniques have been proposed to estimate the speed and position for sensorless AC motor control, such as a motor's physical model based method [1], back-electromotive force (Back-EMF) based method [2], model reference adaptive system (MRAS) based method [3] and extended Kalman filter (EKF) based method [4]. However, all the system uncertainty can not be removed, and it can not be used in the control objects where the distribution is not given.

Recently, due to online learning capabilities, parallel distributed structure and identification of the nonlinear dynamics, neural networks (NNs) are commonly utilized for the control of nonlinear discrete-time systems such as motor control systems [5-8]. A diagonally recurrent NN (DRNN)-based observer is designed to perform the rotor position estimation of the sensorless PMSM [9]. A rotor speed observer based on an Elman NN, which is the feedback NN, is proposed in [10]. A two-layer, online-trained NN stator current observer is used as the adaptive model for the MRAS estimator which requires the rotor magnetic flux information [11]. These studies have attempted to solve the speed and position estimation problem using NN estimator under assuming that there are no NN reconstruction errors [9-11]. In these cases, the estimation result based on the back propagation algorithm may not be convergent or remain bounded to the optimal control.

In this paper, a NN based robust speed estimation strategy of sensorless PMSM is presented. First, a NN adaptive identification method of the sensorless PMSM system with the elimination of the NN reconstruction errors is developed. To learn the unknown nonlinear system, a stable adaptive weight update law is proposed for tuning the nonlinear system estimator. The robust modification term, which is a function of estimation error and an additional tunable parameter, is applied to guarantee asymptotic stability. Next, a NN speed estimation method using the adaptive identified system parameters is studied. Therefore, the convergence of the NN based speed estimator is demonstrated by considering the NN reconstruction errors of PMSM speed control system in contrast to previous works [9-11].

2   Mathematical Model of PMSM and Estimation Strategy

The $d-q$ model cannot be utilized directly in the sensorless motor control because the estimation error of rotor position is not taken into account [2,3]. The mathematical model of sensorless PMSM in the estimated rotating δ-γ frame is derived as (1)

$$
\begin{cases}
L_s \dfrac{di_\delta}{dt} = -\psi_m \omega_k \sin(\Delta\theta_k) - R_s i_\delta + u_\delta + L_s p i_\gamma \omega_s \\[2mm]
L_s \dfrac{di_\gamma}{dt} = -\psi_m \omega_k \cos(\Delta\theta_k) - R_s i_\gamma + u_\gamma - L_s p i_\delta \omega_s \\[2mm]
\dfrac{d\omega_k}{dt} = \dfrac{\psi_m}{J} \sin(\Delta\theta_k) i_\delta + \dfrac{\psi_m}{J} \cos(\Delta\theta_k) i_\gamma - \dfrac{1}{J} T_L(\omega_k) \\[2mm]
\dfrac{d\Delta\theta_k}{dt} = -\omega_k + \omega_s
\end{cases}
\tag{1}
$$

where, ω_k and θ_k are the $d-q$ axis rotor speed and position, ω_s and θ_s are the δ-γ axis rotor speed and position. $\Delta\theta_k$ is a position estimation error, that is, lag theta between $d-q$ frame and δ-γ frame; ψ_m and P represent the flux linkage and the number of magnet pole pairs of the rotor; i_d and i_q are the $d-q$ axis components of the stator current. u_d and u_q are the $d-q$ axis components of the stator voltage, i_δ and i_γ are the δ-γ axis components of the stator current. R_s, J, T_L and L_s are the stator resistance, inertia moment, load torque and stator inductance, respectively.

If the sensorless control is correct, θ_s of δ-γ frame is convergent to ω_k of $d-q$ frame, the position estimation error $\Delta\theta_k$ is also convergent to zero, and all electromagnetic parameters of δ-γ frame can be processed on the $d-q$ frame[3]. Equation (1) can be written as

$$\begin{cases} L_s \dfrac{di_d}{dt} = -R_s i_d + L_s P i_q \omega_k + u_d \\[2mm] L_s \dfrac{di_q}{dt} = -R_s i_q - L_s P i_d \omega_k + u_q - \psi_m \omega_k \\[2mm] \dfrac{d\omega_k}{dt} = \dfrac{\psi_m}{J} i_q - \dfrac{1}{J} T_L(\omega_k) \\[2mm] \dfrac{d\theta_k}{dt} = \omega_k \end{cases} \tag{2}$$

From (1) and (2), we obtain the discrete time system identification model,

$$\begin{cases} i_\delta(t)(R_s + pL_s) = \hat{i}_d(t+1)(R_s + pL_s) = u_d(t) + L_s P i_q(t)\hat{\omega}_k(t) \\[2mm] i_\gamma(t)(R_s + pL_s) = \hat{i}_q(t+1)(R_s + pL_s) = u_q(t) - L_s P i_d(t)\hat{\omega}_k(t) - \psi_m\hat{\omega}_k(t) \end{cases} \tag{3}$$

The control objective is to ensure that the servo drive system tracks a desired target despite of NN reconstruction errors. In the transient state, the speed and position estimation scheme proposed in this paper is shown in Figure 1.

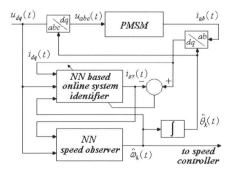

Fig. 1. Control strategy for speed and position estimation of sensorless PMSM

3   NN Based System Identification and Speed Estimation

In order to develop the NN system identifier, the system dynamics (3) are rewritten as

$$x(t+1) = h(x(t), u(t)) \tag{4}$$

The function $h(x(t), u(t))$ in (4) has a NN representation of a compact set. A NN implementation can be made as follows

$$x(t+1) = \eta[v(t)^T \mu(w(t)^T z(t))] + \varepsilon(t) \tag{5}$$

where $w(t) \in \mathfrak{R}^{(n+m) \times l}$ and $v(t) \in \mathfrak{R}^{l \times n}$ are the constant ideal weight matrices. $\mu(\bullet)$ is a tangent sigmoid function. $\eta(\bullet)$ is a linear function. $\varepsilon(t)$ is the bounded NN functional approximation error. $z(t)$ is the NN input and is denoted by

$$z(t) = [x^T(t) \quad u^T(t)]^T \tag{6}$$

where $x(t)$ is the system state and $u(t)$ is the system input. $x(t+1)$ is the NN output. For a simple consideration, $w^T(t)z(t)$ is denoted as $\overline{z}(t)$. From above consideration, (5) can be written as follows

$$x(t+1) = v^T(t)\mu(\overline{z}(t)) + \varepsilon(t) \tag{7}$$

Additionally, bounds of the output layer weights are taken as $\| v(t) \| \leq V_m$ for a constant V_m while NN activation functions are bounded such that $\|\mu(\overline{z}(t))\| \leq \mu_m$ for a constant μ_m.

The NN based system identification scheme is defined as

$$\hat{x}(t+1) = \hat{v}^T(t)\mu(\overline{z}(t)) - b(t) \tag{8}$$

where $\hat{x}(t+1)$ is the estimated system state vector, $\hat{v}(t)$ is the estimation of the ideal weight matrix $v(t)$ and $b(t)$ is the robust modification term defined as

$$b(t) = \frac{\hat{\lambda}(t)\tilde{x}(t)}{(\tilde{x}^T(t)\tilde{x}(t) + C_s)} \tag{9}$$

where $\tilde{x}(t) = x(t) - \hat{x}(t)$ is the system identification error, $\hat{\lambda}(t) \in \mathfrak{R}$ is an additional tunable parameter and $C_s > 0$ is a constant.

Next, the identification error dynamics are written as

$$\tilde{x}(t+1) = x(t+1) - \hat{x}(t+1) = \tilde{v}^T(t)\mu(\overline{z}(t)) + \frac{\hat{\lambda}(t)\tilde{x}(t)}{(\tilde{x}^T(t)\tilde{x}(t)+C_s)} + \varepsilon(t) \tag{10}$$

where $\tilde{v}(t) = v(t) - \hat{v}(t)$.

Assumption 1. The bounded NN functional approximation error term $\varepsilon(t)$ in (7) is assumed to be upper bounded by a function of estimation error such that

$$\varepsilon^T(t)\varepsilon(t) \leq \varepsilon_m(t) = \lambda^* \tilde{x}^T(t)\tilde{x}(t) \tag{11}$$

$$\tilde{\lambda}(t) = \lambda^* - \hat{\lambda}(t) \tag{12}$$

where λ^* is a bounded constant target value such that $\|\lambda^*\| \leq \lambda_m$.

The tuning laws for $\hat{v}(t)$ and $\hat{\lambda}(t)$ are estimated as follows

$$\hat{v}(t+1) = \hat{v}(t) + \alpha_s \mu(\overline{z}(t))\tilde{x}^T(t+1) \tag{13}$$

$$\hat{\lambda}(t+1) = \hat{\lambda}(t) - \gamma_s \frac{\tilde{x}^T(t+1)\tilde{x}(t)}{(\tilde{x}^T(t)\tilde{x}(t)+C_s)} \tag{14}$$

where $\alpha_s > 0$ is a NN learning rate and $\gamma_s > 0$ is a design parameter. The local asymptotic stability of the update laws (13) and (14) and the stability analysis of the system identification scheme are proved from Lyapunov theory.

Theorem 1. Let the NN based identification scheme proposed in (8) be used to identify the system dynamics (4), and let the parameter update law given in (13) and (14) be used for tuning the NN weights and the robust modification term, respectively. In the presence of bounded uncertainties, the state estimation error $\tilde{x}(t)$ is asymptotically stable while the NN parameter estimation errors $\tilde{v}(t)$ and $\tilde{\lambda}(t)$ are bounded, respectively.

Proof. Consider the following positive definite Lyapunov function defined as

$$L(t) = \tilde{x}^T(t)\tilde{x}(t) + tr[\tilde{v}^T(t)\tilde{v}(t)]/\alpha_s + \tilde{\lambda}^2(t)/\gamma_s \tag{15}$$

whose first difference is given by

$$\Delta L = \underbrace{\tilde{x}^T(t+1)\,\tilde{x}(t+1) - \tilde{x}^T(t)\,\tilde{x}(t)}_{\Delta L_1} + \underbrace{tr[\tilde{v}^T(t+1)\,\tilde{v}(t+1) - \tilde{v}^T(t)\,\tilde{v}(t)]/\alpha_s}_{\Delta L_2}$$

$$+ \underbrace{(\tilde{\lambda}^2(t+1) - \tilde{\lambda}^2(t))/\gamma_s}_{\Delta L_3} \tag{16}$$

To solve equation (16), (10) can be rewritten as follows

$$\tilde{x}(t+1) = \psi_1(t) - \psi_2(t) + \frac{\lambda^* \tilde{x}(t)}{(\tilde{x}^T(t)\,\tilde{x}(t) + C_s)} + \varepsilon(t) \tag{17}$$

where

$$\begin{cases} \psi_1(t) = \tilde{v}^T(t)\mu(\tilde{z}(t)) \\[2mm] \psi_2(t) = \dfrac{\tilde{\lambda}(t)\,\tilde{x}(t)}{(\tilde{x}^T(t)\,\tilde{x}(t) + C_s)} \\[2mm] \tilde{\lambda}(t) = \lambda^* - \hat{\lambda}(t) \end{cases} \tag{18}$$

Considering $\| \lambda^* \| \le \lambda_m$ and $\dfrac{\tilde{x}^T(t)\,\tilde{x}(t)}{(\tilde{x}^T(t)\,\tilde{x}(t) + C_s)^2} < \tilde{x}^T(t)\,\tilde{x}(t)$, we can obtain a relation as follows

$$\Delta L \le -[1 - 2(\lambda_m + \lambda_m^2) - 4(\alpha_s\mu_m^2(\lambda_m + \lambda_m^2) + \gamma_s(\lambda_m + \lambda_m^2))]\,\|\,\tilde{x}(t)\,\|^2 \tag{19}$$
$$- (1 - 4\alpha_s\mu_m^2 - 4\gamma_s)(\|\psi_1(t)\|^2 + \|\psi_2(t)\|^2) + 2\|\psi_1(t)\|\,\|\psi_2(t)\|$$

In the next step, we define the change of variables as $\bar{\psi}_1(t) = \psi_1(t)/\beta_1$ and $\bar{\psi}_2(t) = \psi_2(t)/\beta_2$, where β_1 and β_2 are constants. Additionally, the design parameters are selected as $\gamma_s = \alpha_s\mu_m^2$, $\alpha_s \le \beta_1^2/(8\mu_m^2)$ and $\beta_2 = \delta/\beta_1$, where $\delta > 0$ is a constant. Using these relations and new variables, applying the C-S inequality, (19) can be rewritten as

$$\Delta L \le -[1 - 2(\lambda_m + \lambda_m^2) - \beta_1^2(\lambda_m + \lambda_m^2)]\,\|\,\tilde{x}(t)\,\|^2 \tag{20}$$
$$- (1 - \beta_1^2 - \delta/\beta_1^2)\,\|\,\tilde{v}^T(t)\mu(\tilde{z}(t))\,\|^2$$
$$- (1 - \beta_1^2(1 - 1/\delta)) \times \|\,\tilde{\lambda}(t)\,\|^2\,\|\,\tilde{x}(t)/(\tilde{x}^T(t)\,\tilde{x}(t) + C_s)\,\|^2$$

There $\Delta L \le 0$ provided $\beta_1 \le \min\{a_s, b_s, c_s, d_s\}$ and $\delta < 1/4$ where $a_s = \sqrt{(1 + \sqrt{1 - 4\delta})/2}$, $b_s = \sqrt{(1 - \sqrt{1 - 4\delta})/2}$, $c_s = \sqrt{\delta/(1 + \delta)}$ and $d_s = \sqrt{(1 - 2(\lambda_m + \lambda_m^2))/(\lambda_m + \lambda_m^2)}$.

$\tilde{x}(t), \tilde{v}(t)$ and $\tilde{\lambda}(t)$ are bounded by the provided $\tilde{x}(t_0), \tilde{v}(t_0)$ and $\tilde{\lambda}(t_0)$ bounded in the compact set. The upper bound relating to the NN reconstruction error can be reduced by increasing the number of hidden layer neurons[12,13].

The relationship of system input and output for NN is as follows

$$\begin{cases} x(t) = [i_d(t), i_q(t)] \\ u(t) = [u_d(t), u_q(t), \hat{\omega}(t)] \\ y(t) = [\hat{i}_d(t+1), \hat{i}_q(t+1)] \end{cases} \tag{21}$$

NN of $N_{5,5,2}^3$ is selected from (15) and shown in Figure 2.

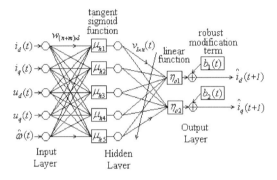

Fig. 2. NN structure used in online system identification

The implementation of NN speed estimator can be described as follows

$$\omega_k(t+1) = f(i_d(t), i_q(t), \omega_k(t)) \tag{16}$$

Because $f(\bullet)$ is a static nonlinear function, any kind of static NN can be used to approximate the nonlinear function. So, we adopt a feedforward multilayer NN in this paper. For the design of the NN speed estimator, a feedforward multilayer NN of $N_{3,13,1}^3$ is designed from the consideration of (16) [6, 9].

4   Simulations and Analyses

In order to verify the proposed control strategy, a control model of pulse width modulation (PWM) inverter-fed PMSM drive system is built in MATLAB/SIMULINK R2012a platform. Simulation studies are carried out with a 10-kHz control frequency. The PMSM electrical and mechanical parameters used in the simulation are listed in Table 1.

Table 1. Parameters for motor and inverter systems in simulations

Quantity	Symbol	Value
Nominal power	$kg \cdot m^2 \quad P_n$	1hp(3-phase)
Stator resistance	R_s	1.5 Ω
Stator self inductance	$L_d = L_q = L_s$	0.05H
Voltage constant	λ_m	0.314V.s/rad
Rated torque	T_e	3.6Nm
Rotor inertia	J	0.003kgm²
Friction coefficient	β_m	0.0009Nm/rad/sec
Nominal speed(electrical)	ω_r	377rad/sec
Number of poles	P	4
Rated current	I	4A
Rated voltage	V	208V
Rated frequency	f	60Hz
Torque constant	K_t	0.95Nm/A
Resolution of the encoder	n	10000p/r

The current estimation results using the conventional current estimation method and proposed system identification method are shown in Figure 3.

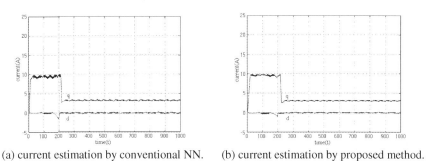

(a) current estimation by conventional NN.   (b) current estimation by proposed method.

Fig. 3. The current estimation using the conventional method and the proposed method

In order to verify the performance of the proposed speed control algorithm, several simulations are carried out on the various operating conditions. In order to evaluate the speed estimation performance, we also give the corresponding simulation results under the PMSM vector control with the incremental encoder as the speed sensor. In order to compare the superiority, the proposed method is compared with the speed estimation results obtained from conventional NN speed estimator under the same simulation conditions. The simulations are realized in low speed and high speed to show the superiority of the proposed method. The simulation results are shown in Figure 4.

In the control of static NN, the accuracy of control results can be improved by increasing the number of hidden layer neurons [21]. The speed estimation results according to the different hidden layer neuron number are shown in Figure 5.

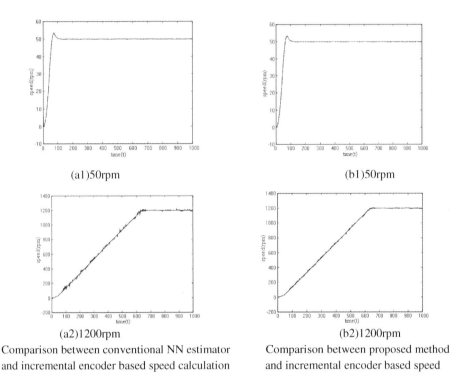

(a1)50rpm

(b1)50rpm

(a2)1200rpm

(b2)1200rpm

Comparison between conventional NN estimator and incremental encoder based speed calculation

Comparison between proposed method and incremental encoder based speed

Fig. 4. Comparison of estimation results between the proposed method and conventional method (actual speed (smooth line), estimated speed(rough line))

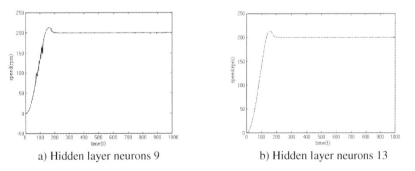

a) Hidden layer neurons 9

b) Hidden layer neurons 13

Fig. 5. Simulations for the number selection of the hidden layer neurons. Actual speed(smooth line) and estimated speed(rough line).

5   Conclusion

In this paper, a novel NN based speed control strategy of sensorless PMSM drive with the consideration of the NN bounded NN functional approximation errors has been developed. The proposed speed estimation strategy is composed with two process: online asymptotic NN system identification and offline NN speed estimation. The results obtained from the Lyapunov theory and the simulations show that the proposed online system identification results converge to the target values of an ideal system. The current error obtained from online system identification does not exceed 0.2%. When the hidden layer has the neurons of thirteen, NN speed estimation result has a good performance. The simulation results show that the proposed control strategy can eliminate the NN reconstruction errors and has the higher robustness and accuracy than the previous method.

References

1. Schroedl, M.: An improved position estimator for sensorless control permanent magnet synchronous motor. In: EPE 1991, Florence, Italy, vol. 3, pp. 418–423 (1991)
2. Babak, N.M., Farid, M.T., Sargos, F.M.: Mechanical sensorless control of PMSM with online estimation of stator resistance. IEEE Trans. Ind. Appl. **40**(2), 457–471 (2004)
3. Kowalska, T.O., Dybkowski, M.: Stator-current-based MRAS estimator for a wide range speed-sensorless induction-motor drive. IEEE Trans. Ind. Electron. **57**(4), 1296–1308 (2010)
4. Bolognani, S., Oboe, R., Zigliotto, M.: Sensorless full-digital PMSM drive with EKF estimation of speed and rotor position. IEEE Trans. Ind. Electron. **46**, 184–308 (1999)
5. Sun, X.D., Chen, L., Yang, Z.B., Zhu, H.Q.: Speed-Sensorless vector control of a bearingless induction motor with artificial neural network inverse speed observer. IEEE/ASME Trans. Mech. **18**(4), 1357–1366 (2013)
6. Shady, M.G., Damian, G., John, W.F.: Stator current model reference adaptive systems speed estimator for regenerating-mode low-speed operation of sensorless induction motor drives. IET Electr. Power Appl. **7**(7), 597–606 (2013)
7. Miroslaw, W., Zbigniew, K., Jarosław, G., Haithem, A., Toliyat, H.A.: Artificial-neural-network-based sensorless nonlinear control of induction motors. IEEE Trans. Ener. Con. **20**(3), 520–528 (2005)
8. Accetta, A., Cirrincione, M., Pucci, M.: TLS EXIN based neural sensorless control of a high dynamic PMSM. Con. Eng. Prac. **20**, 725–732 (2012)
9. Batzel, T.D., Lee, K.Y.: An approach to sensorless operation of the permanent-magnet synchronous motor using diagonally recurrent neural networks. IEEE Trans. Ener. Con. **18**(1), 100–106 (2003)
10. Wang, L.M., Li, X.B.: Position sensorless control for PMLSM using Elman neural network. Rec. IEEE IAS Annu. Meeting **4**, 1–4 (2009)
11. Shady, M.G., Damian, G., John, W.F.: Stator current model reference adaptive systems speed estimator for regenerating-mode low-speed operation of sensorless induction motor drives. IET Electr. Power Appl. **7**(7), 597–606 (2012)
12. Jagannathan, S.: Neural network control of nonlinear discrete-time systems. CRC Press (2006)
13. Dierks, T., Thumati, B.T., Jagannathan, S.: Adaptive dynamic programming based optimal control of unknown affine nonlinear discrete-time systems. In: Proceedings of the IEEE Inter. J. Con. Neu. Net. (2009)

Neural-Network-Based Adaptive Fault Estimation for a Class of Interconnected Nonlinear System with Triangular Forms

Lei Liu, Zhanshan Wang[(✉)], Jinhai Liu, and Zhenwei Liu

College of Information Science and Engineering,
Northeastern University Shenyang, Liaoning, 110819, China
liuleill@live.cn, zhanshan_wang@163.com,
liujinhai@mail.neu.edu.cn, jzlzw@126.com

Abstract. In this paper, a novel fault estimation methodology is proposed for a class of interconnected nonlinear continues-time systems with triangular forms. In the distributed fault estimation architecture, a fault detector is utilized to generate a residual between the subsystem and its detector or observer. Moreover, a threshold for distributed fault detection and estimation in each subsystem is designed. Due to the universal approximation capability of the radial basis function neural networks, it is used to estimate the unknown fault dynamics. The time-to-failure is determined by solving the adaptive law from the current time instant to a failure threshold. Finally, the proposed methods are verified in the simulation.

Keywords: Adaptive fault estimation · Fault detection · Fault threshold · Radial basis function neural networks

1   Introduction

More recently, the investigations on fault detection and estimation (FDE) have received considerable attention because a fault at any given occurrence can be detected which is needed in order to prevent catastrophic failure. Most successful results of fault detection (FD) have been established for linear system [1-2], while some other fruitful FD techniques for nonlinear systems were also presented [3-5]. Typically, a novel unified model-based fault detection and prediction (FDP) scheme was developed for nonlinear multiple input multiple output (MIMO) discrete-time systems in [5].

It is noticed that an efficient FDE scheme is to utilize the adaptive control strategy. Analytical results in [6] guaranteed that the unknown fault dynamics would be

This work was supported by the National Natural Science Foundation of China (Grant Nos. 61473070, 61433004), the Fundamental Research Funds for the Central Universities (Grant Nos. N130504002 and N130104001), and SAPI Fundamental Research Funds (Grant No. 2013ZCX01).

Z. Zeng et al. (Eds.): ISNN 2014, LNCS 8866, pp. 110–120, 2014.
DOI: 10.1007/978-3-319-12436-0_13

estimated by updating the online approximator (OLA) parameters and utilizing the adaptive control law. Two adaptive fault-tolerant control (FTC) schemes were developed in [7] for parametric strict feedback systems to deal with the effects of the actuator failures. Generally speaking, these control techniques in [3-7] can be used to detect and estimate faults for a large kind of nonlinear systems. However, these FDE approaches are not established with respect to large-scale interconnected systems. In fact, the interconnected characters are common phenomena in various practical systems. One passive FTC law and two active FTC laws were designed in [8] to ensure the controlled synchronization of the complex interconnected neural networks in the presence of sensor faults. Recently, designs of distributed fault detection schemes for these interconnected systems have been developed by using multi-block kernel partial least squares [9] and rigorous analysis [10]. Many results were established for nonlinear systems with triangular elements [11-14]. However, the approaches in [11-14] did not extend the strategies into the systems with the interconnected terms. Altogether, few results on adaptive FDE are available on block triangular interconnected system, which motivate us to consider interconnected terms in this paper.

In this paper, a novel fault detector is utilized to generate a residual between the subsystem and its detector. Specifically, it is worth mentioning that the FDE schemes proposed in this paper aim at the interconnected systems which contain block triangular forms. Subsequently, by solving the adaptive law at the current time instant against a failure threshold, the remaining useful life or time-to-failure (TTF) is determined. Finally, a numerical example is used to verify the effectiveness of the proposed method.

2   Problem Description and Preliminaries

2.1   Nonlinear Interconnected System Description

Consider the nonlinear continuous time systems in block-triangular forms with N interconnected subsystems which may be subject to faults occurring at time t_0, in which the dynamics of the $i-th$ subsystem comprised of n_i states are described as

$$\begin{cases} \dot{x}_{i,j} = f_{i,j}\left(\underline{x}_{i,j}\right) + x_{i,j+1}, j = 1,\cdots,n_i-1 \\ \dot{x}_{i,n_i} = f_{i,n_i}\left(\underline{x}_{i,n_i}\right) + b_i u_i + \Delta_i\left(\underline{x}_{i,n_i},\bar{x}_i\right) \\ \qquad + d_i\left(\underline{x}_{i,n_i}\right) + \Psi\left(t-t_0\right)h_i\left(\underline{x}_{i,n_i}\right) \\ y_i = x_{i,1} \end{cases} \tag{1}$$

where $\underline{x}_{i,j} = [x_{i,1},\cdots,x_{i,j}]^T \in R^j$, $i = 1,\cdots,N$, $j = 1,\cdots,$ $n_i - 1$ and $\underline{x}_{i,n_i} = [x_{i,1},\cdots,x_{i,n_i}]^T \in R^{n_i}$ are the local state vectors , u_i is the local control input vector of the $i-th$ subsystem, $\bar{x}_i$ are the vectors of interconnected states which contain the states vectors of other subsystems, $f_{i,j}(\cdot): R^{n_i} \to R$, $j = 1,2,...,n_i$ are unknown smooth functions, b_i is a known constant with $b_i \neq 0$, $\Delta_i(\cdot): R^{n_i} \to R$ is an unknown interconnection function representing the effect of the other subsystems on the $i-th$ subsystem. From a qualitative viewpoint, the term $\Psi(t-t_0)h_i(\cdot)$ represents the local fault, $\Psi(t-t_0)$ stands for the time profiles which reflect the types of fault, and $h_i(\cdot)$ is the description of fault function, $d_i(\cdot)$ stands for the bounded uncertainties in the state equation and includes external disturbances as well as modeling errors and other possibly errors, $d_i(\cdot) \leq \bar{d}_i$ where $\bar{d}_i$ is a known positive constant.

Throughout this paper, the time profiles is chosen similarly to the fault diagnosis procedure in [11] which is described as

$$\Psi(t-t_0) = \begin{cases} 0, & \text{if } t < t_0 \\ 1 - e^{-\bar{k}_i(t-t_0)}, & \text{if } t \geq t_0 \end{cases} \tag{2}$$

where $\bar{k}_i$ is an unknown constant that represents the rate when a fault occurs. Generally speaking, a smaller value of $\bar{k}_i$ indicates that the fault is an incipient fault while a larger value of $\bar{k}_i$ stands for an abrupt type. This time profile definition would capture some of the commonly occurring fault dynamics in a nonlinear system [11].

Furthermore, because only abrupt faults will be considered all over this paper, the time profiles could be represented as

$$\Psi(t-t_0) = \begin{cases} 0, & \text{if } t < t_0 \\ 1, & \text{if } t \geq t_0 \end{cases} \tag{3}$$

2.2   NN Approximation and Some Preliminaries

RBFNN is used to approximate an arbitrary unknown function $f^{NN}(y)$

$$f^{NN}(y) = \theta^T \varphi(y) \tag{4}$$

where $y \in R^q$ is the input variable of the RBFNN, $\theta = [\theta_1, \cdots, \theta_l]^T$ is the weight vector with the RBFNN node number l, $\varphi(y)$ is the smooth basis function vector which is denoted as $\varphi(y) = [\varphi_1(y), \cdots, \varphi_l(y)]^T$.

From equation (4), it yields

$$f^{NN}(y) = \theta^{*T} \varphi(y) + \varepsilon^*(y) \tag{5}$$

where θ^* is the ideal constant weight and $\varepsilon^*(y)$ is the optimal approximation error with its bound $\varepsilon^*(y) \leq \bar{\varepsilon}$, where $\bar{\varepsilon}$ is a known constant.

Assumption 1: There exist a set of known constants $M_{ij} > 0$, $j = 1, \cdots, n_i$ for $\forall X_1, X_2 \in R^{n_i}$, the unknown smooth functions $f_{ij}(X_1)$, $f_{ij}(X_2)$, fault functions $h_{ij}(X_1)$, $h_{ij}(X_2)$ and the unknown interconnection function $\Delta_i(X_3)$, $\Delta_i(X_4)$ satisfy the following Lipschitz conditions

$$\left| f_{ij}(X_1) - f_{ij}(X_2) \right| \leq M_{ij} \left\| X_1 - X_2 \right\|$$
$$\left| h_i(X_1) - h_i(X_2) \right| \leq H_i \left\| X_1 - X_2 \right\|$$
$$\left| \Delta_i(X_3) - \Delta_i(X_4) \right| \leq L_i \left\| X_3 - X_4 \right\|$$

In this paper, $|\cdot|$ denotes the absolute value and $\|\cdot\|$ denotes the 2-norm.

Assumption 2: The state vectors $\underline{x}_{i,n_i} = [x_{i,1}, \cdots, x_{i,n_i}]^T \in R^{n_i}$, $i = 1, \cdots, N$ of each subsystem remain bounded before the occurrence of an unknown fault.

Our objective is to design a fault threshold to detect the fault, and then, to estimate the unknown fault in the case that make sure all the signals are bounded after the occurrence of a fault.

3   Fault Detection and Estimation

Rewritten (1) in the state space form

$$\dot{\underline{x}}_{i,n_i} = A_i \underline{x}_{i,n_i} + K_i x_{i,1} + \sum_{j=1}^{n_i} \left(B_{i,j} f_{i,j}\left(\underline{x}_{i,j} \right) \right) + C_i \Delta_i \left(\underline{x}_{i,n_i}, \bar{x}_i \right)$$
$$+ C_i d_i \left(\underline{x}_{i,n_i} \right) + C_i \Psi\left(t - t_0 \right) h_i \left(\underline{x}_{i,n_i} \right) + C_i b_i u_i \tag{6}$$

where $A_i = \begin{bmatrix} -k_{i1} & \\ \vdots & I_{(n_i-1)\times(n_i-1)} \\ -k_{in_i} & 0 & \cdots & 0 \end{bmatrix}_{n_i \times n_i}$, $K_i = \begin{bmatrix} k_{i1} \\ \vdots \\ k_{in_i} \end{bmatrix}$, $B_{ij} = [\underline{0 \cdots 1}_j, 0 \cdots]^T$, $C_i = \begin{bmatrix} 0 \\ \vdots \\ 0 \\ 1 \end{bmatrix}$.

The FDE for each subsystem is designed as

$$\dot{\hat{\underline{x}}}_{i,n_i} = A_i \hat{\underline{x}}_{i,n_i} + K_i x_{i,1} + \sum_{j=1}^{n_i} \left(B_{i,j} f_{i,j}\left(\hat{\underline{x}}_{i,j} \right) \right) + C_i \Delta_i \left(\hat{\underline{x}}_{i,n_i}, \hat{\bar{x}}_i \right) + C_i b_i u_i \tag{7}$$

Define $e_i = \underline{x}_{i,n_i} - \hat{x}_i$, where $e_i = \left[e_{i,1}, e_{i,2}, \cdots, e_{i,n_i} \right]^T \in R^{n_i}$, $\hat{x}_i = [\hat{x}_{i,1}, \cdots, \hat{x}_{i,n_i}]^T \in R^{n_i}$. Under healthy operating conditions, the residual dynamics are expressed as

$$\dot{e}_i = A_i e_i + \sum_{j=1}^{n_i} B_{i,j} \left(f_{i,j}\left(\underline{x}_{i,j} \right) - f_{i,j}\left(\hat{\underline{x}}_{i,j} \right) \right) + C_i \Delta_i \left(\underline{x}_{i,n_i}, \bar{x}_i \right) - C_i \Delta_i \left(\hat{\underline{x}}_{i,n_i}, \hat{\bar{x}}_i \right) + C_i d_i \left(\underline{x}_{i,n_i} \right) \tag{8}$$

Consider a Lyapunov function $V_i = e_i^T P_i e_i / 2$ for the $i-th$ subsystem. Its time derivative is given by

$$\dot{V}_i = \frac{1}{2} e_i^T \left(P_i A_i + A_i^T P_i \right) e_i + e_i^T P_i C_i d_i \left(\underline{x}_{i,n_i} \right) + e_i^T P_i C_i \left(\Delta_i \left(\underline{x}_{i,n_i}, \overline{x}_i \right) \right)$$
$$- \Delta_i \left(\hat{\underline{x}}_{i,n_i}, \hat{\overline{x}}_i \right) \right) + e_i^T P_i \sum_{j=1}^{n_i} B_{i,j} \left(f_{i,j} \left(\underline{x}_{i,j} \right) - f_{i,j} \left(\hat{\underline{x}}_{i,j} \right) \right) \tag{9}$$

Based on Assumption 1, we have

$$e_i^T P_i \sum_{j=1}^{n_i} B_{i,j} \left(f_{i,j} \left(\underline{x}_{i,j} \right) - f_{i,j} \left(\hat{\underline{x}}_{i,j} \right) \right) \le \left\| e_i^T \right\| \left\| n_i M_i P_i e_i \right\| = n_i e_i^T \left\| M_i P_i \right\| e_i \tag{10}$$

$$e_i^T P_i C_i \left(\Delta_i \left(\underline{x}_{i,n_i}, \overline{x}_i \right) - \Delta_i \left(\hat{\underline{x}}_{i,n_i}, \hat{\overline{x}}_i \right) \right) \le L_i \left\| e_i^T \right\| \left\| P_i \right\| \left\| \tilde{\overline{e}}_i \right\| \tag{11}$$

where $M_i = \max \left\{ M_{i1}, \cdots, M_{ij} \right\}$, $j = 1, \cdots, n_i$ and $\tilde{\overline{e}}_i = \overline{x}_i - \hat{\overline{x}}_i$.

Substituting (10)-(11) into (9) leads to

$$\dot{V}_i \le -\frac{1}{2} e_i^T R_i e_i + L_i \left\| e_i^T \right\| \left\| P_i \right\| \left\| \tilde{\overline{e}}_i \right\| + \overline{d}_i^2 / 2 \tag{12}$$

where $R_i = -\left(P_i A_i + A_i^T P_i + n_i \left\| M_i P_i \right\| I + P_i P_i \right)$.

Then, choose the Lyapunov function for the interconnected system as $V = \sum_{i=1}^{N} V_i$. Construct a positive definite Q whose elements are

$$Q_{ij} = \begin{cases} \lambda_{\min} \left(R_i \right), & i = j \\ -\left\| P_i \right\| L_i - \left\| P_j \right\| L_j, & i \ne j \end{cases} \tag{13}$$

Therefore, one gets

$$\dot{V} \le -\lambda_{\min} \left(Q \right) \tilde{z}^T \tilde{z} / 2 + \sum_{i=1}^{N} \overline{d}_i^2 / 2 \le -cV + \sum_{i=1}^{N} \overline{d}_i^2 / 2 \tag{14}$$

where $\tilde{z} = [e_1, e_2, \cdots, e_N]$, $P = diag \left[P_1, P_2, \cdots, P_N \right]$ and $c = -\lambda_{\min} \left(Q \right) / 2\lambda_{\max} \left(P \right)$.

We can always select a positive constant $\overline{V}_0$ to make sure $V(0) < \overline{V}_0$. Thus, we get

$$V(t) < \overline{V}_0 e^{-ct} + \frac{1}{2} \int_0^t e^{-c(t-\tau)} \sum_{i=1}^{N} \overline{d}_i^2 / 2 \, d\tau \tag{15}$$

This implies that $\left\| \tilde{z} \right\|^2 < \dfrac{\overline{V}_0 e^{-ct}}{\lambda_{\min} \left(P \right)} + \dfrac{1}{2\lambda_{\min} \left(P \right)} \int_0^t e^{-c(t-\tau)} \sum_{i=1}^{N} \overline{d}_i^2 / 2 \, d\tau$.

Therefore, if the detection threshold is chosen as

$$\rho = \frac{\overline{V}_0 e^{-ct}}{\lambda_{\min}(P)} + \frac{1}{2\lambda_{\min}(P)} \int_0^t e^{-c(t-\tau)} \sum_{i=1}^{N} \overline{d}_i^2 / 2 \, d\tau \tag{16}$$

No fault will be detected as long as the interconnected system is working in healthy operating conditions. In other words, when the detection residual exceeds the detection threshold ρ, a fault is declared active in the interconnected system.

In this paper, we assume that the fault only occurs in $i-th$ subsystem. For the $i-th$ subsystem in which fault occurs, considering (3), we know that $\Psi(t-t_0)=1$. Rewrite the fault subsystem as

$$\dot{\underline{x}}_{i,n_i} = A_i \underline{x}_{i,n_i} + K_i x_{i,1} + \sum_{j=1}^{n_i} \left(B_{i,j} f_{i,j} \left(\underline{x}_{i,j} \right) \right) + C_i \Delta_i \left(\underline{x}_{i,n_i}, \overline{x}_i \right)$$
$$+ C_i d_i \left(\underline{x}_{i,n_i} \right) + C_i h_i \left(\underline{x}_{i,n_i} \right) + C_i b_i u_i \tag{17}$$

The following novel fault estimators are developed

$$\dot{\hat{\underline{x}}}_{i,n_i} = A_i \hat{\underline{x}}_{i,n_i} + K_i x_{i,1} + \sum_{j=1}^{n_i} \left(B_{i,j} f_{i,j} \left(\hat{\underline{x}}_{i,j} \right) \right) + C_i \Delta_i \left(\hat{\underline{x}}_{i,n_i}, \hat{\overline{x}}_i \right) + C_i \hat{h}_i \left(\hat{\underline{x}}_{i,n_i} \right) + C_i b_i u_i \tag{18}$$

The derivative of the fault residual is given as

$$\dot{e}_i = A_i e_i + \sum_{j=1}^{n_i} B_{i,j} \left(f_{i,j} \left(\underline{x}_{i,j} \right) - f_{i,j} \left(\hat{\underline{x}}_{i,j} \right) \right) + C_i \Delta_i \left(\underline{x}_{i,n_i}, \overline{x}_i \right)$$
$$- C_i \Delta_i \left(\hat{\underline{x}}_{i,n_i}, \hat{\overline{x}}_i \right) + C_i h_i \left(\underline{x}_{i,n_i} \right) - C_i \hat{h}_i \left(\hat{\underline{x}}_{i,n_i} \right) + C_i d_i \tag{19}$$

We introduce an unknown intermediate variable $h_i \left(\hat{\underline{x}}_{i,n_i} \right)$. And then, it can be approximated by RBFNN

$$h_i \left(\hat{\underline{x}}_{i,n_i} \right) = \theta_i^T \varphi_i \left(\hat{\underline{x}}_{i,n_i} \right) + \varepsilon_i \tag{20}$$

where ε_i is the optimal approximation error.

Adding and subtracting $C_i h_i \left(\hat{\underline{x}}_{i,n_i} \right)$ on the right side of (19), we have

$$\dot{e}_i = A_i e_i + \sum_{j=1}^{n_i} B_{i,j} \left(f_{i,j} \left(\underline{x}_{i,j} \right) - f_{i,j} \left(\hat{\underline{x}}_{i,j} \right) \right) + C_i \Delta_i \left(\underline{x}_{i,n_i}, \overline{x}_i \right) - C_i \Delta_i \left(\hat{\underline{x}}_{i,n_i}, \hat{\overline{x}}_i \right)$$
$$+ C_i h_i \left(\underline{x}_{i,n_i} \right) - C_i h_i \left(\hat{\underline{x}}_{i,n_i} \right) + C_i d_i + C_i \left(\tilde{\theta}_i^T \varphi_i \left(\hat{\underline{x}}_{i,n_i} \right) + \varepsilon_i \right) \tag{21}$$

where $\tilde{\theta}_i = \theta - \hat{\theta}_i$ is the RBFNN parameter estimation error.

Consider the following Lyapunov function $V = \sum_{q=1}^{N} V_q$, where $V_q = e_q^T P_q e_q / 2$ $+ \tilde{\theta}_q^T \tilde{\theta}_q / 2\gamma_{q\theta}$ and $\gamma_{q\theta} > 0$ are known constants. Since only the $i-th$ subsystem is under unhealthy operation, we obtain $\tilde{\theta}_q(t) = 0$ for $q \neq i$. Note that $d_i(\cdot) \leq \overline{d}_i$. Based on Assumption 1, the time derivative of V is

$$\dot{V} \leq \sum_{q=1}^{N}\left\{-\frac{1}{2}e_q^T\left(P_qA_q + A_q^TP_q\right)e_q + n_qM_q\left\|e_q^T\right\|\left\|P_qe_q\right\| + \left\|e_q^TP_qC_q\right\|\right.$$

$$\left.\times\left(\overline{d}_q + L_q\left\|\tilde{\overline{x}}_q\right\| + H_q\left\|e_q\right\|\right)\right\} + e_i^TP_iC_i\tilde{\theta}_i^T\varphi_i\left(\hat{\underline{x}}_{i,n_i}\right) - \frac{1}{\gamma_{i\theta}}\tilde{\theta}_i^T\dot{\hat{\theta}}_i + e_i^TP_iC_i\varepsilon_i \tag{22}$$

By using the properties of 2-norm and the inequality $2ab \leq a^2 + b^2$, one can obtain

$$\left\|e_q^TP_qC_q\right\|\overline{d}_q \leq e_q^T\left(P_qP_q\right)e_q + \overline{d}_q^2/4, \quad n_qM_q\left\|e_q^T\right\|\left\|P_qe_q\right\| \leq e_q^T\left(\left\|n_qM_qP_q\right\|I\right)e_q \tag{23}$$

$$\left\|e_q^TP_qC_q\right\|H_q\left\|e_q\right\| \leq e_q^T\left(\left\|H_qP_q\right\|I\right)e_q, \quad e_i^TP_iC_i\varepsilon_i \leq e_i^T\left(P_iP_i\right)e_i + \overline{\varepsilon}_i^2/4 \tag{24}$$

Choose the following update law as

$$\dot{\hat{\theta}}_i = \gamma_{i\theta}C_i^TP_ie_i\varphi_i - \mu_i\hat{\theta}_i \tag{25}$$

where $\mu_i > 0$ is a known constant.

Substituting (23)-(25) into (22), it gives

$$\dot{V} \leq \sum_{\substack{q=1\\q\neq i}}^{N}\left\{-\frac{1}{2}e_q^TR_qe_q + L_q\left\|P_q\right\|\left\|e_q^T\right\|\left\|\tilde{\overline{x}}_q\right\| + \frac{1}{4}\left|\overline{d}_q\right|^2\right\} - \frac{\mu_i}{2\gamma_{i\theta}}\left\|\tilde{\theta}_i\right\|^2$$

$$-\frac{1}{2}e_i^TR_{if}e_i + \frac{1}{4}\left|\overline{d}_i\right|^2 + L_i\left\|P_i\right\|\left\|e_i^T\right\|\left\|\tilde{\overline{x}}_i\right\| + \frac{1}{4}\overline{\varepsilon}^2 + \frac{\mu_i}{2\gamma_{i\theta}}\left\|\theta\right\|^2 \tag{26}$$

where $R_q = -\left(P_qA_q + A_q^TP_q + n_q\left\|M_qP_q\right\|I + P_qP_q + \left\|H_qP_q\right\|I\right)$ and $R_{if} = -\left(P_iA_i + A_i^TP_i\right.$ $\left. + 2P_iP_i + n_i\left\|M_iP_i\right\|I + \left\|H_iP_i\right\|I\right)$.

Construct a positive definite Q^f whose elements are

$$Q_{qj}^f = \begin{cases} \lambda_{\min}\left(R_q\right), & q = j \neq i \\ \lambda_{\min}\left(R_{if}\right), & q = j = i \\ -\left\|P_q\right\|L_q - \left\|P_q\right\|L_q, & q \neq j \end{cases}$$

Therefore, it yields

$$\dot{V} \leq -\frac{1}{2}\left[e_1, e_2, \cdots, e_N\right]Q^f\left[e_1, e_2, \cdots, e_N\right]^T - \frac{\mu_i}{2\gamma_{i\theta}}\left\|\tilde{\theta}_i\right\|^2 + \frac{1}{4}\overline{\varepsilon}^2 + \frac{\mu_i}{2\gamma_{i\theta}}\left\|\theta\right\|^2 + \sum_{i=1}^{N}\overline{d}_i^2/2 \tag{27}$$

Denote $D = \frac{1}{4}\overline{\varepsilon}^2 + \frac{\mu_i}{2\gamma_{i\theta}}\left\|\theta\right\|^2 + \sum_{i=1}^{N}\overline{d}_i^2/2$, $c^f = -\lambda_{\min}\left(Q^f\right)/-\lambda_{\max}\left(P\right)$ and $c^* = \max$

$\left\{c, \mu_i\right\}$. The time derivative of $\dot{V}$ becomes

$$\dot{V} \le -cV + D \tag{28}$$

By integrating (28) over $[0,t]$, we get

$$V(t) \le V(0)e^{-ct} + \frac{1}{2}\int_0^t e^{-c(t-\tau)}Dd\tau \tag{29}$$

Therefore, the detection residuals and the NN estimation errors are bounded.

Theorem 1: Consider the nonlinear continuous time systems in block-triangular forms with N interconnected subsystems. The fault threshold is defined as (16) to monitor the whole system. After a fault is detected, the RBFNN estimator is initiated and adaptive law (25) is used to update the unknown RBFNN weight vector parameter. Then, the parameter estimation error $\tilde{\theta}_i$ and the FD residual $e_m, m = 1, \cdots, N$ will be bounded after the occurrence of a fault.

In the event of a fault, consider the solution of the RBFNN parameter adaptive law (25) to analytically determine TTF as shown in the following

$$\hat{\theta}_i = e^{-\mu_i(t-t_0)}\hat{\theta}_i(t_0) + \int_{t_0} e^{-\mu_i(t-\tau)}\gamma_i P_i e_i^T(\tau)\varphi_i(\tau)d\tau \tag{30}$$

where t_0 is the instant when a fault occurs and it is determined in the process of fault detection. From the property of the RBFNN's basis function, we know that $\varphi_i(\cdot)$ is bounded. The detection residual is also bounded from the present Theorem 1. Thus, the second term of above equation is bounded if the t is in a finite interval. Suppose that t_f is the time at which the system is failure. The $\hat{\theta}_i(t_f)$ can be represented as

$$\hat{\theta}_i(t_f) = \gamma_i P_i e_i^T(t)\varphi_i(t)\int_{t_0}^{t_f} e^{-\mu_i(t_f-\tau)}d\tau + e^{-\mu_i(t_f-t_0)}\hat{\theta}_i(t_0) \tag{31}$$

By solving the definite integral equation (31), one gets

$$\hat{\theta}_i(t_f) = \gamma_i P_i e_i^T(t)\varphi_i(t)\left(1 - e^{-\mu_i(t_f-t_0)}\right)\Big/\mu_i + e^{-\mu_i(t_f-t_0)}\hat{\theta}_i(t_0)$$

Define $\hat{\theta}_{ij}(t_f)$ and φ_{ij} as the $j-th$ element of $\hat{\theta}_i(t_f)$ and the NN basis function, respectively. It can be rewritten as

$$\hat{\theta}_{ij}(t_f) = \gamma_i P_i e_i^T(t)\varphi_{ij}(t)\Big/\mu_i + e^{-\mu_i(t_f-t_0)}\left(\hat{\theta}_i(t_0) - \gamma_i e_i^T(t)\varphi_{ij}(t)\Big/\mu_i\right)$$

After substituting $\hat{\theta}_{ij}(t_f)$ with $\theta_{ij\,\text{max}}$, it results in

$$e^{-\mu_i(t_f-t_0)} = \frac{\mu_i\theta_{ij\,\text{max}} - \gamma_i P_i e_i^T(t)\varphi_{ij}(t)}{\mu_i\hat{\theta}_{ij}(t) - \gamma_i P_i e_i^T(t)\varphi_{ij}(t)}$$

where $\theta_{ij\,max}$ is the failure limit regarding the parameter of estimated fault function, in terms of maximum value of the RBFNN parameter.

Further mathematical manipulations would give us

$$t_f = \frac{1}{\mu_i} \ln \left| \frac{\mu_i \hat{\theta}_{ij}(t) - \gamma_i P_i e_i^T(t) \varphi_{ij}(t)}{\mu_i \theta_{ij\,max} - \gamma_i P_i e_i^T(t) \varphi_{ij}(t)} \right| + t_o \tag{32}$$

Thus, the equation (32) calculates TTF by applying the estimated fault magnitude at the current time instant and the failure threshold.

4   Simulation Example

To validate the effectiveness of the proposed method, we consider the continues time interconnected systems with 3 subsystems

$$\begin{cases} \dot{x}_{11} = f_{11}(x_{11}) + g_{11}(x_{11})x_{12} \\ \dot{x}_{12} = f_{12}(x_{11},x_{12}) + b_1 u_1 + d_1 + \Delta_1 + h_1(x_{11},x_{12}) \end{cases}$$

$$\begin{cases} \dot{x}_{21} = f_{21}(x_{21}) + g_{21}(x_{21})x_{22} \\ \dot{x}_{22} = f_{22}(x_{21},x_{22}) + b_2 u_2 + d_2 + \Delta_2 \end{cases}$$

$$\begin{cases} \dot{x}_{31} = f_{31}(x_{31}) + g_{31}(x_{31})x_{32} \\ \dot{x}_{32} = f_{32}(x_{31},x_{32}) + b_3 u_3 + d_3 + \Delta_3 \end{cases} \tag{33}$$

where $f_{i1}(\cdot) = 0$, $g_{i1}(\cdot) = 1$, $i = 1,2,3$, $b_1 = 1$, $f_{12}(\cdot) = 2\sin x_{11} - 0.36\sin x_{11}\cos x_{11}$, $b_2 = 0.7$, $b_3 = 1.2$, $d_1 = 0.01\sin x_{11}$, $\Delta_1 = 0.36\sin x_{21}\cos x_{21}$, $f_{22}(\cdot) = 2\sin x_{21} - 3.8$ $\times \sin x_{21}\cos x_{21}$, $d_2 = 0$, $d_3 = 0$, $\Delta_2 = 0.18\sin x_{11}\cos x_{11} + 3.6\sin x_{31}\cos x_{31}$, $f_{32}(\cdot) = 2\sin x_{31} - 1.44\sin x_{31}\cos x_{31}$, $\Delta_3 = 1.44\sin x_{21}\cos x_{21}$ and $h_1(x_{11},x_{12})$ stands for an unknown fault function which will be approximated in the later.

To design an adaptive law conveniently, we consider the case in which abrupt fault only occurs in the first subsystem. This implies that $\Psi(t - t_0) = 1$ when $t \geq t_0$.

Throughout this simulation, we utilize RBFNN to approximate the unknown fault function. The node numbers of the RBFNN are chosen as $l_{1,1} = 12, l_{1,2} = 20$. The centers and widths are selected on a regular lattice in the respective compact sets. The RBFNN $\hat{\theta}_i^T(t)\varphi_i(\hat{\underline{x}})$ contains 12 nodes with the centers μ_i evenly spaced in $[-3,3] \times [-2,2] \times [-1.5,1.5]$ and its width is 5.

The initial conditions are chosen to be $x_{11}(0) = 1$, $x_{12}(0) = -1$, $x_{21}(0) = 0.5$, $x_{22}(0) = -0.5$, $x_{31}(0) = 0.5$, $x_{32}(0) = 1$, $\hat{\theta}_i(0) = 0.2I$ $(i = 1,2,3)$ with I is an identity matrix, $P_i = 0.5I$.

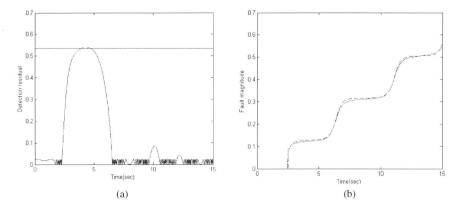

Fig. 1. (a) Detection residual of the subsystem 1 and threshold
(b) Estimated fault magnitudes and actual fault mafnitudes

A fault is initiated at time $t_0 = 2\,\mathrm{s}$ in the first subsystem which is expressed as

$$\Psi\left(t-t_0\right)h_1\left(\underline{x}_{12}\right) = \begin{cases} 0, & if\ t < t_0 \\ 0.03\left(1 - e^{-0.1(t-t_0)}\right)x_{11}, & if\ t \geq t_0 \end{cases}$$

The computer simulation results are given in Fig. 1. To show the performance of the FD estimator, the detection residual of the first subsystem is represented in Fig. 1 (a). We utilize a fault threshold of 0.54 unit magnitude. The residual is below the threshold at the time interval $t \in [0, 2.5]$, but it starts to increase and eventually it reaches the detection threshold at $t = 4.0\mathrm{s}$. The RBFNN is turned on to estimate the unknown fault dynamics by tuning the parameter θ_i online. Moreover, as observed in Fig. 1 (b), it can be concluded that the RBFNN is able to approximate the unavailable fault dynamics with a small error while the fault is detected.

5   Conclusion

In this paper, we focus on the structure of adaptive fault detection and estimation for a class of large-scale nonlinear system. In contrast to various existing schemes, the proposed strategy is designed based on the fact that the input is in block-triangular forms without satisfying the so-called matching condition. A new FD is used to generate a residual between the subsystem and its observer. The local online approximator is initiated while the detection residual of one subsystem exceeds its detection threshold. Finally, a simulation example is employed to illustrate the effectiveness of the proposed schemes.

References

1. Li, X., Zhou, K.: A time domain approach to robust fault detection of linear time-varying systems. Automatica **45**(1), 94–102 (2009)
2. Yang, G., Wang, H.: Fault detection for a class of uncertain state feedback control systems. IEEE Trans. Contr. Syst. Tech. **18**(1), 201–212 (2010)
3. Persis, C., Isidori, A.: A geometric approach to nonlinear fault detection and isolation. IEEE Transactions on Automatic Control **46**(6), 853–865 (2001)
4. Cacccavale, F., Pierri, F., Villani, L.: Adaptive observer for fault diagnosis in nonlinear discrete-time systems. ASME Journal of Dynamic Systems, Measurement, and Control **130**(2), 1–9 (2008)
5. Thumati, B., Jagannathan, S.: A model based fault detection and prediction scheme for nonlinear multivariable discrete-time systems with asymptotic stability guarantees. IEEE Transactions on Neural Networks **21**(3), 404–423 (2010)
6. Thumati, B., Feinstein, M., Sarangapani, J.: A Model-based Fault Detection and Prognostics Scheme for Takagi-Sugeno Fuzzy Systems. IEEE Transactions on Fuzzy Systems (2013). doi:10.1109/TFUZZ.2013.2272584
7. Wang, W., Wen, C.: Adaptive actuators failure compensation for uncertain nonlinear systems with guaranteed transient performance. Automatica **46**(12), 2082–2091 (2010)
8. Wang, Z., Li, T., Zhang, H.: Fault tolerant synchronization for a class of complex interconnected neural networks with delay. International Journal of Adaptive Control and Signal Processing (2013). doi:10.1002/acs.2399
9. Zhang, Y., Zhou, H., Qin, S., Chai, T.: Decentralized fault diagnosis of large-scale processes using multiblock kernel partial least squares. IEEE Transactions on Industrial Informatics **6**(1), 3–10 (2010)
10. Panagi, P., Polycarpou, M.: Distributed Fault Accommodation for a Class of Interconnected Nonlinear Systems with Partial Communication. IEEE Transactions on Automatic Control **56**(12), 2962–2967 (2011)
11. Xu, A., Zhang, Q.: Nonlinear system fault diagnosis based on adaptive estimation. Automatica **40**(7), 1181–1193 (2004)
12. Zhang, X., Polycarpou, M., Parisini, T.: Fault diagnosis of a class of nonlinear uncertain systems with Lipschitz nonlinearities using adaptive estimation. Automatica **46**(2), 290–299 (2010)
13. Zhang, K., Jiang, B., Shi, P.: Fault estimation observer design for discrete-time Takagi-Sugeno fuzzy systems based on piecewise Lyapunov functions. IEEE Transactions on Fuzzy Systems **20**(1), 192–200 (2012)
14. Jiang, B., Gao, Z., Shi, P.: Observer-based integrated robust fault estimation and accommodation design for discrete time systems. International Journal of Control **83**(6), 1167–1181 (2010)

Big Data Matrix Singular Value Decomposition Based on Low-Rank Tensor Train Decomposition

Namgil Lee$^{(\boxtimes)}$ and Andrzej Cichocki

Laboratory for Advanced Brain Signal Processing, RIKEN Brain Science Institute,
Wako-shi, Saitama 351-0198, Japan
{namgil.lee,a.cichocki}@riken.jp

Abstract. We propose singular value decomposition (SVD) algorithms for very large-scale matrices based on a low-rank tensor decomposition technique called the tensor train (TT) format. By using the proposed algorithms, we can compute several dominant singular values and corresponding singular vectors of large-scale structured matrices given in a low-rank TT format. We propose a large-scale trace optimization problem, and in the proposed methods, the large-scale optimization problem is reduced to sequential small-scale optimization problems. We show that the computational complexity of the proposed algorithms scales logarithmically with the matrix size if the TT-ranks are bounded. Numerical simulations based on very large-scale Hilbert matrix demonstrate the effectiveness of the proposed methods.

Keywords: Big data processing · Curse-of-dimensionality · Eigenvalue decomposition · Singular value decomposition · Optimization · Tensor network · Matrix product states

1 Introduction

The singular value decomposition (SVD) is an important matrix factorization technique in wide areas of numerical sciences and engineering. By using the SVD, one can solve linear least squares problems, compute the best low-rank approximation of matrices, compute the pseudo-inverse of matrices, and find a common subspace, just to list a few. The SVD has a wide range of applications in signal processing, multivariate statistics, systems biology, finance, image processing, and so on [1].

In this paper, we propose algorithms for computing K dominant singular values and corresponding singular vectors of large-scale matrices. Standard methods for computing all the singular values and singular vectors of a $P \times Q$ matrix with $P \geq Q$ take $O(PQ^2)$ computational costs, and $O(PK^2)$ for computing $K \leq Q$ singular values [1]. However, if the size of the matrix grows exponentially as $P = Q = 2^N$, the computation of SVD is intractable by desktop computers and standard algorithms. A Monte-Carlo algorithm [2] can be used, but its accuracy is not high enough, and it still suffers from the curse-of-dimensionality for computing singular vectors.

© Springer International Publishing Switzerland 2014
Z. Zeng et al. (Eds.): ISNN 2014, LNCS 8866, pp. 121–130, 2014.
DOI: 10.1007/978-3-319-12436-0_14

On the other hand, the proposed algorithms apply a low-rank tensor approximation technique which is called the tensor train (TT) format [3]. The basic idea underlying the TT format is to reshape matrices and vectors into high-order tensors, i.e., multi-way arrays, and compress them in low-parametric tensor formats. Then, all the numerical operations including the matrix-by-vector multiplication can be performed in a distributed way within a computational cost that scales logarithmically with the matrix size [3]. We refer to [4–6] for modern low-rank tensor approximation techniques, other types of tensor networks, and recent advances of tensor decomposition methods for big data analysis.

In this paper, we show that the singular value decomposition can be efficiently computed by solving a specific type of trace optimization problem. Then, in order to solve the large-scale optimization problem, the K orthonormal singular vectors are efficiently represented in *block TT format* [7–9]. We employ the alternating least squares (ALS) [8,10] and modified alternating least squares (MALS) [10,11] schemes for solving the large-scale optimization problem. In the ALS and MALS schemes, the large-scale optimization problem is reduced to sequential small optimization problems for which any standard optimization algorithms can be applied.

This paper is organized as follows. In Section 2, we propose the trace optimization problem for computing K dominant singular values and corresponding singular vectors. And we describe the TT decompositions of large-scale matrices and vectors. In Section 3, we describe the proposed algorithms with graphical illustration. In Section 4, we present the simulation results demonstrating the effectiveness of the proposed methods in computing the SVD of Hilbert matrix of huge sizes. In Section 5, we provide conclusions and discussions.

2   Tensor Train Formats for Solving Large-Scale Optimization Problem

2.1   Trace Maximization Problem for SVD

We assume that $P \geq Q$ without loss of generality. The $K \leq Q$ dominant singular values and corresponding singular vectors of a matrix $\mathbf{A} \in \mathbb{R}^{P \times Q}$ can be computed by solving the following trace maximization problem:

$$\underset{\mathbf{U},\mathbf{V}}{\text{maximize}} \quad \text{trace}\left(\mathbf{U}^{\mathrm{T}}\mathbf{A}\mathbf{V}\right)$$
$$\text{subject to} \quad \mathbf{U}^{\mathrm{T}}\mathbf{U} = \mathbf{V}^{\mathrm{T}}\mathbf{V} = \mathbf{I}_K . \tag{1}$$

The maximization problem (1) is derived from the eigenvalue decomposition (EVD) of the following $(P+Q) \times (P+Q)$ matrix

$$\mathbf{B} = \begin{bmatrix} \mathbf{0} & \mathbf{A} \\ \mathbf{A}^{\mathrm{T}} & \mathbf{0} \end{bmatrix} \in \mathbb{R}^{(P+Q)\times(P+Q)} . \tag{2}$$

It is known that the largest $K \leq Q$ eigenvalues of $\mathbf{B}$ are same to the K dominant singular values of $\mathbf{A}$, and corresponding eigenvectors are equivalent to

$\mathbf{W}_0 = 2^{-1/2} \left[\mathbf{U}_0^{\mathrm{T}}, \mathbf{V}_0^{\mathrm{T}} \right]^{\mathrm{T}} \in \mathbb{R}^{(P+Q) \times K}$, where $\mathbf{U}_0 = [\mathbf{u}_{01}, \mathbf{u}_{02}, \dots, \mathbf{u}_{0K}] \in \mathbb{R}^{P \times K}$ and $\mathbf{V}_0 = [\mathbf{v}_{01}, \mathbf{v}_{02}, \dots, \mathbf{v}_{0K}] \in \mathbb{R}^{Q \times K}$ denote the left and right dominant singular vectors of $\mathbf{A}$. The EVD of $\mathbf{B}$ can be computed by solving the block Rayleigh quotient maximization problem

$$
\begin{aligned}
\underset{\mathbf{W}}{\text{maximize}} \quad & \text{trace} \left(\mathbf{W}^{\mathrm{T}} \mathbf{B} \mathbf{W} \right) \\
\text{subject to} \quad & \mathbf{W}^{\mathrm{T}} \mathbf{W} = \mathbf{I}_K .
\end{aligned}
\tag{3}
$$

However, the problem (3) incurs larger computational and memory costs than (1) due to the size of $\mathbf{B}$. Proposition 1 shows the equivalence of the solutions of (1) and (3).

Proposition 1. *For $K \leq Q \leq P$, the solution of the maximization problem (1) is equivalent to that of (3).*

Proof. Let $\mathbf{W} = 2^{-1/2}[\mathbf{U}^{\mathrm{T}}, \mathbf{V}^{\mathrm{T}}]^{\mathrm{T}} \in \mathbb{R}^{(P+Q) \times K}$, then we have trace$(\mathbf{W}^{\mathrm{T}} \mathbf{B} \mathbf{W}) = $ trace$(\mathbf{U}^{\mathrm{T}} \mathbf{A} \mathbf{V})$. First, we can show that

$$\max\{\text{trace}(\mathbf{W}^{\mathrm{T}} \mathbf{B} \mathbf{W}) : \mathbf{W}^{\mathrm{T}} \mathbf{W} = \mathbf{I}_K\} \geq \max\{\text{trace}(\mathbf{U}^{\mathrm{T}} \mathbf{A} \mathbf{V}) : \mathbf{U}^{\mathrm{T}} \mathbf{U} = \mathbf{V}^{\mathrm{T}} \mathbf{V} = \mathbf{I}_K\} .$$

Next, we can show that the maximum value of trace$(\mathbf{W}^{\mathrm{T}} \mathbf{B} \mathbf{W})$ is obtained by trace$(\mathbf{U}^{\mathrm{T}} \mathbf{A} \mathbf{V})$ when $\mathbf{U}$ and $\mathbf{V}$ are equal to the K dominant singular vectors.

2.2 Tensor Train Decompositions

An Nth order tensor $\underline{\mathbf{X}}$ is a multi-way array of size $I_1 \times I_2 \times \cdots I_N$, where I_n denotes the size of the nth mode. The $(i_1, i_2, \dots, i_N)$th entry of $\underline{\mathbf{X}}$ is denoted by $x_{i_1, i_2, \dots, i_N}$ or $\underline{\mathbf{X}}(i_1, i_2, \dots, i_N)$. The mode-$n$ unfolding of a tensor $\underline{\mathbf{X}} \in \mathbb{R}^{I_1 \times \cdots \times I_N}$ is denoted by $\mathbf{X}_{(n)} \in \mathbb{R}^{I_n \times I_1 \cdots I_{n-1} I_{n+1} \cdots I_N}$. The mode-$(M, 1)$ contraction of tensors $\underline{\mathbf{A}} \in \mathbb{R}^{I_1 \times I_2 \times \cdots \times I_M}$ and $\underline{\mathbf{B}} \in \mathbb{R}^{I_M \times J_2 \times J_3 \times \cdots J_N}$ is defined by

$$\underline{\mathbf{C}} = \underline{\mathbf{A}} \bullet \underline{\mathbf{B}} \in \mathbb{R}^{I_1 \times I_2 \times \cdots \times I_{M-1} \times J_2 \times J_3 \times \cdots \times J_N} \tag{4}$$

with entries

$$\underline{\mathbf{C}}(i_1, i_2, \dots, i_{M-1}, j_2, j_3, \dots, j_N) = \sum_{i_M=1}^{I_M} \underline{\mathbf{A}}(i_1, i_2, \dots, i_M) \underline{\mathbf{B}}(i_M, j_2, j_3, \dots, j_N) .$$

Tensors and tensor operations can be represented by tensor network diagrams as in Figure 1. The number of edges connected to a vertex shows the order of the tensor, and the connections between vertices represent contraction between tensors. We refer to [12,13] for further notations of tensor operations.

In TT format, a large-scale matrix $\mathbf{A} \in \mathbb{R}^{I_1 I_2 \cdots I_N \times J_1 J_2 \cdots J_N}$ is *tensorized* into a higher-order tensor $\underline{\mathbf{A}} \in \mathbb{R}^{I_1 \times J_1 \times I_2 \times J_2 \times \cdots \times I_N \times J_N}$, and it is represented as contracted products

$$\underline{\mathbf{A}} = \underline{\mathbf{A}}^{(1)} \bullet \underline{\mathbf{A}}^{(2)} \bullet \cdots \bullet \underline{\mathbf{A}}^{(N)}, \tag{5}$$

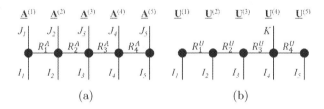

(a) vector, matrix, tensor   (b) $\mathbf{A} \bullet \mathbf{B}$ (c) tensorization of $\mathbf{x}$

Fig. 1. Graphical representations of (a) a vector, a matrix, a 3rd order tensor, (b) contracted product of two 3rd order tensors, and (c) tensorization of a vector

where $\underline{\mathbf{A}}^{(n)} \in \mathbb{R}^{R_{n-1}^A \times I_n \times J_n \times R_n^A}$ are 4th order tensors called TT-cores and R_n^A are called TT-ranks. We suppose that $R_0^A = R_N^A = 1$. Moreover, in TT format, a group of large-scale singular vectors $\mathbf{U} = [\mathbf{u}_1, \mathbf{u}_2, \dots, \mathbf{u}_K] \in \mathbb{R}^{I_1 I_2 \cdots I_N \times K}$ is tensorized and permuted to a higher-order tensor $\underline{\mathbf{U}} \in \mathbb{R}^{I_1 \times I_2 \times \cdots \times I_{n-1} \times K \times I_n \times \cdots \times I_N}$, and it is represented as contracted products

$$\underline{\mathbf{U}} = \underline{\mathbf{U}}^{(1)} \bullet \underline{\mathbf{U}}^{(2)} \bullet \cdots \bullet \underline{\mathbf{U}}^{(N)}, \tag{6}$$

where the TT-cores $\underline{\mathbf{U}}^{(m)} \in \mathbb{R}^{R_{m-1}^U \times I_m \times R_m^U}$, $m \neq n$, are 3rd order tensors and the nth TT-core $\underline{\mathbf{U}}^{(n)} \in \mathbb{R}^{R_{n-1}^U \times K \times I_n \times R_n^U}$ is a 4th order tensor. We suppose that $R_0^U = R_N^U = 1$.

We call (5) as the matrix TT format and (6) as the block-n TT format. Figure 2 shows tensor network diagrams representing a matrix $\mathbf{A} \in \mathbb{R}^{I_1 \cdots I_5 \times J_1 \cdots J_5}$ in matrix TT format, and a group of vectors $\mathbf{U} \in \mathbb{R}^{I_1 \cdots I_5 \times K}$ in block-4 TT format.

$$\begin{array}{cccccc}
\underline{\mathbf{A}}^{(1)} & \underline{\mathbf{A}}^{(2)} & \underline{\mathbf{A}}^{(3)} & \underline{\mathbf{A}}^{(4)} & \underline{\mathbf{A}}^{(5)} & \underline{\mathbf{U}}^{(1)} \quad \underline{\mathbf{U}}^{(2)} \quad \underline{\mathbf{U}}^{(3)} \quad \underline{\mathbf{U}}^{(4)} \quad \underline{\mathbf{U}}^{(5)}
\end{array}$$

(a)   (b)

Fig. 2. Graphical representations of (a) a matrix in matrix TT format and (b) a group of vectors in block-4 TT format

Note that the storage cost for the block TT decomposition is $O(NIR^2 + IR^2K)$, where $I = \max(I_n)$ and $R = \max(R_n^U)$, whereas the original group of vectors has $O(I^N K)$ storage complexity. In the same way, the matrix TT decomposition compresses original data dramatically as well.

3   Algorithms Based on Block TT Decomposition

Let $P = I_1 I_2 \cdots I_N$ and $Q = J_1 J_2 \cdots J_N$. We assume that $\mathbf{A} \in \mathbb{R}^{I_1 I_2 \cdots I_N \times J_1 J_2 \cdots J_N}$ is given in matrix TT format (5). We apply the alternating least squares (ALS) [8] and modified alternating least squares (MALS) [10] methods for solving the large-scale optimization problem (1). We suppose that the left and right singular vectors $\mathbf{U} = [\mathbf{u}_1, \mathbf{u}_2, \ldots, \mathbf{u}_K]$ and $\mathbf{V} = [\mathbf{v}_1, \mathbf{v}_2, \ldots, \mathbf{v}_K]$ in (1) are represented in block TT format. The ALS algorithm for SVD is described in Algorithm 1. The ALS and MALS schemes are depicted graphically in Figure 3 via tensor network diagrams.

Algorithm 1. ALS algorithm for SVD based on block TT format

Data: $\mathbf{A} \in \mathbb{R}^{I_1 I_2 \cdots I_N \times J_1 J_2 \cdots J_N}$ in matrix TT format, $K \geq 2$, $\delta \geq 0$

Result: Left singular vectors $\mathbf{U} \in \mathbb{R}^{I_1 I_2 \cdots I_N \times K}$ and right singular vectors $\mathbf{V} \in \mathbb{R}^{J_1 J_2 \cdots J_N \times K}$ in block-N TT format, TT-ranks $R_1^U, R_2^U, \ldots, R_{N-1}^U$ for $\mathbf{U}$ and $R_1^V, R_2^V, \ldots, R_{N-1}^V$ for $\mathbf{V}$.

1 Initialization: Block-N TT tensors $\underline{\mathbf{U}}$ and $\underline{\mathbf{V}}$ with small TT-ranks and orthogonalized TT-cores ;

2 **repeat**

3   **for** $n = N, N-1, \ldots, 2$ **do** right-to-left half sweep

    // Optimize

4   Fix all the TT-cores except $\underline{\mathbf{U}}^{(n)}$ and $\underline{\mathbf{V}}^{(n)}$, compute

$$\left((\mathbf{U}_{(2)}^{(n)})^{\mathrm{T}}, (\mathbf{V}_{(2)}^{(n)})^{\mathrm{T}} \right) =$$

$$\underset{\mathbf{U}^{(n)}, \mathbf{V}^{(n)}}{\operatorname{argmax}} \left\{ \operatorname{trace}\left((\mathbf{U}^{(n)})^{\mathrm{T}} \overline{\mathbf{A}}_n \mathbf{V}^{(n)} \right) : \left(\mathbf{U}^{(n)} \right)^{\mathrm{T}} \mathbf{U}^{(n)} = \left(\mathbf{V}^{(n)} \right)^{\mathrm{T}} \mathbf{V}^{(n)} = \mathbf{I}_K \right\}$$

    // Separate

5   Compute δ-truncated SVD: $[\mathbf{U}_1, \mathbf{S}_1, \mathbf{V}_1] = \mathrm{SVD}_\delta\left(\mathbf{U}_{(\{1,2\} \times \{3,4\})}^{(n)} \right)$, $[\mathbf{U}_2, \mathbf{S}_2, \mathbf{V}_2] = \mathrm{SVD}_\delta\left(\mathbf{V}_{(\{1,2\} \times \{3,4\})}^{(n)} \right)$;

6   Update $R_{n-1}^U = rank(\mathbf{V}_1)$, $R_{n-1}^V = rank(\mathbf{V}_2)$;

7   Update $\underline{\mathbf{U}}^{(n)} = reshape(\mathbf{V}_1^{\mathrm{T}}, [R_{n-1}^U, I_n, R_n^U])$, $\underline{\mathbf{V}}^{(n)} = reshape(\mathbf{V}_2^{\mathrm{T}}, [R_{n-1}^V, J_n, R_n^V])$;

8   **end**

9   **for** $n = 1, 2, \ldots, N$ **do** left-to-right half sweep

10   Perform left-to-right half sweep in the same way

11   **end**

12 **until** *a stopping criterion is met*;

Specifically, in the ALS, at nth micro-iteration, the singular vectors $\mathbf{U}$ and $\mathbf{V}$ are represented in block-n TT format. We define

$$\underline{\mathbf{U}}^{<n} = \underline{\mathbf{U}}^{(1)} \bullet \underline{\mathbf{U}}^{(2)} \cdots \bullet \underline{\mathbf{U}}^{(n-1)} \in \mathbb{R}^{I_1 \times I_2 \times \cdots \times I_{n-1} \times R_{n-1}},$$

$$\underline{\mathbf{U}}^{>n} = \underline{\mathbf{U}}^{(n+1)} \bullet \underline{\mathbf{U}}^{(n+2)} \cdots \bullet \underline{\mathbf{U}}^{(N)} \in \mathbb{R}^{R_n \times I_{n+1} \times I_{n+2} \times \cdots \times I_N},$$

with $\underline{\mathbf{U}}^{<1} = \underline{\mathbf{U}}^{>N} = 1$. Then, we have $\underline{\mathbf{U}} = \underline{\mathbf{U}}^{<n} \bullet \underline{\mathbf{U}}^{(n)} \bullet \underline{\mathbf{U}}^{>n}$. Note that $\mathbf{U} = \mathbf{U}_{(n)}^{\mathrm{T}}$. Via the mode-$n$ unfolding of $\underline{\mathbf{U}}$, we can derive the following linear equation [9,12]

$$\mathbf{U} = \mathbf{U}_{(n)}^{\mathrm{T}} = \mathbf{U}^{\neq n} \mathbf{U}^{(n)}, \tag{7}$$

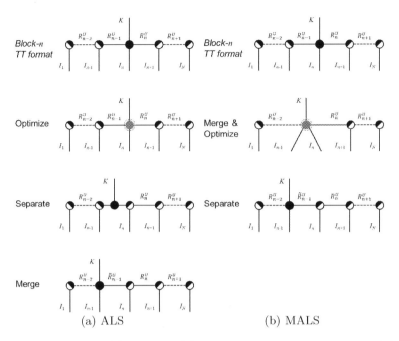

Fig. 3. Illustration of the (a) ALS and (b) MALS schemes based on block TT format during right-to-left half sweep

where
$$\mathbf{U}^{\neq n} = \left((\mathbf{U}^{<n}_{(n)})^{\mathrm{T}} \otimes \mathbf{I}_{I_n} \otimes (\mathbf{U}^{>n}_{(1)})^{\mathrm{T}} \right) \in \mathbb{R}^{I_1 I_2 \cdots I_N \times R^U_{n-1} I_n R^U_n}$$

and $\mathbf{U}^{(n)} = (\mathbf{U}^{(n)}_{(2)})^{\mathrm{T}}$. Similarly, it holds that $\mathbf{V} = \mathbf{V}^{\mathrm{T}}_{(n)} = \mathbf{V}^{\neq n}\mathbf{V}^{(n)}$, where $\mathbf{V}^{\neq n}$ and $\mathbf{V}^{(n)}$ are defined in the same way. The trace in (1) is rewritten by

$$\mathrm{trace}\left(\mathbf{U}^{\mathrm{T}}\mathbf{A}\mathbf{V}\right) = \mathrm{trace}\left((\mathbf{U}^{(n)})^{\mathrm{T}}\overline{\mathbf{A}}_n\mathbf{V}^{(n)}\right), \tag{8}$$

where $\overline{\mathbf{A}}_n = (\mathbf{U}^{\neq n})^{\mathrm{T}}\mathbf{A}\mathbf{V}^{\neq n} \in \mathbb{R}^{R^U_{n-1} I_n R^U_n \times R^V_{n-1} J_n R^V_n}$ has much smaller sizes than $\mathbf{A}$ when the TT-ranks are not large.

Moreover, the orthogonality constraint, $\mathbf{U}^{\mathrm{T}}\mathbf{U} = \mathbf{I}_K$, on $\mathbf{U}$ is equivalent to the orthogonality constraint $(\mathbf{U}^{(n)})^{\mathrm{T}}\mathbf{U}^{(n)} = \mathbf{I}_K$ on the nth TT-core if each core tensor is *orthogonalized* as follows [12]. For $1 \leq m < n$, the mth TT-core $\underline{\mathbf{U}}^{(m)} \in \mathbb{R}^{R^U_{m-1} \times I_m \times R^U_m}$ is called left-orthogonalized if $\mathbf{U}^{(m)}_{(3)}(\mathbf{U}^{(m)}_{(3)})^{\mathrm{T}} = \mathbf{I}_{R^U_m}$. For $n < m \leq N$, the mth TT-core is called right-orthogonalized if $\mathbf{U}^{(m)}_{(1)}(\mathbf{U}^{(m)}_{(1)})^{\mathrm{T}} = \mathbf{I}_{R^U_{m-1}}$. The δ-truncated SVD step of the ALS guarantees the left- and right-orthogonality of the TT-cores, and it determines the block TT-ranks adaptively during iteration. In Figure 3, the half-filled circles represent the orthogonalized TT-cores.

At each iteration, any standard SVD algorithms can maximize (8) and compute the nth TT-cores, which takes $O(IR^2K^2)$ computational complexity, where

$I = \max(\{I_n, J_n\})$ and $R = \max(\{R_n^U, R_n^V\})$. In this step, the matrix $\overline{\mathbf{A}}_n$ don't need to be computed explicitly. Instead, the matrix-by-vector products $\overline{\mathbf{A}}_n^T \mathbf{x}$ and $\overline{\mathbf{A}}_n \mathbf{y}$ can be computed efficiently in a recursive way [10]. At last, the total computational cost for the right-to-left and left-to-right iterations is $O(NIR^2 K^2 + NI^2 R^3 K + NI^2 R^4)$ when $R = max(\{R_n^U, R_n^V, R_n^A\})$, which scales logarithmically with the matrix size.

The MALS scheme is almost similar to the ALS except that it merges two neighboring TT-cores into one supercore as $\underline{\mathbf{U}}^{(n,n+1)} = \underline{\mathbf{U}}^{(n)} \bullet \underline{\mathbf{U}}^{(n+1)} \in \mathbb{R}^{R_{n-1}^U \times I_n \times I_{n+1} \times R_n^U}$ and $\underline{\mathbf{V}}^{(n,n+1)} = \underline{\mathbf{V}}^{(n)} \bullet \underline{\mathbf{V}}^{(n+1)} \in \mathbb{R}^{R_{n-1}^V \times J_n \times J_{n+1} \times R_n^V}$. After optimization of the supercore, it is separated by the δ-truncated SVD, which guarantees orthogonality and TT-rank adaptivity.

4   Simulation

The Hilbert matrix $\mathbf{H} \in \mathbb{R}^{P \times P}$ is a symmetric matrix with entries $h_{i,j} = (i + j - 1)^{-1}, i, j = 1, 2, \ldots, P$. With $P = 2^N$, we consider the rectangular submatrix

$$\mathbf{A} = \mathbf{H}(:, 1 : 2^{N-1}) \in \mathbb{R}^{2^N \times 2^{N-1}}$$

in MATLAB notation. We compared performances of six SVD methods for computing the K dominant singular values and corresponding singular vectors of $\mathbf{A}$. (i) The LOBPCG [14] computes eigenvalues $\mathbf{\Lambda} = \mathrm{diag}(\sigma_1^2, \ldots, \sigma_K^2)$ and eigenvectors $\mathbf{V}$ of the matrix $\mathbf{A}^T \mathbf{A} \in \mathbb{R}^{2^{N-1} \times 2^{N-1}}$, and then computes the left singular vectors by $\mathbf{U} = \mathbf{A} \mathbf{V} \mathbf{\Lambda}^{-1/2}$. (ii) The MATLAB function SVDS applies the Fortran package ARPACK [15] to the matrix $\mathbf{B}$ (2). (iii) The ALS and MALS algorithms for SVD solve the maximization problem (1) as described in the previous section. (iv) The ALS and MALS schemes are also implemented for computing the EVD of $\mathbf{A}^T \mathbf{A}$ by maximizing the block Rayleigh quotient [8].

It is usually difficult to load a large-scale full matrix on the computer memory. Instead, we use an explicit matrix TT representation of $\mathbf{A}$, which can be derived from the matrix TT representation of Toeplitz matrices [16]. In this step, we transform the generating vector of Hilbert matrix into block TT format with $K = 1$ via the DMRG method [17] with relative approximation error 10^{-9}. In Figure 4(a), we can see that the maximum of matrix TT-ranks, $\max\{R_n^A\}$, is bounded by 15 for all $10 \leq N \leq 25$.

Figure 5 compares computational times taken by the six SVD algorithms described in the begining of this section. The tolerance parameter for the relative residual was set at $\epsilon = 10^{-8}$. All the values are averages over 30 repeated Monte Carlo simulations. The two standard algorithms, LOBPCG and SVDS, use the full matrix, which is too large to load on the memory when $N \geq 14$. We can see that the rates of increase of LOBPCG and SVDS are exponential with N. On the other hand, the increase rates of the rest four block TT-based algorithms are polynomial with N. We note that the two block TT-based EVD methods denoted by ALS-EIG and MALS-EIG show larger computational costs than the other two block TT-based SVD methods denoted by ALS-SVD and MALS-SVD, because the EVD methods use $\mathbf{A}^T \mathbf{A}$ whose matrix TT-ranks increase to

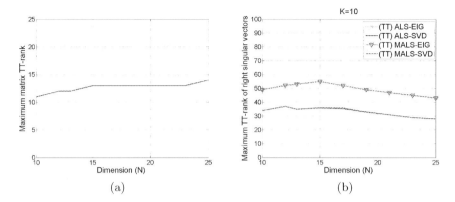

Fig. 4. (a) Maximum value $max(R_n^A)$ of estimated matrix TT-ranks of the rectangular submatrix $\mathbf{A} \in \mathbb{R}^{2^N \times 2^{N-1}}$ of Hilbert matrix, for $10 \leq N \leq 25$. (b) Maximum value $max(R_n^V)$ of block TT-ranks of the group of right singular vectors $\mathbf{V} \in \mathbb{R}^{J_1 J_2 \cdots J_N \times K}$, for $K = 10$ and $10 \leq N \leq 25$.

$\{(R_n^A)^2\}$ and require truncation. Figure 4(b) shows that the maximum of the block TT-ranks of the right singular vectors when $K = 10$ is bounded for all $10 \leq N \leq 25$.

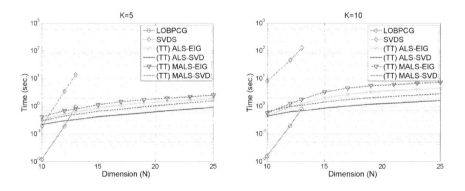

Fig. 5. Computational cost of the SVD algorithms for computing (a) $K = 5$ and (b) $K = 10$ dominating singular values and corresponding singular vectors of the $2^N \times 2^{N-1}$ submatrix $\mathbf{A}$ of Hilbert matrix with $10 \leq N \leq 25$

5   Conclusions and Discussions

We proposed SVD algorithms for very large-scale structured matrices based on TT formats. Unlike previous researches based on TT format for symmetric eigenvalue problems, we proposed a trace maximization problem and efficient

SVD algorithms which can be applied to non-symmetric matrices. Once the large-scale matrix is given in matrix TT format, the proposed methods compute the K singular values and corresponding singular vectors by solving the proposed trace maximization problem. The ALS and MALS schemes reduce the large-scale optimization problem into sequential small optimization problems, and the computational complexity scales logarithmically with the matrix size.

In the simulation experiments, we demonstrated that the proposed algorithms compute $K = 5, 10$ singular values of the $2^{25} \times 2^{24}$ submatrix of Hilbert matrix on desktop computer within a feasible computational time. We are performing further experiments with other kinds of large-scale structured matrices such as Toeplitz matrix and Hankel matrix. Moreover, general tensor networks including the Hierarchical Tucker and PEPS [4–6] are very promising tools for big data analysis.

References

1. Comon, P., Golub, G.H.: Tracking a Few Extreme Singular Values and Vectors in Signal Processing. Proceedings of the IEEE **78**, 1327–1343 (1990)
2. Frieze, A., Kannan, R., Vempala, S.: Fast Monte-Carlo Algorithms for Finding Low-Rank Approximations. In: Proceedings of the 39th Annual IEEE Symposium on Foundations of Computer Science, pp. 370–378 (1998)
3. Oseledets, I.V.: Tensor-Train Decomposition. SIAM J. Sci. Comput. **33**, 2295–2317 (2011)
4. Grasedyck, L., Kressner, D., Tobler, C.: A Literature Survey of Low-Rank Tensor Approximation Techniques. arXiv:1302.7121 (2013)
5. Huckle, T., Waldherr, K., Schulte-Herbrüggen, T.: Computations in Quantum Tensor Networks. Linear Algebra Appl. **438**, 750–781 (2013)
6. Cichocki, A.: Era of Big Data Processing: A New Approach via Tensor Networks and Tensor Decompositions. arXiv:1301.6068 (2014)
7. Lebedeva, O.S.: Tensor Conjugate-Gradient-Type Method for Rayleigh Quotient Minimization in Block QTT-Format. Russian J. Numer. Anal. Math. Modelling **26**, 465–489 (2011)
8. Dolgov, S.V., Khoromskij, B.N., Oseledets, I.V., Savostyanov, D.V.: Computation of Extreme Eigenvalues in Higher Dimensions Using Block Tensor Train Format. Comp. Phys. Comm. **185**, 1207–1216 (2014)
9. Kressner, D., Steinlechner, M., Uschmajew, A.: Low-Rank Tensor Methods with Subspace Correction for Symmetric Eigenvalue Problems. MATHICSE Technical Report 40.2013, EPFL, Lausanne (2013)
10. Holtz, S., Rohwedder, T., Schneider, R.: The Alternating Linear Scheme for Tensor Optimization in the Tensor Train Format. SIAM J. Sci. Comput. **34**, A683–A713 (2012)
11. Schollwöck, U.: The Density-Matrix Renormalization Group in the Age of Matrix Product States. Ann. Physics **326**, 96–192 (2011)
12. Lee, N., Cichocki, A.: Fundamental Tensor Operations for Large-Scale Data Analysis in Tensor Train Formats. arXiv:1405.7786 (2014)
13. Kolda, T.G., Bader, B.W.: Tensor Decompositions and Applications. SIAM Rev. **51**, 455–500 (2009)

14. Knyazev, A.V.: Toward the Optimal Preconditioned Eigensolver: Locally Optimal Block Preconditioned Conjugate Gradient Method. SIAM J. Sci. Comput. **23**, 517–541 (2001)
15. Lehoucq, R.B., Sorensen, D.C., Yang, C.: ARPACK User's Guide: Solution of Large Scale Eigenvalue Problems with Implicitly Restarted Arnoldi Methods. Software Environ. Tools 6. SIAM, Philadelphia (1998). http://www.caam.rice.edu/software/ARPACK/
16. Kazeev, V.A., Khoromskij, B.N., Tyrtyshnikov, E.E.: Multilevel Toeplitz Matrices Generated by Tensor-Structured Vectors and Convolution with Logarithmic Complexity. SIAM J. Sci. Comput. **35**, A1511–A1536 (2013)
17. Holtz, S., Rohwedder, T., Schneider, R.: On Manifolds of Tensors with Fixed TT-Rank. Numer. Math. **120**, 701–731 (2011)

CSP-Based EEG Analysis on Dissociated Brain Organization for Single-Digit Addition and Multiplication

Lihan Wang[1], John Q. Gan[1,2], and Haixian Wang[1(✉)]

[1] Research Center for Learning Science, Southeast University,
Nanjing, 210096 Jiangsu, China
lhwang20@gmail.com, hxwang@seu.edu.cn
[2] School of Computer Science and Electronic Engineering,
University of Essex, Colchester, UK
jqgan@essex.ac.uk

Abstract. This paper presents results of the common spatial pattern (CSP) based electroencephalogram (EEG) analysis on dissociated brain organization for single-digit addition and multiplication. Alpha band EEG oscillations, which have been corroborated to be modulated by arithmetic strategies, are employed for feature extraction based on CSP. Experimental results have confirmed the dissociation between single-digit addition and multiplication. It is indicated that the dissociation originates from different cortical areas across subjects, such as IPS, parieto-occipital, and fronto-parietal regions.

Keywords: Dissociation · Arithmetic strategy · Fronto-parietal network · EEG analysis · Alpha band ERD

1 Introduction

Dissociations among arithmetic operations, initially observed in lesion studies, have led to the postulation of different strategies used in mental arithmetic. Retrieval of answers from long-term memory and procedural solutions, such as counting or transformation, have been identified as two main strategies used by individuals in order to perform simple or more complex calculations at a functional level. Simple multiplications such as 2×4, which are learned by rote at school, are solved by retrieval. On the contrary, simple additions such as 2+4 are usually taught to use procedural strategy. However, the solution of single-digit additions is ambiguous in the domain of numerical cognition. While some studies suggest that retrieval is more likely used in additions, others believe that single-digit additions are solved through manipulations of internal quantity representations just like subtractions [1].

As a matter of fact, the most frequent dissociation, corroborated in normal people by functional magnetic resonance imaging (fMRI) studies, is between multiplications and subtractions rather than multiplications and additions, presumably correlating with the obscureness of additions per se. Only one fMRI study has explored the

© Springer International Publishing Switzerland 2014
Z. Zeng et al. (Eds.): ISNN 2014, LNCS 8866, pp. 131–139, 2014.
DOI: 10.1007/978-3-319-12436-0_15

dissociation between single-digit multiplication and single-digit addition [2]. In the context of the relative short duration (around one second) and considering several phases (number coding, calculation, result production) in the arithmetic operations, electroencephalogram (EEG) recording may have advantage over fMRI due to the better temporal resolution (millisecond). Hence, compared with fMRI, EEG, especially induced EEG, i.e. event-related synchronization (ERS) and desynchronisation (ERD), is more engaged in the analysis of the dissociation between operations. Previous findings have consistently demonstrated the high suitability of the ERD method to uncover stable individual differences in human brain activation patterns during performing cognitively demanding tasks [3].

Although the existing EEG studies have provided some insights into the neural correlate of dissociations among operations, variations in alpha ERD between single-digit addition and multiplication have not been reported so far. Some previous studies were underpinned by the fact that both single-digit addition and multiplication were solved by retrieval. This however is controversial. One of the root causes involves that the traditional EEG analysis has always been implemented across subjects, despite the individual variability such as mathematical abilities and people's brain activation patterns when engaged in performing cognitively demanding tasks [4]. Based on the dissociation postulation derived from the lesions studies and the theoretical models such as encoding-complex model [5], together with the findings that oscillations at alpha frequency band are modulated by procedural strategies [6], we expected that the brain activation patterns in the alpha band are different for single-digit multiplication and single-digit addition. Moreover, finding out in what way the ERD/ERS could be maximized for one class against the other would provide information regarding where the dissociation originates from. Accordingly, the common spatial pattern (CSP) algorithm [7], a popular method in brain-computer interfaces (BCI) using ERD/ERS in the classification of motor imagery, is employed in this study. The algorithm maximizes the variance for one class and at the same time minimizes the variance for the other class. Hence, based on the modulation of EEG oscillations at alpha band by strategy use, the common spatial pattern (CSP) algorithm can be conducted on multiple overlapped time windows to provide evidence for the existence of the dissociation and information on the origination of the dissociation between single-digit addition and multiplication.

2   Materials and Methods

2.1   Subjects

Six right-handed healthy students (5 males and 1 female) studied at Southeast University without known calculation difficulties were recruited as subjects for this experimental study. Exclusion criteria included left handedness, neurological illness, and history of brain injury. All subjects were asked to read and sign an informed consent form before experiments and the study was approved by the Academic Committee of the Research Center for Learning Science, Southeast University, China.

2.2   Experiment Paradigm

A delayed verification paradigm was employed to explore the strategy dissociation between the two operations (Fig. 1). Participants silently produced either the solutions of single-digit multiplication and single-digit addition problems or store operands. On account of the operand-order effect in single-digit multiplications (due to the Chinese multiplication table that contains only smaller-operand-first problems), which has been observed in Mainland Chinese subjects [8], smaller-operand-first problems (e.g., 3×4 or 7×8) were exclusively included. Thus, non-zero problems consisted of 28 possible combinations of operands ranging from 2 to 9 in each operation (i.e., problems ranging from 2×9 to 8×9 in multiplication or 2+9 to 8+9 in addition). Ties (e.g., 3×3 or 3+3) were excluded due to their uniqueness. Consequently, there were 28 problems in each operation, and each problem was repeated 4 times, resulting in 112 trials in each operation.

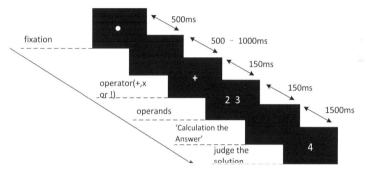

Fig. 1. Schematic depiction of a trial

Previous research has shown strong evidence of interference among arithmetic operations when subjects had to switch among them. To reduce such interference, addition problems and multiplication problems were presented in separate blocks. Such repetitions of a small number of arithmetic problems in each block would result in the use of different cognitive strategies in solving the problems (e.g., calculation at the beginning and memorization for the repeated problems), and the number storage task could serve as the control condition in each block, with all operand combinations in arithmetic operations also used for the storage condition. There are two blocks for each operation (additions and multiplications). Problems were randomly presented within a block, with the constraint that consecutive problems did not have a common operand or the same solution.

2.3   EEG Recording and Preprocessing

The EEG data were recorded by a 60-channel Neuroscan using the international 10-20 system with sampling rate 1000 Hz. Reference electrodes were located at the bilateral mastoids of subjects, and electro-oculographic (EOG) signals were simultaneously recorded by four surface electrodes to monitor ocular movements and eye blinks.

Only the trials with correct response were kept for the further analysis. Prior to the calculation of the spatial filters, all EEG channels were filtered with a pass-band between 8–13Hz. Alpha band has been chosen because alpha band ERD has been an important indicator of procedural strategy. The focus has been put on the interval of 1600ms since the presentation of the operands, i.e., the production of the results. Features were extracted using sliding windows, each of which consisted of 1000 samples, representing 1s of EEG data. Sliding window positions were incremented by 100 samples, each window overlapping the previous by 900 samples.

2.4   Feature Extraction Using CSP

The common spatial pattern (CSP) algorithm is highly successful in constructing spatial filters for detecting ERD/ERS effects, especially in BCI [7]. Given two distributions in a high-dimensional space, the CSP algorithm finds directions (i.e., spatial filters) that maximize the variance for one class and at the same time minimize the variance for the other class. The method used to design such spatial filters is based on the simultaneous diagonalization of two covariance matrices [9]. As a result the CSP algorithm outputs a decomposition matrix and a vector of corresponding eigenvalues. With the projection matrix $W \in R^{C \times C}$ (C is the number of channels), the decomposition (mapping) of a trial $E \in R^{C \times N}$ (N is the number of samples per channel) is given as

$$Z = W^\mathrm{T} E \tag{1}$$

The interpretation of W is two-fold: each column of W is a stationary spatial filter, whereas each column a_j of the matrix $A = (W^{-1})^\mathrm{T}$ can be seen as a spatial pattern, i.e., the time-invariant EEG source distribution vector.

The features used for classification can be obtained by decomposing (filtering) the EEG according to (1). Variance is calculated for each of the CSP channels (band power) and the logarithm is applied to yield a feature vector for each sample point as follows:

$$f_p = \log\left(\frac{\mathrm{var}(z_p)}{\sum_{i=1}^{2m} \mathrm{var}(z_i)}\right) \tag{2}$$

The signals z_p ($p = 1,\ldots, 2m$), which maximize the difference of variance of class 1 versus class 2, are the m first rows and m last rows of Z. In practice m is usually set to 3, i.e., three eigenvectors from both ends. In this study, a projection matrix is generated at each time point, and a linear discriminant analysis (LDA) classifier is used to find a linear separation between the two arithmetic operations. For proper estimation of the classification accuracy, the dataset of each subject is divided into a training set and a testing set. This training/testing procedure is repeated 10 times with different random partitions into training and testing sets (i.e., 10-fold cross-validation).

2.5 Neurophysiological Outcome – Visualization of Dissociation

It is assumed in the current neurophysiological models that the alpha amplitude is inversely related to the activated cortical neuronal populations, and the alpha ERD can be regarded as a correlate of cortical activation [10]. Thus the first column of the CSP pattern matrix $A = (W^{-1})^T$, i.e., the vector which gives maximum ERD of addition against multiplication, has been used to visualize the dissociation between the two difference operations.

3   Results and Discussions

The accuracy in classifying addition and multiplication operations using CSP features is shown in Fig. 2. The accuracy has reached 0.8 for all subjects (above 0.9 for S1 and S3) except for S4 (above 0.7) across the duration of interest. This indicates that the CSP features are good at identifying the arithmetic strategies used in addition and multiplication. To further explore from where such divergence originates and elaborate the development of the dissociation over time, the first pattern, a_1, at a certain time point from each subject, which describes the most influential presumed sources of the dissociation, is displayed in Fig. 3. Since the duration of interest is constituted of converting the stimuli into appropriate internal codes and retrieving or calculating the answer (reporting the answer is omitted due to the delayed verification paradigm), the two stages are discussed separately.

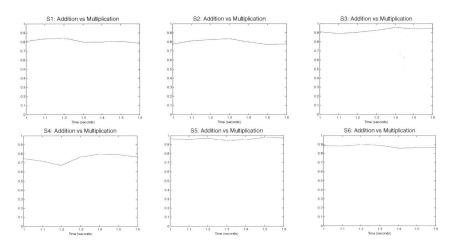

Fig. 2. Accuracy curve of the classification between the two operations. The x-axis represents the time interval since the presentation of the operands. The 1s segment before each time point has been used for feature extraction.

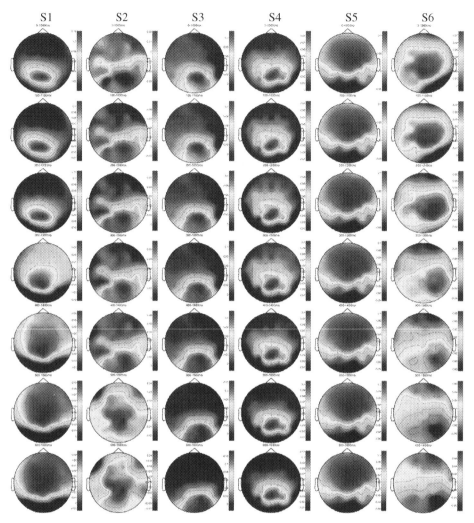

Fig. 3. The visualization of the CSP patterns at seven time points (at 1s, 1.1s, 1.2s, 1.3s, 1.4s, 1.5s and 1.6s since the presentation of the operands). The 1s segment before each time point has been used for feature extraction.

3.1   Focus on the Parieto-Occipital Cortex

Regarding the neurophysiological outcome, the presumed sources for all subjects, except for subjects S5 and S6, disclose the divergence between additions and multiplications in all cortical areas, with a pronounced focus on the parieto-occipital cortex at an earlier stage (the first 4 time points), although the specific location is slightly varied across the individuals. The profile is consistent with the one established in the previous studies [6, 11] which explored the divergence between procedural problems and retrieval problems (but not multiplication vs. addition). The dissimilarity primarily lies in the laterality, with the latter characterized with bilateral parieto-occipital patterns. In view of the group analysis implemented in the previous studies, the average across individuals could result in the obscureness of the individual variability in brain activation areas. However, the subjects in this paper demonstrated different laterality. Higher alpha ERD may reflect the higher engagement of task-related cortical resources, i.e., the parieto-occipital cortex in the current case. The topography focus could be naturally associated with the intraparietal sulcus (IPS) , which is systematically activated whenever numbers are manipulated, and with increasing activation as the task puts greater emphasis on quantity processing [12]. To further inspect the earlier stage of subjects S5 and S6, a smaller time window with 450ms length was employed, which comprised of at least 3 cycles of alpha band signal. This is in considerations of the individual variability in the calculation speed, and the accuracy and the patterns are shown in Fig. 4. With a smaller time window, the pattern demonstrated a similar parieto-occipital pattern as other subjects. Brain activation in 1 second time window could probably involve the frontal cortex, provided that the cognition performance of the subject has a strong reliance on executive processes and working memory at calculation stage. The dissociation pattern on parieto-occipital cortex coincides with the encoding-complex model [5], which posits the number representation depends on the context, e.g., verbal code for memory retrieval of arithmetic facts and the analogue magnitude code for numerical representation.

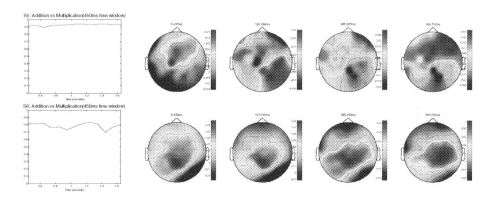

Fig. 4. Accuracy curve and the visualization of the CSP patterns at the first 4 time points of S5 and S6, with 450ms time window used for feature extraction

At the later stage (the last 3 time points), the brain activation began to diverge among the four subjects who exhibited the similar focus at the earlier stage. While two subjects (S3 and S4) preserved the former pattern, the other two subjects (S1 and S2) demonstrated a shift from the parietal cortex to the fronto-parietal region. The pattern for S3 and S4 indicated the parietal cortex was consistently involved throughout the calculation procedure of addition.

3.2   Engagement of the Fronto-Parietal Network

In spite of the high classification accuracy, the pattern in subjects S1, S2, S5 and S6 is divergent from the other two subjects at the later stage, with a focus on fronto-parietal network. Neuroimaging studies have corroborated that contrasted with memory retrieval procedural strategies reliably activated fronto-parietal regions. Then the issue arises with regard to why the dissociation between the two operations could originate, exclusively from IPS (S3, S4) or earlier from IPS and later from fronto-parietal regions (S1, S2, S5 and S6).   Fmri studies have revealed that the recruitment of parietal brain in arithmetic problem solving dynamically changes as a function of training and development. With human brain development and the acquisition of linguistic competencies, there is increasing recruitment of the parietal cortex during calculation, coupled with reduction of activity in bilateral regions of the frontal cortex [12]. The activation of the fronto-parietal cortex (for S1, 2 and 5) indicated a stronger demand for cognition functions such as executive processes and working memory while solving addition. The differential degree of expertise may lead to the differential extent of parietal specialization [13].

4   Conclusion

The present high-resolution EEG study shows the CSP algorithm is able to capture different strategies involved in solving two operations which are learned by different strategies in childhood. Through the alpha band ERD analysis, the findings in this paper suggest that the dissociation between single-digit addition and multiplication indeed exists, and originations develop over the different stages of mental arithmetic. The dissociation originates differently across subjects, which is probably related with the individual trajectory of learning and developmental processes leading to the differential degree of expertise. The EEG analysis based on the CSP algorithm creates an opportunity for elaborating the brain activation dissociation at different stages at individual level, regardless of individual variability. Future research could further explore whether such different engagement correlates with mathematical capability.

Acknowledgments. This work was supported in part by the National Natural Science Foundation of China under Grant 61375118, the Natural Science Foundation of Jiangsu Province under Grant BK2011595, and the Program for New Century Excellent Talents in University of China under Grant NCET-12-0115.

References

1. Fayol, M., Thevenot, C.: The use of procedural knowledge in simple addition and subtraction problems. Cognition **123**, 392–403 (2012)
2. Zhou, X.L., Chen, C.S., Zang, Y.F., Dong, Q., Chen, C.H., Qiao, S.B., Gong, Q.Y.: Dissociated brain organization for single-digit addition and multiplication. Neuroimage **35**, 871–880 (2007)

 3. Neubauer, A.C., Fink, A., Grabner, R.H.: Sensitivity of alpha band ERD to individual differences in cognition. Prog. Brain Res. **159**, 167–178 (2006)
 4. Krause, F., Lindemann, O., Toni, I., Bekkering, H.: Different Brains Process Numbers Differently: Structural Bases of Individual Differences in Spatial and Nonspatial Number Representations. J. Cognitive Neurosci. **26**, 768–776 (2014)
 5. Campbell, J.I., Epp, L.J.: An encoding-complex approach to numerical cognition in Chinese-English bilinguals. Can. J. Exp. Psychol. **58**, 229–244 (2004)
 6. Grabner, R.H., De Smedt, B.: Neurophysiological evidence for the validity of verbal strategy reports in mental arithmetic. Biol. Psychol. **87**, 128–136 (2011)
 7. Blankertz, B., Tomioka, R., Lemm, S., Kawanabe, M., Muller, K.-R.: Optimizing Spatial filters for Robust EEG Single-Trial Analysis. IEEE Signal Processing Magazine **25**, 41–56 (2008)
 8. Zhou, X.: Operation-specific encoding in single-digit arithmetic. Brain Cogn. **76**, 400–406 (2011)
 9. Ramoser, H., Müller-Gerking, J., Pfurtscheller, G.: Optimal spatial filtering of single trial EEG during imagined hand movement. IEEE Transactions on Rehabilitation Engineering: A Publication of the IEEE Engineering in Medicine and Biology Society **8**, 441–446 (2000)
10. Pfurtscheller, G., Lopes da Silva, F.H.: Event-related EEG/MEG synchronization and desynchronization: basic principles. Clinical Neurophysiology: Official Journal of the International Federation of Clinical Neurophysiology **110**, 1842–1857 (1999)
11. De Smedt, B., Grabner, R.H., Studer, B.: Oscillatory EEG correlates of arithmetic strategy use in addition and subtraction. Experimental brain research. Experimentelle Hirnforschung. Expérimentation Cérébrale **195**, 635–642 (2009)
12. Ansari, D.: Effects of development and enculturation on number representation in the brain. Nat. Rev. Neurosci. **9**, 278–291 (2008)
13. Masson, S., Potvin, P., Riopel, M., Foisy, L.M.B.: Differences in Brain Activation Between Novices and Experts in Science During a Task Involving a Common Misconception in Electricity. Mind Brain Educ. **8**, 44–55 (2014)

Lazy Fully Probabilistic Design
of Decision Strategies

Miroslav Kárný$^{(\boxtimes)}$, Karel Macek, and Tatiana V. Guy

Institute of Information Theory and Automation,
Academy of Sciences of the Czech Republic, P.O.Box 18,
182 08 Prague 8, Czech Republic
{school,macek,guy}@utia.cas.cz

Abstract. Fully probabilistic design of decision strategies (FPD) extends Bayesian dynamic decision making. The FPD specifies the decision aim via so-called ideal - a probability density, which assigns high probability values to the desirable behaviours and low values to undesirable ones. The optimal decision strategy minimises the Kullback-Leibler divergence of the probability density describing the closed-loop behaviour to this ideal. In spite of the availability of explicit minimisers in the corresponding dynamic programming, it suffers from the curse of dimensionality connected with complexity of the value function. Recently proposed a lazy FPD tailors lazy learning, which builds a local model around the current behaviour, to estimation of the closed-loop model with the optimal strategy. This paper adds a theoretical support to the lazy FPD and outlines its further improvement.

Keywords: Decision making · Lazy learning · Bayesian learning · Local model

1 Introduction

A decision maker (artificial or human) forms with its environment a closed decision-making (DM) loop and aims to influence the closed-loop behaviour by a sequence of its actions. The behaviour is characterised by a collection of observed, selected and considered variables. The decision maker can only use incomplete knowledge and faces random dynamic changes of the environment. DM understood in this way is wide spread and covers stochastic and adaptive control, fault detection as well as inference tasks like estimation, filtering, prediction, classification, etc. The mentioned DM importance and width have naturally stimulated a search for widely applicable normative DM theories. A long-term development has singled out the Bayesian DM theory [3,4,9,29] as the most promising candidate.

The Bayesian DM provides well-justified solutions of DM tasks but the "curse of dimensionality" [1] limits its applicability and approximations are mostly inevitable. Approximate non-linear estimation and filtering [7,8,27,30] and approximate dynamic programming [4,31,34] are thus unavoidable, permanently-evolving, complements of the basic DM theory. Practically successful techniques

© Springer International Publishing Switzerland 2014
Z. Zeng et al. (Eds.): ISNN 2014, LNCS 8866, pp. 140–149, 2014.
DOI: 10.1007/978-3-319-12436-0_16

mostly rely on local approximations around the current realisation of the behaviour. This applies to learning, with lazy learning being its typical representative [5,20], to adaptive control [13,23] and other techniques like case-based reasoning [10]. Their success or failure strongly depends on a proper specification of the neighbourhood of the current behaviour. The neighbourhood must be narrow to allow a simple and rich modelling containing relevant information learnt within the closed DM loop. To our best knowledge, no established methodology comparable in the width to the underlying DM theory exists. Mostly they either support a subset of DM problems or use a trial and error method.

Fully probabilistic design (FPD) of DM strategies is an extension of the Bayesian DM [12,16,17][1] describes the closed-loop behaviour by a joint probability density (pd) of the involved variables, exactly as the Bayesian DM does. It, however, expresses the DM aims via a decision-maker-adopted ideal pd quantifying desirability of behaviours. The strategy design then reduces to a minimisation of the Kullback-Leibler divergence (KLD, [19]) of the involved pds over the optional strategies. The FPD promises simpler approximations of the unfeasible strategy design as it provides an explicit minimiser in dynamic programming. The rare attempts, e.g. [14], only partially exploited the potential offered by this feature. They are still too much of ad hoc nature and cumbersome. A substantial progress towards an approximate FPD has been recently made, [22]. The proposed *lazy FPD* uses the current ideal pd for weighting the past data records when learning a local model of the optimally closed loop. This treatment overcomes weaknesses of the lazy learning, which: a) serves well to prediction but rarely to dynamic DM; b) is sensitive to a measure quantifying the proximity of behaviours, and c) relies on availability of data records close enough to the current one. The present paper adds a theoretical insight into the technique and improves the lazy FPD using Sanov-type analysis [28]. Section 2 recalls the lazy FPD and Section 3 formulates the addressed problems. Section 4 solves them. Section 5 contains illustrative example and Section 6 offers concluding remarks.

Throughout, x is a set of x-values; all sets are subsets of finite-dimensional spaces; S, O, ... are mappings; $x \in x$ denotes a possible realisation of a random variable X; $\underline{x} \in x$ is a specific realisation of X; probability density (pd) is Radon-Nikodým derivative with respect to a measure d•; pds having different identifiers in arguments are taken as different; $\tau, t \in t \equiv \{1, \ldots, T\}$ label discrete time; $x_m^n = (x_t)_{t=m}^n$ and $x^n = x_0^n$ describe finite sequences.

2 Lazy FPD

The *lazy FPD* selects a decision strategy, which makes a probabilistic description of the closed decision loop close to a pre-specified closed-loop ideal. Instead of the traditional learning of an environment model followed by the strategy optimisation, the lazy FPD uses the currently observed data to estimate, which of simple parametric models provides the closed-loop model near the given ideal.

---

[1] Re-invented in [33], studied in control [11] and used in brain research [32].

The designed strategy is then a marginal of the found closed-loop model. The next text formalises this.

The inspected DM problem deals with sequences of possible realisations x^T of random environment responses $x_\tau \in \boldsymbol{x}_\tau, t \in \boldsymbol{t}$. The realised sequence of responses $\underline{x}^T$ reacts on the realisation $\underline{a}^T$ of actions generated by a randomised strategy, $\mathsf{S}_\tau : \underline{a}^{\tau-1}, \underline{x}^{\tau-1} \to \underline{a}_\tau, \tau \in \boldsymbol{t}$. The action, a_τ, and the environment response, x_τ, forms the data, d_τ, observable at time $\tau \in \boldsymbol{t}$. Pds $\mathsf{S}^T \equiv (\mathsf{S}_t(a_t|a^{t-1}, x^{t-1}))_{\tau=1}^T = (\mathsf{S}_\tau(a_\tau|d^{\tau-1}))_{\tau=1}^T$ describe the strategy. The individual pds in the sequence S^T are decision rules forming the strategy.

Let us consider the current time $t \in \boldsymbol{t}$ delimits the past (when data sequence $\underline{d}^{t-1}$ was observed) and the future, which includes the current inspected DM stage. The current time splits behaviour and all involved pds in their past and future parts. The data considered in the closed DM loop are samples from a closed-loop-describing pd $\mathsf{C}^T = \prod_{\tau=1}^T \mathsf{C}_\tau(d_\tau|d^{\tau-1})$. In the inspected stage, the past and the future closed-loop models are distinguished. The future ideal closed-loop model, given by the joint pd

$$\mathsf{I}_t^T = \mathsf{I}_t(d_t^T|\underline{d}^{t-1}) = \prod_{\tau=t}^T \mathsf{I}_t(d_\tau|\underline{d}^{t-1}, d_t^{\tau-1}), \ t \in \boldsymbol{t}, \tag{1}$$

quantifies the DM aim. Its factors $\mathsf{I}_t(d_\tau|d^{\tau-1})$ for $\tau \geq t$ may differ from the past ideal factors I_τ for $\tau < t$. Notice that the behaviour evolution within the planning periods starts at the realised $\underline{d}^{t-1}$. The future closed-loop model $\mathsf{C}_t^T = \mathsf{C}_t(d_t^T|\underline{d}^{t-1})$ describes the DM loop formed by the environment and the future strategy S_t^T. The strategy making C_t^T close to the future ideal pd $\mathsf{I}_t^T = \mathsf{I}_t(d_t^T|\underline{d}^{t-1})$ (1) is searched for. The lazy FPD uses: i) the observed data realisations $\underline{d}^{t-1}$; ii) the given ideal pd (1); iii) a class of parametric models

$$\mathsf{M}_t(d_t^T|\underline{d}^{t-1}, \theta) = \prod_{\tau=t}^T \mathsf{M}_t(d_\tau|\underline{d}^{t-1}, d_t^{\tau-1}, \theta), \ \theta \in \boldsymbol{\theta}, \tag{2}$$

serving for extrapolation of the past realised closed-loop behaviour $\underline{d}^{t-1}$. Note that the parametric closed-loop models (2) can be simple as the future closed-loop model $\mathsf{C}_t(d_t^T|\underline{d}^{t-1})$ has to be (approximately) valid only for the behaviours prolonging the past $\underline{d}^{t-1}$.

Design concept of the lazy FPD: The lazy FPD uses the data realisation for assigning such a posterior pd $\mathsf{P}(\theta|\underline{d}^{t-1})$ to respective parameters $\theta \in \boldsymbol{\theta}$ in (2) so that the model $\mathsf{C}_t(d_t^T|\underline{d}^{t-1}) = \prod_{\tau=t}^T \mathsf{C}_t(d_\tau|d^{\tau-1})$ describes the closed loop with the desired strategy. Its future-describing factors are predictors

$$\mathsf{C}_t(d_\tau|\underline{d}^{t-1}, d_t^{\tau-1}) \equiv \int_{\boldsymbol{\theta}} \mathsf{M}_t(d_\tau|\underline{d}^{t-1}, d_t^{\tau-1}, \theta)\mathsf{P}(\theta|\underline{d}^{t-1}) \, \mathrm{d}\theta \tag{3}$$

constructed from the parametric model (2) and the posterior pd $\mathsf{P}(\theta|\underline{d}^{t-1})$. The pd $\mathsf{S}_t(a_t|\underline{d}^{t-1}) = \int_{\boldsymbol{x}} \mathsf{C}_t(d_t|\underline{d}^{t-1}) \, \mathrm{d}x_t$ gained from the predictor (3) is the current estimate of the properly tuned decision rule. The action $\underline{a}_t$ is sampled from it and the response $\underline{x}_t$ is observed.

The randomised strategy arising from the lazy FPD cares about the exploration conditioning any successful learning. For a well-peaked $P(\theta|\underline{d}^{t-1})$, the predictors (3) can be approximated by plug in a point estimate of θ into the models $M_t(d_\tau|d^{\tau-1}, \theta)$ (2).

Neither the local model of the environment working in the closed loop nor the future strategy optimal with respect to the future ideal are known. Thus, the parameters $\theta \in \boldsymbol{\theta}$ pointing to the models (2), which guarantee the closeness of the future closed-loop model (3) to the given future ideal pd (1), are unknown. As such, they should be learned in the Bayesian way. The already observed data realisations $\underline{d}^{t-1}$, however, do not origin from the closed loop tuned with respect to the ideal pd I_t^T (1). The lazy FPD faces this serious obstacle by learning the unknown parameter $\theta \in \boldsymbol{\theta}$ via the weighted Bayes rule. It maps a prior pd $P(\theta)$ on the posterior pd, $\forall \theta \in \boldsymbol{\theta}$, as follows

$$P(\theta|\underline{d}^{t-1}) \propto P(\theta) \prod_{\tau=1}^{t-1} M_t^{W_t(\underline{d}^\tau)}(\underline{d}_\tau|\underline{d}^{\tau-1}, \theta) \tag{4}$$

$$W_t(\underline{d}^\tau) \propto I_t(\underline{d}_\tau|\underline{d}^{\tau-1}) \quad \text{and} \quad \propto \text{ denotes proportionality.}$$

After using $\underline{a}_t$ taken from $S_t(a_t|\underline{d}^{t-1}) = \int_{x_t} \int_\theta M_t(d_t|\underline{d}^{t-1}) P(\theta|\underline{d}^{t-1}) \, d\theta \, dx_t$ the response $\underline{x}_t$ is observed and the learning step (4) is repeated for time $t + 1$.

3 Questions Connected with the Lazy FPD

The weights $W_t(\underline{d}^\tau)$ chosen in (4) are intuitively plausible. The weight is the higher the more the realised subsequence $\underline{d}^\tau$ fits the ideal factor $I_t(d_\tau|d^{\tau-1})$ to which closed-loop models (2) with highly probable parameter values should approach. Promising experimental results, partially reported in [22], support this intuition.

The intuition leaves aside the natural questions: i) Is the *use* of the weights W_t in (4) the proper and, ideally, only one? ii) How to *normalise* the weights (4) to get the adequately peaked posterior pd $P(\theta|\underline{d}^{t-1})$? iii) What happens if the processed data realisations indeed come from the properly tuned closed loop describable by the parametric model (2), i.e. what is the asymptotic behaviour under time-invariant circumstances?

The formal inspection of the weighted Bayes rule (4) with a novel choice of the weights presented in the next section forms the paper core and answers the questions above.

4 Answers to the Formulated Questions

The following normalisation of the weights (4) is inspected

$$W_t(d^\tau) = \frac{I_t(d_\tau|d^{\tau-1})}{C_\tau(d_\tau|d^{\tau-1})}, \tag{5}$$

where $C_\tau(d_\tau|d^{\tau-1})$ is the pd describing the realisations of the closed-loop behaviour for $\tau \leq t - 1$. It can be obtained via the standard Bayesian learning using either a specific parametric model or the model (2). The latter option needs a sort of forgetting [18] coping with the approximate nature of the simple models (2), [15]. For a time-invariant ideal pd, it can alternatively be approximated by the predictors (3) obtained when the planning started at times $\tau < t$.

The normalisation (5) has resulted from the Sanov-type analysis [28] of the posterior pd. It is extended here so that it is applicable to the posterior pd obtained in the closed DM loop with the weighted Bayes rule (4). Its idea is often masked by the focus on difficult but technical problems. The common essence is, however, simple. The posterior pd is re-written as

$$P(\theta|\underline{d}^{t-1}) \propto \exp[-(t-1) \times \text{sample mean of a data function depending on } \theta]$$

and a law of large number, ergodic arguments or martingale theory [21] are used to show that this sample mean converges to a function bounded from below. Then, it is easy to see that the posterior pd $P(\theta|\underline{d}^{t-1})$ may concentrate only on $\theta \in \boldsymbol{\theta}$ minimising this function.

The next proposition formalises this way assuming that the time moment $t \in \boldsymbol{t}$ is fixed and the past data d^{t-1} are described by the pd $\prod_{\tau=1}^{t-1} C_\tau(d_\tau|d^{\tau-1})$.

Proposition 1 (On the Weighted Bayesian Learning). *Let*

$$\ln\left(\frac{I_t(d_\tau|d^{\tau-1})}{M_t(d_\tau|d^{\tau-1},\theta)}\right), \quad \tau < t, \tag{6}$$

be essentially bounded for all $\theta \in \boldsymbol{\theta}$. Then, the weighted Bayes rule (4) using the weights (5) provides for $t \to \infty$ the same posterior pd as that obtained by the standard Bayes rule applied to data sampled from the closed-loop described by the ideal pd $I_t(d_\tau|d^{\tau-1})$.

Proof. For any $\theta \in \boldsymbol{\theta}$, the posterior pd obtained from (4) can be given the form

$$P(\theta|\underline{d}^{t-1}) \propto P(\theta) \exp\left[-(t-1)\overbrace{\frac{1}{t-1}\sum_{\tau=1}^{t-1}W_t(\underline{d}^\tau)\underbrace{\ln\left(\frac{I_t(d_\tau|\underline{d}^{\tau-1})}{M_t(d_\tau|\underline{d}^{\tau-1},\theta)}\right)}_{L_\tau\left(\underline{d}^\tau,\theta\right)}}^{\text{the sample mean } \Omega_{t-1}(\underline{d}^{t-1},\theta)}\right], \tag{7}$$

exploiting the fact that the proportionality $\propto$ in (7) defines the same posterior pd even when the right-hand side is multiplied by any positive θ-independent factor. The following innovations N_τ are zero-mean, uncorrelated and essentially bounded due to the assumed bounded-ness of (6), [24],

$$\mathsf{N}_\tau(d_\tau, \underline{d}^{\tau-1}, \theta) \equiv \mathsf{E}_\tau\left[\mathsf{L}_\tau|\underline{d}^{\tau-1}\right] - \mathsf{L}_\tau(d_\tau, \underline{d}^{\tau-1}, \theta) \quad \text{with}$$

$$\mathsf{E}_\tau\left[\mathsf{L}_\tau|\underline{d}^{\tau-1}\right] \equiv \int_{d_\tau} \mathsf{L}_\tau(d_\tau, \underline{d}^{\tau-1}, \theta) \mathsf{C}_\tau(d_\tau|\underline{d}^{\tau-1}) \, dd_\tau$$

$$= \underbrace{\int_{d_\tau} \mathsf{l}_t(d_\tau|\underline{d}^{\tau-1}) \ln\left(\frac{\mathsf{l}_t(d_\tau|\underline{d}^{\tau-1})}{\mathsf{M}_t(d_\tau|\underline{d}^{\tau-1}, \theta)}\right) \, dd_\tau}_{\mathsf{H}_\tau(\underline{d}^{\tau-1}, \theta)} \geq 0, \text{ due to the Jensen inequality, [26].}$$

The decomposition exists due to the essential bounded-ness of $\mathsf{H}_\tau(\underline{d}^{\tau-1}, \theta)$ and splits $\Omega_{t-1}(\underline{d}^{t-1}, \theta)$ into the mean of non-negative terms $\mathsf{H}_\tau(\underline{d}^{\tau-1}, \theta)$ and sample average of innovations $\mathsf{N}_\tau(\underline{d}^\tau, \theta)$, $\tau \leq t-1$, which almost surely converges for $t \to \infty$ to their zero expectation [21]. Thus, the support of the posterior pd concentrates (quickly due to the factor $-(t-1)$) on minimisers $\theta_P \in \boldsymbol{\theta}$ of $1/(t-1) \sum_{\tau=1}^{t-1} \mathsf{H}_\tau(\underline{d}^{\tau-1}, \theta)$: the weighted learning singles out the parametric models as if the data $\underline{d}^{t-1}$ was sampled from the ideally tuned closed loop described by the pd $\prod_{\tau=1}^{t-1} \mathsf{l}_t(d_\tau|\underline{d}^{\tau-1})$ and processed by the usual Bayes rule [2]. $\qquad\square$

Corollary 1 (Asymptotic Optimality of the Lazy FPD). *Let the function (6) be essentially bounded. Then, the predictor of the closed-loop behaviour (3), obtained via the weighted Bayes rule (4) with the weights (5), asymptotically almost surely fulfils the inequality, $\forall\theta \in \boldsymbol{\theta}$:*

$$\int_{d_t} \mathsf{l}_t(d_t|\underline{d}^{t-1}) \ln\left(\frac{\mathsf{l}_t(d_t|\underline{d}^{t-1})}{\mathsf{C}_t(d_t|\underline{d}^{t-1})}\right) \, dd_t \leq \int_{d_t} \mathsf{l}_t(d_t|\underline{d}^{t-1}) \ln\left(\frac{\mathsf{l}_t(d_t|\underline{d}^{t-1})}{\mathsf{M}_t(d_t|\underline{d}^{t-1}, \theta)}\right) \, dd_t. \quad (8)$$

Proof. According to Proposition 1, the support $\boldsymbol{\theta}_P \subset \boldsymbol{\theta}$ of $\mathsf{P}(\theta|\underline{d}_{t-1})$ asymptotically concentrates on minimisers $\theta_P \in \boldsymbol{\theta}$ of

$$\int_{d_t} \mathsf{l}_t(d_t|\underline{d}^{t-1}) \ln\left(\frac{\mathsf{l}_t(d_t|\underline{d}^{t-1})}{\mathsf{M}_t(d_t|\underline{d}^{t-1}, \theta)}\right) \, dd_t.$$

Thus, for any $\theta_P \in \boldsymbol{\theta}_P$ and any $\theta \in \boldsymbol{\theta}$

$$\int_{d_t} \mathsf{l}_t(d_t|\underline{d}^{t-1}) \ln\left(\frac{\mathsf{l}_t(d_t|\underline{d}^{t-1})}{\mathsf{M}_t(d_t|\underline{d}^{t-1}, \theta_P)}\right) \, dd_t \leq \int_{d_t} \mathsf{l}_t(d_t|\underline{d}^{t-1}) \ln\left(\frac{\mathsf{l}_t(d_t|\underline{d}^{t-1})}{\mathsf{M}_t(d_t|\underline{d}^{t-1}, \theta)}\right) \, dd_t.$$

Multiplying this inequality by the posterior pd $\mathsf{P}(\theta_P|\underline{d}^{t-1}) > 0$, integrating over its support $\boldsymbol{\theta}_P$, using the Jensen inequality and taking into account that by definition $\mathsf{P}(\theta_P|\underline{d}^{t-1})$ assigns unit probability to $\boldsymbol{\theta}_P$ give the claim (8). $\qquad\square$

Even when the function (6) is essentially bounded, the values of $\mathsf{W}_\tau(\underline{d}^\tau)$ can be too large. Thus, it is reasonable to limit them from above by $\overline{W} \in (1, \infty)$. Corollary 1 implies that it is always possible to select such $\overline{W}$ that the limitation is almost surely inactive. Then, the asymptotic results hold even when using it.

5   Illustrative Example

The example illustrates that the proposed weighting indeed improves properties of the lazy FPD. A Markov chain with two states $x \in \boldsymbol{x} \equiv \{1, 2\}$ and four actions $a \in \boldsymbol{a} \equiv \{1, 2, 3, 4\}$ is considered. The ideal pd expressing preferability

Table 1. The simulated system

$F(\underline{x}_t|a_t,\underline{x}_{t-1})$	$a_t = 1$	$a_t = 2$	$a_t = 3$	$a_t = 4$
$F(\underline{x}_t = 1|a_t,\underline{x}_{t-1} = 1)$	0.9975	0.0196	0.0196	0.9901
$F(\underline{x}_t = 2|a_t,\underline{x}_{t-1} = 1)$	0.0025	0.9804	0.9804	0.0099
$F(\underline{x}_t = 1|a_t,\underline{x}_{t-1} = 2)$	0.0196	0.9901	0.9967	0.0196
$F(\underline{x}_t = 2|a_t,\underline{x}_{t-1} = 2)$	0.9804	0.0099	0.0033	0.9804

Table 2. The decision rule found

$S(\underline{a}_t|x_{t-1})$	$x_{t-1} = 1$	$x_{t-1} = 2$
$S(\underline{a}_t = 1|x_{t-1})$	0.4607	0.1148
$S(\underline{a}_t = 2|x_{t-1})$	0.0292	0.3241
$S(\underline{a}_t = 3|x_{t-1})$	0.0293	0.4463
$S(\underline{a}_t = 3|x_{t-1})$	0.4808	0.1148

of the state value 1 was selected. The simulated system in given in Table 1. The proposed weighting (5), bounded from above by the value $\overline{W} = 3$, was compared with the standard solution (called un-normalised), which takes the weight W in (4) equal to the ideal pd $I_t(\underline{x}_\tau, \underline{a}_\tau|\underline{x}_{\tau-1})$. The designed strategy is given in Table 2.

Fig. 1 provides samples of simulated closed-loop behaviour when both weighting variants were applied to the same realisation of the underlying random generator. Fig. 2 provides the corresponding time course of weights. The strategy with the proposed weighting (5) reaches the desirable state $\underline{x}_\tau = 1$ in 91% cases while 65% of units occurred when using un-normalised ideal pd as the weight.

The *limited* simulation experience: i) supports the theoretical arguments; ii) shows that the proposed weighting tends to provide (often significant) improvement; iii) indicates that the proposed weighting substantially speeds up the learning of the optimal decision rule while the presented significant difference

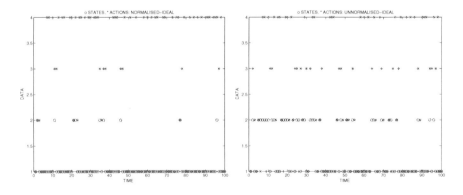

Fig. 1. Simulated behaviour: normalised weight (left), un-normalised weight (right)

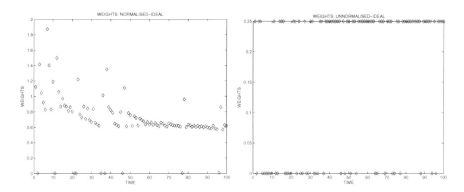

Fig. 2. Time course of the weight: normalised (left) and un-normalised (right)

in quality diminishes in long run; iv) confirms that the used approximation of the past closed-loop model influences visibly the result quality; v) reveals that very high values of the weight W_τ may occur due to the "practical" violation of assumed essential boundedness; vi) shows that the learning of the closed-loop model with a data-dependent forgetting behaves well.

6   Concluding Remarks

The weights are used properly and no other correct way seems to exist. The proposed normalisation of the weights is conceptually unique – the unambiguous *approximate* choice of the numerator in (5) stays open. The asymptotics, when the time-invariance makes its inspection meaningful, is the correct one: when the ideal situation $I_t(\underline{d}_\tau | \underline{d}^{\tau-1}) = C_\tau(\underline{d}_\tau | \underline{d}^{\tau-1})$ occurs, the weight $W_t = 1$ is reached.

Assumption (6) on the logarithmic ratio excludes parametric models that assign zero probability to data realisations, which are accepted as possible by the selected ideal pd. It can be weakened to the requirement on boundedness of the second moments. Algorithmically, it is connected with the upper bound $\overline{W}$ on weights W_t. Sensitivity to specific values of $\overline{W}$ seems to be low.

If almost no past data can be interpreted as coming from the optimally tuned closed loop, then $W_\tau \ll 1$, $\tau \leq t-1$, and the posterior pd becomes flat. This makes the one-step-ahead predictor of the closed-loop behaviour (3) flat, too. This situation enhances the explorative nature of actions generated from it, as desirable.

The choice (5) resembles the trick well-known in Monte Carlo evaluations when a feasible "proposal" pd is used [6]. The past closed-loop model plays its role. The analogy is, however, mechanical and seems to bring no tangible consequences.

Open problems: i) A decision, which of mentioned approximations of $C_\tau(d_\tau | d^{\tau-1})$ is better is to be made or an alternative option found. ii) Closed-loop stability is the major unsolved issue – the approximation of the ideal dynamics $I_t(d_t | d^{t-1})$ by the closed-loop model $C_t(d_t|d^{t-1})$ does not guarantee it; iii)The result guarantees that the one-step-ahead predictor of the closed-loop behaviour approximates the one-step-ahead ideal pd. In truly dynamic cases, the receding horizon strategy [25] can be immediately designed: it suffices to handle blocks od decisions. Other, more efficient ways of coping with DM dynamics have to be developed.

Acknowledgments. The reported research was supported by the project GACR 13-13502S.

References

1. Bellman, R.: Adaptive Control Processes. Princeton U. Press, NJ (1961)
2. Berec, L., Kárný, M.: Identification of reality in Bayesian context. In: Warwick, K., Kárný, M. (eds.) Computer-Intensive Methods in Control and Signal Processing, pp. 181–193. Birkhäuser (1997)
3. Berger, J.: Statistical Decision Theory and Bayesian Analysis. Springer, New York (1985)
4. Bertsekas, D.: Dynamic Programming and Optimal Control. Athena Scientific, US (2001)
5. Bontempi, G., Birattari, M., Bersini, H.: Lazy learning for local modelling & control design. Int. J. of Control **72**(7–8), 643–658 (1999)
6. Cappe, O., Godsill, S., Moulines, E.: An overview of existing methods and recent advances in sequential monte carlo. Proc. of the IEEE **95**(5), 899–924 (2007)
7. Daum, F.: Nonlinear filters: beyond the kalman filter. IEEE Aerospace and Electronic Systems Magazine **20**(8), 57–69 (2005)
8. Doucet, A., Johansen, A.: A tutorial on particle filtering and smoothing: Fifteen years later. In: Handbook of Nonlinear Filtering. Oxford University Press, Oxford (2011)
9. Feldbaum, A.: Theory of dual control. Autom. Remote Control **21**(9) (1960)
10. Gilboa, I., Schmeidler, D.: Case-based decsion theory. The Quaterly Journal of Economics **110**, 605–639 (1995)
11. Guan, P., Raginsky, M., Willett, R.: Online Markov decision processes with Kull-back Leibler control cost. IEEE Trans. on Automatic, Control (2014)
12. Kárný, M.: Towards fully probabilistic control design. Automatica **32**(12), 1719–1722 (1996)
13. Kárný, M.: Adaptive systems: Local approximators? In: Workshop n Adaptive Systems in Control and Signal Processing, pp. 129–134. IFAC, Glasgow (1998)
14. Kárný, M.: On approximate fully probabilistic design of decision making strategies. In: Guy, T., Kárný, M. (eds.) Proceedings of the 3rd International Workshop on Scalable Decision Making, ECML/PKDD 2013. UTIA AV ČR, Prague (2013) iSBN 978-80-903834-8-7
15. Kárný, M.: Approximate bayesian recursive estimation. Information Sciences (2014), doi: 10.1016/j.ins.2014.01.048
16. Kárný, M., Guy, T.V.: Fully probabilistic control design. Systems & Control Letters **55**(4), 259–265 (2006)
17. Kárný, M., Kroupa, T.: Axiomatisation of fully probabilistic design. Information Sciences **186**(1), 105–113 (2012)
18. Kulhavý, R., Zarrop, M.B.: On a general concept of forgetting. Int. J. of Control **58**(4), 905–924 (1993)

19. Kullback, S., Leibler, R.: On information and sufficiency. Annals of Mathematical Statistics **22**, 79–87 (1951)
20. Li, J., Dong, G., Ramamohanarao, K., Wong, L.: Deeps: a new instance-based lazy discovery and classification system. Machine Learning **54**(2), 99–124 (2004)
21. Loeve, M.: Probability Theory. van Nostrand, Princeton, New Jersey (1962) (Russian translation, Moscow 1962)
22. Macek, K., Guy, T., Kárný, M.: A lazy-learning concept of fully probabilistic decision making (2014) (unpublished manuscript)
23. Martín-Sánchez, J., Lemos, J., Rodellar, J.: Survey of industrial optimized adaptive control. Int. J. of Adaptive Control and Signal Processing **26**(10), 881–918 (2013).
24. Peterka, V.: Bayesian system identification. In: Eykhoff, P. (ed.) Trends and Progress in System Identification, pp. 239–304. Pergamon Press, Oxford (1981)
25. Qin, S., Badgwell, T.: A survey of industrial model predictive control technology. Control Engineering Practice **11**(7), 733–764 (2003)
26. Rao, M.: Measure Theory and Integration. John Wiley, NY (1987)
27. Roll, J., Nazin, A., Ljung, L.: Nonlinear system identification via direct weight optimization. Automatica **41**(3), 475–490 (2004)
28. Sanov, I.: On probability of large deviations of random variables. Matematičeskij Sbornik 42, 11–44 (in russian), also in selected translations mathematical statistics and probability. I **1961**, 213–244 (1957)
29. Savage, L.: Foundations of Statistics. Wiley, NY (1954)
30. Schon, T., Gustafsson, F., Nordlund, P.: Marginalized particle filters for mixed linear/nonlinear state-space models. IEEE Tran. on Signal Processing **53**(7), 2279–2289 (2005)
31. Si, J., Barto, A., Powell, W., Wunsch, D. (eds.): Handbook of Learning and Approximate Dynamic Programming. Wiley-IEEE Press, Danvers (2004)
32. Tishby, N., Polani, D.: Information theory of decisions and actions. In: Cutsuridis, V., Hussain, A., Taylor, J. (eds.) Perception-Action Cycle. Springer Series in Cognitive and Neural Systems, pp. 601–636. Springer, New York (2011)
33. Todorov, E.: Linearly-solvable Markov decision problems. In: Schölkopf, B., et al. (eds.) Advances in Neural Inf. Processing, pp. 1369–1376. MIT Press, NY (2006)
34. Zhu, C., Zhu, W.: Feedback control of nonlinear stochastic systems for targeting a specified stationary probability density. Automatica **47**(3), 539–544 (2006)

Memristive Radial Basis Function Neural Network for Parameters Adjustment of PID Controller

Xiaojuan Li[1], Shukai Duan[1(✉)], Lidan Wang[1], Tingwen Huang[2], and Yiran Chen[3]

[1] College of Electronic and Information Engineering,
Southwest University, Chongqing, China
duansk@swu.edu.cn
[2] Department of Electrical and Computer Engineering,
Texas A&M University, Doha, Qatar
[3] Department of Electrical and Computer Engineering,
University of Pittsburgh, Pittsburgh, USA

Abstract. Radial basis function (RBF) based-identification proportional–integral–derivative (PID) can automatically adjust the parameters of PID controller with strong self-organization, self-learning and self-adaptive ability. However, the compound controller has complex weight updating algorithm and large calculation. Memristor, applied well to the investigation of storage circuit and artificial intelligence, is a nonlinear element with memory function. Thus, it can be introduced to RBF neural network as electronic synapse to save and update the synaptic weights. This paper builds a model of memristive RBF-PID (MRBF-PID), and proposes the updating algorithm of weight upon memristance. The proposed MRBF-PID is used for the control of a nonlinear system. Its controlling effect is showed by numerical simulation experiment.

Keywords: Memristor · Radial basis function neural network · PID controller · Simulink model

1 Introduction

Based on the completeness of circuit, Chua proposed theoretically the notion of memristor in 1971 [1]. In 2008, a team at HP Labs announced that the physical model of memristor had been realized, which verified the existence of memristor [2-3], and attracted major research fields' attention. By reason of memristor's non-volatile memristive behavior and dynamic change resistance characteristic, we can primely introduce it into, for example, storage circuit, neural network and pattern recognition etc. Memristor has nanoscale size, which can greatly simplify the circuit by being applied in the large scale integrated circuit.

PID controller is the primary means of control in industrial control systems. In traditional PID controller, once the parameters are identified in the process, they can not be adjusted any more; therefore, systems can not achieve perfect controlling effect. Radial basis function neural network (RBF) is a local approximation network.

© Springer International Publishing Switzerland 2014
Z. Zeng et al. (Eds.): ISNN 2014, LNCS 8866, pp. 150–158, 2014.
DOI: 10.1007/978-3-319-12436-0_17

It has many excellent properties, such as fast convergence, strong generalization ability and simple structure. It can approximate continuous functions with arbitrary precision. RBF-PID is a combination of radial basis function neural network and the traditional PID controller. It boasts not only the ability to dynamically control the parameters of PID system, but also the ability to optimize the system parameters and improve the controlling effect through neural networks' self-learning ability.

This paper is organized as follows. Section 2 gives a brief overview on the PID controller based on RBF network. Section 3 introduces HP Memristor Model. Memristive RBF-PID Model is proposed in Section 4. Numerical simulation results are presented in Section 5. Finally, Section 6 concludes the work.

2   PID Controller Based on RBF Network

2.1   RBF Network

RBF network was proposed in 1988 [5], which possessed good generalization ability and simple network structure. We supposed that RBF network was provided with n input nodes, M hidden layer nodes and one output node. The structure is shown in Fig. 1.

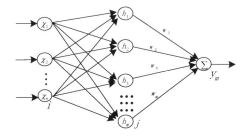

Fig. 1. RBF neural network structure

Generally we select the appropriate hidden layer activation function (Radial Basis Function, namely RBF) in accordance with the need. Gaussian function is used in this paper.

By the structure of RBF network, the input vector of the first layer is $X=[x_1, x_2,..., x_n]^T$. The output vector of hidden layer, namely Radial basis vector, is $H=[h_1,h_2,..., h_m]^T$, where h_j is Gaussian function:

$$h_j = \exp(-\frac{\left\| x - C_j \right\|^2}{2b_j^2}), (j = 1, 2, ..., m) \tag{1}$$

In above equation, m represents the number of the hidden layer neuron. $C_j=[c_{j1},c_{j2},...,c_{ji},...,c_{jn}]^T,(i=1,2,...,n)$ denotes the center vector of jth neuron. b_j denotes the basis width of jth neuron's RBF, which determines the width of basis function around the center. $\|X\text{-}C_j\|$ is the norm of vector $X\text{-}C_j$, denoting the distance between X and C_j.

Assuming that the weight vector of hidden layer to output layer is $W=[w_1,w_2,..,w_m]^T$. Since the mapping from hidden layer to output layer is linear, then the network output is formed by a linearly weighted sum of the number of basic functions in the hidden layer.

$$y_m(k) = W^T H = \sum_0^m w_i h_i \qquad (2)$$

Performance index function of RBF neural network is :

$$J = \frac{1}{2}(y(k) - y_m(k))^2 \qquad (3)$$

In order to minimizing the error objective function between desired output of RBF network and actual output, we use gradient descent to adjust system parameters. Iterative algorithm is as follows:

$$\begin{cases} \Delta w_j = -\dfrac{\partial J}{\partial w_j} = [y(k) - y_m(k)]h_j \\ w_j(k) = w_j(k-1) + \eta \Delta w_j + \alpha[w_j(k-1) - w_j(k-2)] \end{cases} \qquad (4)$$

$$\begin{cases} \Delta b_j = -\dfrac{\partial J}{\partial b_j} = [y(k) - y_m(k)]w_j h_j \dfrac{\|X - C_j\|^2}{b_j^3} \\ b_j(k) = b_j(k-1) + \eta \Delta b_j + \alpha[b_j(k-1) - b_j(k-2)] \end{cases} \qquad (5)$$

$$\begin{cases} \Delta c_{ji} = -\dfrac{\partial J}{\partial c_{ji}} = [y(k) - y_m(k)]w_j \dfrac{x_i - c_{ji}}{b_j^2} \\ c_{ji}(k) = c_{ji}(k-1) + \eta \Delta c_{ji} + \alpha[c_{ji}(k-1) - c_{ji}(k-2)] \end{cases} \qquad (6)$$

In that, $\eta \in (0,1)$ is learning rate. $\alpha \in (0,1)$ is momentum factor. Jacobian matrix algorithm, namely the sensitivity information of object's output towards control input, is:

$$\frac{\partial y(k)}{\partial \Delta u(k)} \approx \frac{\partial y_m(k)}{\partial \Delta u(k)} = \sum_{j=1}^m w_j h_j \frac{c_{ji} - x_1}{b_j^2} \qquad (7)$$

Where $x_1 = \nabla u(k)$.

2.2   RBF-PID

PID controller based on RBF neural network is shown in Fig. 2.

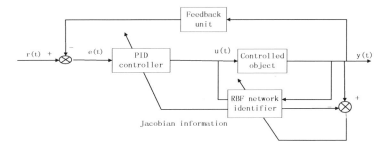

Fig. 2. RBF-PID controller

In the PID controller, its parameters are auto-updated by adjusting the weights of the neural network itself. In order to obtain accurate and optimal performance indicators, we adopt incremental PID controller in this paper. Its control error is:

$$error(k) = rin(k) - y(k) \tag{8}$$

The three inputs of RBF-PID are:

$$xc(1) = error(k) - error(k-1) \tag{9}$$

$$xc(2) = error(k) \tag{10}$$

$$xc(3) = error(k) - 2error(k-1) + error(k-2) \tag{11}$$

The algorithm of Incremental PID controller is:

$$u(k) = u(k-1) + k_p xc(1)(+k_i xc(2) + k_d xc(3) \tag{12}$$

The standard adjusting neural network is:

$$E(k) = \frac{1}{2} error(k)^2 \tag{13}$$

k_p, k_i, k_d are adjusted online in accordance with gradient descent in RBF-PID controller.

$$
\begin{cases}
\Delta k_p = -\eta_p \dfrac{\partial E}{\partial k_p} = -\eta_p \dfrac{\partial E}{\partial y} \dfrac{\partial y}{\partial \Delta u} \dfrac{\partial \Delta u}{\partial k_p} = \eta_p error(k) \dfrac{\partial y}{\partial \Delta u} xc(1) \\[2mm]
\Delta k_i = -\eta_i \dfrac{\partial E}{\partial k_i} = -\eta_i \dfrac{\partial E}{\partial y} \dfrac{\partial y}{\partial \Delta u} \dfrac{\partial \Delta u}{\partial k_i} = \eta_i error(k) \dfrac{\partial y}{\partial \Delta u} xc(2) \\[2mm]
\Delta k_d = -\eta_d \dfrac{\partial E}{\partial k_d} = -\eta_d \dfrac{\partial E}{\partial y} \dfrac{\partial y}{\partial \Delta u} \dfrac{\partial \Delta u}{\partial k_d} = \eta_d error(k) \dfrac{\partial y}{\partial \Delta u} xc(3)
\end{cases} \tag{14}
$$

In above equation, η_p, η_i and η_d denote the learning rate of proportional, integral and differential, respectively, and they are used to adjust the three parameters of PID controller online by RBF neural network.

3   Memristive RBF-PID Model

3.1   HP Memristor Model

In 1971, Prof. Chua, according to completeness of the circuit theory, proposed the notion of memristor. In 2008, HP Labs has proposed the memristor model, in which impurity is linearly drifted. The model is realized by adding metal oxide (TiO_2) between two metal electrodes, including doped layer (TiO_{2-x}) and non-doped layer (TiO_2), in which doped layer contains a part of oxygen vacancies and possesses lesser resistance, and non-doped layer is pure oxide and possesses larger resistance. According to the article [4], the charge-controlled memristor model with boundary conditions is as follow:

$$M = \begin{cases} R_{off}, (q(t) < c_1) \\ R(0) + kq(t) \quad , (c_1 \le q(t) < c_2) \\ R_{on}, (q(t) \ge c_2) \end{cases} \tag{15}$$

where

$$\frac{(R_{on} - R_{off})\mu_v R_{on}}{D^2} = k$$

$$c_1 = \frac{R_{off} - R(0)}{k}, c_2 = \frac{R_{on} - R(0)}{k}$$

The details and the illustrate of parameters see [4].

3.2   Memristve Electronic Synapse

Since memristor appeared, because of its continuous memory function, nanometer size and non-volatile property while power off, it is often applied in artificial intelligence and memory circuit. Neural synapse, as the connection between neurons, needs to be modified and saves its weight constantly in the operational process of network. Due to the similarity between the memristor and synapse, memristive neural network is implemented by applying memristor as electronic synapse. The electronic synapse is introduced in RBF network in this paper. For RBF neural network, the update formulas of weight w, center vector C and basic width b can all be expressed as follows:

$$w(k) = w(k-1) + \Delta w \tag{16}$$

which denotes variation of the parameters in each iteration.

It is known that charge-controlled memristor model has been shown as formula (15). Then we can obtain:

$$M(t) = M(t-1) + kq(t), c_1 \le q(t) \le c_2 \tag{17}$$

Imposing pulse voltage on memristor, whose amplitude is V and duration is t, we acquire the amount of charge flowing through it.

$$q(\Delta t) = \frac{v \Delta t}{M(t)} \tag{18}$$

Updated formula of memristance is:

$$M(t+1) = M(t) + k \frac{v \Delta t}{M(t)} = M(t) + \Delta M(t) \tag{19}$$

While the memristor is used as electronic synapse, the state variable x is considered as synaptic weight in the precious research. Since $x \in (0,1)$, the weight can only vary in the range of $(0,1)$. In actual artificial neural networks, weight's range is $(-1, 1)$. Therefore there are some defects when x is seen as weight. Memristance is as synaptic weight in this article by being linearly mapped in the range of $(-1, 1)$.

$$W = \frac{2}{R_{on} - R_{off}} M + \frac{R_{on} + R_{off}}{R_{off} - R_{on}} \tag{20}$$

Because of   that R_{off} is far greater than R_{on}, the above equation is rewritten as:

$$W \approx -\frac{2}{R_{off}} M + 1 \tag{21}$$

Combining Eq. (21) and (19), the updated formula of weight in RBF neural network based on memristor is as follows:

$$w(t+1) = w(t) - \frac{4}{R_{off}^2} \frac{kv \Delta t}{1 - w(t)} \tag{22}$$

3.3   Memristive RBF-PID Controller

In RBF neural network, the mapping from input layer to hidden layer is nonlinear, while the mapping from hidden to output layer is linear. According to the introduction in the second part, the argument of radial basis function network is Euclidean distance between input and neuron center vector. In order to facilitate memristor as electronic synapse, the Euclidean distance of input vector and weight vector from input layer to hidden is seen as argument of RBF network in this paper, namely the neuron's center vectors are assigned to connecting line connected with input layer, which can make it convenience to adjust weight. What's more, memristor is used inside hidden neuron to adjust basic width of RBF. Structure of memristive RBF network is as follows:

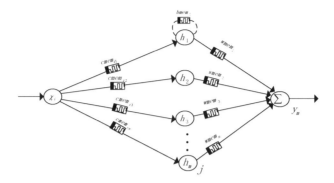

Fig. 3. Structure of memristive RBF network

By combining the above memristive RBF neural network and PID, memristive RBF-PID is constituted. Each time the updated weights are mapped to memristance and saved, they are translated into change of memristance, and accumulation in the process of updating weight is accomplished. We can save computational quantity and reduce the computational complexity with this method.

There are mainly PID control loop and memristive RBF (MRBF) neural network. The structure is same as fig. 2. Based on original RBF-PID, this article introduces memristor to RBF neural network to control PID system adaptively. In our model, MRBF identifies the PID system and obtains identifying information, which is exported to the PID control loop. The parameters of PID controller, i.e. K_p K_i and K_d, are regulated automatically with gradient descent.

4   Numerical Simulation Results

Nonlinear time-vary system is used as controlled object in this paper. The system formula is:

$$y(k) = \frac{a(k)y(k-1)+u(k-1)}{1+y(k-1)} \tag{23}$$

$$a(k) = 0.1(1-e^{-0.1k}) \tag{24}$$

The memristive parameters are set to: R_{on}=40, R_{off}=8000, D=10-8, μ_v=10^{-14}, p=10. The desired system output is square $yd(t)$=0.5square(2πt), and y is actual output. The momentum factor of RBF network is α=0.05, and its learning efficiency is η=0.25. The initial center vector of hidden layer is 3×6 matrix with value 0.1, and its initial basic width is 6×1matrix with value 0.9, and the initialization of weights from hidden to output layer is 6×1matrix with value 0.1. The learning efficiency of proportional unit, integral unit and derivative unit in PID is initialized with η_p=0.15, η_i=0.15 and η_d=0.15, respectively. The results are shown.

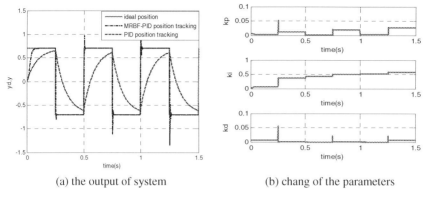

(a) the output of system   (b) chang of the parameters

Fig. 4. The result of simulation

In Fig. 4, The results indicate that MRBF-PID can reach steady state quickly and track input real-time in time-varying difference system compared with conventional PID controller. PID parameters of MRBF-PID adjust automatically with the change of system. When the iterations are 99, the system completes the training. However, the PID parameters of conventional PID controller are fixed with value 0.15 and it is not convergent.

The above results show that the proposed MRBF-PID controller can realize the tracking control for system, and possesses fast response speed and good robustness. It can adjust PID parameters real-time in the process of running to reach perfect control effect.

5   Conclusion and Discussion

This paper reports the application of memristor into RBF-PID network as neural synapse, realizing the weight's preservation and accumulation in the network and the simulation of the related properties. This paper uses memristor mapping for weights of synapse, and realizes the positive and negative polarity of weight. It's a novel attempt in the field of memristive synapse. Memristor is a nanodevice which can simplify circuit, increase the connection density of network, and reduce the physical size of a system in the aspect of hardware realization. Memristor also has important characteristics, such as passive, non-volatile, dependence upon history. As synapse has similar characteristics, it's reasonable to apply memristor in neural network as electronic synapse.

Acknowledgements. This publication was made possible by NPRP grant # NPRP 4-1162-1-181 from the Qatar National Research Fund (a member of Qatar Foundation). The statements made herein are solely the responsibility of the author[s]. The work was supported by Program for New Century Excellent Talents in University ([2013]47), National Natural Science Foundation of China (61372139, 61101233, 60972155), "Spring Sunshine Plan" Research Project of Ministry of Education of China (z2011148), Technology Foundation for Selected Overseas Chinese Scholars, Ministry of Personnel in China (2012-186), Fundamental Research Funds for the Central Universities (XDJK2014A009, XDJK2013B011).

References

1. Chua, L.O.: Memristor-The Missing Circuit Element. IEEE Transactions on Circuit Theory **18**(5), 507–519 (1971)
2. Strukov, D.B., Snider, G.S., Stewart, D.R., Williams, R.S.: The Missing Memristor Found. Nature **453**(7191), 80–83 (2008)
3. Williams, R.S.: How We Found the Missing Memristor. IEEE Spectrum **45**(12), 28–35 (2008)
4. Duan, S., Zhang, Y., Hu, X., et al.: Memristor-based Chaotic Neural Networks for Associative Memory. Neural Computing and Applications, 1–9 (2014)
5. Zhang, Y., Liu, C., Song, X., et al.: Application of RBF Neural Network Controller in the Rectification Column Temperature Control System. In: International Symposium on Computational Intelligence and Design, pp. 72–75 (2013)
6. Wang, L., Fang, X., Duan, S., et al.: PID Controller Based on Memristve CMAC Network. Abstract and Applied Analysis **2013**, 1–6 (2013)
7. Xue, Y., Ye, J., Qian, H., et al.: The Research of Complex BP Neural Network PID Control. In: International Conference on Artificial Intelligence and Computational Intelligence, pp. 55–58. IEEE Press, New York (2009)
8. Zhou, Y., Ding, Q.: Study of PID Temperature Control for Reactor Based on RBF Network. In: 2012 International Conference on Automation and Logistics, pp. 456–460. IEEE Press, New York (2012)

Finite-Time Stability of Switched Static Neural Networks

Yuanyuan Wu[1,2] and Jinde Cao[1,3](✉)

[1] Department of Mathematics, Southeast University, Nanjing 210096, China
[2] College of Electric and Information Engineering, Zhengzhou University
of Light Industry, Zhengzhou 450002, China
[3] Department of Mathematics, Faculty of Science,
King Abdulaziz University, Jeddah 21589, Saudi Arabia
jdcao@seu.edu.cn

Abstract. This paper deals with the finite-time stability problem for switched static neural networks with time-varying delay. Firstly, the concept of finite-time stability is extended to switched static neural networks. Secondly, based on Lyapunov-like functional method, a sufficient criterion is derived, which can guarantee the finite-time stability of the considered systems. Moreover, the obtained conditions can be simplified into linear matrix inequalities conditions for convenient use. Finally, a numerical example is given to show the effectiveness of the proposed results.

Keywords: Finite-time stability · Static neural networks · Switched systems · Time-varying delay · Linear matrix inequalities

1 Introduction

According to whether neuron states (the external states of neurons) or local fields states (the internal states of neurons) are selected as basic variables, neural networks can be classified as static neural networks or local field neural networks [1], and the two models are not always equivalent [2]. Actually, the static neural networks have been widely used to solve various optimization problems, such as some linear variational inequality problems. Compared with the extensive investigation of the stability problem of local field neural networks, the static neural networks have got less attention [3–9].

Considering the switched happened in neural networks, switched neural networks are proposed, and some results have been put forward on the stability analysis of the switched neural networks [10,11]. Furthermore, the stability and passivity analysis are considered for switched neural networks with time-varying delay in [12,13], and the stability analysis is investigated for discrete-time switched neural networks in [14–16]. The global exponential stability is concerned for switched stochastic neural networks with time-varying delays in [17]. The synchronization control is studied for switched neural networks with

© Springer International Publishing Switzerland 2014
Z. Zeng et al. (Eds.): ISNN 2014, LNCS 8866, pp. 159–166, 2014.
DOI: 10.1007/978-3-319-12436-0_18

time delay in [18,19]. As well known, integration and communication delays are unavoidably encountered in neural networks, then it is important and valuable to study the switched neural networks with time delay.

In many practical applications, the dynamical behavior over a fixed finite time interval is paid more attention for a system. For example, the property that the state does not exceed a certain threshold in a finite time interval with a given bound on the initial condition, which is corresponding to the finite-time stability. Now, concepts of finite-time stability have been proposed for several decades [20], and the definition of finite-time stability has been extended to the definition of finite-time bounded by taking the presence of external disturbances into account [21]. Most recently, several valuable results have been proposed for the finite-time problems of neural networks. The finite-time boundedness stability is studied for neural networks with parametric uncertainties in [22], and for uncertain neural networks with Markovian jumps in [23,24]. The problem of finite-time state estimation is investigated for neural networks with time-varying delays in [25]. To the best of the authors' knowledge, there is few work on the finite-time stability of static neural networks.

Motivated by the above analysis, this paper considers finite-time stability of the switched static neural networks (SSNNs) with time-varying delay. Firstly, the concept of finite-time stability is extended to switched static neural networks. Then, employing the proper model transformation and Lyapunov-like functional, a sufficient condition is presented for the finite-time stability of switched static neural networks.

2   Problem Formulation

Consider the following static neural networks with time-varying delay:

$$
\begin{aligned}
\dot{x}(t) &= -Ax(t) + f(Wx(t - \tau(t)) + J), \\
x(t) &= \varphi(t), \quad -\tau \le t \le 0,
\end{aligned}
\tag{1}
$$

where $x(\cdot) = [x_1(\cdot),\ x_2(\cdot), \cdots, x_n(\cdot)]^T \in \mathbb{R}^n$ is the neuron state vector, $f(x(\cdot)) = [f_1(x_1(\cdot)),\ f_2(x_2(\cdot)), \cdots, f_n(s_n(\cdot))]^T \in \mathbb{R}^n$ denotes the neuron activation function, $J = [j_1,\ j_2, \cdots,\ j_n]^T \in \mathbb{R}^n$ is a constant input vector. $A = diag\{a_1,\ a_2, \cdots,\ a_n\} \in \mathbb{R}^{n \times n}$ is a positive diagonal matrix, and $W = [W_1^T,\ W_2^T, \cdots, W_n^T]^T \in \mathbb{R}^{n \times n}$ is the delayed connection weight matrix. $\tau(t)$ is a time-varying delay satisfying $0 \le \tau(t) \le \tau$ and $\dot{\tau}(t) \le \mu$, $\varphi(t)(-\tau \le t \le 0)$ is the initial condition.

As in [12,13], we study the following switched static neural networks composed of the system (1) as the individual subsystems:

$$
\begin{aligned}
\dot{x}(t) &= -A_\beta x(t) + f(W_\beta x(t - \tau(t)) + J), \\
x(t) &= \varphi(t), \quad -\tau \le t \le 0.
\end{aligned}
\tag{2}
$$

Where β is a switching signal, which is unknown a prior and takes its valued in the finite set $\mathbb{E} = 1, 2, \cdots, N$. Throughout this paper, we assume:

Assumption 1. The each neuron activation function in system (2) is assumed to be bounded and satisfy

$$l_i^- \le \frac{f_i(x) - f_i(y)}{x - y} \le l_i^+, \quad \forall x, y \in R, x \ne y, \ i = 1, 2, \cdots, n, \tag{3}$$

where l_i^- and l_i^+ are some known constants. Define $l_i = \max\{|l_i^-|, |l_i^+|\}$, and let $\underline{L} = diag\{l_1^-, l_2^-, \cdots, l_n^-\}, \overline{L} = diag\{l_1^+, l_2^+, \cdots, l_n^+\}$, and $L = diag\{l_1, l_2, \cdots, l_n\}$.

Remark 1. Assume that W_β is invertible and $W_\beta A_\beta = A_\beta W_\beta$ holds, then (2) can be easily transformed to the following local neural networks by $y(t) = W_\beta x(t) + J$. However, in many application, the two models are not equivalent.

Under Assumption 1, there is an equilibrium x^* of (2). For simplicity, let $z(\cdot) = x(\cdot) - x^*$, then system (2) can be transformed into

$$\begin{aligned} \dot{z}(t) &= -A_\beta z(t) + g(W_\beta z(t - \tau(t))), \\ z(t) &= \psi(t), \quad -\tau \le t \le 0. \end{aligned} \tag{4}$$

where $z(\cdot) = [z_1(\cdot), z_2(\cdot), \cdots, z_n(\cdot)]^T$ is the state vector of the transformed system (4), $\psi(t) = \varphi(t) - x^*$ is the initial condition, and the transformed neuron activation functions is $g(W_k z(\cdot)) = [g_1(W_{\beta 1} z(\cdot)), g_2(W_{\beta 2} z(\cdot)), \cdots, g_n(W_{\beta n} z(\cdot))]^T = f(W_\beta z(\cdot) + W_\beta x^* + J) - f(W_\beta x^* + J)$. It is clear that $g_i(\cdot)$ satisfy:

$$l_i^- \le \frac{g_i(x)}{x} \le l_i^+, \quad \forall x \in R, x \ne 0, \ i = 1, 2, \cdots, n. \tag{5}$$

Based on the analysis above, we know that the stability analysis of system (2) on equilibrium is changed into the zero stability problem of system (4). We are now to introduce the notion of the finite-time stability for the system (4) and an employed lemma in this paper.

Definition 1. *Given a positive matrix R, three positive constants c_1, c_2, T with $c_1 < c_2$, the switched static neural networks (4) is said to be finite-time stable with respect to (c_1, c_2, T, R), if for any switched rule,*

$$\sup_{\theta \in [-\tau, 0]} z^T(\theta) R z(\theta) \le c_1 \implies z^T(t) R z(t) < c_2, \forall t \in [0, T].$$

Lemma 1. *(The Jensen Inequality) For any constant matrix $R = R^T \ge 0$, scalar $\tau > 0$ and vector function $x(\cdot) : [-\tau, 0] \to \mathbb{R}^n$ such that the following integrals are well defined, then*

$$-\tau \int_{t-\tau}^t x^T(s) R x(s) ds \le -\left[\int_{t-\tau}^t x(s) ds \right]^T R \left[\int_{t-\tau}^t x(s) ds \right]. \tag{6}$$

3   Main Results

In this section, we will present the finite-time stability criteria for the considered SSNNs (4).

Theorem 1. *Under the Assumption 1, the SSNNs (4) is finite-time stable with respect to* (c_1, c_2, T, R), *if there exists a scalar* $\alpha > 0$, *matrices* $P > 0$, $Q > 0$, $T > 0$, $S > 0$, *and diagonal matrix* $U \geq 0$ *such that the following inequalities hold for* $k = 1, 2, \cdots, N$:

$$\Xi_k = \begin{bmatrix} \Gamma_{11}^k & S & \Gamma_{13}^k \\ * & \Gamma_{22}^k & \Gamma_{23}^k \\ * & * & \Gamma_{33}^k \end{bmatrix} < 0, \tag{7}$$

$$\frac{e^{\alpha T} \vartheta c_1}{\lambda_{min}(\bar{P})} < c_2, \tag{8}$$

where

$$\Gamma_{11}^k = -PA_k - A_k P + Q + W_k^T LTLW_k + \tau^2 A_k SA_k - S - \alpha P,$$
$$\Gamma_{22}^k = -(1 - \mu e^{\alpha \tau})Q - S - 2W_k^T \underline{L} U \overline{L} W_k,$$
$$\Gamma_{13}^k = -\tau^2 A_k S + P, \ \Gamma_{23}^k = W_k^T \underline{L} U + W_k^T \overline{L} U,$$
$$\Gamma_{33} = \tau^2 S - (1 - \mu e^{\alpha \tau})T - 2U,$$
$$\vartheta = \lambda_{max}(\bar{P}) + \tau e^{\alpha \tau} \lambda_{max}(\bar{Q}) + \tau e^{\alpha \tau} \max_{k \in \mathbb{E}}(\lambda_{max}(\bar{T}_k))$$
$$+ 2\tau^3 e^{\alpha \tau} \max_{k \in \mathbb{E}}(\lambda_{max}(\bar{S}_k^1) + \lambda_{max}(\bar{S}_k^2)),$$

with $\bar{P} = R^{-\frac{1}{2}} P R^{-\frac{1}{2}}$, $\bar{Q} = R^{-\frac{1}{2}} Q R^{-\frac{1}{2}}$, $\bar{T}_k = R^{-\frac{1}{2}} W_k^T LTLW_k R^{-\frac{1}{2}}$, $\bar{S}_k^1 = R^{-\frac{1}{2}} A_k SA_k R^{-\frac{1}{2}}$, *and* $\bar{S}_k^2 = R^{-\frac{1}{2}} W_k^T LSLW_k R^{-\frac{1}{2}}$.

Proof. Firstly, we introduce the indicator function $\xi(t) = [\xi_1(t), \xi_2(t), \cdots, \xi_N(t)]^T$, where $\xi_k(t) = 1 (k = 1, 2, \cdots, N)$ when the k-th subsystem is activated, otherwise, it is zero. Then, the SSNNs (4) can be rewritten as

$$\dot{z}(t) = \sum_{k=1}^{N} \xi_k(t)[-A_k z(t) + g(W_k z(t - \tau(t)))]. \tag{9}$$

Obviously, it follows that $\sum_{k=1}^{N} \xi_k(t) = 1$ under any switching rule. Now, choose a Lyapunov-like functional candidate for system (9) as:

$$V(z(t)) = V_1(t) + V_2(t) + V_3(t) + V_4(t), \tag{10}$$

where

$$V_1(t) = z^T(t) P z(t), \quad V_2(t) = \int_{t-\tau(t)}^{t} e^{\alpha(t-s)} z^T(s) Q z(s) ds,$$

$$V_3(t) = \tau \int_{-\tau}^{0} \int_{t+\theta}^{t} e^{\alpha(t-s)} \dot{z}^T(s) S \dot{z}(s) ds d\theta,$$

$$V_4(t) = \sum_{k=1}^{N} \xi_k(t) \int_{t-\tau(t)}^{t} e^{\alpha(t-s)} g^T(W_k z(s)) T g(W_k z(s)) ds.$$

Calculating the derivative of $V_i(i = 1, 2, 3, 4)$ along the solution of system (9) yields

$$\dot{V}_1(t) = \sum_{k=1}^{N} \xi_k(t)\{2z^T(t)P[-A_k z(t) + g(W_k z(t - \tau(t)))]\}, \qquad (11)$$

and

$$\dot{V}_2(t) \leq \alpha V_2(t) + z^T(t)Qz(t) - (1 - \mu e^{\alpha \tau})z^T(t - \tau(t))Qz(t - \tau(t)). \quad (12)$$

Using Lemma 1, it can be deduced that

$$\dot{V}_3(t) \leq \alpha V_3(t) + \tau^2 \sum_{k=1}^{N} \xi_k(t)\{[-A_k z(t) + g(W_k z(t - \tau(t)))]^T S[-A_k z(t)$$
$$+ g(W_k z(t - \tau(t)))]\} - [z(t) - z(t - \tau(t))]^T S[z(t) - z(t - \tau(t))], (13)$$

and it follows that

$$\dot{V}_4(t) \leq \alpha V_4(t) + \sum_{k=1}^{N} \xi_k(t)\{g^T(W_k z(t))Tg(W_k z(t))$$
$$- (1 - \mu e^{\alpha \tau})g^T(W_k z(t - \tau(t)))Tg(W_k z(t - \tau(t))). \qquad (14)$$

From the inequality condition (5), we know that there exists a diagonally matrix $U \geq 0$ such that:

$$\sum_{k=1}^{N} \xi_k(t)2[g(W_k z(t - \tau(t))) - \underline{L}W_k z(t - \tau(t))]^T U$$
$$\times [\overline{L}W_k z(t - \tau(t)) - g(W_k z(t - \tau(t)))] \geq 0, \qquad (15)$$

and

$$g^T(W_k z(t))Tg(W_k z(t)) \leq z^T(t)W_k^T LTLW_k z(t). \qquad (16)$$

Considering (11)-(16), then we can obtain

$$\dot{V}(z(t)) \leq \sum_{k=1}^{N} \xi_k(t)\eta_k^T(t)\Xi_k \eta_k(t) + \alpha V(z(t)), \qquad (17)$$

It is clear that the inequality (7) in Theorem 1 can guarantee

$$\dot{V}(z(t)) \leq \alpha V((z(t)). \qquad (18)$$

Multiplying the above inequality (18) by $e^{\alpha t}$, and then integrating it from 0 to t, with $t \in [0, T]$, we have

$$V(z(t)) \leq e^{\alpha t}V(z(0)) \leq e^{\alpha T}V(z(0)). \qquad (19)$$

According to the definition of $V(z(t))$ in (10), it can be calculated

$$V(z(0)) \leq \vartheta \sup_{\theta \in [-\tau, 0]} z^T(\theta) R z(\theta), \tag{20}$$

where ϑ is defined in Theorem 1 before. On the other hand, we can get

$$V(z(t)) \geq \lambda_{min}(R^{-\frac{1}{2}} P R^{-\frac{1}{2}}) z^T(t) R z(t) \triangleq \lambda_{min}(\bar{P}) z^T(t) R z(t). \tag{21}$$

The inequalities (19)-(21) lead to

$$z^T(t) R z(t) \leq \frac{e^{\alpha T} \vartheta}{\lambda_{min}(\bar{P})} \sup_{\theta \in [-\tau, 0]} z^T(\theta) R z(\theta) \leq \frac{e^{\alpha T} \vartheta c_1}{\lambda_{min}(\bar{P})}. \tag{22}$$

Condition (8) in Theorem 1 implies, for all $t \in [0, T]$, $z^T(t) R z(t) < c_2$. Therefore, the proof follows.

Remark 2. Clearly, Theorem 1 are independent on the switching rule. That is, Theorem 1 holds for any switching rule. Moreover, the conditions in Theorem 1 are not standard linear matrix inequalities (LMIs). Once we have fixed a value of α, the condition (7) can be turned into LMIs and then solved by Matlab LMI toolbox.

On the other hand, we can easily check that the condition (8) in Theorem 1 can be guaranteed by the following LMIs conditions:

$$\gamma_1 I < P < \gamma_2 I, \ Q < \gamma_3 I, \ T < \gamma_4 I, \ S < \gamma_5 I, \tag{23}$$

$$-c_2 \gamma_1 \pi_1 e^{-\alpha T} + c_1[(\gamma_2 + \gamma_3)\pi_1 + (\gamma_4 + \gamma_5)\pi_2 + \gamma_5 \pi_3] < 0, \tag{24}$$

where $\gamma_i (i = 1, 2, 3, 4, 5)$ are positive scalars and $\pi_j (j = 1, 2, 3)$ is defined as

$$\pi_1 = \lambda_{max}(R^{-1}), \ \pi_2 = \max_{k \in \mathbb{E}}(\lambda_{max}(R^{-\frac{1}{2}} W_k^T L L W_k R^{-\frac{1}{2}})),$$

$$\pi_3 = \max_{k \in \mathbb{E}}(\lambda_{max}(R^{-\frac{1}{2}} A_k^T A_k R^{-\frac{1}{2}})).$$

4   A Numerical Example

In this section, a numerical example is given to illustrate the validity of the proposed results. Consider SSNNs (9) with the following parameters:

$$A_1 = \begin{bmatrix} 2.2 & 0 \\ 0 & 2.1 \end{bmatrix}, \ A_2 = \begin{bmatrix} 2.1 & 0 \\ 0 & 1.9 \end{bmatrix}, \ W_1 = \begin{bmatrix} 3.5 & 1.2 \\ -1.0 & 3.5 \end{bmatrix}, \ W_2 = \begin{bmatrix} 3.1 & 0.9 \\ -1.1 & 3.2 \end{bmatrix},$$

The activation functions are defined as $g_1(t) = \tanh(0.3t)$, $g_2(t) = \tanh(0.5t)$, and the time-varying delay is $\tau(t) = 0.8 + 0.3\sin(t)$. It means that

$$\tau = 1.1, \ \mu = 0.3, \ \underline{L} = 0, \ L = \bar{L} = \begin{bmatrix} 0.3 & 0 \\ 0 & 0.5 \end{bmatrix}.$$

The other parameters are chosen as $c_1 = 1$, $c_2 = 10$, $T = 10$ and $R = I$. By Matlab LMI toolbox, solving the inequalities (7), (23) and (24) for $\alpha = 0.01$ gives the feasible solutions:

$$P = \begin{bmatrix} 2.2086 & -0.3598 \\ -0.3598 & 2.0523 \end{bmatrix}, \quad Q = \begin{bmatrix} 2.0215 & -0.0904 \\ -0.0904 & 1.6474 \end{bmatrix},$$

$$T = \begin{bmatrix} 0.9932 & -0.4279 \\ -0.4279 & 0.9769 \end{bmatrix}, \quad S = \begin{bmatrix} 0.3108 & 0.0104 \\ 0.0104 & 0.0033 \end{bmatrix},$$

which satisfy the condition (8). According to Theorem 1, we can conclude that the concerned SSNNs is finite-time stable with respect to $(1, 10, 10, I)$ for any switching rule.

Remark 3. Notice that the conditions (23) and (24) are dependent on the size of c_2, then we can also get the optimal lower bound of c_2 to guarantee the finite-time stability by solving a simple optimal problem. For example, we can obtain the optimal lower bound of c_2 is 3.9351.

5 Conclusions

This paper has studied the problem of finite-time stability for switched static neural networks. The Lyapunov-like function method and LMIs technique are developed to derive a sufficient criterion, which can guarantee the finite-time stability of SSNNs. In the end, a numerical example is provided to show the effectiveness of our proposed theoretical results. In the further investigations, we will search the other analysis technique to obtain the less conservative results, and consider some control problems of SSNNs in finite-time sense.

Acknowledgments. This work is partially supported by Basic and Frontier Technologies Research Program of Henan Province(122300410279) and Doctoral Fund of Zhengzhou University of Light Industry (201BSJJ006).

References

1. Xu, Z.B., Qiao, H., Peng, J., Zhang, B.: A comparative study on two modeling approaches in neural networks. Neural Networks **17**, 73–85 (2004)
2. Qiao, H., Peng, J., Xu, Z.B., Zhang, B.: A reference model approach to stability analysis of neuralnetworks. IEEE Trans. Syst. Man Cybern. PartB **33**, 925–936 (2003)
3. Liang, J.L., Cao, J.D.: A based-on LMI stability criterion for delayed recurrent neural networks. Chaos, Solitons and Fractals **28**, 154–160 (2006)
4. Shao, H.Y.: Delay-dependent stability for recurrent neural networks with time-varying delays. IEEE Trans. on Neural Networks **19**, 1647–1651 (2008)
5. Shao, H.Y.: Delay-dependent approaches to globally exponential stability for recurrent neural networks. IEEE Trans. on Circuits and Systems -II: Express Briefs **55**, 591–595 (2008)

6. Zheng, C.D., Zhang, H., Wang, Z.: Delay-dependent globally exponential stability criteria for static neural networks: an LMI approach. IEEE Trans. Circuits Syst. II 56, 605–609 (2009)
7. Huang, H., Feng, G., Cao, J.D.: State estimation for static neural networks with time-varying delay. Neural Networks **23**, 1202–1207 (2010)
8. Li, X., Gao, H., Yu, X.: A unified approach to the stability of generalized static neural networks with linear fractional uncertainties and delays. IEEE Trans. Systems Man Cybern. Part B 41, 1275–1286 (2011)
9. Xiao, J., Zeng, Z.G., Wu, A.L.: New criteria for exponential stability of delayed recurrent neural networks. Neurocomputing **134**, 182–188 (2014)
10. Huang, H., Qu, Y., Li, H.: Robusts tability analysis of switched Hopfield neural networks with time-varying delay under uncertainty. Phys. Lett. A. 345, 345–354 (2005)
11. Yuan, K., Cao, J.D., Li, H.: Robust stability of switched Cohen CGrossberg neural networks with mixed time-varying delays. IEEE Trans. Syst. Man Cybern. Part B. 36, 1356–1363 (2006)
12. Li, P., Cao, J.: Global stability in switched recurrent neural networks with time-varying delay via nonlinear measure. Nonlinear Dynam. **49**, 295–305 (2007)
13. Hu, M.F., Cao, J.D., Yang, Y.Q., Hu, A.H.: Passivity analysis for switched generalized neural networks with time-varying delay and uncertain output. IMA Journal of Mathematical Control and Information **30**, 407–422 (2013)
14. Hu, M.F., Cao, J.D., Hu, A.H.: Mean square exponential stability for discrete-time stochastic switched static neural networks with randomly occurring nonlinearities and stochastic delay. Neurocomputing **129**, 476–481 (2014)
15. Hou, L., Zong, G., Wu, Y.: Robust exponential stability analysis of discrete-time switched Hopfield neural networks with time delay. Nonlinear Anal. Hybrid Syst. **5**, 525–534 (2011)
16. Arunkumar, A., Sakthivel, R., Mathiyalagan, K., Anthoni, S.M.: Robust stability criteria for discrete-time switched neural networks with various activation functions. Appl. Math. Comput. **218**, 10803–10816 (2012)
17. Wu, X.T., Tian, Y., Zhang, W.B.: Stability analysis of switched stochastic neural networks with time-varying delays. Neural Networks **51**, 39–49 (2014)
18. Yang, X.S., Cao, J.D., Zhu, Q.X.: Synchronization of switched neural networks with mixed delays via impulsive control. Chaos, Solitons and Fractals 44, 817–826 (2011)
19. Yu, W.W., Cao, J.D., Lu, W.L.: Synchronization control of switched linearly coupled neural networks with delay. Neurocomputing **73**, 858–866 (2010)
20. Dorato, P.: Short time stability in linear time-varying systems. In: Proceeding of the IRE International Convention Record Part, vol. 4, pp. 83–87 (1961)
21. Amato, F., Ariola, M., Dorate, P.: Finite-time control of linear systems subject to parameteric uncertainties and disturbances. Automatica **37**, 1459–1463 (2001)
22. Shen, Y.J., Li, C.C.: LMI-based finite-time boundedness analysis of neural networks with parametric uncertainties. Neurocomputing **71**, 502–507 (2008)
23. He, S.P., Liu, F.: Finite-time boundedness of uncertain time-delayed neural network with Markovian jumping parameters. Neurocomputing **103**, 87–92 (2013)
24. Zhang, Y.Q., Shi, P., Nguang, S.K., Zhang, J.H., Karimi, H.R.: Finite-time boundedness for uncertain discrete neural networks with time-delays and Markovian jumps. Neurocomputing (in press)
25. Cheng, J., Zhong, S.M., Zhong, Q.S., Zhu, H., Du, Y.H.: Finite-time boundedness of state estimation for neuralnetworks with time-varying delays. Neurocomputing **129**, 257–264 (2014)

Fixed-Priority Scheduling Policies and Their Non-utilization Bounds

Guangyi Chen[✉] and Wenfang Xie

Department of Mechanical and Industrial Engineering, Concordia University,
Montreal, QC H3G 1M8, Canada
`{guang_c,wfxie}@encs.concordia.ca`

Abstract. Utilization bounds for schedulability are one of the simplest approaches for admission control in real-time systems. Even though they are not resource optimal, they offer the advantages of simplicity and computational efficiency. Liu and Abdelzaher proposed a universal bound that was customized to any fixed-priority scheduling policy by choosing the corresponding load metric. In this paper, we extend Liu and Abdelzaher's work by combining different fixed-priority scheduling policies, and give their non-utilization bounds. We have considered the combination of deadline monotonic scheduling (DMS), shortest-job-first scheduling (SJF), and velocity monotonic scheduling (VMS). The reason why we take this combination is because we can take advantage of the merits of each scheduling policy and at the same time overcome their shortcomings.

Keywords: Fixed priority scheduling · Deadline Monotonic Scheduling (DMS) · Shortest-Job-First scheduling (SJF) · Velocity Monotonic Scheduling (VMS)

1 Introduction

Utilization test is an indirect schedulability test, which does not compute the delays, but rather tests system resource utilization in order to determine the task schedulability. A new task can be admitted only if the utilization is lower than a pre-defined bound. This can be done very efficiently with an O(1) computational complexity. Even though utilization-based schedulability test is simple in terms of concept and computational complexity, it has some limitations. For example, the feasibility condition is sufficient but not necessary (i.e., pessimistic). Also, it imposes unrealistic constrains upon the timing characteristics of tasks (e.g., deadline is equal to period). Fixed-priority scheduling nowadays is one of the most adopted techniques for implementing real-time applications.

We briefly review some of the fixed-priority scheduling methods published in the literature here. In the seminal work of Liu and Layland [1], they derived the well-known 69% utilization bound for rate monotonic scheduling (RMS) on a single processor system, where relative deadlines of periodic tasks are equal to their periods. Since then, many real-time scheduling algorithms have been proposed in the literature.

© Springer International Publishing Switzerland 2014
Z. Zeng et al. (Eds.): ISNN 2014, LNCS 8866, pp. 167–174, 2014.
DOI: 10.1007/978-3-319-12436-0_19

Bini and Buttazzo [2] extended the utilization bounds to multiprocessor real-time systems, and they considered resource constraints and presented a single-stage utilization bound that was less pessimistic than Liu and Layland's work. Kuo and Mok [3] improved the Liu and Layland's bound by collapsing harmonic tasks into one chain. They proved that the bound is a function of the number of the harmonic chains and not the number of individual tasks. This bound was further improved by considering information specific to the task set such as the actual values of task periods [4]. Liu and Abdelzaher [5] derived a general expression for a non-utilization-based schedulability bound. The bound applies to a generalized abstract load metric that is defined for a given system as a function of its particular scheduling policy.

In this paper, we propose to combine different fixed-priority scheduling policies and give their non-utilization bounds. This is based on Liu and Abdelzaher's previous work where they derived non-utilization bounds for arbitrary fixed-priority policies. The combinations of different fixed-priority scheduling policies open a new avenue for real-time admission control, and they can make use of the advantages of existing fixed-priority scheduling policies and at the same time overcome their disadvantages. Simulation results confirm that the proposed method in this paper is a feasible approach for fixed-priority real-time admission control.

2   An Existing Method

Liu and Abdelzaher [5] derived a general expression for a non-utilization-based schedulability bound. Consider a task model in which aperiodic tasks arrive at a system of multiple resources. Each task T_i has a relative end-to-end deadline D_i, which defines its maximum acceptable total latency in the system. The task requires multiple processing stages, collectively denoted by set G_i. The execution time of task T_i on stage $j \in G_i$ is denoted as c_{ij}. It was assumed that a fixed-priority scheduling policy should be used. The priority was assigned monotonically decreasing in some function $x()$ of the task, The value of the function x for task T_i is denoted as x_i. The choice of the function x can be arbitrary as long as it is always positive.

The Universal Feasible Region Theorem: *Consider a system scheduled by a fixed priority scheduling policy, where priorities are monotonically decreasing in some function x. Let x_i be the value of x for some task T_i, which traverses N stages and has an end-to-end deadline D_i. Let the abstract load at stage j be denoted by*

$$M_j = \sum_i \frac{C_{i,j}}{x_i}. \tag{1}$$

T_i meets its end-to-end deadline as long as the following condition is true:

$$\sum_{j \in G_i} \frac{M_j(1 - M_j/2)}{1 - M_j} \le \frac{D_i}{x_i} \tag{2}$$

where G_i is the set of multiple processing stages for task T_i. An admission controller should ensure that

$$\sum_{j \in G_i} \frac{M_j (1 - M_j / 2)}{1 - M_j} \leq \min_i (\frac{D_i}{x_i}) \tag{3}$$

where the minimization is carried out over the set of active tasks, or over the set of tasks in the current busy period.

This theorem quantifies, for the first time in real-time computing literature, a universal schedulable region that is applicable to an arbitrary scheduling policy as a function of its priority definition x.

3   Proposed Method

The above theorem is intended to a single fixed-priority scheduling policy. In this section, we would like to extend it to a combination of existing scheduling policies. Specifically, we will study deadline monotonic scheduling (DMS), shortest-job-first scheduling (SJF), and velocity monotonic scheduling (VMS). Other existing scheduling methods will be left for future research. The main reason why we use a combination of two existing scheduling policies is because we can take advantage of the merits of each scheduling policy and at the same time overcome their shortcomings. It is expected that, by tuning the combination factor α, we can achieve improved scheduling results for fixed-priority tasks.

We briefly give some explanations about DMS, SJF and VMS here. In DMS, tasks are assigned priorities according to their deadlines; the task with the shortest deadline being assigned the highest priority. Deadline-monotonic priority assignment is a priority assignment policy used with fixed priority pre-emptive scheduling. In SJF, a scheduling policy selects the waiting process with the smallest execution time to execute next. SJF is advantageous because of its simplicity and because it maximizes process throughput (in terms of the number of processes run to completion in a given amount of time). It also minimizes the average amount of time each process has to wait until its execution is complete. In VMS, priorities are set proportionally to the ratio of the end-to-end deadline to the number of stages to be traversed. Priority is higher when the deadline lower and when the number of stages to be traversed is more. If two tasks have the same end-to-end deadline but a different number of stages to traverse, the one with more stages should be given a higher priority.

The priority x_i should be different for different scheduling policies. For DMS, we can select $x_i = D_i$, where D_i is the end-to-end deadline. For SJF, we can select $x_i = \sum_j C_{i,j}$, where $C_{i,j}$ is the execution time of job i on stage j. For VMS, we can choose $x_i = \dfrac{D_i}{l_i}$, where $l_i = |G_i|$ is the number of stages task T_i will be processed on.

Since equation (2) holds for any choice of x_i, we can combine two fixed-priority scheduling policies P_1 and P_2 and select x_i based on a linear combination of P_1 and P_2.

For example, P_1 and P_2 can be any of DMS, SJF, and VMS. There are three different combinations by selecting any two scheduling policies, and they are DMS-VMS, DMS-SJF, and SJF-VMS. The priority x_i for the combinations can be given as

$$x_i = \alpha \times x_i^1 + (1-\alpha) \times x_i^2 \qquad (4)$$

where $0 \le \alpha \le 1$, and x_i^1 and x_i^2 are the priorities of the two combined scheduling policies. If we combine DMS with SJF, then

$$x_i = \alpha \times D_i + (1-\alpha) \times \sum_j C_{i,j} \qquad (5)$$

If we combine SJF with VMS, then

$$x_i = \alpha \times \sum_j C_{i,j} + (1-\alpha) \times D_i / l_i \qquad (6)$$

If we combine DMS with VMS, then

$$x_i = \alpha \times D_i + (1-\alpha) \times D_i / l_i \qquad (7)$$

Therefore

$$f(l_i) = \frac{D_i}{x_i} = \frac{1}{\alpha + (1-\alpha)/l_i} \qquad (8)$$

It is easy to know that

$$\min_i f(l_i) = \frac{1}{\alpha + (1-\alpha)/\min_i(l_i)}. \qquad (9)$$

Interestingly, if we choose $\alpha = 1/2$, then we have

$$x_i = D_i \times (1 + 1/l_i)/2 \qquad (10)$$

and

$$f(l_i) = \frac{D_i}{x_i} = \frac{2}{1 + 1/l_i} = \frac{2l_i}{l_i + 1}. \qquad (11)$$

This is a very simple function, which takes very few flops of computation. Fig. 1 shows this function. It can be seen that $f(l_i)$ is monotonically increasing.

Since the universal feasible region theorem proposed in [5] is valid for any x_i, it is also valid for a combination of existing fixed-priority scheduling policies. Theorem 1 states that the priority x_i in this paper is greater than or equal to the priority x_i' for VMS in [5]. Therefore, we have $f(l_i) \le f(l_i')$, which is true for any $\alpha \in [0,1]$.

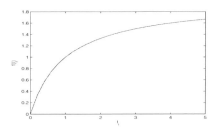

Fig. 1. A plot of the function $f(l_i)$

Theorem 1: *Let D_i be the end-to-end deadline of task T_i, x_i be its priority, $l_i = |G_i|$ be the number of stages to be processed, and x_i' be the priority defined in [5] for VMS. If we combine DMS with VMS, then, for any fixed $\alpha \in [0,1]$,*

$$x_i = D_i \times (\alpha + (1-\alpha)/l_i) = D_i/l_i \times (\alpha(l_i-1)+1) \geq D_i/l_i = x_i' \tag{12}$$

and

$$f(l_i) = \frac{D_i}{x_i} \leq \frac{D_i}{x_i'} = f(l_i') \tag{13}$$

The above discussion only considers a combination of two existing real-time scheduling policies. It is expected that an even better scheduling policy can be achieved by combining several existing scheduling policies. For example,

$$x_i = \alpha_1 x_i^1 + \alpha_2 x_i^2 + \cdots + \alpha_p x_i^P \tag{14}$$

where x_i^p is the priority of the pth scheduling policy and

$$\alpha_p \geq 0 \text{ and } \sum_{p=1}^{P} \alpha_p = 1.0 \tag{15}$$

When p=3, we can combine three existing fixed-priority scheduling policies such as the combination of DMS, SJF and VMS.

4   Experimental Results

In this section, we conducted some experiments in order to simulate the feasibility of the proposed scheduling policies for fixed-priority tasks. We assume that the computation time of different stages is independent, and end-to-end deadlines are selected uniformly from a range. The arrival process is Poisson. The priority is kept constant for all execution stages of the same task. We performed experiments by combining any two of the existing scheduling policies (DMS, SJF and VMS). These combinations are DMS-VMS, DMS-SJF, and SJF-VMS. The number of stages of a task is set to be 10, and the probability that a task selects a stage is $p=0.5$. The average stage load is approximately equal. In our experiments, no deadline misses were observed. This verifies the correctness of the proposed methods in this paper.

We have done experiments for $\alpha \in [0,1]$ with a step of 0.2 and different combinations of existing scheduling policies DMS-VMS, DMS-SJF, and SJF-VMS. Fig. 2 shows the average stage-utilization after admission control for the three combinations of fixed-priority scheduling policies. From the figures, it can be seen that α can balance the trade-off between the two combined scheduling methods. As α increases, the average stage-utilization increases for DMS-VMS and DMS-SJF, but it decreases for SJF-VMS.

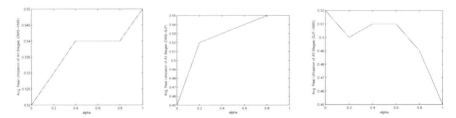

Fig. 2. The average stage-utilization after admission control for the combination of fixed-priority scheduling policies DMS-VMS, DMS-SJF and SJF-VMS, respectively

We have calculated the average stage-utilization for the three combinations of fixed-priority scheduling policies DMS-VMS, DMS-SJF, and SJF-VMS when $\alpha = 0.5$. The input workload is varied from 40% to 200% of a single stage capacity. This means that the sum of computation times of all tasks generated over the duration of the experiment is varied from 40% to 200% of the length of the experiment. Fig. 3 shows the curves of our three combinations. It can be seen that the curves of DMS-VMS, DMS-SJF, and SJF-VMS lie between the two curves of the original scheduling policies. For example, the curve of DMS-VMS lies between the curves of DMS and VMS. In general, SJF is not as good as DMS and VMS, and it is not as good as our combined scheduling policies DMS-VMS, DMS-SJF, and SJF-VMS as well.

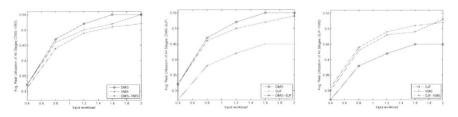

Fig. 3. The average stage-utilization after admission control for the combination of fixed-priority scheduling policies DMS-VMS and the two original scheduling policies DMS and VMS, DMS-SJF and the two original scheduling policies DMS and SJF, and SJF-VMS and the two original scheduling policies SJF and VMS, respectively

We multiply a constant K to the combined priority as $x_i = Kx_i$ for $K=1.0, 10.0$ and 100.0. Fig. 4 shows the curves of our three combinations DMS-VMS, DMS-SJF and SJF-VMS for different values of K. From these figures it can be seen that the constant $K=10.0$ is a good choice for all three combinations DMS-VMS, DMS-SJF and SJF-VMS. In addition, the constant $K=10.0$ is much better than $K=1.0$ in these three figures. This indicates that multiplying a constant to the combined priority is a feasible choice for real-time admission control. It should be noted that we have chosen $\alpha = 0.5$ for this set of experiments.

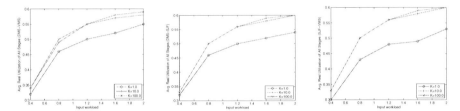

Fig. 4. The average stage-utilization after admission control for the combination of fixed-priority scheduling policies DMS-VMS, DMS-SJF, and SJF-VMS, respectively

We also test the performance of our proposed method by combining three existing fixed-priority policies (DMS, SJF and VMS). The scaling factors satisfy

$$\alpha_p \geq 0 \text{ and } \sum_{p=1}^{3} \alpha_p = 1.0 \tag{16}$$

In this experiment, we set $\alpha_p = 1/3$. Fig. 5 shows the average stage-utilization after admission control for the combination of DMS-SJF-VMS. From the figure it can be seen that DMS-SJF-VMS is better than SJF and VMS. However, it is not as good as DMS. It is expected that by tuning the combination factors, we can achieve even better results.

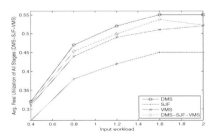

Fig. 5. The average stage-utilization after admission control for the combination of fixed-priority scheduling policies DMS-SJF-VMS

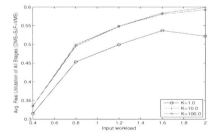

Fig. 6. The average stage-utilization after admission control for the combination of fixed-priority scheduling policies DMS-SJF-VMS

We multiply a constant K to the combined priority as $x_i = Kx_i$ for $K=1.0, 10.0$ and 100.0. Fig. 6 shows the curve of our combination DMS-SJF-VMS for different values of K. From the figure it can be seen that the constant $K=10.0$ is a good choice for DMS-SJF-VMS. In addition, the constant $K=10.0$ is much better than $K=1.0$ in the figure. This indicates that multiplying a constant to the combined priority is a feasible choice for real-time admission control.

5   Conclusions

Understanding the end-to-end temporal behavior of distributed systems is a fundamental concern in real-time computing. Meeting timing requirements is critical to many computer applications. These real-time systems employ a schedulability test to determine whether each task can meet its deadline.

In this paper, we extend Liu and Abdelzaher's non-utilization bound to a combination of existing scheduling methods. For example, we can combine any two of the DMS, SJF, and VMS scheduling policies. From these three scheduling policies, we can have DMS-VMS, DMS-SJF and SJF-VMS as the combined scheduling policies. The main reasons why we combine two different fixed-priority scheduling policies are because we can take advantage of the merits of the two existing scheduling methods, and at the same time reduce their shortcomings. Simulation results demonstrate that this is a feasible method for real-time admission control.

Future work will be done in the following ways. We can derive the non-utilization bounds for periodic tasks, where tighter bounds could be found. In addition, we may try to find the optimal function x in order to improve the derived bounds for real-time scheduling. It would also be interesting to study the scheduling policies for dynamic priority tasks.

References

1. Liu, C.L., Layland, J.W.: Scheduling algorithms for multiprogramming in a hard-real-time environment. Journal of ACM **20**, 46–61 (1973)
2. Bini, E., Buttazzo, G.: A hyperbolic bound for the rate monotonic algorithm. In: 13th Euromicro Conference on Real-time Systems, Delft, Netherlands (2001)
3. Kuo, T.W., Mok, A.K.: Load adjustment in adaptive real-time systems. In: IEEE Real-Time Systems Symposium (1991)
4. Chen, X., Mohapatra, P.: Lifetime behaviour and its impact on web caching. In: IEEE Workshop on Internet Applications (1999)
5. Liu, X., Abdelzaher, T.: On non-utilization bounds for arbitrary fixed priority policies. In: Proceedings of the 12th IEEE Real-Time and Embedded Technology and Applications Symposium (RTAS), San Jose, California (2006)

Modeling Dynamic Hysteresis of Smart Actuators with Fuzzy Tree

Yongxin Guo[1(✉)], Zhen Zhang[1], Jianqin Mao[1], Haishan Ding[2], and Yanhua Ma[3]

[1] School of Automation Science and Electrical Engineering, Beihang University,
Beijing, 100191, China
guoyongxin1979@126.com
[2] China Airborne Missile Academy, Luoyang, 471009, China
chinodhs@ss.buaa.eud.cn
[3] Institute Chemical Physics, Chinese Academy of Sciences, Dalian, 116023, China
yh_ma@126.com

Abstract. The inherent dynamic hysteresis, including rate-dependent and stress-dependent hysteresis, hinders the performance of smart actuators. In this paper, two methods are introduced to model the dynamic hysteresis of smart actuators based on fuzzy tree. The first method is to describe the rate-dependent hysteresis directly with fuzzy tree and the second is to propose a stress-dependent Preisach model by building the relations between compress stress and the density function of the Preisch model with fuzzy tree. Simulation results show that both methods can satisfactorily describe rate-dependent or stress-dependent hysteresis.

Keywords: Smart actuators · Dynamic hysteresis · Rate-dependent · Stress-dependent · Fuzzy Tree · Preisach model

1   Introduction

Smart actuators, including giant magnetostrictive actuators (GMA), piezoelectric actuators (PZT) and shape memory actuators (SMA), are widely used for micro-positioning and micro-vibration control in fields of aerospace, manufacturing, etc. However, smart actuators exhibit dominant hysteresis nonlinearity. What's more, the hysteresis is dynamic, that is, the output displacement of smart actuators depends on the rate of the input signal, or the compress stress that is imposed on the actuators. This poses a significant challenge in the analysis and design of systems with smart actuators.

Dynamic hysteresis models can be classified into physics-based models and phenomenological models. Physics-based models are based on the physical mechanism of systems, and phenomenological models are based on the input-output of systems. The most popular Preisach model [1] [2] and computing intelligence techniques [3] [4] [5] are phenomenological hysteresis models.

Mao et al. [6] proposed a tree-structured-based method (fuzzy tree method for short), and this method is suitable to solve complex nonlinear problems. In this paper,

© Springer International Publishing Switzerland 2014
Z. Zeng et al. (Eds.): ISNN 2014, LNCS 8866, pp. 175–183, 2014.
DOI: 10.1007/978-3-319-12436-0_20

based on the fuzzy tree, two modeling methods are introduced to describe dynamic hysteresis of smart actuators. One is to describe the rate-dependent hysteresis directly with fuzzy tree and another is to propose a stress-dependent Preisach model by building the relations between the compress stress and the density function of the Preisch model with fuzzy tree. Simulation results show that both methods can satisfactorily describe rate-dependent or stress dependent hysteresis.

This paper is organized as follows: In Section 2, the fuzzy tree method and the Preisach model are introduced. In Section 3, approaches of modeling the hysteresis of smart actuators with fuzzy tree are presented. In Section 4, modeling performance is verified by comparison between the outputs of the models and the actuators. Finally, concluding remarks are stated in Section 5.

2   Fuzzy Tree Method and Preisach Model

2.1   Fuzzy Tree

Fuzzy tree is a special type of T-S (Takagi-Sugeno) fuzzy models. T-S models can be described by the following fuzzy rules:

R^l: If x_1 is M_1^l, x_2 is M_2^l, $\cdots$, x_n is M_n^l, then $y_l = (\mathbf{c}_l)^T \hat{\mathbf{x}}$, $l = 1, 2, ..., m$, where the input vector $\mathbf{x} = [x_1, ..., x_n]^T \in R^n$, $\hat{\mathbf{x}} = \left[1, \mathbf{x}^T\right]^T \in R^{n+1}$ is the augmented input vector, $\mathbf{c}_l = [c_0^l, c_1^l, \cdots c_n^l]^T \in R^{n+1}$, M_i^l is the fuzzy set corresponding to the variable x_i, and m is the total rule number.

The membership function of M_i^l is $M_i^l(x_i)$. For an input $\mathbf{x}$, the final output of the fuzzy model is:

$$y(\mathbf{x}) = \sum_{l=1}^{m} \mu_l(\mathbf{x}) y_l = \sum_{l=1}^{m} \mu_l(\mathbf{x})(\mathbf{c}_l)^T \hat{\mathbf{x}} \ , \tag{1}$$

where $\mu_l(\mathbf{x}) = M^l(\mathbf{x}) / \sum_{j=1}^{m} M^j(\mathbf{x})$, $M^l(\mathbf{x}) = \prod_{i=1}^{n} M_i^l(x_i)$, $l = 1, 2, ..., m$.

The main idea of the fuzzy tree method is that the input space is partitioned based on a binary tree, and thus irregular fuzzy subspaces are obtained. Piecewise linear functions defined on the subspaces corresponding to the leaf vertexes of the binary tree are used as the consequent parts of the fuzzy rules. The number of fuzzy rules equals the number of leaf vertexes, so it is *insensitive to the dimension of inputs* and it is suitable to solve high-dimension and complex nonlinear problems.

Let T denotes a binary tree, $r(T)$ denotes the root vertex of T. For each vertex $t \in T$, $l(t)$ and $r(t)$ represent respectively the left and right child of vertex t, and $p(t)$ represents the parent of vertex t. If t has no children, it is called leaf vertex. $\tilde{T}$ denotes the set of all leaf vertexes of T. The depth of vertex t means the number of the ancestors of t. Moreover, the depth of T indicates the maximum depth of the vertexes of T.

In the fuzzy tree method, the partition of the input space is adaptive. At the high density nonlinear part of input data set, the partition will be finer; at the other part, the partition will be relatively rough. The partition corresponds to a binary tree. In each subspace represented by a leaf vertex, the nonlinear function is approximated by an n-dimensional hyper-plane $y_t(\mathbf{x}) = (\mathbf{c}_t)^T \hat{\mathbf{x}}$, where $\hat{\mathbf{x}} = [1, \mathbf{x}^T]^T \in R^{n+1}$. If approximation errors meet the requirements, stop the partition of this subspace. The hyper-plane $g_t(\mathbf{x}) = (\mathbf{c}_t)^T \hat{\mathbf{x}} - \theta_t = 0$ is used as the discriminant function to judge whether the sub-space is divided, where θ_t denotes the gravity center of the corresponding output data in this subspace. The subspace is divided into two smaller subspaces according to $g_t(\mathbf{x}) \leq 0$ and $g_t(\mathbf{x}) \geq 0$, and the fuzzy region is defined near the part $g_t(\mathbf{x}) = 0$.

The following rules can be obtained after the fuzzy partition of the input space:

$$R^l : \text{If } \mathbf{x} \text{ is } N_{t_l}, \text{ then } y_{t_l} = (\mathbf{c}_{t_l})^T \hat{\mathbf{x}} \ , \tag{2}$$

where $t_l \in \tilde{T}$, $\mathbf{c}_{t_l} = \left[c_0^{t_l}, c_1^{t_l}, \cdots, c_n^{t_l} \right]^T$ are the linear parameters, N_{t_l} is a fuzzy set defined on the fuzzy subspace χ_{t_l}, and the corresponding membership function is denoted as $N_{t_l}(\mathbf{x})$. Thus each fuzzy rule is corresponding to a fuzzy subspace which is represented by leaf vertexes.

If the normalized membership function of $N_{t_l}(\mathbf{x})$ is denoted as $\mu_{t_l}(\mathbf{x})$, that is, $\mu_{t_l}(\mathbf{x}) = N_{t_l}(\mathbf{x}) / \sum_{t_l \in \tilde{T}} N_{t_l}(\mathbf{x})$, the output will be the same as that of T-S fuzzy models:

$$\hat{y}(\mathbf{x}) = \sum_{t_l \in \tilde{T}} \mu_{t_l}(\mathbf{x})(\mathbf{c}_{t_l})^T \hat{\mathbf{x}} \ . \tag{3}$$

Membership functions corresponding to each vertex of the binary tree are defined as:

$$N_{r(T)}(\mathbf{x}) \equiv 1 \ , \text{ if } t \text{ is a root vertex} , \tag{4}$$

$$N_t(\mathbf{x}) = N_{p(t)}(\mathbf{x}) \hat{N}_t(\mathbf{x}) , \text{ if } t \text{ is not a root vertex} , \tag{5}$$

where the instrument membership function corresponding to vertex t is defined as:

$$\hat{N}_t(\mathbf{x}) = \frac{1}{1 + \exp[-\alpha_t(\mathbf{c}_{p(t)}^T \hat{\mathbf{x}} - \theta_{p(t)})]} \ , \tag{6}$$

where $\theta_{p(t)}$ is the gravity center of the corresponding output data on the parent vertex of t. It is defined as:

$$\theta_{p(t)} = \frac{\sum_{i=1}^M N_{p(t)}(\mathbf{x}^i)(\mathbf{c}_{p(t)}^T \hat{\mathbf{x}}^i)}{\sum_{i=1}^M N_{p(t)}(\mathbf{x}^i)} \ , \tag{7}$$

$|\alpha_t|$ is the width of the fuzzy region. For a left vertex, $\alpha_t = -\alpha$, for a right vertex, $\alpha_t = \alpha$, α is a positive number. $\mathbf{x}^i$ is an input sample, $i = 1, 2, \cdots, M$.

Suppose $\tilde{T} = \{t_1, t_2, \cdots, t_L\}$, denotes $\mathbf{c}_{\tilde{T}} = \left[\mathbf{c}_{t_1}^{\mathrm{T}}, \mathbf{c}_{t_2}^{\mathrm{T}}, \cdots, \mathbf{c}_{t_L}^{\mathrm{T}}\right]^{\mathrm{T}}$, the linear parameters $\mathbf{c}_{\tilde{T}}$ in fuzzy rules are solved by the recursive least square method:

$$\begin{cases} \mathbf{c}_{\tilde{T}}^{i+1} = \mathbf{c}_{\tilde{T}}^i + \mathbf{S}_{i+1}\tilde{\mathbf{X}}^{i+1}(y^{i+1} - (\tilde{\mathbf{X}}^{i+1})^{\mathrm{T}}\mathbf{c}_{\tilde{T}}^i) \\ \mathbf{S}_{i+1} = \mathbf{S}_i - \dfrac{\mathbf{S}_i\tilde{\mathbf{X}}^{i+1}(\tilde{\mathbf{X}}^{i+1})^{\mathrm{T}}\mathbf{S}_i}{1 + (\tilde{\mathbf{X}}^{i+1})^{\mathrm{T}}\mathbf{S}_i\tilde{\mathbf{X}}^{i+1}} \end{cases} , i = 0, 1, \cdots, M-1 , \tag{8}$$

where $\mathbf{c}_{\tilde{T}}^0 = 0$, $\mathbf{S}_0 = \lambda\mathbf{I}$, λ is a positive number, which is large enough. $\mathbf{I}$ is an identity matrix, and

$$\tilde{\mathbf{X}}^i = \left[\frac{N_{t_1}(\mathbf{x}^i)}{\sum\limits_{t_l \in \tilde{T}} N_{t_l}(\mathbf{x}^i)}(\hat{\mathbf{x}}^i)^{\mathrm{T}}, \frac{N_{t_2}(\mathbf{x}^i)}{\sum\limits_{t_l \in \tilde{T}} N_{t_l}(\mathbf{x}^i)}(\hat{\mathbf{x}}^i)^{\mathrm{T}}, \cdots, \frac{N_{t_L}(\mathbf{x}^i)}{\sum\limits_{t_l \in \tilde{T}} N_{t_l}(\mathbf{x}^i)}(\hat{\mathbf{x}}^i)^{\mathrm{T}} \right]^{\mathrm{T}} . \tag{9}$$

2.2   Preisach Model

The Preisach model [1] is a weighted superposition of delayed relay operators. For a pair of thresholds (β, α) with $\beta \leq \alpha$ and the initial configuration $\zeta \in \{-1, 1\}$, for the input $u \in C([0, T])$ and $t \in [0, T]$, the output of the delayed relay operator $f = \hat{\gamma}_{\beta\alpha}[u, \zeta]$ is defined as:

$$f(t) \triangleq \begin{cases} -1 & \text{if } u(t) < \beta \\ 1 & \text{if } u(t) > \alpha \\ f(t^-) & \text{if } \beta \leq u(t) \leq \alpha \end{cases} , \tag{10}$$

where $t^- \triangleq \lim\limits_{\varepsilon > 0, \varepsilon \to 0}(t - \varepsilon)$ and $f(0^-) = \zeta$

The Preisach plane is defined as:

$$P = \{(\beta, \alpha) \in P \mid \beta \leq \alpha\} . \tag{11}$$

For $u \in C([0, T])$ and a Borel measurable initial configuration ζ_0 of all operators, $\zeta_0 : P \to \{-1, 1\}$, the Preisach model Γ is defined as:

$$\Gamma[u, \zeta_0](t) = \iint_P \mu(\beta, \alpha)\hat{\gamma}_{\beta,\alpha}[u, \zeta_0(\beta, \alpha)](t)d\beta d\alpha , \tag{12}$$

where the weighting function $\mu(\beta, \alpha)$ is called the Preisach density function. The Preisach model is a static one, and thus it cannot describe dynamic hysteresis.

3   Modeling Dynamic Hysteresis with Fuzzy Tree

The fuzzy tree method can be used to describe dynamic hysteresis of smart actuators. The first method is to describe the rate-dependent hysteresis directly with fuzzy tree, another method is to propose a stress-dependent Preisach model by building the relations between compress stress and the density functions of the Preisch model with fuzzy tree.

3.1   Method I: Modeling Rate-Dependent Hysteresis with Fuzzy Tree

The discrete model based on fuzzy tree is:

$$\hat{y}(k+1) = f(x(k), x(k-1), \cdots, x(k-m+1);$$
$$y(k), y(k-1), \cdots, y(k-n+1)) \tag{13}$$

where $x(k)$ and $y(k)$ are respectively the input and output of the actuator at time k, with orders m and n, and $\hat{y}(k+1)$ represents the output of the model at time $k+1$. The essential of the model is that it predicts the output at the next time using the input and output information at current and historical times.

In order to describe the rate-dependent hysteresis, the input signals should excite all the modes in the frequency range sufficiently. Gaussian random signal, random binary signal, pseudo random binary signal, and sinusoidal scanning signal can be used for excite the actuators for model identification. The details of the identification process can be found in [7].

3.2   Method II: Modeling Stress-Dependent Hysteresis with Fuzzy-Tree-Based Preisach Model

The compressive stress opposed on smart actuators is equivalent to the mechanical load on smart actuators. Then the static Preisach model can be modified to describe stress-dependent hysteresis by including mechanical load m into the density function. The stress-dependent Preisach model can be written as:

$$y(t) = \Xi[u, \zeta_0, m](t) = \iint_P \mu(\beta, \alpha, m)\, \gamma_{\beta,\alpha}[u, \zeta_0(\beta, \alpha)](t)\, d\beta d\alpha . \tag{14}$$

Due to the great variation of mechanical load, the density function in (14) would be severely influenced by the variable m. To overcome this drawback, another form of (14) is proposed as:

$$y(t) = \Xi[u, \zeta_0, m](t) = \iint_P \mu(\beta, \alpha, g(m))\, \gamma_{\beta,\alpha}[u, \zeta_0(\beta, \alpha)](t)\, d\beta d\alpha , \tag{15}$$

where $g(m)$ is a function of the mechanical load m. If the amplitude of $g(m)$ has a small variation, then the power series expansion of the density function in (15) with respect to $g(m)$ is:

$$\mu\left(\beta,\alpha,g\left(m\right)\right)=\mu_0\left(\beta,\alpha\right)+g\left(m\right)\mu_1\left(\beta,\alpha\right)+\cdots \ . \tag{16}$$

Keep the first two terms of (16) and substitute it into (15):

$$\begin{aligned}
y(t)&=\Xi[u,\zeta_0,m](t)=\iint_P\mu\left(\beta,\alpha,g\left(m\right)\right)\gamma_{\beta,\alpha}\left[u,\zeta_0\left(\beta,\alpha\right)\right](t)d\beta d\alpha\\
&=\iint_P\mu_0\left(\beta,\alpha\right)\gamma_{\beta,\alpha}\left[u,\zeta_0\left(\beta,\alpha\right)\right](t)d\beta d\alpha\\
&\quad+\iint_P g\left(m\right)\mu_1\left(\beta,\alpha\right)\gamma_{\beta,\alpha}\left[u,\zeta_0\left(\beta,\alpha\right)\right](t)d\beta d\alpha
\end{aligned} \tag{17}$$

$\mu_0\left(\beta,\alpha\right)$ in (17) should coincide with $\mu\left(\beta,\alpha\right)$ of the Preisach model in (12), then (17) can be rewritten as:

$$y\left(t\right)=\tilde{y}\left(t\right)+\iint_{\alpha\geq\beta}g\left(m\right)\mu_1\left(\beta,\alpha\right)\gamma_{\beta,\alpha}\left[u,\zeta_0\left(\beta,\alpha\right)\right]\left(t\right)d\beta d\alpha \ , \tag{18}$$

where $\tilde{y}(t)$ stands for the part with no mechanical load. The second term in (18) represents the variation part with different stresses.

Denote $g(m_j)$ as the value of $g(m)$ with mechanical load m_j. Each $g(m_j)$ can be identified with the method in [8]. Then $g(m)$ can be identified with fuzzy tree to describe the relations between the load m and the density function of stress-dependent Preisach model.

4   Modeling Performance

In this section, the above two modeling methods are used to describe the dynamic hysteresis loops of GMAs. The modeling results are compared with experimental data to reveal the performance. The two GMAs used in experiments are manufactured by Beihang University. The output displacements of the GMAs are measured by eddy current sensor. The dSPACE system DS1103 was used for data acquisition, and the sampling frequency is 10 kHz. The power amplifier that drove the GMA at different frequencies worked in voltage mode, and the one that drove the GMA with different mechanical load worked in current mode.

4.1   Performance of Method I

The hysteresis loops measured in experiments at different frequencies and those simulated with the identified fuzzy tree are shown in Fig. 1. It is obvious that the fuzzy tree can describe the rate-dependent hysteresis. Table 1 lists the root mean square error, which is defined as:

$$\text{RMSE} = \sqrt{\frac{\sum_{k=1}^{M} \left(\hat{y}(k) - y(k) \right)^2}{M}} \quad , \tag{19}$$

where $y(k)$ is the GMA's output measured in experiment at time k, $\hat{y}(k)$ is simulated with the fuzzy tree, M is the number of data.

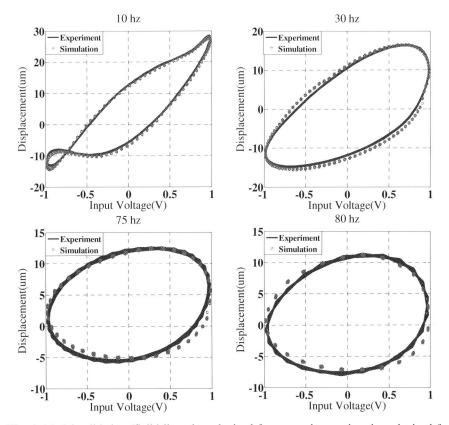

Fig. 1. Model validation (Solid line: data obtained from experiment; dot: data obtained from model)

Table 1. RMSE of method I

Frequency	5Hz	10Hz	15Hz	20Hz	25Hz
RMSE/μm	0.5977	0.7537	0.7753	1.0747	1.5480
Frequency	30Hz	75Hz	80Hz	95Hz	
RMSE/μm	0.8085	0.8966	1.0873	0.6455	

4.2   Performance of Method II

The hysteresis loops with different mechanical load measured in experiments and those simulated with the identified stress-dependent Preisach model based on fuzzy-tree are shown in Fig. 2. It is obvious that the Preisach model with fuzzy tree can describe the stress-dependent hysteresis. Table 2 lists the root mean square error.

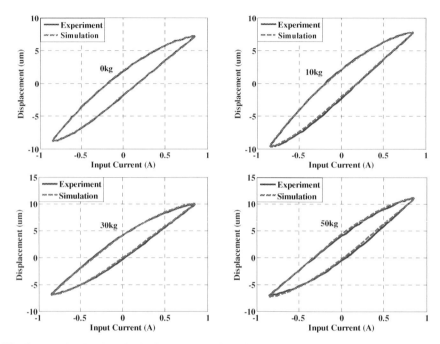

Fig. 2. Model validation (Solid line: data obtained from experiment; dashed: data obtained from model)

Table 2. RMSE of method II

Load	0kg	10kg	20kg	30kg	40kg	50kg
RMSE/μm	0.0850	0.1495	0.1736	0.1534	0.2103	0.2418

5   Conclusion

The hysteresis in engineering practices always has dynamic effects, such as the rate-dependent and stress-dependent hysteresis effects in smart actuators. The fuzzy tree method is insensitive to the dimension of inputs, and it is suitable to describe high dimensional and complex nonlinear systems.

In this paper, taking giant magnetostrictive actuators as the study object, we introduced two modeling methods to describe dynamic hysteresis of smart actuators based on the fuzzy tree. The first method is modeling the rate-dependent hysteresis directly with fuzzy tree. Another method is proposing a stress-dependent Preisach model by building the relations between compress stress and the density functions of the Preisch model with fuzzy tree. Simulation results show that both methods can satisfactorily describe rate-dependent or stress dependent hysteresis.

It should be noted that the methods introduced in this paper can also be used to model other smart actuators, such as piezoelectric actuators and shape memory actuators, etc.

Acknowledgments. This work is supported by National Natural Science Foundation of China (91016006, 91116002) and the Fundamental Research Funds for the Central High Education Institutions (30420111109, 30420120305), State Key Laboratory of Alternate Electrical Power System with Renewable Energy Sources (LAPS13019).

References

1. Tan, X.B., Baras, J.S.: Modeling and Control of Hysteresis in Magnetostrictive Actuator. Automatica **40**, 1469–1480 (2004)
2. Bergqvist, A., Engdahl, G.: A Stress-dependent Magnetic Preisach Hysteresis Model. IEEE Trans. Magn. **27**(6), 4796–4798 (1991)
3. Dong, R., Tan, Y., Chen, H., Xie, Y.: A Neural Networks Based Model for Rate-dependent Hysteresis for Piezoceramic Actuators. Sensor Actuat. A-Phys. **143**(3), 370–376 (2008)
4. Deng, L., Tan, Y.: Diagonal Recurrent Neural Network with Modified Backlash Operators for Modeling of Rate-dependent Hysteresis in Piezoelectric Actuators. Sensor Actuat. A- Phys. **148**(1), 259–270 (2008)
5. Lei, W., Mao, J.Q., Ma, Y.H.: A New Modeling Method for Nonlinear Rate-dependent Hysteresis System Based on LS-SVM. In: 10th IEEE International Conference on Control, Automation, Robotics and Vision, pp. 1442–1446. IEEE Press, Hanoi (2008)
6. Mao, J.Q., Zhang, J.G., Yue, Y.F., Ding, H.S.: Adaptive Tree-Structured -based Fuzzy Inference Systems. IEEE Trans. Fuzzy Syst. **13**(1), 1–12 (2005)
7. Mao, J.Q., Ding, H.S.: Intelligent Modeling and Control for Nonlinear Systems with Rate-dependent Hysteresis. Sci. China Inform. Sci. **52**(4), 656–673 (2009)
8. Ma, Y.H., Mao, J.Q.: On Modeling and Tracking Control for a Smart Structure with Stress-dependent Hysteresis Nonlinearity. Acta. Automatica Sinica. **36**(11), 1611–1619 (2010)

Neurodynamics-Based Model Predictive Control for Trajectory Tracking of Autonomous Underwater Vehicles

Xinzhe Wang$^{(\boxtimes)}$ and Jun Wang

School of Control Science and Engineering,
Dalian University of Technology, Dalian 116023, Liaoning, China
wxzagm@dlut.edu.cn, jwang@mae.cuhk.edu.hk

Abstract. This paper presents a model predictive control (MPC) method based on a recurrent neural network for control of autonomous underwater vehicles (AUVs) in a vertical plane. Both kinematic and dynamic models are considered in the trajectory tracking control of the AUV. A one-layer recurrent neural network called the simplified dual neural network is applied for real-time optimization to compute optimal control variables. Simulation results are discussed to demonstrate the effectiveness and characteristics of the proposed model predictive control method.

Keywords: Model predictive control · Autonomous underwater vehicles · Simplified dual neural network

1   Introduction

Autonomous underwater vehicles (AUVs) have attracted much attention in recent years. There has been a considerable interest over the last few years for marine vehicle motion control; e.g., set-point control, trajectory tracking, and path-following control. The trajectory tracking control refers to the problem of steering a vehicle to follow a given route. The trajectory tracking control of underwater vehicles is one of the most important parts in AUV control [1].

For AUV trajectory tracking control, A recurrent neuro-fuzzy system is used to model the inverse dynamics of the AUV and then utilized as a feedforward controller to compute the nominal torque in [2], and a PD feedback controller is constructed as a feedback controller to compute the error torque to minimize the system error along the desired trajectory. In [3], an adaptive switching control method is proposed for position trajectory-tracking and path-following control of AUV. In this method, motion parameters are estimated and backstepping control algorithm is used, then switching algorithm is designed for tracking control

This research is supported by the project (61273307) of the National Nature Science Foundation of China and the Fundamental Research Funds for the Central Universities (DUT12RC(3)97).

© Springer International Publishing Switzerland 2014
Z. Zeng et al. (Eds.): ISNN 2014, LNCS 8866, pp. 184–191, 2014.
DOI: 10.1007/978-3-319-12436-0_21

and following control. In [4], an observer is designed for velocity estimation, then the tracking control is implement based on the estimated velocity. In [5], model prediction control method and genetic algorithm are combined for controller designing. Line of sight tracking of AUV is realized by using the proposed controller. In [6], A vision system is implemented to obtain the deviation between AUV and the tracked target. Trajectory tracking control is divided into path planning, attitude control and position control that are implemented based on the vision information.

Model predictive control (MPC) is an optimization-based advanced control method and entails extensive online computation of real-time solutions to formulated optimization problems [7]. For large-scale and realtime optimization problems, recurrent neural networks emerged as promising computational models for real-time optimization problems. For example, in [8], a one layer general projection neural network is presented for solving convex optimization problems. In [9], another one-layer neural network was presented for pseudoconvex optimization problems. These recurrent neural network models are shown to perform well in terms of convergence property and model complexity. Some studies on MPC based on recurrent neural networks were carried out. In [10], the simplified dual network is applied for solving real-time quadratic optimizations in various MPC approaches. In [11], a two-layer recurrent neural network is applied for solving reformulated minimax optimization problems of robust MPC approaches. These neurodynamics-based MPC approaches are developed to improve the computational efficiency and control performance substantially.

2   Problem Formulation

In this section, the kinematic and vertical dynamic models of the Taipan-2 AUV are presented, and the formulation of driving the vehicle in the vertical plane to track a trajectory is stated. The mathematical model of an AUV in six DOF can be described as follows:

$$
\begin{aligned}
\dot{\eta} &= J\left(\eta\right)\nu \\
M\dot{v} + C\left(v\right)v + D\left(v\right)v + g\left(\eta\right) + \tau_d &= \tau \\
y &= \eta
\end{aligned}
\tag{1}
$$

where $\eta = [x\ y\ z\ \phi\ \theta\ \varphi]^T$ denotes the vehicle location and orientation in the earth-fixed frame, $\nu = [u\ v\ w\ p\ q\ r]^T$ is the vehicle's velocity and angular rate vector expressed in the body-fixed frame, y is the output of the system, $J\left(\eta\right)$ is the kinematic transformation matrix expressing the transformation from the body-fixed frame to earth-fixed frame. In this paper, we consider the AUV kinematic model in the vertical plane only which can be expressed as follows:

$$
\begin{aligned}
\dot{x} &= u\cos\theta \\
\dot{z} &= -u\sin\theta \\
\dot{\theta} &= q
\end{aligned}
\tag{2}
$$

In this research, for the control design purpose, we simplify the full model by neglecting the stable roll motion. Then the simplified vertical plane dynamic model can be written as:

$$\dot{u} = \frac{F_u - d_u}{m_u}$$
$$\dot{q} = \frac{\Gamma_q - m_{pr}pr - d_q}{m_q} \tag{3}$$

where F_u is the force along the x axis, Γ_q is the torque acting on the pitch angle θ; $m_u = m - X_{\dot{u}}, m_q = I_{yy} - M_{\dot{q}}, m_{pr} = -I_{zz}, d_u = -X_{uu}u\,|u| + m(qw - vr + z_g(pr)), d_q = -M_{qq}q\,|q| - M_{uq}uq - M_{uw}uw + (z_g mg - z_b bg)\sin\theta + mz_g(wq - vr)$; X, Z, and M represent the dynamic derivative coefficients of the vertical plane dynamics of Taipan-2; the terms m, b and I are the mass, buoyancy, and moments of inertia of the vehicle, respectively; z_g and z_b are the location of the center of gravity and the center of buoyancy along the z_B axis with respect to the axis of propulsion. All the coefficients involved here are listed in Table 1.

Table 1. Hydrodynamic dimensional coefficients of the Taipan AUV

$X_{uu} = -4.00kg\ m^{-1}$	$X_{\dot{u}} = -5.070kg$
$Z_{uq} = -37.327kg\ rad^{-1}$	$Z_{ww} = -350.00kg\ m^{-1}$
$Z_{uu\delta_s} = -4.4913kg\ m^{-1}\ rad^{-1}$	$Z_{uw} = -40.750kg\ m^{-1}$
$Z_{uu\delta_b} = 4.4913kg\ m^{-1}\ rad^{-1}$	$Z_{\dot{w}} = -50.700kg$
$M_{uw} = 10.280kg$	$M_{uq} = -34.192kgmrad^{-1}$
$M_{qq} = -200.00kg\ m^2\ rad^{-2}$	$M_{\dot{q}} = -18.020kg\ m^2\ rad^{-1}$
$M_{uu\delta_s} = -16.874kg\ rad^{-1}$	$M_{uu\delta_b} = -8.4729kg\ rad^{-1}$
$z_g = 0.01757m$	$I_{yy} = 10.900kg\ m^2$
$z_b = 0.00316m$	$m = 50.7kg$
$g = 9.81m\ s^{-1}$	$b = 50.9kg$

For the Taipan-2 AUV, considering both its kinematics and dynamics in the vertical plane, we define the state vector $\mathbf{x} = [x\ z\ \theta\ u\ q]^T$ and the input vector $\mathbf{u} = [F_u\ \Gamma_q]^T$.

By using Euler discretization, the AUV model can be transformed into a discrete-time model in the following form:

$$x(k+1) = f(x(k)) + g(x(k))u(k)$$
$$y(k) = Cx(k) \tag{4}$$

where $x(k) \in \Re^n$ is the state vector; $u(k) \in \Re^m$ is the input vector; $y(k) \in \Re^p$ is the output vector; $f(\cdot)$ and $g(\cdot)$ are nonlinear functions, and $C \in \Re^{p \times n}$. The system is subject to the constraints:

$$u_{\min} \leq u(k) \leq u_{\max},$$
$$\Delta u_{\min} \leq \Delta u(k) \leq \Delta u_{\max},$$
$$x_{\min} \leq x(k) \leq x_{\max}, \tag{5}$$
$$y_{\min} \leq y(k) \leq y_{\max},$$

MPC is an iterative optimization technique: at each sampling time k, measure or estimate the current state, then obtain the optimal input vector by solving a real-time optimization problem. For model (4), the following cost function is commonly used in MPC for calculation:

$$J(k) = \sum_{j=1}^{N} \| r(k+j|k) - y(k+j|k) \|_Q^2$$
$$+ \sum_{j=0}^{N_u-1} \| \Delta u(k+j|k) \|_R^2 \qquad (6)$$

where $r(k+j|k)$ denotes the reference vector for output, $y(k+j|k)$ denotes the predicted output vector, and $\Delta u(k+j|k)$ denotes the input increment vector, $\Delta u(k+j|k) = u(k+j|k) - u(k-1+j|k)$, N and N_u are prediction horizon and control horizon ($N > N_u > 0$), respectively. Q and R are appropriate weighting matrices, $\|\cdot\|$ denotes the Euclidean norm of the corresponding vector. The first term in (6) represents the error between the predicted output and the reference output while the second term considers the control energy. Hence with appropriate N, N_u, Q and R, the cost function (6) can also guarantee closed-loop stability. According to model (4), future state $x(k+j|k)$, $j = 1, 2, ..., N$ at sampling instant k can be predicted by using the optimal input obtained at previous time instant, i.e., $u(k+j|k-1), j = 1, 2, ..., N_u$:

Define the following vectors:

$$\bar{x}(k) = \left[x(k+1|k) \ldots x(k+N|k) \right]^T \in \Re^{Nn}$$
$$\bar{u}(k) = \left[u(k|k) \ldots u(k+N_u-1|k) \right]^T \in \Re^{N_u m}$$
$$\bar{y}(k) = \left[y[k+1|k] \ldots y(k+N|k) \right]^T \in \Re^{Np} \qquad (7)$$
$$\bar{r}(k) = \left[r(k+1|k) \ldots r(k+N|k) \right]^T \in \Re^{Nn}$$
$$\Delta\bar{u}(k) = \left[\Delta u(k|k) \ldots \Delta u(k+N_u-1|k) \right]^T \in \Re^{N_u m}$$

Then the predicted output $\bar{y}(k)$ can be expressed in the following form:

$$\bar{y}(k) = \tilde{C}\left(G\Delta\bar{u}(k) + \tilde{f} + \tilde{g} \right) \qquad (8)$$

hence, the original optimization problem (6) becomes:

$$\min \left\| \bar{r}(k) - \tilde{C}\left(G\Delta\bar{u}(k) + \tilde{f} + \tilde{g} \right) \right\|_Q^2 + \| \Delta\bar{u}(k) \|_R^2$$
$$s.t. \ \Delta\bar{u}_{\min} \leq \Delta\bar{u}(k) \leq \Delta\bar{u}_{\max},$$
$$\bar{u}_{\min} \leq \bar{u}(k-1) + \tilde{I}\Delta\bar{u}(k) \leq \bar{u}_{\max}, \qquad (9)$$
$$\bar{x}_{\min} \leq \tilde{f} + \tilde{g} + G\Delta\bar{u}(k) \leq \bar{x}_{\max},$$
$$\bar{y}_{\min} \leq \tilde{C}\left(\tilde{f} + \tilde{g} + G\Delta\bar{u}(k) \right) \leq \bar{y}_{\max},$$

Problem (6) can be rewritten as a time-varying quadratic programming (QP) problem:

$$\min \tfrac{1}{2}\Delta\bar{u}^T W \Delta\bar{u} + c^T \Delta\bar{u},$$
$$s.t. \ \ l \leq E\Delta\bar{u} \leq h. \tag{10}$$

The solution to the QP problem (10) gives optimal control increment vector $\Delta\bar{u}(k)$ whose first element $\Delta u(k)$ can be used to calculate the optimal control input

3   Neurodynamic Optimization

In recent years, various neural network models have been developed as goal-seeking solvers for QP problems. The essence of neural computation lies in its parallel and distributed information processing. In particular, the simplified dual network showed superior performances in MPC applications. This neural network model is applied for solving (10), whose dynamic equations can be described as:

State equation

$$\varepsilon \frac{dz}{dt} = -Ev + h(Ev - z).$$

Output equation

$$v = W^{-1}(E^T z - c). \tag{11}$$

where z is the state vector, v is the output vector and h is an activation function defined as

$$h(x_i) = \begin{cases} l_i, \ x_i < l_i; \\ x_i, \ l_i \leq x_i \leq h_i; \\ h_i, \ x_i > h_i. \end{cases} \tag{12}$$

The simplified dual network has a single-layer structure with totally $3N_u m + 2Np$ neurons. The MPC scheme for formation control of multi-robot systems based on the simplified dual network is summarized as follows:

1. Let k=1. Set control time terminal T, prediction horizon N, Control horizon N_u, sample period t, weight matrices Q and R.
2. Calculate process model matrices G, $\tilde{f}$,$\tilde{g}$, $\tilde{C}$ and neural network matrices W, c, E.
3. Solve the convex quadratic minimization problem (10) by using the simplified dual neural network to obtain the optimal control action Δu_k.
4. Calculate the optimal input vector $\bar{u}(k)$ and implement the first element $u(k\,|k)$.
5. If $k < T$, set $k = k$+1, go to step 2;otherwise end.

4   Simulation Resultss

In this section, simulation results are discussed to demonstrate the effectiveness of the proposed MPC scheme for the AUV control based on both its kinematics and dynamics.The AUV is supposed to move in the vertical plane to track a sine trajectory with the orientation of tangential direction. $x = t, z = -sin(t)$, $\theta = atan(cos(t))$ The initial inputs are $(F_u, \Gamma_q) = (5, 1)$ and the initial position and orientation are $(x\ z\ \theta\ u\ q) = (0.1, -0.05, \pi/4, 0, 0)$, the output matrix is $C = I$. Both prediction horizon N and the control horizon N_u are 10, $Q = 10I$, $R = I$ and the sample time $Ts = 0.01s$. For the discrete-time model (4), we have the following $g(x)$ and $f(x)$:

$$g(x) = Ts \begin{bmatrix} 0 & 0 \\ 0 & 0 \\ 0 & 0 \\ 0.01793 & 0 \\ 0 & 0.3458 \end{bmatrix} \in \Re^{5 \times 2}$$

$$f(x) = \begin{bmatrix} x(k) \\ z(k) \\ \theta(k) \\ u(k) \\ q(k) \end{bmatrix}^T + Ts \begin{bmatrix} u(k)\cos\theta(k) \\ -u(k)\sin\theta(k) \\ q(k) \\ u_1 u(k)|u(k)| \\ q(k)(q_1|q(k)| + q_2 u(k)) + q_3 \sin\theta(k) \end{bmatrix}^T$$

Coefficients involved in $f(x)$ are $u_1 = -0.0717$, $q_1 = -6.9156$, $q_2 = -1.1823$, $q_3 = -0.2476$. Figs. 1-3 illustrated the practical positions and the target positions of x axis, z axis and angle around the y axis. Figs. 4-5 shown the input incremental and the control inputs. The results show that with a proper input, the AUV can reach tracking a given nonlinear route with a satisfactory precision.

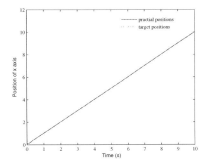

Fig. 1. Practical and target x

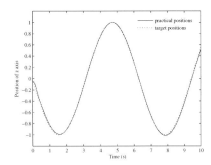

Fig. 2. Practical and target along z

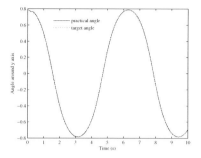

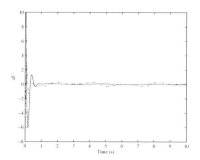

Fig. 3. Practical and target angle around y axis

Fig. 4. Incremental of input action

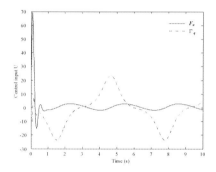

Fig. 5. Input action

5   Conclusions

This paper presents an MPC approach to steering a class of autonomous underwater vehicles in the vertical plane to track a given route. Based on an AUV model in the vertical plane, the MPC problem is formulated as a time-varying quadratic optimization problem which can be repeatedly solved by using a single-layer globally convergent recurrent neural network called the simplified dual neural network. Simulation results show that the proposed method is able to control the AUV in the vertical plane with a good performance. The three-dimensional control of AUVs deserves further investigations.

References

1. Herman, P.: Decoupled PD set-point controller for underwater vehicles. Ocean Engineering **36**(6), 529–534 (2009)
2. Wang, J., Lee, C.S.: Self-adaptive recurrent neuro-fuzzy control of an autonomous underwater vehicle. IEEE Transactions on Robotics and Automation **36**(2), 283–295 (2003)

3. Aguiar, A., Hespanha, J.: Trajectory-tracking and path-following of underactuated autonomous vehicles with parametric modeling uncertainty. IEEE Transactions on Automatic Control **52**, 1362–1379 (2007)
4. Antonelli, G.: On the use of adaptive/integral actions for six-degrees-of-freedom control of autonomous underwater vehicles. IEEE Journal of Oceanic Engineering **32**, 300–312 (2007)
5. Naeem, W., Sutton, R., Chudleum, J., Dalgleish, F.R., Tettow, S.: An online genetic algorithm based model predictive control autopilot design with experimental verification. International Journal of Control **78**, 1076–1090 (2005)
6. Tang, S.H., Li, S., Wu, Q.X., Li, Y.P., Zhang, Q.F.: A dynamic positioning method based on vision for under water vehicle. Ocean Engineering **24**(2), 112–122 (2005)
7. Qin, S.J., Badgwell, T.A.: A survey of industrial model predictive control technology. Control Engineering Practice **11**(7), 734–764 (2003)
8. Xia, Y., Wang, J.: A general projection neural network for solving optimization and related problems. IEEE Transactions on Neural Networks **3**, 2334–2339 (2003)
9. Pan, Y., Wang, J.: Model predictive control for nonlinear affine systems based on the simplified dual neural network. In: Proc. of IEEE International Symposium on Intelligent Control, pp. 683–688. IEEE Press, Saint Petersburg (2009)
10. Pan, Y., Wang, J.: A neurodynamic optimization approach to nonlinear model predictive control. In: Proc. of the 2010 IEEE International Conference on Systems, Man, and Cybernetics, pp. 1597–1602. IEEE Press, Istanbul (2010)
11. Shan, Y., Yan, Z., Wang, J.: Model predictive control of underwater gliders based on a one-layer recurrent neural network. In: Proc. of 6th International Conference on Advanced Computational Intelligence, pp. 238–333. IEEE Press, Hangzhou (2013)

Feedback-Dependence of Correlated Firing in Globally Coupled Networks

Jinli Xie[1(✉)], Zhijie Wang[2], and Jianyu Zhao[1]

[1] School of Electrical Engineering, University of Jinan, Jinan 250022, Shandong, China
{cse_xiejl,cse_zjy}@ujn.edu.cn
[2] College of Information Science and Technology, Donghua University,
Shanghai 201620, China
wangzj@dhu.edu.cn

Abstract. It is known that, in feed-forward nets, the degree of neural correlation generally increases with firing rate. Here, we study the correlations of neurons that are part of a homogeneous global feedback network, under the influence of partially correlated external input. By using numerical simulations of a network of noisy leaky integrate-and-fire neurons with delayed and smoothed spike-driven feedback, we obtain a non-monotonic relationship between the correlation coefficient and the strength of inhibitory feedback connections. This non-monotonic relationship can be explained by the interplay between the mean rate and the regularity of firing activity caused by the inhibitory feedback connections. We also show that this non-monotonic relationship is robust in both sub-threshold and supra-threshold dynamic regimes, for low and moderate internal noise levels, as well as when the network is heterogeneous. Our results point to a potent functional role for feedback as a modulator of correlated activity in neural networks.

Keywords: Feedback · Correlation · Oscillation · Heterogeneity

1 Introduction

The properties of the correlations of spiking activity of neurons are critical for understanding sensory and cortical processing [1-3]. The mechanisms underlying correlations and the influence of neural connectivity on these correlations continues to attract the attention of theorists and computational neuroscientists [1,4-7]. It has been observed, both in experimental systems and in computational models of varying complexity, that the correlations within a population of neurons are influenced not only by the magnitude of correlations of the external stimuli, but also by the direct interactions and dynamic states within the population [4,8-10].

Inhibition is beneficial for controlling the timing and probability of action potential generation, as well as for generating high frequency activity in cortex [8]. We expect global inhibitory feedback to couple the activity of cells, thereby increasing the correlations. On the other hand, this feedback reduces the firing rate, which would likely

© Springer International Publishing Switzerland 2014
Z. Zeng et al. (Eds.): ISNN 2014, LNCS 8866, pp. 192–199, 2014.
DOI: 10.1007/978-3-319-12436-0_22

lead to the opposite effect [1]. Interestingly, our computational study of networks of stochastic leaky integrate-and-fire (LIF) neurons reveals that the correlation coefficient is a non-monotonic function of the strength of the feedback gain: the correlation coefficient drops with weak feedback gain, and rises after the strength of feedback gain exceeds a threshold. We discuss this non-monotonic relationship for networks in different dynamic regimes.

To understand the increase of the correlation coefficient when the feedback gain is sufficiently strong, we study the effect of gain on oscillations, which in turn affects correlations. We calculate the power spectral densities of spike trains as a function of feedback gain and analyze peaks in the gamma range. Gamma oscillations have been shown to enhance correlated activity of neurons [8,11]. Therefore, we quantify gamma oscillations by the degree of these spectral peaks [12,13] and evaluate the influence of inhibitory feedback on this coherence.

Since parameter heterogeneity often reduces correlations [14,15], we further investigate the robustness of the non-monotonic relationship in the presence of heterogeneity. Together our findings point to the importance of the regime of firing, feedback strength, oscillation characteristics, and network heterogeneity to understand the degree of pairwise correlations.

2   Model and Numerical Methods

In this paper, we investigate firing correlations in a network with inhibitory feedback. The network has two interacting layers. The N excitatory LIF neurons, which receive external stimuli, provide excitatory input to an inhibitory LIF neuron population, which feeds its output back to all the excitatory neurons via a global delayed inhibitory feedback loop.

Each excitatory neuron receives an input $I_i(t)$ from a sensory neuron composed of the following components [4]:

$$I_i(t) = \mu_E + \eta_i(t) + \sigma\left[\sqrt{1-c}\xi_i(t) + \sqrt{c}\xi_c(t)\right] + I_g \tag{1}$$

where μ_E denotes the base current and $\eta_i(t)$ is an internal Gaussian white noise of zero-mean and intensity D_E. The next term stands for the external noise, and consists of two Gaussian low-pass filtered (0-150Hz, eight-order Butterworth filter) noise processes of unit variance: $\xi_i(t)$ is a noise specific to each neuron and $\xi_c(t)$ is shared by all neurons. These two noise processes are scaled by the input correlation coefficient c to determine the degree of shared input of all the excitatory neurons. We set $c = 0.6$ throughout our study.

The last term in Eq. 1 represents the inhibitory feedback generated by the inhibitory neuron, which is calculated by the convolution of a delayed α function and the spike train y_I of the inhibitory neuron:

$$I_g(t) = G \int_{\tau_D}^{\infty} \alpha(\tau) y_I(t - \tau) d\tau \tag{2}$$

Here τ_D is half the transmission delay around the feedback loop. We use the spike train cross-correlogram (CCG) [16,17] to compute the pairwise spike correlations:

$$CCG_{ij}(\tau) = \frac{\sum_{k=1}^{M} \sum_{t=0}^{L} y_i^k(t) y_j^k(t + \tau)}{M(L - |\tau|)\sqrt{\lambda_i \lambda_j}} \tag{3}$$

where M is the number of trials or realizations, L is the duration of every trial, and λ_i and λ_j are the firing rates of neurons i and j, respectively. The term $L - |\tau|$ is used to correct for the degree of overlap. The auto-correlograms (ACG), are calculated similarly as the CCG, but by letting $i = j$. The pairwise spike correlation (C_{ij}) of the two neurons i and j is estimated by the ratio of the area of the CCG within a certain range of lags defining a window T to the geometric mean area of the ACG over the same window. All CCGs and ACGs are corrected by subtracting the shift predictor SPT. When T is large enough, C_{ij} saturates to a steady value, which is defined here as the pairwise correlation coefficient of neurons i and j. Finally, the correlation coefficient Cor of the network reported in our results below is obtained by averaging the pairwise correlation coefficients over all pairs of excitatory neurons. The parameter values chosen in our work are: $N = 100$, $\tau_D = 4ms$, $\sigma = 0.2$, $T = 100$.

3　Results

We now explore the relationship between the correlation coefficient of the network and the inhibitory feedback gain using the numerically generated spike trains of excitatory neurons in the first layer. First, a sub-threshold base current $\mu_E = 0.9$ with low internal noise $D_E = 0.112$ is chosen. In this case, spikes are solely induced by noise. As shown in Fig. 1 (top), Cor first drops to a lower level for small values of the feedback G, but later rises after G exceeds some threshold value. Afterwards, a stable and relatively high level of pairwise correlation is maintained with further increases in G. The non-monotonic curve thus reveals a minimum correlation coefficient of the network for a moderate value of the feedback strength.

We then consider the supra-threshold regime, where μ_E is raised to $\mu_E = 1.2$. The curve of Cor vs G in Fig. 1 (bottom) remains non-monotonic as in the sub-threshold regime. Moreover, since correlation is proportional to firing rate in feedforward networks [1], over the whole range of the values of feedback gain selected in the simulations, the correlated activity measured by Cor in the supra-threshold regime is significantly higher than in sub-threshold regime.

To further explain the non-monotonic relationship between the correlation coefficient of the network and the strength of inhibitory feedback, we calculate the spike train power spectrum and the firing rate of a single excitatory neuron in the network for different values of G. The spike train power spectrum of neuron i is determined by $S(\omega)=\langle \tilde{y}_i \tilde{y}_i^* \rangle$ where $\tilde{y}_i$ is the Fourier transform of the spike train, and $\tilde{y}_i^*$ denotes the complex conjugate of $\tilde{y}_i$.

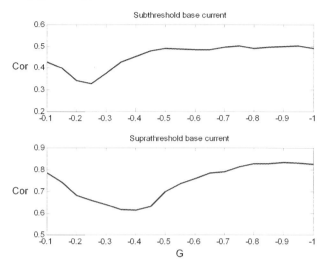

Fig. 1. The correlation coefficient versus the inhibitory feedback gain

The intensity of the power spectrum can be quantified by the degree of the coherence [12,13]:

$$\beta = \frac{h_p}{\Delta\omega} \cdot \omega_p \qquad (4)$$

where h_p and $\Delta\omega$ are, respectively, the height and width-at-half-maximum of the averaged spectrum peak at frequency ω_p.

We first consider the case where the base current is sub-threshold and the internal noise intensity is small. The correlation coefficient of the network, the coherence of the spike trains, and the mean firing rate R are plotted as a function of feedback gain in Fig. 2. As previously discussed, the curve of the correlation coefficient of the network is non-monotonic in the sub-threshold regime (Fig. 2 top). Because of the inhibitory feedback loop, the firing rate of the neuron decreases as G increases (Fig. 2 middle). Further, the coherence increases monotonically with G (Fig. 2 bottom). And the variation of the coherence with G, i.e. the slope of the plot, goes through a maximum as a function of G. Here $\mu_E = 0.92$, $D_E = 0.11$.

The following story emerges from inspection of Fig. 2. When G is weak, β remains very small, and the network has no apparent gamma oscillations. The drop of Cor at small G is mainly due to the effect of the sharp decline of the firing rate [1]. However, an increase in G leads to growth in β, but only a small change in R. The relatively flat values of R imply that the effect of firing rate on Cor is almost unchanged over this range. Therefore, Cor is associated closely with the increase of β.

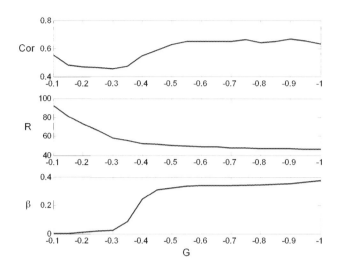

Fig. 2. Correlation coefficient, mean firing rate and coherence vs inhibitory feedback gain

An induced gamma oscillation modulates the activity of the excitatory neurons, leading to an improvement of the correlation coefficient of the network. In fact, the change in β in Fig. 2 (bottom) parallels very closely the increase in Cor at roughly the same values of G in Fig. 2 (top).

The correlation coefficient and coherence as a function of G are shown for larger internal noise in Fig. 3 (Left: sub-threshold regime, $\mu_E = 0.9$, $D_E = 0.246$; Right: supra-threshold regime, $\mu_E = 1.2$, $D_E = 0.246$). The non-monotonic relationship between Cor and G is also observed when the internal noise is high, in either regime. Further, we find a qualitatively different dependence of the coherence on the gain. It exhibits a minimum at a moderate feedback gain, in both regimes. Hence, in this higher noise regime, the initial decrease of the spectral coherence represents reduced periodic activity, and the associated changes in correlations are linked to both coherence and firing rate drops.

As shown by previous studies, heterogeneity can desynchronize networks [15], because it can translate into variability in intrinsic action potential frequency, which weakens correlations in the network. Here we study the effect of the heterogeneity on the relationship between Cor and G. We achieve this goal by distributing either G or the firing threshold for each cell according to Gaussian statistics. Fig. 4 shows the results of computer simulations for the correlation coefficient of the network with delayed inhibitory feedback, when either G (circles, solid black line) or v_T (squares, solid gray line) is Gaussian-distributed. Fig. 4 also shows the projection onto the Cor -vs- G and R - G planes, with the same symbols as in the three-dimensional curves. In Fig. 5, the heterogeneous cases are compared to the homogenous case, the curves correspond to the two-dimensional projected curves obtained from Fig. 4. Solid black lines show Cor vs. G (top) and R vs. G (bottom) when the network is homogeneous. Owing to the disordering effect of the heterogeneity of the network on the correlations, both *Cor* and *R* in the heterogeneous cases are weaker than in the homogeneous case. However, in spite of this moderate amount of heterogeneity, the non-monotonic relationship between the correlation coefficient of the network and the strength of the inhibitory feedback gain is robust as in the homogeneous case. Here $\mu_E = 0.9$, $v_T = 1$, $D_E = 0.112$, $\sigma_E = 0.03$.

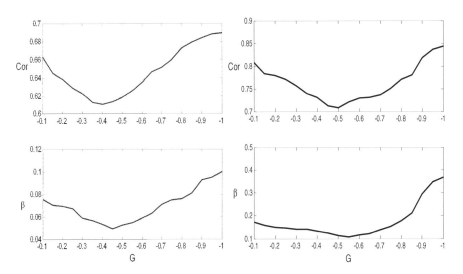

Fig. 3. Correlation coefficient and spectral coherence of spike trains as a function of inhibitory feedback gain in the sub-threshold regime and the supra-threshold regime

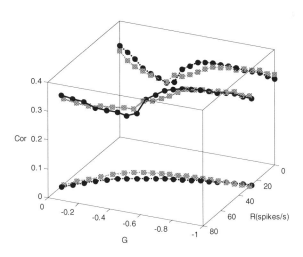

Fig. 4. Relationship between correlation coefficient, inhibitory feedback gain and mean firing rate in a heterogeneous network

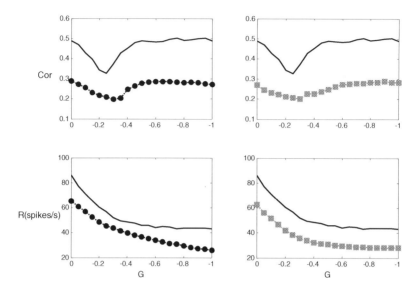

Fig. 5. The heterogeneous networks are compared to the homogeneous ones

4   Discussion

Correlations are at the center of many controversies about the coding and processing of neural information. In the present paper, we have shown that the relationship between the network-averaged pairwise correlation coefficient of spiking activity and the strength of the global inhibition in that network is non-monotonic. We find that, as the magnitude of the inhibitory feedback increases, the firing rate first decreases, causing a drop in correlation. However, beyond a certain magnitude, the feedback coupling increases the correlation. In particular, the inhibitory feedback begins to elicit asynchronous network oscillations. This spectral coherence counteracts the drop in firing rate, leading to an increase in correlation. This is true in both the sub-threshold and supra-threshold regimes, i.e. when each cell can (supra) or can't (sub) fire periodically in the absence of feed-forward and feedback inputs (i.e. when it has sufficient bias or not). The non-monotonic relationship also holds at low or high internal noise in either regime. The structure of our model also enables us to predict that this will be the case if the external noise intensity changes. The effects presented here are also likely to occur in non-sensory systems characterized by closed neural loops functioning as oscillators, such as the basal ganglia.

The non-monotonic relationship is not only robust over a range of noise intensities and biases, but also when the network is heterogeneous. Gaussian-distributed firing thresholds or feedback gains yield moderately smaller correlation values, but the correlation-vs-gain non-monotonicity is preserved. Apart from heterogeneity, different forms of plasticity in the feedback pathway would, to lowest order, have an effect similar to distributed gain values, and thus we may expect that plasticity will also reduce correlation – this will be explored in future work.

Acknowledgements. This work was supported by the National Natural Science Foundation of China under Grant No. 61203375, and the Doctoral Foundation of University of Jinan under Grant No. XBS1240.

References

1. de La Rocha, J., Doiron, B., Shea-Brown, E., Josic, K., Reyes, A.: Correlation between Neural Spike Trains Increases with Firing Rate. Nature **448**, 802–806 (2007)
2. Gutnisky, D.A., Dragoi, V.: Adaptive Coding of Visual Information in Neural Populations. Nature **452**, 220–224 (2008)
3. Ecker, A.S., Berens, P., Keliris, G.A., Bethge, M., Logothetis, N.K., Tolias, A.S.: Decorrelated Neuronal Firing in Cortical Microcircuits. Science **327**, 584–587 (2010)
4. Lindner, B., Doiron, B., Longtin, A.: Theory of Oscillatory Firing Induced by Spatially Correlated Noise and Delayed Inhibitory Feedback. Phys. Rev. E **72**, 061919 (2005)
5. Renart, A., de la Rocha, J., Bartho, P., Hollender, L., Parga, N., Reyes, A., Harris, K.D.: The Asynchronous State in Cortical Circuits. Science **327**, 587–590 (2010)
6. Tchumatchenko, T., Geisel, T., Volgushev, M., Wolf, F.: Signatures of Synchrony in Pairwise Count Correlations. Front. Comput. Neurosci. **4**, 1 (2010a)
7. Tchumatchenko, T., Malyshev, A., Geisel, T., Volgushev, M., Wolf, F.: Correlations and Synchrony in Threshold Neuron Models. Phys. Rev. Lett. **104**, 058102 (2010b)
8. Galan, R.F., Fourcaud-Trocme, N., Ermentrout, G.B., Urban, N.N.: Correlation-Induced Synchronization of Oscillations in Olfactory Bulb Neurons. J. Neurosci. **26**, 3646–3655 (2006)
9. Marinazzo, D., Kappen, H.J., Gielen, S.C.A.M.: Input-Driven Oscillations in Networks with Excitatory and Inhibitory Neurons with Dynamic Synapses. Neural Comput. **19**, 1739–1765 (2007)
10. Ostojic, S., Brunel, N., Hakim, V.: How Connectivity Background Activity and Synaptic Properties Shape the Cross Correlation between Spike Trains. J. Neurophysiol. **29**, 10234–10253 (2009)
11. Masuda, N., Doiron, B.: Gamma Oscillations of Spiking Neural Populations Enhance Signal Discrimination. PLoS Comput. Biol. **3**, e236 (2007)
12. Pakdaman, K., Tanabe, S., Shimokawa, T.: Coherence Resonance and Discharge Time Reliability in Neurons and Neuronal Models. Neural Networks **14**, 895–905 (2001)
13. Lindner, B., Schimansky-Geier, L., Longtin, A.: Maximizing Spike Train Coherence or Incoherence in the Leaky Integrate-and-Fire Model. Phys. Rev. E **66**, 031916 (2002)
14. Borgers, C., Epstein, S., Kopell, N.J.: Background Gamma Rhythmicity and Attention in Cortical Local Circuits: A Computational Study. Proc. Nat. Acad. Sci. USA **102**, 7002–7007 (2005)
15. Bartos, M., Vida, I., Jonas, P.: Synaptic Mechanisms of Synchronized Gamma Oscillations in Inhibitory Interneurons Networks. Nat. Rev. Neurosci. **8**, 45–56 (2007)
16. Chacron, M.J., Bastian, J.: Population Coding by Electrosensory Neurons. J. Neurophysiol. **99**, 1825–1835 (2008)
17. Kohn, A., Smith, M.A.: Stimulus Dependence of Neuronal Correlation in Primary Visual Cortex of the Macaque. J. Neurosci. **25**, 3661–3673 (2005)

Wilcoxon-Norm-Based Robust Extreme Learning Machine

Xiao-Liang Xie[✉], Gui-Bin Bian, Zeng-Guang Hou, Zhen-Qiu Feng,
and Jian-Long Hao

State Key Laboratory of Management and Control for Complex Systems,
Institute of Automation, Chinese Academy of Sciences, Beijing, China
{xiaoliang.xie,guibin.bian,zengguang.hou,
zhenqiu.feng,jianlong.hao}@ia.ac.cn

Abstract. It is known in statistics that the linear estimators using the rank-based Wilcoxon approach in linear regression problems are usually insensitive to outliers. Outliers are the data points that differ greatly from the pattern set by the bulk of the data. Inspired by this, Hsieh *et al* introduced the Wilcoxon approach into the area of machine learning. They investigated four new learning machines, such as Wilcoxon neural network (WNN) etc., and developed four descent gradient based backpropagation algorithms to train these learning machines. The performances of these machines are better than the ordinary nonrobust neural networks. However, it is hard to balance the learning speed and the stability of these algorithms which is inherently the drawback of gradient descent based algorithms. In this paper, a new algorithm is used to train the output weights of single-layer feedforward neural networks (SLFN) with its input weights and biases being randomly chosen. This algorithm is called Wilcoxon-norm based robust extreme learning machine or WRELM for short.

Keywords: Extreme learning machine · Wilcoxon neural network · Wilcoxon-norm based robust extreme learning machine

1 Introduction

It is said that the modern age of neural network began with the work of McCulloch and Pitts in 1943. Since then, some popular and powerful artificial neural networks (ANN) have been proposed, such as self organizing map (SOM) [1], radial basis function neural networks (RBF) [2], and support vector machines (SVMs) [3] etc. Several learning algorithms have been proposed in the literature for training the aforementioned learning machines [1]-[5]. Among these machines, one simple structure is multilayer perceptron artificial neural networks (MLP). Some offline algorithms have been introduced to learn the weights and biases of MLP. One well known gradient descent based batch learning algorithm is the back propagation [4]. In order to improve the convergence speed of BP algorithm, several improvements were made in [5,6]. It has been approved that a

© Springer International Publishing Switzerland 2014
Z. Zeng et al. (Eds.): ISNN 2014, LNCS 8866, pp. 200–209, 2014.
DOI: 10.1007/978-3-319-12436-0_23

single-hidden layer feedforward neural network (SLFN) with additive hidden nodes and with a nonpolynomial activation function can approximate any function in a compact set [7]. Huang *et al* further rigorously proved that SLFNs with randomly assigned input weights and hidden neurons' biases and with almost any nonzero activation functions can universally approximate any continuous function on any compact input sets [8]. Based on this concept, the extreme learning machine (ELM) algorithm was proposed for batch learning [8,9]. Later, Liang *et al* developed an online sequential learning machine algorithm for SLFNs with additive or RBF hidden nodes in a unified framework which is referred to as online sequential extreme learning machine (OS-ELM) [10]. Both the ELM algorithm and OS-ELM algorithm are based on the principle of least square error minimization, so the performances of these algorithms are easily affected by outliers. In other words, these algorithms are not robust. Inspired by different mechanisms, two robust algorithms were proposed, namely least trimmed squares (LTS) [21] and rank-based Wilcoxon neural networks [20]. In this paper, performances of both robust and nonrobust algorithms will be compared with the algorithm we proposed.

This paper is organized as follows. Section 2 reviews the Wilcoxon neural network proposed by Hsieh and discusses some related problems. Section 3 illustrates the basic background of ELM and the introduced WRELM. The experimental results are conducted in Section 4. Finally, some conclusions are included in Section 5.

2   Wilcoxon SLFN

2.1   Wilcoxon Norm

The Wilcoxon norm of a vector will be used as the objective function or dispersion function for Wilcoxon learning machines. In order to define the Wilcoxon norm of a vector, a score function is introduced. A score function is a nondecreasing function $\phi : [0,1] \to \Re^1$ which satisfies $\int_0^1 \phi(u)du = 0$ and $\int_0^1 \phi^2(u)du = 1$.

The score $a_\phi(\cdot)$ associated with the score function ϕ is defined by

$$a_\phi(i) = \phi\left(\frac{i}{N+1}\right), \quad i = 1, 2, \cdots, N$$

where N is a fixed positive integer. Hence $a_\phi(1) \leqslant a_\phi(2) \leqslant \ldots \leqslant a_\phi(N)$. It can be shown that the following function is a pseudonorm (seminorm) on $\Re^N$:

$$\|e\|_W = \sum_{i=1}^{N} a(R(e_i))e_i = \sum_{i=1}^{N} a(i)e_{(i)} \tag{1}$$

where $e = [e_1, \ldots, e_N]^T \in \Re^N$, $R(e_i)$ denotes the rank of e_i among $e_1, \ldots, e_N$, $e_{(1)} \leq \ldots \leq e_{(N)}$ are the ordered values of $e_1, \ldots, e_N$, $a(i) = \phi[i/(N+1)]$, and $\phi(u) = \sqrt{12}(u - 0.5)$. We call $\|e\|_W$ defined in (1) the Wilcoxon norm of the vector e.

2.2   Wilcoxon Neural Network

Consider the single-hidden layer Wilcoxon neural network with $n+1$ nodes in its input layer, m nodes in its hidden layer, and p nodes in its output layer.

Let the input vector be $x = [x_1, x_2, \ldots, x_n, 1]^T \in \Re^{n+1}$, and let v_{ij} denote the connection from the ith input node to the jth hidden node. The input u_j and output r_j of the jth hidden node are respectively given by

$$u_j = \sum_{i=1}^{n+1} v_{ji} x_i, \quad r_j = f(u_j), \quad \text{for } j = 1, 2, \cdots, m$$

where f is the activation function of hidden nodes.

Let w_{kj} denote the connection weight from the output of the jth hidden node to the kth output node. Then, the output of kth output node t_k and final output y_k are respectively given by

$$t_k = \sum_{j=1}^{m} w_{kj} r_j, \quad y_k = t_k + b_k, \quad \text{for } k = 1, 2, \cdots, p$$

where b_k is the bias of the kth output node.

Suppose we are given the training data set, $\{(\mathbf{x}_i, d_i)\}_1^N$ with $\mathbf{x}_i \in \Re^{n+1}$ and $d_i \in \Re^p$, where N is the number of training data, $\mathbf{x_i} = [x_{1i}, \ldots, x_{ni}, 1]^T$ is the ith input vector, and d_i is the desired output for the input $\mathbf{x_i}$. In the WNN, the approach is to choose network weights ($\mathbf{v}$ and $\mathbf{w}$) that minimize the Wilcoxon norm of the total residuals

$$D(\mathbf{v}, \mathbf{w}) = \sum_{k=1}^{p} \sum_{i=1}^{N} a(R(e_{i,k})) e_{i,k} = \sum_{k=1}^{p} \sum_{i=1}^{N} a(i) e_{(i),k} \tag{2}$$

where $e_{i,k} = d_{i,k} - t_{i,k}$, $R(e_{i,k})$ denotes the rank of the residual $e_{i,k}$ among $e_{1,k}, \ldots, e_{N,k}$ and $e_{(1),k} \leqslant \cdots \leqslant e_{(N),k}$ are the ordered values of $e_{1,k}, \ldots, e_{N,k}$.

The neural network used above is the same as that used in traditional artificial neural network, except the bias terms at the output node. The main reason is that the Wilcoxon norm is a pseudonorm instead of the usual norm. $\|e\|_W = 0$ implies that $e_1 = \cdots = e_N$. So, without the bias terms, the resulting predictive function with small Wilcoxon norm of total residuals may deviate from the desired function by constant offsets. The bias term b_k is estimated by the median of the residuals at the kth output node, i.e., $b_k = \text{med}_{1 \leqslant i \leqslant N} \{d_{ki} - t_{ki}\}$.

In [20], the proposed gradient descent based algorithm can effectively train WNN, however, there are some practical issues involved in its application. Firstly, the synaptic weights are initially set to small random values, so that the nodes are not saturated. Secondly, the learning rate parameter severely affects the speed of convergence. In this paper, we use an algorithm in linear regression to train WNN which will be discussed in the following section.

3   Wilcoxon-Norm-Based Robust Extreme Learning Machine

In this section, a brief description of the ELM algorithm developed by Huang *et al* is given first. Then the WRELM algorithm is introduced.

3.1   ELM Algorithm

In supervised batch learning, the learning algorithms use a finite number of input-output samples for learning networks' parameters. For N arbitrary distinct samples $(x_i, y_i) \in \Re^n \times \Re^p$, standard SLFNs with m hidden neurons and activation function (or radial basis function) $g(x)$ are modeled as

$$\sum_{j=1}^{m} \mathbf{w}_j \, G(\mathbf{a}_j, b_j, \mathbf{x}_i) = \mathbf{y}_i, \quad \text{for } i = 1, \cdots, N \tag{3}$$

where $\mathbf{a}_j$ and b_j are the learning parameters of hidden neurons and $\mathbf{w}_j$ is the weight connecting the jth hidden node to output neurons. For additive hidden neuron with the activation function $g(x)$ (e.g., sigmoid or threshold), $G(\mathbf{a}_j, b_j, \mathbf{x})$ is given by [10]

$$G(\mathbf{a}_j, b_j, \mathbf{x}) = g(\mathbf{a}_j \cdot \mathbf{x} + b_j), \quad b_j \in \Re.$$

For RBF hidden neuron with Gaussian activation function $g(x)$, $G(a_i, b_i, x)$ is given by $G(\mathbf{a}_j, b_j, \mathbf{x}) = g\left(\frac{\|\mathbf{x} - \mathbf{a}_j\|}{2b_j^2}\right), \quad b_j \in \Re.$

Equation (3) can be written compactly as $H \cdot W = Y$

H is called the hidden layer output matrix of the network [8]. The ith column of H is the ith hidden node's output vector with respect to inputs $\mathbf{x}_1, \mathbf{x}_2, \ldots, \mathbf{x}_N$.

By minimizing the objective function $\|H \cdot W - Y\|_2^2$, we can get the estimation of output weights of hidden layer

$$W = \arg\min_{W} \|H \cdot W - Y\|_2^2 = H^+ Y \tag{4}$$

where H^+ is the P-M pseudo inverse of H.

3.2   Description of the Proposed WRELM

Like ELM algorithm, if the weights and biases of the input layer of WNN are randomly chosen, the dimension of the parameters to be learned in WNN could be greatly reduced. Based on the above principle, the WRELM algorithm is proposed.

After the input weights and the hidden layer biases are arbitrarily chosen, single-layer Wilcoxon neural network can be simply considered as a linear system

$$y_{i,k} = b_k + H_i \cdot \mathbf{w}_k + e_{i,k}, \text{ for } i = 1, \cdots, N, \ k = 1, \cdots, p$$

where H_i is the ith row of hidden layer output matrix H, and $\mathbf{w}_k \in \Re^{m \times 1}$ to be learned is the weight connecting the hidden neurons to the kth output neuron, and $e_{i,k}$ is a random variable with density f_k and distribution function F_k. In its general form, Jaeckel's rank dispersion function can be stated as

$$D_R(e_k) = \sum_{i=1}^{n} e_{i,k}\, a[R(e_{i,k})] \tag{5}$$

where $a(1) \leqslant a(2) \leqslant \ldots \leqslant a(N)$ is a set of scores generated by $a(i) = \varphi(i/(n+1))$ and $e_{i,k} = y_{i,k} - H_i \cdot \mathbf{w}_k$. One usually used score function is $\phi(u) = \sqrt{12}(u - 0.5)$. Some other forms of score functions can been found in [12,13,18]. It is easy to prove that $D_R(e)$ is an even $(D_R(e) = D_R(-e))$ and location free $(D_R(e) = D_R(e - \gamma I))$ dispersion function. Jaeckel shows that $D_R(e)$ is a nonnegative continuous, and convex function of $W = [\mathbf{w}_1, \ldots, \mathbf{w}_p]$ which attains its minimum with bounded W if X has full rank [13].

We denote the rank based estimator of $\mathbf{w}_k$ by $\tilde{\mathbf{w}}_k$ which is

$$\tilde{\mathbf{w}}_k = \arg\min_{\mathbf{w}_k} D_R\left(Y_k - H \cdot \mathbf{w}_k\right) = \arg\min_{\mathbf{w}_k} \|Y_k - H \cdot \mathbf{w}_k\|_W \tag{6}$$

where $\|\cdot\|_W$ is a pseudonorm defined in (1).

In order to minimize $D_R\left(Y_k - H \cdot \mathbf{w}_k\right)$, we need to compute its partial derivative with respect to $\mathbf{w}_k$ which exists almost everywhere [14]

$$\nabla D_R = \frac{\partial D_R}{\partial \mathbf{w}_k} = -S\left(Y_k - H \cdot \mathbf{w}_k\right) = -H^T a\left(R\left(Y_k - H \cdot \mathbf{w}_k\right)\right).$$

Thus $\tilde{\mathbf{w}}_k$ is the solution to the following R-normal equations

$$H^T a\left(R\left(Y_k - H \cdot \mathbf{w}_k\right)\right) = 0_N \tag{7}$$

Let $\mathbf{w}_{k0}$ denote the true parameters which satisfy R-normal equations and the scale factor

$$\tau_k = \left(\sqrt{12} \int_{-\infty}^{+\infty} f_k^2(x)\, dx\right)^{-1}, \quad k = 1, \cdots, p$$

where f_k is the p.d.f. of the noise e_k. If the following requirements are satisfied, the dispersion function $D_R(\cdot)$ can be approximated by a quadratic function $Q(\cdot)$ [15]

$$Q(Y_k - H \cdot \mathbf{w}_k) = \frac{1}{2\tau_k}\left(\mathbf{w}_k - \mathbf{w}_{k0}\right)^T H^T H \left(\mathbf{w}_k - \mathbf{w}_{k0}\right)$$

$$- \left(\mathbf{w}_k - \mathbf{w}_{k0}\right)^T S\left(Y_k - H \cdot \mathbf{w}_{k0}\right) + D\left(Y_k - H \cdot \mathbf{w}_{k0}\right).$$

1. The density f_k is absolutely continuous and its Fisher information $I(f_k) = \int_{-\infty}^{+\infty} \left[f_k'(x)\right]^2 / f_k(x)dx < \infty$.
2. $\lim\limits_{N \to \infty} N^{-1} X^T X = \Sigma$, where X is an $N \times m$ design matrix and Σ is a $m \times m$ positive definite matrix.

3. $\lim\limits_{N \to \infty} \max\limits_{1 \leq i \leq N} x_{iq}^2 / \sum_{j=1}^{N} x_{jq}^2 \to 0$ for all $q = 1, \cdots, m$.

The following estimate minimizes $Q(\cdot)$ [16]

$$\tilde{\mathbf{w}}_k(t+1) = \tilde{\mathbf{w}}_k(t) + \tau_k(t)(H^T H)^{-1} H^T a(R(Y_k - H \cdot \tilde{\mathbf{w}}_k(t)))$$

The scale factor $\tau_k(t)$ needs to be estimated. One estimate of $\int_{-\infty}^{+\infty} f_k^2(x)dx$ is by Schuster who first obtained a kernel type of estimate of $f_k(x)$ [17]

$$\tilde{f}_k(x) = \frac{1}{Nh} \sum_{i=1}^{N} K\left(\frac{x - e_{i,k}}{h}\right)$$

where h is the kernel bandwidth and $K(\cdot)$ is a uniform kernel function

$$K(x) = \begin{cases} 1, & x \in [-1/2, 1/2] \\ 0, & \text{otherwise} \end{cases}$$

Then $\delta_k = \int_{-\infty}^{+\infty} f_k^2(x)dx$ can be estimated by

$$\hat{\delta}_k = 1/N^2 h \sum_{i=1}^{N} \sum_{j=1}^{N} I\left(|e_{i,k} - e_{j,k}| < h/2\right).$$

A modified version of the above estimate, $\hat{\delta}_{k,c}$ is proposed in [14] to ease the computation [22]

$$\hat{\delta}_{k,c} = \frac{1}{Nc} + \frac{1}{N(N-1)h} \sum_{i=1}^{N} \sum_{j \neq i}^{N} K\left(\frac{e_{i,k} - e_{j,k}}{h}\right)$$

where c is a fixed constant. When $h = c/\sqrt{N}$, the modified $\hat{\delta}_{k,c}$ is consistent of δ_k [14].

4   Illustrative Examples

In this section, we compare the performances of five neural networks for both artificial regression problem and two other real world benchmark nonlinear regression examples. In those examples, in order to test the generalization capability of the learned machines, the machines are tested by another set of training data without noise or outliers. The five learning machines compared here are two nonrobust neural networks, namely standard ANN and ELM in [8], three robust neural networks, namely LTS in [21], original Wilcoxon neural network in [20] and WRELM. All these illustrative simulations are conducted in Matlab 6.5 running on a Pentium IV 2.0 GHz and 512 MB RAM personal computer.

For a fair comparison, each neural network's hidden layer nodes have the same activation functions, and the same number of hidden nodes.

4.1 Artificial Problem

In this simulation, the true function is given by the Hermite function [20]

$$y = 1.1 \cdot (1 - x + 2x^2) \cdot e^{-x^2/2}, \qquad x \in [-5, 5].$$

A training data set (x_i, y_i) with 100 data is generated, where x_i's are uniformly randomly distributed in the interval $[-5, 5]$. The gross error model used for modeling outliers is $D_\varepsilon = 0.85 \cdot G + 0.15 \cdot H$, where $G \backsim N(0, 1)$ and $H \backsim N(0, 0.1)$. For all the machines concerned, the number of hidden layer nodes is 20 and the activation functions of the hidden nodes are sigmoid functions.

The learning rate η used in BP training algorithm of ANN is 0.008, in LTS algorithm 0.003, and in BP algorithm of WNN 0.001. The number of training epochs for ANN is 8,000, and for LTS is 80,000, and for WNN is 8,000. The trainings of the aforementioned three algorithms are time consuming, while the WRELM is trained only 5 epochs in neglectable time.

The simulation results are shown in Fig. 1. For highly corrupted data as shown in Fig. 1(a), LTS, WNN and WRELM are robust to outliers and they are not affected by outliers. While the performances of least square based ANN and ELM as shown in Fig. 1(b) are severely affected by outliers. In this example, the performance of WRELM is almost as good as other two robust algorithms, but WRELM converges pretty faster.

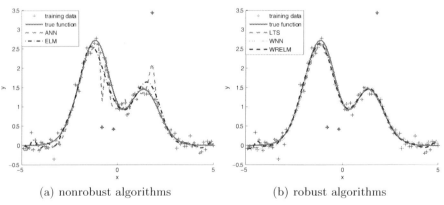

(a) nonrobust algorithms   (b) robust algorithms

Fig. 1. Simulation results of some non-robust and robust algorithms in Example 1

4.2 Real World Benchmark Regression Problems

1) Fuel Consumption Prediction of Automobiles [19]: In this example, a regression benchmark problem is studied, namely, auto-mpg. This problem is to predict city-cycle fuel consumption of different models of car by 3 multivalued discrete and 4 continuous input attributes and one continuous output attributes. The dataset contains 392 data. In our simulation, about 3/4 of the total data are randomly chosen to form the training data set and the remaining data to form the testing data set. The corrupted training data set is formed by keep the normalized input attributes unchanged but with 5% randomly chosen output attribute values replaced by random values from a uniform distribution defined on [-100,100]. The testing data set remains unchanged. For simplicity, the eight input attributes are normalized to the range [-1,1].

In this simulation, the learning rate in BP algorithm of WNN is 0.01, in BP algorithm of ANN is 0.001, and in LTS algorithm is 0.001, they are all chosen by trial and error. Fig. 2 shows that compared to ELM and ANN, the three robust algorithms LTS, WNN and WRELM achieves good generalization performance when there exists outliers. It can be further seen from Fig. 2(b) that WRELM algorithm archives least RMSE for the testing data in just a few training epochs.

2) Abalone Age Prediction [19]: This problem has 4177 cases predicting the age of abalone from physical measurements. The age of abalone is determined by cutting the shell through the cone, staining it, and counting the number of rings through a microscope which is a boring and time-consuming task. Other

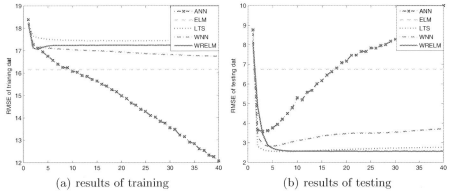

(a) results of training   (b) results of testing

Fig. 2. Performance comparison of the concerned algorithms in Example 2 (the training epochs for ANN and LTS x25, and for WNN x50)

Table 1. Performance Comparison for the Methods in Example3

Algorithms	Time(Seconds)		Training		Testing		η	Epochs	#Nodes
	Mean	Dev	Mean	Dev	Mean	Dev			
ANN	28.9195	0.4107	6.8668	0.0002	2.5747	0.0734	0.0002	600	20
ELM [8]	**0.0142**	0.0001	**6.8065**	0.0566	2.5983	0.0114	-	1	20
LTS [21]	283.9649	55.9718	7.2171	0.0037	2.2730	0.0207	0.0002	5,000	20
WNN [20]	24.9874	0.6430	7.0637	0.0699	2.0355	0.0151	0.001	500	20
WRELM	1.3656	0.0001	7.0506	0.0674	**2.0028**	0.0149	-	5	20
ANN	36.8601	1.0196	6.4117	0.0020	2.4861	0.1030	0.0002	600	30
ELM [8]	**0.0237**	0.0001	**6.3159**	0.0001	2.4635	0.0010	-	1	30
LTS [21]	356.6765	145.3539	6.6625	0.0023	2.1135	0.0025	0.0002	5,000	30
WNN [20]	32.2266	1.1770	6.5274	0.0002	1.9847	0.0002	0.001	500	30
WRELM	1.3780	0.0001	6.5198	0.0002	**1.9727**	0.0004	-	5	30

8 measurements, which are easier to obtain, are used to predict the age. The 8 measurements are sex, length, diameter, height, whole weight, viscera weight, shell weight and rings. For simplicity, the eight input attributes are normalized to the range [-1,1]. In this regression problem, about 75% of the total data are randomly chosen to form the training data set with 10% of the total training data are corrupted by keep the input attributes unchanged but output values are replaced by random values from a uniform distribution defined on [0,50].

Table 1 summarizes the results for this benchmark regression problem in terms of training time, training RMSE and testing RMSE for each network with different number of nodes. We run each of the concerned five algorithms 10 times. The learning rate of BP algorithm of ANN, LTS and WNN are chosen by trial and error in consideration of converge speed and stability. The input weights and biases of hidden layer nodes of ELM and WRELM are of the same at each simulation, and they are chosen randomly in range [-1,1]. From Table 1, we can see that although the training time of ELM is neglectable and the RMSE of training uncorrupted data set of ELM is the smallest among the five algorithms, its RMSE of testing data is pretty large. In other words, the ELM network is not robust to outliers. WRELM algorithm has fastest convergence speed with smallest RMSE of testing data among the other four algorithms.

5   Conclusion

In this paper, a robust ELM-like learning machine was proposed, which called Wilcoxon-norm based robust extreme learning machine or WRELM for short. Like ELM algorithm, after the input weights and the hidden layer biases are chosen randomly, single-layer WNN can be simply considered as a linear system, so the output weights can be tuned by well studied robust linear regression methods. Based on this principle, the new robust algorithm called WRELM was introduced. Performance of WRELM was compared with ANN, ELM, LTS, and WNN on both artificial regression problem and some real world benchmark regression problems. The results indicate that WRELM algorithm, like WNN algorithm and LTS algorithm, is robust to outliers, but with no additional vital parameters, such as learning rate in gradient based algorithms, to been decided. The WRELM algorithm can converge fast and is stable with good generalization capability.

Acknowledgments. This research is supported in part by the National Natural Science Foundation of China (Grant #61203318, #61203342, and #61333016) and the National Hi-Tech R&D Program (863) of China (Grant #2013AA013803).

References

1. Kohonen, T.: Self-organizing formation of topologically correct feature maps. Biological Cybernetics **43** (1982)
2. Powell, M.J.D.: Radial basis functions for multivariable interpolation: a review. In: Mason, J.C., Cox, M.G. (eds.) Algorithms for Approximation, pp. 143–167. Clarendon Press, Oxford (1987)

3. Cortes, C., Vapnik, V.: Support vector networks. Machine Learning **20**, 273–297 (1995)
4. Werbos, P.: Beyond regression: new tools for prediction and analysis in the behavioral sciences. Ph.D. dissertation, Harvard Univ., Cambridge, MA (1974)
5. Gibb, J.: Back propagation family album. Technical Report C/TR96-05, Macquarie University (August 1996)
6. Riedmiller, M., Braun, H.: A direct adaptive method for faster backpropagation learning: the RPROP algorithm. In: Proc. of the IEEE Int. Conf. on Neural Netw., San Francisco, CA (April 1993)
7. Leshno, M., Lin, V.Y., Pinkus, A., Schocken, S.: Multilayer feedfor-ward networks with a nonpolynomial activation function can approximate any function. Neural Netw. **6**, 861–867 (1993)
8. Huang, G.B.: Extreme learning machine: a new learning scheme of feedforward neural networks. In: Proc. Int. Joint Conf. Neural Netw. (IJCNN 2004), Budapest, Hungary, July 25–29, vol. 2, pp. 985–990 (2004)
9. Huang, G.B., Zhou, H.M., Ding, X.J., Zhang, R.: Extreme learning machine for regression and multiclass classification. IEEE Transactions on Systems, Man, and Cybernetics, Part B: Cybernetics **42**(2), 513–529 (2012)
10. Liang, N.Y., Huang, G.B., Saratchandran, P., Sundararajan, N.: A fast and accurate online sequential learning algorithm for feedforward network. IEEE Trans. on Neural Netw., **17**(6) (November 2006)
11. Hawkins, D.M.: Identification of Outliers. Chapman & Hall, London (1980)
12. Jureckova, J.: Asymptotic linearity of a rank statistic in regression parameter. Ann. Math. Statist. **40**, 1889–1900 (1969)
13. Jaeckel, L.A.: Estimating regression coefficients by minimizing the dispersion of the residuals. Ann. Math. Statist. **43**, 1449–1458 (1972)
14. Hettmansperger, T.P.: Statistical inference based on ranks. Wiley, New York (1984)
15. Hettmansperger, T.P., McKean, J.W.: Robust non-parametric statistics. Wiley, New York (1998)
16. Hettmansperger, T.P., McKean, J.W.: Robust nonparametric statistical methods. Arnold, London (1998)
17. Scuster, E.: On the rate of convergence of an estimate of a functional of a probability density. Scandinavian Actuarial Journal **1**, 103–107 (1974)
18. Choi, Y.H., Ozturk, O.: A new class of score generating functions for regression models. Statistics & Probability Letters **57**, 205–214 (2002)
19. Asuncion, A., Newman, D.J.: UCI Machine Learning Repository (2007). http://www.ics.uci.edu/~mlearn/MLRepository.html
20. Hsieh, J.G., Lin, Y.L., Jeng, J.H.: Preliminary study on Wilcoxon learning machines. IEEE Trans. on Neural Netw. **19**(2), 201–211 (2008)
21. Rusiecki, A.: Robust LTS backpropagation learning algorithm. In: Sandoval, F., Prieto, A.G., Cabestany, J., Graña, M. (eds.) IWANN 2007. LNCS, vol. 4507, pp. 102–109. Springer, Heidelberg (2007)
22. Qing, C.Y., Annpey, P., Biao, X.: Rank regression in stability analysis. Journal of Biophamaceutical Statistics **13**, 463–479 (2003)

Modeling

An Artificial Synaptic Plasticity Mechanism for Classical Conditioning with Neural Networks

Caroline Rizzi Raymundo$^{(\boxtimes)}$ and Colin Graeme Johnson

School of Computing, University of Kent, Canterbury, Kent CT2 7NF, UK
{cr519,C.G.Johnson}@kent.ac.uk

Abstract. We present an *artificial synaptic plasticity* (ASP) mechanism that allows artificial systems to make associations between environmental stimuli and learn new skills at runtime. ASP builds on the classical neural network for simulating associative learning, which is induced through a conditioning-like procedure. Experiments in a simulated mobile robot demonstrate that ASP has successfully generated conditioned responses. The robot has learned during environmental exploration to use sensors added after training, improving its object-avoidance capabilities.

Keywords: Synaptic plasticity · Classical conditioning · Artificial neural networks

1 Introduction

Natural environments change often, which makes adaptation an essential survival skill for most organisms. Like animals, robotic systems may also find themselves in ever-changing environments. The need for more effective artificial intelligence, capable to overcome environmental changes, has led researchers to find inspiration in nature's solutions for adaptation, such as animals' reflexive behaviors [2,5], and the human brain [1] and its hormonal mechanisms [7,9].

In nature, the ability to learn new behaviors by means of associations between external stimuli is known to be essential for adaptation in a variety of animals, including humans [4]. However, to date, most works on adaptive systems seem to ignore this fact, seeking to adapt only the system's native behaviors and overlooking the need of a system that autonomously learn new ones [1,2,5,7,9].

In this paper, we propose a mechanism for allowing artificial systems to autonomously learn new skills based on environmental feedback and on its pre-existing skills. Our approach consists in an *artificial synaptic plasticity* (ASP) mechanism that builds on the classical artificial neural network (ANN) [3] for simulating associative learning. The system learns at runtime, through a procedure analogous to *classical conditioning* [8], to associate different environmental stimuli and use newly available information to solve problems in ways it was not trained for. We have evaluated ASP in a simulated mobile robot, which has successfully expressed conditioned responses.

© Springer International Publishing Switzerland 2014
Z. Zeng et al. (Eds.): ISNN 2014, LNCS 8866, pp. 213–221, 2014.
DOI: 10.1007/978-3-319-12436-0_24

This paper is organized as follows: Section 2 introduces ASP's biological inspiration, followed by its implementation in Section 3. We present experimental results in Section 4 and conclude in Section 5.

2   Biological Background

2.1   Classical Conditioning

Classical conditioning, first documented by Pavlov [8], is an important form of learning that involves the association of a behavioral response with an event that normally does not trigger that response. In his most famous experiment, Pavlov conditioned a dog to salivate on the ringing of a bell, after repeatedly ringing the bell whenever he presented food to the dog.

Pavlov argued that some reflexes are "hard-wired" and, therefore, do not need to be learned. For example, dogs do not need to learn to salivate when they smell food. This kind of reflex, which is native and automatic, is called *unconditioned response* (UR) and is triggered by an *unconditioned stimulus* (US). In Pavlov's dog example, the smell of food is an US that triggers salivation as an UR.

By contrast, the ringing of a bell is considered a *neutral stimulus* (NS), because it naturally produces no salivation in dogs. After pairing the bell sound with food smell, association occurs and it becomes a *conditioned stimulus* (CS), being able to trigger salivation by itself as a *conditioned response* (CR). Unlike URs, a CR can be extinguished after learned, for example, the dog will diminish its salivation response to the bell if food is repeatedly presented on the absence of the bell sound and vice-versa.

2.2   Synaptic Plasticity

At neural level, associative learning happens when a neuron is simultaneously excited by a strong and a weak electrical stimulus. This process gives rise to a phenomenon known as *long-term potentiation* (LTP), which strengthens the communication between two neurons [4]. LTP takes place at the *synapse*, which is the structure that connects two neurons and allows neural communication. However, LTP is not the only process that affects synapses' efficiency. *Long-term depression* (LTD) is a process similar to LTP, but instead of strengthening, it weakens synapses capability to transmit signals between neurons.

Synapses' ability to change their strength in signal transmission according to neural activity level, called *synaptic plasticity* or *Hebbian plasticity*, is known to play an important role in classical conditioning [4]. As a simplified example of this relation, a weak electrical stimulus could come from a CS, such as the ringing of a bell for Pavlov's dog, whereas a strong electrical stimulation could come from an US, such as the smell of food for Pavlov's dog. The target neuron (i.e., the one receiving these stimulations), in turn, could be a neuron that meaningfully contributes for triggering the dog's salivation response.

The pairing of both weak (from the bell) and strong (from the food) electrical stimulus generates LTP, which makes the target neuron more responsive to the weak stimulus. In the future, the weak stimulus will be able to activate the target neuron by itself, allowing the bell's sound to trigger the salivation response. If CS and US are repeatedly presented in the absence of each other, LTD occurs, leading the dog to no longer respond to the bell sound.

3   Artificial Synaptic Plasticity

We propose to simulate the neural mechanism of classical conditioning in the classical feedforward ANN [3]. As discussed in Section 2, the neural mechanism of classical conditioning consists in strengthening the signaling efficiency of synapses, which are represented as weights in the classical ANN. We argue that it is possible to generate an *artificial synaptic plasticity* (ASP) for artificial systems by gradually changing the ANN's weights according to the activity coincidence of its inputs.

In feedforward ANNs, the input value of a neuron i, known as *net input*, is given by Equation 1, where x_j is the output of neuron j, w_{ij} is the weight that connects neurons j and i, and b_i is the *bias* of neuron i.

$$net_i = b_i + \sum_j x_j w_{ij} \,. \tag{1}$$

The association process induced by ASP takes place after the ANN's training phase, during the system's operational cycle. Each input of the ANN is considered an external stimulus, which may be an *artificial conditioned stimulus* (ACS) or an *artificial unconditioned stimulus* (AUS). Therefore, the ANN's inputs are divided into two groups: the AUS group, depicted by the vector $\boldsymbol{u}$, of size p; and the ACS group, depicted by the vector $\boldsymbol{c}$, of size q. Together, these two stimuli vectors compose the input vector $\boldsymbol{x}$ of the ANN, with size $p + q$. A particular ACS cannot be an AUS at the same time and vice-versa. Therefore

$$\boldsymbol{x} = [u_1, u_2, ..., u_p, c_1, c_2, ..., c_q]^T \,;$$
$$net_i = b_i + \sum_{k=1}^{p} u_k w_{ik} + \sum_{j=1}^{q} c_j w_{ij} \,.$$

Consider, for now on, that the variables $k \in \{1, 2, ..., p\}$ and $j \in \{1, 2, ..., q\}$ are reserved for indexing AUS and ACS elements, respectively. Also, for the following explanation, we assume that the ANN's inputs are normalized in the range [0,1] and that the higher the value of an input, the higher its influence for generating the behavior of interest.

ASP's methodology consists in gradually changing the first-layer weights of ACSs, so that they become able to activate neurons of the second layer by themselves with the same pattern that AUSs would. Thus, each weight w_{ij} connecting ACS inputs to the second layer should be updated by a delta Δw_{ij}, so that $w_{ij}(t+1) = w_{ij}(t) + \Delta w_{ij}(t)$, where $w_{ij}(t)$ is the value of w_{ij} at time t. Note that this updating rule excludes weights w_{ik} that are related to AUS inputs.

The value of Δw_{ij} should consider not only the amount by which a given pattern is associated, determined by LTP, but also the amount by which the same pattern is extinguished/dissociated, determined by LTD. The variables Δa (association amount) and Δd (dissociation amount) control the level of association, reinforcing it (if $\Delta a > \Delta d$) or diminishing it (if $\Delta d > \Delta a$), so that

$$\Delta w_{ij} = \alpha_j(\Delta a_{ij} - \Delta d_{ij}),\tag{2}$$

where $\alpha_j \in [0,1]$ is the constant that determines the rate at which the ANN associates or dissociates stimulus c_j, and is called as the *association rate* (AR) of stimulus c_j. Hence, $\alpha_j = 0$ means that no association will occur, and the closer α_j is to 1 the faster the system associates c_j.

The value of w_{ij} cannot be increased/decreased indefinitely, because the ANN's outcome could be very different from the outcome produced by AUSs, diverging from the concept of classical conditioning. To avoid that, w_{ij} must be kept in a range $[w'_{ij}, w''_{ij}]$, where w'_{ij} is the initial value of w_{ij} and w''_{ij} is the desired conditioned value of w_{ij}. Hence, the closer w_{ij} is from a w''_{ij}, the closer it is from a complete association. Analogously, the closer w_{ij} is from w'_{ij}, the closer it is from a complete dissociation. Therefore

$$\Delta a_{ij} = (w''_{ij} - w_{ij}) \times assoc_j,\tag{3}$$

$$\Delta d_{ij} = (w_{ij} - w'_{ij}) \times dissoc_j,\tag{4}$$

where the variables $assoc_j$ (association factor) and $dissoc_j$ (dissociation factor), both in the interval $[0,1]$, dictate the degree of synaptic-activity coincidence between c_j and $\boldsymbol{u}$. We will return to these variables later on.

According to our definition, the value of w''_{ij} should allow c_j to activate neuron i of the second layer with the same pattern that vector $\boldsymbol{u}$ would, which implies in Equation 5 (remember that $\{x \in \boldsymbol{x} \mid 0 < x < 1\}$). The constant $s_{jk} \in [0,1]$ is the *sensitivity* of stimulus c_j to stimulus u_k. The matrix that maps the sensitivity between vectors $\boldsymbol{c}$ and $\boldsymbol{u}$ should be calibrated according to the designer's judgment, depending on the purpose of the associative learning and the architecture of the ANN. If correctly calibrated, the SM (sensitivity matrix) can prevent the system from learning "superstitions", i.e., patterns that are no more than random coincidences.

$$w''_{ij} = \sum_k s_{jk}w_{ik}.\tag{5}$$

In biological synaptic plasticity, association between a pair of CS and US occur when their values are simultaneously high, and it is analogous for ASP. Therefore, the higher the values of an ACS c_j and an AUS u_k, the higher the association between both (i.e., the higher the association factor $assoc_j$). However, c_j may be associated with more than one AUS, at different sensitivity values. Thereafter, it is more correct to state that $assoc_j$ is proportional to the average signal strength of $\boldsymbol{u}$ weighted by the respective sensitivities. This implies that

$$assoc_j = c_j \sum_k s'_{jk}u_k,\tag{6}$$

where $s'_{jk} = s_{jk}/\sum_k s_{jk}$. Analogously, the dissociation (i.e., the extinction of an association) of an ACS c_j with an AUS u_k should occur when these stimuli are no longer paired. Therefore, the smaller the value of c_j and the higher the mean of $\boldsymbol{u}$ weighted by the respective sensitivities, the higher the dissociation between both (i.e., the higher the dissociation factor $dissoc_j$):

$$dissoc_j = (1 - c_j) \sum_k s'_{jk} u_k \,. \tag{7}$$

Also, according to classical conditioning, the higher the ACS and the smaller the AUS, the higher the dissociation; however, we have omitted it in the first version of ASP and considered it as future work. By replacing Equations 3, 4, 6 and 7 in Equation 2 and simplifying, we find

$$\Delta w_{ij} = \alpha_j \left[c_j(w''_{ij} - w_{ij}) - (w_{ij} - w'_{ij})(1 - c_j) \right] \sum_k s'_{jk} u_k \,, \tag{8}$$

where w''_{ij} is given by Equation 5. If a particular pair of associated stimuli, say c_j and u_k, have high input values at the same time, a net-input extrapolation may occur. This is because after being associated with u_k, c_j is able to mimic the effect of u_k in the ANN. Therefore, if both inputs are high, the neural network will receive a total input twice as high as it would if association had not occurred. So, the net input value must be restricted according to Equation 9, where $\underline{v}$ and $\overline{v}$ are, respectively, the minimum and maximum values that v can assume.

$$b_i + \sum_j \underline{x_j w'_{ij}} \; < \; net_i \; < \; b_i + \sum_j \overline{x_j w'_{ij}} \,. \tag{9}$$

4   Experimental Evaluation

In this section, we evaluate ASP in a multi-stimulus association case. A robot is equipped with distance and touch sensors, but is trained (by means of the *backpropagation algorithm* [3]) to recognize and avoid obstacles using only touch sensors. Environmental exploration provides a natural conditioning, since it is probable that at least one distance sensor will measure high proximity to obstacles whenever a collision occur. By means of ASP, the robot is expected to gradually associate collision with proximity at runtime and eventually start to use information from the distance sensors to avoid obstacles before colliding.

4.1   Experimental Setup

We have used the robot simulator Webots [6] to simulate the Pioneer 2 robot[1], which is equipped with 16 distance sensors plus 5 custom touch-sensors. Figure 1a shows the architecture of the ANN that controls the robot's movements, whose inputs come from the sensors depicted in Figure 1b. The first five inputs,

---

[1] http://www.mobilerobots.com

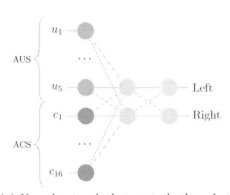

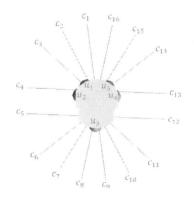

(a) Neural network that controls the robot. Inputs are divided into unconditioned (vector u) and conditioned (vector c) stimuli. Outputs provide the speeds for the left and right wheels in radians per second.

(b) Aerial view of robot's sensors disposition. Red lines represent distance-sensor rays and blue spheres represent the contact area of touch sensors.

Fig. 1. Robot controller setup

representing AUS, come from the touch sensors and assume only binary values, where one means that a collision has been detected and zero means the opposite. The last 16 inputs, representing ACS, come from the distance sensors and assume integer values from 0 to 1024, where the higher the input value the closer the robot is to an obstacle. The robot's maximum detection range is 0.5 meters.

The robot was placed in a $4m^2$ box with a narrow and curved corridor leading to two dead ends, where it was initialized in three different positions and evaluated for five ARs: 0, 0.001, 0.01, 0.1 and 1. AR zero represents the execution of the pure ANN implementation, i.e., when there is no associative learning at all. In order to investigate performance variation deriving from noise error (simulated by Webots), we have executed each setup combination (3 initial positions and 5 ARs) 30 times, each for 5 minutes.

The SM has been configured to prevent the robot from associating random coincidences. For example, if the robot occurs to be near the left wall while touching the wall at its front, it may associate its left distance sensors with its frontal touch sensors, which is a mistake. Therefore, for this particular experiment, the SM should map the disposition of the robot's sensors, so that distance sensors are associated with the nearest touch sensor. Table 1 depicts the SM used in this experiment. Some distance sensors, such as c_3, are close to two touch sensors and, thus, have their sensitivity divided between them. By contrast, c_6 and c_{11} are relatively far from all touch sensors, so they have no sensitivity mapping.

Table 1. Sensitivity matrix (zeroed cells were omitted)

Touch	Distance Sensors															
Sensors	c_1	c_2	c_3	c_4	c_5	c_6	c_7	c_8	c_9	c_{10}	c_{11}	c_{12}	c_{13}	c_{14}	c_{15}	c_{16}
u_1	0.2	0.4	0.4	-	-	-	-	-	-	-	-	-	-	-	-	-
u_2	-	-	0.4	0.4	0.2	-	-	-	-	-	-	-	-	-	-	-
u_3	-	-	-	-	-	-	0.1	0.4	0.4	0.1	-	-	-	-	-	-
u_4	-	-	-	-	-	-	-	-	-	-	-	0.2	0.4	0.4	-	-
u_5	-	-	-	-	-	-	-	-	-	-	-	-	-	0.4	0.4	0.2

4.2   Results

Figure 2 counts all detected collisions during a complete run for each initial position of the robot. The number of collisions when ASP is used with AR 0.01 is about 70% smaller than when it is not used. The difference is even greater for AR 1.0, when the number of collisions is about 96% smaller than without ASP.

Outcomes for AR 0.001, however, are not as good as the results observed for the other ARs. This is because, when under AR 0.001, the robot got trapped in a corner from where it could not easily escape due to the disposition of its touch sensors. The trapping was persistent in runs starting from position 3 (causing the high collision count) and more occasional for runs starting from positions 1 and 2 (causing the high standard deviations). Despite trapping, results for the other ARs (0.01, 0.1 and 1.0) present a consistent decrease in number of collisions, with low standard deviation. Also, the lines are very close to each other, suggesting that outcomes are the same regardless the robot's start position.

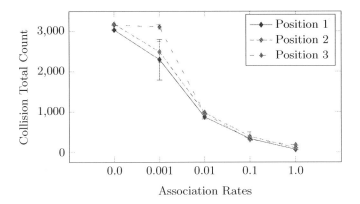

Fig. 2. Count of all detected collisions during an complete run

The robot's behavioral changes were beyond our expectations. Because of locomotion difficulty in the dead ends of the narrow corridor, the robot had to move more "carefully" in order to make a turn without touching the walls. As consequence, the more we increased the AR, the more time the robot spent making the turn in the dead ends, and the less it collided. This "cautiousness" is a positive collateral-effect that was neither deliberately designed, nor predicted.

The explored area is another unexpected and positive collateral-effect. When running without ASP, the robot's vision-range was limited to touch, so it could not perceive alternative (and perhaps better) paths to avoid an obstacle. As consequence, the robot kept doing laps in a small space. By contrast, when using ASP, the increased vision-range of the distance sensors improved the robot's space-awareness, which doubled the robot's explored area.

In addition to this experiment, we have also performed experiments for evaluating ASP regarding the dissociation process, i.e., when the robot forgets the association learned. Videos of all experiments are available on-line[2].

5   Conclusion and Future Work

We have presented an artificial mechanism of synaptic plasticity for generating associative learning in artificial systems through a conditioning-like process. Experiments with ASP successfully generated a conditioned response, allowing the robot to learn a new skill at runtime: use its distance sensors to avoid obstacles before bumping into them. This improved the robot's locomotion efficiency, allowing it to explore a bigger area and preventing physical damage.

For future work, we plan to use ASP for triggering and memorizing artificial emotions in computer systems. Neuroscientific findings indicate that emotions are essential for intelligent behavior and fast decision making in humans [4]. As for animals, emotions may be also valuable for artificial intelligence [5,7,9]. We hypothesize that ASP could be used for generating emotive conditioned responses, so that the system could adapt its behavior according to its past "emotional experiences" in relation to a particular place, entity or object.

In addition, we plan to test ASP with a real mobile robot in a physical environment in order to provide more realistic evaluation on its efficacy.

Acknowledgments. The first author is financially supported by CAPES, a Brazilian research-support agency (process number 0648/13-2).

References

1. Alnajjar, F., Murase, K.: A Simple Aplysia-Like Spiking Neural Network to Generate Adaptive Behavior in Autonomous Robots. Adaptive Behavior **16**(5), 306–324 (2008)
2. Arkin, R.C., Ali, K., Weitzenfeld, A., Cervantes-Pérez, F.: Behavioral models of the praying mantis as a basis for robotic behavior. Robotics and Autonomous Systems **32**(1), 39–60 (2000)
3. Callan, R.: The Essence of Neural Networks. The Essence of Computing Series. Prentice Hall Europe, London (1999)

---

[2] http://carolrizzi.github.io/ASP

4. Kolb, B., Whishaw, I.Q.: An Introduction to Brain and Behavior, 2nd edn. Worth Publishers, New York (2004)
5. Likhachev, M., Arkin, R.C.: Robotic Comfort Zones. In: Proceedings of SPIE on Sensor Fusion and Decentralized Control in Robotic Systems III, Boston, vol. 4196, pp. 27–41 (2000)
6. Michel, O.: Webots: Professional Mobile Robot Simulation. International Journal of Advanced Robotic Systems **1**(1), 39–42 (2004)
7. Neal, M., Timmis, J.: Timidity: A Useful Emotional Mechanism for Robot Control? Informatica **27**(2), 197–204 (2003)
8. Pavlov, I.P.: Conditioned Reflexes: An Investigation of the Physiological Activity of the Cerebral Cortex. Oxford University Press (1927)
9. Timmis, J., Neal, M., Thorniley, J.: An Adaptive Neuro-Endocrine System for Robotic Systems. In: IEEE Workshop on Robotic Intelligence in Informationally Structured Space, pp. 129–136. IEEE Press, Nashville (2009)

An Application of Dynamic Regression Model and Residual Auto-Regressive Model in Time Series

Yu-zhen Lu, Ming-hui Qu[✉], and Min Zhang

Dalian Maritime University, Dalian, China
shanyanqu@sina.com

Abstract. This paper, by using the dynamic regression model (ARIMAX) models and predicts the tourist date from 1979 to 2004 in Zhejiang Province, and makes stationary test and white noise test of the residual date generated by the above analysis. The innovation point of this paper is that it is suitable to establish dynamic regression by cointegration test and proves the data of the residual data validation is stationary. Further testing and analysis of residual data, finds that the residual data can establish auto-regression model. This method has made full use of data information. Thus the paper presents that the prediction effect of the combination of the dynamic regression model and the residual auto regressive model is superior to that of the prediction model of the ARMA model. This combination model has better adaptability, greatly improves the predicted effect of the model.

Keywords: The dynamic regression model (ARIMAX) · Cointegration test · Residual autoregressive model · Time series analysis

1 Introduction

The so-called time series analysis is a list of ordered date which is recorded in accordance with the time.This method makes prediction and tries to control the future development of things or system by observing, studying and finding the developing law of the time series.This is how the time series analysis works. Time series analysis has been widely used in daily life and production, such as weather forecast, yield estimation, stock movements, etc.

For a set of data $\{X_t\}$,ARIMA is the preferred model [1]. The model was put forward by Box and Jenkins in1976, which was developed from unary time series to binary. However, this model is a stable multi-series model, which shows strict requirements for the stability of each component in the stationary time series. In this model, both the input time series and studied time series should be steady, which, however is very difficult to achieve in many cases.

The concept Cointegration wasput forward by Engle and Granger in 1987 [2]. The condition of stationary was historically relaxed. It requires only a certain linear combination of data setsbe stationary, in other words,that is to insure their auto-regression

© Springer International Publishing Switzerland 2014
Z. Zeng et al. (Eds.): ISNN 2014, LNCS 8866, pp. 222–231, 2014.
DOI: 10.1007/978-3-319-12436-0_25

residual data be stationary.Cointegration theory has greatly promoted the development of multivariate time series. The later Johansen cointegration test, proposed by Johansen and Juselius [3], is a method testing thecointegration relationship under the VAR (vector auto-regression) system.

In 2005,Yinyin Wu, using the method of Gibbs mined abnormal points in the ARIMAX model [4].Yuanzheng Wang andYajing Xu made the ARIMAX model a skilled use in multiple stationary time seriesin 2007 [5]. In 2008, Hui Zeng and others applied the residual auto-regressive model to solve the forecasting analysis and prediction of the non-stationary time series of economic [6]. In this paper, by the combining of the dynamic regression model and the residual auto-regressive model, we update the ARMA model and improve its prediction accuracy.

2   Dynamic Regression Model and Residual Auto-Regressive Model

2.1   Dynamic Regression Model

Assuming that the output variable (the dependent variable sequence) $\{Y_t\}$ and the input variable sequences (independent variable sequence) $\{X_{1t}\}, \{X_{2t}\}, \cdots, \{X_{kt}\}$ are all stationary, the establishing of regression model of the output and input sequences is the first thing to do:

$$Y_t = \mu + \sum_{i=1}^{k} \frac{\Theta_i(B)}{\Phi_i(B)} B^{l_i} X_{it} + \varepsilon_t \tag{1}$$

In formulary (1), $\Phi_i(B)$ is the i-th polynomial in regression coefficients of input variables; $\Theta_i(B)$ is the i-th polynomial in moving mean coefficients of input variables; l_i is the i-th hysteretic order of input variables; $\{\varepsilon_t\}$ is regression residual sequence. Sequences $\{Y_t\}$ and sequences $\{X_{1t}\}, \{X_{2t}\}, \cdots, \{X_{kt}\}$, are all stationary, and Linear combination of stationary sequence remains stable, so the residual sequence is a stationary sequence, we have

$$\varepsilon_t = Y_t - \left(\mu + \sum_{i=1}^{k} \frac{\Theta_i(B)}{\Phi_i(B)} B^{l_i} X_{it} \right) \tag{2}$$

If necessary, we can use ARMA model to extract relevant information of residual sequence $\{\varepsilon_t\}$, then a model shown as following:

$$\begin{cases} Y_t = \mu + \sum_{i=1}^{k} \frac{\Theta_i(B)}{\Phi_i(B)} B^{l_i} X_{it} + \varepsilon_t \\ \varepsilon_t = \frac{\Theta(B)}{\Phi(B)} a_t \end{cases} \tag{3}$$

This model is called dynamic regression model and abbreviated ARIMAX. In formulary(3), $\Phi(B)$ is the polynomial in regression coefficients of residual sequence; $\Theta(B)$ is the polynomial in moving mean coefficients of residual sequence; a_t is mean zero white noise sequence.

2.2   Engle-Granger Two Step Cointegration Check

EG check works like this: estimating coefficient with the ordinary least square regression among independent variables, and testing tin stability of regression residual, if the residual data is stationary, then the two variables have cointegration relationship. Otherwise, they have no cointegration relationship.Therefore the EG test hypothesis can be represented as:

H_0:There is not cointegration relationship between binary non-stationary series

H_1:There is cointegration relationship between binary non-stationary series

Due to the fact that the cointegration relationship is mainly determined by inspecting the stability of the regression residual series, thus the above assumption is equivalent to:

H_0:Regression residual sequence $\{\varepsilon_t\}$ is non-stationary

H_1:Regression residual sequence $\{\varepsilon_t\}$ is stationary

2.3   Residual Auto-Regressive Model

The construction thought of Residual autoregressive model is firstly extracting main deterministic information sequence by decomposing method of certainty factors:

$$x_t = X_t + S_t + \varepsilon_t \tag{4}$$

In formulary (4), T_t is trend effect imitation (this paper usesa dynamic regression model), S_t is seasonal effect imitation.

Considering the factor that the extraction ofdeterministic information by the decomposition method may not be sufficient enough, we need totest autocorrelation of residual sequence. If the test result shows no significant autocorrelation between the residual sequences, then the deterministic regression model has fullyextracted information and the analysis can be stopped.Otherwise, the result will show that the deterministic regression model is inadequate for information extraction.Then we can consider fitting regression model on the residual sequence and further extracting the information:

$$\varepsilon_t = \phi_1 \varepsilon_{t-1} + \cdots + \phi_p \varepsilon_{t-p} + a_t \tag{5}$$

Then, we can get the following model shown as following:

$$\begin{cases} x_t = X_t + S_t + \varepsilon_t \\ \varepsilon_t = \phi_1 \varepsilon_{t-1} + \cdots + \phi_p \varepsilon_{t-p} + a_t \\ E(a_t) = 0, Var(a_t) = \sigma^2, Cov(a_t, a_{t-i}) = 0, \forall i \geq 1 \end{cases} \tag{6}$$

Formulary (6) is called residual auto-regressive model.

3   Modeling Process

3.1   Process of Establishing Dynamic Regression Model

- Plotting data on a graph, and defining image direction;
- Determining the order of model and establishing dynamic regression model;
- Cointegration test and white noise teston the regression residuals, confirming whether to continue modeling.

3.2   Process of Establishing Residual Auto-Regressive Model

- Processing abnormal points with residual data;
- Determining autocorrelation coefficient and partial autocorrelation coefficient of the residual data
- Setting up residual auto-regressive model;
- Model test;
- Forecasting application and comparing effect.

4   The Example Analysis and Contrast

In this paper, based on the number of inbound tourism in zhejiang province in 1979-1979 (table 1) [7], and by using the SAS software, we set up data files and predict the number of inbound tourism in 2005-2007.

Table 1. Zhejiang inbound tourist arrivals in 1979-1979 (ten thousand people)

year	People(Ten thousand)	year	People(Ten thousand)
1979	9.31	1994	61.27
1980	13.89	1995	67.27
1981	17.17	1996	72.9
1982	18.03	1997	81.15
1983	18.43	1998	81.96
1984	21.31	1999	94.78
1985	27.29	2000	112.59
1986	29.4	2001	146.95
1987	33.01	2002	204.18
1988	39.27	2003	181.8
1989	29.41	2004	276.67
1990	49.62	2005	348.01
1991	55.41	2006	426.83
1992	68.54	2007	511.18
1993	72.84		

According to the data in the table 1 we draw a sequence diagram, as shown in Fig 1. The sequence seen from the diagram is exponential growth trend. It proves that sequence is non-stationary.

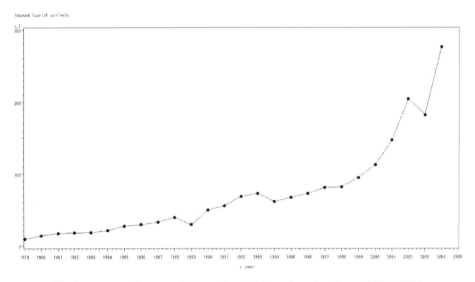

Fig. 1. sequence diagram of the number of inbound tourism from 1997 to 2004

Data, on the number of inbound tourism on Fig. 1, shows remarkable autocorrelation. According to the sequence of ACF figure, data shows has 3 order truncation and establish dynamic regression model (ARIMAX) by using SAS software:

$$X_t = \beta_1 X_{t-1} + \beta_2 X_{t-2} + \beta_3 X_{t-3} + \varepsilon_t \tag{7}$$

Then residual data can be shown as following:

$$\varepsilon_t = X_t - (\beta_1 X_{t-1} + \beta_2 X_{t-2} + \beta_3 X_{t-3}) \tag{8}$$

We need to getfitting data and the residual data,then map the dynamic regression model (Fig 2) and residual sequence diagrams (Fig 3)

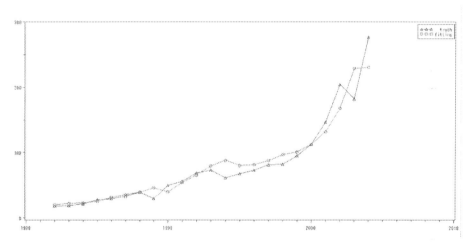

Fig. 2. Fitting effect of dynamic regression mode

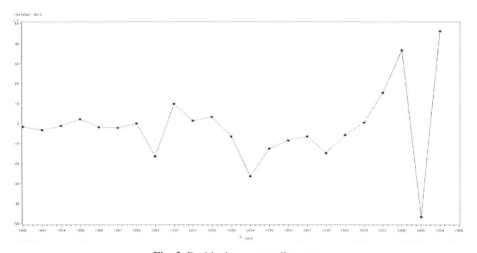

Fig. 3. Residual sequence diagrams

Observing the Fig 3, residual data in 2003 is found to be abnormal point. According to the relevant information in 2003, it is found that in 2003 we were in a special period of SARS. In order to prevent SARS, most provinces in China's implemented a strict entrance number control, thus the number of inbound tourism was greatly reduced.To improve the accuracy of forecast data, the data in 2003 is regarded as the abnormal points.

For abnormal point ε_{t+1} [8], We can use $\hat{\varepsilon}_t$ instead of ε_{t+1}, that is :

$$\hat{\varepsilon}_t = 2\varepsilon_t - \varepsilon_{t-1} \tag{9}$$

According to the measures for the handling of abnormal points, here we use $2\times36.61-15.25 = 59.97$ to replace -46.92, then we have a new residual sequence diagrams (Fig 4)

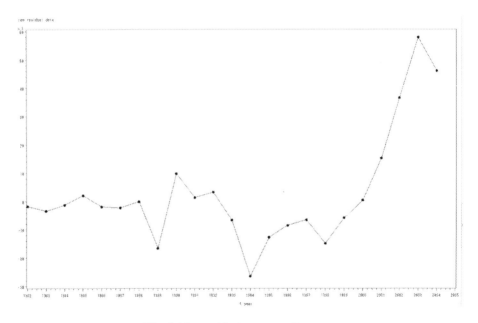

Fig. 4. New residual sequence diagrams

Using SAS program, stationary test (Fig 5) and white noise test (Fig 6) were carried out on the residual data.

As you can see the residual data in Fig 5, the autocorrelation coefficient of residual data after 2 order fell on confidence intervaland gradually tends to zero. It shows that residual data is stable [9]. The dynamic regression model can be established.

```
                                  Autocorrelations
 Lag    Covariance    Correlation    -1 9 8 7 6 5 4 3 2 1 0 1 2 3 4 5 6 7 8 9 1     Std Error
  0      412.216       1.00000                          |********************|            0
  1      313.877       0.76144                        . |******************            0.204124
  2      201.407       0.48859                        . |**********  .              0.299970
  3      59.234902     0.14370                        . |***       .               0.331475
  4     -18.557710     -.04502                        .    *|       .               0.334061
  5     -52.366883     -.12704                        .  ***|       .               0.334314
  6     -82.496655     -.20013                        . ****|       .               0.336319
  7     -114.511       -.27779                        .******|      .               0.341245
  8     -134.837       -.32710                        .*******|     .               0.350541
  9     -120.427       -.29215                        .******|      .               0.363036
 10     -66.151472     -.16048                        .  ***|       .               0.372703
 11     -15.783907     -.03829                        .    *|       .               0.375571
 12      3.255281      0.00790                        .     |       .               0.375734
 13     -4.750348      -.01152                        .     |       .               0.375741
 14     -38.885391     -.09433                        .   **|       .               0.375756
 15     -26.666097     -.06469                        .    *|       .               0.376741
 16     -31.103488     -.07545                        .   **|       .               0.377204
 17     -12.181099     -.02955                        .    *|       .               0.377832
 18     -12.305470     -.02985                        .    *|       .               0.377928
```

                    ".″ marks two standard errors

Fig. 5. Residual autocorrelation test

```
                    Autocorrelation Check for White Noise

 To       Chi-            Pr >
 Lag     Square    DF    ChiSq   ------------------Autocorrelations-------------------

  6       17.58     6    0.0074   0.700    0.359    0.087   -0.056   -0.088   -0.145
```

Fig. 6. White noise test

According to the dynamic regression model parameter, we establish X_t model of the number of inbound tourism:

$$X_t = 0.8518074X_{t-1} + 0.3360194X_{t-2} + 0.0481506X_{t-3} + \varepsilon_t \qquad (10)$$

The next task is to check if the residual data is the data for white noise sequence [10]. If residual data is white noise sequence, we may think the model is reasonable and suitable for prediction. Otherwise, it means that there is useful information in the residual sequence that has not been put forward, we should further transform the model. The value of Prwith a value of 0.0074 in Fig 6 is less than 0.05, and the instructions for the extraction of information is not complete, so we need to consider improving the original model using the residual data model.

As is shown in Fig. 7, the parameter estimated value is less than 0.0001 and has a very significant effect, and the produced residual sequence via examining is proved to be white noise sequence.White noise testing Pr values in Fig. 7 are respectively 0.9089, 0.8985, 0.9666, all figures are greater than 0.05. Residual data is white noise,

The ARIMA Procedure

Conditional Least Squares Estimation

Parameter	Estimate	Standard Error	t Value	Approx Pr > |t|	Lag
AR1,1	0.89604	0.15103	5.93	<.0001	1

Autocorrelation Check of Residuals

To Lag	Chi-Square	DF	Pr > ChiSq				Autocorrelations		
6	1.54	5	0.9089	0.044	0.147	-0.032	-0.037	0.134	-0.007
12	5.60	11	0.8985	0.031	-0.218	-0.180	-0.061	0.101	0.069
18	8.00	17	0.9686	0.057	-0.178	0.034	-0.014	-0.016	0.024

Fig. 7. Parameter estimation and White noise test

and the establishment model is reasonable, soAR(1) model can be established for data sets $\{\varepsilon_t\}$:

$$\varepsilon_t = 0.89604\varepsilon_{t-1} + a_t \qquad (11)$$

By the studying of dynamic regression model of the data and the residual data of residual autoregressive model, the combination model can be represented as:

$$\begin{cases} X_t = 0.8518074X_{t-1} + 0.3360194X_{t-2} + 0.0481506X_{t-3} + \varepsilon_t \\ \varepsilon_t = 0.89604\varepsilon_{t-1} + a_t \end{cases} \qquad (12)$$

Then:

$$x_t = 0.8518074X_{t-1} + 0.3360194X_{t-2} + 0.0481506X_{t-3} + 0.89604\varepsilon_{t-1} + a_t \quad (13)$$

By using combination model, we can forecast the data of 2005, 2006 and 2007 data and compare the predicting data of the ARMA model [7] (table 2).

Table 2. The predicted results contrast table

year	truth	predict data of the ARMA model	Relativeerror	predict data of the combination model	relativeerror
2005	348.01	318.11	8.60%	348.21	0.05%
2006	426.83	397.09	7.00%	435.75	2.08%
2007	511.18	482.64	5.60%	527.81	3.25%

5   Summary

As shown from the analysis results of ARMA models prediction,the relative error of the three years' data from 2005 to 2007 are 8.60%, 7.00% and 5.60%. While,that of the combined model prediction are 0.05%, 2.08% and 3.25%. Combination model greatly improves the accuracy of prediction and shows combination of the dynamic regression model and residual auto-regression model has better adaptability. But the relative errors of combination model prediction areslightly declines, thus this combination model can only be adapted to the short-term forecast. There is also room for improvement for long-term forecasts.

References

1. Wang, Y.: Application of Time Series Analysis, 3rd edn. China Renmin University Press, Beijing (2012)
2. Wang, L.M., Wan, L., Yang, N.: Application of Time Series Analysis. Fudan University Press, Shanghai (2009)
3. Zhong, Z.W., Lei, L.: Johansen and Juselius Cointegration Should Pay Attention to Several Issues. Statitcs and Information Forum. **23**, 80–85 (2008)
4. Wu, Y.: Mining Method with Gibbs ARIMAX Model Outliers. Journal of Guizhou University of Technology **37**, 8–11 (2005)
5. Wang, Y.Z., Xu, Y.J.: ARIMAX Model using Multiple Stationary Time Series. J. Statistics and Decision **18**, 132–134 (2008)
6. Zeng, H., Zheng, C.P.: Forecast Analysis of Non-stationary Time Series for Economic Modelby MATLAB Based on Residual Auto-regressive Model. Journal of Jiamusi University **26**, 70–74 (2008)
7. Yu, X.F., Chen, Y., Huang, H.: Forecast Tourist Arrivals in Zhejiang Province. J. Market Modernization **576**, 129–130 (2009)
8. Xiao, Z.H., Guo, M.Y.: Time Series Analysis and SAS Application. Wuhan University Press, Hubei (2009)
9. Yuan, J.: SAS Statistical Analysis from Entry to the Master. People Post Press, Hebei (2009)
10. Liu, L., Tang, H.P., Zhang, L.J.: Non-steady Time-Series Analysis of Finance Investment in Education based on ARMA Model. Journal of Beijing Normal University **46**(2), 194 (2010)

Reweighted l_2-Regularized Dual Averaging Approach for Highly Sparse Stochastic Learning

Vilen Jumutc$^{(\boxtimes)}$ and Johan A.K. Suykens

ESAT-STADIUS, KU Leuven, Kasteelpark Arenberg 10, 3001 Leuven, Belgium
{vilen.jumutc,johan.suykens}@esat.kuleuven.be

Abstract. Recent advances in dual averaging schemes for primal-dual subgradient methods and stochastic learning revealed an ongoing and growing interest in making stochastic and online approaches consistent and tailored towards sparsity inducing norms. In this paper we focus on the reweighting scheme in the l_2-Regularized Dual Averaging approach which favors properties of a strongly convex optimization objective while approximating in a limit the l_0-type of penalty. In our analysis we focus on a regret and convergence criteria of such an approximation. We derive our results in terms of a sequence of strongly convex optimization objectives obtained via the smoothing of a sub-differential and non-smooth loss function, *e.g.* hinge loss. We report an empirical evaluation of the convergence in terms of the cumulative training error and the stability of the selected set of features. Experimental evaluation shows some improvements over the l_1-RDA method in the generalization error as well.

Keywords: Stochastic learning · l_0 penalty · Regularization · Sparsity

1 Introduction

In this paper we investigate an interplay between l_2-Regularized Dual Averaging (RDA) approach [18] in the context of stochastic learning and parsimony concepts arising from the application of sparsity inducing norms, like the l_0-type of penalty. Learning with $\|x\|_0$ pseudonorm regularization is a NP-hard problem [10] and is feasible only via the reweighting schemes [3], [5], [16] while lacking a proper theoretical analysis of convergence in the online and stochastic learning cases. Some methods, like [7], consider an embedded approach where one has to solve a sequence of QP-problems, which might be very computationally- and memory-wise expensive while still missing some proper convergence criteria.

There are many important contributions of the parsimony concept to the machine learning field, e.g. understanding the obtained solution or simplified and easy to extract decision rules. Many methods, such as Lasso and Elastic Net, were studied in the context of stochastic and online learning in several papers [15], [18], [4] but we are not aware of any l_0-norm sparsity inducing approaches which were applied in the context of Regularized Dual Averaging and stochastic optimization.

© Springer International Publishing Switzerland 2014
Z. Zeng et al. (Eds.): ISNN 2014, LNCS 8866, pp. 232–242, 2014.
DOI: 10.1007/978-3-319-12436-0_26

In many existing iterative reweighting schemes [5], [9] the analysis is provided in terms of the Restricted Isometry (RIP) or the Null Space Properties (NSP) [8]. In this paper we are trying to provide a supplementary analysis and sufficient convergence criteria for learning much sparser linear Pegasos-like [14], [13] models from random observations. We use the l_2-Regularized Dual Averaging approach and a sequence of strongly convex reweighted optimization objectives to accomplish this goal. The solution of every optimization problem at iteration t in our approach is treated as a hypothesis of a learner which is induced by an expectation of a non-smooth loss function (*e.g.* hinge loss) $f(w) \triangleq \mathbb{E}_\xi[l(w, \xi)]$, where the expectation is taken *w.r.t.* the random sequence of observations $\xi = \{\xi_\tau\}_{1 \leq \tau \leq t}$. We regularize it by a re-weighted l_2-norm at each iteration t. This approach in case of satisfying the sufficient conditions will converge to a global optimal solution *w.r.t.* our objective and the loss function which is generating a sequence of stochastic sub-gradients endowing our dual space E^* [12].

This paper is structured as follows. Section 2 describes our reweighted l_2-RDA method. Section 2.3 gives an upper bound on a regret for the sequence of strongly convex optimization objectives under the setting of stochastic learning. Section 3 presents our numerical results and Section 4 concludes the paper.

2   Proposed Method

2.1   Problem Definition

In the Regularized Dual Averaging approach for stochastic learning developed by Xiao [18] we approximate the expected loss function $f(w) \triangleq \mathbb{E}_\xi[l(w, \xi)]$ on a particular random question-answer sequence $\{\xi_\tau\}_{1 \leq \tau \leq t}$, where $\xi_\tau = (x_\tau, y_\tau)$ and $y_\tau \in \{-1, 1\}$. In this particular setting the loss function is regularized by a general convex penalty and hence we are minimizing the following optimization objective:

$$\min_w \quad \phi(w)$$

$$\text{s.t.} \quad \phi(w) \triangleq \frac{1}{t} \sum_{\tau=1}^{t} f(w, \xi_\tau) + \Psi(w), \tag{1}$$

where $\Psi(w)$ can be either a strongly convex $\|\cdot\|_2$ norm or a non-smooth sparsity promoting $\|\cdot\|_1$ norm.

In our particular setting we are dealing with the squared l_2 norm and $\Psi(w) \triangleq \lambda\|w\|_2^2$. For promoting additional sparsity we add to the l_2-norm the reweighted $\|\Theta_t^{1/2} w\|_2^2$ term such that we have $\Psi_t(w) \triangleq \lambda\|w\|_2^2 + \|\Theta_t^{1/2} w\|_2^2$. At every iteration t we will be solving a separate λ-strongly convex instantaneous optimization objective conditioned on a diagonal reweighting matrix Θ_t.

To solve problem in Eq.(1) we split it into a sequence of separated optimization problems which should be cheap to compute and hence should have a closed form solution. These problems are interconnected through the sequence of dual variables $\tilde{g}_\tau \in \partial f(w, \xi_\tau), \tau \in \overline{1,t}$ which are averaged $w.r.t.$ to the current iterate t. Because we are working with the non-smooth hinge loss the reweighted l_2-regularization is imposed via a composite smoothing term which is being gradually increased with every iteration t.

According to a simple dual averaging scheme [12], [18] we can solve Eq.(1) with the following sequence of iterates w_{t+1}:

$$w_{t+1} = \arg\min_w \{\sum_{\tau=1}^{t} \langle \tilde{g}_\tau, w \rangle + t\Psi_t(w) + \beta_t h(w)\}, \tag{2}$$

where $h(w)$ is an auxiliary strongly convex smoothing term and $\{\beta_t\}_{t \geq 1}$ is a non-negative and either constant or increasing input sequence, which in case of non-strongly convex $\Psi_t(w)$ function entirely determines the convergence properties of the algorithm. In our reweighted l_2-RDA approach we use a zero β_t-sequence[1] such that we omit the auxiliary smoothing term $h(w)$ which is not necessary since our $\Psi_t(w)$ function is already smooth and λ-strongly convex. Hence the solution for every iterate w_{t+1} in our approach is given by

$$w_{t+1} = \arg\min_w \{\langle \hat{g}_t, w \rangle + \|\Theta_t^{1/2} w\|_2^2 + \lambda \|w\|_2^2\}, \tag{3}$$

where for derivations we do average stochastic sub-gradients as $\hat{g}_t = \frac{1}{t}\sum_{\tau=1}^{t} \tilde{g}_\tau$. We will explain the details regarding recalculation of Θ_t in the next subsection.

2.2   Algorithm

In this subsection we will outline our main algorithmic scheme. It consists of a simple initialization step, computation and averaging of the subgradient $\tilde{g}_\tau$, evaluation of the iterate w_{t+1} and finally recalculation of the reweighting matrix Θ_{t+1}. In Algorithm 1 we do not have any explicit sparsification mechanism for the iterate w_{t+1} except for the auxiliary function "Sparsify" which utilizes an additional hyperparameter ε to truncate the final solution w_t or any other w below the desired number precision as follows:

$$w^{(i)} := \begin{cases} 0, & \text{if } |w^{(i)}| \leq \varepsilon, \\ w^{(i)}, & \text{otherwise}, \end{cases} \tag{4}$$

where $w^{(i)}$ is i-th component of the vector w. In general we do not restrict ourselves to a particular choice of the loss function $f(w_t, \xi_t)$ but as it was mentioned before we stick to the hinge loss for the completeness. In comparison with the simple l_2-RDA approach [18] we have one additional hyperparameter ϵ, which enters the closed form solution for w_{t+1} and should be tuned or adjusted $w.r.t.$ the iterate t as described in [3] and highlighted in [2].

In Algorithm 1 we perform an optimization $w.r.t.$ to the intrinsic bias term b, which doesn't enter our decision function

$$\hat{y} = \text{sign}(w^T x), \tag{5}$$

---

[1] we assume $\beta_0 = \lambda$ and $\beta_t = 0, t \geq 1$ for completeness.

Algorithm 1. Stochastic Reweighted l_2-Regularized Dual Averaging

---

Data: $\mathcal{S}, \lambda > 0, k \geq 1, \epsilon > 0, \varepsilon > 0, \delta > 0$

1 Set $w_1 = 0, \hat{g}_0 = 0, \Theta_0 = diag([1, \ldots, 1])$
2 **for** $t = 1 \rightarrow T$ **do**
3 $\quad$ Select $\mathcal{A}_t \subseteq \mathcal{S}$, where $|\mathcal{A}_t| = k$
4 $\quad$ Calculate $\tilde{g}_t \in \partial f(w_t, \mathcal{A}_t)$
5 $\quad$ Compute the dual average $\hat{g}_t = \frac{t-1}{t} \hat{g}_{t-1} + \frac{1}{t} \tilde{g}_t$
6 $\quad$ Compute the next iterate $w_{t+1}^{(i)} = -\hat{g}_t^{(i)} / (\lambda + \Theta_t^{(ii)})$
7 $\quad$ Recalculate the next Θ by $\Theta_{t+1}^{(ii)} = 1/((w_{t+1}^{(i)})^2 + \epsilon)$
8 $\quad$ **if** $\|w_{t+1} - w_t\| \leq \delta$ **then**
9 $\quad\quad |$ Sparsify(w_{t+1}, ε)
10 $\quad$ **end**
11 **end**
12 **return** Sparsify(w_{T+1}, ε)

---

but is appended to the final solution w. The trick is to append every input x_t in the subset $\mathcal{A}_t$ with an additional feature column which will be set to 1. This will alleviate the decision function with an offset in the input space. Empirically we have verified that sometimes this design has a crucial influence on the performance of a linear classifier.

2.3 Theoretical Guarantees

In this subsection we will provide the theoretical guarantees for the upper bound on the regret of the function $\phi_t(w) \triangleq f(w, \xi_t) + \Psi_t(w)$, such that for any $w \in \mathbb{R}^n$ we have:

$$\mathbf{R}_t(w) = \sum_{\tau=1}^{t} (\phi_\tau(w_\tau) - \phi_\tau(w)). \tag{6}$$

In this case we are interested in the guaranteed boundedness of the sum generated by this function applied to the sequences $\{\xi_1, \ldots, \xi_t\}$ and $\{\Theta_1, \ldots, \Theta_t\}$. From [12] and [18] we know that a particular gap function defined as $\delta_t = \max_w \{ \sum_{\tau=1}^{t} (\langle \tilde{g}_\tau, w_\tau - w \rangle + \Psi_t(w_t) - \Psi_t(w)) \}$ is an upper bound for the regret

$$\delta_t \geq \sum_{\tau=1}^{t} (\phi_\tau(w_\tau) - \phi_\tau(w)) = \mathbf{R}_t(w) \tag{7}$$

due to the convexity of $f(w, \xi_t)$ [1]. In the next theorem we will provide the sufficient conditions for the boundedness of δ_t if the imposed regularization is given by the reweighted λ-strongly convex term $\|\Theta_t^{1/2} w\|_2^2 + \lambda \|w\|_2^2$. Due to the page limitations the proof of the following theorem is not included hereafter but provided online[2].

---

[2] ftp://ftp.esat.kuleuven.be/pub/stadius/vjumutc/proofs/proofs_rl2rda.pdf

Theorem 1. *Let the sequences $\{w_t\}_{t \geq 1}$, $\{\hat{g}_t\}_{t \geq 1}$ and $\{\Theta_t\}_{t \geq 1}$ be generated by Algorithm 1. Assume $\|\Theta_{t+1}^{1/2}w\|_2 \geq \|\Theta_t^{1/2}w\|_2$ for any $w \in \mathbb{R}^n$, $\Psi_t(w_t) \leq \Psi_1(w_1)$, $\|g_t\|_* \leq G$, where $\|\cdot\|_*$ stands for the dual norm and constant $\lambda > 0$ is given for all $\Psi_t(w)$. Then:*

$$\mathbf{R}_t(w) \leq \frac{G^2}{2\lambda}(1 + \log(t)). \tag{8}$$

Our intuition is related to the asymptotic convergence properties of an iterative reweighting procedure discussed in [7] where with each iterate of Θ_t our approximated norm becomes $\|\Theta_t w\|_2 \simeq \|w\|_p$ with $p \to 0$ thus in a limit applying the l_0-type of a penalty. This implies $p_{t+1} \leq p_t$ and $\|w\|_{p_{t+1}} \geq \|w\|_{p_t}$. In the next theorem we will relax the sufficient conditions on $\Psi_t(w_t)$ and Θ_t. This will introduce into the bound a new term which governs the accumulation of an error w.r.t. these conditions.

Theorem 2. *Let the sequences $\{w_t\}_{t \geq 1}$, $\{g_t\}_{t \geq 1}$ and $\{\Theta_t\}_{t \geq 1}$ be generated by Algorithm 1. Assume $\|\Theta_t^{1/2}w\|_2 - \|\Theta_{t+\tau}^{1/2}w\|_2 \leq \nu_1/\tau$ and $\Psi_{t+\tau}(w_{t+\tau}) - \Psi_t(w_t) \leq \nu_2/\tau$ for some $\tau \geq 1$, $\nu_1, \nu_2 \geq 0$ and $w \in \mathbb{R}^n$, $\|g_t\|_* \leq G$, where $\|\cdot\|_*$ stands for the dual norm and constant $\lambda > 0$ is given for all $\Psi_t(w)$. Then:*

$$\mathbf{R}_t(w) \leq \log(t)(\lambda\nu_1 + \nu_2) + \frac{G^2}{2\lambda}(1 + \log(t)). \tag{9}$$

The above bound boils down to the bound in Theorem 1 if we set ν_1, ν_2 to zero.

3   Simulated Experiments

3.1   Experimental Setup

For all methods in our experiments we use a 2-step procedure for tuning hyper-parameters. This procedure consists of Coupled Simulated Annealing [17] initialized with 5 random sets of parameters for the first step and the simplex method [11] for the second step. After CSA converges to some local minima we select a tuple of hyperparameters which attains the lowest cross-validation error and start the simplex procedure to refine our selection. On every iteration step for CSA and simplex method we proceed with a 10-fold cross-validation. In l_1-RDA and our reweighted l_2-RDA we are promoting additional sparsity with a slightly modified cross-validation criteria. We introduce an affine combination of the validation error and obtained sparsity in proportion 90% : 10% where sparsity is calculated as $\sum_i I(|w^{(i)}| > 0)/d$.

All experiments with large-scale UCI datasets [6] were repeated 50 times (iterations) with the random split to training and test sets in proportion 90% : 10%. Every iteration all methods are evaluated with the same test set to provide a consistent and fair comparison in terms of the generalization error and obtained p-values of a pairwise two-sample t-test. In the presence of 3 or more classes we perform binary classification where we learn to classify the first class versus all others. For CT slices[3] dataset we performed a binarization of an output y_i by the median value. For URI dataset we took only "Day0" subset as a probe. For evaluation of the Algorithm 1 for UCI datasets we set $T = 1000$, $k = 1$, $\delta = 10^{-5}$ and other hyperparameters λ, ϵ and ε were determined using the cross-validation tuning procedure described above. For extremely sparse datasets with $d \gg n$, like Dexter and URI we increased k by 10 times. Information on all public UCI datasets one can find in [6].

3.2 Numerical Results

In this subsection we will provide an outlook on the performance of l_1-RDA, our reweighted l_2-RDA and Pegasos [14] methods. We provide the results of the Pegasos approach for the completeness and a fair comparison in terms of the affected generalization error $w.r.t.$ the obtained sparsity. In Table 1 one can see generalization errors with standard deviations (in brackets) for different UCI datasets.

Table 1. Performance

Dataset	Generalization (test) errors					
	(re)l_2-RDA		l_1-RDA		Pegasos	
Pen Digits	0.0745^{**}	(±0.02)	0.1043	(±0.04)	**0.0573**	(±0.02)
Opt Digits	0.0680^{**}	(±0.03)	0.0554	(±0.03)	**0.0356**	(±0.01)
Semeion	0.0619^{*}	(±0.03)	**0.0414**	(±0.02)	0.0549	(±0.02)
Spambase	0.1228^{*}	(±0.02)	0.1205	(±0.02)	**0.0989**	(±0.02)
Shuttle	0.0744^{*}	(±0.02)	0.0734	(±0.02)	**0.0488**	(±0.02)
CT slices	0.0643^{*}	(±0.02)	0.0845	(±0.13)	**0.0478**	(±0.01)
Magic	**0.2242**	(±0.01)	0.2259	(±0.02)	0.2254	(±0.01)
CNAE-9	$\mathbf{0.0109}^{**}$	(±0.01)	0.0172	(±0.02)	0.0448	(±0.02)
Covertype	$\mathbf{0.2670}^{*}$	(±0.01)	0.2715	(±0.03)	0.2791	(±0.01)
Dexter	0.0922^{*}	(±0.02)	0.0956	(±0.01)	**0.0765**	(±0.01)
URI	0.0458^{**}	(±0.01)	0.0623	(±0.03)	**0.0388**	(±0.01)

In Table 1 one can find asterisk symbols next to the results of our method ((re)l_2-RDA). These symbols indicate p-values < 0.05 of a pairwise two-sample t-test on generalization errors. Here p-values are reflecting the statistical significance of having the null-hypothesis true: the equivalence of normal distributions from which the test errors are drawn. By having two asterisk symbols we assume strong presumption against null hypothesis $w.r.t.$ both competing methods, and by having one asterisk symbol - to at least one of them. Analyzing Table 1 we can conclude that for the majority of UCI datasets we are doing equally good $w.r.t.$ l_1-RDA method and the significance of the obtained difference is quite high. One can see that for some datasets our reweighted l_2-RDA approach is doing better than Pegasos as well. This phenomenon could be understood from the underlying sparsity pattern which is likely to be very sparse for some datasets, for instance CNAE-9.

---

[3] Originally it is a regression problem.

3.3 Sparsity and Stability

In this subsection we will provide some of the findings which highlight the enhanced sparsity of the reweighted l_2-RDA approach as well as the consistency and stability for the selected set of features (dimensions). In Table 2 one can observe the evidence of an additional sparsity promoted by the reweighting procedure which in some cases significantly reduce the number of non-zeros in the obtained solution. We do not provide any results for the Pegasos-based approach because it consists of a generic l_2-norm penalty and a projection step which all together do not provide sparse solutions. In Table 2 we provide the statistical significance of the given result by an asterisk symbol. By analyzing the results on immediately imply that in cases where we are performing equally good or slightly worse the p-values are quite high. Next we perform several

Table 2. Sparsity $\sum_i I(|w^{(i)}| > 0)/d$

Dataset	(re)l_2-RDA		l_1-RDA	
Pen Digits	0.12*	($\pm$0.06)	**0.09**	($\pm$0.11)
Opt Digits	**0.16***	($\pm$0.09)	0.24	($\pm$0.07)
Semeion	**0.13***	($\pm$0.08)	0.19	($\pm$0.05)
Spambase	0.35	($+$0.07)	**0.34**	($\pm$0.08)
Shuttle	**0.32**	($\pm$0.17)	0.32	($\pm$0.10)
CT slices	0.26*	($\pm$0.08)	**0.21**	($\pm$0.05)
Magic	**0.22***	($\pm$0.05)	0.34	($\pm$0.15)
CNAE-9	**0.02***	($\pm$0.01)	0.03	($\pm$0.03)
Covertype	**0.06***	($\pm$0.03)	0.09	($\pm$0.06)
Dexter	**0.08***	($\pm$0.07)	0.17	($\pm$0.06)
URI	**0.0012***	($\pm$0.0011)	0.0027	($\pm$0.0007)

series of experiments with UCI datasets to reveal the consistency and stability of our algorithm *w.r.t.* the selected sparsity patterns. For every dataset first we tune the hyperparameters with all available data. We run our reweighted l_2-RDA approach and l_1-RDA [18] method 100 times in order to collect frequencies of every feature (dimension) being non-zero in the obtained solution. In Figure 1 we present the corresponding histograms. As we can see our approach results in much more sparser solutions which are quite robust *w.r.t.* a sequence of random observations. l_1-RDA approach lacks these very important properties being relatively unstable under the stochastic setting.

In the next experiment we adopted a simulated setup from [4] and created a toy dataset of sample size 10000, where every input vector a is drawn from a normal distribution $\mathcal{N}(0, I_{d \times d})$ and the output label is calculated as follows $y = \text{sign}(a^T w_* + \epsilon)$, where $w_*^{(i)} = 1$ for $1 \leq i \leq \lfloor d/2 \rfloor$ and 0 otherwise and the noise is given by $\epsilon \sim \mathcal{N}(0, 1)$. We run each algorithm for 100 times and report the mean F1-score reflecting the performance of sparsity recovery. F1-score is defined as $2\frac{\text{precision} \times \text{recall}}{\text{precision} + \text{recall}}$, where

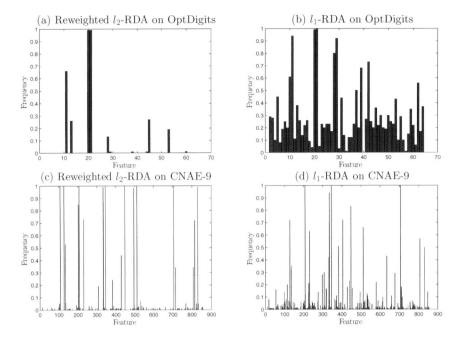

Fig. 1. Frequency of being non-zero for the features of Opt Digits and CNAE-9 datasets. In the left subfigures (a,c) we present the results for the reweighted l_2-RDA approach, while the right subfigures (b,d) correspond to l_1-RDA method.

$$\text{precision} = \frac{\sum_{i=1}^{d} I(\hat{w}^{(i)} \neq 0, w_*^{(i)} = 1)}{\sum_{i=1}^{d} I(\hat{w}^{(i)} \neq 0)}, \qquad \text{recall} = \frac{\sum_{i=1}^{d} I(\hat{w}^{(i)} \neq 0, w_*^{(i)} = 1)}{\sum_{i=1}^{d} I(w_*^{(i)} = 1)}.$$

Figure 2 shows that the reweighted l_2-RDA approach selects irrelevant features much less frequently as in comparison to l_1-RDA approach. As it was empirically verified before for UCI datasets we perform better both in terms of the stability of the selected set of features and the robustness to the stochasticity and randomness.

The higher the F1-score is, the better the recovery of the sparsity pattern. In Figure 3 we present an evaluation of our approach and l_1-RDA method *w.r.t.* to ability to identify the right sparsity pattern as the number of features increases. We clearly do outperform l_1-RDA method in terms of F1-score for $d \leq 300$. In conclusion we want to point out some of the inconsistencies that we've discovered comparing our F1-scores with [4]. Although the authors in [4] use a batch-version of the accelerated l_1-RDA method and a quadratic loss function they obtain very low F1-score (0.67) for the feature vector of size 100. In our experiments all F1-scores were above 0.7. For the dimension of size 100 our method obtains F1-score ≈ 0.95 while authors in [4] have only 0.87.

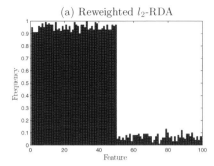
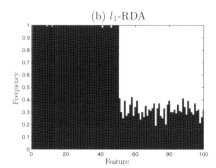

Fig. 2. Frequency of being non-zero for the features of our toy dataset ($d = 100$). Only the first half of features do correspond to the encoded sparsity pattern. In the left subfigure (a) we present the results for the reweighted l_2-RDA approach, while the right subfigure (b) corresponds to l_1-RDA method.

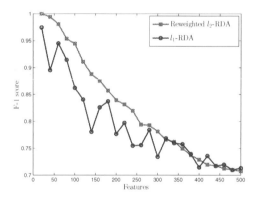

Fig. 3. F1-score as the function of the number of features. We ranged the number of features from 20 to 500 with the step size of 20.

4   Conclusion

In this paper we presented a novel and promising approach, namely Reweighted l_2-Regularized Dual Averaging. This approach helps to approximate very efficient l_0-type of a penalty using a proven and reliable simple dual averaging scheme. Our method is suitable both for online and stochastic learning, while our numerical and theoretical results mainly consider only stochastic setting. We provided theoretical guarantees of the boundedness of the regret under different conditions and demonstrated the empirical convergence of the cumulative training error (loss). Experimental results validate the usefulness and promising capabilities of the proposed approach in obtaining much sparser and consistent solutions while keeping the convergence of Pegasos-like approaches at hand.

For the future we consider to improve our algorithm in terms of the accelerated convergence discussed in [4], [12], [18] and develop some further extensions towards online and stochastic learning applied to the huge-scale[4] data.

Acknowledgments. EU: The research leading to these results has received funding from the European Research Council under the European Union's Seventh Framework Programme (FP7/2007-2013) / ERC AdG A-DATADRIVE-B (290923). This paper reflects only the authors' views, the Union is not liable for any use that may be made of the contained information. Research Council KUL: GOA/10/09 MaNet, CoE PFV/10/002 (OPTEC), BIL12/11T; PhD/Postdoc grants Flemish Government: FWO: projects: G.0377.12 (Structured systems), G.088114N (Tensor based data similarity); PhD/Postdoc grants IWT: projects: SBO POM (100031); PhD/Postdoc grants iMinds Medical Information Technologies SBO 2014 Belgian Federal Science Policy Office: IUAP P7/19 (DYSCO, Dynamical systems, control and optimization, 2012-2017)

References

1. Boyd, S., Vandenberghe, L.: Convex Optimization. Cambridge University Press, New York (2004)
2. Candès, E., Wakin, M., Boyd, S.: Enhancing sparsity by reweighted l1 minimization. Journal of Fourier Analysis and Applications **14**(5), 877–905 (2008)
3. Chartrand, R., Yin, W.: Iteratively reweighted algorithms for compressive sensing. In: IEEE International Conference on Acoustics, Speech and Signal Processing, ICASSP 2008, pp. 3869–3872 (March 2008)
4. Chen, X., Lin, Q., Peña, J.: Optimal regularized dual averaging methods for stochastic optimization. In: Bartlett, P.L., Pereira, F.C.N., Burges, C.J.C., Bottou, L., Weinberger, K.Q. (eds.) NIPS, pp. 404–412 (2012)
5. Daubechies, I., DeVore, R., Fornasier, M., Güntürk, C.S.: Iteratively reweighted least squares minimization for sparse recovery. Comm. Pure Appl. Math. **63**(1), 1–38 (2010)
6. Frank, A., Asuncion, A.: UCI machine learning repository (2010). http://archive.ics.uci.edu/ml
7. Huang, K., King, I., Lyu, M.R.: Direct zero-norm optimization for feature selection. In: ICDM, pp. 845–850 (2008)
8. Lai, M.J., Liu, Y.: The null space property for sparse recovery from multiple measurement vectors. Applied and Computational Harmonic Analysis **30**(3), 402–406 (2011)
9. Lai, M.J., Xu, Y., Yin, W.: Improved iteratively reweighted least squares for unconstrained smoothed l_q minimization. SIAM J. Numerical Analysis **51**(2), 927–957 (2013)
10. Lázaro, J.L., De Brabanter, K., Dorronsoro, J.R., Suykens, J.A.K.: Sparse LS-SVMs with l_0-norm minimization. In: ESANN, pp. 189–194 (2011)
11. Nelder, J.A., Mead, R.: A simplex method for function minimization. Computer Journal **7**, 308–313 (1965)
12. Nesterov, Y.: Primal-dual subgradient methods for convex problems. Mathematical Programming **120**(1), 221–259 (2009)

---

[4] Both in terms of dimensions and number of samples.

13. Shalev-Shwartz, S., Singer, Y.: Logarithmic regret algorithms for strongly convex repeated games. Tech. rep., The Hebrew University (2007)
14. Shalev-Shwartz, S., Singer, Y., Srebro, N.: Pegasos: Primal Estimated sub-GrAdient SOlver for SVM. In: Proceedings of the 24th International Conference on Machine Learning, ICML 2007, New York, NY, USA, pp. 807–814 (2007)
15. Shalev-Shwartz, S., Tewari, A.: Stochastic methods for l1 regularized loss minimization. In: Proceedings of the 26th Annual International Conference on Machine Learning, ICML 2009, pp. 929–936. ACM, New York (2009)
16. Wipf, D.P., Nagarajan, S.S.: Iterative reweighted l_1 and l_2 methods for finding sparse solutions. J. Sel. Topics Signal Processing **4**(2), 317–329 (2010)
17. Xavier-De-Souza, S., Suykens, J.A.K., Vandewalle, J., Bollé, D.: Coupled simulated annealing. IEEE Trans. Sys. Man Cyber. Part B **40**(2), 320–335 (2010)
18. Xiao, L.: Dual averaging methods for regularized stochastic learning and online optimization. J. Mach. Learn. Res. **11**, 2543–2596 (2010)

Perceptual Learning Model on Recognizing Chinese Characters

Jiawei Chen, Yan Liu, Xiaomeng Li, and Liujun Chen[✉]

School of Systems Science, Beijing Normal University,
Beijing 100875, People's Republic of China
chenlj@bnu.edu.cn

Abstract. Perceptual learning is the improvement in performance on a variety of simple sensory tasks through practice. Based on the perceptual model, with the lateral interaction applied to the neurons of the middle layer, a neural network is developed to simulate the transition of perceptual mode from global perception to local perception in the process of Chinese characters learning. Using some Chinese characters with the same structure to train the network, the components and radicals of the Chinese characters can be extracted through the local perceptual mode. The perceptual learning process under the damage of neural connections is also simulated, and the result are coincident with the somatosensory cortex changes experiment on owl monkeys. It is a self-organization process, in which the lateral interaction among the neurons are the core mechanism.

Keywords: Perceptual learning · Lateral interaction · Neural network · Self organization

1 Introduction

Sensory perception is a learned trait. The brain strategies we use to perceive the world are constantly modified by experience. With practice, we subconsciously become better at identifying familiar objects or distinguishing fine details in our environment [1]. The perception of Chinese characters reflects the characteristics of learning. Researches on the students whose mother language is not Chinese showed that the global strategy is commonly used at the first stage of Chinese learning [2]. Chinese characters are most pictures symbolizing an idea or object, so beginners often try to setup a relationship between the characters with some pictures that already exist in the brain. The relationship is important for students to remember these Chinese characters, while they often lost some strokes. For the learners at intermediate and advanced level, the radicals of Chinese characters and their position play an important role in the process

The paper was supported by MOE Youth Fund Project of Humanities and Social Sciences (Project No.11YJC840006) and Fundamental Research Funds for the Central Universities (Fund number: 2013YB76).

© Springer International Publishing Switzerland 2014
Z. Zeng et al. (Eds.): ISNN 2014, LNCS 8866, pp. 243–251, 2014.
DOI: 10.1007/978-3-319-12436-0_27

of perception [3]. While for the learners whose mother language is Chinese, they can quickly separate every components of a Chinese character, instead of learn it as a whole.

Therefore, Chinese characters can be perceived by two basic mode: global to local (global perception) and local to global (local perception). The simulation results of perceptual model [4] and the behavior experiments both show that the local perception is more efficient than global perception. The transition from global perception to local perception corresponds to the change of the weight distribution in the neural system. How does the weight distribution transfer from one state to the other one in the actual learning process? This is a self-organizing process, while what is the neural dynamic mechanisms in the process?

Lateral interaction is one of neural mechanism which has been applied in many models with self-organization features. H. Kohonen proposed SOM model in 1982 [5], which well simulated some self-organizing processes in brain. Erwin et al. showed that SOM model can explain the formation process of the rhesus monkey's primary visual cortex [6]. Xing et al. simulated children's acquisition of Chinese characters, and the simulation results are well consistent with the behavior experiment in consistency and regulation [7]. In this paper, lateral interaction is applied to the perceptual model [4], and a perceptual learning model is developed. The model can simulate the transition of perceptual mode in the Chinese characters learning process, and the result are coincident with behavior and physiological experiments.

2   The Neural Network Model

2.1   Network Structure

The perceptual learning model is composed of three layers, including the input layer X, the middle layer Y and the output Z as shown in Fig. 1. The middle layer includes 4 receptive fields RF_1, RF_2, RF_3 and RF_4. The input layer, the output layer and each receptive field are all composed of 40×40 neurons. The neurons of the input layer X_{mn} connects to the corresponding neurons $Y_{i,mn}$ in the 4 receptive fields RF_i with the weights of $u_{i,mn}$. And the 4 neurons $Y_{i,mn}$ of the middle layer transfer the integrated signal to the neurons Z_{mn} of the output layer with the weights of $v_{i,mn}$.

The receptive fields of the middle layer compete with each other. When the input $X_{mn} \neq 0$, the 4 neurons connected with X_{mn} in every receptive field compete with each other and there is only one neuron $Y_{c,mn}$ is activated. In each receptive field, there are lateral interaction between neurons within neighborhood.

2.2   The Adjustment of the Weights from the Input Layer to the Middle Layer

The lateral interaction is applied to the middle layer, so there are two factors that change the weight U from the input layer to the middle layer. One factor is the

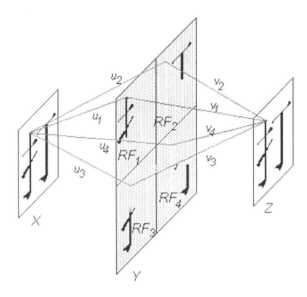

Fig. 1. The schematic diagram of the perceptual learning neural network model. The input and the output layer are 40×40 neurons. The middle layer is composed of 4 receptive fields, each of which is a 40×40 neurons. The 4 receptive fields in the middle layer compete with each other. While in each receptive field, there are lateral interactions between neurons.

input from the outside, and the other factor is the competition and cooperation in the middle layer. The weight adjustment on the input from the outside is according to the Pseudo-Hebb learning rule with the equations as follows,

$$
\begin{cases}
\theta_i(t) = \alpha \cdot \min\{c_i(t-1), ..., c_i(t-h)\} + (1-\alpha) \cdot \max\{c_i(t-1), ..., c_i(t-h)\} \\
c_i(t) = \sum\limits_{m,n} u_{i,mn}(t) \cdot x_{mn} - \theta_i(t) \\
\begin{cases} Y_i(t) = \tanh(k \cdot c_i(t)), \text{for } c_i(t) \geq 0 \\ Y_i(t) = 0, \text{for } c_i(t) < 0 \end{cases} \\
u^e_{i,mn}(t) = u^e_{i,mn}(t-1) + \eta \cdot (Y_i(t) \cdot x_{mn} \cdot u_{i,mn}(0) - Y_i^2(t) \cdot u^e_{i,mn}(t))
\end{cases}
\tag{1}
$$

where the last equation shows the weight adjustment on the input from the outside, and the superscript e denotes external.

The interaction among the receptive field of the middle layer is determined by both the input from the outside and the current state of the system. The interaction can be divided into two stages, the first stage is competition and the second stage is cooperation. Each neuron X_{mn} in the input layer connects to four neurons in the middle layer $Y_{1,mn}$, $Y_{2,mn}$, $Y_{3,mn}$, $Y_{4,mn}$, and the corresponding weights are $u_{1,mn}$, $u_{2,mn}$, $u_{3,mn}$, $u_{4,mn}$. If the input from outside $x_{mn} \neq 0$, and $x_{mn} \cdot u_{c,mn} \geq x_{mn} \cdot u_{i,mn}$ for $\forall i$, then it is defined that $Y_{c,mn}$ wins out, which is the competition process.

The winner neuron stimulates its neighborhood neurons to enhance the weight from the input layer to the middle layer, which is the cooperation process. The radius of neighborhood is $N_{c,mn}$ and its size changes with time according to the equation, $N_{c,mn} = \beta \cdot x_{mn} \cdot u_{c,mn} \cdot S$, where $S = 40$, $\beta(t)$ is the parameter denotes the size of the neighborhood. The distance of the winner neurons determines the magnitude of the lateral interaction. Here the interaction obeys Gaussian distribution, that if $m'n'$ is the winner neuron, the change of the weight of mn is:

$$\Delta u_{c,mn}^{I(m'n')} = x_{m'n'} \cdot u_{c,m'n'} \cdot \frac{1}{\sqrt{2\pi}\sigma} \exp(-\frac{d^2}{2\sigma^2}) \qquad (2)$$

where $d = \sqrt{(m-m')^2 + (n-n')^2}$. So based on the lateral interaction, the weights $u_{c,mn}$ of all the winner neurons are

$$\Delta u_{c,mn}^I = \sum_{m',n'} \Delta u_{c,mn}^{I(m'n')}$$

The total change of weight is the sum of the weight change from the interlayer and the intra-layer, that is

$$u_{i,mn}(t) = u_{i,mn}(t-1) + \Delta u_i(t) = u_{i,mn}(t-1) + \Delta u_{i,mn}^e(t) + \Delta u_{i,mn}^I(t)$$

Considering the normalization of the weight from the input layer to the middle layer,

$$u_{i,mn}(t) = \frac{u_{i,mn}(t)}{\sum_j u_{j,mn}(t)}$$

the output of the middle layer is $y_{i,mn}(t) = x_{mn} \cdot u_{i,mn}(t)$.

2.3 The Adjustment of the Weights from the Middle Layer to the Output Layer

Similar with the perceptual model [4], the weights from the middle layer to the output layer are adjusted by the perceptron rule. Each training process includes three steps of computing the output, adjusting the weight and normalization with the equations as follows,

$$\begin{cases} z_{mn}(t) = hardlim(v_{mn}(t) \cdot y_{mn}(t) - \rho) \\ v_{mn}(t) = v_{mn}(t-1) + \xi \cdot (x_{mn}(t) - z_{mn}(t)) \cdot y'_{mn}(t) \\ v_{i,mn}(t) = v_{i,mn}(t)/\sum_j v_{j,mn}(t) \end{cases} \qquad (3)$$

where $v_{mn} = (v_{1,mn}, v_{2,mn}, v_{3,mn}, v_{4,mn})$, $y_{mn} = (y_{1,mn}, y_{2,mn}, y_{3,mn}, y_{4,mn})'$.

3 The Results of Simulation

3.1 Initial State

This paper discuss how perceptual learning modifies the perceptual mode. Based on the perceptual model [4], suppose the initial state is the global perception, and the initial distribution of the weights $u_{i,mn}$ is random, i.e. $u_{i,mn} = rand(0 \sim 1), \forall i, m, n$. Then the weights are normalized as $u_{i,mn} = u_{i,mn}/\sum_j u_{j,mn}$ $\forall i, m, n$, as shown in Fig. 2.

Fig. 2. The initial weights $u_{i,mn}(0)$ normalized by $\sum_i u_{i,mn} = 1$

3.2 The Model is Trained by Single Chinese Character

A Chinese character is used to train the network. The parameters in the eq.1-eq.4 are $\theta_i(0) = 50$, $\eta = 0.5$, $h = 5$, $k = 0.1$, $\alpha = 0.2$, $\rho = 0.3$, $\sigma = N_{c,mn}(t)$, $\xi = 0.5$, $\beta = 0.1$. The weights u_i, the outputs of each receptive field and the output layer are shown in Fig. 3.

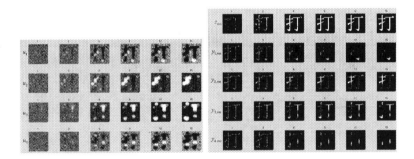

Fig. 3. One Chinese character is used to train the perceptual learning model. (a) the outputs of the middle layer and the output layer; (b) The weights U.

At the beginning of training, the output of each neuron in the middle layer is mainly the basic outline of the Chinese character. As the training carrying on, each receptive field turns to sensitive to some part of the character and then realize the local perceptual mode. From the results of the perceptual model we know that the local perceptual mode is more efficient than the global perceptual mode. In fact, here the input signal can reappear in the output layer only after 9 times of training.

3.3   The Model Is Trained by Multiple Chinese Character

Some Chinese characters with the same structure are used to train the network. The perceptual mode changes from global mode to local mode, and the partial feature of the Chinese characters can be extracted. Ten Chinese characters with left-right structures, as shown in Fig. 4a, are selected to train the network, randomly picking one each time. The parameters are the same with above except $\beta = 0.35$. The changes of weights U in 40 times of training are shown in Fig. 4b. The outputs of the middle layer and the output layer are shown in Fig. 4c. Similarly, another 10 Chinese characters with top-bottom structures are used to training the network, and the results are shown in Fig. 5.

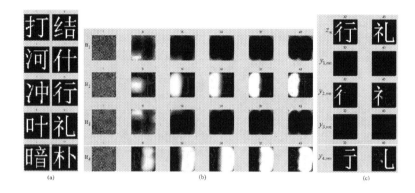

Fig. 4. (a) Training Sampleten Chinese characters with left-right structures. (b) Randomly choose Chinese characters to train the network. The weights U changes with time. (c) Two outputs of the middle layer and the output layer.

After training, the weights of the receptive fields change from the random distribution to the block distribution. So each receptive field turns to sensitive to some part of the character and then realize the local perceptual mode. If the input Chinese character is with left-right structure or up-down structure, the components and radicals of the Chinese characters can be extracted through the local perceptual mode.

3.4   The Perceptual Learning Model Under the Damage of Neural Connections

In certain brain regions, the damage of neural connections will injure its function. So we also simulate the perceptual learning process under the damage of neural connections. How do the weights redistribute in this process is discussed. The parameters are the same with above except $\beta(t) = 0.35 - (0.35 - 0.1) \cdot t/40$. For the first 10 times, the connection in the network is intact. At $t = 11$, the connections are damaged as $u_{4,mn} = 0, \forall m, n$, and then training the model 30 times, the changes of the weights are shown as Fig. 6.

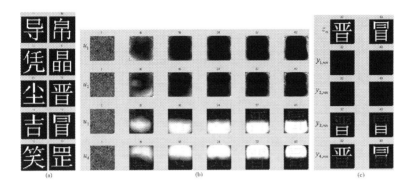

Fig. 5. (a) Training Sampleten Chinese characters with top-bottom structures. (b) Randomly choose Chinese characters to train the network. The weights U changes with time. (c) Two outputs of the middle layer and the output layer.

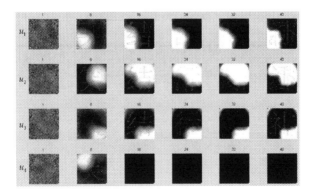

Fig. 6. Before and after the damage of neural connections, the changes of the weights U during the training. For the first 10 times, the connection in the network is intact and the weights change from the random distribution to the block distribution. At $t = 11$, the connections between the input layer to the receptive field RF_4 are damaged, so the sensitive area of RF_4 will be covered by the sensitive area of RF_1 and RF_2, whose sensitive areas are near to the sensitive area of RF_4 before damaged.

When the connections between the input layer to the receptive field RF_4 are damaged, the sensitive area of RF_4 will be covered by the sensitive area of RF_1 and RF_2, whose sensitive areas are near to the sensitive area of RF_4 before damaged. This results are coincident with the somatosensory cortex changes experiment on owl monkeys [8].

4 Conclusion and Discussion

Based on the perceptual model, with the lateral interaction applied to the neurons in the middle layer, perceptual learning network is developed to simulate the transition of perception mode in the Chinese characters learning process. Train the model with one or several Chinese characters, the weight distribution evolves from global perception to local perception. Using some Chinese characters with the same structure to train the network, the components and radicals of the Chinese characters can be extracted through the local perceptual mode. The perceptual learning process under the damage of neural connections is also simulated, and the result are coincident with the somatosensory cortex experiment on owl monkeys.

Another method based on NMF (non-negative matrix factorization) [9] is trained using self-made Chinese character, and all the meaningful features are extracted. But when training this model with actual Chinese character which is used above, not all the extracted sections are meaningful. On the other hand, there is no any self organization in NMF, which is just an algorithm. The evolvement of the weights between input layer and middle layer in our perceptual learning model is characterized by self-organization. Although the weights are only local adjusted with random distribution initially, the whole weights of the network tends to a block distribution to perceive the local information of the input signal.

References

1. Tsodyks, M., Gilbert, C.: Neural networks and perceptual learning. Nature **14**, 775–781 (2004)
2. Jiang, X., Zhao, G.: A Survey on the Strategies for Learning Chinese Characters among CSL Beginners. Language Teaching and Linguistic Studies **4**, 10–17 (2001)
3. Feng, L., Lu, H., Xu, C.: The Role of Information about Radical Position in Processing Chinese Characters by Foreign Students. Language Teaching and Linguistic Studies **3**, 66–72 (2005)
4. Chen, J., Liu, Y., Chen, Q., Chen, L., Fang, F.: A Neural Network Model for Chinese Character Perception. In: The 5th International Conference on Natural Computation, pp. 319–323 (2009)
5. Kohonen, T.: The Self-Organizing Map, 2nd edn. Springer (1997)
6. Erwin, E., Obermayer, K., Schulten, K.: Models of Orientation and Ocular Dominance Columns in the Visual Cortex: A Critical Comparison. Neural Comp. **7**, 425–468 (1995)
7. Xing, H., Shu, H., Li, P.: A self-organizing connectionist model of character acquisition in Chinese. In: Proceedings of the Twenty-fourth Annual Conference of the Cognitive Science Society. Lawrence Erlbaum, Mahwah (2002)

8. Merzenich, M.M., Nelson, R.J., Stryker, M.P., Cynader, M.S., Schoppmann, A., Zook, J.M.: Somatosensory cortical map changes following digit amputation in adult monkeys. Journal of Comparative Neurology **224**, 591–605 (1984)
9. Lee, D.D., Seung, H.S.: Learning the parts of objects by non-negative matrix factorization. Nature **401**, 788–791 (1999)

Hierarchical Solving Method
for Large Scale TSP Problems

Jingqing Jiang[1,2(✉)], Jingying Gao[2], Gaoyang Li[3], Chunguo Wu[3], and Zhili Pei[1,2]

[1] College of Computer Science and Technology,
Inner Mongolia University for Nationalities, Tongliao, China
jiangjingqing@yahoo.com.cn
[2] College of Mathematics, Inner Mongolia University for Nationalities, Tongliao, China
[3] College of Computer Science and Technology, Jilin University, Changchun, China

Abstract. This paper presents a hierarchical algorithm for solving large-scale traveling salesman problem (TSP), the algorithm first uses clustering algorithms to large-scale TSP problem into a number of small-scale collections of cities, and then put this TSP problem as a generalized traveling salesman problem (GTSP), convert solving large-scale TSP problem into solving GTSP and several small-scale TSP problems. Then all the sub-problems will be solved by ant colony algorithm and At last all the solutions of each sub-problem will be merged into the solution of the large-scale TSP problem by solution of GTSP. Experimental part we uses the traditional ant colony algorithm and new algorithm for solving large-scale TSP problem, numerical simulation results show that the proposed algorithm for large-scale TSP problem has a good effect, compared with the traditional ant colony algorithm, the solving efficiency has been significantly improved.

Keywords: Traveling salesman problem · Ant colony algorithm · Affinity propagation clustering · K-means algorithm · Generalized traveling salesman problem

1   Introduction

Traveling Salesman Problem (TSP) is one of the most typical combinatorial optimization problems in computer science and operations research. Given a set of cities and the distance between them, the goal is to find a lowest cost loop (this loop is called a Hamiltonian circuit) that starts from a certain city, visits each city and the city only be visited once, finally returns to the start city. The algorithm to solve TSP problem can be divided into two categories: exact algorithms and heuristic algorithms (or approximate algorithms). The exact algorithm ensures to find the optimal solution in a finite number of steps. Today, people can solve TSP problem from hundreds to thousands

Supported by the National Natural Science Foundation of China under Grant No. 61163034, 61373067 and Inner Mongolia Natural Science Foundation under Grant No. 2013MS0910, 2013MS0911.

Z. Zeng et al. (Eds.): ISNN 2014, LNCS 8866, pp. 252–261, 2014.
DOI: 10.1007/978-3-319-12436-0_28

of cities with the exact solution. Among the exact solutions, Branch and bound [1], branch and cut [2], cutting-plane or facet-finding algorithms [3], branch and price [4] and the Concorde algorithm [5] are the most typical exact algorithms for TSP problem. TSP problem is a well-known NP-hard combinatorial problems, namely it is difficult to obtain the optimal solution for the large-scale TSP problem [6]. Therefore, the heuristic algorithm is usually used to obtain the quasi-optimal solution. The LK algorithm with a simple local search presented by Lin and Kernighan is the representation of the heuristic algorithm for the large-scale TSP problem [7-10]. In recent years, more and more swarm intelligence optimization algorithms are being used to solve the TSP problem. Ant colony system proposed by Dorigo is considered the most representative method [11]. But the existed swarm intelligence optimization algorithms are so limited to the scale that cannot be applied to solve the large-scale TSP problem.

This paper presents a hierarchical method based on an ant colony algorithm for solving large scale TSP problem. The idea of this algorithm is using clustering algorithm to large-scale TSP problem. The large-scale TSP problem is clustered into some small-scale TSP problems. The normal TSP problem is converted into a generalized traveling salesman problem (GTSP) [12]. So, solving large-scale TSP problem is transformed into solving GTSP and several small-scale TSP problems.

2   Description of the Algorithm

This paper presents a hierarchical solving algorithm which can effectively solve the large-scale TSP. The hierarchical solving is refers to using clustering algorithm to large-scale TSP problem firstly, and then the large-scale TSP problem is divided into some city groups. So the original large-scale TSP problem is transformed into GTSP problem. This algorithm needs to obtain an optimal solution of the GTSP problem. However, due to the large scale of the TSP problem, the city number in each clustered group is still very large. Hence we need to use the clustering algorithm to the city group again. The second clustering equals to cluster the center of the city group. If the number of city (center) group is still very large after clustering the center, we continue to cluster layer by layer till the final number of the clustering center is so small that ant colony system algorithm can handle effectively. The process of clustering will stop when the average number of cluster and the final size of each group are less than 40. The procedure is shown in Figure 1. On the first layer, the clustering algorithm is used to large-scale TSP problem. The large scale cities are clustered into multiple city groups. And then the center of each group is regarded as a city. If the number of clusters is greater than 40, then perform clustering on the second layer. This process continues till the number of city group is less than 40. Finally GTSP problems algorithm is applied to obtain the connection order between the groups on the last layer, which is the m-th layer. Solving the shortest loop of the cities in each group on the last layer and the shortest loop is used as the connection order between the groups on the higher layer. This process continues till the connection order between the cities on the first layer is solved.

2.1   Selection of Clustering Algorithm

Ant colony algorithm for small-scale TSP problems (such as 40 cities TSP problem) has a high performance. If the large-scale TSP problem can be divided into several sub-problems and the number of cities in each sub-problem is less than 40, ant colony algorithm can be used to solve each small sub-problem respectively. Finally the solution of each small sub-problem can be merged as the solution of original large-scale TSP problems. This method will greatly improve the performance of the algorithm for the large-scale TSP problem. How to decompose a large-scale TSP problem into sub-problems becomes the key of the algorithm.

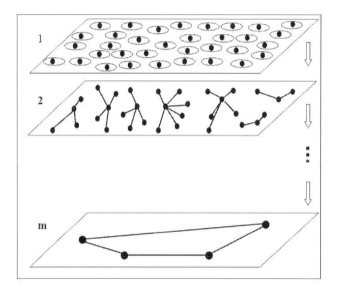

Fig. 1. Schematic diagram of hierarchical clustering for large scale TSP

This article chooses two different clustering algorithms to cluster the cities. One is the affinity propagation (AP) clustering algorithm [13], and the other is the K-means clustering algorithm [14]. Both these algorithms have some advantages and disadvantages: AP clustering algorithm need not to pre-define the number of the cluster. But the computational complexity of AP algorithm is high, which is $O(N \cdot n^2)$. N is the number of the iterations and n is the number of the cities. Therefore, AP clustering algorithm often requires long time when n is relatively large. While K-means algorithm need to set the clustering number but the computational complexity is relatively low, which is $O(N \cdot k \cdot n)$. N is the number of iterations, n is the number of city, and k is the number of clusters. The computation time of K-means clustering algorithm for big data is smaller than AP algorithm. So, integrated the advantages and disadvantages mentioned above, this paper uses the AP clustering algorithm for TSP problem with less than 3000 cities and the K-means clustering algorithm for more than 3000 cities.

2.2   Determine the Connection Order between Groups

In order to obtain the desired optimal solution of GTSP problem on the last layer, we improve the ACS algorithm proposed in literature [11]. The principle of the improved ACS algorithm is almost the same as the algorithm in literature [11]. But there is one difference: the next city selected by ant colony algorithm in Literature [11] is unvisited, While the selected city in improved ACS algorithm is not only unvisited but also not in the same group of the visited cities. For example, in Figure 2, when an ant in the first group selects the next city, the next city cannot be selected from the first group. It can be selected from the second or the third group. Assuming that the ant selects the next city from the second cluster, then the other city can only be selected from the third cluster, and then return to the start city. Figure 2 shows the shortest circuit obtained by improved ACS which connection order is $1 \rightarrow 2 \rightarrow 3 \rightarrow 1$. The three large ellipses are obtained by clustering algorithm.

2.3   Determine the Boundary City of the Cluster

The boundary city of each group is determined according to the connection order between groups. That is to determine the nearest two cities in adjacent groups. Set group a and group b ($a, b = 1, 2, \cdots, c$ and $a \neq b$) are two adjacent groups according to the connection order. $u_i^a \in V_a$ is the city in group a and $u_j^b \in V_b$ is the city in group b. We can determine the boundary city of group a and b with equation (1).

$$\{u_k^a, u_l^b\} = \arg \min_{u_i^a \in V_a, u_j^b \in V_b} d(u_i^a, u_j^b) \tag{1}$$

Where u_k^a is the boundary city in group a and u_l^b is the boundary city in group b. $d(\cdot, \cdot)$ represents the Euclidean distance between the two cities. As shown in Figure 3, the black point is the boundary city of each group.

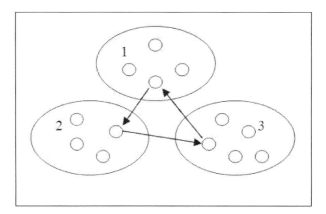

Fig. 2. Connection order between the cluster and the cluster

2.4   Solve the Optimal Path within the Group

Using the above method in section 2.3 determines the boundaries cities of each group. Each group has two border cities. We will solve the shortest path in each group, in which the endpoints are the boundary cities. For convenience this paper changes the shortest path problem with fixed endpoints into a shortest Hamilton circuit problem. The latter is solved by the traditional ACS algorithm. The procedure is as follows: when an ant selects the next city, if the current city is one of two boundary cities and the other boundary city has not been visited, and then jump directly to the other boundary city. As shown in Figure 4 the black points are the boundary cities. The ant travels from the fifth city, through the sixth, seventh to the eighth city (boundary city). And then skips to the third city (boundary city). And then travels back to the original city. Finally set up an optimal path (solid lines).

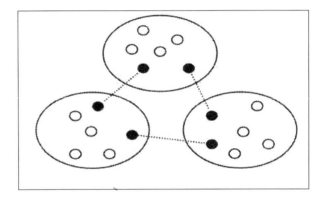

Fig. 3. Schematic diagram of Border City

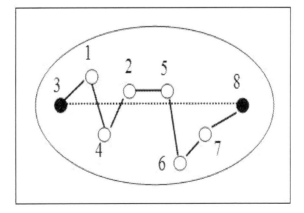

Fig. 4. Schematic diagram of optimal path

2.5   Merge the Path

After solved the shortest path in each group, we merge the shortest path according to the connection order between groups to set up a feasible solution of large-scale TSP problem. As shown in Figure 5, the solid line plus the dot line constitute a feasible solution of original TSP problem.

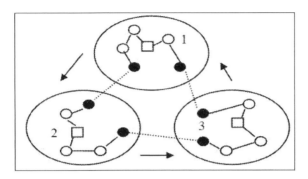

Fig. 5. Quasi-optimal path of TSP problem

3   Experimental Results

To evaluate the performance of the proposed algorithm, computational experiments are conducted to compare the performance of traditional Ant Colony System algorithm with the proposed algorithm. The test problems are a standard collection of TSPs selected from the Library of Traveling Salesman Problems [15]. In our simulation experiments, all programs are implemented in Java software on a workstation with 4G memories, Intel (R) Xoon (R) CPU. Because ACS algorithm cannot directly obtain the solution of large-scale TSP problem, we use the original ACS algorithm to ten TSP problems out of TSP standard database which have relatively small scales and the two experimental data in the literature [16]. They are Pr107、pr136、Pr299、Pcb442、U574、D657、Rat783、Pr1002、D2103、Pr2392、Block107、Block364. In the name of the test problems, the letter represents the type of the problem and the number represents the scale of the problem. We set parameters of the ACS algorithm as follows: ant number $m = 10$, real number $\alpha = 1$, $\beta = 4$, $q_0 = 0.8$, $\gamma = 0.1$, $\rho = 0.1$, and the iteration number is 1000. We execute 10 times for each problem. Table 1 shows the experimental results of the ACS algorithm.

The problem Block107 and Block364 in table 1 are a kind of machining data. We use Concorde to obtain the optimal solution of them. The error is obtained according to the equation (2).

$$Err = (Ave - opt) / opt \times 100\% \qquad (2)$$

And we use the proposed algorithm to the 12 standard test problems mentioned above and six large-scale TSP problems. These six large-scale TSP problems are Rl11849、Brd14051、D15112、D18512、Pla33810、Pla85900. We set parameters of the proposed Algorithm as follows: ant number $m = 10$, real number $\alpha = 1$, $\beta = 4$, $q_0 = 0.8$, $\gamma = 0.1$, $\rho = 0.1$, and the iteration number is 500. We execute 10 times for each problem. Table 2 shows the experimental results of the proposed algorithm, in which the error is obtained according to the equation (2). Figure 6 shows the errors of the two algorithms (black is the error of the ACS algorithm, gray is the error of the proposed algorithm), Figure 7 shows the run time of the two algorithms (dot line is the run time of the ACS algorithm, solid line is the run time of the proposed algorithm). N is the label of the TSP problem.

Table 1. Experimental results of ACS algorithm

N	TSP	Optimal	Best	Average	Error (%)	Time (s)
1	Pr107	44303	44524	45042	1.668	18
2	Pr136	96772	98243	99152	2.459	30
3	Pr299	48191	49586	51642	7.161	140
4	Pcb442	50778	53822	54901	8.120	309
5	U574	36905	40833	41655	12.870	517
6	D657	48912	56011	57244	17.034	682
7	Rat783	8806	10315	10853	23.245	956
8	Pr1002	259045	297530	299256	15.522	1574
9	D2103	80450	88525	89225	10.907	6837
10	R2392	378032	459847	461623	22.112	9367
11	Block107	2769	2826	2854	2.059	18
12	Block364	3011	3166	3189	5.148	208

Table 3 shows the results of the proposed algorithm for solving randomly generated TSP problem. The horizontal and the vertical ordinate of cities in the randomly generated TSP problem are integer numbers in the interval $[0, n]$. n is the scale of TSP problem. Ratio is given by equation (3).

$$ratio = \frac{best}{n\sqrt{n}} \quad (3)$$

Where $best$ is the best solution of the TSP problem obtained by the proposed algorithm. This ratio represents the quality of the optimal solution of TSP problem. Johnson et al. [17] obtained the $ratio = 0.7124 \pm 0.0002$ in 1996. This indicates that the obtained ratio is closer to 0.7124 ± 0.0002, the solution of the TSP problem is better.

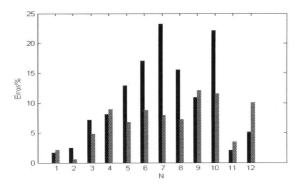

Fig. 6. Errors comparison

Table 2. Experimental results of the proposed algorithm

N	TSP	C	Optimal	Best	Average	Error(%)	Time(s)
1	Pr107	6	44303	44663	45251	2.169	4
2	Pr136	8	96772	96922	97335	0.581	5
3	Pr299	13	48191	50273	50498	4.787	17
4	Pcb442	13	50778	54838	55326	8.957	32
5	U574	18	36905	39228	39411	6.790	54
6	D657	18	48912	52665	53234	8.836	71
7	Rat783	23	8806	9453	9505	7.938	147
8	Pr1002	27	259045	275911	277768	7.228	136
9	D2103	40	80450	89712	90169	12.080	258
10	Pr2392	55	378032	419476	421630	11.532	247
11	Block107	7	2769	2865	2874	3.467	4
12	Block364	13	3011	3315	3339	10.100	26
13	Rl11849	300	923288	1052300	1056325	14.409	582
14	Brd14051	350	469385	507713	507876	8.200	714
15	D15112	400	1573084	1707900	1714300	8.977	845
16	D18512	500	645238	702630	706110	9.434	1018
17	Pla33810	900	66048945	75055791	75317000	14.032	2853
18	Pla85900	2100	142382641	159269135	159543976	12.052	13470

C is the cluster number

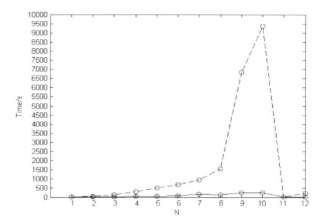

Fig. 7. Comparison of time between two algorithms

Table 3. The results of randomly generated TSP problem

TSP	City num-ber	Trial number	Average ratio	Time
R1000	1000	10	0.784	105
R10000	10000	10	0.785	404
R100000	100000	5	0.783	16847
R400000	400000	5	0.782	62404
R900000	900000	5	0.781	143014
R2500000	2500000	1	0.782	425925

4   Conclusions

TSP problem is a well-known NP-hard combinatorial problem. Ant colony system algorithm for large scale TSP problem has some shortcomings, such as slower convergence speed and longer runtime. Especially when the scale of TSP problem is relatively large, ant colony system algorithm cannot solve the problem in expected time. This paper presents a hierarchical algorithm based on clustering algorithm for solving large-scale TSP problem. First of all, through clustering algorithm to decompose large-scale traveling salesman problem into several small scale traveling salesman problem. And then, traditional ant colony algorithm is used to solve each small traveling salesman problem. Finally, the optimal path of sub-problems merged into a quasi-optimal solution of the TSP problem. The experimental results show that the proposed algorithm for traveling salesman problem, especially for cities distribution has clustering characteristics, will obtain good results. With the scale of the TSP problem increases, the advantage of the presented algorithm becomes more significant.

References

1. Fischetti, M., Lodi, A., Toth, P.: Exact Methods for the Asymmetric Traveling Salesman Problem. The Traveling Salesman Problem and Its Variations **12**, 169–205 (2002)
2. Naddef, D.: Polyhedral Theory and Branch and Cut Algorithms for the Symmetric TSP. The Traveling Salesman Problem and Its Variations **12**, 29–116 (2002)
3. Gomory, R.E.: Outline of an Algorithm for Integer Solutions to Linear Programs. Bulletin of the American Mathematical Society **64**(5), 275–278 (1958)
4. Barnhart, C., Johnson, E.L., Nemhauser, G.L., Savelsbergh, M.W.P., Vance, P.H.: Branch-and-price: Column Generation for Solving Huge Integer Programs. Operations Research **46**(3), 316–329 (1998)
5. Applegate, D.L., Bixby, R.E., Chvatal, V., Cook, W.J.: The Traveling Salesman Problem: A Computational Study. Princeton University Press, New Jersey (2006)
6. Papadimitriou, C.H., Steglitz, K.: Combinatorial Optimization: Algorithm and Complexity. Prentice Hall of India Private Limited (1997)
7. Lin, S., Kernighan, B.W.: An Effective Heuristic Algorithm for the Traveling Salesman Problem. Operations Research **21**(2), 498–516 (1973)
8. Helsgaun, K.: An Effective Implementation of the Lin-Kernighan traveling salesman heuristic. European Journal of Operational Research **126**(1), 106–130 (2000)
9. Applegate, D., Cook, W., Rohe, A.: Chained Lin-Kernighan for Large Traveling Salesman Problems. INFORMS Journal on Computing **15**(1), 82–92 (2003)
10. Hung, D.N., Ikuo, Y., Kunihito, Y., Moritoshi, Y.: Implementation of an Effective Hybrid GA for Large-scale Traveling Salesman Problems. IEEE Transactions on Systems, Man, and Cybernetics Part B: Cybernetics **37**(1), 92–99 (2007)
11. Dorigo, M., Gambardella, L.M.: Ant Colony System: A Cooperative Learning Approach to the Traveling Salesman Problem. IEEE Transactions on Evolutionary Computation **1**(1), 53–66 (1997)
12. Henry-Labordere, A.L.: The Record Balancing Problem: A Dynamic Programming Solution of a Generalized Traveling Salesman Problem. RAIROB **2**, 43–49 (1969)
13. Frey, B.J., Dueck, D.: Clustering by Passing Messages Between Data Points. Science **315**(5814), 972–976 (2007)
14. MacQueen, J.B.: Some Methods for Classification and Analysis of Multivariate Observations. In: Proceedings of 5th Berkeley Symposium on Mathematical Statistics and Probability, pp. 281–297. University of California Press, Berkeley (1967)
15. TSPLIB, http://www.iwr.uni-heidelberg.de/groups/comopt/software/TSPLIB95
16. Wu, C.G., Liang, Y.C., Lee, H.P., Lu, C.: Generalized Chromosome Genetic Algorithm for Generalized Traveling Salesman Problems and Its Applications for Machining. Physical Review E **70**(1), 1–13 (2004)
17. Johnson, D.S., McGeoch, L.A., Rothberg, E.E.: Asymptotic Experimental Analysis for the Held-Karp Traveling Salesman Bound. In: Proceedings of the Seventh Annual ACM-SIAM Symposium on Discrete Algorithms, pp. 341–350. Society for Industrial and Applied Mathematics (1996)

Gaussian Process Learning:
A Divide-and-Conquer Approach

Wenye Li[✉]

Macao Polytechnic Institute, Rua de Luís Gonzaga, Macao SAR, China
wyli@ipm.edu.mo
http://staff.ipm.edu.mo/~wyli

Abstract. The Gaussian Process (GP) model is used widely in many hard machine learning tasks. In practice, it faces the challenge from scalability concerns. In this manuscript, we proposed a domain decomposition method in GP learning. It is shown that the GP model itself has the inherent capability of being trained through *divide-and-conquer*. Given a large GP learning problem, it can be divided into smaller problems. By solving the smaller problems and merging the solutions, it is guaranteed to reach the solution to the original problem. We further verified the efficiency and the effectiveness of the algorithm through experiments.

Keywords: Gaussian process · Domain decomposition · Machine learning

1 Introduction

Recently considerable researches have been devoted to the study of the Gaussian Process (GP) model. With flexible non-parametric nature, the GP model provides a powerful statistical tool and has been routinely used to solve many difficult machine learning problems [1–3].

Unfortunately in practice, the GP model faces the challenge of computational scaling. The model requires the solution of a dense positive definite linear system $Ac = y$. The computational cost is $O\left(n^3\right)$ (suppose A is an $n \times n$ matrix) by direct solvers, which is prohibitive for large tasks. To deal with the difficulty, one line of work suggests to use sparse or low-rank approximations [4–7]. These approximated methods significantly reduce the computation, but they generally do not provide the optimal solution.

To seek a quick yet optimal solution, people have explored the conjugate gradient method [8,9]. The method modifies the components of the solution successively, until convergence is achieved. If $rank\left(A\right) = r$, it converges in $r + 1$ steps. The computational complexity is $O\left(rn^2\right)$. While this represents an improvement for some problems, unfortunately the rank of the matrix may not be small. Unfortunately in GP learning, The matrix A is typically full rank and again the complexity becomes $O\left(n^3\right)$.

© Springer International Publishing Switzerland 2014
Z. Zeng et al. (Eds.): ISNN 2014, LNCS 8866, pp. 262–269, 2014.
DOI: 10.1007/978-3-319-12436-0_29

In this paper, we consider a different iterative approach, a domain decomposition method. We show the GP model itself has the capability of being trained in a *divide-and-conquer* manner. That is, for a large GP problem, we can divide the problem into smaller ones. By solving these smaller GP tasks and merging the solutions, it is guaranteed to reach the optimal solution. This line of research has attracted considerable research in GP models recently [10,11].

A word on notation: A lower-cased k and ϕ denote a kernel function, and a capital letter denotes a matrix (e.g. K and Φ). A bold lower-cased letter denotes a vector (e.g. $\mathbf{c}$, $\mathbf{f}$), and a corresponding normal letter with subscript i refers to the i-th entry (e.g. c_i). $\mathbf{0}$ denotes a vector of all 0's. Furthermore, let A be an $n \times n$ matrix and $\mathbf{c}$ be an $n \times 1$ vector. If $s_1, s_2 \subseteq \{1, \cdots, n\}$, then A_{s_1, s_2} is a matrix obtained from A by keeping only the rows with indices in s_1 and the columns with indices in s_2, and $\mathbf{c}_{s_1}$ is a vector obtained from $\mathbf{c}$ by keeping the elements with indices in s_1.

2   Gaussian Process Model

Given a set of input vectors $X = \{\mathbf{x}_1, \cdots, \mathbf{x}_n\}$ and a set of observed output scalars $\mathbf{y} = (y_1, \cdots, y_n)^T$. We are concerned with the problem of seeking a function f which explains the relationship between each pair of $\mathbf{x}$ and y:

$$y = f(\mathbf{x}) + \varepsilon,$$

where ε is an additive i.i.d. Gaussian noise with a mean 0 and a variance σ_n^2. With this function, we are able to give the output vector $\mathbf{f}_*$ corresponding with a set of testing inputs $X_* = \{\mathbf{x}_1^*, \cdots, \mathbf{x}_m^*\}$.

In Gaussian processes, under some prior, the joint distribution of the observed target values and the function values for X_* is given by

$$\begin{bmatrix} \mathbf{y} \\ \mathbf{f}_* \end{bmatrix} \sim \mathcal{N} \left(\mathbf{0}, \begin{bmatrix} \Phi_{X,X} + \sigma_n^2 I & \Phi_{X,X_*} \\ \Phi_{X_*,X} & \Phi_{X_*,X_*} \end{bmatrix} \right).$$

Here Φ_{X,X_*} denotes the $n \times m$ matrix of the covariances evaluated at all pairs of training and testing points. The i,j-th entry is given by $\phi(\mathbf{x}_i, \mathbf{x}_j^*)$ where ϕ is a predefined symmetrical positive definite function. The similarity holds for the entries of $\Phi_{X,X}$, $\Phi_{X_*,X}$ and Φ_{X_*,X_*}.

The key predictive equation for $\mathbf{f}_*$ for GP learning is

$$\mathbf{f}_* | X, \mathbf{y}, X_* \sim \mathcal{N}\left(\bar{\mathbf{f}}_*, cov(\mathbf{f}_*)\right)$$

where

$$\bar{\mathbf{f}}_* \doteq E\left[\mathbf{f}_* | X, \mathbf{y}, X_*\right] = \Phi_{X_*,X}\left[\Phi_{X,X} + \sigma_n^2 I\right]^{-1} \mathbf{y} \tag{1}$$

and

$$cov(\mathbf{f}_*) = \Phi_{X_*,X_*} - \Phi_{X_*,X}\left[\Phi_{X,X} + \sigma_n^2 I\right]^{-1} \Phi_{X,X_*}.$$

Another way to look at the solution in equation (1) is to see it as a linear combination of n kernel functions:

$$\bar{f}(\mathbf{x}^*) = \sum_{i=1}^{n} c_i \phi(\mathbf{x}_i, \mathbf{x}^*), \quad \mathbf{x}^* \in X_* \tag{2}$$

where

$$\mathbf{c} = (c_1, \cdots, c_n)^T = \left(\Phi_{X,X} + \sigma_n^2 I\right)^{-1} \mathbf{y}. \tag{3}$$

We need to solve a linear system to train a GP model. The computational cost is $O(n^3)$ by direct solvers, where n is the number of input points. For large problems (e.g. $n > 10,000$), the cost becomes expensive.

3   Domain Decomposition in GP

To provide scalability to the GP model, we propose a *divide-and-conquer* method in GP learning. Algorithm (1) uses a sub-routine GP_naive, which has three input arguments. The first is a training input set X. The second is the observed output $\mathbf{y}$. The last is a testing input set X_*. It has two outputs. The first is a function $\bar{f}$ by equation (2). The second is a vector of predicted function values by equation (1).

Algorithm 1. GP Learning by *Divide-and-Conquer*.

1: $s = \{1, \cdots, m\}$.
2: Divide s into subsets: $s = s_1 \cup \cdots \cup s_\ell$.
3: $t = 0, \bar{f} = 0$.
4: **repeat**
5:     $t = t + 1$.
6:     **for** $j = 1$ to ℓ **do**
7:        $\left(\bar{f}_{s_j}^t, \bar{\mathbf{f}}_{s-s_j}^t\right) = GP\_naive\left(X_{s_j}, \mathbf{y}_{s_j}, X_{s-s_j}\right)$.
8:        $\mathbf{y}_{s_j} = 0, \ \mathbf{y}_{s-s_j} = \mathbf{y}_{s-s_j} - \bar{\mathbf{f}}_{s-s_j}^t$.
9:        $\bar{f} = \bar{f} + \bar{f}_{s_j}^t$.
10:    **end for**
11: **until** $\bar{f}$ converges
12: **return** $\bar{f}$

This algorithm divides the training data X into ℓ different subsets and then solves the whole problem iteratively. At step 7, it treats one subset X_{s_j} as the training set and the rest X_{s-s_j} as the testing set, and returns a solution $\bar{f}_{s_j}$. It is expected that the input-output relationship cannot be explained fully by $\bar{f}_{s_j}$ for all the data, so it updates the observations $\mathbf{y}$, extracting the part of the observations that has been explained by $\bar{f}_{s_j}$ in step 8. Then the updated observations will be used in the next round training process. Accordingly, the solution $\bar{f}_{s_j}$ is added to $\bar{f}$ in step 9.

4   Justification

To show the correctness of algorithm (1), one needs to observe the actual computation of the algorithm, which involves solving a linear system in equation (3). Denote $A = \Phi_{X,X} + \sigma_n^2 I$, and equation (3) becomes the problem of solving $\mathbf{c} = A^{-1}\mathbf{y}$. Then we re-write the actual computation of algorithm (1) as in algorithm (2). This algorithm is a variant of block Gauss-Seidel method in solving linear systems [12], and we omit the discussion here.

Algorithm 2. Computation in GP Learning.

1: $s = \{1, \cdots, m\}$
2: Divide s into subsets: $s = s_1 \cup \cdots \cup s_\ell$.
3: $t = 0, \mathbf{c} = \mathbf{0}$.
4: **repeat**
5:     $t = t + 1, \mathbf{c}^t = \mathbf{0}$.
6:     **for** $j = 1$ to ℓ **do**
7:         Solve $\mathbf{c}_{s_j}^t$ by $A_{s_j, s_j} \mathbf{c}_{s_j}^t = \mathbf{y}_{s_j}$.
8:         $\mathbf{y}_{s_j} = \mathbf{0}, \mathbf{y}_{s-s_j} = \mathbf{y}_{s-s_j} - A_{s-s_j, s_j} \mathbf{c}_{s_j}^t$.
9:         $\mathbf{c}_{s_j} = \mathbf{c}_{s_j} + \mathbf{c}_{s_j}^t$.
10:    **end for**
11: **until** $\mathbf{c}$ converges
12: **return** $\mathbf{c}$

4.1   Preliminaries and Definitions

Algorithm (2) can be analyzed as successive alternating projections in Hilbert space, based on the work of [13–15].

Lemma 1. *For any $n \times n$ positive definite matrix A, there exist n points*[1] $\mathbf{x}_1, \cdots,$ $\mathbf{x}_n$ *in some space R^d and a kernel function k defined on R^d, such that $A_{ij} = k(\mathbf{x}_i, \mathbf{x}_j)$.*

With this lemma, we can see that the problem of finding a solution to $A\mathbf{c} = \mathbf{y}$ has been changed to the problem of finding a function $f = \sum_{j=1}^n c_j k(\mathbf{x}_j, \cdot)$ such that $f(\mathbf{x}_i) = y_i$ for all $\mathbf{x}_i$.

Furthermore, given a kernel function $k(\mathbf{x}, \mathbf{x}')$ and a finite set of distinct points $X = \{\mathbf{x}_1, \cdots, \mathbf{x}_n\}$ in R^d, let $\mathcal{H}_K$ denote the function space induced by k:

$$\left\{ \sum_{j=1}^n a_j k(\mathbf{x}_j, \cdot) : a_1, \cdots, a_n \in R \right\}$$

endowed with the inner product

$$\langle f, g \rangle = \sum_{i,j=1}^n a_i b_j k(\mathbf{x}_i, \mathbf{x}_j)$$

---

[1] The $\mathbf{x}_1, \cdots, \mathbf{x}_n$ here have a slightly different meaning from those in section 2.

where

$$f = \sum_{i=1}^{n} a_i k\left(\mathbf{x}_i, \cdot\right) \text{ and } g = \sum_{j=1}^{n} b_j k\left(\mathbf{x}_j, \cdot\right).$$

Given $X_i \subseteq X$ $(1 \leq i \leq \ell)$, let $\mathcal{H}_i$ denote the subspace of functions in $\mathcal{H}_K$ associated with X_i:

$$\mathcal{H}_i = \left\{ f \in \mathcal{H}_K : f = \sum_{\mathbf{x} \in X_i} c_{\mathbf{x}} k\left(\mathbf{x}, \cdot\right), \text{ where } c_{\mathbf{x}} \in R \right\}.$$

4.2   Interpolation Operator and Orthogonal Projection

The definitions of orthogonal projection and interpolation operator in a Hilbert space are as follows.

Definition 1. *(Orthogonal Projection) Let $\mathcal{V}$ be a closed subspace of a Hilbert space $\mathcal{H}$. The linear operator $P : \mathcal{H} \to \mathcal{V}$ is called the orthogonal projection onto $\mathcal{V}$ if for any $f \in \mathcal{H}$ and any $v \in \mathcal{V}$*

$$\langle v, f - Pf \rangle = 0,$$

where $\langle \cdot, \cdot \rangle$ denotes the inner product in $\mathcal{H}$.

Definition 2. *(Interpolation Operator) Let $X_1, \cdots, X_\ell$ be subsets of X, such that $\cup_{i=1}^{\ell} X_i = X$. Given $f \in \mathcal{H}_K$, define interpolation operators $P_i : \mathcal{H}_K \to \mathcal{H}_i, i = 1, \cdots, \ell$ by*

$$P_i f = \sum_{\mathbf{x} \in X_i} c_{\mathbf{x}} K\left(\mathbf{x}, \cdot\right)$$

and

$$(P_i f)\left(\mathbf{z}\right) = f\left(\mathbf{z}\right) \text{ for all } \mathbf{z} \in X_i.$$

With the definition, we can see P_i is the orthogonal projection onto $\mathcal{H}_i$.

Lemma 2. *Let $X_i \subseteq X, (1 \leq i \leq \ell)$ be a finite set of distinct points in R^d and P_i denote the interpolation operator defined above. Then P_i is the orthogonal projection from $\mathcal{H}_K$ onto $\mathcal{H}_i$.*

Now algorithm (2) can be seen as a version of alternating projections. Each execution of step 7 and step 8 corresponds to an orthogonal projection from $\mathcal{H}_K$ onto subspace $\mathcal{H}_i$.

4.3   A Domain Decomposition Approach

We study the orthogonal projection onto the intersection $\mathcal{U} \cap \mathcal{V}$ of the two closed subspaces $\mathcal{U}$ and $\mathcal{V}$ of a general Hilbert space $\mathcal{H}$. Denote the projection by $P_u \wedge P_v$, and based on von Neumann's alternating projection theorem [16] we have:

$$\lim_{t \to \infty} \left(P_u P_v\right)^t f = \left(P_u \wedge P_v\right) f,$$

where the convergence is in the norm of $\mathcal{H}$ and $f \in \mathcal{H}$. The theorem generalizes to any finite number of subspaces. From [17], the algorithm is linearly converged.

Based on the justification, we have a domain decomposition approach. It generates a sequence $\{f_{\ell t+i}\}$, where $t = 1, \cdots$ and $i = 1, \cdots, \ell$ via

$$f_0 = f \ and \ f_{\ell t+i} = f_{\ell t+i-1} - P_i f_{\ell t+i-1}.$$

And the sequence of approximations is given by:

$$\bar{f}^0 = 0 \ and \ \bar{f}^{\ell t+i} = \bar{f}^{\ell t+i-1} + P_i f_{\ell t+i-1}.$$

The algorithms (1) and (2) exactly follow this approach.

5   Results

We compared the domain decomposition approach, and the conjugate gradient method in GP learning. Three datasets from CMU text mining group were used. The 20-newsgroups dataset has about $19,000$ pages in 20 classes. The webkb dataset has about $8,300$ pages in 7 classes. The 7-sectors dataset has about $4,600$ pages in 7 classes. In each experiment, the matrix $K_{X,X}$ was computed from the dataset X with a Gaussian kernel. Then we got the matrix $A = K + \sigma_n^2 I$. The vector $\mathbf{y}$ was set with document's class labels. We solved $A\mathbf{c} = \mathbf{y}$ for the parameters $\mathbf{c}$. For domain decomposition, we set the size of each subsystem to be $1,000$.

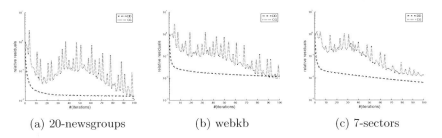

| (a) 20-newsgroups | (b) webkb | (c) 7-sectors |

Fig. 1. Comparison of the domain decomposition (DD) method and the conjugate gradient (CG) method

Figure (1)(a)-(c) depicts the results. In each sub-figure, the horizontal axis is the iterations and the vertical axis depicts the relative residuals after each iteration. We can see that the domain decomposition (DD) approach reaches an acceptable residual in much fewer iterations than the conjugate gradient (CG) method. In each iteration the complexities of the methods are similar, and the DD method actually runs quicker than the CG method.

6   Conclusion

In this paper, we studied a *divide-and-conquer* method in training Gaussian Process and related models [18,19]. It divides a large learning problem into small problems and solves these small problems iteratively. The process converges with at least a linear rate.

We justified the algorithm as alternating projections in Hilbert space. This *divide-and-conquer* property is inherent in GP model. It equips the model with the ability in handling large problems.

Acknowledgments. The work is partially supported by Macao Polytechnic Institute (RP/ESAP-01/2014) and The Science and Technology Development Fund, Macao SAR, China (044/2010/A).

References

1. Williams, C., Rasmussen, C.: Gaussian Processes for Regression. In: Advances in Neural Information Processing Systems'8. MIT Press (1996)
2. MacKay, D.: Introduction to Gaussian Processes. Technical report, Cambridge university (1997)
3. Seeger, M.: Gaussian Processes for Machine Learning. International Journal of Neural Systems **14**, 69–106 (2004)
4. Csató, L., Opper, M.: Sparse On-line Gaussian Processes. Neural Computation **14**, 641–668 (2002)
5. Lawrence, N., Seeger, M., Herbrich, R.: Fast sparse gaussian process methods: The informative vector machine. In: Advances in Neural Information Processing Systems'15, pp. 609–616. MIT Press (2003)
6. Tipping, M.: Sparse Bayesian Learning and the Relevance Vector Machine. Journal of Machine Learning Research **1**, 211–244 (2001)
7. Williams, C., Seeger, M.: Using the Nyström Method to Speed Up Kernel Machines. In: Advances in Neural Information Processing Systems'13, pp. 682–688. MIT Press (2001)
8. Rifkin, R.: Everything Old is New Again: A Fresh Look at Historical Approaches in Machine Learning. PhD thesis, Massachusetts Institute of Technology (2002)
9. Yang, C., Duraiswami, R., Davis, L.: Efficient Kernel Machines Using the Improved Fast Gauss Transform. In: Advances in Neural Information Processing Systems'17, pp. 1561–1568. MIT Press (2005)
10. Chalupka, K., Williams, C., Murray, I.: A Framework for Evaluating Approximation Methods for Gaussian Process Regression. Journal of Machine Learning Research **14**, 333–350 (2013)
11. Bo, L., Sminchisescu, C.: Greedy Block Coordinate Descent for Large Scale Gaussian Process Regression. In: Proceedings of the 24th Conference in Uncertainty in Artificial Intelligence, pp. 43–52. AUAI Press (2008)
12. Golub, G., van Loan, C.: Matrix Computations. John Hopkins Studies in the Mathematical Sciences. 3rd edn. Johns Hopkins University Press (1996)
13. Faul, A., Powell, M.: Proof of Convergence of an Iterative Technique for Thin Plate Spline Interpolation in Two Dimensions. Advances in Computational Mathematics **11**, 183–192 (1999)

14. Schaback, R., Wendland, H.: Numerical Techniques Based on Radial Basis Functions. In: Curve and Surface Fitting, pp. 359–374. Vanderbilt University Press (2000)

15. Li, W., Lee, K.H., Leung, K.S.: Large-scale RLSC Learning Without Agony. In: Proceedings of the 24th Annual International Conference on Machine Learning, pp. 529–536. ACM (2007)

16. von Neumann, J.: Mathematical Foundations of Quantum Mechanics. Princeton University Press (1955)

17. Smith, K., Solomon, D., Wagner, S.: Practical and Mathematical Aspects of the Problem of Reconstructing Objects from Radiographs. Bulletin of the American Mathematical Society, 1227–1270 (1977)

18. Li, W., Lee, K.H., Leung, K.S.: Generalized Regularized Least-Squares Learning with Predefined Features in a Hilbert Space. In: Advances in Neural Information Processing Systems'19, pp. 881–888. MIT Press (2007)

19. Li, W., Leung, K.S., Lee, K.H.: Generalizing the Bias Term of Support Vector Machines. In: Proceedings of the 20th International Joint Conference on Artificial Intelligence, pp. 919–924. AAAI (2007)

A Kernel ELM Classifier for High-Resolution Remotely Sensed Imagery Based on Multiple Features

Wei Yao[1,2], Zhigang Zeng[1(✉)], Cheng Lian[1], and Huiming Tang[3]

[1] School of Automation, Huazhong University of Science and Technology, Wuhan, China
wyao@mail.scuec.edu.cn, Zgzeng_hust@163.com,
liancheng@hust.edu.cn
[2] School of Computer Science, South-Central University for Nationalities, Wuhan, China
[3] Faculty of Engineering, China University of Geosciences, Wuhan, China
hmtang_cug@163.com

Abstract. Better interpretation about the contents in high-resolution remote sensing images can be obtained by using multiple features of various types. In order to process large image data sets with high feature dimensions, the very efficient algorithm of kernel extreme learning machine is employed to in our study to build image classifiers. In order to avoid the overflow problem, the classification strategy is improved by training classifiers on different features independently and then fusing the classification results. The effectiveness of the proposed classification approaches are shown by the experimental results achieved on a realistic remote sensing image data set.

Keywords: Classification · Kernel extreme learning machine · Remote sensing · Multiple features

1 Introduction

For decades, remote sensing images have been very important sources for people to learn about the earth. The knowledge in remote sensing images that can be extracted using image classification approaches is generally referred as land use and land cover (LULC) information. LULC information is important in many aspects including environment protection, city management and planning, resource survey, natural disaster prevention and so on.

Thanks to the development of remote sensing techniques in recent years, remote sensing images with much higher spatial resolution and much more spectral bands become available. And this may lead to more accurate classifications. However, higher feature dimensions also lead to two major problems. First, the computation burden will be largely increased. Therefore the processing time will be lengthened and the efficiency becomes a problem. Furthermore, the data sizes of high resolution remote sensing images are usually very large as compared to the storage and processing abilities of common computers. So the overflow problem may also happen during the processing of remote sensing image data.

© Springer International Publishing Switzerland 2014
Z. Zeng et al. (Eds.): ISNN 2014, LNCS 8866, pp. 270–277, 2014.
DOI: 10.1007/978-3-319-12436-0_30

In order to avoid these problems in remote sensing image classifications, algorithms which are followed to establish image classifiers should be highly efficient. There are many successful applications of support vector machine (SVM) in tasks of remote sensing image classification [1-3]. Extreme learning machine (ELM), which is "extremely" fast, is also a guaranteed choice for obtaining LULC information [4, 5]. In our study, an improved version of ELM, the kernel ELM (k-ELM) [6], is employed. Highly efficient classification algorithms can tackle the efficiency problem of image classification, while the classification strategies should also be optimized to avoid the overflow problem. Therefore, we develop a multi-feature classifier fusion approach.

2   Classifier for Multiple Features

2.1   Multiple Features

Image classification accuracy is heavily relied on the feature dimension of the image data set. Since the spectral features have its limitation, it's quite necessary to include some higher level image features in the classification processes. In our study, we made use of texture features and shape features, in addition to the original spectral features.

A. Texture features based on Gray-Level Co-Occurrence Matrix (GLCM)
The GLCM technique is a standard technique for extracting texture features from remote sensing images [7]. There are several different texture measures, which can represent the gray-level difference between neighboring pixels. In order to extract GLCM features, first there has to be a base image. As for remote sensing images with multiple spectral bands, there are many ways to build different base images. Then moving windows are used to define the neighborhoods of pixels in the image, and pixels with different gray levels in the window are counted to form the GLCM. In order to capture multi-scale characteristics, the GLCM texture measures should be computed using moving windows of different sizes. Therefore, by selecting different base images, different moving window sizes and different textural directions, many sets of texture measures can be calculated. So it's quite convenient to extract high dimension texture features from remote sensing images.

B. Morphological Profiles (MPs)
Morphology is another way to express the spatial information of the remote sensing images [8]. There are two fundamental morphological operators, known as erosion and dilation. These two operators are applied to an image with a basic shape, called a structuring element (SE). When applied to an image, the erosion operator can produce output images showing where the SE fits the objects in the image, and the dilation operator can produce output images showing where the SE hits the objects in the image [9]. Therefore, morphological operators can produce features that are related to the shapes of the objects in images. The erosion and dilation operators can be combined into another two operators, which are more widely used for extracting shape

features of images. These two operators are known as opening and closing. The opening and closing operators, as well as the erosion and dilation operators, can be performed with different SEs. The dimension of morphological features is determined by the number of SEs and the number of morphological operators that have been applied to the images.

2.2   Principles of k-ELM

K-ELM is a development from ELM, which is a very fast learning algorithm for single-layer feed-forward networks (SLFNs). The forward function of a SLFN can be expressed as

$$y_k = W_o \cdot g(W_i \cdot x_k + b), \quad k = 1, 2, ..., K \tag{1}$$

where x_k is the kth input in a sample set with the size of K, and y_k is the corresponding output of the network. W_i and W_o are two matrix containing the input and output connection weights of the network, while b is a vector representing the biases of the hidden-layer neurons. For an M-input N-output SLFN with L hidden-layer neurons, the size of W_i, W_o and b are $L \times M$, $L \times N$ and $L \times 1$ respectively. The Sigmoid-type activation function of the hidden-layer neurons is denoted as $g()$. Following the learning algorithm of ELM, both W_i and b will be randomly assigned in previous and only W_o need to be tuned using the training samples $\{x_k, t_k\}$, where t_k represents the desired output. And this process can be implemented in a very simple way that

$$W_o = TH^+ = T (H^T H)^{-1} H^T \tag{2}$$

where H^+ is the Moore–Penrose generalized inverse of the hidden-layer output matrix H [10] and

$$H = \left[g(W_i \cdot x_1 + b), \ g(W_i \cdot x_2 + b), ..., \ g(W_i \cdot x_K + b) \right]_{L \times K}$$
$$T = [t_1, t_2, ..., t_K]$$

Then the output of the SFLN can be calculated following

$$y = T(H^T H)^{-1} H^T \cdot g(W_i \cdot x + b) \tag{3}$$

In k-ELM, the 'kernel trick' [11] is made use of. The activation functions of the hidden-layer neurons will be replaced by kernel functions. The kernel function for a SLFN can be defined as

$$G(x, x) = g^T (W_i \cdot x + b) \cdot g(W_i \cdot x + b) \tag{4}$$

Furthermore, according to the ridge regression theory, a regularization term is added to improve the generalization ability of the learning process and the forward function can be rewritten as

$$y = T(H^T H + \frac{I}{c})^{-1} H^T g(W_i \cdot x + b) = T\left(\begin{bmatrix} G(x_1, x_1) & \cdots & G(x_1, x_K) \\ \vdots & \vdots & \vdots \\ G(x_K, x_1) & \cdots & G(x_K, x_K) \end{bmatrix} + \frac{I}{c}\right)^{-1} \begin{bmatrix} G(x_1, x) \\ \vdots \\ G(x_K, x) \end{bmatrix} \quad (5)$$

where c is the regularization parameter and I is a unity matrix.

In k-ELM, the activation functions and the number of hidden-layer neurons need not to be known, while only the kernel function needs to be defined in previous. The random feature mappings in ELM, which are implemented by the activation functions and input connection weights, are replaced by kernel mappings. Therefore, as a learning algorithm, k-ELM is more stable than basic ELM.

2.3   Classifier Based on k-ELM

In order to build a k-ELM based classifier for remote sensing images, the numbers of input neurons and output neurons in the SFLN will be set equally to the feature dimension and class number of the image. The features of a given sample (usually a pixel) will be assigned to the input neurons, and the outputs can be calculated as a vector following (5). The class label of the sample can be predicted as the index of the output node which has the highest output.

A k-ELM classifier can also be probabilistic. The outputs in (5) can be rescaled into the range of [0, 1] as

$$p_n = \frac{y_n - y_{min}}{y_{max} - y_{min}}, n = 1, 2, ..., N \quad (6)$$

where y_n is the nth element of the vector y in (5), namely the output of the nth output neuron. y_{min} and y_{max} are the minimum and maximum values among the N outputs. Then p_n can be considered as the probability that the input sample may belong to the nth class. It's quite convenient to build soft classifiers using k-ELM, and this is beneficial to our classification tasks.

3   Feature Fusion Based on k-ELM

Both the diversity and the dimension of image features are important conditions for generating accurate classification. The traditional approach for the integration of multiple features is known as vector stacking (VS), which concatenate the multiple features and feed them into a classifier as a vector. This approach may suffer a lot from the aforementioned efficiency and overflow problems. In order to chase high classification accuracies, feature dimensions are increased with the help of various feature extraction techniques. Then the efficiency and overflow problems will be even more serious, which is known as the curse of dimensionality. Besides the application of fast learning algorithms, such as k-ELM, new classification strategies are also required.

Instead of classifying the image in a very high dimension feature space, we divide the features into sub-sets and classify the image based on different feature sets. Then we fuse the classification results achieved on different features. Two fusion strategies [12], as illustrated in Fig. 1, are implied in our study.

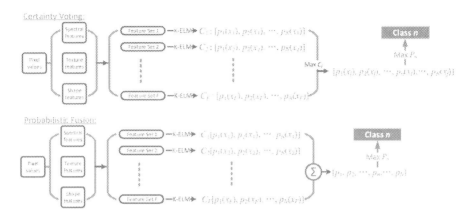

Fig. 1. Classifications of k-ELM classifiers following different strategies

A. Certainty voting

Multiple k-ELM classifiers are established on different feature sets. After training, these classifiers can give their individual predictions about the class label of an un-known sample, which is usually a pixel in the image. The classifications of these classifiers are not necessarily the same. If a pixel is identified as the same class by all the classifiers, it's considered as a reliable classification. If a pixel is an unreliable one, its class label will be decided by comparing the classification certainty degree of different sub feature space classifiers. This process is known as certainty voting.

The sub-features of a target sample can be express as x_f. Assuming the original feature space has been divided into F sub spaces, there will be F different classifiers. The certainty degree of these k-ELM classifiers for an unknown sample can be defined as

$$C_f = \sum_{n=1}^{N-1} \frac{1}{n}[q_n(x_f) - q_{n+1}(x_f)], \ f = 1, 2, ..., F \tag{7}$$

where q_n represent the probabilistic outputs p_n in a descending order. The class label of the sample will be decided according to the classifier achieving highest certainty degree on this sample.

B. Probabilistic fusion

Certainty voting is a 'winner takes all' strategy, while probabilistic fusion is more moderate. In probabilistic fusion, the finial classification is calculated based on all the classifiers' outputs. The certainty degree of each k-ELM classifier is considered as the weight of the probabilistic output. Subsequently, the weighted probabilistic outputs of the k-ELM classifiers are fused for the final classification. This process can be expressed as

$$p_n = \frac{1}{F} \sum_{f=1}^{F} C_f \, p_n(x_f), \, n = 1, 2, ..., N \tag{8}$$

where $p_n(x_f)$ is the probability calculated from the fth set of features, denoted as x_f, following (5) and (6). And the class label with the highest probability will be assigned to the sample.

4   Experiments

The aim of our study is to obtain LULC information about a landslide affected region along Yangtze River, from a high resolution Worldview-II [13] remote sensing image. Before classification, the image has been enhanced using techniques such as pansharpening [14]. The enhanced image of the region is illustrated in Fig.2. There are totally 28 features. Pixel values in the four spectral bands of the image are considered as spectral features. The dimension of the GLCM texture features is 16, eight for contrast and eight for energy. An eight-dimension morphological profile has also been produced.

The objects in the image can be categorized into seven basic LULC types, namely water body, forests, bare fields, rocks, roofs, roads and landslides. Some parts in the image can be classified via visual interpretation. These classified regions are also illustrated in Fig. 2. Training samples and testing samples are selected from these regions out of the same image. In our experiments, there are 2330 pixels in the training set and 1664 pixels in the testing set.

The k-ELM classifiers are trained on the training set and their performances are evaluated on the testing set. The accuracies reported in table 1 confirmed the previous arguments that classification accuracies are highly relied on the variety and dimension of features. The highest accuracy can be obtained when all the 28 features are involved in the classification process. The best sub feature set is the combination between spectral features and shape features.

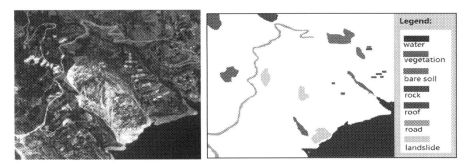

Fig. 2. The enhanced image of the study region (left) and the visual interpretation results (right)

In order to verify the effectiveness of the classification approaches proposed, k-ELM classifiers based on the texture feature set and the spectral-shape combinational feature set are fused, following the strategies of certainty voting and probabilistic fusion, respectively. As reported in table 2, the results show that the classification fusion approaches can produce classifications as accurate as, and even more accurate than the vector stacking approach. The classification accuracies and processing time of an ELM classifier and a SVM classifier are also listed in table as benchmarks.

Table 1. Comparisons of classification results achieved on different features

Features	Training Accuracy	Testing Accuracy
Spectral+Shape+Texture	99.48%	96.88%
Spectral	89.31%	85.28%
Shape	91.85%	88.34%
Texture	56.14%	49.28%
Spectral+Shape	98.28%	96.39%
Spectral+Texture	96.70%	94.41%
Shape+Texture	96.05%	92.79%

Table 2. Comparisons among different feature fusion strategies

Fusion Strategy	Training Time	Testing Accuracy
ELM classifier	277.0ms	94.29%
SVM classifier	423.2ms	96.27%
Vector stacking	282.5ms	96.88%
Certainty Voting	276.6ms	94.95%
Probabilistic Fusion	276.9ms	97.06%

5   Conclusion

The performance of remote sensing image classification is heavily relied on image features. However, high resolution and high dimension advanced remote sensing images may also bring efficiency and overflow problems. The efficiency problems can be eased using faster learning algorithm for the classifier, while the overflow problems can be avoid by using the multiple features separately. Therefore, we came up with multi-classifier fusion approaches based on k-ELM in our study. The proposed classification approaches are testified on a worldview-II image data set, and the results show that fusion approaches can obtain comparably accurate classifications, while the risk of overflow is much lower.

Acknowledgements. This work was supported by the Natural Science Foundation of China under Grant 61203286 and 61125303, National Basic Research Program of China (973 Program) under Grant 2011CB710606, the Program for Science and Technology in Wuhan of China under Grant 2014010101010004, the Program for Changjiang Scholars and Innovative Research Team in University of China under Grant IRT1245.

References

1. Pal, M., Foody, G.M.: Feature Selection for Classification of Hyperspectral Data by SVM. IEEE Transactions on Geoscience and Remote Sensing **48**(5), 2297–2307 (2010)
2. Giacco, F., Thiel, C., Pugliese, L., Scarpetta, S., Marinaro, M.: Uncertainty Analysis for the Classification of Multispectral Satellite Images Using SVMs and SOMs. IEEE Transactions on Geoscience and Remote Sensing **48**(10), 3769–3779 (2010)
3. Byun, H., Lee, S.-W.: Applications of Support Vector Machines for Pattern Recognition: A Survey. Springer, Heidelberg (2002)
4. Chang, N.-B., Han, M., Yao, W., Chen, L.-C., Xu, S.: Change detection of land use and land cover in an urban region with SPOT-5 images and partial Lanczos extreme learning machine. Journal of Applied Remote Sensing **4**(1), 043551 (2010)
5. Huang, G.-B., Wang, D., Lan, Y.: Extreme learning machines: a survey. International Journal of Machine Learning and Cybernetics **2**(2), 107–122 (2011)
6. Guang-Bin, H., Hongming, Z., Xiaojian, D., Rui, Z.: Extreme Learning Machine for Regression and Multiclass Classification. IEEE Transactions on Systems, Man, and Cybernetics, Part B: Cybernetics **42**(2), 513–529 (2012)
7. Huang, X., Zhang, L., Li, P.: An adaptive multiscale information fusion approach for feature extraction and classification of IKONOS multispectral imagery over urban areas. IEEE Geoscience and Remote Sensing Letters **4**(4), 654–658 (2007)
8. Soille, P.: Morphological Image Analysis. Springer, Berlin (2004)
9. Benediktsson, J., Palmason, J., Sveinsson, J.: Classification of hyperspectral data from urban areas based on extended morphological profiles. IEEE Transactions on Geoscience and Remote Sensing **43**(2), 480–491 (2005)
10. Serre, D.: Matrices: Theory and Applications. Springer, New York (2002)
11. Mariéthoz, J., Bengio, S.: A kernel trick for sequences applied to text-independent speaker verification systems. Pattern Recognition **40**(8), 2315–2324 (2007)
12. Huang, X., Zhang, L.: An SVM Ensemble Approach Combining Spectral, Structural, and Semantic Features for the Classification of High-Resolution Remotely Sensed Imagery. IEEE Transactions on Geoscience and Remote Sensing **51**(1), 257–272 (2013)
13. http://worldview2.digitalglobe.com/
14. Yao, W., Han, M.: Improved GIHSA for image fusion based on parameter optimization. International Journal of Remote Sensing **31**(10), 2717–2728 (2010)

Early-Stopping Regularized
Least-Squares Classification

Wenye Li$^{(\boxtimes)}$

Macao Polytechnic Institute, Rua de Luís Gonzaga, Macao SAR, China
wyli@ipm.edu.mo
http://staff.ipm.edu.mo/~wyli

Abstract. We study optimization of the regularized least-squares classification algorithm, and proposes an early-stopping training procedure. Different from previous empirical training methods which separate model selection and parameter learning into two stages, the proposed method performs the two processes simultaneously and thus reduces the training time significantly. We carried out a series of evaluations on text categorization tasks. The experimental results verified the effectiveness of our training method, with comparable classification accuracy and significantly improved running speed over conventional training methods.

Keywords: Machine Learning · Kernel Methods · Iterative Regularization

1 Introduction

Supervised learning has been widely studied recently [1]. It is used to predict an object's category that is unknown. We have some training objects' category information at hand. Some external mechanisms are assumed which provide the correct class labels. We are interested in the task of seeking a computer algorithm that takes labeled training data as input and outputs a *classifier* which performs well in predicting the labels of the unlabeled objects.

Many methods have been developed. Among them, the support vector machines (SVM) algorithm [2,3] and a family of related kernel methods have attracted much attention. Although with significant empirical success, such algorithms face a challenge of scalability. To address the issue, this paper studies the speed-up of an alternative to SVM, the regularized least-squares classification (RLSC) algorithm. We propose an early-stopping training method for RLSC. The method performs iteratively, with each iteration involving limited computation. Comparing with conventional training procedures, the proposed method saves significant computation yet without losing classification accuracy.

2 Background

2.1 Support Vector Machines

Consider the binary classification setting for given data $X = \{\mathbf{x}_1, \cdots, \mathbf{x}_m\}$ that are vectors in a space $\mathcal{R}^d$ with their class labels $Y = \{y_1, \cdots, y_m\}$ where each

© Springer International Publishing Switzerland 2014
Z. Zeng et al. (Eds.): ISNN 2014, LNCS 8866, pp. 278–285, 2014.
DOI: 10.1007/978-3-319-12436-0_31

$y_i \in \{-1, +1\}$. SVM is based on the principle of finding a hyperplane that separates the two classes by the maximal margin.

Besides the geometric viewpoint, SVM can also be explained by Tikhonov regularization theory [4,5].

$$\min_{f \in \mathcal{H}_K} \frac{1}{m} \sum_{i=1}^{m} [1 - y_i f(\mathbf{x}_i)]_+ + \gamma \|f\|_K^2$$

Here $[1 - yf(\mathbf{x})]_+ = \max(1 - yf(\mathbf{x}), 0)$ is a hinge function used to measure the empirical loss. K is a kernel function. $\|f\|_K$ is the norm of f in $\mathcal{H}_K$, where the feature space $\mathcal{H}_K$ is the completion of $\{g | g(\mathbf{x}) = \sum_{c_u} c_u k_{\mathbf{u}}(\mathbf{x}), \mathbf{u} \in \mathcal{R}^d, c_u \in \mathcal{R}\}$. γ is a positive regularization parameter trading off the empirical loss and the *complexity* of the solution. By minimizing the loss and the complexity simultaneously, we wish to find a stable solution that fits the training data well.

The non-differentiability of the hinge loss brings much difficulty in computation. To get the solution, we need to solve a quadratic program. Although special techniques have been proposed [6], the inherent nature still poses a challenge when SVM is applied to large scale problems.

2.2   RLSC: An Alternative to SVM

An alternative is to replace the non-differentiable hinge loss by a differentiable squared loss: $V(y, f(\mathbf{x})) = (y - f(\mathbf{x}))^2$. Hence,

$$\min_{f \in \mathcal{H}_K} \frac{1}{m} \sum_{i=1}^{m} (y_i - f(\mathbf{x}_i))^2 + \gamma \|f\|_K^2.$$

By the representer theorem [7–9], the solution to this model also possess the form $f(\mathbf{x}) = \sum_{i=1}^{m} c_i k(\mathbf{x}, \mathbf{x}_i)$ and each c_i comes from a linear system:

$$(K + \gamma m I) \mathbf{c} = \mathbf{y}, \tag{1}$$

where K is an $m \times m$ kernel matrix with $K_{ij} = k(\mathbf{x}_i, \mathbf{x}_j)$ $(1 \leq i, j \leq m)$ and vector $\mathbf{y} = (y_1, \cdots, y_m)^T$. This leads to the regularized least-squares classification algorithm (RLSC) [10–12].

RLSC and SVM only differ in the loss function, and have similar empirical performances [10,11]. On the other hand, different from SVM which solves a quadratic program, RLSC only needs to solve a system of linear equations that is usually less demanding in computation, and hence provides a potentially faster alternative to SVM.

2.3   Model Selection and Parameter Learning

When applying kernel machines in practice, a necessary step is to determine the value of γ. This model selection process is often done by grid search. The given data is split into a training set and a validation set. People try different regularization parameter values, learn parameters, and evaluate its performance on the validation set. We repeat the processes by enumerating all values in a given set of regularization parameters, and return a classifier with the best result on the validation set.

One shortcoming of the training procedure is that the computation of parameters cannot be re-used. Each time given a new value of γ, the parameter learning has to start from scratch. It is time consuming as we need to try different γ tens of times. The situation becomes even worse when the kernel selection problem is also a concern, where we often need to train and evaluate hundreds of classifiers or more.

3   Early Stopping RSLC

Different from SVM and standard RLSC which use a regularization parameter for model selection, we study a new method to train RLSC. For simplicity, we call the standard RLSC training algorithm derived from solving equation (1) as *RLSC-Tik*, and call our algorithm as *RLSC-ES*.

Instead of training and evaluating many classifiers, RLSC-ES uses an iterative scheme. After each iteration, RLSC-ES returns a classifier with less empirical error on the training set. To avoid over-fitting, the classifier is evaluated until no improvement can be made on the validation set. By RLSC-ES, the model selection and parameter learning processes are merged, which reduces the computation significantly.

3.1   Algorithm

Start from equation (1). The linear equation can be regarded as a regularized alternative to $K\mathbf{c} = \mathbf{y}$. Practically, the condition number [13] of the kernel matrix K is usually high, which makes the solution sensitive to the changes of $\mathbf{y}$. Suppose $\mathbf{c}^*$ is the solution to $K\mathbf{c} = \mathbf{y}$. A small perturbation in $\mathbf{y}$ could bring a large change to $\mathbf{c}^*$ and thus make the solution unstable. To lessen the effect from this *ill-posedness*, regularization techniques are often used, among which Tikhonov regularization is well-known [4]. As in equation (1), the addition of γmI improves the well-posedness of K by decreasing its condition number which helps to improve the stability of $\mathbf{c}^*$.

Besides Tikhonov regularization, a number of other techniques are possible, for example, Landweber iterative regularization [14]. Starting from an initial guess of $\mathbf{c}$, the Landweber method computes the next approximated solution to $\mathbf{c}$ recursively. Repeating the process, the solution converges to a least-squares solution of $K\mathbf{c} = \mathbf{y}$. The iterative method exhibits a "self-regularization property" in the sense that the early termination of the iterative process has a regularizing effect. The iteration index plays the role of the regularizing parameter γ, and the stopping rule plays the role of model selection.

Motivated by Landweber regularization, we resort to the following technique. In literature it belongs to a category called "early stopping methods" [15]. It is a form of regularization used when a model, such as a neural network, is trained by on-line gradient descent. The data is split into a training set and a validation set. Gradient descent is applied to the training set. After each iteration through the new training set, the network is evaluated on the validation set. The model with the best performance in validation is kept for subsequent testing.

This technique is simple but efficient to deal with the problem of over-fitting. Early stopping effectively limits the scope of a solution model, imposes a regularization, and thus effectively lowers the VC dimension in learning [2].

Algorithm 1. Early-Stopping RLSC

---

1. $s = \{1, \cdots, m\}$
2. Divide s into subsets $s_j, j = 1, \cdots, \ell$.
3. $t = 0, \mathbf{c} = \mathbf{0}, \mathbf{cbest} = \mathbf{0}$
4. **repeat**
5. $t = t + 1, \mathbf{c}^t = \mathbf{0}$
6.   **for** $j = 1$ to ℓ **do**
7.   Solve $\mathbf{c}^t_{s_j}$ by $K_{s_j, s_j} \mathbf{c}^t_{s_j} = \mathbf{y}_{s_j}$.
8. $\mathbf{y} = \mathbf{y} - K_{s, s_j} \mathbf{c}^t_{s_j}$
9. $\mathbf{c}_{s_j} = \mathbf{c}_{s_j} + \mathbf{c}^t_{s_j}$
10.   **end for**
11.   **if** c performs better on validation set **then**
12.   **cbest** = **c**
13.   **end if**
14. **until** a certain number of iterations reaches
15. **return cbest**

The detailed algorithm is shown in Algorithm (1). It is based on a variant of block Gauss-Seidel method [13]. The learning algorithm and the convergence property can be understood as alternating projections in a reproducing kernel Hilbert space [16], which is omitted in this paper.

To speed up the convergence, we divide the matrix K into a number of blocks along the diagonal and solve the equations with the blocks one by one. Step 6 to step 10 form one iteration of block Gauss-Seidel method. After each iteration, the empirical error on the training set decreases. We evaluate the performance of the resulting classifier on the validation set and records the best solution so far (steps 11 to 13). After a fixed number of iterations, the best solution is kept and used for future classification.

The major computation comes from the modification of the residual in step 8. The total computation of $K_{s, s_j} \mathbf{c}^t_{s_j}, 1 \leq j \leq \ell$ roughly requires $O(m^2)$ by direct matrix-vector multiplications. For linear kernels, this can be reduced to $O(dm)$, which is generally smaller than $O(m^2)$. For other kernels, we may resort to fast matrix-vector multiplication methods [17], which may simplify the computation to a linear complexity in certain applications.

The major memory consumption comes from the storage of K_{s, s_j} in step 8. The memory requirement is $O(m^2)$ if we store the whole coefficient matrix K. It is also possible to store only one K_{s,s_j} at a time, which requires only an $O\left(\frac{m^2}{\ell}\right)$ storage. When the fast matrix-vector multiplication methods are applicable or linear kernels are used on sparse data such as in text categorization, the memory requirement can be substantially reduced.

The most desirable property of this training procedure is that it performs model selection and parameter learning simultaneously. It uses the number of iterations for model selection. It shares the idea as in iterative regularization methods [14], which provide admissible *filter factors*. In our algorithm, moving into the next iteration is equivalent to changing to a different model. After each iteration, the parameters are also automatically updated from the previous parameters with limited computation. In this way it avoids the heavy computation of repeating model selection and parameter learning in separate stages.

4   Experiments

4.1   Accuracy

To evaluate the performance of RLSC-ES, we carried out a series of empirical studies. Our experiment was on the 20-newsgroups data set, which collects news postings from twenty categories. Each category has roughly $1,000$ documents. The documents are represented by *TFIDF* weights. We compared the classification accuracy of SVM, RLSC-Tik, and RLSC-ES. All algorithms used a linear kernel. Thirty values which cover a reasonably large range of regularization parameters were tried, and the one with the best performance on the validation set was chosen.

We treated each class as the positive category; while treating all others as the negative. We carried out ten runs. In each run we randomly selected 30% of both positive and negative documents as the training set, selected 10% as validation set, and used the rest 60% as testing set.

Both the classification precision and the $F_{1.0}$ values averaged from ten runs were recorded. Here

$$F_{1.0} = \frac{2 \times \#\,(TP)}{2 \times \#\,(TP) + \#\,(FP) + \#\,(FN)}$$

where $\#\,(TP)$ is the number of documents a classifier correctly assigns to the positive category (true positives), $\#\,(FP)$ is the number of documents a classifier incorrectly assigns to the category (false positives), and $\#\,(FN)$ is the number of documents that belong to the positive category but are not assigned to the category by the classifier (false negatives).

Table (1) depicts the results on twenty different positive categories. It is quite evident that SVM and RLSC-ES have similar accuracy.

4.2   Speed

Knowing the comparable classification accuracies, we'd like to investigate the speed of the algorithms on the text corpus. We recorded their running time with

Table 1. Classification accuracy ($F_{1.0}$ and Precision) on 20-newsgroups

	SVM	RLSC-Tik	RLSC-ES
alt.*	0.759/0.982	0.683/0.979	0.751/0.981
comp.gra*	0.696/0.974	0.666/0.973	0.707/0.974
comp.os*	0.713/0.974	0.682/0.972	0.712/0.973
comp.ibm*	0.641/0.969	0.586/0.966	0.639/0.967
comp.mac*	0.736/0.977	0.701/0.975	0.736/0.977
comp.win*	0.757/0.979	0.755/0.979	0.778/0.980
misc.*	0.744/0.978	0.679/0.974	0.727/0.976
rec.autos	0.818/0.983	0.796/0.981	0.821/0.983
rec.motor*	0.880/0.988	0.867/0.987	0.881/0.988
rec.base*	0.881/0.988	0.876/0.988	0.888/0.989
rec.hoc*	0.931/0.993	0.930/0.993	0.934/0.993
sci.crypt	0.875/0.988	0.854/0.986	0.870/0.987
sci.ele*	0.686/0.974	0.654/0.971	0.692/0.973
sci.med	0.824/0.984	0.815/0.983	0.835/0.985
sci.space	0.858/0.987	0.836/0.985	0.854/0.986
soc.rel*	0.796/0.980	0.966/0.978	0.790/0.979
talk.guns*	0.808/0.983	0.780/0.981	0.811/0.983
talk.mid*	0.904/0.991	0.895/0.990	0.906/0.991
talk.misc*	0.721/0.981	0.631/0.977	0.721/0.980
talk.rel*	0.548/0.977	0.453/0.975	0.575/0.977

training data sets of different sizes (from $1,000$ to $18,000$). The time was recorded on a conventional linux workstation. Only the training time was counted; while the input/output time was neglected.

For SVM, we used an implementation of *SVMperf* that is specially optimized for linear kernels [18]. We recorded the time for ten different runs and calculated the average for problems with the same training size. For RLSC-Tik, we implemented it in MATLAB and used Cholesky factorization to solve linear systems. We recorded the time for training sizes from $1,000$ to $10,000$. For larger sizes, we didn't finish the training on our workstation and used a theoretically estimated running time. RLSC-ES was also implemented in MATLAB.

We observed significantly improved speed by RLSC-ES for large-scale tasks. From figure (1), RLSC-Tik is the fastest when the training set is less than $3,000$. When a problem gets larger than $7,000$, it becomes the slowest. SVMperf also reported excellent performance. RLSC-ES is the fastest with more than $3,000$ documents. For a problem with $18,000$ documents, RLSC-ES finished training with tens of seconds. Comparatively, SVM requires hundreds of seconds and RLSC-Tik needs thousands of seconds, partially because these two algorithms need to evaluate dozens of regularization parameters to get a solution.

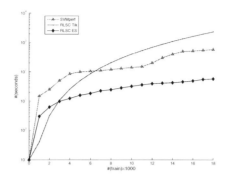

Fig. 1. Running time in log-scale on 20-newsgroups with different-sized training sets

5   Conclusion

Motivated by the recent research in kernel-based methods [19,20], in this paper we investigate the speed-up of regularized least-squares classification. We present an iterative approach with ideas tracing back to modern iterative regularization. The training method allows us to perform model selection and parameter learning simultaneously and provides a faster solution.

One noteworthy point is to decide when to stop the training process. In this paper we proposed to use cross validation to determine this iteration index and has acquired good empirical results. In future work, new criteria will be investigated with the hope of seeking more direct theoretical guidance.

Acknowledgments. The work is partially supported by Macao Polytechnic Institute (RP/ESAP-01/2014) and The Science and Technology Development Fund, Macao SAR, China (044/2010/A).

References

1. Bishop, C.: Pattern Recognition and Machine Learning. Springer (2008)
2. Vapnik, V.: Statistical Learning Theory. John Wiley and Sons (1998)
3. Schölkopf, B., Smola, A.: Learning with Kernels. MIT Press (2002)
4. Tikhonov, A., Arsenin, V.: Solutions of Ill-Posed Problems. Winston and Sons (1977)
5. Poggio, T., Girosi, F.: Regularization Algorithms for Learning That Are Equivalent to Multilayer Networks. Science **247**, 978–982 (1990)
6. Platt, J.: Fast Training of Support Vector Machines Using Sequential Minimal Optimization. In: Advances in Kernel Methods – Support Vector Learning. MIT Press (1999)
7. Kimeldorf, G., Wahba, G.: Some Results on Thebycheffian Spline Functions. J. Math. Anal. Appl. **33**, 82–95 (1971)
8. Girosi, F., Jones, M., Poggio, T.: Regularization Theory and Neural Networks Architectures. Neural Comput. **7**, 219–269 (1995)

9. Schölkopf, B., Herbrich, R., Smola, A.J.: A Generalized Representer Theorem. In: Helmbold, D.P., Williamson, B. (eds.) COLT 2001 and EuroCOLT 2001. LNCS (LNAI), vol. 2111, pp. 416–426. Springer, Heidelberg (2001)

10. Fung, G., Mangasarian, O.: Proximal Support Vector Machine Classifiers. In: Proceedings of KDD 2001 (2001)

11. Rifkin, R.: Everything Old is New Again: A Fresh Look at Historical Approaches in Machine Learning. PhD thesis. MIT (2002)

12. Suykens, J., Gestel, T., Brabanter, J., B.D.Moor, Vandewalle, J.: Least Squares Support Vector Machines. World Scientific Publishing (2002)

13. Golub, G., Loan, C.V.: Matrix Computations. John Hopkins Press (1996)

14. Engl, H., Hanke, M., Neubauer, A.: Regularization of Inverse Problems. Kluwer Academic Publishers (1996)

15. Bauer, F., Pereverzev, S., Rosasco, L.: On Regularization Algorithms in Learning Theory. Journal of Complexity **23**, 52–72 (2007)

16. Li, W., Lee, K.H., Leung, K.S.: Large-scale RLSC Learning Without Agony. In: Proceedings of the 24th Annual International Conference on Machine Learning, pp. 529–536. ACM (2007)

17. Sun, X., Pitsianis, N.: A Matrix Version of the Fast Multipole Method. SIAM Review **43**, 289–300 (2001)

18. Joachims, T.: Training Linear SVMs in Linear Time. In: Proceedings of SIGKDD 2006. ACM (2006)

19. Li, W., Lee, K.H., Leung, K.S.: Generalized Regularized Least-Squares Learning with Predefined Features in a Hilbert Space. In: Advances in Neural Information Processing Systems, vol. 19, pp. 881–888. MIT Press (2006)

20. Li, W., Leung, K.S., Lee, K.H.: Generalizing the Bias Term of Support Vector Machines. In: Proceedings of the 20th International Joint Conference on Artificial Intelligence, pp. 919–924. AAAI (2007)

Z-Type Model for Real-Time Solution
of Complex ZLE

Long Jin[1,2,3], Hongzhou Tan[1,2], Ziyi Luo[1,2,3], Zhan Li[1],
and Yunong Zhang[1,2,3]($\boxtimes$)

[1] School of Information Science and Technology, Sun Yat-sen University (SYSU),
Guangzhou 510006, China
zhynong@mail.sysu.edu.cn
[2] SYSU-CMU Shunde International Joint Research Institute, Shunde 528300, China
ynzhang@ieee.org
[3] Key Laboratory of Autonomous Systems and Networked Control,
Ministry of Education, Guangzhou 510640, China
jinlongsysu@foxmail.com

Abstract. A Z-type model for real-time solution of complex ZLE (i.e.,
complex-valued Zhang linear equation or termed complex-valued time-
varying linear equation) is proposed and analyzed in this paper. Different
from conventional G-type model, such a Z-type model utilizes adequately
the first-order time-derivative information of time-varying coefficients,
and eliminates a predefined vector-valued error function rather than a
scalar-valued error function to zero. The state vector of such a Z-type
model globally and exponentially converges to the unique theoretical
time-varying solution-pair of complex ZLE. Computer-simulation results
further verify and illustrate the effectiveness, efficiency and novelty of
the proposed Z-type model.

Keywords: Z-type model · G-type model · Complex ZLE (complex
Zhang linear equation) · Global exponential convergence

1 Problem Formulation and Solvers

In mathematics, the problem of complex ZLE (or termed complex-valued time-
varying linear equation) can be generally formulated as

$$A(t)x(t) = b(t), \tag{1}$$

where nonsingular coefficient matrix $A(t) \in \mathbb{C}^{n \times n}$ and vector $b(t) \in \mathbb{C}^n$ are
smooth and differentiable at any time instant $t \in [0, +\infty)$ (here notations $\mathbb{C}^{n \times n}$
and $\mathbb{C}^n$ denote $n \times n$- and n-dimensional complex-valued sets, respectively),
$x(t) \in \mathbb{C}^n$ is the unknown complex time-varying vector to be obtained in real
time. Considering that any complex matrix/vector can be seen as the combi-
nation of its real and imaginary parts [i.e., notated by subscripts $_{\text{re}}$ and $_{\text{im}}$,
respectively], we can rewrite problem (1) evidently and equivalently as

$$[A_{\text{re}}(t) + jA_{\text{im}}(t)][x_{\text{re}}(t) + jx_{\text{im}}(t)] = b_{\text{re}}(t) + jb_{\text{im}}(t), \tag{2}$$

© Springer International Publishing Switzerland 2014
Z. Zeng et al. (Eds.): ISNN 2014, LNCS 8866, pp. 286–293, 2014.
DOI: 10.1007/978-3-319-12436-0_32

where
$$A(t) = A_{\text{re}}(t) + jA_{\text{im}}(t) \in \mathbb{C}^{n \times n} \text{ with } A_{\text{re}}(t), A_{\text{im}}(t) \in \mathbb{R}^{n \times n},$$
$$x(t) = x_{\text{re}}(t) + jx_{\text{im}}(t) \in \mathbb{C}^{n} \text{ with } x_{\text{re}}(t), x_{\text{im}}(t) \in \mathbb{R}^{n},$$
$$b(t) = b_{\text{re}}(t) + jb_{\text{im}}(t) \in \mathbb{C}^{n} \text{ with } b_{\text{re}}(t), b_{\text{im}}(t) \in \mathbb{R}^{n},$$

and $j = \sqrt{-1}$ denotes the imaginary unit.

By considering that real or imaginary parts of the left-side and right-side of equation (2) are equivalent, the following time-varying coupled linear equations can be further refined from equation (2) as

$$\begin{cases} A_{\text{re}}(t)x_{\text{re}}(t) - A_{\text{im}}(t)x_{\text{im}}(t) = b_{\text{re}}(t) \\ A_{\text{re}}(t)x_{\text{im}}(t) + A_{\text{im}}(t)x_{\text{re}}(t) = b_{\text{im}}(t) \end{cases} \tag{3}$$

or

$$U(t)y(t) = w(t) \tag{4}$$

with

$$U(t) = \begin{bmatrix} A_{\text{re}}(t) & -A_{\text{im}}(t) \\ A_{\text{im}}(t) & A_{\text{re}}(t) \end{bmatrix}, y(t) = \begin{bmatrix} x_{\text{re}}(t) \\ x_{\text{im}}(t) \end{bmatrix}, w(t) = \begin{bmatrix} b_{\text{re}}(t) \\ b_{\text{im}}(t) \end{bmatrix}.$$

From the above problem reformulation procedure, we can evidently conclude that complex ZLE problem (1) can be solved via solving the time-varying coupled linear equations depicted in (3) or equivalently (4). It is emphasized here that, to ensure the existence of the unique complex time-varying solution at any time instant $t \in [0, +\infty)$ for problem (1), the following condition has to be satisfied according to Cramer's law:

$$\det(U(t)) \neq 0, \tag{5}$$

where operator $\det(\cdot)$ denotes the determinant of a square matrix. This condition guarantees the non-singularity of the coefficient matrix and the uniqueness of the theoretical solution. Besides, it is worth pointing out here that Zhang problem solving is the real-time solution of a time-varying problem related to division (or generalized division), where the divisor or generalized divisor (or say, denominator or generalized denominator) may be time-varying as well and may pass through zero, which originally causes the notorious division-by-zero problem that can now be conquered. By considering that the determinant of $U(t)$ may vary and pass through zero in practical applications, the complex-valued time-varying linear equation solving is potentially a Zhang problem (i.e., complex-valued Zhang linear equation) solving; and finding solutions of such Zhang problems (e.g., ZLE) is a future research direction.

1.1   Expanded Z-Type Model

In order to monitor the solving process of complex ZLE (1) via time-varying coupled linear equation (4), we can firstly consider the vector-valued Z-type error function (or say, Z function) [1,2]:

$$e(t) = U(t)y(t) - w(t) \in \mathbb{R}^{2n},$$

of which each element can be positive or negative (and even lower-unbounded). Expanding classical Z-type model [1] leads to the following complex Z-type model depicted in implicit dynamics:

$$U(t)\dot{y}(t) = -\dot{U}(t)y(t) - \gamma\Phi(U(t)y(t) - w(t)) + \dot{w}(t), \qquad (6)$$

where $\Phi(\cdot) : \mathbb{R}^{2n} \to \mathbb{R}^{2n}$ denotes an activation-function processing array [1]. Note that $\phi(\cdot)$ indicates a scalar-valued processing unit of array $\Phi(\cdot)$, and it should be a monotonically-increasing odd activation function. Besides, two kinds of such activation functions are discussed in the ensuing section. State vector $y(t) \in \mathbb{R}^{2n}$ of Z-type model (6), starting from initial condition $y(0) \in \mathbb{R}^{2n}$, is proved to globally converge to theoretical solution-pair $y^*(t) = [y_1^*(t), \cdots, y_n^*(t),$ $\cdots, y_{2n}^*(t)]^{\mathrm{T}} \in \mathbb{R}^{2n}$, which constitutes the theoretical solution to problem (1); i.e., complex time-varying theoretical solution $x^*(t) = x_{\mathrm{re}}^*(t) + jx_{\mathrm{im}}^*(t)$ of complex ZLE (1) is

$$[y_1^*(t), y_2^*(t), \cdots, y_n^*(t)]^{\mathrm{T}} + j[y_{n+1}^*(t), y_{n+2}^*(t), \cdots, y_{2n}^*(t)]^{\mathrm{T}}.$$

1.2   Expanded G-Type Model

For comparison, an expanded G-type model is developed correspondingly to solve complex ZLE (1). According to the authors' previous researches [1,2], the G-type model can be generalized as follows:

$$\dot{y}(t) = -\gamma U^{\mathrm{T}}(t)\Phi(U(t)y(t) - w(t)). \qquad (7)$$

2   Convergence Analyses and Results

In this section, three theorems on convergence properties of Z-type model (6) are presented for real-time solution of complex ZLE (1). The analyses include the situations of using linear or power-sigmoid activation functions.

Theorem 1. *Consider complex ZLE problem (1). If a monotonically-increasing odd activation-function array $\phi(\cdot)$ is used, then state vector $y(t) \in \mathbb{R}^{2n}$ of Z-type model (6), starting from any initial state $y(0) \in \mathbb{R}^{2n}$, converges to theoretical solution-pair vector $y^*(t) = [x_{\mathrm{re}}^{*\mathrm{T}}(t), x_{\mathrm{im}}^{*\mathrm{T}}(t)]^{\mathrm{T}}$ of equation (2), which constitutes time-varying theoretical solution $x^*(t) = x_{\mathrm{re}}^*(t) + jx_{\mathrm{im}}^*(t)$ to problem (1).*

Proof. For Z-type model (6), we can define a Lyapunov function (or say, Lyapunovian) candidate $v(t) = \|e(t)\|_2^2/2 \geqslant 0$, and its time-derivative is thus

$$\dot{v}(t) = \frac{\mathrm{d}v(t)}{\mathrm{d}t} = e^{\mathrm{T}}(t)\frac{\mathrm{d}e(t)}{\mathrm{d}t} = -\gamma e^{\mathrm{T}}(t)\Phi(e(t)) = -\gamma\sum_{i=1}^{2n} e_i(t)\phi(e_i(t)).$$

With $\phi(\cdot)$ being a monotonically-increasing odd activation function, we have

$$e_i(t)\phi(e_i(t)) \begin{cases} > 0, \text{ if } e_i(t) > 0, \\ = 0, \text{ if } e_i(t) = 0, \\ > 0, \text{ if } e_i(t) < 0, \end{cases}$$

which guarantees the negative-definiteness of $\dot{v}(t)$. By Lyapunov theory, $e(t)$ globally converges to zero. In view of $e(t) = U(t)y(t) - w(t) = U(t)(y(t) - y^*(t))$ and $U(t)$ being nonsingular for any time instant t, we obtain that $y(t) \to y^*(t)$, as $t \to +\infty$. That is, state vector $y(t)$ of Z-type model (6) globally converges to exact time-varying theoretical solution-pair vector $y^*(t) = [x_{re}^{*T}(t), x_{im}^{*T}(t)]^T$ of complex ZLE (2); therefore the exact theoretical solution $x^*(t) = x_{re}^*(t) + jx_{im}^*(t)$ to complex ZLE problem (1) is achieved. The proof is thus complete. $\qquad \square$

Theorem 2. *Consider complex ZLE problem (1). If the array of linear activation function $\phi(e_i) = e_i$ is used, then state vector $y(t) \in \mathbb{R}^{2n}$ of Z-type model (6), starting from any initial state $y(0) \in \mathbb{R}^{2n}$, globally exponentially converges to theoretical solution-pair vector $y^*(t) = [x_{re}^{*T}(t), x_{im}^{*T}(t)]^T$ of equation (2), which constitutes time-varying theoretical solution $x^*(t) = x_{re}^*(t) + jx_{im}^*(t)$ to (1).*

Proof. Let us review Z-type model (6), which can be rewritten as

$$U(t)\dot{\tilde{y}}(t) = -\dot{U}(t)\tilde{y}(t) - \gamma\Phi(U(t)\tilde{y}(t)), \qquad (8)$$

where $\tilde{y}(t) = y(t) - y^*(t)$ denotes the difference between state vector $y(t)$ and time-varying theoretical solution-pair vector $y^*(t)$. In addition, the relation between $\tilde{y}(t)$ and $e(t)$ is

$$\|\tilde{y}(t)\|_2 \leqslant \frac{1}{\sqrt{\lambda_{min}}}\|U(t)(y(t) - y^*(t))\|_2 \leqslant \frac{1}{\sqrt{\lambda_{min}}}\|e(t)\|_2,$$

where $\lambda_{min} > 0$ denotes the minimal eigenvalue of the positive-definite matrix $U^T(t)U(t)$ all over the time $t \in [0, +\infty)$. If linear activation function $\phi(e_i) = e_i$ is used, then it follows from the corresponding Z-type design formula [1] that $e(t) = e(0)\exp(-\gamma t)$. Thus, one can have

$$\|y(t) - y^*(t)\|_2 \leqslant \frac{1}{\sqrt{\lambda_{min}}}\|e(0)\|_2 \exp(-\gamma t),$$

which implies that state vector $y(t)$ of Z-type model (6) globally converges to time-varying theoretical solution-pair vector $y^*(t)$ with exponential convergence rate γ and thus theoretical solution $x^*(t) = x_{re}^*(t) + jx_{im}^*(t)$ to problem (1) is obtained. The proof is thus complete. $\qquad \square$

Theorem 3. *In addition to Theorems 1 and 2, if we use the array of power-sigmoid activation function*

$$\phi(e_i) = \begin{cases} e_i^p, & \text{if } |e_i| \geqslant 1, \\ \frac{1+\exp(-\xi)}{1-\exp(-\xi)} \cdot \frac{1-\exp(-\xi e_i)}{1+\exp(-\xi e_i)}, & \text{if } |e_i| < 1, \end{cases}$$

with suitable design parameters (e.g., odd integer $p \geqslant 3$ and $\xi \geqslant 2$), then state vector $y(t) \in \mathbb{R}^{2n}$ of Z-type model (6), starting from any initial state $y(0) \in \mathbb{R}^{2n}$, superiorly converges to theoretical solution-pair vector $y^*(t) = [x_{re}^{*T}(t), x_{im}^{*T}(t)]^T$ of equation (2), which constitutes time-varying theoretical solution $x^*(t) = x_{re}^*(t) + jx_{im}^*(t)$ to problem (1), as compared with the situation of using the array of linear activation function presented in Theorem 2.

Proof. 1) For error range $|e_i| \geqslant 1$

Power activation function $\phi(e_i) = e_i^p$ with $p \geqslant 3$ is used specifically for this error range. It can be generalized from [1] that, if we use the power-sigmoid activation function over error range $|e_i(t)| \geqslant 1$, superior convergence can be achieved for Z-type model (6), as compared with the situation of using the linear activation function with exponential convergence rate γ.

2) For error range $|e_i(t)| < 1$

Bipolar-sigmoid activation function $\phi(e_i) = ((1 + \exp(-\xi))/(1 - \exp(-\xi))) \cdot (1 - \exp(-\xi e_i))/(1 + \exp(-\xi e_i))$ is used specifically for such an error range. Review Lyapunov function candidate $v(t) = e^{\mathrm{T}}(t)e(t)/2$ and its time-derivative equation once again. Over error range $|e_i(t)| < 1$, we have $|\phi(e_i)| \geqslant |e_i|$ and then

$$\dot{v}_{\mathrm{ps}}(t) = -\gamma \sum_{i=1}^{2n} e_i(t)\phi(e_i(t)) \begin{cases} < -\gamma \sum_{i=1}^{2n} e_i^2(t) = \dot{v}_{\mathrm{li}}(t), \text{ if } 0 < |e_i(t)| < 1, \\ = -\gamma \sum_{i=1}^{2n} e_i^2(t) = \dot{v}_{\mathrm{li}}(t), \text{ if } e_i(t) = 0, \end{cases}$$

which implies that, if the power-sigmoid activation function is used over error range $0 < |e_i(t)| < 1$, superior convergence can also be achieved for Z-type model (6), as compared to the situation of using the linear activation function with exponential convergence rate γ.

Summarizing the above analyses of the two sub-cases, if we use the array of power-sigmoid activation function to construct Z-type model (6), superior convergence is achieved, as compared to the situation of using the array of linear activation function. The proof is now complete.   □

3   Simulative Verification and Comparison

In this section, two illustrative computer-simulation examples are presented to illustrate the characteristics of the presented Z-type model (6). Note that both Z-type model (6) and G-type model (7) are activated by the power-sigmoid activation function array with design parameters $p = 3$ and $\xi = 4$.

Example 1. Let us consider the following coefficients of equation (1):

$$\begin{cases} A(t) = \begin{bmatrix} \cos(12t) + j\sin(8t) & \sin(12t) + j\cos(8t) \\ -\sin(12t) + j\cos(8t) & \cos(12t) - j\sin(8t) \end{bmatrix}, \\ b(t) = \begin{bmatrix} \sin(60t) + j\sin(100t) \\ \cos(60t) + j\cos(100t) \end{bmatrix}. \end{cases} \tag{9}$$

It can be seen from Fig. 1(a) that, with randomly-generated initial state $y(0) \in \mathbb{R}^4$, state vector $y(t) \in \mathbb{R}^4$ of Z-type model (6) globally converges to the theoretical time-varying solution-pair vector $[x_{\mathrm{re}}^{*\mathrm{T}}(t), x_{\mathrm{im}}^{*\mathrm{T}}(t)]^{\mathrm{T}}$ of problem (9) in a rather short time. In contrast, as shown in Fig. 1(b), state vector $y(t) \in \mathbb{R}^4$ of G-type model (7) does not fit well with the theoretical solution-pair vector $[x_{\mathrm{re}}^{*\ \mathrm{T}}(t), x_{\mathrm{im}}^{*\ \mathrm{T}}(t)]^{\mathrm{T}}$ with quite large computational errors. In summary, the exact solution $x^*(t)$ to problem (9) is achieved by Z-type model (6) with much better accuracy. In addition, Fig. 2(a) shows that computational error $\|x(t) - x^*(t)\|_2$

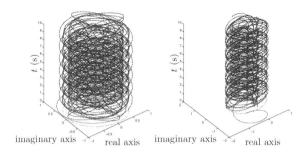

(a) $x_1(t)$ and $x_2(t)$ of Z-type model (6) solving (9)

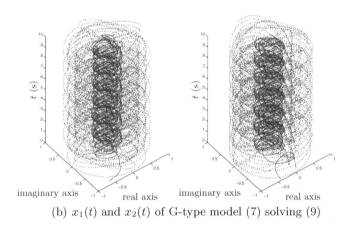

(b) $x_1(t)$ and $x_2(t)$ of G-type model (7) solving (9)

Fig. 1. Trajectories of Z-type model (6) and G-type model (7) with $\gamma = 10$ for solving complex ZLE (9), where red-dotted curves denote theoretical-solution trajectories of (9) and blue-solid curves denote solutions of Z-type model (6) and G-type model (7)

synthesized by Z-type model (6) decreases to zero rapidly, while Fig. 2(b) shows once again that computational error $\|x(t) - x^*(t)\|_2$ synthesized by G-type model (7) is rather larger. This illustrates the efficacy of the presented Z-type model (6) for solving complex ZLE (9).

Example 2. Let us consider the following coefficients of equation (1):

$$\begin{cases} A(t) = \begin{bmatrix} \sin(100t) + j\exp(\sin(100t)) & \exp(\cos(100t)) - j\cos(100t) \\ \exp(\sin(100t)) + j\cos(100t) & -\sin(100t) + j\exp(\cos(100t)) \end{bmatrix}, \\ b(t) = \begin{bmatrix} \cos(100t) + j\sin(100t) \\ \cos(1000t) + j\sin(1000t) \end{bmatrix}. \end{cases} \tag{10}$$

It can be observed from Fig. 3 that, with randomly-generated initial state $y(0) \in \mathbb{R}^4$, residual error $\|A(t)x(t) - b(t)\|_2$ synthesized by Z-type model (6)

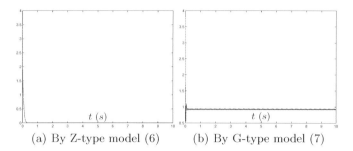

(a) By Z-type model (6)   (b) By G-type model (7)

Fig. 2. Computational error $\|x(t) - x^*(t)\|_2$ synthesized by Z-type model (6) and G-type model (7) with parameter $\gamma = 10$ for solving complex ZLE (9)

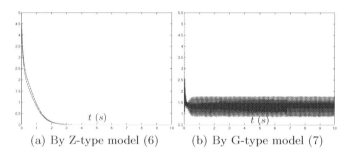

(a) By Z-type model (6)   (b) By G-type model (7)

Fig. 3. Residual error $\|A(t)x(t) - b(t)\|_2$ synthesized by Z-type model (6) and G-type model (7) with parameter $\gamma = 1$ for solving complex ZLE (10)

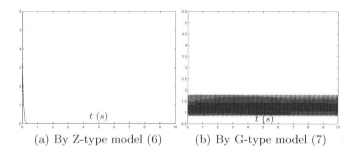

(a) By Z-type model (6)   (b) By G-type model (7)

Fig. 4. Residual error $\|A(t)x(t) - b(t)\|_2$ synthesized by Z-type model (6) and G-type model (7) with parameter $\gamma = 10$ for solving complex ZLE (10)

for solving complex ZLE (10), decreases to zero within around 3 seconds, while residual error $\|A(t)x(t) - b(t)\|_2$ synthesized by G-type model (7) is once again rather large with drastic oscillation as time t goes on. Besides, the convergence process can be accelerated by increasing design parameter γ for Z-type model (6) for real-time solution of (10), which is illustrated in Fig. 4(a); i.e., residual error $\|A(t)x(t) - b(t)\|_2$ synthesized by Z-type model (6) diminishes to zero within about 0.5 second. However, as shown in Fig. 4(b), residual error $\|A(t)x(t)-b(t)\|_2$ synthesized by G-type model (7) is still large in the steady-state situation. This illustrates the efficacy of the presented Z-type model (6) for solving complex ZLE (1) once more.

4   Conclusions

A Z-type model for complex-valued ZLE problems solving has been proposed, developed and analyzed in this paper. Computer-simulation examples with comparison to the G-type model have further illustrated the effectiveness, efficiency and novelty of the proposed Z-type model. Before ending this paper, it is worth clarifying the contributions as follows. First, by reformulating complex-valued ZLE (1) as (3) or (4), the solving process of complex-valued ZLE (1) can be consistent with that of the conventional real-valued linear equations. Second, by considering that the determinant of $U(t)$ may vary and pass through zero, this work provides a direction to the general division-by-zero problem solving.

References

1. Zhang, Y., Ge, S.S.: Design and Analysis of A General Recurrent Neural Network Model for Time-Varying Matrix Inversion. IEEE Trans. Neural Netw. **16**, 1477–1490 (2005)
2. Zhang, Y., Li, Z., Guo, D., Chen, K., Chen, P.: Superior Robustness of Using Power-Sigmoid Activation Functions in Z-Type Models for Time-Varying Problems Solving. In: Proceedings of the International Conference on Machine Learning and Cybernetics, pp. 759–764. IEEE Press, New York (2013)

A Dynamic Generation Approach for Ensemble of Extreme Learning Machines

Hualong Yu[1(✉)], Yulong Yuan[1], Xibei Yang[1], and Yuanyuan Dan[2]

[1] School of Computer Science and Engineering, Jiangsu University of Science and Technology,
Zhenjiang 212003, Jiangsu, People's Republic China
yuhualong@just.edu.cn, hongshi89@126.com, yangxibei@hotmail.com
[2] School of Environment and Chemical Engineering,
Jiangsu University of Science and Technology, Zhenjiang 212003,
Jiangsu, People's Republic China
danyuanyuan@126.com

Abstract. Extreme learning machine (ELM) as one new learning algorithm has been proposed for single hidden-layer feed-forward neural network (SLFN). In contrast with the popular back-propagation (BP) algorithm, ELM often has obviously faster learning speed and stronger generalization performance. However, ELM lacks stability as the weights and biases between the input layer and the hidden layer are randomly assigned, and meanwhile, it often suffers from overfitting as the learning model will approximate all training instances well. In this article, a dynamic generation approach for ensemble of extreme learning machine (DELM) is proposed to overcome the problems above. Specifically, cross-validation and one target function are embedded into the learning phase. Experimental results on several benchmark datasets indicate that DELM is robust and accurate.

Keywords: Extreme learning machine · Ensemble learning · Cross-validation · Neural network · Generalization performance

1 Introduction

Extreme learning machine (ELM) as one new algorithm was proposed by Huang et al. [1] for single-hidden layer feed-forward network (SLFN). Unlike gradient decent-based back-propagation (BP) algorithm [2], ELM doesn't need to iteratively tune the parameters between the input layer and the hidden layer, but denotes their values randomly, then calculates the weights connecting the hidden layer and the output layer by least-square method [3]. It has been found that ELM often provides similar or better generalization performance at a much faster learning speed than those traditional classifiers, including BP neural network (BPNN), support vector machine (SVM) and least-square support vector machine (LS-SVM) et al. [1,3-4]. In recent years, ELM has been widely applied to solve various real-world problems, such as face recognition [5], human action recognition [6], sales forecasting [7-8], credit scoring [9], bioinformatics [10] et al.

© Springer International Publishing Switzerland 2014
Z. Zeng et al. (Eds.): ISNN 2014, LNCS 8866, pp. 294–302, 2014.
DOI: 10.1007/978-3-319-12436-0_33

ELM promotes the learning speed and generalization performance of SLFN, however, it still suffers from two problems as following:

- ELM lacks stability as the weights and bias between the input layer and the hidden layer are assigned randomly [11].
- ELM is apt to be overfitting as it will approximate all training instances well [12].

Ensemble learning could help alleviate the problems above to some extent [11-15]. The concept of ensemble learning was firstly proposed by Hansen and Salamon [13]. They found that a collection of different neural networks trained on the same task can significantly improve the generalization performance and robustness of the prediction. Krogh and Vedelsby [14] indicated that the generalization performance of ensemble is closely related with two factors: the average generalization error of each component learner and the average diversity among all component learners. That means a successful ensemble classier should be both accurate and diverse. Based on this theory, Zhou et al. [15] further found that selecting some both accurate and diverse components can often acquire better performance than using all. As for ensemble of ELM, Lan et al. [11] integrated several online sequential ELM (OS-ELM) and used their average value of outputs as the final measurement of network performance. They found the results of ensemble performs more stable than each OS-ELM. Sun et al. [7] proposed an ensemble algorithm of ELM named ELME which also integrates the average value of outputs belonging to multiple ELMs to predict sales amount. ELME can be seen as one simple weighted Bagging [16]. Liu and Wang [12] presented an ensemble based on ELM (EN-ELM) algorithm. EN-ELM puts cross-validation and prior information about the generalization performance, i.e., the norm of output weight matrix, of single ELM into the learning phase, and finally selects the top half component learners to make decision.

In this article, we combine the ideas of ref.[12] and ref.[14] to present an novel ensemble algorithm of ELM named DELM. Specifically, DELM dynamically generates component learners to guarantee each new inserted component can promote the quality of the current ensemble. First, the training set is averagely divided into K subsets, and then K pairs of training and validation sets are obtained so that each training set consists K-1 subsets. Next, for the first pair of training and validation sets, one initial ELM is randomly generated and is put into the first sub-ensemble, then random ELMs are sequentially generated and are put into ensemble one by one, if the new inserted ELM can reduce the value of the target function, it will be reserved, otherwise, it will be removed. The learning process stops until there are M ELMs in the sub-ensemble. The other pairs of training and validation sets implement the same learning procedure. At last, we use the $K \times M$ ELMs to constitute ensemble and to make decision for those unseen testing instances. It is worth noting that the target function in DELM is expressed as the difference between the average generalization error and the average diversity of all component learners, which is the same measurement to evaluate the quality of ensemble learning presented in the ref. [14].

The remaining of this article is organized as follows. Section 2 first briefly introduces ELM algorithm, and then presents the proposed DELM ensemble learning algorithm. Experimental results and discussions are given in Section 3. At last, Section 4 summarizes the contributions of this article.

2   Methods

2.1   Extreme Learning Machine

As we know, in supervised learning, the learning algorithms always use a finite number of input-output instances for training. Suppose there are N arbitrary distinct training instances $(x_i, t_i) \in R^n \times R^m$, where x_i is one $n{\times}1$ input vector and t_i is one $m{\times}1$ target vector. If an SLFN with L hidden nodes can approximate these N samples with zero error, it then implies that there exist β_i, a_i and b_i, such that:

$$f_L(x_j) = \sum_{i=1}^{L} \beta_i G(a_i, b_i, x_j) = t_j, j = 1, \dots, N \tag{1}$$

where a_i and b_i are learning parameters between the input layer and the hidden layer, β_i is the weight vector connecting the ith hidden node to the output node. Then Eq. (1) can be written compactly as:

$$H\beta = T \tag{2}$$

where

$$H(a_1, \dots, a_L, b_1, \dots, b_L, x_1, \dots, x_N) = \begin{bmatrix} G(a_1, b_1, x_1) & \cdots & G(a_L, b_L, x_1) \\ \vdots & \ddots & \vdots \\ G(a_1, b_1, x_N) & \cdots & G(a_L, b_L, x_N) \end{bmatrix} \tag{3}$$

$$\beta = \begin{bmatrix} \beta_1^T \\ \cdot \\ \cdot \\ \cdot \\ \beta_L^T \end{bmatrix}_{L \times m} \quad \text{and} \quad T = \begin{bmatrix} t_1^T \\ \cdot \\ \cdot \\ \cdot \\ t_N^T \end{bmatrix}_{N \times m} \tag{4}$$

Here, $G(a_i, b_i, x_j)$ denotes activation function which is used to calculate the output of the ith hidden node for the jth training instance. H is called hidden layer output matrix of the network, where its ith column denotes the ith hidden node's output vector with respect to inputs $x_1, x_2, \dots, x_N$ and its jth line represents the output vector of the hidden layer with respect to the input x_j. Fig. 1 gives the basic structure of one SLFN.

In SLFN, the number of hidden nodes, L, will always be less than the number of training samples, N, and hence, the training error cannot be made exactly zero but can approach a nonzero training error ε. The hidden node parameters a_i and b_i need not be tuned during training and may simply be assigned with random values according to any continuous sampling distribution [1, 3]. Eq. (2) then becomes a linear system and the output weights β are estimated as:

$$\hat{\beta} = H^\dagger T \tag{5}$$

where $H^\dagger$ is the Moore-Penrose generalized inverse of the hidden layer output matrix H. $H^\dagger = (H^T H)^{-1} H^T$ if $H^T H$ is nonsingular or $H^\dagger = H^T (HH^T)^{-1}$ if HH^T is nonsingular. Here, $\hat{\beta}$ is the minimum-norm least squares solution of Eq. (2) [1]. The ELM algorithm can then be summarized as:

ELM Algorithm[1]: Given a training set $\aleph = \{(x_i, t_i) | x_i \in R^n, t_i \in R^m, i = 1,2, ..., N\}$, activation function $G(x)$, and hidden node number L.

Step1: Assign random hidden nodes by randomly generating parameters (a_i, b_i) according to any continuous sampling distribution, $i=1, 2,..., L$;
Step 2: Calculate the hidden layer output matrix H;
Step 3: Calculate the output weight β by $\beta = H^\dagger T$.

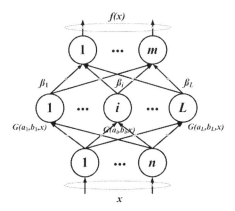

Fig. 1. The basic network structure of single-hidden layer feed-forward network (SLFN)

It has been proved that training SLFN with randomly generated additive or RBF nodes with a wide range of activation functions by ELM algorithm can universally approximate any continuous target functions in any compact subset of the euclidean space R^n [1, 3]. In contrast with back propagation (BP) algorithm [2], ELM can usually produce better generalization performance with obviously faster training speed [1, 3-4].

2.2   Dynamic Ensemble of Extreme Learning Machine

ELM promotes learning speed and generalization performance of SLFN, however, it is still instable and apt to be overfitting [11-12]. Integrating multiple ELMs could help avoid both problems above [7, 11-12]. Krogh and Vedelsby [14] indicated that a sufficient and necessary condition the ensemble outperforms its component members is that component learners should be simultaneously accurate and diverse:

---

[1] The source codes of ELM are available at the homepage of Huang: http://www.ntu.edu.sg/home/egbhuang/.

$$E = \bar{E} - \bar{A} \qquad (6)$$

where E is ensemble generalization error, $\bar{E}$ and $\bar{A}$ are average of generalization errors and diversities of all component learners, respectively. To generate successful ensemble learning models, two aspects should be considered simultaneously.

DELM uses Eq.(6) as target function to dynamically construct ensemble. To calculate the average generalization error and diversity, cross-validation is adopted, where training set is used to train ELM, validation set is used to calculate the generalization error of each ELM and the diversity of multiple ELMs. Specifically, average diversity is detected by using disagreement measurement [17]. Disagreement measurement first runs each component learner on the validation set and acquires the corresponding binary prediction sequence, then compares the difference between every two sequences by counting their mismatched bits. Indeed, the average percentage of mismatched bits can be seen as their diversity. The average generalization error can be detected in the similar way. In DELM, each sub-ensemble, i.e., the component learners in each fold of cross-validation, sequentially generates ELMs to gradually decrease the value of the target function, i.e., Eq.(6), until the sub-ensemble consists M members. It is clear that each sub-ensemble might be overfitting, but when integrating all sub-ensembles to make decision, the problem can be avoided. DELM algorithm can be summarized as:

DELM Algorithm: Given a training set $\aleph = \{(x_i, t_i)|x_i \in R^n, t_i \in R^m, i = 1,2, \dots, N\}$, activation function $G(x)$, hidden node number L, fold number K, and number of component learners in each sub-ensemble M.

Step1: Divide the original training set $\aleph$ into K folds randomly;
Step2: for $i=1:K$

    2.1: Set the ith fold as validation set and the other folds as training set;
    2.2: Generate an ELM randomly, calculate its target function value E_1 on the validation set, and put the ELM into sub-ensemble$_i$;
    2.3: Set flag=1;
    2.4: while (flag<M)
        2.4.1: Generate an ELM randomly, put it into sub-ensemble$_i$, and calculate the new target function value;
        2.4.2: if the new target function value is lower than E_{flag}
            2.4.2.1 flag=flag+1;
            2.4.2.2 Set E_{flag} as the new target function value;
            2.4.3: else, remove the new ELM from sub-ensemble$_i$;
Step 3: Integrate all $K \times M$ ELMs to make decision for unseen testing instances.

At the testing phase, we use majority voting to make the final decision. Suppose there is one C classes problem, h_m^k denotes the kth ELM in the mth sub-ensemble, and $v_{m,c}^k$ is set to one if h_m^k predicts the testing instance as class c, otherwise, $v_{m,c}^k$ is set to zero. Then the class label l of an testing instance can be calculated by the following formula:

$$l = \underset{c}{\text{argmax}} \sum_{k=1}^{K} \sum_{m=1}^{M} v_{m,c}^k \qquad (7)$$

Obviously, DELM takes advantage of the target function to maximize the efficiency and cross-validation to guarantee the completeness of training instances. Also, the dynamic generation procedure of DELM can be seen as one process of selective ensemble [15].

3   Experiments

3.1   Datasets and Initial Parameters Settings

The experiments are carried out on six benchmark datasets which come from UCI machine learning repository [18]. The detailed information of these datasets are summarized in Table 1.

Table 1. Datasets used in the experiments

Dataset	Instance number	Feature number	Class number
Pima	768	8	2
Balance	625	4	3
Segmentation	2310	19	7
Spambase	4597	57	2
Waveform	5000	40	3
letter	20000	16	26

In the experiments, we compare the proposed DELM algorithm with ELM [1], ELME [7] and EN-ELM [12] algorithms. All experiments are implemented in the environment of Matlab 2013a. ELM adopts sigmoid function as activation function [3]. The number of hidden nodes are provided with different settings for different datasets, which will be given in next sub-section. To guarantee the impartiality of experimental results, 50 times' 5-fold cross-validation is implemented for each algorithm and the results are provided in the form of mean ± standard deviation. In addition, for three ensemble learning algorithms, they share one identical parameter, i.e., the number of component learners which participate in voting is 50. Due to both EN-ELM and DELM adopt internal cross-validation, we set the number of folds as 5, too. That means the parameter K in EN-ELM should be assigned as 20 because only top half learners are used to vote, and the parameter M in DELM should be set as 10.

3.2   Results and Discussions

Table 2 presents the comparision results of four learning algorithms, where the running time denote the average value all over the 50 random runs. On the same dataset, different learning algorithms share the same hidden node number.

From Table 2, we observe that three ensemble algorithms of ELM perform more accuracte and robust than ELM, although ELM is one faster learner. EN-ELM has similar classification performance with ELME, but consumes more training time as there exists much information interchange during the generation of component learners. As for our proposed DELM, it often performs best on each dataset, not only acquires the highest classificaion accuracy, but also performs more stable which can be observed by the standard deviation of classification accuracy. DELM is both robust and accurate, however, it requires more training time as the continuous computation of the target function, as well the generation of lots of useless component learners. DELM can dynamically extract accurate and diverse ELMs to construct ensemble and to maximize the effeciency of ensemble.

Table 2. Comparision of the experimental results on six datasets

Dataset	Algorithm	Classification Accuracy (%)	Running time (s)	Hidden node number
Pima	ELM	76.30±0.70	0.06	50
	ELME	76.84±0.66	2.50	50
	EN-ELM	77.48±0.67	8.85	50
	DELM	77.80±0.31	33.57	50
Balance	ELM	96.02±0.66	0.04	50
	ELME	97.19±0.42	2.18	50
	EN-ELM	96.74±0.71	7.84	50
	DELM	97.35±0.47	29.89	50
Segmentation	ELM	93.56±0.24	0.31	100
	ELME	94.22±0.19	13.77	100
	EN-ELM	94.12±0.22	46.71	100
	DELM	94.95±0.16	142.33	100
Spambase	ELM	90.21±0.26	0.69	100
	ELME	90.81±0.13	30.48	100
	EN-ELM	90.90±0.14	104.26	100
	DELM	91.42±0.11	278.13	100
Waveform	ELM	84.06±0.31	0.74	100
	ELME	86.23±0.25	32.89	100
	EN-ELM	86.09±0.30	110.89	100
	DELM	87.17±0.19	301.68	100
letter	ELM	82.04±0.19	7.6	200
	ELME	83.06±0,09	365	200
	EN-ELM	83.00±0.20	1205	200
	DELM	83.47±0.08	2718	200

We also investigate the effects of two important parameters in DELM, i.e., the number of component learners and the hidden node number. Both parameters are assigned from 10 to 100 with an increment of 10. When one parameter changes, all other parameters use the initial settings. Taking Balance dataset as an example, the change of classification accuracy with the change of two parameters are depicted in Fig.2 and Fig.3.

From Fig.2, we observe that the classification accuracy increases rapidly until there are 40 component learners. Further injecting component learners could not obviously promote the classification accuracy. When there are 40-100 component learners in ensemble, the classification accuracies lie on a narrow range of [0.973, 0.974]. Fig.3 shows when hidden node number is undersize or oversize, DELM both performs poorly. A small hidden node number could not accurately approximate the targets and a large hidden node number is apt to overfit the training instances, consequently decreasing the generalization performance of the classifier. Therefore, these parameters should be pre-designed carefully according to the characteristics of data in real-world applications.

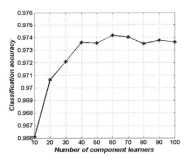

Fig. 2. Change of classification accuracy with the increase of component learners in DELM on Balance dataset

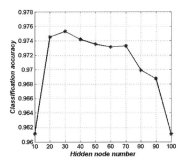

Fig. 3. Change of classification accuracy with the increase of hidden node number in DELM on Balance dataset

4   Conclusions

In this article, a novel ensemble algorithm of ELM named DELM is proposed. DELM integrates cross-validation and an target function to dynamically generate component learners of ensemble. The adoption of the target function guarantees the component learners are both accurate and diverse. The experimental results indicate that DELM is not only accurate, but also robust. DELM outperforms ELM and two previously proposed ensemble learning algorithms of ELM, though it needs to consume more time resources. In future work, we wish to accelerate DELM by introducing some prior knowledge. Also, the possibility of applying DELM to solve regression problems will be investigated.

Acknowledgements. This work was supported in part by National Natural Science Foundation of China under grant No.61305058, No.61100116, Natural Science Foundation of Jiangsu Province of China under grant No.BK20130471, No.BK2011492, China Postdoctoral Science Foundation under grant No.2013M540404 and Jiangsu Planned Projects for Postdoctoral Research Funds under grant No.1401037B.

References

1. Huang, G.B., Zhu, Q.Y., Siew, C.K.: Extreme learning machine: theory and applications. Neurocomputing **70**, 489–501 (2006)
2. Rumelhart, D.E., Hinton, G.E., Williams, R.J.: Learning representations by back-propagation errors. Nature **323**, 533–536 (1986)
3. Huang, G.B., Wang, D.H., Lan, Y.: Extreme learning machine: a survey. International Journal of Machine Learning and Cybernetics **2**, 107–122 (2011)
4. Huang, G.B., Zhou, H., Ding, X., Zhang, R.: Extreme learning machine for regression and multiclass classification. IEEE Transactions on Systems, Man, and Cybernetics, Part B: Cybernetics **42**, 513–529 (2012)
5. Choi, K., Toh, K.A., Byun, H.: Realtime training on mobile devices for face recognition applications. Pattern Recognition **44**, 386–400 (2011)
6. Minhas, R., Baradarani, A., Seifzadeh, S., Wu, Q.M.J.: Human action recognition using extreme learning machine based on visual vocabularies. Neurocomputing **73**, 1906–1917 (2010)
7. Sun, Z.L., Choi, T.M., Au, K.F., Yu, Y.: Sales forecasting using extreme learning machine with applications in fashion retailing. Decision Support Systems **46**, 411–419 (2008)
8. Chen, F.L., Ou, T.Y.: Sales forecasting system based on gray extreme learning machine with Taguchi method in retail industry. Expert Systems with Applications **38**, 1336–1345 (2011)
9. Li, F.C., Wang, P.K., Wang, G.E.: Comparison of the primitive classifiers with extreme learning machine in credit scoring. In: 16th International Conference on Industrial Engineering and Engineering Management, pp. 685–688. IEEE Press, New York (2009)
10. Li, L.N., Ouyang, J.H., Chen, H.L., Liu, D.Y.: A Computer Aided Diagnosis System for Thyroid Disease Using Extreme Learning Machine. Journal of Medical Systems **36**, 3327–3337 (2012)
11. Lan, Y., Soh, Y.C., Huang, G.B.: Ensemble of online sequential extreme learning machine. Neurocomputing **72**, 3391–3395 (2009)
12. Liu, N., Wang, H.: Ensemble based extreme learning machine. IEEE Signal Processing Letters **17**, 754–757 (2010)
13. Hansen, L.K., Salamon, P.: Neural network ensembles. IEEE Transactions on Pattern Analysis and Machine Intelligence **12**, 993–1001 (1990)
14. Krogh, A., Vedelsby, J.: Neural network ensembles, cross validation, and active learning. Advances in Neural Information Processing Systems **7**, 231–238 (1995)
15. Zhou, Z.H., Wu, J., Tang, W.: Ensembling neural networks: Many could be better than all. Artificial Intelligence **137**, 239–263 (2002)
16. Breiman, L.: Bagging Predictors. Machine Learning **24**, 123–140 (1996)
17. Tang, E.K., Suganthan, P.N., Yao, X.: An analysis of diversity measures. Machine Learning **65**, 247–271 (2006)
18. Blake, C., Keogh, E., Merz, C.J.: UCI Repository of Machine Learning Databases. Department of Information and Computer Science, University of California. http://archive.ics.uci.edu/ml/

Saliency Detection: A Divisive Normalization Approach

Ying Yu[1(✉)], Jie Lin[2], and Jian Yang[1]

[1] School of Information Science and Engineering, Yunnan University, Kunming 650091, China
`yuying.mail@163.com, nxryang@126.com`
[2] Department of Information Management, Yunnan Normal University, Kunming, China
`linjie@ynnu.edu.cn`

Abstract. Saliency detection for images has become a valuable tool in applications like object segmentation, adaptive compression, and object recognition. In this paper, we propose a method for saliency detection that outputs full resolution saliency maps of the input images. The key idea is to exploit a computational process of divisive normalization that simulates the similar feature suppression in human primary visual cortex, and thereby is capable of generating visual saliency. The method, which only employs low-level features of color and luminance, is simple and computationally efficient. We compare our method with five state-of-the-art saliency detection algorithms by use of psychophysical patterns and natural images. Experimental results show that our method outperforms these five algorithms both on the psychophysical ground-truth evaluation and on the eye fixations prediction task.

Keywords: Saliency detection · Visual attention · Divisive normalization

1   Introduction

Visual saliency refers to the perceptual quality that makes an object or location stand out or pop out relative to its neighbors and thereby attract our attention. Typically, visual attention is either driven by fast, pre-attentive, bottom-up visual saliency, or controlled by slow, task-dependent, top-down cues [1].

This paper is primarily concerned with the automatic detection of bottom-up visual saliency, which has already attracted intensive investigations in the area of computer vision in relation to robotics, cognitive science and neuroscience. One of the most influential algorithms of saliency detection was proposed by Itti et al. [2], which is designed based on the biological model of human early visual system. Itti et al.'s model (denoted ITTI) is able to detect salient objects and predict human fixations. However, it is ad-hoc designed and suffers from over-parameterization.

Some recent approaches compute visual saliency in an information theoretic way. These information theory-based approaches include the attention model based on information maximization (denoted AIM) [3], and the graph-based visual saliency approach (denoted GBVS) [4]. While these approaches show good performance in saliency detection, they are computationally expensive for some real-world systems. Another kind of saliency algorithms are implemented in the frequency domain, which

© Springer International Publishing Switzerland 2014
Z. Zeng et al. (Eds.): ISNN 2014, LNCS 8866, pp. 303–311, 2014.
DOI: 10.1007/978-3-319-12436-0_34

are not at all biologically motivated, but they have fast computational speed. These algorithms include the so-called spectral residual approach [5], the saliency algorithm using binary spectrum of discrete cosine transform [6], and the approach using phase spectrum of quaternion Fourier transform (denoted PQFT) [7].

Most of current saliency detection methods generate saliency maps that have low resolution, or are expensive to compute. Moreover, some methods produce higher saliency values at object edges instead of generating maps that uniformly cover the whole object. Recently, a so-called frequency-tuned saliency algorithm (denoted FT) was proposed [8], which computes saliency maps by use of the Euclidean distance in the CIE LAB space between a given position's value and the mean value of the whole image. Although this algorithm is simple to implement and can generate full resolution saliency maps, it is not based on any biological model, and thereby often fails to detect salient regions of images.

In this paper, we introduce a method for salient region detection that employs a computational process referred to as divisive normalization. The method computes chromatic saliency values of all pixels of an image by use of color and luminance features. The divisive normalization simulates intra-cortical suppression among neurons that are tuned to similar features, and thereby has biological plausibility. Our method offers three advantages over existing approaches: computational efficiency, full resolution, and uniformly highlighted salient regions with clear-cut shapes. The saliency map can be effectively used in accurate object segmentation, which is important for object recognition. We provide an objective and visual comparison of the accuracy of the saliency maps against five state-of-the-art methods using a group of psychophysical patterns as well as a 120 images dataset.

The rest of this paper is organized as follows. Section 2 describes the proposed method for salient region detection and its biological plausibility. Section 3 presents the experiments and evaluates the consistency of the method with psychophysical patterns and human fixations. Finally, conclusions are given in Section 4.

2   Proposed Method

In human visual pathway, there exists a "color double-opponent" system. In the center of the receptive fields, neurons are excited by one color and inhibited by another, while the converse is true in the surround. Such spatial and chromatic opponency exists for the green/red and blue/yellow color pairs [9]. Recently, Li hypothesized that human primary visual cortex (V1) creates a bottom-up saliency map of the visual space and the contextual influence is necessary for saliency computation [10]. The dominant contextual influence in V1 is "iso-feature suppression", i.e., nearby neurons tuned to similar features are linked by intra-cortical inhibitory connections [11]. Li's hypothesis can explain "why a red flower is salient among green leaves". In this section, we propose a computational process of divisive normalization to simulate the iso-feature suppression in human visual system.

We consider an $M{\times}N$ color image, where M and N are the number of rows and columns respectively. To begin with, the input image is transformed into the CIE1976

LAB color space. The LAB space is preferred over other color spaces because it is perceptually uniform and similar to human psycho-visual space. In the LAB space, the dynamic range of luminance channel $L*$ is $0 \sim 100$. The range of green/red opponent channel $a*$ is $-128 \sim 127$, where negative and positive values indicate green and red respectively. The range of blue/yellow opponent channel $b*$ is $-128 \sim 127$, where negative and positive values indicate blue and yellow respectively.

For the input image, the LAB transformation generates three biologically plausible channels: a luminance channel $\mathbf{L}$, a green/red opponent channel $\mathbf{A}$, and a blue/yellow opponent channel $\mathbf{B}$. Next, we decompose channel $\mathbf{A}$ into a pair of sub-channels $\mathbf{A}_-$ and $\mathbf{A}_+$, which are obtained by setting all positive and negative entries of matrix $\mathbf{A}$ to zeroes, respectively. Similarly, we decompose channel $\mathbf{B}$ into another pair of sub-channels $\mathbf{B}_-$ and $\mathbf{B}_+$, which are produced by setting all positive and negative entries of matrix $\mathbf{B}$ to zeroes, respectively. From the definition of LAB color space, we can consider $\mathbf{A}_-$, $\mathbf{A}_+$, $\mathbf{B}_-$ and $\mathbf{B}_+$ as four color channels: green, red, blue and yellow.

Then, we need to compute the energy of each color channel. In this work, the energy of each color channel is defined as the summation of the absolute value of all coefficients in each color matrix, which can be formulated as

$$E_g = \sum_{x=1}^{M} \sum_{y=1}^{N} \left| \mathbf{A}_-(x, y) \right|, \tag{1a}$$

$$E_r = \sum_{x=1}^{M} \sum_{y=1}^{N} \left| \mathbf{A}_+(x, y) \right|, \tag{1b}$$

$$E_b = \sum_{x=1}^{M} \sum_{y=1}^{N} \left| \mathbf{B}_-(x, y) \right|, \tag{1c}$$

$$E_y = \sum_{x=1}^{M} \sum_{y=1}^{N} \left| \mathbf{B}_+(x, y) \right|, \tag{1d}$$

where E_g, E_r, E_b and E_y are the energy of green, red, blue and yellow channels, respectively. The divisive normalization that is performed on a matrix $\mathbf{X}$, is defined as

$$\tilde{\mathbf{X}} = \frac{\mathbf{X}}{\sum_{x} \sum_{y} \left| \mathbf{X}(x, y) \right|}, \tag{2}$$

where $\tilde{\mathbf{X}}$ is the divisive normalized matrix. It should be noted that each coefficient of matrix $\mathbf{X}$ is divided by a summation of all coefficients in a way that models a surround inhibition. Conforming to equation (2), we perform divisive normalization on each color matrix to simulate the iso-feature suppression, which can be written as

$$\tilde{\mathbf{A}}_-(x, y) = \mathbf{A}_-(x, y) / E_g, \tag{3a}$$

$$\tilde{\mathbf{A}}_+(x, y) = \mathbf{A}_+(x, y)/E_r \, , \tag{3b}$$

$$\tilde{\mathbf{B}}_-(x, y) = \mathbf{B}_-(x, y)/E_b \, , \tag{3c}$$

$$\tilde{\mathbf{B}}_+(x, y) = \mathbf{B}_+(x, y)/E_y \, . \tag{3d}$$

It is worth stating that the energy of each divisive normalized color matrix is equal to 1. This means that through divisive normalization, the coefficients of a low-energy channel are relatively magnified, whereas those of a high-energy channel are relatively suppressed. As a result of divisive normalization, the iso-feature suppression in V1 can be simulated. In fact, human cannot perceive those color features of very weak energy. Therefore, for the color channels of which the energy is less than 3% of possible maximum, we need to suppress them in some proper way or simply set them to zeros after divisive normalization. Afterwards, we recombine these four divisive normalized color matrices into two color opponent channels as follows:

$$\tilde{\mathbf{A}} = \tilde{\mathbf{A}}_- + \tilde{\mathbf{A}}_+ \, , \tag{4a}$$

$$\tilde{\mathbf{B}} = \tilde{\mathbf{B}}_- + \tilde{\mathbf{B}}_+ \, . \tag{4b}$$

In order to compute visual saliency within a uniform energy scale among all color channels, the luminance channel is also subjected to divisive normalization:

$$\tilde{\mathbf{L}} = \frac{\mathbf{L}}{\displaystyle\sum_{x=1}^{M}\sum_{y=1}^{N}\mathbf{L}(x, y)} \, . \tag{5}$$

To this end, we can form a divisive normalized image by use of $\tilde{\mathbf{L}}$, $\tilde{\mathbf{A}}$, and $\tilde{\mathbf{B}}$, which can be considered as the result of the iso-feature suppression.

The saliency value at a particular coordinate position in a given channel is defined as the absolute difference of the pixel intensity value to the mean intensity value of the channel. The final saliency map $\mathbf{S}$ is also of dimension $M{\times}N$. The final saliency value for a given position is computed as the Euclidean norm of saliencies over different divisive normalized channels in the LAB color space. This calculation process can be formulated as

$$\mathbf{S}(x, y) = \sqrt{\left[\tilde{\mathbf{L}}(x, y) - m_{\tilde{\mathbf{L}}}\right]^2 + \left[\lambda \cdot (\tilde{\mathbf{A}}(x, y) - m_{\tilde{\mathbf{A}}})\right]^2 + \left[\lambda \cdot (\tilde{\mathbf{B}}(x, y) - m_{\tilde{\mathbf{B}}})\right]^2} \, , \tag{6}$$

where $m_{\tilde{\mathbf{L}}}$, $m_{\tilde{\mathbf{A}}}$ and $m_{\tilde{\mathbf{B}}}$ are the mean values of the divisive normalized channel $\tilde{\mathbf{L}}$, $\tilde{\mathbf{A}}$, and $\tilde{\mathbf{B}}$, respectively. The parameter λ is used to adjust the computation weights of saliency values of the two color channels. Considering the disparity in dynamic range between the luminance channel and two color channels in the LAB space, we set λ to 2.55 (i.e., $(128+127)/100$). Finally, the resulting saliency map $\mathbf{S}$ is normalized in the gray-scale interval [0, 255] for visibility.

Our divisive normalization-based saliency method is referred to as **DN** in this paper. Note that our method does not downscale the input image to a lower resolution like other approaches [2][4][7]. In addition, our method does not require prior training bases in contrast to [3]. Also no parameter in our method requires tuning. Fig. 1 gives a visual comparison between the proposed saliency maps of some natural images and the maps generated by five state-of-the-art approaches: ITTI, GBVS, AIM, PQFT, and FT. It can be seen that our proposed DN method can generate full-resolution, uniformly highlighted salient regions, whereas PQFT, AIM, GBVS, and ITTI cannot. Note that FT, which is not motivated by any biological model, fails to highlight salient objects, particularly for those images with dark or bright regions.

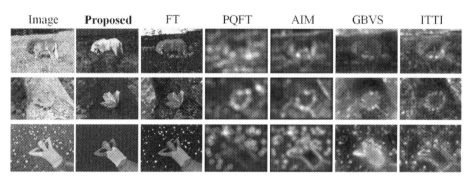

Fig. 1. Visual comparison of the generated saliency maps of natural images

3   Experimental Validation

In this section we compare the capability of psychophysical consistency and fixation prediction of our proposed DN method with five state-of-the-art saliency algorithms: the original Itti et al.'s saliency model (ITTI) [2], Harel et al.'s graph-based visual saliency (GBVS) [4], Bruce and Tsotsos's attention model based on information maximization (AIM) [3], Guo and Zhang's phase spectrum of quaternion Fourier transform (PQFT) [7], and Achanta et al.'s frequency-tuned saliency approach (FT) [8]. These experiments can provide an objective evaluation as well as visual comparison of all saliency maps. All of the saliency approaches are based on the original Matlab implementations available on the author's websites.

3.1   Psychophysical Consistency

In this subsection, we show the consistency of the proposed DN method with some psychophysical patterns. For each psychophysical pattern, we present comparisons with our method against five saliency algorithms mentioned above.

Fig. 2 shows two psychophysical patterns of red-green and green-red color pop-out. It can be seen that, a red bar among green bars as in the first pattern and a green bar among red bars as in the second pattern are salient objects. This is a fundamental task for the test of color saliency. It can be seen that the salient bars can pop out relative to the distracters in the saliency maps generated by our DN method. However, other five methods cannot detect the salient objects correctly.

Fig. 3 shows a group of psychophysical patterns of color pop-out. In each pattern, one salient object, which has a unique color, is present. It can be seen that the extent of visual saliency of the targets gradually decrease from the first image to the fourth one. This is a somewhat difficult task. For all four patterns, the disparity between saliency value of target and distracters of our proposed DN method is consistent with perception, but other 5 algorithms cannot correctly highlight these salient objects conforming to visual perception. This means that our proposed DN method is very sensitive to the variation of color and is consistent with human visual perception.

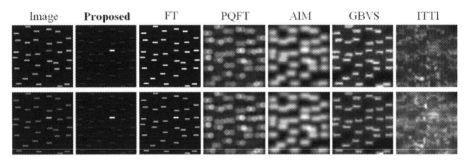

Fig. 2. Responses to psychophysical patterns of red-green and green-red color pop-out

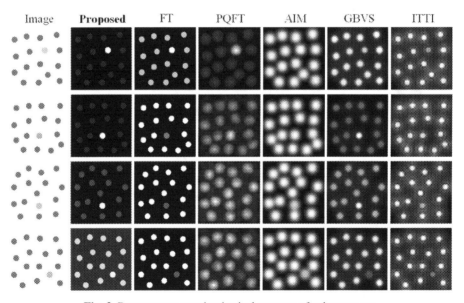

Fig. 3. Responses to psychophysical patterns of color pop-out

3.2   Eye Fixation Prediction

In this subsection, we validate the proposed saliency maps by use of the data set of 120 color images from an urban environment and corresponding human eye-fixation data from 20 subjects provided by Bruce and Tsotsos [3]. These color images consist of indoor and outdoor scenes, of which some have very salient items, and others have no particular regions of interest. In order to quantify the consistency of a particular saliency map with a set of fixations of the image, we employ an objective evaluation metric that is referred to as receiver operating characteristic (ROC) area under the curve (AUC). Note that a number of published papers employed ROC-AUC score to evaluate a saliency map's ability to predict human eye fixations (e.g., [3][7]).

Following Tatler et al.'s approach [12], we compute the ROC-AUC score conforming to the following procedure. For one image, the positive point set is composed of the fixated locations from all subjects on that image, whereas the negative point set is composed of the non-fixated locations of the image. Each saliency map is binarized by a particular threshold and thereby considered as a binary classifier. At a particular threshold level, a binary saliency map can be divided into the target (white) region and the background (black) region. The true positive rate (TPR) is the proportion of the positive points that fall in the target region of the binary saliency map. The false positive rate (FPR) can be calculated in the same way by using the negative point set. Varying the threshold yields an ROC curve of TPRs versus FPRs, of which the area beneath provides a good measure of the capability of the saliency map to accurately predict where human eye fixations occurred on an image. Since the AUC is a portion of the area of the unit square, its value will always be between 0 and 1.0. Chance level is 0.5, and perfect prediction is 1.0.

In this test, we resize all images to 384×288px before computing the saliency maps. We compare our saliency maps generated from the proposed DN method to five state-of-the-art saliency approaches. An important note about these experiments is that the ROC-AUC score is sensitive to the number of fixations we use in calculation. Former fixations are more likely to be driven by bottom-up manner, whereas later fixations are more likely to be influenced by top-down cues [12]. We calculate the ROC-AUC scores for each image with respect to all fixations, and repeat the process but use only the first two fixation points. Table 1 lists the ROC-AUC score averaged over all 120 images for each saliency method. As expected, the ROC-AUC scores with only the first two fixations are higher than those with all fixations. It can be seen that in both tests our proposed DN method has the best capability for predicting eye fixations.

Fig. 4 gives the saliency maps for 7 sample images from the image data set, which provides a qualitative comparison of all saliency methods. A fixation density map,

Table 1. The ROC-AUC performance of all six methods

Method	**Proposed**	FT	PQFT	AIM	GBVS	ITTI
All fixations	**0.7765**	0.7163	0.7751	0.7706	0.7127	0.7062
First 2 fixations	**0.7884**	0.7204	0.7846	0.7777	0.7267	0.7182

| Image | Fixation | **Proposed** | FT | PQFT | AIM | GBVS | ITTI |

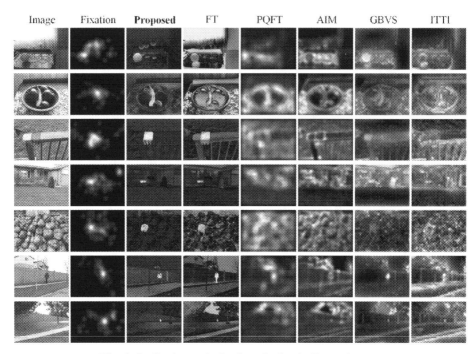

Fig. 4. Qualitative analysis of results for the Bruce data set

generated for each image by convolution of the fixation map for all subjects with a Gaussian filter, serves as ground truth. Analyzing the qualitative results, we can see that the proposed DN method is more predictive of human fixations. The regions highlighted by our proposed method overlap to a surprisingly large extent with those image regions looked at by humans in free viewing. In addition, our method generates clear-cut and uniformly highlighted salient regions as compared to other methods. Once again, we notice that FT is vulnerable by some dark or bright regions (e.g., the first two images). PQFT, AIM, GBVS and ITTI cannot generate full-resolution saliency maps, and are easily distracted by branches (e.g., the last two images).

We record the computational time cost per image in a standard desktop computing environment. Table 2 shows each method's Matlab runtime measurements averaged over the data set. It can be noticed that, our proposed DN method use only 0.0853s on the average to compute a saliency map, which is significantly faster than AIM, GBVS, and ITTI. Although our DN method is slightly slower than FT and PQFT, it is more predictive of human fixations and can detect salient regions more accurately. All six saliency approaches are implemented in the Matlab R2012a environment on such a computer platform as Intel 3.3 GHz CPU with 8 GB of memory.

Table 2. Computational time cost per image for all six methods

Method	**Proposed**	FT	PQFT	AIM	GBVS	ITTI
Time (s)	**0.0853**	0.0744	0.0793	5.0766	2.5957	1.1842

4   Conclusions

This paper proposed a saliency detection method using low level features of color and luminance. We manifested that the color pop-out saliency of an image can be generated by using a divisive normalization approach performed on the features of color and luminance. The proposed method is simple, fast, and provides full resolution saliency maps with well-defined boundaries, which are important for salient object segmentation. Experiments in this paper showed that the proposed method outperforms state-of-the-art saliency algorithms when evaluated by the ability to detect color pop-out saliency or predict human fixations. Actually, ROC-AUC score cannot fully reflect the advantage of our DN method over other saliency algorithms. Our future work will focus on applying the proposed DN method to salient object segmentation, which is important for object recognition.

Acknowledgements. This research was supported by the National Natural Science Foundation of China (Grant No. 61263048), by the Scientific Research Foundation of Yunnan Provincial Department of Education (2012Y277), by the Scientific Research Project of Yunnan University (2011YB21), and by the Young and Middle-Aged Backbone Teachers' Cultivation Plan of Yunnan University (XT412003).

References

1. Itti, L., Koch, C.: Computational Modeling of Visual Attention. Nat. Rev. Neurosci. **2**(3), 194–203 (2001)
2. Itti, L., Koch, C., Niebur, E.: A Model of Saliency-Based Visual Attention for Rapid Scene Analysis. IEEE Trans. Patt. Anal. Mach. Intell. **20**(11), 1254–1259 (1998)
3. Bruce, N.D., Tsotsos, J.K.: Saliency, Attention, and Visual Search: An Information Theoretic Approach. J. Vis. **9**(3), 1–24 (2009)
4. Harel, J., Koch, C., Perona, P.: Graph-Based Visual Saliency. In: Proc. NIPS (2006)
5. Hou, X., Zhang, L.: Saliency Detection: A Spectral Residual Approach. In: Proc. CVPR (2007)
6. Yu, Y., Wang, B., Zhang, L.: Bottom-Up Attention: Pulsed PCA Transform and Pulsed Cosine Transform. Cogn. Neurodyn. **5**(4), 321–332 (2011)
7. Guo, C., Zhang, L.: A Novel Multiresolution Spatiotemporal Saliency Detection Model and Its Applications in Image and Video Compression. IEEE Trans. Image Process. **19**(1), 185–198 (2010)
8. Achanta, R., Hemami, S., Estrada, F., Susstrunk, S.: Frequency-Tuned Salient Region Detection. In: Proc. CVPR (2009)
9. Engel, S., Zhang, X., Wandell, B.: Colour Tuning in Human Visual Cortex Measured with Functional Magnetic Resonance Imaging. Nature **388**(6637), 68–71 (1997)
10. Li, Z.: A Saliency Map in Primary Visual Cortex. Trends Cogn. Sci. **6**(1), 9–16 (2002)
11. Li, Z., Dayan, P.: Pre-attentive Visual Selection. Neural Netw. **19**, 1437–1439 (2006)
12. Tatler, B.W., Baddeley, R.J., Gilchrist, I.D.: Visual Correlates of Fixation Selection: Effects of Scale and Time. Vision Res. **45**(5), 643–659 (2005)

A Novel Neural Network Based Adaptive Control for a Class of Uncertain Strict-Feedback Nonlinear Systems

Baobin Miao and Tieshan Li[✉]

Navigational College, Dalian Maritime University, Dalian, 116026, China
tieshanli@126.com

Abstract. In this paper, a novel robust adaptive tracking control approach is presented for a class of strict-feedback single input single output nonlinear systems. In the controller design process, all unknown functions at intermediate steps are passed down, and only one neural network is used to approximate the lumped unknown function of the system at the last step. Although some similar design themes have been proposed, the approach presented in this paper is more reasonable and simpler. The most contribution in this paper is that a new concept named "filter technique" is proposed for how to avoid generating new unknown functions when derivation of virtual control law in the backstepping based control methods. So the neural network is just used to approximate the finite or less unknown functions and the good capabilities in function approximation of neural network are guaranteed. Stability analysis shows that the uniform ultimate boundedness of all the signals in the closed-loop system can be guaranteed, and the steady state tracking error can be made arbitrarily small by appropriately choosing control parameters. Simulation results demonstrate the effectiveness of the proposed scheme.

Keywords: Filter technique · Neural network · Adaptive control · Nonlinear systems

1 Introduction

In the past years, backstepping based nonlinear adaptive control has been paid considerable attentions and a great deal of progress had been achieved for the adaptive control of strict-feedback nonlinear systems with linearly parameterized uncertainty [1]. Although significant progresses have been made by combining backstepping methodology with neural network technologies, there are still some problems that need to be solved for practical implementations. The main drawback of the aforementioned control design methods is the problem of complexity [2]. That is, the complexity of the designed controller grows drastically as the system order increase, this phenomenon is caused by four reasons. The first reason is that the repeated differentiations of the virtual control laws in the traditional backstepping approach. The second reason is that neural network is used to approximate major terms of unknown functions.

© Springer International Publishing Switzerland 2014
Z. Zeng et al. (Eds.): ISNN 2014, LNCS 8866, pp. 312–320, 2014.
DOI: 10.1007/978-3-319-12436-0_35

The third reason is that with an increase of neural network nodes, the number of parameters to be estimated will increase significantly. The last reason is that the use of multiple approximators. All of above reasons make the complexity growing problem harder for implementation.

For the first reason behind the complexity growing problem, in [3], the authors addressed a modification that obviated the repeated differentiations of the command derivatives by introducing command filter technique in the backstepping design. A dynamic surface control technique was proposed to solve the complexity growing problem in [4]. In addition, the virtual control laws were modeled as portions of unknown functions that were approximated during operation in [5]. The point is that there is not only one question in the typical adaptive backstepping control, a new concept named "filter technique" is proposed for how to avoid generating new unknown functions when derivation of virtual control law. That is, the virtual control law a_i include the elements of $[x_1, x_2, \ldots, x_i]$, while in the backstepping based control design process, it needs to calculate the derivative of the virtual control law, so it will generate new terms of unknown functions, i.e., $\dfrac{\partial a_i}{\partial x_j} \dot{x}_j$, where x_j for $j \le i$ is an element of the state vector. In order to solve this problem, the most contribution in this paper, a first-order filtering of the synthetic input is introduced at each step of the traditional backstepping approach. So the neural network is just used to approximate the finite or less unknown functions and the good capabilities in function approximation of neural network are guaranteed. Consequently, the second reason behind the complexity growing problem is removed.

In many practical applications on the control of uncertain nonlinear systems, neural network based control methods are shown to be more efficient compared with other modern control techniques and many remarkable results have been obtained. The most useful property of neural network is their ability to approximate arbitrary linear or nonlinear mapping through learning. Although there are significant advantages by employing neural network to control uncertain nonlinear systems, these neural network based schemes suffered from some limitations. For example, many approximators are still used to construct virtual control laws and actual control law in these methods. That is, for solving the uncertainty of nonlinear systems, every virtual control laws and actual control law are constructed by at least one neural network to approximate the unknown functions. So, if the uncertain nonlinear systems order is more than three, the computational burden grows due to the adaptive computation of these approximators. Although some novel themes had been proposed for solving this problem in [4], I think there exist some problems in their literatures: (1) the virtual control laws are composed by parts of unknown function, it is unreasonable. (2) the control signal u include the derivative of a_n, which requires the second derivative of a_{n-1}, which requires the third derivative of a_{n-2}, and so on, i.e., the repeated differentiations of virtual control laws and generating new unknown functions. Another central issue within approximation based adaptive control schemes is that the number of adaptation laws depends on the number of the neural network nodes. With an increase of neural network nodes to improve approximation accuracy, the number of parameters to be estimated will increase significantly. As a result, the on-line learning time will become prohibitively large. To solve this problem, the norm of the ideal weighting vector in neural network is considered as the estimation parameter instead of the elements of weighting vector. Thus, the number of adaptation laws is reduced considerably. So, both the third and the last reasons behind the complexity growing problem are removed.

In this paper, a neural network approximation based adaptive control approach is presented for a class of uncertain strict-feedback nonlinear systems. The most contributions are that a new concept named "filter technique" is proposed for how to avoid generating new unknown functions and a reasonable controller design procedure about using one approximator is proposed. In addition, by using a first-order filter, both problems of the repeated differentiations of the virtual control laws and generating new unknown functions are solved. These features guarantee that the computational burden of the algorithm can drastically be reduced and that the algorithm is convenient to implement in applications. Stability analysis shows that all the closed-loop system signals are uniformly ultimately bounded, and the steady state tracking error can be made arbitrarily small by appropriately choosing control parameters. Theoretical result is illustrated by simulation results.

2   Problem Formulation and Preliminaries

Consider a class of uncertain nonlinear dynamical systems in the following form:

$$\begin{cases} \dot{x}_i = x_{i+1} + f_i(\bar{x}_i) & 1 \le i \le n-1 \\ \dot{x}_n = u + f_n(\bar{x}_n) \\ y = x_1 \end{cases} \tag{1}$$

where $\bar{x}_i = [x_1,\ldots,x_i]^T \in R^i$, $i = 1,\ldots,n$, $u \in R$ and $y \in R$ are system state variables, system input, and output, respectively; $f_i(\bar{x}_i)$, $i = 1,\ldots,n$, are unknown smooth nonlinear functions.

The control objective is to design an adaptive controller for the system (1), such that all the close-loop system signals remain uniformly ultimately bounded, and the system output y follows the reference signal $y_r(t)$.

Notation 1. $\|\cdot\|$ stands for Frobenius norm of matrices and Euclidean norm of vectors, i.e., given a matrix B and a vector Q, the Frobenius norm and Euclidean norm are given by $\|B\|^2 = tr(B^T B) = \sum_{i,j} b_{ij}^2$ and $\|Q\|^2 = \sum_i q_i^2$.

3   Controller Design and Stability Analysis

In the following part, for the purpose of simplicity, the time variable t and the state vector $\bar{x}_i$ will be omitted from the corresponding functions.

Step 1: Let $z_1 = x_1 - y_r$, the derivative of z_1 is

$$\dot{z}_1 = x_2 + f_1 - \dot{y}_r \tag{2}$$

The virtual control law a_2^0 is chosen as follows:

$$a_2^0 = -k_1 z_1 + \dot{y}_r \tag{3}$$

where k_1 is a positive real constant which will be specified later.

Introduce a new state variable a_2 and let a_2^0 pass through a first-order filter with time constant e_2 to obtain a_2

$$e_2 \dot{a}_2 + a_2 = a_2^0 \tag{4}$$

Define the filter error as follows

$$p_2 = a_2 - a_2^0 \tag{5}$$

Let $z_2 = x_2 - a_2$ and consider the (3)-(5), we can get

$$\dot{z}_1 = z_2 + p_2 + f_1 - k_1 z_1 \tag{6}$$

Consider the following Lyapunov function candidate

$$V_1 = \frac{1}{2} z_1^2 + \frac{1}{2} p_2^2 \tag{7}$$

Differentiating V_1 yields

$$\dot{V}_1 = -k_1 z_1^2 + z_1 z_2 + z_1 p_2 + z_1 f_1 + p_2 \dot{p}_2 \tag{8}$$

It is worth noting that

$$\dot{p}_2 = -\frac{p_2}{e_2} + B_2 \left(z_1, z_2, p_2, y_r, \dot{y}_r, \ddot{y}_r \right) \tag{9}$$

where $B_2 (\cdot)$ is a continuous function and has a maximum value M_2 [3].

Using the facts that

$$z_1 p_2 \leq z_1^2 + \frac{p_2^2}{4} \tag{10}$$

$$p_2 B_2 \le \frac{p_2^2 B_2^2}{2} + \frac{1}{2} \tag{11}$$

Apparently, if we choose $k_1 - 1 \ge a_0$, $\frac{1}{e_2} \ge \frac{1}{4} + \frac{M_2^2}{2} + a_0$ and consider the (9)-(11), the (8) can be rewritten as

$$\dot{V}_1 \le -a_0 z_1^2 - a_0 p_2^2 + z_1 z_2 + z_1 f_1 + \frac{1}{2} \tag{12}$$

Step i ($2 \le i \le n-1$): Let $z_i = x_i - a_i$, the derivative of z_i is

$$\dot{z}_i = x_{i+1} + f_i - \dot{a}_i \tag{13}$$

The virtual control law a_{i+1}^0 is chosen as follows:

$$a_{i+1}^0 = -k_i z_i + \dot{a}_i - z_{i-1} \tag{14}$$

where k_i is a positive real constant which will be specified later.

Introduce a new state variable a_{i+1} and let a_{i+1}^0 pass through a first-order filter with time constant e_{i+1} to obtain a_{i+1}

$$e_{i+1} \dot{a}_{i+1} + a_{i+1} = a_{i+1}^0 \tag{15}$$

Define the filter error as follows

$$p_{i+1} = a_{i+1} - a_{i+1}^0 \tag{16}$$

Let $z_{i+1} = x_{i+1} - a_{i+1}$ and consider the (14)-(16), we can get

$$\dot{z}_i = z_{i+1} + p_{i+1} + f_i - k_i z_i - z_{i-1} \tag{17}$$

Consider the following Lyapunov function candidate

$$V_i = V_{i-1} + \frac{1}{2} z_i^2 + \frac{1}{2} p_{i+1}^2 \tag{18}$$

Differentiating V_i yields

$$\dot{V}_i = \dot{V}_{i-1} + z_i z_{i+1} - z_i z_{i-1} + z_i p_{i+1} + z_i f_i - k_i z_i^2 + p_{i+1} \dot{p}_{i+1} \tag{19}$$

It is worth noting that

$$\dot{p}_{i+1} = -\frac{p_{i+1}}{e_{i+1}} + B_{i+1}\left(\bar{z}_{i+1}, p_2, \ldots, p_{i+1}, y_r, \dot{y}_r, \ddot{y}_r\right) \tag{20}$$

where $B_{i+1}(\cdot)$ is a continuous function and has a maximum value M_{i+1}.

Using the facts that

$$z_i p_{i+1} \le z_i^2 + \frac{p_{i+1}^2}{4} \tag{21}$$

$$p_{i+1} B_{i+1} \le \frac{p_{i+1}^2 B_{i+1}^2}{2} + \frac{1}{2} \tag{22}$$

Apparently, if we choose $k_i - 1 \ge a_0$, $\frac{1}{e_{i+1}} \ge \frac{1}{4} + \frac{M_{i+1}^2}{2} + a_0$, the (19) can be rewritten as

$$\dot{V}_i \le -a_0 \sum_{l=1}^{i} z_l^2 - a_0 \sum_{l=1}^{i} p_{l+1}^2 + \sum_{l=1}^{i} z_l f_l + z_i z_{i+1} + \frac{i}{2} \tag{23}$$

Step n: Let $z_n = x_n - a_n$, the derivative of z_n is

$$\dot{z}_n = u + f_n - \dot{a}_n \tag{24}$$

Consider the following Lyapunov function candidate

$$V_n = V_{n-1} + \frac{1}{2} z_n^2 + \frac{1}{2} \tilde{\theta}^2 \tag{25}$$

where $\tilde{\theta} = \theta - \hat{\theta}$. Differentiating V_n yields

$$\dot{V}_n = \dot{V}_{n-1} + z_n \left(u + f_n - \dot{a}_n \right) - \tilde{\theta} \dot{\hat{\theta}} \tag{26}$$

Choose the actual control law as

$$u = -k_n z_n - z_{n-1} + \dot{a}_n + \left(-\hat{\theta} \right)^{\frac{1}{2}} \tag{27}$$

one has

$$\dot{V}_n \le -a_0 \sum_{l=1}^{n-1} z_l^2 - a_0 \sum_{l=1}^{n-1} p_{l+1}^2 + \sum_{l=1}^{n} z_l f_l + \frac{n-1}{2} - k_n z_n^2 + z_n \left(-\hat{\theta} \right)^{\frac{1}{2}} - \tilde{\theta} \dot{\hat{\theta}} \tag{28}$$

Given a compact set $\Omega \subset R^n$, and let W^* and ε be such that for any $Z \in \Omega$,

$$\sum_{l=1}^{n} z_l f_l = W^{*T} S(Z) + \varepsilon \tag{29}$$

where $|\varepsilon| \le \varepsilon^*$.

the (28) can be rewritten as

$$\dot{V}_n \le -a_0 \sum_{l=1}^{n} z_l^2 - a_0 \sum_{l=1}^{n-1} p_{l+1}^2 + \varepsilon^* + \frac{n-1}{2} + \frac{\|S(Z)\|^2}{4} + \tilde{\theta} - \tilde{\theta} \dot{\hat{\theta}} \tag{30}$$

The adaptive law is chosen as

$$\dot{\hat{\theta}} = -k_0\hat{\theta} + 1 \tag{31}$$

where k_0 is a positive constant.

It is worth noting that

$$k_0\tilde{\theta}\hat{\theta} \leq -\frac{k_0}{2}\tilde{\theta}^2 + \frac{k_0}{2}\theta^2 \tag{32}$$

Choose $\dfrac{k_0}{2} \geq a_0$, one has

$$\dot{V}_n \leq -2a_0V_n + d \tag{33}$$

where $d = \dfrac{\|S(Z)\|^2}{4} + \varepsilon^* + \dfrac{n-1}{2} + \dfrac{k_0}{2}\theta^2$.

From (33), one has

$$V_n(t) \leq \frac{d}{2a_0} + \left(V_n(t_0) - \frac{d}{2a_0}\right)e^{-2a_0(t-t_0)} \tag{34}$$

It follows that, for any $\mu_1 > (b_0/a_0)^{1/2}$, there exists a constant $T > 0$ such that $z_1(t) \leq \mu_1$ for all $t \geq t_0 + T$, and the tracking error can be made small, since μ_1 can arbitrarily be made small if the design parameters are appropriately chosen.

4   Simulation Examples

In this section, an example will be used to test the effectiveness of the proposed controller. Consider the following nonlinear system:

$$\begin{cases} \dot{x}_1 = x_2 + x_1^2 \\ \dot{x}_2 = x_3 + x_1^2 + x_2^2 \\ \dot{x}_3 = u + x_1x_3 \end{cases} \tag{35}$$

The reference signal is given as $y_r = 0.5(\sin(t) + \sin(1.5t))$. The design paremeters of the above controller are $k_1 = 10$, $k_2 = 5$, $k_3 = 5$, $e_2 = 0.01$, $e_3 = 0.01$, $k_0 = 10$. The simulations are run with the initial conditions $x(0) = [2,0,0]^T$ and $\hat{\theta}(0) = -1$. The simulation results are shown in Figs. 1-2.

As it can be seen from the simulation results, good tracking accuracy and the stability of the closed-loop system are guaranteed under the proposed controller.

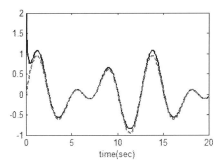

Fig. 1. Output tracking performance (y—solid line and yr—dashed line)

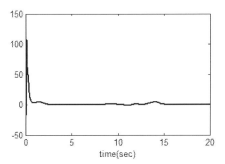

Fig. 2. The control law u

5   Conclusion

In this paper, the tracking control problem has been considered for a class of uncertain nonlinear systems with a strict-feedback structure. In the controller design process, only one neural network approximator is used to address the lumped unknown function of the system. In addition, by using the "filter technique", both problems of the repeated differentiations of the virtual control laws and generating new unknown functions are solved. By this approach, the structure of the controller can be simplified observably, and the computational burden can be reduced drastically. The main feature of the control scheme proposed in this paper is simplicity. In particular, no matter how many neural network nodes are used, there is only one parameter to be updated online. The proposed controller is derived in the sense of Lyapunov function, thus the system can be guaranteed to be asymptotically stable. Simulation results demonstrate the effectiveness of the proposed scheme.

Acknowledgments. This work was supported in part by the National Natural Science Foundation of China (Nos. 51179019, 61374114, 61001090), the Natural Science Foundation of Liaoning Province (No. 20102012), the Program for Liaoning Excellent Talents in University (LNET) (Grant No.LR 2012016) , the Applied Basic Research Program of Ministry of Transport of P. R. China (Nos. 2011-329-225-390 and 2013-329-225-270), and the National

Fundamental Research 973 Program of China under Grant 2011CB302801, the Macau Science and Technology Development Foundation under Grant 008/2010/A1, and Multiyear Research Grants.

References

1. Kanellakopoulos, I., Kokotovic, P.V., Morse, A.S.: Systematic design of adaptive controllers for feedback linearizable systems. IEEE Trans. Autom. Control **36**, 1241–1253 (2008)
2. Sun, G., Wang, D., Li, T.S., Peng, Z.H., Wang, H.: Single neural network approximation based adaptive control for a class of uncertain strict-feedback nonlinear systems. Nonlinear Dynamics **72**, 175–184 (2013)
3. Farrell, J.A., Polycarpou, M., Sharma, M., Dong, W.J.: Command filtered backstepping. IEEE Trans. Autom. Control **54**, 1391–1395 (2009)
4. Swaroop, D., Hedrick, J.K., Yip, P.P., Gerdes, J.C.: Dynamic surface control for a class of nonlinear systems. IEEE Trans. Autom. Control **45**, 1893–1899 (2000)
5. Ge, S.S., Wang, C.: Direct adaptive NN control of a class of nonlinear systems. IEEE Transactions on Neural Networks **13**, 214–221 (2002)

Rough Sets Theory Approch to Extenics Risk Assessment Model in Social Risk

Guangyao Gu, Yaodong Le$^{(\boxtimes)}$, Fajie Wei, and Chen Li

School of Economics and Management, Beihang University, Beijing, 100191, China
guguangyao@126.com, {leyaodong,lichenbuaa}@163.com,
weifajie@buaa.edu.cn

Abstract. Social risks in large-scale projects are ranked according to the results of risk assessment, and the order helps assign the resource priority for the high risk events. The common method of risk evaluation is mostly based on probability and consequences of risk events, but ignores the management properties of them. This paper combines rough sets theory and extenics theory, and then builds the extenics risk assessment model based on rough sets theory. This model eliminates some valueless indicators and evaluates their weights using the rough sets theory and calculates the risk grades by extenics evaluation. In the last part, this paper analyzes an airport construction project to prove the practicality and effectiveness of the model.

Keywords: Risk assessment model · Rough sets theory · Extenics · Social risk

1   Introduction

Large-scale projects, such as nuclear power plant construction projects, airport construction projects, etc. have a significant impact on local economy and environment. Recent years, more and more group unexpected incidents happened because of the phenomena of land requisition and demolishing, barbaric construction and ignorance the ecological environment in large-scale projects[1][2]. The research for risk assessment model of social risk is necessary.

Risk assessment is a process which sorts the risk events according to the value-at-risk. The common method of risk evaluation is based on Probability and Consequences, which is denoted as $R = f(P,C) = P \times C$. The higher the score, the greater the risk[3]. This method just considers the natural properties of risk, but not considers the management properties, which means it ignores the management resources and their effects devoting to the project. The risk evaluation indexes are not comprehensive.

Many researchers do a lot of researches on the selection of risk evaluation indexes in recent years. Haimes(1993) said the common method in analyzing risk is insufficient[4]. Ward(1999) stated the risk assessment model should consider the attributes of the response measures and the response time can be used in addition to the probability and consequences of risk[5]. Li Shaoming(2006) came up with a new model called Risk Management Priority Evaluation Model, which contained four indexes:

© Springer International Publishing Switzerland 2014
Z. Zeng et al. (Eds.): ISNN 2014, LNCS 8866, pp. 321–329, 2014.
DOI: 10.1007/978-3-319-12436-0_36

probability, urgency, controllability and management efficiency[3]. However, the current study exist two disadvantages: the first one is that the selection of indexes is unreasonable, most models focus on the objectivity and accuracy of risk evaluation methods but ignore the comprehensiveness and necessity of index system; the second one is that most evaluation methods are suitable for common two indexes, not apply to multi-index system.

Extenics evaluation is a kind of method for multi-index models and it's widely used in risk assessment in recent years[6]. The attribute reduction theory has good performance on selection of indexes and evaluation of index weight on the basis of rough sets[7]. This paper combines these two theories and builds the extenics risk assessment model based on rough sets theory. This model reduces risk evaluation indexes and evaluates their weights by rough sets theory and calculates the risk grade by extenics evaluation. Finally, this paper analyzes an airport construction project to prove the practicality and effectiveness of the model.

2   Basic Concepts of Models

2.1   Rough Sets Theory

The rough sets theory introduced by Pawlak (1982) is a mathematical tool for approximate reasoning for decision support and is particularly well suited for classification of objects[8][9].

Definition 1. Let $S = (U, R, V, f)$ be an information system, $r \in R$, If satisfy the following equations:

$$ind(R - \{r\}) = ind(R) \qquad (1)$$

where attribute r is redundant in attribute R, otherwise, attribute r is necessary in attribute R, and the set consists of all necessary attributes in R called the core of R, in symbols $red(R)$.

Definition 2. Let $S = (U, R, V, f)$ be an information system, $P, Q \subseteq R$, the importance of attribute R is defined as:

$$U_r = \frac{\| Pos_p(Q) \| - \| Pos_{p-|r|}(Q) \|}{|U|} \qquad (2)$$

2.2   Extenics Evaluation

Professor Cai Wen proposed the extenics in 1983. It is a new theory which calculates the evaluation system[6].

Definition 3. Define an ordered triad $R = (N, c, v)$ as the basic element for describing things, called matter-element, where N represents the matter, c, the characteristics, v is the N's measure about the characteristic, the expression $v = c(N)$ describes the relation between quality and quantity. A matter has many characteristic-elements, which can be described by n-dimensional matter-elements

$$R = \begin{bmatrix} N, & c_1 & v_1 \\ & c_2 & v_2 \\ & \cdots & \cdots \\ & c_n & v_n \end{bmatrix} = \begin{bmatrix} R_1 \\ R_2 \\ \cdots \\ R_n \end{bmatrix} \tag{3}$$

The dynamic matter-element $R_i = (N, c_i, v_i), i = 1, 2, ..., n$ describes the change of matter N with time.

Definition 4. The concept of distance in real variable functions has been generalized.
The distance on real axis between point x and a given real interval $X_0 = <a, b>$ is defined as

$$\rho(x, X_0) = |x - \frac{a+b}{2}| - \frac{1}{2}(b-a) \tag{4}$$

The formulas of dependent function can be defined as

$$K(x) = \frac{\rho(x, X_0)}{D(x, X_0, X)} \tag{5}$$

where $D(x, X_0, X) = \begin{cases} \rho(x, X) - \rho(x, X_0), & x \notin X_0 \\ -1, & x \in X_0 \end{cases} \tag{6}$

3   Establish the Risk Assessment Model

Extension assessment method is based on a combination of qualitative and quantitative analysis. This method uses extension dependence function to evaluate correlation degree about evaluation indexes to risk level, and then sorts the risk events due to the result.

How to determine index weight is the key to extension assessment method, and the quality of selection of indexes decides the final result directly. This paper chooses rough sets theory to evaluate index weight, and establishes a multi-index risk assessment model with extension assessment method. The steps are shown as follows:

Step 1, establish extension assessment model according to extenics. Define Q as matter-element, R as basic risk field, in symbols $R = (R_1, R_2, ..., R_n)$, where n is the number of risk events, C as initial evaluation risk index system, in symbols

$C = (c_1, c_2, ..., c_l)$, where l is the number of risk indexes, U as risk grade field, in symbols $U = (u_1, u_2, ... u_m)$, where m is the number of risk grades.

Step2, determine sutra field and controlled field of risk grade. Define

$$R_j = (U_j, C, V_j) = \begin{bmatrix} u_j, & c_1 & V_{1j} \\ & c_2 & V_{2j} \\ & ... & ... \\ & c_l & V_{lj} \end{bmatrix} = \begin{bmatrix} u_j, & c_1 & <a_{1j}, b_{1j}> \\ & c_2 & <a_{2j}, b_{2j}> \\ & ... & ... \\ & c_l & <a_{lj}, b_{lj}> \end{bmatrix} \tag{7}$$

Where U_j represents risk grade, $j = 1, 2, ..., m$; c_k represents the evaluation indexes of risk grade U_j; V_{kj} represents u_j's value range on c_k, the value range $<a_{kj}, b_{kj}>$ is sutra field of u_j. Define

$$R_U = (U, C, V_U) = \begin{bmatrix} U, & c_1 & V_{1U} \\ & c_2 & V_{2U} \\ & ... & ... \\ & c_l & V_{lU} \end{bmatrix} = \begin{bmatrix} U, & c_1 & <a_{1U}, b_{1U}> \\ & c_2 & <a_{2U}, b_{2U}> \\ & ... & ... \\ & c_l & <a_{lU}, b_{lU}> \end{bmatrix} \tag{8}$$

Where U represents risk grade field, V_{kU} represents U's value range on C_k, the value range $<a_{kU}, b_{kU}>$ is controlled field of U.

Step 3, acquire data. Build evaluation Criteria and use expert evaluation method to get each risk event R_p's values about revaluation indexes C_k. It establishes R_p's Matter-element model,

$$Q_p = (R_p, C, V_{lp}) = \begin{bmatrix} R_p, & c_1 & v_{1p} \\ & c_2 & v_{2p} \\ & ... & ... \\ & c_l & v_{lp} \end{bmatrix} \tag{9}$$

Step 4, reduct initial evaluation index system through the attribute reduction theory which is on the basis of rough sets. Use formula (2) to evaluate weight of index c_k, in symbols α_k

$$\alpha_k = U_{c_k} / \sum_{k=1}^{l} U_{c_k} \tag{10}$$

And then establish the preferred evaluation risk index system.

Step 5, use extension dependence function to evaluate the correlation degree of risk event R_p's evaluation index c_k to risk grade u_j according to formula (6),

$$K(c_k, u_j) = \frac{\rho(v_{kp}, V_{kj})}{\rho(v_{kp}, V_{kU}) - \rho(v_{kp}, V_{kj})} \tag{11}$$

And then the correlation degree of risk event R_p to risk grade u_j is evaluated by

$$K(R_p, u_j) = \sum_{k=1}^{l} \alpha_k K(c_k, u_j) \tag{12}$$

And finally, get one risk event's risk grade.

$$K_{j_0} = \max_{j_0 \in \{1,2,\dots,m\}} K(R_p, u_j) \tag{13}$$

Where j_0 represents risk event R_p's risk grade, and the degree is K_{j_0}.

4   Case Study

4.1   Establish Initial Evaluation Risk Index System

We identify 10 social risk events from one airport construction project in Beijing[2], as is shown in Table 1.

Table 1. Risk events

Symbols	Risk event	Symbols	Risk event
R_1	Policy planning risk	R_6	Social environment risk
R_2	Public participation risk	R_7	Economic risk
R_3	Land requisition risk	R_8	Project management risk
R_4	Project technical risk	R_9	Public opinion risk
R_5	Natural environment risk	R_{10}	Project special risk

The 10 risk events above make up a risk field $R = (R_1, R_2, \dots, R_{10})$. We define a five-level risk degree field $U = \{u_1, u_2, u_3, u_4, u_5\}$, including "very low", "low", "medium", "high" and "very high". Their corresponding scores are "0~2", "2~4", "4~6", "6~8" and "8~10".

In this case, we establish the initial evaluation indexes on the basis of other related researches, including probability, consequence, urgency, controllability, management costs and management efficiency.

Get sutra field and controlled field of risk degree according to formula(9)(10),

$$R_j = (U_j, C, V_j) = \begin{bmatrix} u_j, & c_1 & <a_{1j}, b_{1j}> \\ & c_2 & <a_{2j}, b_{2j}> \\ & c_3 & <a_{3j}, b_{3j}> \\ & c_4 & <a_{4j}, b_{4j}> \\ & c_5 & <a_{5j}, b_{5j}> \\ & c_6 & <a_{6j}, b_{6j}> \end{bmatrix} \qquad R_U = (U, C, V_U) = \begin{bmatrix} U, & c_1 & <0,10> \\ & c_2 & <0,10> \\ & c_3 & <0,10> \\ & c_4 & <0,10> \\ & c_5 & <0,10> \\ & c_6 & <0,10> \end{bmatrix}$$

where $j = 1,2,3,4,5$, and when j is 1,2,3,4,5, $<a_{1j}, b_{1j}>$, $<a_{2j}, b_{2j}>$, $<a_{3j}, b_{3j}>$, $<a_{4j}, b_{4j}>$, $<a_{5j}, b_{5j}>$ are $<0,2>$, $<2,4>$, $<4,6>$, $<6,8>$, $<8,10>$.

4.2   Data Acquisition

Get scores and grades of 6 evaluation indexes on 10 risk events through expert evaluation method, the result is shown in Table 2.

Table 2. Scoring and grade of evaluation indexes for each risk event

	c_1	c_2	c_3	c_4	c_5	c_6
R_1	5.5(3)	3(2)	4.5(3)	3.5(2)	3.5(2)	4.5(3)
R_2	7(4)	7(4)	1(1)	1.5(1)	6.5(4)	1.5(1)
R_3	5.5(3)	3.5(2)	5(3)	3(2)	2.5(2)	2(1)
R_4	5(3)	4.5(3)	4.5(3)	3.5(2)	1(1)	6(3)
R_5	5(3)	4.5(3)	4.5(3)	5(3)	1.5(1)	5(3)
R_6	3(2)	3.5(2)	2.5(2)	6.5(4)	7(4)	6.5(4)
R_7	5.5(3)	7(4)	2(1)	1.5(1)	7.5(4)	1.5(1)
R_8	6.5(4)	3(2)	3.5(2)	4.5(3)	5(3)	5.5(3)
R_9	7(4)	7.5(4)	3.5(2)	5(3)	4.5(3)	4.5(3)
R_{10}	3(2)	3(2)	5(3)	3(2)	3(2)	4.5(3)

4.3   Prefer Evaluation Index System and Determine Their Weights

Reduct the indexes through formula (1), the result is:

$U / ind(C) = \{\{R_1\}, \{R_2\}, \{R_3\}, \{R_4\}, \{R_5\}, \{R_6\}, \{R_7\}, \{R_8\}, \{R_9\}, \{R_{10}\}\}$

$U / ind(C - \{c_1\}) = \{\{R_1, R_{10}\}, \{R_2, R_7\}, \{R_3\}, \{R_4\}, \{R_5\}, \{R_6\}, \{R_8\}, \{R_9\}\}$

$U / ind(C - \{c_2\}) = \{\{R_1\}, \{R_2\}, \{R_3\}, \{R_4\}, \{R_5\}, \{R_6\}, \{R_7\}, \{R_8, R_9\}, \{R_{10}\}\}$

$U / ind(C - \{c_3\}) = \{\{R_1\}, \{R_2\}, \{R_3\}, \{R_4\}, \{R_5\}, \{R_6\}, \{R_7\}, \{R_8\}, \{R_9\}, \{R_{10}\}\}$

$U / ind(C - \{c_4\}) = \{\{R_1\}, \{R_2\}, \{R_3\}, \{R_4, R_5\}, \{R_6\}, \{R_7\}, \{R_8\}, \{R_9\}, \{R_{10}\}\}$

$$U / ind(C - \{c_5\}) = \{\{R_1\}, \{R_2\}, \{R_3\}, \{R_4\}, \{R_5\}, \{R_6\}, \{R_7\}, \{R_8\}, \{R_9\}, \{R_{10}\}\}$$
$$U / ind(C - \{c_6\}) = \{\{R_1, R_3\}, \{R_2\}, \{R_4\}, \{R_5\}, \{R_6\}, \{R_7\}, \{R_8\}, \{R_9\}, \{R_{10}\}\}$$
$$U / ind(C) \neq U / ind(C - \{c_1\}) \neq U / ind(C - \{c_2\}) = U / ind(C - \{c_3\}) \neq U / ind(C - \{c_4\})$$
$$= U / ind(C - \{c_5\}) \neq U / ind(C - \{c_6\})$$

According to Definition 2, index c_3 and c_5 can be deleted, the preferred evaluation risk index system is $C_f = \{c_1, c_2, c_4, c_6\}$. And we can get their weights through formula(2).

$$U_{c_1} = \frac{\| Pos_C(C) \| - | Pos_{C - \{c_1\}}(C) \|}{|U|} = 1 - \frac{6}{10} = \frac{2}{5} \text{ , Similarly , } U_{c_2} = U_{c_3} = U_{c_4} = \frac{1}{5} .$$

And then use formula (10) to get weights: $\alpha_1 = \frac{2}{5}$, $\alpha_2 = \frac{1}{5}$, $\alpha_4 = \frac{1}{5}$, $\alpha_6 = \frac{1}{5}$

4.4   Determine Risk Grade

For an example of a risk event R_1, we can calculate the correlation degree of R_1's evaluation index c_1 to risk grade u_1 through formula (11),

$$K(c_1, u_1) = \frac{|5.5 - (0 + 2) / 2| - (2 - 0) / 2}{[|5.5 - (0 + 10) / 2| - (10 - 0) / 2] - [|5.5 - (0 + 2) / 2| - (2 - 0) / 2]} = -0.4375$$

Similarly, the results of correlation degrees of 4 indexes to 5 risk grades are show in below:

$$K(c_1, u_1) = \{K(c_1, u_1), K(c_2, u_1), K(c_3, u_1), K(c_4, u_1)\} = \{-0.4375, -0.25, -0.3, -0.3571\}$$
$$K(c_1, u_2) = \{K(c_1, u_2), K(c_2, u_2), K(c_3, u_2), K(c_4, u_2)\} = \{-0.25, 0.5, 0.1667, -0.1\}$$
$$K(c_1, u_3) = \{K(c_1, u_3), K(c_2, u_3), K(c_3, u_3), K(c_4, u_3)\} = \{0.125, -0.25, -0.125, 0.125\}$$
$$K(c_1, u_4) = \{K(c_1, u_4), K(c_2, u_4), K(c_3, u_4), K(c_4, u_4)\} = \{-0.1, -0.5, -0.4167, -0.25\}$$
$$K(c_1, u_5) = \{K(c_1, u_5), K(c_2, u_5), K(c_3, u_5), K(c_4, u_5)\} = \{-0.3571, -0.625, -0.5625, -0.4375\}$$

Use formula (12) to get
$$K(R_1, u_1) = 0.4 \times (-0.4375) + 0.2 \times (-0.25) + 0.2 \times (-0.3) + 0.2 \times (-0.3571) = -0.3564$$

Similarly, $K(R_1, u_2) = 0.0133$ $K(R_1, u_3) = 0$, $K(R_1, u_4) = -0.4$, $K(R_1, u_5) = -0.4679$. We use formula (13) to get the result: $K_{j_0} = \max\{-0.3564, 0.0133, 0, -0.4, -0.4679\}$.

It means the risk grade of risk event R_1 is "low", the correlation degree is 0.0133. The result of all the risk events' grades and correlation degrees are shown in Table 3.

4.5   Results Analyze

We can get the result that the urgency and management costs are redundant indexes in social risk assessment. The preferred evaluation risk index system includes

Table 3. Grade of risk events

	Risk grade	Correlation degree
R_1	low	0.0133
R_2	high	0
R_3	low	0.0333
R_4	medium	0.1
R_5	medium	0.225
R_6	low	0.0667
R_7	Very low	-0.1
R_8	medium	-0.05
R_9	high	0.1667
R_{10}	low	0.38

probability, consequence, controllability and management efficiency, their weights are 0.4, 0.2, 0.2, 0.2. Use extension dependence function to analyze the data shown in Table 3, the result tells us that the grade of R_2 and R_9 are "high", R_4, R_5 and R_8 are "medium", R_1, R_3, R_6 and R_{10} are "low", and R_7 is "very low". The order is $R_9 > R_2 > R_5 > R_4 > R_8 > R_{10} > R_6 > R_3 > R_1 > R_7$, so we should invest our resources to solve public opinion risk and public participation risk firstly, and manage economic risk at last.

5   Conclusion

Risk assessment is a major process of risk management. Because of the contradiction between limited resources and numerous risk events, we should sort risk events and utilize resources rationally to improve the efficiency of risk management. So we need not only to consider the probability and consequences of risk events, but also the management properties, such as management costs and management efficiency.

A well performing risk assessment model uses reasonable evaluation indexes and scientific method to measure risk events' grade. This paper prefers the initial evaluation risk index system through rough sets theory, and calculates the grade of risk events combined with extension assessment method. Finally, this paper analyzes an airport construction project to prove the practicality and reliability of the model.

References

1. Ulrich, B., Joost, V.L.: The risk society and beyond: critical issues for social theory. Sage (2000)
2. Holzmann, R., Jørgensen, S.: Social Risk Management: A new conceptual framework for Social Protection, and beyond. International Tax and Public Finance **8**(4), 529–556 (2001)

3. Li, S.M.: A Summarize on the Project Risk Sequencing. Sci./Tech. Information Development & Economy **1**, 158–160 (2006)
4. Haimes, Y.Y.: Risk modeling, assessment, and management, vol. 40. John Wiley & Sons (2005)
5. Ward, S.C.: Assessing and managing important risks. International Journal of Project Management **17**(6), 331–336 (1999)
6. Cai, W.: Extension theory and its application. Chinese Science Bulletin **44**(17), 1538–1548 (1999)
7. Baccarini, D., Richard, A.: The risk ranking of projects: a methodology. International Journal of Project Management **19**(3), 139–145 (2001)
8. Pawlak, Z.: Rough sets. International Journal of Computer & Information Sciences **11**(5), 341–356 (1982)
9. Ayyub, B.M.: Risk analysis in engineering and economics. CRC Press (2014)

Kernel Parameter Optimization for KFDA Based on the Maximum Margin Criterion

Yue Zhao and Jinwen Ma[✉]

Department of Information Science, School of Mathematical Sciences
and LMAM, Peking University, Beijing 100871, China
jwma@math.pku.edu.cn

Abstract. Kernel parameters optimization is one of the most challenging problems on kernel Fisher discriminant analysis (KFDA). In this paper, a simple and effective KFDA kernel parameters optimization criterion is proposed on the basis of the maximum margin criterion (MMC) that maximize the distances between any two classes. Actually, this MMC-based criterion is applied to the kernel parameters optimization on KFDA and KFDA with Locally Linear Embedding affinity matrix (KFDA-LLE). It is demonstrated by the experiments on six real-world multiclass datasets that, in comparison with two other criteria, our MMC-based criterion can detect the optimal KFDA kernel parameters more accurately in the cases of both RBF kernel and polynomial kernel.

Keywords: Kernel parameter optimization · Maximum margin criterion · Feature extraction · Kernel Fisher discriminant analysis (KFDA) · Affinity matrix

1 Introduction

Fisher Discriminant Analysis (FDA) is a popular method for dimensionality reduction. It tries to make the between-class scatter be maximized while the within-class scatter be minimized in the lower dimensional space [1]. Kernel Fisher discriminant analysis (KFDA) is the kernelized version of FDA [2],[3]. The main idea of KFDA is to map the input space to a higher dimensional feature space where the corresponding sample classes can be linearly separated so that we can conduct the FDA method in this projected feature space [4]. In this case, kernel function is utilized to implement the inner product of two data in the kernel space [5]. In fact, the parameters of the kernel function determine the data distribution in the projected feature space, which directly influences the performance of KFDA. Therefore, kernel parameters optimization is a critical problem in KFDA.

In order to solve this kernel parameters optimization problem, two kinds of optimization approaches have been suggested. One approach is the leave-one-out or k-fold cross validation, the other is criterion based optimization method [6]. In fact, the leave-one-out validation is rather time-consuming because of the large

© Springer International Publishing Switzerland 2014
Z. Zeng et al. (Eds.): ISNN 2014, LNCS 8866, pp. 330–337, 2014.
DOI: 10.1007/978-3-319-12436-0_37

repetition computations [7]. On the other hand, there are many clustering criteria based on the between-class scatter and the within-class scatter such as Hartigan criterion [8], McClain and Rao criterion [9], and Friedman and Rubin criterion [10]. However, these criteria are not so useful in kernel parameters optimization. For example, Friedman and Rubin suggested a criterion for determining the optimal number of clusters by using $F = tr(Sw^{-1}Sb)$. But it is is too complex to be applied to the kernel parameters optimization since we cannot derive any explicit expressions for the optimal kernel parameters.

In this paper, we propose a new criterion for KFDA kernel parameters optimization based on the maximum margin criterion (MMC), which has been already used for feature extraction. The main idea of the MMC is to maximize the distances between classes in the kernel space [11]. That is, a pattern in the kernel space should be close to those in the same class but be far from those in different classes. By employing a simple formula, the MMC for kernel parameters optimization can be expressed as the weighted sum of kernel functions. Thus, the optimal solution can be easily calculated. Actually, this MMC-based criterion does not depend on the classifiers or the dimensionality after the feature extraction. It is demonstrated by the experiments on six real-world multiclass datasets that, in comparison with two other possible criteria, the MMC-based criterion can detect the optimal KFDA kernel parameters more accurately in the cases of both RBF kernel and polynomial kernel.

The rest of the paper is organized as follows. We present the MMC-based approach for kernel parameter optimization in two FDA methods in Section 2. Section 3 contains its experimental results on six real-world datasets in comparison with two other clustering criteria. Section 4 makes a brief conclusion, along with directions for further research.

2   MMC-Based Approach for Kernel Parameters Optimization

For clarity, we introduce some mathematical notations used throughout the paper. $x_i \in \mathbb{R}^d$ and $z_i \in \mathbb{R}^r (1 \le r < d)$ are the i-th input data and its corresponding low dimensional projection or embedding $(i = 1, 2, \cdots, n)$, where n is the number of samples, d is the dimensionality of the input data and r is the reduced dimensionality. $y_i \in \{1, 2, \cdots, c\}$ are the associated class labels, and c is the number of classes. n_l is the number of samples in class l, thus $\sum_{l=1}^{c} n_l = n$. X is defined as the matrix of collection of all samples, i.e., $X = (x_1, x_2, ..., x_n)$.

2.1   Maximum Margin Criterion

As defined in [11], the maximum margin criterion (MMC) can be given as

$$J = \frac{1}{2} \sum_{i=1}^{c} \sum_{j=1}^{c} \frac{n_i}{n} \frac{n_j}{n} d(C_i, C_j),$$

where $d(C_i, C_j)$ denotes the difference between class C_i and class C_j. One way to measure this difference is that

$$d(C_i, C_j) = dis(\mu_i, \mu_j) - tr(S_i) - tr(S_j),$$

where $dis(\mu_i, \mu_j)$ is the Euclidean distance between μ_i and μ_j that are the mean vectors of class C_i and class C_j, respectively. And $tr(S_i)$ means the scatter of the class C_i, where S_i is the covariance matrix of the class C_i.

In FDA, the within-class scatter matrix S_w and the between-class scatter matrix S_b are defined as follows.

$$S_w = \sum_{l=1}^{c} \sum_{i:y_i=l} (x_i - \mu_l)(x_i - \mu_l)^T; S_b = \sum_{l=1}^{c} n_l(\mu_l - \mu)(\mu_l - \mu)^T,$$

where μ is the mean of all samples.

Applying S_w and S_b in FDA, the maximum margin criterion can be written as

$$J = tr(S_b) - tr(S_w). \tag{1}$$

Since $tr(S_b)$ measures the overall variance of the class mean vectors, a large $tr(S_b)$ implies that the class mean vectors scatter in a large space. On the contrary, a small $tr(S_w)$ implies that every class has a small spread. Thus, a large J indicates that patterns are close to each other if they are from the same class but are far from each other if they are from different classes. It is clear that the maximum margin criterion does not depend on the classifiers or the dimensionality after feature extraction.

2.2   MMC for KFDA

We now apply the maximum margin criterion to kernel Fisher discriminant analysis (KFDA). Let $\phi : z \in \mathbb{R}^d \rightarrow \phi(z) \in \mathbb{F}$ be a nonlinear mapping from the input space to a higher dimensional feature space $\mathbb{F}$. KFDA is conducting FDA in the feature space $\mathbb{F}$ [4]. Generally, the kernelization of a conventional method makes use of a kernel function $\kappa(.)$ which serves as the inner product in the higher dimensional space, i.e., $\kappa(x, y) = < \phi(x), \phi(y) >$.

In KFDA, we need to calculate the top r generalized eigenvectors associated with the top r generalized eigenvalues of the generalized eigenvalue problem: $S_b^\phi \varphi = \lambda S_w^\phi \varphi$, where S_b^ϕ and S_w^ϕ are the between-class scatter matrix and within class scatter matrix in kernel space, respectively. Then, the MMC for KFDA is changed as

$$J_1 = tr(S_b^\phi) - tr(S_w^\phi). \tag{2}$$

According to [12], we have

$$S_w^\phi = \frac{1}{2} \sum_{i=1}^{n} \sum_{j=1}^{n} W_{i,j}^w (\phi(x_i) - \phi(x_j))(\phi(x_i) - \phi(x_j))^T;$$

$$S_b^\phi = \frac{1}{2} \sum_{i=1}^{n} \sum_{j=1}^{n} W_{i,j}^{'b} (\phi(x_i) - \phi(x_j))(\phi(x_i) - \phi(x_j))^T.$$

where, $W_{ij}^w = \begin{cases} 1/n_l, & \text{if } y_i = y_j = l, \\ 0, & \text{if } y_i \neq y_j, \end{cases}$ and $W_{ij}^b = \begin{cases} 1/n - 1/n_l, & \text{if } y_i = y_j = l, \\ 1/n, & \text{if } y_i \neq y_j, \end{cases}$.

So, we have

$$tr(S_w^\phi) = \frac{1}{2} \sum_{i=1}^n \sum_{j=1}^n W_{i,j}^w [K(x_i, x_i) - 2K(x_i, x_j) + K(x_j, x_j)];$$

$$tr(S_b^\phi) = \frac{1}{2} \sum_{i=1}^n \sum_{j=1}^n W_{i,j}^b [K(x_i, x_i) - 2K(x_i, x_j) + K(x_j, x_j)].$$

Let θ denote the set of kernel parameters, MMC for optimizing KFDA kernel parameters is

$$J_1(\theta) = \frac{1}{2} \sum_{i=1}^n \sum_{j=1}^n W_{ij} [K(x_i, x_i) - 2K(x_i, x_j) + K(x_j, x_j)], \qquad (3)$$

where $W_{ij} = \begin{cases} 1/n - 2/n_l, & \text{if } y_i = y_j = l, \\ 1/n, & \text{if } y_i \neq y_j, \end{cases}$.

By maximizing J, the average margin is maximized. The optimal kernel parameters set is

$$\theta^* = \text{argmax}_\theta J_1(\theta).$$

This is an optimization problem of a unary function and can be solved by using the optimal toolboxes in MATLAB.

2.3   MMC for KLFDA-LLE

In [13], we already established a new local fisher discriminant analysis method with LLE affinity matrix (LFDA-LLE) considering the local information of the data set. The local between-class scatter matrix and local within-class scatter matrix are defined as,

$$\widetilde{S}_w = X(D - \widetilde{W})X^T; \widetilde{S}_b = X(\widetilde{W} - B)X^T,$$

where D is the $n \times n$ identity matrix and $B = (B_{ij})_{n \times n}$ with $B_{ij} = 1/n$. Here, $\widetilde{W}$ is the reconstruct matrix in the locally linear embedding (LLE) method satisfying $\sum_j \widetilde{W}_{ij} = 1$ and $\widetilde{W}_{ij} = 0$ if x_j is not the k-th or less nearest neighbor of x_i, where x_{i_j} is the j-th neighbor of x_i. $\widetilde{W}$ can preserve the local structure of the input data. The KLFDA-LLE method is better than KFDA because of its locality preserving property.

We can reexpress $\widetilde{S}_w$ and $\widetilde{S}_b$ as

$$\widetilde{S}_w = \frac{1}{2} \sum_{i,j=1}^n P_{i,j}^w (x_i - x_j)(x_i - x_j)^T, \qquad (4)$$

$$\widetilde{S}_b = \frac{1}{2} \sum_{i,j=1}^n P_{i,j}^b (x_i - x_j)(x_i - x_j)^T, \qquad (5)$$

where $P_{i,j}^w = \widetilde{W}_{i,j}$ and $P_{i,j}^b = 1/n - \widetilde{W}_{i,j}$. The theoretical derivations of Eq.(4) and Eq.(5) are based on the fact that $\sum_j \widetilde{W}_{ij} = 1$.

Similar to Section 2.2, applying the MMC on LFDA-LLE, we have

$$J_2(\theta) = \frac{1}{2}\sum_{i=1}^{n}\sum_{j=1}^{n} P_{ij}[K(x_i, x_i) - 2K(x_i, x_j) + K(x_j, x_j)], \tag{6}$$

where $P = P^b - P^w$.

3 Experimental Results

To test the performance of our MMC-based approach to kernel parameters optimization for both KFDA and KFDA-LLE, we implement it on six real-world datasets. For comparison, we choose two other criteria to learn the optimal kernel parameters. The first compared criterion is the trace S_w criterion $(O = tr(S_w))$, which has been one of the most popular indices suggested for use in clustering context [14]. The second one is the statistical index of cluster recovery proposed by Hartigan [8], i.e., $H = log(SSB/SSW)$, where SSB and SSW are the sums of squared distances between and within the groups, respectively.

Our experimental procedure is given as follows. Firstly, we compute the optimal kernel parameters for RBF kernel $(\kappa(x, y) = exp\{-\|x - y\|^2/\sigma^2\})$ and polynomial kernel $(\kappa(x, y) = (x^T y + 1)^b)$. We then extract the features through KFDA and KFDA-LLE on the training data, respectively. Finally, we conduct the one-nearest-neighbor classification for the test data to evaluate these kernel parameter optimization criteria.

We use the Iris, Wine, Seeds, Wisconsin Breast Cancer (WBC), Wisconsin Diagnostic Breast Cancer (WDBC) and Landsat satellite (LS) datasets selected from UCI Machine Learning Repository [15]. Actually, there are 16 missing values in WDBC and we just set them as zero. For simplicity, the first and second classes of the Landsat satellite dataset which are called red soil and cotton crop are used. Some basic numbers are listed in Table 1. For clarity, the numbers of training and test sample points are denoted as $n_{training}$ and n_{test}. The number of training points of each dataset is about 60% of the number of the total dataset.

Table 1. The basic numbers of five real-world datasets in the experiments

Dataset	c	d	r	n	$n_{training}$	n_{test}
Iris	3	4	2	150	90	60
Wine	3	13	4	178	106	72
Seeds	3	7	2	210	126	84
WBC	2	9	2	699	420	279
WDBC	2	30	5	569	341	228
LS	2	36	6	1551	930	621

We firstly implement the MMC-based approach to learn the optimal kernel parameters for both KFDA and KFDA-LLE on the six datasets. In the same way, we also implement the criteria O and H based approaches on the six datasets. The optimization problem is solved using optimal function in MATLAB. The kernel parameter interval of polynomial kernel and RBF kernel are set as $[0, 10]$ and $[0, 5000]$, respectively. The kernel parameters of polynomial kernel and RBF kernel learned by four criteria for each dataset are listed in Table 2 and Table 3, respectively.

Table 2. The parameters of polynomial kernel learned by four criteria

Dataset	Criterion O	Criterion H	MMC for KFDA	MMC for KFDA-LLE
Iris	3.71×10^{-5}	5.30×10^{-5}	5.69	10.00
Wine	6.03×10^{-5}	5.90×10^{-5}	6.50×10^{-5}	6.60×10^{-5}
Seeds	5.02×10^{-5}	2.68	6.27	5.78
WBC	3.04×10^{-5}	1.41	1.38	5.10×10^{-5}
WDBC	6.01×10^{-5}	5.99	5.50×10^{-5}	0.14
LS	6.01×10^{-5}	5.20×10^{-5}	0.66	4.74

Table 3. The parameters of RBF kernel learned by four criteria

Dataset	Criterion O	Criterion H	MMC for KFDA	MMC for KFDA-LLE
Iris	5000.00	5000.00	2.95	1.27
Wine	5000.00	4821.90	71.62	5.42
Seeds	5000.00	5000.00	6.65	2.00
WBC	5000.00	5000.00	29.56	9.80
WDBC	5000.00	5000.00	5000.00	109.78
LS	5000.00	5000.00	466.32	95.26

It should be noted that for Criterion O, the optimal kernel parameters of RBF kernel are the maximum of the interval, which means the larger kernel parameter the smaller $tr(S_w)$.

Next, we implement the KFDA and KFDA-LLE approaches for dimensionality reduction. Moreover, we use the 1-nearest-neighbour classifier for supervised classification. In order to test and compare the performances of these optimization approaches, we implement the above procedure for each dimensionality reduction approach on a couple of randomly selected training and test sets with the fixed numbers $n_{training}$ and n_{test} for 30 times. Table 4 lists the average Classification Accuracy Rates (CARs) of the kNN($k = 1$) classifiers using feature extracted through polynomial kernel with kernel parameters listed in Table 2 on each dataset. Table 5 lists the average Classification Accuracy Rates (CARs) of the kNN($k = 1$) classifiers using feature extracted through RBF kernel with kernel parameters listed in Table 3 on each data set.

Table 4. The average classification accuracy rates (CARs) of the kNN($k = 1$) classifiers with KFDA and KFDA-LLE using polynomial kernel with parameters learned by four criteria

Dataset	Criterion O	Criterion H	MMC for KFDA	MMC for KFDA-LLE
Iris	0.8033	0.8111	0.9815	0.9900
Wine	0.8236	0.8292	0.8403	0.8772
Seeds	0.8083	0.9381	0.9952	0.9679
WBC	0.9613	0.9946	0.9728	0.9566
WDBC	0.6069	0.6272	0.9302	0.9680
LS	0.9714	0.9726	0.9988	1.0000

Table 5. The average classification accuracy rates (CARs) of the kNN($k = 1$) classifiers with KFDA and KFDA-LLE using RBF kernel with parameters learned by four criteria

Dataset	Criterion O	Criterion H	MMC for KFDA	MMC for KFDA-LLE
Iris	0.8967	0.8978	0.9522	0.9689
Wine	0.6778	0.6778	0.9628	1.0000
Seeds	0.8262	0.8262	0.9488	0.9948
WBC	0.9498	0.9498	0.9677	0.9785
WDBC	0.8904	0.8904	0.8904	0.9693
LS	0.9934	0.9934	1.0000	1.0000

Table 4 indicates that when the polynomial kernel is used, the maximum margin criterion is better than Criterion O and Criterion H on all the datasets except for the WBC dataset. This exceptional result may be related to the data structure of the WBC dataset.

As for the RBF kernel, we can observe from Table 5 that, the maximum margin criterion obtains improved classification results for all the six datasets, especially for the Wine data and the Seeds data. Moreover, the KFDA-LLE with MMC is much better than the other criteria because the local information is taken into consideration. Comprehensively, the experimental results of the RBF kernel are better than those of polynomial kernel. It is consistent with the fact that the RBF kernel is applied more frequently than the polynomial kernel in practice. Experimental results show that the maximum margin criterion is effective for kernel parameter optimization for both RBF kernel and polynomial kernel.

4   Conclusions

We have established a simple and effective criterion for KFDA kernel parameters optimization based on the maximum margin criterion (MMC). We apply this criterion to the kernel parameters optimization of both KFDA and KFDA-LLE for feature extraction. It is demonstrated by the experiments on six real-world multiclass datasets that, the MMC-based approach can detect the optimal KFDA kernel parameters more accurately in the cases of both RBF kernel and polynomial kernel than two other existing criteria. In the future, we will extend the maximum margin criterion to the other kernel-based learning methods.

Acknowledgments. This work was supported by the Natural Science Foundation of China for Grant 61171138 and BGP Inc., China national petroleum corporation.

References

1. Fukunnaga, K.: Introduction to Statistical Pattern Recognition, 2nd edn. Academic Press, Boston (1990)
2. Baudat, G., Anouar, F.: Generalized Discriminant Analysis Using a Kernel Approach. Neural Computation **12**, 2385–2404 (2000)
3. Müller, K., Mika, S., Rätsch, G., Tsuda, K., Schölkopf, B.: An Introduction to Kernel-Based Learning Algorithms. IEEE Transactions on Neural Networks **12**(2), 181–201 (2001)
4. Tenenbaum, J.B., Silva, V., Langford, J.C.: A Global Geometric Framework for Nonlinear Dimensionality Reduction. Science **290**, 2319–2323 (2000)
5. Taylor, J.S., Cristianini, N.: Kernel Methods for Pattern Analysis. Cambridge University Press, London (2004)
6. Liu, J., Zhao, F., Liu, Y.: Learning Kernel Parameters for Kernel Fisher Discriminant Analysis. Pattern Recognition Letters **34**, 1026–1031 (2013)
7. Huang, J., Chen, X., et al.: Kernel Parameter Optimization for Kernel-based LDA methods. In: International Joint Conference on Neural Networks, pp. 3840–3846. IEEE Press, Hong Kong (2008)
8. Hartigan, J.A.: Clustering Algorithms. Wiley, New York (1975)
9. Millgan, G., Cooper, M.: An Examination of Procedures for Determining The Number of Clusters in A Data Set. Pyschometrika **50**(2), 159–179 (1985)
10. Friedman, H.P., Rubin, J.: On Some Invariant Criteria for Grouping Data. Journal of the American Statistical Association **62**, 1159–1178 (1967)
11. Li, H., Jiang, T., Zhang, K.: Efficient and Robust Feature Extraction by Maximum Margin Criterion. IEEE Transactions on Neural Networkd **17**(1), 157–165 (2006)
12. Sugiyama, M.: Dimensionality Reduction of Multimodal Labeled Data by Local Fisher Discriminant Analysis. Journal of Machine Learning Research **8**, 1027–1061 (2007)
13. Zhao, Y., Ma, J.: Local Fisher Discriminant Analysis with Locally Linear Embedding Affinity Matrix. In: Guo, C., Hou, Z.-G., Zeng, Z. (eds.) ISNN 2013, Part I. LNCS, vol. 7951, pp. 471–478. Springer, Heidelberg (2013)
14. Orloci, L.: An Agglomerative Method for Classification of Plant Communities. Journal of Ecology **55**, 193–206 (1967)
15. UCI Machine Learning Repository, http://mlearn.ics.uci.edu/databases

A Polynomial Time Solvable Algorithm to Binary Quadratic Programming Problems with Q Being a Seven-Diagonal Matrix and Its Neural Network Implementation

Shenshen Gu$^{(\boxtimes)}$, Jiao Peng, and Rui Cui

School of Mechatronic Engineering and Automation, Shanghai University,
149 Yanchang Road, Shanghai, China
gushenshen@shu.edu.cn

Abstract. In this paper, we consider the binary quadratic programming problems (BQP). The unconstrained BQP is known to be NP-hard and has many practical applications like signal processing, economy, management and engineering. Due to this reason, many algorithms have been proposed to improve its effectiveness and efficiency. In this paper, we propose a novel algorithm based on the basic algorithm proposed in [1], [2], [3] to solve problem BQP with Q being a seven-diagonal matrix. It is shown that the proposed algorithm has good performance and high efficiency. To further improve its efficiency, the neural network implementation is realized.

Keywords: Binary quadratic programming · NP-hard · Neural network

1 Introduction

We consider in this paper the following unconstrained binary quadratic programming problem:

$$\min_{x \in \{0,1\}} \frac{1}{2} x^T Q x + c^T x, \tag{1}$$

where $Q = (q_{ij})_{n \times n}$ is a symmetric matrix with zero elements in the main diagonal, $c \in \mathbb{R}^n$. There is no loss of generality in assuming the zero diagonal because $x_i^2 = x_i (1 \leqslant i \leqslant n)$. The above problem, termed also as the pseudo-Boolean programming problem is a classical combinational optimization problem and is well known to be NP-hard (see [4]).

There exist many real-world applications of binary quadratic programming in the field of signal processing, including financial data analysis [5], molecular conformation problem [6] and cellular radio channel assignment [7]. Many combinational optimization problems, such as the Max-cut problem (see, e.g.,

This work was supported by the Specialized Research Fund for the Doctoral Program of Higher Education (SRFDP) under Grant 20113108120010.

Z. Zeng et al. (Eds.): ISNN 2014, LNCS 8866, pp. 338–346, 2014.
DOI: 10.1007/978-3-319-12436-0_38

[7][8]), are special cases of problem BQP. Various exact solution methods of a branch-and-bound framework for solving problem BQP and its variants have been proposed in the literature (see, e.g., [3][7][10][11][12][13][14] and the references therein).

We focus in this paper on the unconstrained binary quadratic programming problems, where Q is a seven-diagonal matrix, denoted by $BQP7$ problems. Identifying polynomially solvable subclasses of binary quadratic programming problems not only offers theoretical insight into the complicated nature of the problem, but also provides useful information for designing efficient algorithms for finding optimal solution to problem BQP. More specifically, the properties of the polynomially solvable subclasses provide hints and facilitate the derivation of efficient relaxations for the general form of BQP.

The rest of the paper is organized as follows: First, in section 2, a new algorithm is proposed to solve the problem $BQP7$. Then neural network implementation of this algorithm is addressed in section 3. To illustrate the effectiveness and efficiency of this new algorithm, the computational experiments are performed in section 4. Finally the conclusion is given.

2 A New Algorithm Proposed to Solve $BQP7$

Here, we proposed a novel algorithm to solve $BQP7$. In our algorithm, each time when we set x_i to 0 or 1, only a maximum of eight states of $f(x)$ existed. This can lead to limited calculation, and finally make the algorithm effective and efficient.

Consider $BQP7$, the special case of binary quadratic problems, where Q in Equ. (1) is a seven-diagonal symmetric matrix with zero diagonal elements:

$$
Q = \begin{pmatrix}
0 & q_{12} & q_{13} & q_{14} & \cdots & 0 & 0 & 0 & 0 \\
q_{12} & 0 & q_{23} & q_{24} & \cdots & 0 & 0 & 0 & 0 \\
q_{13} & q_{23} & 0 & q_{34} & \cdots & 0 & 0 & 0 & 0 \\
q_{14} & q_{24} & q_{34} & 0 & \cdots & 0 & 0 & 0 & 0 \\
\cdots & \cdots & \cdots & \cdots & \cdots & \cdots & \cdots & \cdots & \cdots \\
0 & 0 & 0 & 0 & \cdots & 0 & q_{n-3,n-2} & q_{n-3,n-1} & q_{n-3,n} \\
0 & 0 & 0 & 0 & \cdots & q_{n-3,n-2} & 0 & q_{n-2,n-1} & q_{n-2,n} \\
0 & 0 & 0 & 0 & \cdots & q_{n-3,n-1} & q_{n-2,n-1} & 0 & q_{n-1,n} \\
0 & 0 & 0 & 0 & \cdots & q_{n-3,n} & q_{n-2,n} & q_{n-1,n} & 0
\end{pmatrix}
$$

For each x_i, it can be set to either 0 or 1. That is to say, if we use enumeration method, there will be 2^n possibilities. With the increasing of the number of dimension, calculation will grow exponentially. In our proposed algorithm, we first set x_i, x_{i-1} and x_{i-2} to 0 or 1, generating eight states of x ($x = (x_1, ..., x_n)$) and the corresponding eight states of $f(x)$, for every two adjacent states of $f(x)$ (of which the high bit of vector x is different, like $state$ (0) and $state$ (1), $state$ (2) and $state$ (3) in Table 1), only two polynomial terms are different, one is the term contains variable x_{i-3} and the other is the constant term. When we further set $x_{i-3} = 0$ or $x_{i-3} = 1$, the different two terms can be compared. Thus, we eliminate the bad state and keep the good one for next calculations. We can apply the similar process for the remaining seven states. Therefore, each time when we assign 0 or 1 to x_i, there will be only eight states of $f(x)$. The flow chart of our algorithm is shown in Fig. 1 and the procedures of our algorithm are given in detail as follows:

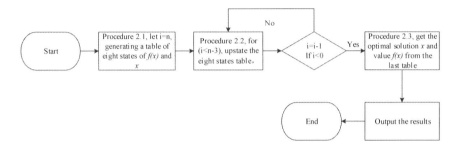

Fig. 1. The flow chart of the new algorithm

Since Q is a seven-diagonal matrix, $f(x)$ takes the following form:

$$f(x) = q_{12}x_1x_2 + q_{13}x_1x_3 + q_{14}x_1x_4 + q_{23}x_2x_3 + \ldots + q_{i-3,i}x_{i-3}x_i + q_{i-2,i}x_{i-2}x_i$$
$$+ q_{i-1,i}x_{i-1}x_i + \ldots + q_{n-1,n}x_{n-1}x_n + c_1x_1 + \ldots + c_ix_i + \ldots + c_nx_n$$

Procedure 2.1: Generate eight states of $f(x)$ by assigning 0 or 1 to x_i, x_{i-1}, x_{i-2}.

Step 1: $x_i = 0$ or $x_i = 1$
Obviously, when we set $x_i = 0$, one state of $f(x)$ will be generated, i.e., *state* (0). And another state of $f(x)$ named *state* (1) will be obtained when we set $x_i = 1$. That is to say, totally two states of $f(x)$ can be generated in this case.

Step 2: $x_{i-1} = 0$ or $x_{i-1} = 1$
In this case, x_{i-1} can be either 0 or 1. By applying the consequences of step 1, we will obtain four states of x, namely, $(0,0)$, $(0,1)$, $(1,0)$ and $(1,1)$. It means there are also four states of $f(x)$, denoted by *state* (0), *state* (1), *state* (2) and *state* (3) respectively.

Step 3: $x_{i-2} = 0$ or $x_{i-2} = 1$
Similarly, in this case, Based on the previous states obtained from step 2, there will be eight states of $f(x)$, denoted by *state* (0) to *state* (7). Its corresponding states of x denoted by $(0,0,0)$, $(0,0,1)$, $(0,1,0)$, $(0,1,1)$, $(1,0,0)$, $(1,0,1)$, $(1,1,0)$ and $(1,1,1)$ respectively, which is summarized in Table 1.

Table 1. Different states of $f(x)$

states	(x_{i-2}, x_{i-1}, x_i)	$f(x_1,.....,x_n)$
state (0)	(0,0,0)	$q_{12}x_1x_2 + \ldots + q_{i-4,i-3}x_{i-4}x_{i-3} + q_{i+1,i+2}x_{i+1}x_{i+2} + \ldots + q_{n-1,n}x_{n-1}x_n$ $+c_1x_1 + \ldots + c_{i-4}x_{i-4} + c_{i+1}x_{i+1} + \ldots + c_nx_n + c_{i-3}x_{i-3}$
state (1)	(0,0,1)	$q_{12}x_1x_2 + \ldots + q_{i-4,i-3}x_{i-4}x_{i-3} + q_{i+1,i+2}x_{i+1}x_{i+2} + \ldots + q_{n-1,n}x_{n-1}x_n$ $+c_1x_1 + \ldots + c_{i-4}x_{i-4} + c_{i+1}x_{i+1} + \ldots + c_nx_n + (c_{i-3} + q_{i-3,i})x_{i-3} + c_i$
state (2)	(0,1,0)	$q_{12}x_1x_2 + \ldots + q_{i-4,i-3}x_{i-4}x_{i-3} + q_{i+1,i+2}x_{i+1}x_{i+2} + \ldots + q_{n-1,n}x_{n-1}x_n$ $+q_{n-1,n}x_{n-1}x_n + c_1x_1 + \ldots + c_{i-5}x_{i-5} + c_{i+1}x_{i+1} + \ldots + c_nx_n$ $+(c_{i-4} + q_{i-4,i-1})x_{i-4} + (c_{i-3} + q_{i-3,i-1})x_{i-3} + c_{i-1}$
state (3)	(0,1,1)	$q_{12}x_1x_2 + \ldots + q_{i-4,i-3}x_{i-4}x_{i-3} + q_{i+1,i+2}x_{i+1}x_{i+2} + \ldots + q_{n-1,n}x_{n-1}n$ $+c_1x_1 + \ldots + c_{i-5}x_{i-5} + c_{i+1}x_{i+1} + \ldots + c_nx_n$ $+(c_{i-4} + q_{i-4,i-1})x_{i-4} + (c_{i-3} + q_{i-3,i-1} + q_{i-3,i})x_{i-3} + c_{i-1} + c_i$
state (4)	(1,0,0)	$q_{12}x_1x_2 + \ldots + q_{i-4,i-3}x_{i-4}x_{i-3} + q_{i+1,i+2}x_{i+1}x_{i+2} + \ldots + q_{n-1,n}x_{n-1}x_n$ $+c_1x_1 + \ldots + c_{i-6}x_{i-6} + c_{i+1}x_{i+1} + \ldots + c_nx_n + (c_{i-5} + q_{i-5,i-2})x_{i-5}$ $+(c_{i-4} + q_{i-4,i-2})x_{i-4} + (c_{i-3} + q_{i-3,i-2})x_{i-3} + c_{i-2}$
state (5)	(1,0,1)	$q_{12}x_1x_2 + \ldots + q_{i-4,i-3}x_{i-4}x_{i-3} + q_{i+1,i+2}x_{i+1}x_{i+2} + \ldots + q_{n-1,n}x_{n-1}x_n$ $+c_1x_1 + \ldots + c_{i-4}x_{i-4} + c_{i+1}x_{i+1} + \ldots + c_nx_n + (c_{i-5} + q_{i-5,i-2})x_{i-5}$ $+(c_{i-4} + q_{i-4,i-2})x_{i-4} + (c_{i-3} + q_{i-3,i-2} + q_{i-3,i})x_{i-3} + c_{i-2} + q_{i-2,i} + c_i$
state (6)	(1,1,0)	$q_{12}x_1x_2 + \ldots + q_{i-4,i-3}x_{i-4}x_{i-3} + q_{i+1,i+2}x_{i+1}x_{i+2} + \ldots + q_{n-1,n}x_{n-1}x_n$ $+c_1x_1 + \ldots + c_{i-6}x_{i-6} + c_{i+1}x_{i+1} + \ldots + c_nx_n + (c_{i-5} + q_{i-5,i-2})x_{i-5}$ $+(c_{i-4} + q_{i-4,i-2} + q_{i-4,i-1})x_{i-4} + (c_{i-3} + q_{i-3,i-2} + q_{i-3,i-1})x_{i-3}$ $+c_{i-2} + q_{i-2,i-1} + c_{i-1}$
state (7)	(1,1,1)	$q_{12}x_1x_2 + \ldots + q_{i-4,i-3}x_{i-4}x_{i-3} + q_{i+1,i+2}x_{i+1}x_{i+2} + \ldots + q_{n-1,n}x_{n-1}x_n$ $+c_1x_1 + \ldots + c_{i-6}x_{i-6} + c_{i+1}x_{i+1} + \ldots + c_nx_n + (c_{i-5} + q_{i-5,i-2})x_{i-5}$ $+(c_{i-4} + q_{i-4,i-2} + q_{i-4,i-1})x_{i-4} + (c_{i-3} + q_{i-3,i-2} + q_{i-3,i-1} + q_{i-3,i})x_{i-3}$ $+c_{i-2} + q_{i-2,i-1} + q_{i-2,i} + c_{i-1} + q_{i-1,i} + c_i$

Procedure 2.2   Set $x_{i-3} = 0$ and $x_{i-3} = 1$. This is the key procedure of the whole algorithm.

Step 1: $x_{i-3} = 0$

For *state* (0) and *state* (1) in Table 1, they are only different in the linear term coefficient of x_{i-3} and the constant coefficient. Therefore we set $x_{i-3} = 0$, calculate $f(x)$ of *state* (0) and *state* (1) respectively. Compare two results, eliminate the bad state and preserve the good one as the new *state* (0). Similarly, the new *state* (1) can be generated from *state* (2) and *state* (3) by setting $x_{i-3} = 0$, the new *state* (2) is from *state* (4) and *state* (5) and the new *state* (3) is from *state* (6) and *state* (7).

Step 2: $x_{i-3} = 1$

By applying the same updating process, when we set $x_{i-3} = 1$, calculate $f(x)$ of *state* (0) and *state* (1). Compare the results and choose the good one as the new *state* (4). Similarly, we can obtain the new *state* (5) from *state* (2) and *state* (3) by setting $x_{i-3} = 1$; the new *state* (6) is from *state* (4) and *state* (5); and the new *state* (7) is from *state* (6) and *state* (7). Through the above two steps, we will get a new table of eight states of $f(x)$ and its corresponding states of x.

Procedure 2.3   Set the remaining x_i to 0 or 1.

Through applying Procedure 2.1 and Procedure 2.2, we can generate a table of eight states of $f(x)$ and its corresponding states of x for each time when we set $x_i = 0$ or $x_i = 1$. After every variable in x is being set to 0 or 1, we find out that the final table contains only constant coefficient of $f(x)$ and its corresponding states of x, choose the optimal value and the corresponding sequence of x is the optimal solution.

3   Implementation of Neural Network

To further improve the efficiency of our algorithm, a subset of neural networks is implemented in this section. As we all known that a Hopfield neural network is characterized by an energy function, which is uniquely specified by the weights on the connections between the neurons and by the thresholds of the neurons. That is to say, if we write $f(x)$ as the same form of energy function, it can always be represented by a neural network. Then a subset of neural networks can be transformed to logic gate circuits through the next algorithm. The energy function is of the following form:

$$E = -[\frac{1}{2}\sum_{i=1}^{n}\sum_{j=1}^{n}T_{ij}x_ix_j] - [\sum_{i}^{n}I_ix_i] + K \tag{2}$$

where $T_{ij} \in \mathbb{R}$ is the weight associated with the connection from neuron i to j, x_i is the activation value of the neuron i, $I_i \in \mathbb{R}$ is the threshold of neuron i, and K is a constant.

Neural network for multi-input AND, OR, NAND, NOR and XOR gates can be constructed from the basis set [17], Table 2 gives the basis set of logic gates:

Table 2. Basis set of the neural network

Gate	T and I
AND	$T_{ij} = (1 - \delta(i,j)) \times ((A + B) \times connected(i,j) - B \times inputs(i,j))$ $I_i = -(2A + B) \times output(i)$ $K = 0$
OR	$T_{ij} = (1 - \delta(i,j)) \times (A + B) \times connected(i,j) - B \times inputs(i,j))$ $I_i = -B \times output(i) - A \times input(i)$ $K = 0$
NAND	$T_{ij} = (1 - \delta(i,j)) \times (-A - B) \times connected(i,j) - B \times inputs(i,j))$ $I_i = (2A + B) \times output(i) + (A + B) \times input(i)$ $K = 2A + B$
NOR	$T_{ij} = (1 - \delta(i,j)) \times (-A - B) \times connected(i,j) - B \times inputs(i,j))$ $I_i = B$ $K = B$
NOT	$T_{ij} = -2J \times (1 - \delta(i,j))$ $I_i = J$ $K = J$

By applying a linear time algorithm [18], a subset of neural networks defined by binary quadratic programs can be transformed to the instance of the logic simulation problem. We extend the 2-input logic gate circuits in [18] to multi-input logic gate circuits. Firstly, rewrite the function $f(x)$ in the same form as the energy function, then the corresponding neural network graph G_f can be developed.

Consider the following example:

$$f(x) = -27x_1x_2 - 32x_1x_3 + 39x_1x_4 - 36x_2x_3 + 35x_2x_4 + 6x_2x_5 + 9x_3x_4 - 20x_3x_5$$
$$- 30x_3x_6 - 32x_4x_5 - 42x_4x_6 - 13x_4x_7 - 60x_5x_6 - 31x_5x_7 + 42x_5x_8 - 31x_6x_7$$
$$+ 42x_6x_8 + 42x_7x_8 - 8x_1 - 9x_2 + 16x_3 + 3x_4 + 7x_5 + 15x_6 - 10x_7 - 18x_8 \tag{3}$$

It can be rewritten as:

$$f(x) = -[27x_1x_2 + 32x_1x_3 - 39x_1x_4 + 36x_2x_3 - 35x_2x_4 - 6x_2x_5 - 9x_3x_4 + 20x_3x_5$$
$$+ 30x_3x_6 + 32x_4x_5 + 42x_4x_6 + 13x_4x_7 + 60x_5x_6 + 31x_5x_7 - 42x_5x_8 + 31x_6x_7$$
$$- 42x_6x_8 - 42x_7x_8] - [8x_1 + 9x_2 - 16x_3 - 3x_4 - 7x_5 - 15x_6 + 10x_7 + 18x_8]$$

The corresponding neural network graph G_f is shown in Fig. 2. Each circle corresponds to a neuron. The name of the neuron is written in the upper half and its threshold is indicated in the lower half. The link between neurons x_i and x_j is denoted by T_{ij}. Every neuron corresponds to a signal in the logic circuit. A combinational logic circuit is constructed through the following procedure.

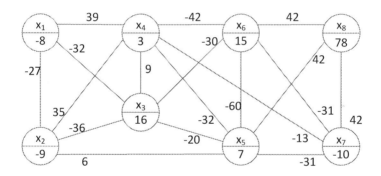

Fig. 2. The neural network G_f of function $f(x)$

Let x_n be the primary output neuron of G_f, which has equal edge weights incident to x_n. The threshold of x_n and the weights on its edges can be uniquely determined according to Table 2:

Case 1: $w > 0$. From Table 2, only in G_{AND} and G_{OR}, the edges incident on the primary output neuron have positive edge weights. Therefore, gate x_n could be an AND or OR gate. Again, from Table 2, if $w < -I_n$, then the gate is AND type with $A + B = w$ and $-(2A + B) = I_n$, i.e., $A = -(w + I_n)$, $B = 2w + I_n$. If $w > -I_n$, then the gate is OR type with $A + B = w$ and $-B = I_n$, i.e., $A = w + I_n$, $B = -I_n$.

Case 2: $w < 0$. The gate is NAND or NOR type. From Table 2, if $-w < I_n$, then the gate is NAND type with $-A - B = w$ and $2A + B = I_n$, i.e., $A = w + I_n$,

$B = -2w - I_n$. If $-w > I_n$, then the gate is NOR type with $-A - B = w$ and $B = I_n$, i.e., $A = -w - I_n$.

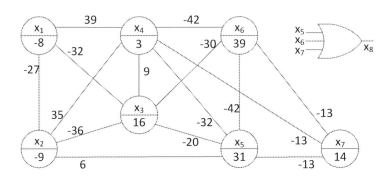

Fig. 3. Transform of G_f to $G_f \setminus x_8$

After gate x_n being determined, let C_n be the circuit by deleting gate x_n from the original circuit and let E_{C_n} and G_{C_n} be the energy function and neural network graph. The deleting rules are as follows:

If the edge linked to x_n is more than two, since the weights on the edges are equal [18]. Let x_k be the vertices connected to x_n, I_k be their thresholds, and T_k be the weight between the connected circles. If x_n is an AND gate, I_k unchanged and $T_k + B$ is the new edge weight. Similarly, for OR type, I_k and T_k are modified to $I_k + A$ and $T_k + B$, respectively; for NAND type, $I_k - A - B$ and $T_k + B$, respectively; and for NOR type, $I_k - B$ and $T_k + B$, respectively.

Consider the above example Equ. (3), the specific implementation process is shown in Fig. 3 and Fig. 4. The final logic circuit is shown in Fig. 5. We finally obtain the primary input x_1 and x_2 of the circuit. Therefore, the logic circuit corresponding to $f(x)$ has x_1 and x_2 two inputs. Logic simulation of the circuit with $x_1 = 1$ and $x_2 = 1$ yields $x_3 = 1, x_4 = 1, x_5 = 1, x_6 = 0, x_7 = 1, x_8 = 1$, substituting these values into $f(x)$ gives the minimum value of the quadratic function Equ. 3.

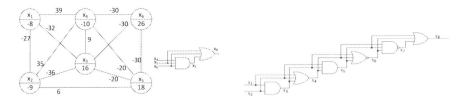

Fig. 4. Transform of $G_f \setminus x_8$ to $G_f \setminus x_7$ **Fig. 5.** The logic circuit of function $f(x)$

4   Computational Results

In order to show the efficiency and effectiveness of the algorithm, the algorithm was coded by C and run on a Inter (R) Core (TM) PC. Numerical results of computational time are represented in Table 3.

Table 3. Computation time of $BQP7$

n	10	20	30	40	50	60	`70	80	90	100
Ave	0.00279	0.00289	0.00330	0.00419	0.00539	0.00820	0.00828	0.01177	0.01290	0.01541

All instances are generated randomly. For the dimension n of Q from $n = 10$ to $n = 100$, ten tests have been performed and each test is the average computational time (s) of 100 iterations. Also, we demonstrate that our algorithm is effective and efficient with the polynomial feature. Fig. 6 shows the average computational time for different dimensional $BQP7$. The curve in the figure is approximately polynomial, which meanwhile directly reveals the effectiveness and high efficiency of our algorithm.

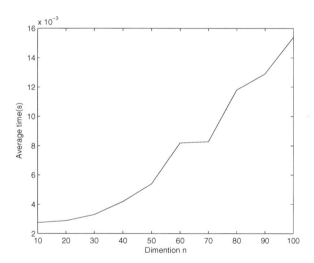

Fig. 6. Average computational time for different dimensional $BQP7$

5   Conclusions

We proposed a new algorithm for solving binary quadratic programming problems with Q being a seven-diagonal matrix. To further improve the efficiency, neural network implementation is realized. The illustration and analysis proved that the novel algorithm can solve the problem in polynomial time. And the computational time results shows that with the increase of dimension its effectiveness and efficiency is still considerable.

References

1. Hammer, P.L., Rudeanu, S.: Boolean Methods in Operations Research and Related Areas. Springer, Heidelberg (1968)
2. Crama, Y., Hansen, P., Jaumard, B.: The basic algorithm for pseudo-Booleab programming revisited. Discrete Appl. Math. **29**, 171–185 (1990)
3. Li, D., Sun, X.L.: Nonlinear Integer Programming. Springer, New York (2006)
4. Garey, M.R., Johnson, D.S.: Computers and Intractability: A Guide to the Theory of NP-Completeness. WH Freeman. Co., New York (1979)
5. Mcbride, R.D., Yormark, J.S.: An implicit enumeration algorithm for quadratic integer programming. Manage. Sci. **26**, 282–296 (1980)
6. Phillips, A.T., Rosen, J.B.: A quadratic assignment formulation of the molecular conformation problem. J. Global Optim. **4**, 229–241 (1994)
7. Chardaire, P., Sutter, A.: A decomposition method for quadratic zero-one programming. Manage. Sci. **41**, 704–712 (1995)
8. Delorme, C., Poljak, S.: Laplacian eigenvalues and the maximum cut problem. Math. Program **62**, 557–574 (1993)
9. Goemans, M.X., Williamson, D.P.: Improved approximation algorithms for maximum cut and satisfiability problems using semidefinite programming. J. Assoc., Comput, Mach., 42, 1115–1145 (1995)
10. Chardaire, P., Sutter, A.: A decomposition method for quadratic zero-one programming. Manage, Sci. **41**, 704–712 (1995)
11. Helmberg, C., Rendl, F.: Solving quadratic (0,1)-problems by semidefinite programs and cutting planes. Math. Program **82**, 291–315 (1998)
12. Rendl, F., Rinaldi, G., Wiegele, A.: Solving max-cut to optimality by intersecting semidefinite and polyhedral relaxations. Lecture Notes Comput, Sci. **4513**, 295–309 (2007)
13. Pardalos, P.M., Rodgers, G.P.: Computational aspects of a branch-and-bound algorithm for quadratic zero-one programming. Computing **45**, 131–144 (1990)
14. Barahona, F.: Jünger, M., Reinelt, G.: Experiments in quadratic 0–1 programming. Math. Program **44**, 127–137 (1989)
15. Gu, S.: Polynomial time solvable algorithm to binary quadratic programming problems with Q being a tri-diagonal or five-diagonal matrix. In: Proceedings of 2010 International Conference on Wireless Communication and Signal Processing (2010)
16. Gu, S.: A Polynomial time solvable algorithm to linearly constrained binary quadratic programming problems with Q being a tri-diagonal. Advance in Information Science and Service Sciences 3(6) (July 2011)
17. Chakradhar, S.T., Bushnell, M.L., Agrawal, V.D.: Toward massively parallel automatic test generation, pp. 981–994. IEEE Trans, Computer-Aided (1990)
18. Chakradhar, S.T., Bushnell, M.L.: A solvable class of quadratic 0–1 programming. Discrete Applied Mathemetics **36**, 233–551 (1992)

A New Nonlinear Neural Network
for Solving QP Problems

Yinhui Yan[(⊠)]

Shenzhen Airlines Co., Ltd., Shenzhen, China
A15601@shenzhenair.com

Abstract. In this paper, a new nonlinear neural network is proposed to solving quadratic programming problems subject to linear equality and inequality constraints without any parameter tuning. This nonlinear neural network is proved to be stable in the sense of Lyapunov under certain conditions. Simulation results are further presented to show the effectiveness and performance of this neural network.

Keywords: Nonlinear neural network · Lyapunov stability · Quadratic programming

1 Introduction

Quadratic programming (QP) studies problem of optimizing (minimizing or maximizing) a quadratic function of several variables subject to linear equality or linear inequality constraints on these variables. It has been successfully applied to various fields [1] such as transportation, energy, telecommunications, and manufacturing. Traditional approaches to solve QP problems [2–8] include interior point method, active set method, augmented Lagrangian method, conjugate gradient method and gradient projection method etc. However traditional methods usually require much computational time and can not meet real-time requirements in practical applications.

In 1986, based on a gradient method, Hopfield and Tank [9] in their paper proposed a new approach to solve LP problem by using recurrent neural network. The main advantage of this method is that it can be implemented by using analog electronic circuits, possibly on a VLSI (very large-scale integration) circuit, which can operate in parallel. In contrast with traditional approaches which may involve an iterative process and require long computational time, this model can potentially provide an optimal solution in real time. After their pioneer work [9,10], numerous neural network models have been developed to solve optimization problems, such as the Lagrangian neural network [11], the deterministic annealing neural network [12], the projection neural network [13], the delayed projection neural network [14], the dual neural network [15,16] and the primal-dual neural network [17]. In 1988, Kennedy and Kan [18] developed a neural network for solving nonlinear programming problems based on Karush-Kuhn-Tucker (KKT) optimal conditions. By using a penalty parameter its solution

© Springer International Publishing Switzerland 2014
Z. Zeng et al. (Eds.): ISNN 2014, LNCS 8866, pp. 347–357, 2014.
DOI: 10.1007/978-3-319-12436-0_39

usually approximates the optimal solution. Only when the penalty parameter is very large, it is same as the exact solution. Later Maa and Shanblatt [19] extended this penalty based method by using two-phase model and ensured that the model converges to the optimal solution. However, their model is more complex and still requires careful parameter selection. To overcome these drawbacks, Xia [20] proposed a primal and dual model to solving this problem. Zhang and Constantinides [11] invented a lagrangian neural network based on the idea of lagrangian multiplier. In this model slack variables are introduced as new variables to deal with inequality constraints, this may lead to high dimension thus require more computation. Unlike previous approaches using a fixed parameter, Wang etc. [12] used a time-variant temperature to design a deterministic annealing neural network to resolve the linear programs. In International Symposium on Mathematical Programming 2000, Nguyen [21] presented a novel recurrent neural network model to solve linear optimization problem. Compared with Xia's model, Nguyen's model not only retains the advantages of Xia's model but also have a more intuitive economic interpretation and much faster convergence. The most interested thing for this model is its nonlinear dynamic structure and high convergence speed. This paper will extend the Nguyen's neural network model to solving quadratic programming problems. For the background and details of neural networks, we refer to [22–31].

The rest of this paper is organized as follows: Section 2 presents a nonlinear neural network to solving quadratic problem and the convergence property of this neural network. Section 3 studies the stability of the proposed dynamical neural network and proves that this neural network is stable in the sense of Lyapunov under certain conditions. Section 4 demonstrates the power and effectiveness of the proposed neural network. In the end, Section 5 gives a summary of this paper and points out some future research directions.

2   Model Description

Consider the QP Problem

$$
\begin{aligned}
&\text{Find } \mathbf{x} \text{ which minimizes :} \quad && \tfrac{1}{2}\mathbf{x}^T\mathbf{Q}\mathbf{x} + \mathbf{e}^T\mathbf{x}, \\
&\text{subject to} && \mathbf{D}\mathbf{x} = \mathbf{b}, \\
& && \mathbf{A}\mathbf{x} \geq \mathbf{c}, \\
& && \mathbf{x} \geq 0,
\end{aligned}
\tag{1}
$$

where $\mathbf{x}$ and $\mathbf{e}$ are n-dimensional vectors, $\mathbf{Q}$ is an $n \times n$ symmetric positive definite matrix, $\mathbf{D} \in \mathbb{R}^{p \times n}$, $\mathbf{A} \in \mathbb{R}^{m \times n}$, $\mathbf{b} \in \mathbb{R}^{p \times 1}$, $\mathbf{c} \in \mathbb{R}^{m \times 1}$. We call this problem as the primal QP problem.

The lagrangian function of this minimization problem can be written as

$$
\mathcal{L}(\mathbf{x}, \mathbf{y}, \mathbf{z}) = \frac{1}{2}\mathbf{x}^T\mathbf{Q}\mathbf{x} + \mathbf{e}^T\mathbf{x} - \mathbf{y}^T(\mathbf{D}\mathbf{x} - \mathbf{b}) - \mathbf{z}^T(\mathbf{A}\mathbf{x} - \mathbf{c}),
\tag{2}
$$

where $\mathbf{z} \in \mathbb{R}_+^p = \{\mathbf{z} \in \mathbb{R}^p | \mathbf{z} \geq 0\}$, $\mathbf{y} \in \mathbb{R}^m$ are Lagrangian multipliers. According to the Karush-Kuhn-Tucker (KKT) conditions [32,33], $\mathbf{x}^\star$ is a solution of (1) if and only if there exist $\mathbf{y}^\star \in \mathbb{R}^m$, $\mathbf{z}^\star \in \mathbb{R}_+^p$ so that $(\mathbf{x}^\star, \mathbf{y}^\star, \mathbf{z}^\star)$ satisfies the following conditions:

$$\mathbf{Qx}^\star + \mathbf{e} - \mathbf{D}^T\mathbf{y}^\star - \mathbf{A}^T\mathbf{z}^\star \geq 0,$$

$$\mathbf{x}^{\star T}\left(\mathbf{Qx}^\star + \mathbf{e} - \mathbf{D}^T\mathbf{y}^\star - \mathbf{A}^T\mathbf{z}^\star\right) = 0,$$

$$\mathbf{b} - \mathbf{Dx}^\star = 0,$$

$$\mathbf{c} - \mathbf{Ax}^\star \leq 0,$$

$$\mathbf{z}^{\star T}\left(\mathbf{c} - \mathbf{Ax}^\star\right) = 0. \tag{3}$$

We propose a recurrent neural network for solving the primal and dual problem as follows:

$$\dot{\mathbf{x}} = -\mathbf{Q}(\mathbf{x} + k\dot{\mathbf{x}}) - \mathbf{e} + \mathbf{D}^T(\mathbf{y} + k\dot{\mathbf{y}}) + \mathbf{A}^T(\mathbf{z} + k\dot{\mathbf{z}}), \mathbf{x} \geq 0, \tag{4a}$$

$$\dot{\mathbf{y}} = \mathbf{b} - \mathbf{D}(\mathbf{x} + k\dot{\mathbf{x}}), \tag{4b}$$

$$\dot{\mathbf{z}} = -\mathbf{A}(\mathbf{x} + k\dot{\mathbf{x}}) + \mathbf{c}, \mathbf{z} \geq 0, \tag{4c}$$

where k is a positive constant. The architecture of the proposed neural network model is shown in Fig. 1. The proposed neural network consists of two layers of

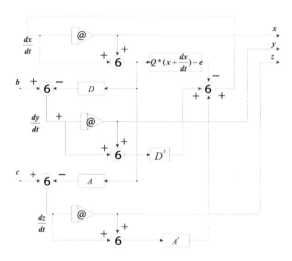

Fig. 1. Block diagram of the neural network (4a, 4b, and 4c)

neurons, i.e., primal neurons and dual neurons. The outputs from one layer are the inputs to the other layer. The inputs of the primal neurons are composed of the dual neuron's outputs and their derivatives, while the inputs of the dual neurons are composed of the primal neuron's outputs and their derivatives. Due to the involvement of these derivatives, this neural network model is a nonlinear dynamic system. The convergence property of the system is stated by the following theorem.

Theorem 1: If the neural network whose dynamics guided by the differential equations (4a, 4b, and 4c) converges to a steady state $(\mathbf{x}^\star, \mathbf{y}^\star, \mathbf{z}^\star)$, then $\mathbf{x}^\star$ will be the optimal solution of the primal QP problem and the Lagrangian multipliers $\mathbf{y}^\star$ and $\mathbf{z}^\star$ the optimal solution of the dual of the QP problem.

Proof. Let $\mathbf{x}_i$ be the ith component of $\mathbf{x}$, then the equation (4a) can be written as

$$\frac{d\mathbf{x}_i}{dt} = \{(-\mathbf{Q}(\mathbf{x} + k\frac{d\mathbf{x}}{dt}) - \mathbf{e}) + \mathbf{D}^T(\mathbf{y} + \frac{d\mathbf{y}}{dt}) + \mathbf{A}^T(\mathbf{z} + k\frac{d\mathbf{z}}{dt})\}_i \quad \text{if } \mathbf{x}_i > 0, \quad (5)$$

$$\frac{d\mathbf{x}_i}{dt} = \max\{\{(-\mathbf{Q}(\mathbf{x} + k\frac{d\mathbf{x}}{dt}) - \mathbf{e}) + \mathbf{D}^T(\mathbf{y} + \frac{d\mathbf{y}}{dt}) + \mathbf{A}^T(\mathbf{z} + k\frac{d\mathbf{z}}{dt})\}_i, 0\} \quad \text{if } \mathbf{x}_i = 0. \quad (6)$$

Note that (6) is to ensure that $\mathbf{x}$ will bounded from below by 0. Let $\mathbf{x}^\star$, $\mathbf{y}^\star$ and $\mathbf{z}^\star$ be the limit of $\mathbf{x}(t)$, $\mathbf{y}(t)$ and $\mathbf{z}(t)$ respectively. In other words

$$\lim_{t \to \infty} \mathbf{x}(t) = \mathbf{x}^\star \tag{7}$$

$$\lim_{t \to \infty} \mathbf{y}(t) = \mathbf{y}^\star \tag{8}$$

$$\lim_{t \to \infty} \mathbf{z}(t) = \mathbf{z}^\star \tag{9}$$

By the definition of convergence, we have $\frac{d\mathbf{x}^\star}{dt} = 0$, $\frac{d\mathbf{y}^\star}{dt} = 0$ and $\frac{d\mathbf{z}^\star}{dt} = 0$. From Eqns. (5) and (6) we conclude that

$$0 = \{(-\mathbf{Q}(\mathbf{x}^\star) - \mathbf{e} + \mathbf{D}^T\mathbf{y}^\star + \mathbf{A}^T\mathbf{z}^\star)\}_i \quad \text{if} \quad \mathbf{x}_i^\star > 0 \tag{10}$$

$$0 = \max\{\{(-\mathbf{Q}(\mathbf{x}^\star) - \mathbf{e} + \mathbf{D}^T\mathbf{y}^\star + \mathbf{A}^T\mathbf{z}^\star)\}_i, 0\} \quad \text{if} \quad \mathbf{x}_i^\star = 0 \tag{11}$$

In other words:

$$\left(-\mathbf{Q}(\mathbf{x}^\star) - \mathbf{e} + \mathbf{D}^T\mathbf{y}^\star + \mathbf{A}^T\mathbf{z}^\star\right)_i \leq 0 \tag{12}$$

$$\mathbf{x}_i^\star \left(-\mathbf{Q}(\mathbf{x}^\star) - \mathbf{e} + \mathbf{D}^T\mathbf{y}^\star + \mathbf{A}^T\mathbf{z}^\star\right)_i = 0 \tag{13}$$

or

$$\left(\mathbf{Q}(\mathbf{x}^\star) + \mathbf{e} - \mathbf{D}^T\mathbf{y}^\star - \mathbf{A}^T\mathbf{z}^\star\right) \geq 0 \tag{14}$$

$$\mathbf{x}^{\star T} \left(-\mathbf{Q}(\mathbf{x}^\star) - \mathbf{e} + \mathbf{D}^T\mathbf{y}^\star + \mathbf{A}^T\mathbf{z}^\star\right) = 0 \tag{15}$$

Similarly, from Eqns. (4b) and (4c), we have:

$$\mathbf{D}\mathbf{x}^\star - \mathbf{b} = 0 \tag{16}$$

$$\mathbf{A}\mathbf{x}^\star - \mathbf{c} \geq 0 \tag{17}$$

$$\mathbf{z}^{\star T}(\mathbf{A}\mathbf{x}^\star - \mathbf{c}) = 0 \tag{18}$$

By KKT conditions in (3) and conditions provided in (15-18) we have shown that $x^\star$ and $(y^\star, z^\star)$ are the optimal solutions for the problem (1) and its dual problem respectively. This concludes the proof.

3   Stability Analysis

It's easy to prove that the differential equations (4a), (4b) and (4c) are equivalent to the following second order differential equations:

$$(\mathbf{I} + k\mathbf{Q} + k^2\mathbf{D}^T\mathbf{D} + k^2\mathbf{A}^T\mathbf{A})\ddot{\mathbf{x}} +$$
$$(\mathbf{Q} + 2k\mathbf{D}^T\mathbf{D} + 2k\mathbf{A}^T\mathbf{A})\dot{\mathbf{x}} +$$
$$(\mathbf{D}^T\mathbf{D} + \mathbf{A}^T\mathbf{A})\mathbf{x} - (\mathbf{D}^T\mathbf{b} + \mathbf{A}^T\mathbf{c}) = 0. \tag{19}$$

Suppose the entity $\mathbf{D}^T\mathbf{D} + \mathbf{A}^T\mathbf{A}$ is non-singular, we introduce a transformation $\mathbf{x} = \mathbf{u} + (\mathbf{D}^T\mathbf{D} + \mathbf{A}^T\mathbf{A})^{-1}(\mathbf{D}^T\mathbf{b} + \mathbf{A}^T\mathbf{c})$, then we have $\dot{\mathbf{x}} = \dot{\mathbf{u}}$ and $\ddot{\mathbf{x}} = \ddot{\mathbf{u}}$. By this transformation, the ordinary differential equation (19) can be written as

$$(\mathbf{I} + k\mathbf{Q} + k^2\mathbf{D}^T\mathbf{D} + k^2\mathbf{A}^T\mathbf{A})\ddot{\mathbf{u}} + (\mathbf{Q} + 2k\mathbf{D}^T\mathbf{D} + 2k\mathbf{A}^T\mathbf{A})\dot{\mathbf{u}}$$
$$+(\mathbf{D}^T\mathbf{D} + \mathbf{A}^T\mathbf{A})\mathbf{u} = 0 \tag{20}$$

Now we would like to study the stability of the equation (20).

Generally we study the stability of the following second order ordinary differential equation

$$\mathbf{L}\ddot{\mathbf{u}} + \mathbf{M}\dot{\mathbf{u}} + \mathbf{N}\mathbf{u} = 0. \tag{21}$$

where $\mathbf{L}$, $\mathbf{M}$ and $\mathbf{N}$ are all positive definite.

First we consider the simplified second order ordinary differential equation

$$\ddot{\mathbf{u}} + \mathbf{M}\dot{\mathbf{u}} + \mathbf{N}\mathbf{u} = 0. \tag{22}$$

where $\mathbf{M}$ and $\mathbf{N}$ are both positive definite.

Theorem 2: If the coefficient matrices $\mathbf{M}$ and $\mathbf{N}$ of the system (22) are both positive definite, then this dynamic system is global asymptotic stable.

Proof. If we set $\mathbf{u}_1 = \mathbf{u}$, $\mathbf{u}_2 = \dot{\mathbf{u}}$, we have the system

$$\begin{cases} \dot{\mathbf{u}}_1 = \quad\quad \mathbf{u}_2, \\ \dot{\mathbf{u}}_2 = -\mathbf{M}\mathbf{u}_2 - \mathbf{N}\mathbf{u}_1. \end{cases}$$

In order to show the global asymptotic the stability of (22), we only need to show the real parts of the eigenvalues of $\mathbf{P}$ are negative, where $\mathbf{P} = \begin{pmatrix} 0 & \mathbf{I} \\ -\mathbf{N} & -\mathbf{M} \end{pmatrix}$.

Suppose $\lambda \in \mathbb{C}^n$ be an eigenvalue of $\mathbf{P}$ with the corresponding non-zero eigenvector $\mathbf{v} = (\mathbf{v}_1, \mathbf{v}_2)$, by the definition of eigenvector, we have

$$\begin{pmatrix} 0 & \mathbf{I} \\ -\mathbf{N} & -\mathbf{M} \end{pmatrix} \begin{pmatrix} \mathbf{v}_1 \\ \mathbf{v}_2 \end{pmatrix} = \begin{pmatrix} \mathbf{v}_2 \\ -\mathbf{N}\mathbf{v}_1 - \mathbf{M}\mathbf{v}_2 \end{pmatrix} = \lambda \begin{pmatrix} \mathbf{v}_1 \\ \mathbf{v}_2 \end{pmatrix}.$$

Since $\mathbf{N}$ is positive definite, $\mathbf{P}$ is non-singular. This concludes that λ can not be an eigenvalue of $\mathbf{P}$. Since $\lambda \neq 0$ and $\mathbf{v}_2 = \lambda\mathbf{v}_1$, we claim that $\mathbf{v}_1 \neq 0$ and

$\mathbf{v}_2 \neq 0$. Without loss of generality we may assume that $\mathbf{v}_1^* \cdot \mathbf{v}_1 = 1$, where $*$ denotes complex conjugate transpose. Using this assumption, we can write $\lambda^2 = \mathbf{v}_1^* \lambda^2 \mathbf{v}_1 = \mathbf{v}_1^* \lambda \mathbf{v}_2 = \mathbf{v}_1^*(-\mathbf{N}\mathbf{v}_1 - \mathbf{M}\mathbf{v}_2) = -\mathbf{v}_1^*\mathbf{N}\mathbf{v}_1 - \lambda\mathbf{v}_1^*\mathbf{M}\mathbf{v}_1$, where we have used the identity $\lambda\mathbf{v}_1 = \mathbf{v}_2$ and $\lambda\mathbf{v}_2 = -\mathbf{N}\mathbf{v}_1 - \mathbf{M}\mathbf{v}_2$. Since $\mathbf{N}$ is positive definite, the entity $\beta = \mathbf{v}_1^*\mathbf{N}\mathbf{v}_1$ is positive real. Similarly, the entity $\alpha = \mathbf{v}_1^*\mathbf{M}\mathbf{v}_1$ is positive because of the positive definiteness of $\mathbf{M}$. Substitute these scalars into the equation $\lambda^2 = -\mathbf{v}_1^*\mathbf{N}\mathbf{v}_1 - \lambda\mathbf{v}_1^*\mathbf{M}\mathbf{v}_1$, we have a quadratic equation of λ, i.e.,

$$\lambda^2 + \alpha\lambda + \beta = 0.$$

Note that every eigenvalue of $\mathbf{P}$ satisfies the above equation. The solution of the above equation is

$$\lambda_{1,2} = \frac{1}{2}(-\alpha \pm \sqrt{\alpha^2 - 4\beta}).$$

If $\alpha^2 - 4\beta \geq 0$, the real parts of $\lambda_{1,2}$ are:

$$Re\{\lambda_{1,2}\} = \frac{1}{2}(-\alpha \pm \sqrt{\alpha^2 - 4\beta}).$$

Recall that α and β are positive, we claim that $Re\{\lambda_{1,2}\}$ are negative.

If $\alpha^2 - 4\beta < 0$, the real parts of $\lambda_{1,2}$ are:

$$Re\{\lambda_{1,2}\} = -\frac{1}{2}\alpha.$$

Since α is positive, it's obviously that the real parts of $\lambda_{1,2}$ are negative.

In all cases, we conclude that the real parts of $\lambda_{1,2}$ are always negative.

Theorem 3: For the second order ordinary differential equation

$$\mathbf{L}\ddot{\mathbf{x}} + \mathbf{M}\dot{\mathbf{x}} + \mathbf{N}\mathbf{x} = 0,$$

if its coefficient matrices $\mathbf{L}$, $\mathbf{M}$ and $\mathbf{N}$ are all positive definite, then it is asymptotic stable.

Proof. Note that for the positive definite matrix $\mathbf{L}$, we have a decomposition such that $\mathbf{L} = \mathbf{L}^{\frac{1}{2}}\mathbf{L}^{\frac{1}{2}}$ and $\mathbf{L}^{-1} = \mathbf{L}^{-\frac{1}{2}}\mathbf{L}^{-\frac{1}{2}}$, where $\mathbf{L}^{\frac{1}{2}}$ and $\mathbf{L}^{-\frac{1}{2}}$ are positive definite. Now we define the transformation $\tilde{\mathbf{x}} = \mathbf{L}^{\frac{1}{2}}\mathbf{x}$ or $\mathbf{x} = \mathbf{L}^{-\frac{1}{2}}\tilde{\mathbf{x}}$, using this transformation we have

$$\mathbf{L}\mathbf{L}^{-\frac{1}{2}}\ddot{\tilde{\mathbf{x}}} + \mathbf{M}\mathbf{L}^{-\frac{1}{2}}\dot{\tilde{\mathbf{x}}} + \mathbf{N}\mathbf{L}^{-\frac{1}{2}}\tilde{\mathbf{x}} = 0 \qquad (23)$$

Premultiplying $\mathbf{L}^{-\frac{1}{2}}$ to both hands of equation, we get

$$\ddot{\tilde{\mathbf{x}}} + \mathbf{L}^{-\frac{1}{2}}\mathbf{M}\mathbf{L}^{-\frac{1}{2}}\dot{\tilde{\mathbf{x}}} + \mathbf{L}^{-\frac{1}{2}}\mathbf{N}\mathbf{L}^{-\frac{1}{2}}\tilde{\mathbf{x}} = 0 \qquad (24)$$

This system is exactly of the form used in *Theorem 2*, but instead of $\mathbf{M}$ and $\mathbf{N}$ we now have $\mathbf{L}^{-\frac{1}{2}}\mathbf{M}\mathbf{L}^{-\frac{1}{2}}$ and $\mathbf{L}^{-\frac{1}{2}}\mathbf{N}\mathbf{L}^{-\frac{1}{2}}$. If the later system is asymptotic stable, it implies that (1) is asymptotic stable, since the two systems differ only by a non-singular transformation. Therefore the global asymptotic stability of (6) follows from *Theorem 2*.

Theorem 4: If the matrix $\mathbf{A}^T\mathbf{A}$ or $\mathbf{D}^T\mathbf{D}$ is non-singular and $k > 0$, then the neural network described by differential Eqns. (4a, 4b and 4c) is asymptotic stable.

Proof. Since $\mathbf{A}^T\mathbf{A}$ or $\mathbf{D}^T\mathbf{D}$ is symmetric and non-singular, the matrices $\mathbf{I} + k\mathbf{Q} + k^2\mathbf{D}^T\mathbf{D} + k^2\mathbf{A}^T\mathbf{A}$, $\mathbf{Q} + 2k\mathbf{D}^T\mathbf{D} + 2k\mathbf{A}^T\mathbf{A}$ and $\mathbf{A}^T\mathbf{A} + \mathbf{D}^T\mathbf{D}$ are positive definite. By *Theorem 3*, we conclude that the dynamical system of (4a, 4b and 4c) is asymptotic stable in the sense of Lyapunov.

4   Simulation Results

To demonstrate the behavior and properties of the proposed nonlinear neural network model, one example with four different initial vectors is simulated. The simulation is conducted with MATLAB. We use the Euler method to solve the neural system of ordinary differential equations (4a, 4b and 4c).
Consider the following quadratic programming problem:

$$
\begin{aligned}
\text{Minimize} \quad & 0.4x_1 + 1.25x_1^2 + x_2^2 - x_1x_2 + 0.5x_3^2 + 0.5x_4^2, \\
\text{subject to} \quad & -0.5x_1 - x_2 + x_4 \geq -0.5, \\
& x_1 + 0.5x_2 - x_3 = 0.4, \\
& \mathbf{x} \geq 0.
\end{aligned}
\tag{25}
$$

We tested the proposed neural network guided by (4a, 4b and 4c) with four different initial vectors(four combination for feasible and infeasible vectors) for the primal and dual problems:
case 1: $\mathbf{x}_0 = (1, 1, 1.1, 2)^T$(feasible) and $(\mathbf{y}_0, \mathbf{z}_0) = (-1, 1)$(feasible),
case 2: $\mathbf{x}_0 = (1, 1, 1.1, 2)^T$(feasible) and $(\mathbf{y}_0, \mathbf{z}_0) = (1, -1)$(infeasible),
case 3: $\mathbf{x}_0 = (1, 2, -1, -2)^T$(infeasible) and $(\mathbf{y}_0, \mathbf{z}_0) = (-3, 1)$(feasible),
case 4: $\mathbf{x}_0 = (-1, 2, 4, 3)^T$(infeasible) and $(\mathbf{y}_0, \mathbf{z}_0) = (1, -1)$(infeasible),
and the transient behaviors of $\mathbf{x}(t)$ are depicted in Fig. 2, Fig. 3, Fig. 4, Fig. 5 respectively.

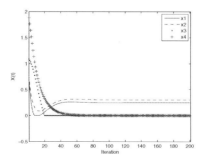

Fig. 2. Transient behavior of $\mathbf{x}(t)$ for case 1

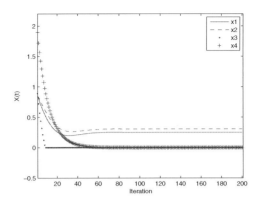

Fig. 3. Transient behavior of $\mathbf{x}(t)$ for case 2

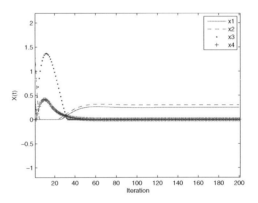

Fig. 4. Transient behavior of $\mathbf{x}(t)$ for case 3

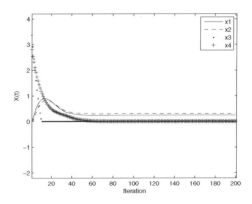

Fig. 5. Transient behavior of $\mathbf{x}(t)$ for case 4

It can be seen that after about 80 iterations the vector $\mathbf{x}$ will converge to the optimal solution $\mathbf{x}^\star = (0.2483, 0.3034, 0, 0)^T$ for all cases.

5 Conclusions

This paper presents a new nonlinear neural network to solving quadratic programming problems. It's proved that this novel neural network is stable in the sense of Lyapunov under certain conditions. Numerical simulation results show the effectiveness and efficiency this neural network. Future research direction include application the proposed neural network to solving the K-Winners-Take-All (KWTA) problem [34–36] based on linear programming or quadratic programming formulations, assignment problem [37,38] and maximum flow problem [39,40], extension the nonlinear model to convex programming and more general optimization problems.

References

1. Franklin, N.J.: Methods of Mathematical Economics: Linear and Nonlinear Programming. Society for Industrial and Applied Mathematics, Fixed-Point Theorems (2002)
2. Luenberger, D.G., Ye, Y.: Linear and Nonlinear Programming. Springer (2008)
3. Bazaraa, M.S., Sherali, D.H., Shetty C.M.: Nonlinear Programming: Theory and Algorithms. Wiley-Interscience (2006)
4. Surhone, L.M., Tennoe, M.F., Henssonow, S.F.: Quadratic Programming. Betascript Publishing (2010)
5. Gärtner, B., Matousek, J.: Approximation Algorithms and Semidefinite Programming. Springer (2012)
6. Griva, I., Nash, S.G., Sofer, A.: Linear and Nonlinear Optimization, 2nd edn. Society for Industrial Mathematics (2008)
7. Hertog, D.: Interior Point Approach to Linear, Quadratic and Convex Programming: Algorithms and Complexity. Springer (1994)
8. Lee, G.N.: Quadratic Programming and Affine Variational Inequalities: A Qualitative Study. Springer (2010)
9. Tank, D.W., Hopfield, J.J.: Simple Neural Optimization Networks: An A/D Converter, Signal Decision Circuit, and a Linear Programming Circuit. IEEE Transactions on Circuits and Systems **33**(5), 533–541 (1986)
10. Hopfield, J.J., Tank, D.W.: Computing with Neural Circuits: A Model. Science **233**, 625–633 (1986)
11. Zhang, S., Constantinides, A.G.: Lagrange Programming Neural Networks. IEEE Transactions on Circuits and Systems II **39**(7), 441–452 (1992)
12. Wang, J.: A Deterministic Annealing Neural Network for Convex Programming. Neural Networks **5**(4), 962–971 (1994)
13. Ding, K., Huang, N.J.: A New Class of Interval Projection Neural Networks for Solving Interval Quadratic Program. Chaos, Solitons and Fractals **3**, 149–164 (2008)
14. Yang, Y., Cao, J.: Solving Quadratic Programming Problems by Delayed Projection Neural Network. IEEE Transactions on Neural Networks **17**(6), 1630–1634 (2006)

15. Zhang, Y., Wang, J.: A Dual Neural Network for Convex Quadratic Programming Subject to Linear Equality and Inequality Constraints. Physics Letters A **298**, 271–278 (2002)

16. Hu, X., Wang, J.: An Improved Dual Neural Network for Solving a Class of Quadratic Programming Problems and Its K-Winners-Take-All Application. IEEE Transactions on Neural Networks **19**(2), 2022–2031 (2009)

17. Xia, Y., Feng, G., Wang, J.: Primal-Dual Neural Network for Online Resolving Constrained Kinematic Redundancy in Robot Motion Control. IEEE Transactions on Systems, Man and Cybernetics **35**(1), 54–64 (2005)

18. Kennedy, M.P., Chua, L.O.: Neural Networks for Nonlinear Programming. IEEE Transactions on Circuits and Systems **35**(5), 554–562 (1988)

19. Maa, C.Y., Schanblatt, M.A.: A Two-Phase Optimization Neural Network. IEEE Transactions on Neural Network **3**(6), 1003–1009 (1992)

20. Xia, Y.: A New Neural Network for Solving Linear Programming Problems and Its Application. IEEE Transactions on Neural Networks **7**(2), 525–529 (1996)

21. Nguyen, K.V.: A Nonlinear Neural Network for Solving Linear Programming Problems. In: International Symposium on Mathematical Programming, ISMP 2000, Atlanta, GA, USA (2000)

22. Taylor, J.G.: Mathematical Approaches to Neural Networks. North-Holland (1993)

23. Harvey, R.L.: Neural Network Principles. Prentice Hall (1994)

24. Veelenturf, L.: Analysis and Applications of Artifical Neural Networks. Prentice Hall (1995)

25. Rojas, R., Feldman, J.: Neural Networks A Systematic Introduction. Springer (1996)

26. Mehrotra, K., Mohan, C.K., Ranka, S.: Elements of Artificial Neural Networks. MIT Press (1997)

27. Haykin, S.: Neural Networks: A Comprehensive Foundation, 2nd edn. Prentice Hall (1998)

28. Michel, A., Liu, D.: Qualitative Analysis and Synthesis of Recurrent Neural Networks. CRC Press (2001)

29. Hagan, M.T., Demuth, H.B., Beale, M.H.: Neural Network Design. Martin Hagan (2002)

30. Gurney, K.: An Introduction to Neural Networks. CRC Press (2003)

31. Graupe, D.: Principles of Artificial Neural Networks, 2nd edn. World Scientific Pub Co., Inc. (2007)

32. Boyd, S., Vandenbeghe, L.: Convex Optimization. Cambridge University Press, Cambridge (2004)

33. Bertsekas, D.P., Tsitsiklis, J.N.: Parallel and Distributed Computation: Numerical Methods. Prentice-Hall, Englewood Cliffs (1989)

34. Maass, W.: Neural Computation with Winner-Take-All as the Only Nonlinear Operation. Advances in Neural Information Processing Systems **12**, 293–299 (2000)

35. Marinov, C.A., Calvert, B.D.: Performance Analysis for a K-Winners-Take-All Analog Neural Network: Basic Theory. IEEE Transactions on Neural Networks **14**(4), 766–780 (2003)

36. Liu, S.B., Wang, J.: A Simplified Dual Neural Network for Quadratic Programming with Its KWTA Application. IEEE Transactions on Neural Networks **17**(6), 1500–1510 (2006)

37. Wang, J.: Analogue Neural Network for Solving the Assignment Problem. Electronics Letters **28**(11), 1047–1050 (1992)

38. Hu, X., Wang, J.: Solving the Assignment Problem with the Improved Dual Neural Network. In: Liu, D., Zhang, H., Polycarpou, M., Alippi, C., He, H. (eds.) ISNN 2011, Part I. LNCS, vol. 6675, pp. 547–556. Springer, Heidelberg (2011)
39. Effati, S., Ranjbar, M.: Neural Network Models for Solving the Maximum Flow Problem. Applications and Applied Mathematics **3**(3), 149–162 (2008)
40. Nazemi, A., Omidi, F.: A Capable Neural Network Model for Solving the Maximum Flow Problem. Journal of Computational and Applied Mathematics **236**(14), 3498–3513 (2012)

Untrained Method for Ensemble Pruning and Weighted Combination

Bartosz Krawczyk[✉] and Michał Woźniak

Department of Systems and Computer Networks, Wrocław University
of Technology, Wybrzeże Wyspiańskiego 27, 50-370 Wrocław, Poland
{bartosz.krawczyk,michal.wozniak}@pwr.edu.pl

Abstract. The combined classification is an important area of machine learning and there are a plethora of approaches methods for constructing efficient ensembles. The most popular approaches work on the basis of voting aggregation, where the final decision of a compound classifier is a combination of discrete individual classifiers' outputs, i.e., class labels. At the same time, some of the classifiers in the committee do not contribute much to the collective decision and should be discarded. This paper discusses how to design an effective ensemble pruning and combination rule, based on continuous classifier outputs, i.e., support functions. As in many real-life problems we do not have an abundance of training objects, therefore we express our interest in aggregation methods which do not required training. We concentrate on the field of weighted aggregation, with weights depending on classifier and class label. We propose a new untrained method for simultaneous ensemble pruning and weighted combination of support functions with the use of a Gaussian function to assign mentioned above weights. The experimental analysis carried out on the set of benchmark datasets and backed up with a statistical analysis, prove the usefulness of the proposed method, especially when the number of class labels is high.

Keywords: Machine learning · Classifier ensemble · Classifier combination · Ensemble pruning · Weighted fusion · Untrained aggregation

1 Introduction

For a given classification task, we may often have more than a single classifier available. What is interesting, the number of misclassified objects by all individual classifiers is typically small. From this we can conclude, that even if individual classifiers do not have high quality, their union could form a reasonably good compound classifier. The considered approach is called a multiple classifier system (MCS), combined classifier or classifier ensemble and is considered as one of the most vital fields in the contemporary machine learning [10].

During the ensemble design process, we must take into consideration several important aspects, such as which classifiers to use, how to select the proper

© Springer International Publishing Switzerland 2014
Z. Zeng et al. (Eds.): ISNN 2014, LNCS 8866, pp. 358–365, 2014.
DOI: 10.1007/978-3-319-12436-0_40

topology, or what would be the best method for combining their outputs. In this work, we focus on two crucial steps: ensemble pruning and classifier combination.

For most considered problems, we can create / collect a huge number of classifiers. However, for ensemble to work properly it should be formed by mutually complementary models of high individual quality. Adding new classifiers that do not exploit a new area of competence do not improve the ensemble, only increases the computational cost and reduces its robustness. The problem lies on how to select a useful subgroup from a large pool of classifiers at hand.

However, one should note that these methods require specific criteria to evaluate the selected subgroup of classifiers, such as accuracy, AUC or diversity. Such criteria do not often lead to a satisfactory results (as using accuracy may lead to large and similar ensembles, while diversity will not take into account the individual quality of models) and selecting a proper metric for a given problem is not a trivial task.

When having selected a number of competent classifiers, one need to design a combination rule in order to establish a collective decision of the ensemble. Such a mechanism should be able to exploit the individual strengths of classifiers in the pool, while at the same time minimizing their drawbacks. In literature, two methods for classifier combination can be distinguished: methods that make decisions on the basis of discrete outputs (class labels) returned by the individual classifiers and methods that work with continuous outputs (supports returned by the individual classifiers).

The former group consists of mainly voting algorithms [2], where majority voting is still the most popular method used so far. Other works in this area suggest to train the weights for controlling the level of importance assigned to each vote.

The latter group of combination methods is based on discriminants, or support functions. In general the support function is a measure of support given in favor of a distinguished class, as neural network output, *posterior* probability or fuzzy membership function. There are many approaches dealing with this problem as [7], in which the optimal projective fuser was presented, or [8] employing a probabilistic approach. Several analytical properties of aggregating methods were discussed e.g. in [9]. Basically, the aggregating methods, which do not require a learning procedure, use simple operators as the maximum, minimum, sum, product, or average value. Other works suggest to use a trained combiner in order to efficiently establish weights [6]. However although this is an efficient method, such an approach requires an extensive computational time and additional training dataset - both of which are not often available in real-life applications.

In this work, we introduce a novel method for simultaneous ensemble pruning and weighted combination. We propose novel weighted aggregation operators which do not require learning and have embedded pruning procedure that do not require any criterion to work. We work on modification of two popular operators: average of supports and maximum of supports. Their main drawback lies in lack of robustness to weak and irrelevant classifiers, and in minimizing the influence of other ensemble members. By using a Gaussian function to estimate the weights for the entire ensemble, we achieve a smooth method for reducing, but not eliminating the influence of weaker classifiers. At the same time by adjusting a threshold on the value of weights, we are able to prune the ensemble by discarding incompetent learners.

2   Classifier Combination Methods

As in this work we concentrate on weighted combination of continuous outputs, therefore let us assume that each individual classifier makes a decision on the basis of the values of support functions.

2.1   Weighted Aggregation

Let $\Pi = \left\{ \Psi^{(1)}, \Psi^{(2)}, ..., \Psi^{(n)} \right\}$ be the pool of n individual classifiers and $F_{i,k}(x)$ stands for a support function that is assigned to class i ($i \in \mathcal{M} = \{1, ..., M\}$) for a given observation x and which is used by the classifier $\Psi^{(k)}$ from the pool Π.

The combined classifier $\Psi(x)$ uses the following decision rule

$$\Psi(x) = i \quad if \quad F_i(x) = \max_{k \in M} F_k(x), \tag{1}$$

where $F_k(x)$ is the weighted combination of the support functions of the individual classifiers from Π for the class k .

In this work, we assume that weights are dependent on classifier and class number. Weight $w_{i,k}$ is assigned to the k-th classifier and the i-th class. For a given classifier, weights assigned for different classes could be different. In our previous works, we have shown that this approach leads to a significant improvement over traditional methods [6]. With this, we can formulate our combination scheme as follows:

$$F_i(x) = \sum_{k=1}^{n} w_{i,k} F_{i,k}(x) \text{ and } \forall i \in \mathcal{M} \sum_{k=1}^{n} w_{i,k} = 1. \tag{2}$$

3   Untrained Ensemble Pruning and Weighted Combination

In this work, we propose new untrained aggregation operators which could exploit the competencies of the individual classifiers. The simple operators as maximum or average usually behave reasonably well but their work could be spoil by very imprecise estimators of the support functions used by only a few classifiers from a pool. Therefore we propose the modifications of the mentioned above operators which take into consideration all available support functions returned by the individual classifiers from the pool, but the functions which have the similar values to maximum or average have the strongest impact in the final value of the common support function calculated by using eq. 2. Additionally, we should notice that there may be some irrelevant classifiers in the pool and that for a large pool of classifiers most of the weights will become very small (in order to satisfy the condition from eq. 2). To deal with this problem, we propose to embed an ensemble pruning algorithm to eliminate incompetent classifiers. Then we normalize the weights for a reduced number of learners, thus increasing their level of influence over the ensemble decision. We propose to implement the pruning threshold ϕ, in order to discard all classifiers with assigned weights $w_{i,k} \leq \phi$.

The proposed operators are called NP-AVG and NP-MAX and can be calculated according to the Alg. 1. The only difference is the calculation of the $\overline{F}_i(x)$. For NP-AVG it is calculated according to

$$\overline{F}_i(x) = \frac{\sum\limits_{k=1}^{N} F_{i,k}}{N}, \tag{3}$$

and for NP-MAX using the following formulae

$$\overline{F}_i(x) = \max_{k \in \mathcal{M}} F_{i,k}. \tag{4}$$

Algorithm 1. General framework for ensemble pruning and weight calculation

---

Require: Π - pool of n elementary classifiers
  $F_{i,k}(x)$ - support function value for each class i returned by each individual classifier k from Π
  ϕ - pruning threshold
Ensure: $w_{i,k}(x)$ - weights assigned to each support function $F_{i,k}(x)$ which could be used in eq.2
  1: **for** $i := 1$ **to** M **do**
  2:     $w := 0$
  3:     Calculate $\overline{F}_i(x)$ according to eq. 3 for NP-AVG or according to eq. 4 for NP-MAX
  4:     **for** $k := 1$ **to** n **do**
  5:         $w_{i,k}(x) = \frac{1}{\sigma\sqrt{2\pi}} \exp\left(\frac{-(F_{i,k}(x) - \overline{F}_i(x))}{2\sigma^2}\right)$
  6:         $w := w + w_{i,k}(x)$
  7:     **end for**
  8:     **for** $k := 1$ **to** n **do**
  9:         **if** $w_{i,k} \leq \phi$
 10:         discard the k-th classifier
 11:     **end for**
 12:     **return** pruned pool of p classifiers
 13:     **for** $k := 1$ **to** p **do**
 14:         $w := \frac{w_{i,k}(x)}{w}$
 15:     **end for**
 16: **end for**

The only parameters of the proposed operators is σ which equivalent of standard deviation in normal distribution, and a pruning threshold ϕ.

4   Experimental Investigations

The aims of the experiment was to check the performance of the two proposed aggregation operators N-AVG and N-MAX and to compare them with several popular methods for aggregating classifiers.

4.1   Datasets

In total we chose 10 well known datasets from the UCI Repository [4]. For datasets with missing values (autos, cleveland and dermatology), instances without full set of features available were removed.

4.2   Set-up

As a base classifier, we have decided to use neural network (NN) - realized as a multi-layer perceptron, trained with back-propagation algorithm, with number of neurons depending on the considered dataset: in the input layer equal to the number of features, in the output layer equal to the number of classes and in the hidden layer equal to half of the sum of neurons in previously mentioned layers. Each model was initialized with random starting values and their training process was stopped prematurely after 200 iterations, in order to assure the initial diversity of the pool and that we are working on weak classifiers.

The pool of classifiers used for experiments was homogeneous and consisted of 30 neural networks.

As a reference methods we decided to use popular classifier combination algorithms: majority voting (MV), maximum of support (MAX), average of supports (AVG) and product (PRO).

For a pairwise comparison, we use a 5x2 combined CV F-test [1]. For assessing the ranks of classifiers over all examined benchmarks, we use a Friedman ranking test [3] and Shaffer post-hoc test [5]. For all statistical analysis, we use the significance level $\alpha = 0.05$.

4.3   Results

Firstly, we need to establish the level of influence of value of pruning threshold ϕ on the quality of the ensemble. A grid search was performed for $\phi \in [0; 0.5]$ with step $= 0.05$. The best parameter values according to the final accuracy and the avg. size of the ensemble after pruning are given in Table 1. If $\phi = 0$, then no pruning was applied. We use the established values of this parameter for further comparisons.

Results of the experiments, presented according to the accuracy and reduction rate of the examined methods, are given in Table 2. Outputs of Shaffer post-hoc test over accuracy are given in Table 3.

Table 1. Selecting the value of pruning threshold ϕ, and is influence on the size of the ensemble. Numbers in brackets stands for a standard deviation in the ensemble size.

Dataset	Best ϕ value		Avg. size of the ensemble	
	NP-AVG	NP-MAX	NP-AVG	NP-MAX
Autos	0.00	0.00	30 (0.00)	30 (0.00)
Car	0.3	0.25	21 (2.45)	19 (3.03)
Cleveland	0.00	0.00	30 (0.00)	30 (0.00)
Dermatology	0.15	0.10	19 (3.23)	16 (2.09)
Ecoli	0.10	0.15	17 (4.23)	18 (2.78)
Flare	0.2	0.15	13 (1.28)	12 (2.03)
Lymphography	0.00	0.00	30 (0.00)	30 (0.00)
Segment	0.2	0.15	20 (4.02)	18 (3.11)
Vehicle	0.05	0.05	17 (2.26)	17 (1.84)
Yeast	0.15	0.05	12 (3.72)	11 (2.39)

Table 2. Comparison of the classifier combination methods, with respect to their accuracy [%]. Small numbers under accuracies stand for indexes of methods, from which the considered one is statistically superior. Last row stands for the avg. rank after the Friedman test.

Dataset	MV[1]	MAX[2]	AVG[3]	PRO[4]	NP-AVG[5]	NP-MAX[6]
Autos	62.34 –	65.84 1,3,4	64.23 1,4	63.05 –	67.54 ALL	66.32 1,3,4
Car	89.12 4,5	89.23 3,4,5	88.43 4,5	85.31 –	87.74 4	91.03 ALL
Cleveland	52.38 –	57.23 1,5,7	57.43 1,5,7	55.64 1	55.02 1	57.14 1,5,7
Dermatology	93.23 –	95.75 1,5,7	95.05 1,5	92.87 –	94.67 1	95.83 1,5,7
Ecoli	71.02 –	77.43 1,3,4,5	75.36 1,4,5	71.61 –	79.62 ALL	77.60 1,3,4,5
Flare	74.31 2,4,5	72.69 –	75.72 1,2,4,5,6,7	73.12 2	73.90 2,4,5	77.12 ALL
Lymphography	82.27 ALL	80.32 5	80.87 5	79.32 –	81.12 2,5	80.32 5
Segment	86.23 4,5	86.74 4,5	87.54 1,2,4,5,7	85.62 4	86.89 4,5	91.21 ALL
Vehicle	66.43 –	74.03 1,3,4,5,7	72.63 1,4,5,7	67.90 1	70.12 1,4,5	73.87 1,3,4,5,7
Yeast	43.41 –	52.36 1,3,4,5,7	49.78 1,4,5	45.02 1	50.11 1,4,5	57.98 ALL
Avg. rank	4.51	3.21	5.72	6.48	7.62	2.78

Let's present the conclusions derived from the experiments. The proposed operators behaved reasonably well and outperformed, with statistical significance, all of the traditional methods for 5 out 10 data sets. Modifications of the average operator N-AVG was significantly better than the original one in 3 out 10 experiments, while N-MAX (and N-AVG as well) was not significantly better than the original maximum operator. The Shaffer test confirmed that the combination rule which takes into consideration additional information (coming e.g. individual classifier accuracy) can outperform untrained operators. This confirmed our intuition, because the trained combination rule usually behave better than untrained one, what was confirmed in the literature. This test also showed that N-MAX is a slightly better than N-AVG, and what is interesting it can outperform most of the traditional untrained approaches except maximum operator. Analyzing characteristics of the used data benchmark sets we can suppose that proposed operators work well especially for the classification task where the number of possible classes is a quite high, but additional computer experiments should be carried out to confirm this dependency. Each of the proposed operators outperform majority voting for almost all data sets. We can conclude, that in the case of an absence of additional learning examples (which can be used to train the combination rule) the untrained aggregation is a better choice than voting methods. This observation is also known and confirmed by other researches as [11]. Our proposed methods allow to establish efficient weighted combination rules with a low computational complexity. Trained fusers require an additional processing time, which increases the complexity of the ensemble. Our methods, due to their low complexity, seem as an attractive proposition for real-life problems with limitations on processing time, e.g., ensembles for data streams.

Table 3. Shaffer test for comparison between the proposed combination methods and reference fusers. Symbol '=' stands for classifiers without significant differences, '+' for situation in which the method on the left is superior and '-' vice versa.

hypothesis	p-value
NP-AVG vs MV	+ (0.0423)
NP-AVG vs MAX	= (0.3895)
NP-AVG vs AVG	= (0.4263)
NP-AVG vs PRO	+ (0.0136)
NP-MAX vs MV	+ (0.0262)
NP-MAX vs MAX	= (0.4211)
NP-MAX vs AVG	+ (0.0249)
NP-MAX vs PRO	+ (0.0097)
NP-AVG vs NP-MAX - (0.0314)	

5   Conclusions

The paper presented two novel untrained aggregation operators which could be used in the case of the absence of additional learning material to train the combination rule. Otherwise the trained combination rule should be advised. The proposed methods could be valuable alternatives for the traditional aggregating operators which do not required learning and should be used in the mentioned above case instead of voting methods, of course in the case that we can access to the support function values of individual classifiers. The computer experiments confirmed that performances of the proposed methods are satisfactory compared to the traditionally untrained operators, especially for tasks when the number of possible classes is high. Therefore, we are going to continue the work on the proposed models, especially we would like to carried out the wider range of computer experiments which would define precisely the type of the classification tasks when the N-AVG and N-MAX could be used.

Acknowledgments. This work was supported by the Polish National Science Center under the grant no. DEC-2013/09/B/ST6/02264.

References

1. Alpaydin, E.: Combined 5 x 2 cv f test for comparing supervised classification learning algorithms. Neural Computation **11**(8), 1885–1892 (1999)
2. Biggio, B., Fumera, G., Roli, F.: Bayesian Analysis of Linear Combiners. In: Haindl, M., Kittler, J., Roli, F. (eds.) MCS 2007. LNCS, vol. 4472, pp. 292–301. Springer, Heidelberg (2007)
3. Demšar, J.: Statistical comparisons of classifiers over multiple data sets. J. Mach. Learn. Res. **7**, 1–30 (2006)
4. Frank, A., Asuncion, A.: UCI machine learning repository (2010), http://archive.ics.uci.edu/ml
5. García, S., Fernández, A., Luengo, J., Herrera, F.: Advanced nonparametric tests for multiple comparisons in the design of experiments in computational intelligence and data mining: Experimental analysis of power. Inf. Sci. **180**(10), 2044–2064 (2010)
6. Jackowski, K., Krawczyk, B., Woźniak, M.: Improved adaptive splitting and selection: the hybrid training method of a classifier based on a feature space partitioning. Int. J. Neural Syst. 24(3) (2014).
7. Rao, N.S.V.: A Generic Sensor Fusion Problem: Classification and Function Estimation. In: Roli, F., Kittler, J., Windeatt, T. (eds.) MCS 2004. LNCS, vol. 3077, pp. 16–30. Springer, Heidelberg (2004)
8. Rokach, L., Maimon, O.: Feature set decomposition for decision trees. Intell. Data Anal. **9**(2), 131–158 (2005)
9. Wozniak, M.: Experiments on linear combiners. In: Pietka, E., Kawa, J. (eds.) Information Technologies in Biomedicine. AISC, vol. 47, pp. 445–452. Springer, Berlin / Heidelberg (2008)
10. Woźniak, M., Graña, M., Corchado, E.: A survey of multiple classifier systems as hybrid systems. Information Fusion **16**, 3–17 (2014)
11. Wozniak, M., Zmyslony, M.: Designing combining classifier with trained fuser - analytical and experimental evaluation. Neural Network World **20**(7), 925–934 (2010)

An Improved Learning Algorithm with Tunable Kernels for Complex-Valued Radial Basis Function Neural Networks

Xia Mo[1], He Huang[1([⊠])], and Tingwen Huang[2]

[1] School of Electronics and Information Engineering, Soochow University,
Suzhou 215006, People's Republic of China
cshhuang@gmail.com
[2] Texas A&M University at Qatar, Doha 5825, Qatar

Abstract. In this paper, as an extension of real-valued orthogonal least-squares regression with tunable kernels (OLSRTK), a complex-valued OLSRTK is presented which can be used to construct a suitable sparse regression model. In order to enhance the real-valued OLSRTK, the random traversal process and method of filtering center are adopted in complex-valued OLSRTK. Then, the complex-valued OLSRTK is applied to train complex-valued radial basis function neural networks. Numerical results show that better performance can be achieved by the developed algorithm than by the original real-valued OLSRTK.

Keywords: Complex-valued radial basis function neural networks · Random traversal process · Repeat weighted boosting search · Filtering center

1   Introduction

The complex-valued radial basis function neural networks (CVRBFNNs) were initially introduced in [1]. Since then, it has been extensively studied and many successful applications have been found in various fields such as image recognition [2], image segmentation [12], circuit components modeling [3], signal processing [4,5,6,7,8] and pattern classification [13,14] etc. As is well known now, the selection of centers of neurons in hidden layer plays an important role in the learning process of CVRBFNNs. And the connection weights between neurons in output and hidden layers can be efficiently computed by least-squares method after centers selected. Many learning algorithms have been available in literature [1], [9,10] which can be efficiently applied to determine the centers of hidden neurons. Among them, the so-called K-means clustering algorithm is a popular one, where the number of centers is fixed in advance. This may lead to an unreasonable structure of CVRBFNNs. The orthogonal least-squares

The work was supported by the National Natural Science Foundation of China under Grant No. 61005047, and the Natural Science Foundation of Jiangsu Province of China under Grant No. BK2010214. Also, this publication was made possible by NPRP grant #4-1162-1-181 from the Qatar National Research Fund (a member of Qatar Foundation). The statements made herein are solely the responsibility of the author[s].

Z. Zeng et al. (Eds.): ISNN 2014, LNCS 8866, pp. 366–373, 2014.
DOI: 10.1007/978-3-319-12436-0_41

(OLS) method for CVRBFNNs was introduced by S. Chen in [1]. Based on it, one can establish a sparse regression model. It is known that the centers obtained by OLS algorithm must be chosen from samples. Nevertheless, in some situations, it should be more suitable that the centers come from outside of samples. In fact, a combination of some samples can be chosen as a center. The complex-valued gradient descent learning algorithm was proposed in [9]. It was not applicable to problems of which the gradient was difficult to calculate. In real domain, it was solved by the real-valued orthogonal least-squares regression with tunable kernels (OLSRTK), where a combination of some samples was chosen as a candidate center without gradient calculation [10]. Furthermore, the center's parameters of each individual regressor were tuned in [11] by incrementally maximizing the training error reduction ratio (ERR) based on a guided random search algorithm. This algorithm was named as repeating weighted boosting search (RWBS) [11].

It is found that although the real-valued OLSRTK has some benefits, two aspects are ignored by them. Firstly, some training samples may be ignored when using RWBS to generate centers randomly from samples. Secondly, the decline rule, which centers should obey in OLS, is broken in RWBS.

Motivated by these, an improved complex-valued OLSRTK is presented in this paper in order that it can be applied in complex domain. Two adjustments are adopted to resolve the two issues mentioned above. Specifically, for the first one, a random traversal process (RTP) is added; for the second one, a method of filtering center (FC) is utilized to ensure that the better center obeys the decline rule in OLS. Then, the improved algorithm is employed to train CVRBFNNs to deal with some practical problems. Two examples are given to illustrate the effectiveness of the developed algorithm. The experimental results show that better performance can be achieved by our algorithm.

2   Preliminaries

Consider a problem of approximating N pairs of training samples $\{x_t, y_t\}_{t=1}^{N}$, with $x_t \in C^m, y_t \in C$ and $y = [y_1, \cdots, y_N]^T$ by using the following model

$$y(x) = \hat{y}(x) + e(x) = \sum_{i=1}^{M} p_i(x)\theta_i + e(x) \tag{1}$$

where M is the number of regressors; $x \in \{x_t\}_{t=1}^{N}$; θ_i is the connection weight; $e(x)$ is the modeling error; and $p_i(\cdot)$ denotes a regressor. Generally, the regressor uses the Gaussian function that maps $C^m \to R$, resulting in inaccurate phase approximation. To overcome the above limitation, a Gaussian like fully complex function $sech(\cdot)$ $(C^m \to C)$[9] was used. Here we take the variant of origin $sech(\cdot)$ [9]:

$$p_i(x) = sech\big(\sigma_i(x - u_i)^T(x - u_i)\big) \tag{2}$$

where the amplitude of samples are normalized to one; the constant scalar σ_i is the spread of the i th regressor and $u_i \in C^m$ is the i th kernel center. Let $\theta = [\theta_1, \cdots, \theta_M]^T$, $e = [e(x_1), \cdots, e(x_N)]^T$ and $P_{N \times M} = [p_1, \cdots, p_M]$ with regression vector $p_i = [p_i(x_1), \cdots, p_i(x_N)]^T$, Eq. (1) can be rewritten as

$$y = P\theta + e \tag{3}$$

Make an orthogonal decomposition of P as $P = WA$, where $A_{M \times M}$ is an upper triangular matrix with the unit diagonal elements and $W_{N \times M} = [w_1, \cdots, w_M]$ with $w_i \in C^N$ satisfies $w_m^H w_n = 0$, if $m \neq n$. Eq. (3) can be rewritten as

$$y = WA\theta + e = Wg + e \tag{4}$$

where $g = [g_1, \cdots, g_M]^T$. The sum of squares of y_t is

$$y^H y = \sum_{i=1}^{M} w_i^H w_i g_i^2 + e^H e \tag{5}$$

According to Eq. (5), an ERR of candidate center x_i with regard to w_i can be defined as

$$J(x_i) = (w_i)^H w_i g_i^2 / (y^H y) \tag{6}$$

with $g_i = (w_i)^H y / ((w_i)^H w_i)$. Define the orthogonal coefficients

$$\alpha_{ji} = \left(w_{ols}^{(j)}\right)^H p_i / \left(\left(w_{ols}^{(j)}\right)^H w_{ols}^{(j)}\right) \tag{7}$$

Here $1 \leq i \leq M$, $1 \leq j < k$ and $w_{ols}^{(j)}$ is the orthogonal column vector corresponding to jth center. In OLS algorithm, the selection of the kth center is as follows. For $1 \leq i \leq M$, $i \neq i_1, \cdots, i_{k-1}$ where i_s, $1 \leq s \leq k - 1$ records the index of sth center in samples. We use Eq. (6) and Eq. (7) to calculate

$$w_i^{(k)} = p_i - \sum_{j=1}^{k-1} \alpha_{ji}^{(k)} w_{ols}^{(j)} \tag{8}$$

$$J^{(k)}(x_i) = \left(w_i^{(k)}\right)^H w_i^{(k)} \left(g_i^{(k)}\right)^2 / (y^H y) \tag{9}$$

$$i_k = max_i \{J^{(k)}(x_i)\} \tag{10}$$

Then the kth center is chosen as x_{i_k} corresponding to the maximum of $J^{(k)}(x_i)$. Its orthogonal column vector and ERR are $w_{ols}^{(k)} = w_{i_k}^{(k)}$ and $J^{(k)}(x_{i_k})$ respectively. The decline law in OLS is

$$\begin{cases} J^{(1)}(x_{i_k}) < J^{(1)}\left(x_{i_{(k-1)}}\right) < \cdots < J^{(1)}(x_{i_2}) < J^{(1)}(x_{i_1}) \\ \quad\quad\quad\quad\quad\quad \vdots \\ J^{(k-1)}(x_{i_k}) < J^{(k-1)}\left(x_{i_{(k-1)}}\right) \end{cases} \tag{11}$$

3   Improved Algorithm with RTP and FC

3.1   RTP

In the training procedure of real-valued OLSRTK, the candidate centers are selected from samples randomly by RWBS. However, some training samples may be skipped in RWBS. The RTP is added in order that all samples can be learned. RWBS is to repeat weighted boosting search (WBS) NG times. The description of WBS is as follows.

Suppose an optimization problem is $max_{x \in X} J(x)$. A population of P_s samples is selected randomly from sample set X. Let $x_{best} = arg\ max\ J(x)$ and $x_{worst} = arg\ min\ J(x)$ with $x \in \{x_1, \cdots, x_{P_s}\}$. The $(P_s + 1)$th point is obtained by computing a combination of $x_1, \cdots, x_{P_s}$:

$$x_{P_s+1} = \sum_{i=1}^{P_s} \delta_i x_i \tag{12}$$

where the distribution weightings $\delta_i > 0$ and $\sum_{i=1}^{P_s} \delta_i = 1$. Then x_{P_s+2} is defined as

$$x_{P_s+2} = x_{best} + \left(x_{best} - x_{P_s+1}\right) \tag{13}$$

According to their ERR $J\left(x_{P_s+1}\right)$ and $J\left(x_{P_s+2}\right)$, x_{worst} is replaced by the better one between x_{P_s+1} and x_{P_s+2}. The WBS process is stopped until the above iteration is repeated for NB times, or the following condition is satisfied:

$$\left\| x_{P_s+1} - x_{P_s+2} \right\| < \varepsilon_B \tag{14}$$

where ε_B is a small positive scalar. For example, suppose that $X = \{x_1, \cdots, x_6\}$, $P_s = 3$, NB = 2 and NG = 1. In the first iteration, one randomly selects 3 samples, e.g., $\{x_2, x_4, x_5\}$ from X to generate x_{P_s+1} and x_{P_s+2}. If the termination condition is not satisfied, one goes to the second iteration and then randomly selects 3 points, e.g., $\{x_1, x_3, x_4\}$ from X for further calculation. Then, the WBS and RWBS are stopped. We can see that x_6 is ignored in the process. Hence, we cannot make sure that the center selected by RWBS is the most appropriate one.

RTP is to make sure that all samples are involved in the selection procedure. Suppose that the sample set is $\{x_t\}_{t=1}^N, x_t \in C^m$ and we are going to determine the kth center. Hence, $k - 1$ centers have been selected. Some of the selected centers may come from samples and we assume that the number is d. Others may be generated by Eq. (12) and Eq. (13), and the number is $(k - 1) - d$. Define $avail\_NumSam$ to record the number of samples which are not selected as centers, that is $avail\_NumSam = N - d$. ε is a small positive scalar. The detailed RTP is as follows:

```
For k=1:N    (outer loop process)
   Store samples which are not selected as centers in
   variable: temp_sample. Set values for Ps and NB
   respectively, then set δi(0) = 1/Ps;
   Compute NG = fix |(avail_NumSam−Ps)/(Ps−1) + 1|;
   For n=1:NG    (RWBS process)
     If n==1
        Select Ps samples randomly from temp_sample and
        delete them from temp_sample. Then calculate
```

```
   their ERR: J^(k)(x_i), 1 ≤ i ≤ P_s;
Else
   x_1 = x_best^{n-1} is regarded as an candidate center. Select
   the rest (P_s - 1) samples randomly from temp_sample and
   delete them from temp_sample. Then calculate their
   ERR: J^(k)(x_i), 1 ≤ i ≤ P_s;
End If
For t=1:NB  (WBS process)
   Update the distribution weightings as follows:
   First, normalize ERR:
```

$$\bar{J}^{(k)}(x_i) = J^{(k)}(x_i) / \sum_{i=1}^{P_s} J^{(k)}(x_i), \ 1 \le i \le P_s,$$ calculate $\eta_t = \sum_{i=1}^{P_s} \delta_i(t-1)\bar{J}^{(k)}(x_i)$, then compute:

$$\delta_i(t) = \begin{cases} \delta_i(t-1)\beta^{1-\bar{J}^{(k)}(x_i)}, for\ \beta_t \le 1 \\ \delta_i(t-1)\beta_t^{\bar{J}^{(k)}(x_i)}, for\ \beta_t > 1 \end{cases}, with\ \beta_t = \frac{\eta_t}{1-\eta_t}$$

```
   Use Eq. (12) and Eq. (13) to generate x_{P_s+1} and
   x_{P_s+2}. If either of both obeys the decline rule,
   then find i_* = arg max J^(k)(x_i), i = P_s + 1, P_s + 2. And If
   J^(k)(x_{i_*}) > J^(k)(x_worst), then x_worst is replaced by x_{i_*};
   If ‖x_{P_s+1} - x_{P_s+2}‖ < ε_B, exit WBS;
End WBS
Obtain local optimal center x_best^n;
End RWBS
If n==NG && ((NG-1)*(P_s-1)+P_s)< avail_NumSam
   n=n+1;
   tp_P_s = avail_NumSam - ((NG - 1) * (P_s - 1) + P_s);
   x_best^{n-1} with the remainder tp_P_s samples are as candidate
   centers. Then calculate their ERR;
   P_s = tp_P_s + 1;
   Implement WBS process again with the last P_s samples;
   Obtain local optimal center x_best^n;
End If
Obtain center x_(k) = x_best^n and its ERR: J^(k)(x_(k));
If 1 - (J^(1)(x_(1)) + ⋯ + J^(k)(x_(k))) < ε  , exit outer loop;
End outer loop
```

3.2   FC

The second stage of training CVBRFNNs is to determine the centers from the candidates obtained by RWBS. However, according to Eq. (11), the decline law may be broken in RWBS. Suppose that $x_{(1)}$ is the first selected center, its ERR $J^{(1)}(x_{(1)})$ is the largest one among the candidate centers. Then we need to select the second center $x_{(2)}$. Its ERR is $J^{(2)}(x_{(2)})$. However, if $x_{(2)}$ is from new points x_{P_s+1}, x_{P_s+2} and it has not been involved in the selection of determining the first center $x_{(1)}$. Hence, its $J^{(1)}(x_{(2)})$ might be larger than $J^{(1)}(x_{(1)})$. This may break the decline rule in OLS.

The method of FC is to delete new point which breaks the rule. Therefore we can ensure that all centers selected in the training obey the decline rule. Assume that it is in the calculation of the kth center with $k \geq 2$, at the nth loop in RWBS with $1 \leq n \leq NG$. The value of $flag\_P_{s1}$ is to show whether x_{P_s+1} obeys the decline law or not. If $flag\_P_{s1}$ equals to $k - 1$, it means that x_{P_s+1} obeys it. Otherwise, we cannot consider x_{P_s+1}. Similarly, $flag\_P_{s2}$ is used for x_{P_s+2}. The idea of FC is as follows:

```
For t=1: NB     (WBS process)
   Update the distribution weightings, Use Eq. (12) and
      Eq. (13) to generate x_{P_s+1} and x_{P_s+2};
   Set flag_P_{s1} = 0, flag_P_{s2} = 0;
   Calculate J^{(s)}(x_{P_s+1}) and J^{(s)}(x_{P_s+2}), with s = 1,···,k − 1;
   For m=1:k-1
        If J^{(m)}(x_{P_s+1}) < J^{(m)}(x_{(m)}), flag_P_{s1} = flag_P_{s1} + 1;
        If J^{(m)}(x_{P_s+2}) < J^{(m)}(x_{(m)}), flag_P_{s2} = flag_P_{s2} + 1;
   End
   If flag_P_{s1} == k − 1, calculate J^{(k)}(x_{P_s+1});
   Otherwise, set J^{(k)}(x_{P_s+1}) = 0;
   If flag_P_{s2} == k − 1 , calculate J^{(k)}(x_{P_s+2});
   Otherwise, set J^{(k)}(x_{P_s+2}) = 0;
   Then find i_* = arg max J^{(k)}(x_i), i = P_s + 1, P_s + 2. And if
   J^{(k)}(x_{i_*}) > J^{(k)}(x_{worst}), then x_{worst} is replaced by x_{i_*};
   If ‖x_{P_s+1} − x_{P_s+2}‖ < ε_B, exit WBS;
End WBS
Find local optimal center x^n_{best}.
```

4 Applications

4.1 Modeling

Consider a second order band-stop FIR digital filter. Its amplitude-frequency response function is described by

$$H(e^{j\omega}) = \sum_{n=0}^{L-1} h(n)e^{-j\omega n} \tag{15}$$

where $L - 1$ denotes the order of filter. Its normalized amplitude-frequency response is

$$H(e^{j\omega}) = \begin{cases} 1, \omega \in (0, 0.25\pi) \\ 0, \omega \in (0.25\pi, 0.75\pi) \\ 1, \omega \in (0.75\pi, \pi) \end{cases} \tag{16}$$

According to the frequency transforming equation $\omega = 2\pi\Omega/f_s$, Eq. (16) corresponds to a band-pass filter when $\omega \in [0, 0.25\pi]$ and $\omega \in [0.75\pi, \pi]$. And it corresponds to a stop-band filter when $\omega \in [0.25\pi, 0.75\pi]$. Totally 200 samples are produced, where the input vector is $x = [e^{-j\omega 0}, e^{-j\omega 1}, e^{-j\omega 2}]^T$ and ω is randomly distributed in $[0, \pi]$. 100 samples are used for training and the others for testing. Tab. 1 shows the averaged experimental results for 100 times between real-valued OLSRTK and improved algorithm. It can be seen from Tab. 1 that our algorithm achieved better performance.

Table 1. Results of real-valued OLSRTK and our algorithm in Modeling

Methods	Training MSE	Testing MSE	Number of neuron
Real-valued OLSRTK [10]	0.0568	0.0798	14.56
Our algorithm	0.0518	0.0682	16.36

4.2   Classification

Two UCI datasets Iris and BLOGGER are further tested to show the performance of the developed algorithm. In Iris dataset, there are three classes of samples. We use 0, 1 and 2 to represent the corresponding desired output, respectively. In BLOGGER dataset, there are two classes of samples. We use 0 and 1 to represent the corresponding desired output, respectively. Since both of them are real-valued, it is necessary to transform them into complex ones:

$$\begin{cases} z_{in} = e^{1i \cdot x}, x \in R^l \\ z_{out} = e^{1i \cdot (\pi/2) \cdot y}, y \in R \end{cases} \tag{17}$$

where x is the input vector and y is the desired output. z_{in} is adopted as the input vector of CVRBFNNs and z_{out} is the output of CVRBFNNs. The result is given in Tab. 2. It is also verified that better performance can be guaranteed by our algorithm.

Table 2. Results of real-valued OLSRTK and our algorithm in Classification

Methods	Iris	BLOGGER
Real-valued OLSRTK [10]	88.99%	77.2%
Our algorithm	89.78%	78.12%

References

1. Chen, S., Grant, P.M., Mclaughlin, S., Mulgrew, B.: Complex-valued Radial Basis Function Networks. In: 3rd International Conference on Artificial Neural Networks, pp. 148–152. IET Press, London (1993)

2. Pande, A., Goel, V.: Complex-valued Neural Network in Image Recognition: A Study on the Effectiveness of Radial Basis Function. World Academy of Science, Engineering and Technology **20**, 220–225 (2007)
3. Li, M.Y., He, S.B., Li, X.D.: Modeling the Nonlinear Power Amplifier with Memory Using Complex-valued Radial Basis Function Networks. In: International Conference on Microwave and Millimeter Wave Technology, pp. 100–103. IEEE Press, New York (2008)
4. Hong, X., Chen, S.: Modeling of Complex-valued Wiener Systems Using B-spline Neural Network. IEEE Transactions on Neural Network **22**, 818–825 (2011)
5. Cha, I., Kassam, S.A.: Channel Equalization Using Adaptive Complex Radial Basis Function Networks. IEEE Journal on Selected Areas in Communications **13**, 122–131 (1995)
6. Gan, Q., Saratchandran, P., Sundararajan, N., Subramanian, K.R.: A Complex Valued Radial Basis Function Network for Equalization of Fast Time Varying Channel. IEEE Transactions on Neural Network **10**, 958–960 (1999)
7. Deng, J., Sundararajan, N., Saratchandran, P.: Communication Channel Equalization Using Complex-valued Minimal Radial Basis Function Neural Networks. IEEE Transactions on Neural Network **13**, 687–696 (2002)
8. Botoca, C., Budura, G.: Symbol Decision Equalizer Using a Radial Basis Functions Neural Network. In: 7th WSEAS International Conference on Neural Networks, pp. 79–84. WSEAS Press, Stevens Point (2006)
9. Savitha, R., Suresh, S., Sundararajan, N.: A Fully Complex-valued Radial Basis Function Network and Its Learning Algorithm. International Journal of Neural Systems **19**, 253–267 (2009)
10. Chen, S., Wang, X.X., Brown, D.J.: Orthogonal Least Squares Regression with Tunable Kernels. Electronics Letters **41**, 484–486 (2005)
11. Chen, S., Wang, X.X., Harris, C.J.: Experiments with Repeating Weighted Boosting Search for Optimization in Signal Processing Application. IEEE Transactions on Systems, Man and Cybernetics, Part B: Cybernetics **35**, 682–693 (2005)
12. Ceylan, M., Yacar, H.: Blood Vessel Extraction from Retinal Images Using Complex Wavelet Transform and Complex-valued Artificial Neural Network. In: 36th International Conference on Telecommunications and Signal Processing, pp. 822–825. IEEE Press, New York (2013)
13. Savitha, R., Suresh, S., Sundararajan, N., Kim, H.J.: A Fully Complex-valued Radial Basis Function Classifier for Real-valued Classification Problems. Neurocomputing **78**, 104–110 (2012)
14. Amin, M.F., Murase, K.: Signal-layered Complex-valued Neural Network for Real-valued Classification Problems. Neurocomputing **72**, 945–955 (2009)

Combined Methodology Based on Kernel Regression and Kernel Density Estimation for Sign Language Machine Translation

Mehrez Boulares[✉] and Mohamed Jemni

Research Laboratory of Technologies of Information and Communication and Electrical Ingineering (LaTICE), Ecole Supérieure des Sciences et Techniques de Tunis, Tunis, Tunisia
mehrez.boulares@gmail.com, Mohamed.jemni@fst.rnu.tn

Abstract. The majority of current researches in Machine Translation field are focalized essentially on spoken languages. The aim is to find a most likely translation for a given source sentence based on statistical learning techniques which are applied to very big parallel corpora. In this work, we focused on gesture languages especially on Sign Language in order to present a new methodological foundation for Sign Language Machine Translation. Our approach is based on Kernel Regression combined to Kernel Density Estimation method applied to Sign Language n-grams. The translation process is modelled as an n-gram to n-gram mapping with the consideration of the n-gram positions in the source and the target phrases. For doing so, we propose a new feature mapping process (Weighted Sub n-gram Feature Mapping) which is a modified version of the String Subsequence Kernel SSK feature mapping. The Weighted Sub n-gram aims to generate feature vectors mapping of both source and target n-gram. Afterwards, to learn the function that map source n-grams to target n-grams, we used and compared four learning techniques (Gaussian Process Regressor, K-Nearest Neighbors Regressor, Support Vector Regressor with Gaussian Kernel and Kernel Ridge Regression) for the purpose to choose the efficient one which minimizes the SSE (Sum of Squared Error). Even so, to find solution to the pre-image problem, we rely on the De-Bruijn Multi Graph search applied on n-grams target. For the purpose to obtain the best translation, we relied on the search of the most frequently observed bilingual n-gram alignment in term of the maximization of the translation probability. For unknown n-grams, we used kernel ridge regression for the purpose to predict the probability through learning the Density Estimation function of the bilingual n-grams alignments. We obtained encouraging experimental results on a small-scale reduced-domain corpus.

Keywords: Kernel ridge regression · String kernel · De-Bruijn · Kernel density estimation · Sign language · Gaussian process for regression · KNN-Regressor · SVR Gaussian kernel · ASL signing space

© Springer International Publishing Switzerland 2014
Z. Zeng et al. (Eds.): ISNN 2014, LNCS 8866, pp. 374–384, 2014.
DOI: 10.1007/978-3-319-12436-0_42

1   Introduction

Machine learning technology has progressively extended its application domain to reach Machine Translation field. However, in this field spoken languages are mainly used, there are few works that treat gesture languages such as Sign Language and this is due to the lack of sign language parallel corpora. In this work, we focused on the American Sign Language (ASL) as the primary means of communication for about one-half million deaf people in the US [3]. Especially, we are concentrated on the translation from English gloss text (without taking into account the signing space) to English gloss text that includes signing space information.

The study [6] has shown that signing space information improves the translation understanding. A phenomenon in which signers use special hand movements to indicate the location and movement of invisible objects (representing entities under discussion) in space around their bodies as shown in figure 1. Signing space information are frequent in ASL and are necessary for conveying many concepts. In other words, the translation process that integrates spatial information is more understandable than classical translation. This is due to also that deaf people have many difficulties related to the creation of a mental image that reflects the real meaning of the translated textual information.

Fig. 1. An Example of the sentential use of space in ASL. Nominal (cat, dog) are first associated with spatial loc_i through indexation. The direction of the movement of the verb (BITE) indicates the grammatical role of subject and object.

In this context, we propose a new methodological foundation that aims to use Kernel Regression methods in machine translation processing to generate translation that includes signing space information. For this purpose, we built a small-scale parallel corpora composed of both English gloss texts without signing space information and English gloss texts with signing space information in order to apply our methodology using Kernel Regression technique. The translation process is modeled as an n-gram to n-gram mapping with the consideration of the n-gram positions in the source and the target phrases.

For doing so, we rely on our feature mapping process in order to generate feature vectors mapping of both source and target n-grams. To learn the function that map source n-gram to target n-grams, we used and compared four learning techniques (Gaussian Process Regressor, K-Nearest Neighbors Regressor, Support Vector Regressor with Gaussian Kernel and Kernel Ridge Regression) for the purpose to choose the efficient one(that minimizes the SSE). As a solution to the pre-image problem, we rely on the De-Bruijn Multi Graph search applied on n-grams target. In fact, we are based on the search of the translation that maximizes the probability score of n-grams translation. For unknown n-grams, we used ridge regression for the purpose to predict the probability through learning the Density Estimation function of the bilingual n-grams. The remainder of this paper is organized as follows. In section 2, we present some related works. Section 3 is dedicated to discuss our approach. Section 4 introduces our experimental results. Finally, we conclude by a conclusion and some perspectives.

2   Related Work

Regression techniques can be used to model the relationship between strings. However, there is the work of [4] which is based on a general regression technique for learning transductions using a string-to-string mapping. Wang [17] applies a string to string mapping approach to machine translation by using ordinary least squares regression and n-gram string kernels on a small subset of the Europarl corpus. This work uses the pre-image model as a score to the standard statistical machine translation systems such as phrased-based search [9]. Furthermore, this approach loses some of main advantages of the regression approach, as proposed and cited in [4].Wang and Shawe-Taylor [18] used also later the L2 regularized least squares regression in machine translation. Although the translation quality they achieved still not better than Moses [8], which is accepted to be the state-of-the-art, they show the feasibility of the approach. Ergun Bicici [1] uses L2 regularized regression for sparse regression estimation of target features and graph decoding to find translation results. Serrano and al. [12] work is based on the learning of the translation mapping by linear regression applied to constrained hotel front desk requests domain (corpora). Once the target feature vector is obtained, they use a multi-graph search to find all the possible target strings. We noticed that the majority of existing works use mainly regression or statistical techniques on spoken languages corpora. Except, some works such as Daniel Stein [14] work that uses statistical approach on sign language machine translation on small-sized corpora.

The Hung-Yu Su [7] work, relies on improving structural statistical machine translation for Sign Language with small corpus using thematic role templates as Translation Memory. Furthermore, we also observed that there are no studies that combine the statistical and regression approaches in Sign language machine translation. For this purpose, we follow the works of Cortes and al. [4], Wang and Shawe-Taylor [18], and P. Koehn [9] in order to derive benefit from statistical and regression approaches. Furthermore, we present a new methodological foundation of Hybrid technique which consists to combine the Kernel Regression and the Kernel Density Estimation approaches applied to small-size sign language corpora taking into a count the signing space informations.

3   Our Approach

3.1   Problem Formulation

Machine translation deals with the problem of mapping sentences x from a source language X^* to a target language Y^*. Due to the complexity of the translation problem, the relationships between these two languages cannot be properly enumerated as a set of rules [12]. Let X and Y correspond to the token sets used to represent source and target N-Gram, then a training sample of m inputs can be represented as: (X_1, Y_1). . . $(X_m, Y_m) \in X^* \times Y^*$, where (x_i, y_i) corresponds to a pair of source and target language token string. Input N-Gram in X^* are mapped via φ_x to feature space F_x and the output string are mapped to F_y via the mapping φ_y. The mapping can be defined implicitly by a positive symmetric Kernel K_x and K_y associated with the mappings φ_x and φ_y. Our goal is to find a mapping f: $X^* \rightarrow Y^*$ that can convert a given set of source string to a set of target string that share the same meaning in the target language. In other words, a regression technique can be used to learn and to estimate the mapping g from X^* to F_y.

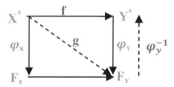

Fig. 2. The String-to-string mapping presented in Cortes work [4]

Afterwards, given the target feature vector F_y obtained from the translation mapping, we compute its pre-image set φ_y^{-1}. Figure 2 depicts a general scheme of the whole translation process.

3.2   Our Feature Mapping Approach

Standard learning systems (like neural networks or decision trees) operate on input data after they have been transformed into feature vectors d_1. . . $d_n \in D$ living in an m dimensional space. However, many techniques can be applied to these data points in order to analyze, to classify, to cluster, to interpolate or to derive useful informations in order to make predictions. There are many cases, however, where the input data cannot readily be described by explicit feature vectors: for example bio sequences, images, graphs and text documents. For such datasets, the construction of a feature extraction module can be as complex and expensive as solving the entire problem [11]. It is also possible to lose some important information during the feature extraction process.

In general, the effectiveness of a system is closely related to the accuracy and the performance of the feature extraction process. We may cite that Kernel Methods can be considered as an efficient alternative in the feature extraction process. However, the key idea behind the Kernel Methods is to build function that gives the inner product between the mapped data points in a higher dimensional space. Afterwards, Support Vector Machine (SVM), Perceptron, PCA or Nearest Neighbor can be directly applied to this feature space in order to predict new data. In our context, we focused on the feature mapping process in String Kernels such as String Subsequence Kernels (SSK), N-Gram String Kernels or p-spectrum Kernel [10]. In fact, SSK is based on the work [16] mostly motivated by bioinformatics applications. It maps strings to a feature vector indexed by all k tuples of characters. The feature mapping φ used in SSK is for a string "s" [11]:

$$\varphi_u(s) = \Sigma_{i:u=s[i]} \lambda^{l(i)}. \tag{1}$$

The p-spectrum kernel maps strings according to the occurrences of common substrings of length p. Even so, the n-grams Kernel maps strings (documents) into high dimensional feature vectors; each entry of the vector represents occurrence or non-occurrence of a contiguous subsequence by a number. Unfortunately, these features mappers are not enough to generate a unique feature vector mapping. Recall that even in the theoretical situation this would not be possible since several n-grams can be built from different reordering of the same token counts.

These n-grams have the same feature vector and therefore, we have no way to discriminate among them. If we take the example shown in figure 3, without taking into account the weighting values assigned to each value of the table, we can deduce that all of the three rows has the same values but at different places:
$(\varphi_1 = (\lambda^2, \lambda^2, 1, 1, 1, 1), \varphi_2 = (1, 1, \lambda^2, \lambda^2, 1, 1), \varphi_3 = (1, 1, 1, 1, \lambda^2, \lambda^2)$). If we apply cosine similarity for example, we conclude that:
cosine (φ_1, φ_2) =cosine (φ_1, φ_3) = cosine(φ_2, φ_3).

	[I-ENTER-ROOM]$	$ [TABLE]	[LEAF]LOC(a)$	$[FALLING]	[DOG]LOC(a)$	$[CAT]LOC(b)
φ ([I-ENTER-ROOM]$[TABLE])	λ^2	$2\lambda^2$	-1	-1	-1	-1
φ ([LEAF]LOC(a)$[FALLING])	-1	-1	$3\lambda^2$	$4\lambda^2$	-1	-1
φ ([DOG]LOC(a)$[CAT]LOC(b))	-1	-1	-1	-1	$5\lambda^2$	$6\lambda^2$

Fig. 3. From N-Gram alignment (according to 2-gram position = 1) to feature vector mapping: a simple demonstration

Obviously, this problem is solved by applying (2) which multiply the λ values by the "p" position of the token in the columns (see definition 1):

ω ([I-ENTER-ROOM]$)$_1$=1, ω ($[TABLE])$_2$=2, etc.... In this manner, $(\varphi_1 = (\lambda^2, 2\lambda^2, 1, 1, 1, 1), \varphi_2 = (1, 1, 3\lambda^2, 4\lambda^2, 1, 1), \varphi_3 = (1, 1, 1, 1, 5\lambda^2, 6\lambda^2))$, consequently cosine $(\varphi_1, \varphi_2)! =$ cosine (φ_1, φ_3) ! = cosine (φ_2, φ_3).

Definition 1(Weighted Sub n-gram Feature Mapping). Let C be a corpus and Σ be a finite word in C. An N-Gram is a finite contiguous sequence of words from Σ, separated by spaces designed by ($) symbol. Let n of n-gram is a positive number that define the count of contiguous words in a sequence separated by $ with n ∈ {2..4}. For n-gram G, we denote by |G| the length of the n-gram $G = g1...g_{|G|}$. The n-gram G[i:j] is the sub n-gram Gi...Gj of G. We say that h is a subn-gram of G, if there exist indices $i=(i1...i_{|h|})$ with $1 \leq i1 \leq ... \leq i_{|h|} \leq |G|$, such that $h_j = G_{ij}$, for $j = 1... |h|$ or h = G[i] for short. The length l(i) of the sub n-gram in G is $i_{|h|}$-i1 + 1. We denote by $\omega(h)_p$ the weight of h according to his position $p \in \{1...m\}$ with m is the count of different word in C that start and end with space symbol ($). We now define the feature spaces mapping φ for an n-gram G is given by:

$$\varphi_h(G) = \sum_{i:h=G[i]} \omega(h)_p * \lambda^{l(i)}. \tag{2}$$

In summary, our feature mapping solution (2) is a weighted version of the feature mapping process used in SSK. Firstly, we generate all the n-grams (n=2...4) in the corpora according to their positions in the phrases as shown in the example in figure 3 (I-ENTER-ROOM$TABLE, DOG$CAT, LEAF$FALLING: are all 2-GRAMs at position 1 in the three phrases). Secondly, for a "p" position n-gram, we split all thengrams according to space symbol separator which is replaced by the "$" symbol as is mentioned in figure 3 (in the table).Finally, we apply (2) to generate the feature mapping vectors. These three steps are applied both for source and target phrases with n=2...4 and positions p=1...4.

3.3   Denoising Data in Machine Translation

We consider that our data are obtained by applying (2) in order to extract the feature mapping of both source and target n-gram. Let δ be a data and let δ_M with M∈ {1...m} be a set of m data where data are the count of n-gram source or target. We denote by k the count of the entire different token obtained by splitting the n-grams according to the space symbol separator which is replaced by the "$" symbol as is mentioned in figure 3 (in the table columns). By applying (3) respectively on the source and the target n-gram, we obtain a set of data points X for source and Y for target $\{x_1...x_m\} \in X$ and $\{y_1...y_m\} \in Y$. In other words, δ is considered as a data point with x and y coordinate that symbolize the alignment between the source and the target n-gram.

$$\delta_M = \sum_{j=1}^{k} \sum_{i:h=G[i]} \omega(h)_p * \lambda^{l(i)}. \tag{3}$$

However, denoising is a technique used to reconstruct patterns eliminating noise from data. In our context, denoising means that we must choose the best translation from the source to the target n-gram in term of the most frequently observed alignment. For this purpose, we introduced the Kernel Density Estimation in order to choose the best n-gram translation (that maximizes the alignment probability) δ (x,y) according to the position in both source and target corpus sentences. Let $(\delta_1 (x_1,y_1), \delta_2 (x_2,y_2)...\delta_n (x_n,y_n))$ be an independent and identically distributed sample drawn from some distribution with an unknown density f. We are interested in estimating the shape of this function f. Its kernel density estimator is:

$$f_h(x) = \frac{1}{nh} \sum_{i=1}^{n} K\left(\frac{x - x_i}{h}\right) \tag{4}$$

Where $K(\bullet)$ is the kernel and $h > 0$ is a smoothing parameter called the bandwidth. In our context, we have used a Gaussian kernel.

3.4   Learning the Regression Function

Now in the machine translation process, given $S = \{(x_i, y_i): x_i \in X, y_i \in Y, i = 1 \dots m\}$, on a set of training n-gram samples$\in$ C (bilingual corpus), i.e. bilingual n-gram pairs $S = \{(x_i, y_i): x_i \in X, y_i \in Y, i = 1\dots m\}$. We denote by $\delta(x) \in \delta_M$with$M \in \{1\dots m\}$ in (3) of source n-gram feature mapping and we denote by $\delta(y)$in a similar way for the target n-gram. Let £: $x \to \delta(x)$ and ¥: $y \to \delta(y)$ are the feature mapping respectively of source$\delta(x)$ and target$\delta(y)$ n-gram.Once the feature mapping is defined, the training problem is restated as a regression problem where a source n-gram to target n-gram mapping must to be found, i.e. finding the translation mapping h:

$$¥(y) = f(x) = h(£(x)) \tag{5}$$

In other words, we try to learn the regression function h that can be a linear or a nonlinear function in the real practice. For this purpose, we used and compared some linear and nonlinear regression methods through kernel ridge regression (6), support vector Regressor with Gaussian kernel (nonlinear kernel) [2], Gaussian process Regressor [2] and K-Nearest neighbors Regressor.

$$\min ||W M_£ - M_¥||_F^2 + v||W||_F^2 \text{with } h \text{ be W for a linear regression} \tag{6}$$

$$W = M_¥(K_£ + vI)^{-1} M_£^T \tag{7}$$

The solution to Eq. (9) is found by differentiating the expression and equaling it to zero to obtain the explicit solution of the ridge regression problem with (8) as the prediction formula:

$$¥(y) = M_¥(K_£ + vI)^{-1} K_£(x) \tag{8}$$

Our aim is to apply the appropriate regression method to the entire n-gram corpus at different positions $p \in [1\dots4]$ in order to find the best estimation of the translation. In fact, table 1 show that the Kernel Ridge Regression has the minimum SSE (Sum of Squared Error) value compared to Gaussian Process Regressor, SVR (Gaussian Kernel) and KNN Regressor. This is due to that linear models can often outperform fancier nonlinear models, especially in situations with small numbers of training data or a low signal-to-noise ratio [15].

Table 1. A part of the SSE (Sum of Squared Error) results of the n-grams regression (2-gram) with 300 corpus phrases

SSE / 2-gram	2-gram			
	P1	**P2**	**P3**	**P4**
SSE-Kernel Ridge Regression	$9.29263e^3$	$1.52454899e^4$	$8.167e^3$	$5.793e^2$
SSE-Gaussian Process Regressor	$1.5353374e^5$	$1.13597150e^5$	$3.7071e^4$	$1.5906e^3$
SSE-SVR (Gaussian Kernel Regressor	$2.21779207e^6$	$1.21887605e^6$	$1.867892e^6$	$2.17080e^5$
SSE-KNN Regressor	$1.25842897e^7$	$2.26991483e^7$	$6.578504e^6$	$3.97199e^5$

3.5   Regression on the Density Estimation of the Bilingual n-Gram Alignment

Our approach is based on the estimation of the most appropriate translation according to the different n-gram positions in the corpus sentences. In fact, we rely on the Kernel Density Estimation process applied to bilingual n-gram at different positions for the purpose to find the most commonly observed translation in term of probability. However, to choose the best n-gram translation, we used (9):

$$argmax_{n=[2...4]}\left\{\prod_{p=1}^{m} \theta\left(s_{n_p}, t_{n_p}\right)\right\} \tag{9}$$

Picking $\theta(s_{n_p}, t_{n_p})$ to be the density estimation probability of the source n-gram s_{n_p} translated to the n-gram target t_{n_p} and let p be the set of n-gram positions in sentences with $p \in [1...4]$. Unfortunately, we cannot compute the probability of the bilingual n-gram translation with unknown input n-grams. For this reason, we used Regression method through Kernel Ridge Regression in order to predict the unknown translation probability.

3.6   The Pre-image Problem

The pre-image problem consists of determining the predicted output: given $z \in F_Y$, the problem is to find $y \in Y^*$ such that $\varphi_y(y) = z$, see figure 2. In fact, based on the work of [4] in pre-image resolution, we used the same idea which relies on the De Bruijn graph, except that we are based on words instead of letters. We associate a vertex to each one of the entire different token obtained by splitting the n-grams to different words that could start or end with "$" symbol as is mentioned in figure 3 (in the table columns). The edges between all vertexes are weighted using (2). Each sentence of the target corpora has a De Bruijn graph where the sum of all his edges weight is represented by (3). Note that there is an overlap between resembling De Bruijn graphs which makes possible to generate new links. For a new regression value, we employ a graph search method in order to find the best eulerian path (the nearest value) in term of the sum of edges weight (3).

3.7   Our Algorithm

Now we will describe each step of the algorithm in details:

1- **Words Decomposition**: All of the aligned bilingual phrases are decomposed in order to obtain two set of words. The generated words contain a "$" symbol that can be placed before or after the words according to their original phrases location for the purpose to represent the "space" symbol location.

2- **Feature Mapping**: Let $p \in [1...4]$ be the positions of the n-grams used for both source and target phrases. For each p in $[1...4]$, we generate all the n-gram with n=2 to 4 from the aligned bilingual corpora for both source and target phrases. We can now apply (2) formula in order to generate the feature mapping to all of the bilingual n-grams.

3- **Data Denoising**: In our context, denoising means that we must choose the best translation from the source to the target n-gram in term of the most frequently observed alignment. For this purpose, we apply (3) for each of the feature mapping of the source and the target n-grams in order to obtain a set of data points X for source and Y for target. For data denoising, we apply (4) then we choose the most frequently observed alignment.

4- **Choosing the learning method and learning the regression function**: To learn the mapping function, we apply and we compute the SSE of the Kernel Ridge Regression (6) (7), the Gaussian Process Regressor [2], the SVR (Gaussian Kernel) [2] and the KKN Regressor [2]. Afterwards, we choose automatically the method that has the best approximation results (minimum of SSE).For unknown input n-grams we used the same regression method (to learn the Density Estimation function of the aligned bilingual n-grams) in order to predict the unknown n-gram translation probability.

5- **Solving the pre-image problem**: Now for a new input phrase, we apply step 2 with (3) formula in order to predict the regression value (using (8) in the case of Kernel Ridge Regression) which is used in De Bruijn graphs to find the best eulerian path (the nearest value).

6- **Choosing the best translation**: This step aims to choose the best translation from the generated 2-grams, 3-grams and 4-grams in term of maximizing the translation probability using (9).

4   Experiment Evaluation

We have carried out experiments on 300 ASL bilingual phrases corpus. The proposed system was trained on these phrases and tested on 100 different ASL phrases. We applied and compared bleu scores of the translation results with Kernel Ridge Regression, Gaussian Process Regressor, SVR (Gaussian kernel) and K-NN Regressor. As shown in table 2, using Kernel Ridge Regression we obtained encouraging result as 87% score using the Bleu metric [5].

Table 2. Results from 300 training bilingual ASL phrases with 100 different testing phrases

	Kernel Ridge Regression	GP Regressor	SVR	K-NN Regressor
BLEU$\approx$	87%	68%	61%	49%

5   Conclusion and Future Works

The research described in this paper has laid a methodological foundation for future research in Sign Language Machine Translation based on kernel regression and kernel density estimation methods applied to ASL n-grams (with signing space informations). Our aim is to provide an efficient approximation function (in term of the minimum of SSE comparing Gaussian Process Regressor, K-Nearest Neighbors Regressor, Support Vector Regressor with Gaussian Kernel and Ridge Regression) that learns the mapping between the source and the target n-grams.

For feature mapping, we proposed a new mapping process (Weighted Sub n-gram Feature Mapping) which is based on the String Subsequence Kernel SSK feature mapping. Even so, to find solution to the pre-image problem, we relied on the De-Bruijn Multi Graph search applied on n-grams target. We are based on the search of the translation that maximizes the translation probability score of n-grams translation. For unknown n-grams, we used ridge regression for the purpose to predict the probability through learning the Density Estimation function of the aligned bilingual n-grams. We report encouraging experimental results on a small-scale reduced-domain corpus with 87% BLEU score. The ultimate goal of our future research is to build a big American Sign Language Corpora (with signing space informations) in order to improve the translation quality in term of blue score. We also plan to enlarge the phrases length for the purpose to treat longer phrases (to reach 8 n-gram position and 5-grams).

References

1. Biçici, E., Yuret, D.: L1 regularization for learning word alignments in sparse feature matrices. In: Proceedings of the Computer Science Student Workshop (2010)
2. Bishop, C.M.: Pattern Recognition and Machine Learning. Springer (2006). ISBN: 978-0-387-31073-2
3. Charles, A., Rebecca, S.: Reading optimally builds on spoken language implication for deaf readers. Learning research and development center University of Pittsburgh (2000)
4. Cortes, C., Mehryar, M., Jason, W.: A general regression framework for learning string-to-string mappings. In: Bakir, G.H., Hofmann, T., Sch, B. (eds.) Predicting Structured Data, pp. 143–168. The MIT Press (September 2007)
5. Finch, A., Hwang, Y.-S., Sumita, E.: Using machine translation evaluation techniques to determine sentence-level semantic equivalence. In: IWP 2005 (2005)
6. Huenerfauth, M., Lu, P.: Effect of spatial reference and verb inflection on usability of sign language animations. Springer-Verlag Univ. Access Inf. Soc. (2011). doi 10.1007/s10209-011-0247-7

7. Hung-Yu, S., Chung-Hsien, W.: Improving structural statistical machine translation for sign language with small corpus using thematic role templates as translation memory. IEEE Transactions on Audio Speech, and Language Processing **17**(7), 1305–1315 (2009)

8. Koehn, P., Hoang, H.: Factored translation models. In: Proc. of EMNLP-CoNLL 2007 (2007)

9. Koehn, P., Och, F.J., Marcu, D.: Statistical phrase-based translation. In: Proc. of HAACL-HLT 2003, pp. 48–54 (2003)

10. Leslie, C., Eskin, E., Stafford, W.: The spectrum kernel: a string kernel forsvm protein classification. In: Pacific Symposium on Biocomputing, pp. 566–575 (2002)

11. Lodhi, H., Saunders, C., Shawe-Taylor, J., Nello, C., Watkins, C.: Text Classification using String Kernels. Journal of Machine Learning Research **2**, 419–444 (2002)

12. Serrano, N., Andres-Ferrer, J., Casacuberta, F.: On a kernel regression approach to machine translation. In: Iberian Conference on Pattern Recognition and Image Analysis, pp. 394–401 (2009)

13. Scott, D.W.: Multivariate Density Estimation: Theory, Practice, and Visualization. John Wiley & Sons, New York (1992)

14. Stein, D., Schmidt, C., Hermann, N.: Analysis, preparation, and optimization of statistical sign language machine translation. Machine Translation **26**(4), 325–357 (2012)

15. Trevor, H., Tibshirani, R., Friedman, J.: The Elements of Statistical Learning: Data Mining, Inference and Prediction, 2nd edn. Springer (2009)

16. Watkins, C.: Dynamic alignment kernels. Advances in Large Margin Classifiers, pp. 39–50 (2000)

17. Zhuoran, W., Shawe-Taylor, J., Sandor, S.: Kernel regression based machine translation. In Human Language Technologies. In: The Conference of the North American Chapter of the Association for Computational Linguistics, pp. 185–188 (2007)

18. Zhuoran, W., Shawe-Taylor, J.: Kernel regression framework for machine translation: UCL system description for WMT 2008 shared translation task. In: Proceedings of the Third Workshop on Statistical Machine Translation, pp. 155–158 (2008)

Adaptive Intelligent Control for Continuous Stirred Tank Reactor with Output Constraint

Dong-Juan Li[1(✉)] and Yan-Jun Liu[2]

[1] School of Chemical and Environmental Engineering, Liaoning University of Technology,
Jinzhou 121001, Liaoning, China
ldjuan@126.com
[2] College of Science, Liaoning University of Technology, Jinzhou 121001, Liaoning, China
liuyanjun@live.com

Abstract. For a class of continuous stirred tank reactor with the output constraint and the uncertainties, an adaptive control approach is proposed based on the approximation property of the neural networks. The considered systems can be viewed as a class of pure-feedback systems. It is proven that all the signals in the closed-loop system are bounded and the system output is not violated by using Lyapunov stability analysis method. A simulation example is given to verify the effectiveness of the proposed approach.

Keywords: Continuous stirred tank reactor · Adaptive control · The neural networks · Barrier Lyapunov function

1   Introduction

Recently, the adaptive control of uncertain nonlinear systems has attracted much attention. Based on the approximation of the fuzzy logic systems and the neural networks, some significant works were obtained for nonlinear systems with completely unknonwn functions. In [1], chen et al desiged an adaptive output feedback neural control for nonlinaer SISO systems with time-delay. Based on the small gain theory, an adaptive fuzzy output feedback control was given in [2] for a class of nonlinear SISO systems with unmodelled dynamics. In [3], a robust adaptive neural network control algorithm was developed for uncertain nonlinear systems with unknown gain function and control direction.

At present, the adaptive control for continuous stirred tank reactor (CSTR) has obtained many interests. An adaptive tracking control method was studied in [4] to be applied in CSTR. In [5], H. G. Zhang et al provided an adaptive fuzzy sliding control to a general class of nonlinear systems and this method is used to control CSTR to verify the effectiveness. In [6, 7], two adaptive fuzzy control approaches were proposed for CSTR and experiment results were given for validating the effectiveness. For SISO and MIMO CSTR with dead-zone, two adaptive neural network control approaches were presented in [8,9].

© Springer International Publishing Switzerland 2014
Z. Zeng et al. (Eds.): ISNN 2014, LNCS 8866, pp. 385–392, 2014.
DOI: 10.1007/978-3-319-12436-0_43

A common restriction in the above approaches is that output constraint problem is not considered. To this end, two adaptive control were studied in [10] for nonlinear systems with output constraint by using Barrier Lyapunov function and the output constraint is not violated. Subsequently, the full state constraint problem was solved in [11] for a class of nonlinear systems. In [12], an adaptive neural output feedback control algorithm was proposed for a class of nonlinear systems with output constraint and unknown function. The results in [10-12] can be only to control a class of strict-feedback systems. It is known that the pure-feedback systems are more complex than strict-feedback systems.

In this paper, an adaptive neural network control problem is solved for a class of CSTR with uncertainties. The neural networks are used to approximate uncertain function of systems. The considered reactor can be viewed as a class of nonlinear pure-feedback systems with output constraint. To control this class of systems, the novel Barrier Lyapunov function is chosen and the mean-value theorem is correctly used to decompose the systems. Based on decomposed systems, a stable controller is designed. Using Lyapunov stability, it is proven that all the signals in the closed-loop system are bounded and the system output is not violated. A simulation example is illustrated to validate the feasibility of the approach.

2   Problem Description

Consider the following continuous stirred tank reactors:

$$
\begin{cases}
V\dfrac{dC_A}{dt'} = \left(C_{AF} - C_A\right)F - VK_0\exp\left(E_a / RT\right)C_A \\
V\rho C_P\dfrac{dT}{dt'} = F\rho C_P\left(T_F - T\right) - hA\left(T - T_c\right) \\
\qquad\qquad - \left(\Delta H\right)VK_0 \times \exp\left(-\dfrac{E_a}{RT}\right)C_A
\end{cases}
\tag{1}
$$

where the meanings of some notations in system (1) were defined in [8].

Define the variables as follows in [8]. The equation (1) can be expressed as

$$
\begin{cases}
\dot{x}_1 = -x_1 + D_a\left(1 - x_1\right)\exp\left[x_2 / \left(1 + \left(x_2/\varphi\right)\right)\right] \\
\dot{x}_2 = -x_2\left(1 + \delta\right) + BD_a\left(1 - x_1\right) \\
\qquad \times \exp\left[x_2 / \left(1 + \left(x_2/\varphi\right)\right)\right] + \delta u(t) \\
y = x_1
\end{cases}
\tag{2}
$$

where x_1 and x_2 denotes the dimensionless reactant concentration and mixture temperature, respectively; u is the dimensionless coolant flow rate; control target of process is to use the coolant flow rate u control reactant temperature x_2; $y = x_1$ is the output of system be constrained in $|x_1| \le k_{c_1}$ with k_{c_1} being a constant.

Define

$$f_1(x_1, x_2) = -x_1 + D_a(1 - x_1)\exp\left[x_2 / (1 + x_2/\varphi)\right]$$
$$f_2(x_1, x_2) = -x_2(1 + \delta) + BD_a(1 - x_1)\exp\left[x_2 / (1 + x_2/\varphi)\right]$$

So, we rewrite (2) as

$$\begin{cases} \dot{x}_1 = f_1(x_1, x_2) \\ \dot{x}_2 = f_2(x_1, x_2) + \delta u(t) \\ y = x_1 \end{cases} \tag{3}$$

In this paper, the our goal is to construct an adaptive NN scheme such that $y = x_1$ follows the reference signal $y_d(t)$ to a small set and all the signals in the closed-loop are retained to be bounded and output constraint is not violated.

For the system (3), as follow assumptions and lemma are given.

Assumption 1: There exist constants $\bar{a} > \underline{a} > 0$ satisfying $\underline{a} \le \partial f_1(x_1, x_2)/\partial x_2 \le \bar{a}$.

Assumption 2: There exist constants $\bar{\delta} > \underline{\delta} > 0$ satisfying $\underline{\delta} \le \delta \le \bar{\delta}$.

Assumption 3: For $k_{c_1} > 0$, there exists constants $0 < A_0 \le k_{c_1}$ and A_1 satisfying $|y_d(t)| \le A_0$ and $|\dot{y}_d(t)| \le A_1$.

Lemma 1 [4]: Suppose $\forall (x, u) \in R^n \times R$, $f(x, u)$ is continuously differentiable and exists positive definite constant d satisfying $\partial f(x, u)/\partial u > d > 0$. So, exists a continuously function $u^* = u(x)$ satisfying $f(x, u^*) = 0$.

Lemma 2 [12] : For any positive constant k_{b_1} , for all of e_1 ,exists $|e_1| < k_{b_1}$ satisfy-

ing $\log \dfrac{k_{b_1}^2}{k_{b_1}^2 - e_1^2} < \dfrac{e_1^2}{k_{b_1}^2 - e_1^2}$.

Because unknown functions are contained in the system (3), in the design process of controller, in which the unknown term can not be utilized directly. Due to the inherent approximation ability of neural network, it has been successfully applied to the modeling and control problem of uncertain nonlinear systems. The approximation performance of neural network can be seen in [4].

3   Controller Design and Stability Analysis

An adaptive neural network controller is designed to stable the system in the subsection A. The detailed design procedure is given in the following.

Step 1: Define the tracking error $e_1 = x_1 - y_d(t)$ and $\dot{e}_1$ is

$$\dot{e}_1 = f_1(x_1, x_2) - \dot{y}_d(t) \tag{4}$$

By Assumption 1, we have $\partial\{f_1(x_1, x_2) + \mu_1\}/\partial x_2 \geq \underline{a} > 0$, which $\mu_1 = -\dot{y}_d(t)$. According to lemma 1, we get

$$f_1(x_1, \alpha_1^*) + \mu_1 = 0 \tag{5}$$

Using the mean value theorem, there is $\lambda (0 < \lambda < 1)$ such that

$$f_1(x_1, x_2) = f_1(x_1, \alpha_1^*) + g_{1\lambda}(x_2 - \alpha_1^*) \tag{6}$$

where $g_{1\lambda} := f_2(x_1, x_{2\lambda})$, $x_{2\lambda} = \lambda x_2 + (1 - \lambda)\alpha_1^*$. According to Assumption 1, we know $\underline{a} \leq h_{1\lambda} \leq \bar{a}$. Using (5) and (6), (4) can be rewritten as

$$\dot{e}_1 = g_{1\lambda}(x_2 - \alpha_1^*) \tag{7}$$

By using the neural networks, α_1^* can be approximated as

$$\alpha_1^* = \theta_1^{*T} \xi_1(Z_1) + d_1 \tag{8}$$

where $Z_1 = [x_1, y_d, \dot{y}_d]^T \in \Omega_1 \subset R^3$, θ_1^* denotes the ideal constant weight and $|d_1| \leq d_1^*$ is the approximation error with constant.

Let $\hat{\theta}_1$ be the estimate of θ_1^* and let $\tilde{\theta}_1 = \hat{\theta}_1 - \theta_1^*$. Defined $e_2 = x_2 - \alpha_1$ and α_1 is the virtual control input to be defined as

$$\alpha_1 = -c_1 e_1 - k_1 e_1 \left(k_{b_1}^2 - e_1^2\right) + \hat{\theta}_1^T \xi_1(Z_1) \tag{9}$$

where $k_1 > 0$ is the design parameter and let $k_{b_1} = k_{c_1} - A_0$. Using (9), (7) becomes

$$\dot{e}_1 = g_{1\lambda}\left[e_2 + \tilde{\theta}_1^T \xi_1(Z_1) - d_1 - k_1 e_1\left(k_{b_1}^2 - e_1^2\right) - c_1 e_1\right] \tag{10}$$

Consider the Lyapunov function candidate as follows

$$V_1 = \frac{1}{2g_{1\lambda}} \log \frac{k_{b_1}^2}{k_{b_1}^2 - e_1^2} + \frac{1}{2} \tilde{\theta}^T \Gamma^{-1} \tilde{\theta} \tag{11}$$

where $\Gamma_1 = \Gamma_1^T > 0$ is a design parameter. Its time derivative is

$$\dot{V}_1 = \frac{e_1 \dot{e}_1}{\left(k_{b_1}^2 - e_1^2\right) g_{1\lambda}} - \frac{\dot{g}_{1\lambda}}{2g_{1\lambda}^2} \log \frac{k_{b_1}^2}{k_{b_1}^2 - e_1^2} + \tilde{\theta}^T \Gamma^{-1} \dot{\tilde{\theta}} \tag{12}$$

Thus, we have

$$-\frac{\dot{g}_{1\lambda}}{2g_{1\lambda}^2} \log \frac{k_{b_1}^2}{k_{b_1}^2 - e_1^2} \leq \frac{|\dot{g}_{1\lambda}|}{2g_{1\lambda}^2} \log \frac{k_{b_1}^2}{k_{b_1}^2 - e_1^2} \leq \frac{a_d}{a^2} \log \frac{k_{b_1}^2}{k_{b_1}^2 - e_1^2} \leq \frac{a_d}{a^2} \frac{e_1^2}{k_{b_1}^2 - e_1^2} \tag{13}$$

Basis on (10) and (13), (12) can be repressed as

$$\dot{V}_1 \leq \frac{e_1 e_2 + \tilde{\theta}_1^T \xi_1(Z_1) e_1 - d_1 e_1 - k_1 e_1^2 \left(k_{b_1}^2 - e_1^2\right)}{k_{b_1}^2 - e_1^2} - \frac{c_1 e_1^2}{k_{b_1}^2 - e_1^2} + \frac{a_d}{a^2} \frac{e_1^2}{k_{b_1}^2 - e_1^2} + \tilde{\theta}^T \Gamma^{-1} \dot{\tilde{\theta}} \tag{14}$$

Design the adaptation laws as follows

$$\dot{\hat{\theta}}_1 = \Gamma_1 \left[\frac{-\xi_1(Z_1) e_1}{k_{b_1}^2 - e_1^2} - \varsigma_1 \hat{\theta}_1 \right] \tag{15}$$

where $\varsigma_1 > 0$ is the design constant. Substituting (15) into (14) leads to

$$\dot{V}_1 \leq \frac{e_1 e_2 - d_1 e_1}{k_{b_1}^2 - e_1^2} - k_1 e_1^2 - c_1^* \frac{e_1^2}{k_{b_1}^2 - e_1^2} - \varsigma_1 \tilde{\theta}_1^T \hat{\theta}_1 \tag{16}$$

where c_1 is selected to satisfy $c_1^* = c_1 - a_d / 2\bar{a}^2 > 0$. Using Young's inequality, we obtain

$$-\frac{d_1 e_1}{k_{b_1}^2 - e_1^2} \leq \frac{1}{2} d_1^2 + \frac{1}{2} \frac{e_1^2}{\left(k_{b_1}^2 - e_1^2\right)^2} \tag{17}$$

$$-\varsigma_1 \tilde{\theta}_1^T \hat{\theta}_1 = -\varsigma_1 \tilde{\theta}_1^T \left(\tilde{\theta}_1 + \theta_1^*\right) \leq -\frac{\varsigma_1 \left\| \tilde{\theta}_1 \right\|^2}{2} + \frac{\varsigma_1 \left\| \theta_1^* \right\|^2}{2} \tag{18}$$

Substituting (17) and (18) into (16) leads to

$$\dot{V}_1 \le \frac{e_1 e_2}{k_{b_1}^2 - e_1^2} - \frac{c_1^* e_1^2}{k_{b_1}^2 - e_1^2} - \frac{\varsigma_1 \left\| \tilde{\theta}_1 \right\|^2}{2} + \frac{\varsigma_1 \left\| \theta_1^* \right\|^2}{2} + \frac{1}{2} d_1^2 \tag{19}$$

Step 2: Define the variable $e_2 = x_2 - \alpha_1$ and $\dot{e}_2$ is

$$\dot{e}_2 = f_2(x_1, x_2) + \delta u - \dot{\alpha}_1 \tag{20}$$

where $\dot{\alpha}_1 = \frac{\partial \alpha_1}{\partial x_1} f_1(x_1, x_2) + \frac{\partial \alpha_1}{\partial y_d} \dot{y}_d + \frac{\partial \alpha_1}{\partial \hat{\theta}_1} \dot{\hat{\theta}}_1$.

The unknown function is approximated by RBF neural network

$$\left[-f_2(x_1, x_2) + \dot{\alpha}_1 \right] / \delta = \theta_2^{*T} \xi_2(Z_2) + d_2 \tag{21}$$

where $Z_2 = \left[x_1, x_2, y_d, \dot{y}_d, \hat{\theta}_1 \right]^T$; θ_2^* indicate the ideal constant weight, $|d_2| \le d_2^*$ is approximation error and $d_2^* > 0$. Let $\hat{\theta}_2$ be the estimate of θ_2^* and build the actual controller as follows

$$u = -c_2 e_2 - \frac{e_1}{k_{b_1}^2 - e_1^2} + \hat{\theta}_2^T \xi_2(Z_2) \tag{22}$$

Then, Substituting (22) into (20), function $\dot{e}_2$ be written as follow

$$\dot{e}_2 = \delta \left[-c_2 e_2 - \frac{e_1}{k_{b_1}^2 - e_1^2} + \tilde{\theta}_2^T \xi_2(Z_2) - d_2 \right] \tag{23}$$

Consider the Lyapunov function candidate as follows

$$V_2 = V_1 + \frac{1}{2\delta} e_2^2 + \frac{1}{2} \tilde{\theta}_2^T \Gamma_2^{-1} \tilde{\theta}_2 \tag{24}$$

Based on (23), its time derivative is

$$\dot{V}_2 = \dot{V}_1 + \tilde{\theta}_2^T \Gamma_2^{-1} \dot{\hat{\theta}}_2 + e_2 \left[-c_2 e_2 - \frac{e_1}{k_{b_1}^2 - e_1^2} + \tilde{\theta}_2^T \xi_2(Z_2) - d_2 \right] \tag{25}$$

Design the adaptation laws as follows

$$\dot{\hat{\theta}}_2 = \Gamma_2 \left[-\xi_2(Z_2) e_2 - \varsigma_2 \hat{\theta}_2 \right] \tag{26}$$

where $\varsigma_2 > 0$ is the design constant. Then, Substituting (26) into (25), we get

$$\dot{V}_2 = \dot{V}_1 + e_2 \left[-c_2 e_2 - \frac{e_1}{k_{b_1}^2 - e_1^2} - d_2 \right] - \varsigma_2 \tilde{\theta}_2^T \hat{\theta}_2 \tag{27}$$

Utilizing Young's inequality, we obtain

$$-d_2 e_2 \le \frac{1}{2} d_2^2 + \frac{1}{2} e_2^2 \tag{28}$$

$$-\varsigma_2 \tilde{\theta}_2^T \hat{\theta}_2 = -\varsigma_2 \tilde{\theta}_2^T \left(\tilde{\theta}_2 + \theta_2^* \right) \le -\frac{\varsigma_2 \left\| \tilde{\theta}_2 \right\|^2}{2} + \frac{\varsigma_2 \left\| \theta_2^* \right\|^2}{2} \tag{29}$$

Then, substituting (19), (28) and (29) into (27), we get

$$\dot{V}_2 \le -k_1 e_1^2 + \frac{1}{2} d_1^2 + \frac{1}{2} d_2^2 - c_1^* \frac{e_1^2}{k_{b_1}^2 - e_1^2} - c_2^* e_2^2 - \frac{\varsigma_1 \left\| \tilde{\theta}_1 \right\|^2}{2} + \frac{\varsigma_1 \left\| \theta_1^* \right\|^2}{2} - \frac{\varsigma_2 \left\| \tilde{\theta}_2 \right\|^2}{2} + \frac{\varsigma_2 \left\| \theta_2^* \right\|^2}{2}$$

$$\le -c_1^* \log \frac{e_1^2}{k_{b_1}^2 - e_1^2} - c_2^* e_2^2 - \frac{\varsigma_1 \left\| \tilde{\theta}_1 \right\|^2}{2} - \frac{\varsigma_2 \left\| \tilde{\theta}_2 \right\|^2}{2} + \frac{1}{2} d_1^2 + \frac{1}{2} d_2^2 + \frac{\varsigma_1 \left\| \theta_1^* \right\|^2}{2} + \frac{\varsigma_2 \left\| \theta_2^* \right\|^2}{2} \tag{30}$$

where c_2 is chosen to satisfy $c_2^* = c_2 - 1/2 > 0$.

Let $p = \min\{ 2c_1^* \underline{a}, 2c_2^* \delta, \varsigma_1 \lambda_{\min}(\Gamma_1), \varsigma_2 \lambda_{\min}(\Gamma_2) \}$ and $q = \frac{1}{2} d_1^2 + \frac{1}{2} d_2^2 + \frac{\varsigma_1 \left\| \theta_1^* \right\|^2}{2}$

$+ \frac{\varsigma_2 \left\| \theta_2^* \right\|^2}{2}$, (30) can be expressed as follows

$$\dot{V}_2 \le -p V_2 + q \tag{31}$$

The stability of the closed-loop system is pointed out by the following theorem.

Theorem 1: Consider the system (1) and (2), Under the condition of hypothesis 1-3, the control scheme can guarantee that all the signals are bounded and the system outputs converge to a small neighbourhood of zero.

Proof: The process of proof can be found in [9-10].

4 Conclusion

In this article, an adaptive NN controller has been proposed for a general class of continuous stirred tank reactors. Consider the tank reactor is a kind of pure feedback nonlinear systems and with output constraints. Pure feedback system is decomposed by using the mean value theorem. Based on backstepping technique, the adaptive controller and adaptive law have been designed. We have shown that tracking error converges to zero within the neighborhood, all closed loop signals remain bounded and output-constraints are not violated.

Acknowledgments. The work was supported by the National Natural Science Funds of China under grant 61104017, 61473139; the Foundation of Educational Department of Liaoning Province L2013243; the Program for Liaoning Excellent Talents in University under Grant LR2014016.

References

1. Chen, W.S., Li, J.M.: Adaptive Neural Tracking Control for Unknown Output Feedback Nonlinear Time-delay Systems. Acta Automatica Sinica **31**(5), 799–803 (2005)
2. Tong, S.C., He, X.L., Zhang, H.G.: A Combined Backstepping and Small-Gain Approach to Robust Adaptive Fuzzy Output Feedback Control. IEEE Transactions on Fuzzy Systems **17**, 1059–1069 (2009)
3. Foster, I., Kesselman, C.: The Grid: Blueprint for a New Computing Infrastructure. Morgan Kaufmann, San Francisco (1999)
4. Ge, S.S., Hang, C., Zhang, C.T.: Nonlinear adaptive control using neural networks and its application to CSTR systems. Journal of Process Control **9**, 313–323 (1999)
5. Zhang, H.G., Cai, L.L.: Nonlinear adaptive control using the Fourier integral and its application to CSTR systems. IEEE Transactions on Systems, Man, and Cybernetics, Part B: Cybernetics **32**, 367–372 (2002)
6. Salehi, S., Shahrokhi, M.: Adaptive fuzzy approach for H^∞ temperature tracking control of continuous stirred tank reactors. Control Engineering Practice **16**, 1101–1108 (2008)
7. Salehi, S., Shahrokhi, M.: Adaptive fuzzy backstepping approach for temperature control of continuous stirred tank reactors. Fuzzy Sets and Systems **160**, 1804–1818 (2009)
8. Li, D.J.: Adaptive Neural Network Control for a Class of Continuous Stirred Tank Reactor Systems (2013). doi: 10.1007/s11432-013-4824-7
9. Li, D.J.: Neural network control for a class of continuous stirred tank reactor process with dead-zone input. Neurocomputing **131**, 453–459 (2014)
10. Tee, K.P., Ge, S.S., Tay, E.H.: Barrier Lyapunov functions for the control of the output-constrained nonlinear systems. Automatica **45**(4), 918–927 (2009)
11. Tee, K.P., Ren, B.B., Ge, S.S.: Control of nonlinear systems with time-varying output constraints. Automatica **47**(11), 2511–2516 (2011)
12. Ren, B.B., Ge, S.S., Tee, K.P., Lee, T.H.: Adaptive neural control for output feedback nonlinear systems using a barrier Lyapunov function. IEEE Transactions on Neural Networks **21**(8), 1339–1345 (2010)

Applications

Modeling and Application of Principal Component Analysis in Industrial Boiler

Wenbiao Wang[(✉)], Lan Chen, Xinjie Han, Zhanyuan Ge, and Siyuan Wang

IT College of Dalian Maritime University, Dalian, China
{wwb201,chenlan198898}@163.com, dl_wsy@sina.com

Abstract. An identification model based on principal component analysis which can reflect thermal efficiency is proposed, in order to improve the operation efficiency of boiler. It can monitor the thermal efficiency online and estimate the key influential parameters. The monotonic relationship between thermal efficiency and SPE statistic is verified by large numbers of historical data. When the boiler's operation efficiency decreases, the influential parameters can be directly got by contribution plot method, which guide operators in real-time to adjust these and maintain boiler efficient operation. The practice shows that this method is feasible.

Keywords: Industrial boiler · Principal component analysis · Operation optimization · Parameters adjusted

1 Introduction

Huge number and low operating efficiency are the present situation of China's industrial boiler system. There are many parameters, some of which are automatically controlled by appropriate algorithm, but economical operation index is often ignored [1,3]. The control of relatively complex combustion systems is usually depended on artificial experience in the actual production. This may lead to unstable operation of boilers resulting in lower thermal efficiency, and a lot of energy is wasted. With wide application of computer technology, large amounts of process data are collected. However, these data which contain information of process conditions are not well exploited. Finding useful information in such a large amounts of data has been noticed in industrial community. Especially with the rapid development of data mining and data-driven technology, the combination of data analysis theory and industrial applications is constantly promoted [4-8].

Thermal efficiency is an economical operation index of boilers. Many factors influencing it exist, such as operational parameters, equipment status, climate conditions, and process is complex [3]. In the practical applications, mature control methods are mostly designed based on feedback regulation. The best operational efficiency of boiler is difficult to be determined as the changes of external factors, so designing a suitable controller for boiler efficiency is impossible. The application of data analysis method is more feasible which can find a better mode from the mass of historical data guiding manual operation. The multivariate statistical methods such as principal component

© Springer International Publishing Switzerland 2014
Z. Zeng et al. (Eds.): ISNN 2014, LNCS 8866, pp. 395–404, 2014.
DOI: 10.1007/978-3-319-12436-0_44

analysis(PCA), partial least square(PLS), canonical correlation analysis(CCA) and so on are the most commonly used.

The factors that influence thermal efficiency usually have a certain correlation. Using these variables directly not only makes a higher computational complexity but also makes a non-optimal result. PCA is a typical statistical analysis theory [9-13]. It reduces the number of variables monitored, eliminates their correlation, and simplifies the complexity of characteristic analysis of original process. Another advantage is that completely free parameter limits, and final result is only influenced by data. Therefore, PCA is used in this paper. An optimization method based on PCA for the operation of boiler is proposed. PCA model is established by efficient samples of historical data. The relationship between thermal efficiency and squared prediction error (SPE, also called Q) statistic is verified by large numbers of historical data. The running status of boiler can be on-line monitored and when boiler's operation efficiency decreases, the key influential parameters can be got by contribution plot method. It guides operators to make reasonable adjustments and the level of boiler operation is improved.

2   PCA Theory and Modeling

2.1   PCA Algorithm

PCA is a method of data dimensionality reduction. It reduces the dimensionality under the premise of keeping process information as much as possible. Uncorrelated characteristic signals' data matrix is obtained.

Suppose A is an $n \times m$ data matrix. It is normalized to develop matrix X according to formula (1).

$$x_{ij} = \frac{a_{ij} - b_j}{\sigma_j} \left(i = 1, 2, \ldots, n, j = 1, 2, \ldots, m \right) \qquad (1)$$

In the formula, a_{ij} and x_{ij} represent each element of matrix X and A, respectively. b_j and σ_j represent mean and standard deviation of column, respectively. Covariance matrix S of matrix A can be obtained according to formula (2).

$$S = \text{cov}(x) \approx \frac{1}{n-1} X^T X \qquad (2)$$

The principal component analysis of matrix X is actually equivalent to the vector analysis of covariance matrix S. If the eigenvalues of S are arranged as follows $\lambda_1 \geq \lambda_2 \geq \cdots \geq \lambda_m$, eigenvectors $p_1, p_2, \cdots, p_m$ corresponding to these eigenvalues are the load vectors of X. Doing eigenvalue decomposition of S, and arranging them by descending order, X can be written as formula (3).

$$X = t_1 p_1^T + t_2 p_2^T + \cdots + t_k p_k^T + E = TP^T + E = \hat{X} + E(k < m) \qquad (3)$$

Among it, $P_{m \times k}$ composes to the former k eigenvectors of S. Each column of $T_{n \times k}$ is called principal component. k is the number of principal component. E is residual matrix. X' changes are reflected in the former several load vectors' directions. X' projection on the last several load vectors will be very small which can be ignored.

PCA divides original variable space into two subspaces--principal component subspace(PCS) and residual subspace(RS). Any sample can be decomposed into projections on PCS and RS. The formula is shown in (4).

$$\begin{cases} X = \hat{X} + \tilde{X} \\ \hat{X} = PP^T X \in R_p \equiv span\{P\} \\ \tilde{X} = (I - PP^T)X \in R_y \equiv span\{R\} \end{cases} \tag{4}$$

Among it, $\hat{X}$ is projection on PCS, $\tilde{X}$ is projection on RS. They are orthogonal and statistically independent.So PCA has natural advantage applied to process monitoring.

Usually, SPE and T^2 statistics are used to monitor samples [9]. If both of them are lower than their control limits, it shows that the status described by monitor samples is the same status described by PCA model, and vice versa.

SPE statistic and its control limit SPE_α are shown in formulas (5) and (6), respectively.

$$SPE = \left\| (I - PP^T)x \right\|^2 \tag{5}$$

$$SPE_\alpha = \theta_1 \left[\frac{C_\alpha \sqrt{2\theta_2 h_0^2}}{\theta_1} + 1 + \frac{\theta_2 h_0 (h_0 - 1)}{\theta_1^2} \right]^{1/h_0} \tag{6}$$

There, $\theta_i = \sum_{j=k+1}^{m} \lambda_j^i (i = 1,2,3)$, $h_0 = 1 - \frac{2\theta_1\theta_3}{3\theta_2}$, C_α is the threshold when standard normal distribution is at confidence level α. Principal component number of PCA model is k, and λ_j is the eigenvalue of covariance matrix S.

T^2 statistic and its control limit T^2_α are shown in formulas (7) and (8), respectively.

$$T^2 = x^T P\Lambda^{-1}P^T x \tag{7}$$

$$T^2_\alpha = \frac{k(n^2 - 1)}{n(n - 1)} F_{k,n-k;\alpha} \tag{8}$$

Among it, $\Lambda = diag\{\lambda_1, \cdots \lambda_k\}$. $F_{k,n-k;\alpha}$ is the distribution critical value of F, when the freedom is k and n-k, the significant level is α.

2.2   Determination of Principal Component Number

In PCA model, cumulative percent variance(CPV)is commonly used to determine the number of principal component according to the percentage of cumulative sum of principal component variances. Principal component variances of data matrix are equivalent to the eigenvalues of covariance matrix, so cumulative contribution rate of the former k principal component means the sum of former k eigenvalues dividing by the total of all eigenvalues of covariance matrix. It represents the proportion of data changes explained by the former k principal component to all changes. Cumulative contribution rate of the former k principal component is expressed as formula(9).

$$CPV = \frac{\sum_{i=1}^{k} \lambda_i}{\sum_{i=1}^{m} \lambda_i} \tag{9}$$

Generally, when the rate of former k principal component is more than 85%, k is regarded as the number of principal component. Such as in Fig. 1, the rate of former 4 is over 85%, so the number is 4.

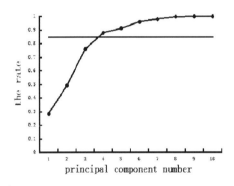

Fig. 1. Determination of principal component number by CPV

2.3   Contribution Plot Method

To estimate parameters which leading to SPE or T^2 statistic over its control limit, when monitor samples in industrial process by PCA model. Contribution plot method is the common separation method proposed by Miller P [12]. It gives the contribution of each process variable for detection statistics.

Contribution plot based on SPE is defined as follows:

$$Cont_i^{SPE} = [\xi_i^T (I - PP^T)x]^2 \tag{10}$$

Among it, $Cont_i^{SPE}$ represents each contribution values for SPE statistic, ξ_i represents i column of unit matrix I_m.

Contribution plot based on T^2 is defined as follows:

$$Cont_i^{T^2} = x^T D^{\frac{1}{2}} \xi_i \xi_i^T D^{\frac{1}{2}} x \tag{11}$$

Among it, $D = P^T \Lambda^{-1} P$, ξ_i represents i column of unit matrix I_m.

The largest contribution values for two statistics in contribution plot are considered to be the main reason variables. With this, the results can be fed back to operator guiding how to adjust parameters.

3   Experiment and Analysis

3.1   Experimental Design

Hot water boiler is used in this paper whose parameters are shown in Table 1. Its data of parameters are collected when boiler runs in normal operating conditions and stored in database.

Table 1. Parameters table of hot water boiler

serial number	measuring point	unit	serial number	measuring point	unit
1	effluent temperature	°C	7	furnace negative pressure	Pa
2	effluent pressure	MPa	8	exhaust smoke temperature	°C
3	water feeding pressure	MPa	9	blast valve position	%
4	water feeding temperature	°C	10	air-induced valve position	%
5	effluent flow	t/h	11	grate valve position	%
6	furnace temperature	°C	12	coal supply valve position	%

Selecting 300 samples from database whose thermal efficiency is high for composing matrix $A_{300 \times 12}$. A is processed by PCA. According to CPV principle, 6 is principal component number. Here the value of CPV is 91.19% greater than 85%. When $\alpha = 0.01$, control limit of SPE and T^2 are 6.4854 and 17.5329, respectively. Flow chart of project design is shown in Fig. 2.

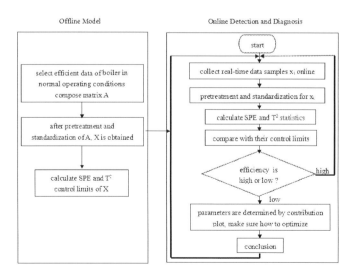

Fig. 2. Flow chart of project design

3.2   Statistics and Analysis of Thermal Efficiency

Formula (8) shows that the control limit of T^2 largely depends on the principal component number k and samples' number. For different number of samples, the control limit is different even if the same k. This has little significance for practical guidance. SPE statistic and its control limit are analyzed in this paper. For better achieving the experiment purpose, T^2 statistic is considered.

Over 100 samples are selected arbitrarily from database. Thermal efficiency is calculated by the amount of inputs and outputs, and they are arranged from low to high. As known to us, the first 20 samples' thermal efficiency is less than 62%, and others are higher than 70%. Values of two statistics are calculated of each sample to test the performance of PCA model. The result is shown in Fig. 3.

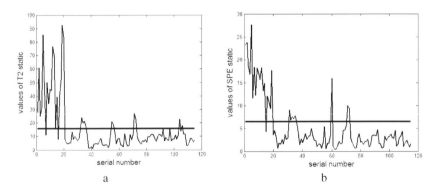

a   b

Fig. 3. T2 and SPE statistics of samples

It shows the relationship between statistics and their control limits. Straight lines are control limits of T^2 and SPE statistics, respectively. Values of the first 20 samples are above control limits, and others are below. It shows that PCA model can better identify boiler efficiency. b also shows that values of SPE statistic decrease with boiler's thermal efficiency increasing in general. Making a further analysis to verify.

Historical data of one year is selected and grouped according to different thermal efficiency whose interval is 1%. SPE statistics of each group are calculated. The curve of boiler thermal efficiency and SPE statistic is shown in Fig. 4.

SPE values decrease with boiler's thermal efficiency increasing. When boiler's thermal efficiency is over 70%, SPE is lower than limit, and the smaller of SPE, the higher of thermal efficiency. When boiler's thermal efficiency is over 80% considered to be higher, the values of SPE are lower than 2. In conclusion, samples' SPE values reflect the size of boiler's thermal efficiency. When SPE values are larger than its preset value, boiler's thermal efficiency is determined to be low. Due to the limitation of experiment, relevant conclusions need to be further analyzed and proved.

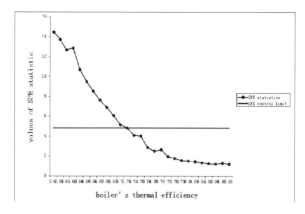

Fig. 4. The cure of boiler thermal efficient and SPE statistics

3.3   Experimental Analysis

Collecting online data at different times. Select 7 sets of data A1~A7 arranged by values of thermal efficiency from low to high. The samples are recorded in Table 2. Their SPE and T^2 statistics are calculated and recorded in Table 3.

Table 2. Parameter values of samples collected online

serial number	parameters of samples	A1	A2	A3	A4	A5	A6	A7
1	effluent temperature	93.23	97.39	98.59	96.22	98.60	98.79	101.51
2	effluent pressure	0.42	0.54	0.39	0.55	0.43	0.54	0.55
3	water feeding pressure	0.49	0.61	0.47	0.64	0.50	0.61	0.62
4	water feeding temperature	50.41	59.40	51.83	53.43	47.43	51.65	48.58
5	effluent flow	452.95	445.49	449.83	454.69	445.14	450.17	453.47
6	furnace temperature	695.89	656.39	765.41	684.61	770.75	707.18	818.52
7	furnace negative pressure	-0.93	0.81	0.01	-1.04	1.97	7.292	0.23
8	exhaust smoke temperature	93.287	94.91	96.41	91.96	96.07	94.68	98.32
9	blast valve position	52.14	51.45	49.36	50.58	49.77	52.03	51.27
10	air-induced valve position	77.08	78.71	79.98	79.34	77.72	74.19	78.88
11	grate valve position	44.39	35.71	39.84	33.68	34.56	32.87	32.87
12	coal supply valve position	47.84	37.80	44.24	35.60	38.50	33.34	34.84

Table 3. Statistical values of each sample

	A1	A2	A3	A4	A5	A6	A7
boiler's thermal efficiency (%)	54.2	58.9	66.5	70.12	75.17	80.14	82.15
SPE	26.094	12.798	8.259	3.659	1.008	0.371	2.151
T^2	22.073	7.963	15.43	11.51	9.18	5.64	5.356

From this table, boiler's thermal efficiency of A1 is 54.2%, and SPE and T^2 statistics are all over their theoretical control limits. T^2 statistics of A2~A7 do not exceed the control limit. SPE statistics of A4~A7 do not exceed the control limit. SPE and T^2 statistics present a downward trend with the improving of boiler's thermal efficiency.

For example, making an analysis for A3 using contribution plot of SPE and T^2 statistics. SPE statistic is over control limit, and T^2 statistic is not exceed control limit. Therefore just calculate the contribution plot of SPE is enough. The result is shown in Fig.5. The contribution value of water feeding temperature is the maximum, followed by the value of coal supply valve position.

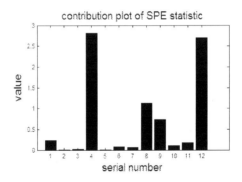

Fig. 5. Contribution plot of SPE statistic

The parameter of water feeding temperature is uncontrollable, so it is ignored. Because of blast valve position is constant, larger coal supply valve position causes lower air-coal ratio leading to insufficient combustion of coal, so boiler's thermal efficiency should be reduced. Through above analysis, the value of coal supply valve position is adjusted to about 40 initially. Two sets of samples $A3_1$ and $A3_2$ are collected after running a long time considering the delay and hysteretic nature of boiler. It is shown in Table 4. Their SPE and T^2 statistics are calculated and recorded in Table 5.

Table 4. Parameter values after adjustment

serial number	parameters of samples	A3	$A3_1$	$A3_2$
1	effluent temperature	99.595	99.826	104.04
2	effluent pressure	0.398	0.439	0.53
3	water feeding pressure	0.468	0.508	0.597
4	water feeding temperature	50.83	49.739	53.212
5	effluent flow	449.83	444.097	440.451
6	furnace temperature	765.41	754.948	796.701
7	furnace negative pressure	0.010	1.042	-0.694
8	exhaust smoke temperature	96.412	98.206	101.331
9	blast valve position	49.363	49.363	49.363
10	air-induced valve position	79.977	66.435	68.808
11	grate valve position	39.835	38.368	38.136
12	coal supply valve position	44.24	39.78	39.84

Table 5. Statistical values of each sample after adjustment

	A3	$A3_1$	$A3_2$
boiler's thermal efficiency(%)	67.7	74.8	75.19
SPE	8.059	5.689	3.219
T^2	16.43	7.584	6.791

It shows that after reducing coal supply valve position, SPE and T^2 statistics are reduced and SPE statistics are less than control limit. Thermal efficiency of boiler is enhanced. It verifies the correctness of conclusion.

4   Conclusion

An operation optimization method of boiler based on PCA is proposed in this paper. An identification model which can reflect thermal efficiency is established based on historical data. It is used to monitor boiler's operating conditions online. When boiler's operation efficiency decreases, influential parameters are directly obtained. This can guide operators work and improve the boiler's efficiency. Through an application of hot water boiler, it shows that the method guides operation correctly and significantly improves the operating efficiency of boiler.

Acknowledgment. This work was financially supported by "the Fundamental Research Funds for the Central Universities".

References

1. Xia, H.Y.: The Study of Determining the Optimal Target Value of Circulating Fluidized Bed Boiler Operation Parameter. Shanxi University, Shanxi (2011)
2. Ghosh, K., Ramteke, M., Srinivasan, R.: Optimal Variable Selection for Effective Statistical Process Monitoring. Computers and Chemical Engineering **60**, 260–276 (2014)
3. Shah, S., Adhyaru, D.M.: Boiler Efficiency Analysis using Direct Method. In: International Conference on Current Trends in Technology, pp. 1–5. IEEE Press, New York (2011)
4. Laurí, D., Rossiter, J.A., Sanchis, J., Martínez, M.: Data-driven Latent-variable Model-based Predictive Control for Continuous Processes. Process Control **20**, 1207–1219 (2010)
5. Wang, H.: Understanding Data Driven Control for Industrial Processes. Control Engineering of China **20**(3), 197–200 (2013)
6. MacGregor, J., Cinar, A.: Monitoring, Fault Diagnosis, Fault-tolerant Control and Optimization: data driven methods. Computers and Chemical Engineering **47**, 111–120 (2012)
7. Yin, S., Ding, S.X., Haghani, A., Hao, H., Zhang, P.: A Comparison Study of Basic Data-driven Fault Diagnosis and Process Monitoring Methods on the Benchmark Tennessee Eastman process. Process Control **22**, 1567–1581 (2012)
8. Liu, Q., Chai, T.Y., Qin, S.J., Zhao, L.J.: Progress of Data-driven and Knowledge-driven Process Monitoring and Fault Diagnosis for Industry Process. Control and Decision **25**(6), 801–807, 813 (2010)
9. Hou, R.R., Wang, H.G., Xiao, Y.C., Xu, W.L.: Incremental PCA based Online Model Updating for Multivariate Process Monitoring. In: Proceedings of the 10th World Congress on Intelligent Control and Automation, pp. 3422–3427. IEEE Press, Beijing (2012)
10. Godoy, J.L., Vega, J.R., Marchetti, J.L.: Relationships between PCA and PLS-regression. Chemometrics and Intelligent Laboratory Systems **130**, 182–191 (2014)
11. Alaei, H.K., Salahshoor, K., Alaei, H.K.: A New Integrated On-line Fuzzy Clustering and Segmentation Methodology with Adaptive PCA Approach for Process Monitoring and Fault Detection and Diagnosis. Soft Computer **17**, 345–362 (2013)
12. Lau, C.K., Ghosh, K., Hussain, M.A., Che-Hassan, C.R.: Fault Diagnosis of Tennessee Eastman Process with Multi-scale PCA and ANFIS. Chemometrics and Intelligent Laboratory Systems **15**(120), 1–14 (2013)
13. Kong, X.G., Guo, J.Y., Lin, A.J.: Fault Diagnosis for Batch Processes based on Two-dimensional Principal Component Analysis. Journal of Computer Applications **33**(2), 350–352 (2013)
14. Dunia, R., Edgar, T.F., Nixon, M.: Process Monitoring using Principal Components in Parallel Coordinates. AIChE Journal **59**(2), 445–456 (2013)

An Orientation Column-Inspired Contour Representation and Its Application in Shape-Based Recognition

Hui Wei[⊠] and Wentao Ge

Laboratory of Cognitive Model and Algorithm, School of Computer Science,
Fudan University, Shanghai 200433, China
{weihui,wge13}@fudan.edu.cn

Abstract. Recognizing an object from its background in a real-world image is always a very challenging task. During the recognition process, shape (or contour) information of an object is useful. In this paper, we build a bio-inspired contour detection model which can organize the edge information into a structured data form. Biological primary visual cortex, which can be simulated by computer, is specialized in detecting orientation of edge to producing a set of line segments. Then we propose the concept of route that indicates a continuous part of the contour. The set of line segments is divided into several routes which are the basic processing units of following recognition steps.

Keywords: Orientation column · Line segments · Route · Line context · Object recognition

1 Introduction

The shape of a given object, which is generally stable, plays an important role in object recognition. For the recognition utilizing the shape of an object, there are usually three steps: contour detection, representation of the contour and shape comparison. Many researchers built a lot of computational models to simulate human vision for object recognition [1–4]. Some biologically inspired algorithms were also proposed in the past concerning subjective contours [5,6].

The motivation of this paper is to design a recognition method based on line segments, which are the contour detection results of our bio-inspired model [7]. Our method can effectively eliminate background interference when recognizing objects in real-world images.

Concretely, the bio-inspired model is based on orientation selectivity of the biological primary visual cortex(V1). The contour of natural image can be detected and output as a set of line segments in the model. Then a graph-based algorithm is used to divide the set of line segments into several routes, which are our basic processing units of recognition. After that, we improve our previous method [8] that is similar to the well-known shape-context descriptor [9] for shape comparison. Once the background interference presents, the performance is not ideal when the recognition is based on either shape-context or our previous method. Consider

© Springer International Publishing Switzerland 2014
Z. Zeng et al. (Eds.): ISNN 2014, LNCS 8866, pp. 405–413, 2014.
DOI: 10.1007/978-3-319-12436-0_45

a template object O_t and a real-world image I_r. In this paper, we generate some routes in I_r using a graph-based algorithm, and compare the line-context of each route to that of O_t. Finally we can get several best-matched routes, which indicate the contour of target object, from the real-world image with complex background.

The rest of this paper is organized as follows: Section 2 introduces the bio-inspired model and the line-detection algorithm. Section 3 shows the graph-based algorithm of dividing line segments into some routes. Section 4 describes the shape comparison using improved line-context descriptor. Experimental results are presented in section 5, and section 6 is the conclusion.

2   The Bio-inspired Model and Line-Detection Algorithm

2.1   Theoretical Basis in Biology

Hubel and Wiesel won the 1981 Nobel Prize for finding that simple cells in the V1 area of primary visual cortex respond strongly to special orientations [10]. They proposed the concept of orientation column which consists of some simple cells. The orientation selectivity of simple cells in each column vary systematically, and adjacent cells have approximate orientation preferences [11]. An array of orientation columns can be regarded as a bio-inspired representation platform. The whole framework of the image representation platform is shown in Fig. 1, the left part shows several orientation columns and their receptive fields. Neighbouring columns have overlapping receptive fields to guarantee that no information is missed. In the middle part of Fig. 1, many GC/LGN cells under the column layer perform edge detection, produce edge images and send the information up to the simple cell layer. In biological theory, the LGN cells are the relays of ganglion cells. In the right part of Fig. 1, the larger hexagon consists of 19 small green hexagons denote the orientation column that can detect the orientations range from 0° to 180°. In the receptive field of the column, the orientations of the red curve can be detected. Finally, four corresponding simple cells are activated and four red line segments are output to represent the orientation information.

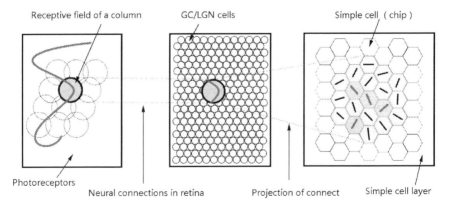

Fig. 1. Framework of the hierarchical image representation platform

2.2   Line-Detection Algorithm

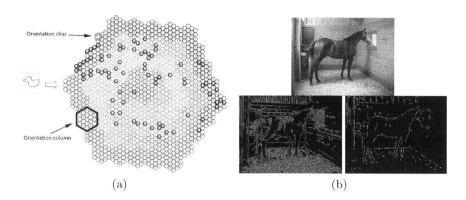

(a)   (b)

Fig. 2. (a) The array of orientation columns and the activated results of an image of duck. The contour of duck is represented by some activated simple cells (also called orientation chip here). (b) A real-world image, result of Canny detector and result of our bio-inspired model.

Based on the aforementioned biological theories, the computational model of orientation columns is shown in Fig. 2(a). Each small hexagon is a simple cell (also called orientation chip in the model) and the larger black hexagonal frame indicates an orientation column containing nineteen orientation chips. Each of these nineteen orientation chips can detect one unique orientation so that an orientation column can detect the orientations range from 0°to 180°. When processing an image by this bio-inspired model, each column detects the orientations in its own receptive field and the corresponding orientation chips will be activated. Finally, each column outputs several line segments with these orientations. Fig. 2(a) also shows a shape of a duck represented by some activated cells. This bio-inspired model provides a scheme, with an image as the input while a set of line segments as the output, to represent an object. Fig. 2(b). shows the line-detection result of a real-world image. The result of the traditional Canny edge detection algorithm [12] can also be seen. Obviously, the line-detection result of bio-inspired model is more cleaner than that of Canny method.

3   A Graph-Based Algorithm of Dividing the Set of Line Segments into Some Routes

3.1   The Graph and its Adjacent Matrix

After detecting line segments by the bio-inspired model, we utilize an undirected graph $G = (V, E)$ to represent an image. All the line segments are treated as vertices $v_i \in V$, and edges $(v_i, v_j) \in E$ correspond to pairs of neighboring vertices. Each edge $(v_i, v_j) \in E$ has a corresponding weight $w((v_i, v_j))$, which is a non-negative measure of the distance between two line segments represented by v_i and v_j.

We want to reaffirm that the goal to generate routes is to get some relatively continuous parts of the contour of image. That is to say, two adjacent vertices in the graph indicate that the two corresponding line segments are contiguous on the contour of object. Based on this principle, two factors are considered to calculate the distances between line segments.

The first is the included angle θ (range from 0 to $\frac{\pi}{2}$) of two line segments. Two line segments with large included angle (closed to the right-angle) are not considered contiguous on the contour. On the contrary, the smaller included angle means that two line segments are more reasonable to be contiguous. For the purpose of normalization, the first factor is finally denoted by $FAC_1 = \frac{2\theta}{\pi}$ which is ranged from 0 to 1.

In order to clearly state the definition of the second factor, consider two line segments denoted by L_m and L_n (m_a, m_b; n_a, n_b are their end points). D_{large} is the largest value in the set of point-to-point distances: $\{|m_a n_a|, |m_a n_b|, |m_b n_a|, |m_b n_b|\}$. The distance between the other two end points, which have nothing to do with D_{large}, is denoted by D_{small}. We define the second factor as $FAC_2 = \frac{D_{small}}{D_{large}}$ (range from 0 to 1). There are three situations in Fig. 3 that show how this factor works. The two line segments in situation I are obviously adjacent because they are close to each other. The lines in situation II are too far away from each other to be adjacent. In situation III, the two lines are almost parallel, they belong to different contours although the distance and included angle (FAC_1) are both very small. So they also can't be adjacent. The value of FAC_2 in situation I is small while the values in situation II and situation III are large. So the value of FAC_2 has the ability to measure the level of adjacency of line segments.

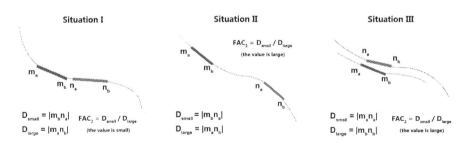

Fig. 3. Three situations show how FAC_2 works

The distance d_{ij} between two line segments (L_i and L_j) is finally defined as follow:

$$d_{ij} = \begin{cases} 0 & \text{if } i = j \\ \alpha * FAC_1 + \beta * FAC_2 & \text{if } i \neq j \end{cases} \tag{1}$$

where α and β are parameters adjusting the importance of FAC_1 and FAC_2. It is important to emphasize that, the smaller distance between two line segments means that they are more reasonable to be contiguous on the contour.

Based on equation (1), the distance between any two line segments in the set can be calculated. Assuming that the size of the set of line segments is n, a $n \times n$ distance matrix is generated. Then we connect each vertex with its two nearest neighbors by edges. That is to say, for each vertex, there are two edges which connect itself to its two nearest neighbors. Finally we build an adjacent matrix $A_{n \times n}$, the value of element a_{ij} is 1 if there exists a edge between v_i and v_j, otherwise the value is 0. All elements in $A_{n \times n}$ are 1's or 0's (1 for adjacent and 0 for not adjacent).

3.2 Generating Routes by Depth-First Search

In graph theory, a connected component (or just component) of an undirected graph is a subgraph in which any two vertices are connected to each other by paths. In our case, one connected component is a subset of line segments that belong to a continuous part on the contour of image. So the connected components of graph is equivalent to the routes we mentioned before.

Fig. 4. On the top row, the detected lines are overlaid in different colors on the original images. The bottom row shows the corresponding route maps. Different routes are drew in different colors, and we draw those routes, which contain more than eight line segments, wider than others.

It is straightforward to compute the connected components of a graph in linear time (in terms of the numbers of the vertices and edges of the graph) using the classical depth-first search algorithm. So we can divide the set of line segments into several routes which are the basic processing units for subsequent steps. Fig. 4 shows the results of some images from ETHZ [13] data set. We highlight those routes, which contain more than eight line segments, by increasing the widths of them. The reason is that, the more line segments a route contains, the more important it is in the final step of recognition.

4   Shape Comparison Based on Route and Line-Context

Given a template object and some real-world images, the goal of this paper is to recognize a target object in every image. Shape-context is a well-known method which can be used for shape description and comparison. The key points underlying this technique are counting the pixels in a series of fan-shaped areas and producing a 2D histogram to be used as a representation of pixels distributing through the shape. In the environment of line segments, we use an improved method which relied on line-context rather than shape-context.

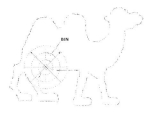

Fig. 5. Line-context descriptor system

Table 1. Line-context matrix of a line segment

SECTOR	RING = 1						...	RING = 4
	$0°<\delta \leq 10°$	$10°<\delta \leq 20°$	$20°<\delta \leq 30°$	...	$80°<\delta \leq 90°$	...		
$0°$-$45°$	1	0	2	...	1	...		
$45°$-$90°$	2	3	1	...	2	...		
$90°$-$135°$	0	2	1	...	1	...		...
...				...		...		
$315°$-$360°$	1	0	2	...	1	...		

Fig. 5 shows the system of line-context descriptor in the environment of line segments. For each line segment, we treated it as the center line and establish a polar coordinate system with four-layer concentric rings divided into eight sections. Each of the 32 (4 × 8) sectors is called a BIN. In each BIN_{ij} (the j_{th} sector in i_{th} layer), we calculate the included angles between the center line and every other line segments. Then we record the numbers of line segments in 9 intervals according to the values of included angles. For each line segment in the set, we finally get a 3-D matrix which is shown in Table. 1.

For comparing the similarities of different contours, we firstly define the similarity of two line segments. Consider the p_{th} line segment l_p^X on the first shape X and the q_{th} line segment l_q^Y on the second shape Y. Let $C_{pq} \equiv \text{C}(\ l_p^X\ ,\ l_q^Y\ )$ denote the cost of matching these two lines:

$$C_{pq} \equiv C(l_p^X\ ,\ l_q^Y) = \frac{1}{IJK} \sum_{i=1}^{I=8} \sum_{j=1}^{J=4} \sum_{k=1}^{K=9} \frac{\left|INTER_{i,j,k}^X - INTER_{i,j,k}^Y\right|}{max(INTER_{i,j,k}^X\ ,\ INTER_{i,j,k}^Y)} \quad (2)$$

Where X, Y denote two shapes, and I, J, K denote the number of layers, the number of sections and the number of slope intervals of a bin respectively.

$INTER_{i,j,k}$ is the number of line segments in the k_{th} slope interval, which belongs to j_{th} section in i_{th} layer. Given the set of cost C_{pq} between all pairs of line segments l_p^R on route R and l_q^T on template T, we want to minimize the average cost of matching,

$$H(\pi) = \frac{\sum_p C(l_p^R , l_{\pi(p)}^T)}{P} \tag{3}$$

subject to the constraint that the matching be one-to-one. The denominator P is the number of line segments in route R. The result $\pi(p)$, which can be calculated using Hungarian algorithm, is a permutation such that (3) is minimized.

The classical shape-context method has a drawback: the recognition sometimes fails in the presence of large amounts of background noises. In this paper, we compare the line-context of one route to that of the template object each time. After comparing all the routes to the template, we choose those best-matched routes (with the lowest cost value of equation (3)) that indicate the most possible area of the target object.

5   Experiment

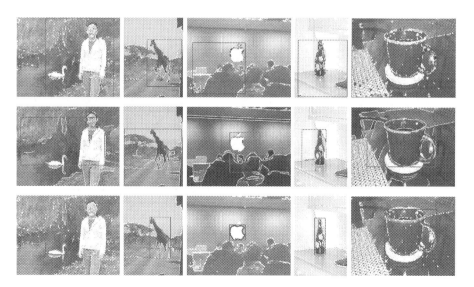

Fig. 6. Comparison of different methods. The last row shows our results and the multi-colored curves are actually some best-matched routes in the images. The red frame indicates the area of target object.

We experimented on the ETHZ [13] data set, which consists of 5 different categories (each contains a template object). The advantage of our route-based method is effectively eliminating background interference. Some results of a contrast experiment in Fig. 6 show the advantage of our method. In the first row

Fig. 7. More recognition results of our method

of Fig. 6, each picture shows the result of directly comparing the shape-context descriptor of template to that of the whole image where all sample points are the midpoints of line segments. In the second row, we compared the line-context descriptor of template to that of the whole image. And the results of our route-based method are shown in the last row. Every route was compared to the template and some best-matched routes represented by colorful curves in the last row of Fig. 6 were chosen. Comparing to the method in this paper, the first two methods cannot precisely detect the target objects because of the background interference. Fig. 7 shows more recognition results of our method.

We compared the time performance of this method to our previous work [8] whose model of line detection and module of shape comparison are different from this paper. Fig. 8 shows the time comparisons on 40 samples of ETHZ data set.

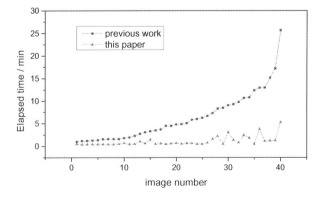

Fig. 8. The performance comparison between this paper and previous work [8]

6 Conclusion

In this paper, we propose the concept of route which is our basic processing unit of recognition. After utilizing a bio-inspired model to detect line segments, we generate some routes using a graph-based algorithm. Finally, we measure the similarities between every route and the template object, and choose the best-matched routes which indicate the area of target object. The whole method, which based on bio-inspired model and route, can effectively eliminate background interference in real-world images. In addition, the time performance is also greatly improved.

References

1. Hinton, G.E., Osindero, S., Teh, Y.W.: A Fast Learning Algorithm for Deep Belief Nets. Neural Computation **18**(7), 1527–1554 (2006)
2. Jhuang, H., Serre, T., Wolf, L., Poggio, T.: A Biologically Inspired System for Action Recognition. In: IEEE 11th International Conference on Computer Vision, pp. 1–8. IEEE Press, New York (2007)
3. Krizhevsky, A., Sutskever, I., Hinton, G.E.: Imagenet Classification with Deep Convolutional Neural Networks. In: Advances in Neural Information Processing Systems, pp. 1097–1105 (2012)
4. Serre, T., Wolf, L., Bileschi, S., Riesenhuber, M., Poggio, T.: Robust Object Recognition with Cortex-like Mechanisms. IEEE Transactions on Pattern Analysis and Machine Intelligence **29**(3), 411–426 (2007)
5. Ullman, S.: Filling-in the Gaps: The Shape of Subjective Contours and a Model for Their Generation. Biological Cybernetics **25**(1), 1–6 (1976)
6. Carpenter, G.A., Grossberg, S.: A Massively Parallel Architecture for a Self-organizing Neural Pattern Recognition Machine. Computer Vision, Graphics, and Image Processing **37**(1), 54–115 (1987)
7. Wei, H., Li, H.: Shape Description and Recognition Method Inspired by the Primary Visual Cortex. Cognitive Computation **6**(2), 164–174 (2014)
8. Wei, H., Xiao, J.: A Shape Recognition Method based on Graph-and Line-contexts. In: 2013 IEEE 25th International Conference on Tools with Artificial Intelligence (ICTAI), pp. 235–241. IEEE Press, New York (2013)
9. Belongie, S., Malik, J., Puzicha, J.: Shape Matching and Object Recognition using Shape Contexts. IEEE Transactions on Pattern Analysis and Machine Intelligence **24**(4), 509–522 (2002)
10. Hubel, D.H., Wiesel, T.N.: Sequence Regularity and Geometry of Orientation Columns in the Monkey Striate Cortex. Journal of Comparative Neurology **158**(3), 267–293 (1974)
11. Swindale, N.: The Development of Topography in the Visual Cortex: A Review of Models. Network: Computation in Neural Systems **7**(2), 161–247 (1996)
12. Canny, J.: A Computational Approach to Edge Detection. IEEE Transactions on Pattern Analysis and Machine Intelligence **8**(6), 679–698 (1986)
13. Ferrari, V., Jurie, F., Schmid, C.: From Images to Shape Models for Object Detection. International Journal of Computer Vision **87**(3), 284–303 (2010)

Region Based Image Preprocessor for Feed-Forward Perceptron Based Systems

Keith A. Greenhow[✉] and Colin G. Johnson

School of Computing, University of Kent, Canterbury, Kent CT2 7NF, UK
{kg246,C.G.Johnson}@kent.ac.uk

Abstract. In this paper, we investigate the notion that there may be alternate methods, beyond typical rectilinear interpolations such as Bilinear Interpolation, that have a greater suitability for use in visual/image preprocessors for Artificial Neural Networks. We present a novel method for down-sampling image data in preparation for a Feed-Forward Perceptron system assisted by a neural *usefulness* metric, inspired by those common to pruning algorithms. This new method achieves greater accuracy compared to the same system using by Bilinear Interpolation, and has a reduced computational time.

Keywords: Artificial neural networks · Preprocessor · Image processing · Salience · Relevance assessment

1   Introduction

When building a high-throughput or real-time image processing system that uses an Artificial Neural Network (ANN), it is common place to down-sample the input image as part of the preprocessing or data preparation stage. The typical reasons behind these choices are either that it is computationally cheaper to resize the image than it is to have the larger ANN input layer, or that there is an inability to learn/solve the task due to relevant signals getting lost in the noise of the full image; as seen in the following extracts:

> "The size of the captured image is 640×480, but due to the limited computational power, it is used after resizing to 26×20 using the OpenCV library."   [1, p. 1447]

> "...the DVS128 retina has 16,384 elements, ... [so] a relevant method of reducing the data flow must therefore be employed, without losing information relevant to the line-following task."   [2, p. 8]

Our issue with this practice is that both a considerable amount of information is lost in the process (however, most is validly disposed of as it shows no/little predictive value for the ANN), or that different regions of the image may benefit from being presented at different resolutions. Such losses or inefficiencies may reduce the effectiveness of the system or outright prevent it from working at a given image resolution. This can be a greater problem when the developer has

© Springer International Publishing Switzerland 2014
Z. Zeng et al. (Eds.): ISNN 2014, LNCS 8866, pp. 414–422, 2014.
DOI: 10.1007/978-3-319-12436-0_46

little (or no) a priori knowledge of what areas in the source images will prove useful to the task required. Typically, they have to search multiple different resolutions to find the suitable values so that the ANN can solve the problem.

Our hypothesised model is to automatically explore a large range of possible resolutions and quantify how *useful* each pixel is, at each resolution. Using this information, we generate a recommendation of which regions of the image are resized to what resolutions and which regions have no need to be processed at all as they serve no predictive benefit to the specified task.

2   Brief Overview

The process of constructing/training an ANN using our model is separated into three stages: Salience Heat Mapping (SHM, section 4), Region Selection and Optimisation (RSO, section 5), and traditional ANN training using the bespoke preprocessor constructed in the first two stages.

The objective of SHM is to generate a heat map of the input scene, which quantifies the predictive value of the different points in that scene. This information can then be passed to the selection stage, RSO, where it can be used to find which areas (both location and size) of the scene appear to have enough predictive value to warrant preprocessing at run-time. The final objective is to construct a preprocessor that can take advantage of this information to optimally pre-process only the areas of the scene that provide suitable predictive value, and at suitable resolutions; details in section 6.

3   Measuring an Input Units '*Worth*'

One of the first problems to solve was to find a suitable metric to determine the usefulness of the individual datum generated by the preprocessor. As each datum has a 1:1 relationship with each input unit of the ANN, determining the usefulness of each input unit can be used as a representation of the data's usefulness, allowing us to use concepts that are common in pruning algorithms.

One metric we looked at for suitability is that of 'Relevance Assessment' [3,4]. In their work, they describe a method for measuring a unit's functionality, or *relevance*, as its contribution to minimising error [3, p. 1]. We can reinterpret this as being a measure of how much a unit contributes to the overall accuracy of the ANN against the given data-set.

For input and hidden units, the approximate relevance, $\hat{\rho}_i$, is computed during an additional feed-back phase (similar to back-propagation) as

$$\hat{\rho}_i = \sum_{j}^{N_i} |w_{ij}\,\hat{\rho}_j| \, , \tag{1}$$

where N_i is the set of neurons that use unit i as an input and w_{ij} is the weight between them. For output units, using the linear error function ($|t_i - o_i|$), the approximate relevance becomes

$$\hat{\rho}_i = -o_i \times \mathrm{sgn}\,(t_i - o_i) \, , \tag{2}$$

where t_i is the target value for the current pattern, and $\mathrm{sgn}\,(x)$ the sign of x.

The issue with this definition is that the output units relevance is dependant on the latest input received by the ANN and the approximated relevance varies considerably with time. Mozer and Smolensky attempted to rectify this issue by an "exponentially-decaying time average of the derivative" [3, p. 4],

$$\hat{\rho}_i(t) = 0.8\hat{\rho}_i(t-1) + 0.2\varrho(t) \, , \qquad (3)$$

where $\varrho(t)$ is the approximated relevance from equations (1) or (2) at time t. This makes the computed relevance dependant on the order in which the training set was presented. When applied to our system, this produced relevancies that would react very slowly to the learning of the ANN. If the weighting was adjusted towards newer values, to decrease the response time to learning, then the variability would destabilise the system.

Other metrics like 'Sensitivity Analysis' [5], solve the temporal noise problem by parsing the entire data-set then computing an overall value. This still leave the sensitivities generated dependant on the patterns in the data-set; such that a poor choice of patterns can result in useless sensitivities.

To resolve these issues, we defined 'salience' as the amount of contribution by each input unit has to the output function of an ANN. The method we use to quantify this is derived from Relevance Assessment mentioned above.

In our model, we are not interested in the contribution of the inputs to the ANN's predictive accuracy, but rather to the ANNs output function. We found that disregarding the exponentially-decaying time average and defining the relevance of each output unit to be exactly 1 instead produced suitable results, reduced memory costs and avoided the previously aforementioned issues:

$$\hat{\rho}_i = \begin{cases} 1 & \text{when unit } i \text{ is an output unit,} \\ \sum_j^{N_i} |w_{ij}\,\hat{\rho}_j| & \text{otherwise.} \end{cases} \qquad (4)$$

Finally, the units salience, S_i is determined by normalising the approximate salience using the hyperbolic tangent, to improve suitability,

$$S_i = \tanh(\hat{\rho}_i) \, . \qquad (5)$$

It is also possible to include a scaling factor prior to normalising if the approximated salience is too small or large. We chose this function due to the constraining effects of back-propagation on the weights. Additionally, our interest in the salience tends toward the order of magnitude as opposed to the absolute magnitude.

4   Salience Heat Mapping

For the first stage of the process, SHM, the developer is required to specify three groups of information:

- The specific task with associated training and test sets
- An ANN topology and training regime (input layer size will be inferred from the dimensionality of the preprocessor output and output layer size can be inferred from the requirements of the task)

– Values for res_{min}, res_{max} and res_{rep} that specify the minimum and maximum limits of search for resolution saliency (inclusive) and how many times to repeat the process for averaging, respectively.

In this stage, an ANN using the topology specified will be generated for each possible resolution (as all possible permutations in the range $res_{min} \ldots res_{max}$, inclusive) and trained against the training set for 5 epochs.[1] The quality of that resolution is quantified by Cohen's kappa (κ, the agreement between the test set defined class and the ANN predicted class) and input's normalised computed salience. The salience can be viewed as a heat map of each potential pixel at each resolution. The salience maps are resized (by Bilinear Interpolation[2]) such that all sides have a length of $res_g \times res_{min} \times res_{max}$, where res_g is a scaling value to provide suitable granularity of the final heat map.[3] The rescaled saliency maps which have $\kappa > 0$ can be combined into a single heat map by generating an average heat map, weighted by κ. This process is then repeated res_{rep} times. The res_{rep} resultant saliency maps are then averaged together, weighted by mean contributing κ (average of only κ values greater than 0) to produce the final Salience Heat Map.

5   Region Selection and Optimisation

The second stage is to take the generated Salience Heat Map and select regions that would provide the greatest contribution to predictive accuracy. This is done in two parts; a region selection process followed by a performance optimising process. Figure 1 shows an example heat map generated by SHM and the regions selected by RSO.

Fig. 1. A heat map showing the saliency at various locations of the input image space. The bright regions have been determined to be of greater salience for the task (Section 7.2 and A). The red rectangles identify regions determined to be of high predictive value for use in the preprocessor.

---

[1] An epoch consists of a complete showing of the training or test set in a random order.

[2] In this paper we only use Bilinear Interpolation as we found that all other traditional interpolation techniques either produced lower accuracies or had negligible difference in accuracy at the cost of increased run-time.

[3] $res_g = 5$ seems to produce adequate results (found by trial and error).

Prior to performing the selection process, it is beneficial to transform the heat map produced prior such that the values $min \ldots max$ are normalised to the range 0..1 if not already.

As part of the selection process, the salience of each possible region (each pixel at all valid permutations of resolutions) is approximated by performing a down sample on the Salience Heat Map of the area covered by that region. The region with the highest predicted salience is then selected. The area on the Salience Heat Map covered by the region is then set to 0 to prevent redundant selections. This is repeated until there are no potential regions remaining with a predictive salience greater than some threshold value.[4]

The way the Region Preprocessor functions requires a new Bilinear Interpolation call for each region provided (Section 6). This can have a significant impact on performance, due to overheads incurred in the set-up of each interpolation process. The goal of the optimisation step is to combine regions at the same resolution into single regions and then filter out the selected pixels, as the runtime demand of the additional pixels is typically less than the overhead of multiple interpolation calls.

First, all the regions are grouped together by resolution. At each resolution, the minimum and maximum x and y coordinates of the regions are determined and a new region that covers the bounding box is define, r_{bound}. r_{bound} is then attached to a list of all the coordinates of the regions at this resolution, such that each entry in the data structure takes the form $\{r_{bound}, pixels = [(x_0 - \underline{x}, y_0 - \underline{y}), (x_1 - \underline{x}, y_1 - \underline{y}), \ldots, (x_n - \underline{x}, y_n - \underline{y})]\}$ (where $\underline{x}$ is the minimum value for x at the current resolution, similarly for $\underline{y}$).

6 Regioned Preprocessor

With the salient regions selected and optimised, the design of the preprocessor is the last stage. For each region (representative of each resolution), the source image is cropped and resized (by Bilinear Interpolation) so that the resultant image has the same size as the r_{bound}'s w and h properties and the top left pixel corresponds to the pixel at coordinate $r_{round} \rightarrow (x, y)$ at resolution of $r_h \times r_v$. This can be seen in the sample figures in Appendix A. Each point in the pixels part of the map is then extracted from the resized image and return as a value from the preprocessor. Care needs to be taken to make sure that the order the pixel are returned in is consistent between runs for each ANN, but do not need to be the same between different ANNs in the population.

7 Preliminary Results

We have used our model, as described above, to generate some preliminary results. With this, we hope to demonstrate that our model is worth further investigation and improvement as a candidate replacement for traditional interpolation techniques in preprocessors for ANN based image processors.

---

[4] In our experimentation, we found a value 75% of the max value suitable.

7.1   Test Methodology

Both of the tests, ours and the control, were run for 10 epochs. Each test was repeated 500 times to produce a reliable average accuracy. The control test used a traditional Bilinear Interpolation preprocessor set to down-sample to a 6×5 image and then normalise the pixels in preparation for the ANN.

We believe that, for our method to be considered successful it must have an accuracy that is at least comparable to the Bilinear preprocessor whilst having a significantly reduced time. We consider the rate of learning to solve the task as of lower priority. The reasoning behind this is that we intend for our algorithm to be used to improve runtime after learning has occurred, so that the processor can be freed up to perform other tasks on a real-time system.

7.2   Task

The task we chose to perform was to detect the presence of a specific object in a complex scene. The object position would always be centred vertically, but could be positioned at either 25% or 75% horizontally, or not be present in the scene at all. To add to the difficulty of the task, an object with no rotational symmetry was used, and randomly orientated. See Appendix A.

7.3   Test and Training Set

The training set used in this preliminary experiment consisted of 600 randomly generated images. Each image was generated from one of 17 possible background images and, where relevant, the object to be identified was overlaid onto the image. Each image was in an RGB format, providing 3 colour channels. 200 images had the object on the left, 200 had the object on the right and the remaining 200 were just the background. The scale of the object was not altered, but the orientation was randomly assigned. To add suitable complexity to the images (especially those with no object present) the backgrounds had their brightness/contrast level altered by small random amounts.

The test set was produced by the same method, but with 50 of the each to the three types (object on left, object on right, no object).

7.4   Test Environment

The test-rig used for our model was implemented to support both the traditional preprocessors and our Regioned Preprocessor. The ANN was to predict which of 3 possible classes the input image belongs to. For this task, the ANN topology was structured with 3 layers sized, $I \to 10 \to 3$ where I equals the number of outputs provided by the related preprocessor ($6 \times 5 \times 3 = 90$ for Bilinear and $3 \times 3 = 9$ for Regioned).

The processing units (hidden and output units) implemented the model

$$y_i = \text{norm}\left(b_i + \sum_{j}^{N_i} w_{ij} y_j\right), \tag{6}$$

where b_i is the bias value for unit i, and norm (x) is the logistic function. The task of normalising the input units was considered part of the preprocessing stage, so the inputs units are implemented as the function $f(x) \rightarrow x$.

To minimise the effect of background processes, non-essential processes where shut down with the test where run in parallel on a multi-core CPU. Each CPU core was assigned one test thread at any one time. Additionally, to mitigate disk IO delays, all the test and training set images were preloaded into RAM.

7.5   Results

On this task our Regioned preprocessor outperformed the Bilinear preprocessor in processing speed. Our Regioned method took a mean processing time of 0.6 ms compared to Bilinear's 1.6 ms, a reduction of about 65% run-time. It can also be seen (Fig 2) that the two methods eventually surpass an accuracy of 90%. We also found that the ANN is able to learn how to solve the task at a faster rate (probably due to the reduced noise in the input data) before converging and with significantly less variation in the accuracy of the population.

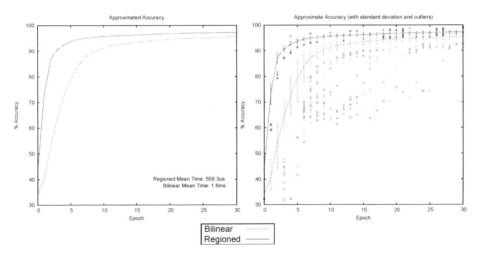

Fig. 2. These graph shows the accuracy of the ANNs at correctly classifying the values of the test set after each epoch of learning. Outliers are defined as any value outside $1.5 \times$ IQR. Positive and negative standard deviations are displayed independently.

8   Conclusions

We have shown that, when compared against Bilinear Interpolation on the given task, our algorithm is able to learn to solve the task in fewer epochs and perform the preprocessing computations in less time. This can be particularly relevant to the construction of image processing systems that are required to run in real-time, e.g. a robotic controller or environment monitoring system.

9   Future Work

Our current implementation uses running averages and integer data types during the selection of regions (Sec. 4). Changing the process to use floating-point and using single step averages would reduce the rounding errors that accumulate.

We would also like to test our preprocessor on a greater variety of different tasks to determine generality and suitability to these different types of tasks.

A   Samples Images

Each image is divided into two sections. The left shows the original image, plus the regions being processed in red. The right section shows the output after processing. A.1 shows one complex backgrounds used. Backgrounds like A.2 where selected to increase the possibility of false-positives. Figures A.3 and A.4 show the same background with modified contrast/brightness.

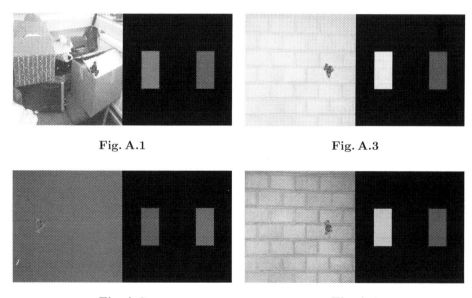

Fig. A.1   Fig. A.3

Fig. A.2   Fig. A.4

References

1. Shibata, K., Utsunomiya, H.: Discovery of pattern meaning from delayed rewards by reinforcement learning with a recurrent neural network. In: The 2011 International Joint Conference on Neural Networks (IJCNN), pp. 1445–1452 (2011)
2. Davies, S., Patterson, C., Galluppi, F., Rast, A., Lester, D., Furber, S.: Interfacing real-time spiking i/o with the spinnaker neuromimetic architecture. Australian Journal of Intelligent Information Processing System **11**, 7–11 (2010)
3. Mozer, M.C., Smolensky, P.: Using relevance to reduce network size automatically. Connection Science **1**(1), 3–16 (1989)

4. Mozer, M.C., Smolensky, P.: Skeletonization: a technique for trimming the fat from a network via relevance assessment. Advances in Neural Information Processing Systems **1**, 107–115 (1989)
5. Engelbrecht, A., Cloete, I., Zurada, J.: Determining the significance of input parameters using sensitivity analysis. In: Mira, J., Sandoval, F. (eds.) From Natural to Artificial Neural Computation. LNCS, vol. 930, pp. 382–388. Springer, Berlin Heidelberg (1995)

Image Denoising with Signal Dependent Noise Using Block Matching and 3D Filtering

Guangyi Chen[1], Wenfang Xie[1], and Shuling Dai[2]

[1] Department of Mechanical and Industrial Engineering, Concordia University, Montreal,
QC H3G 1M8, Canada
{guang_c,wfxie}@encs.concordia.ca
[2] State Key Lab. of Virtual Reality Technology and Systems, Beihang University,
No. 37, Xueyuan Rd., Haidian District, Beijing 100191, People's Republic of China
sldai@yeah.net

Abstract. In this paper, we propose a new method for image denoising. We use block matching 3D filtering (BM3D) to denoise the noisy image, and then denoise the noisy residual and merge this denoised residual into the denoised image. We can perform another BM3D to this merged image if the noise-level is still higher than a threshold. Our method performs similarly as the BM3D for Gaussian white noise, and it outperforms the BM3D, Poisson-Gaussian BM3D (PGBM3D), and Bivariate shrinking (BivShrink) for nearly all cases in our experiments for signal dependent noise. The method does not assume the noise to be Gaussian alone, and it works well for a mixture of Gaussian and signal-dependent noise. However, the computational complexity of the new method is twice and at most three-times that of the standard BM3D for image denoising.

Keywords: Image denoising, Block matching 3D filtering (BM3D), signal-dependent noise.

1 Introduction

Digital images are often contaminated by different types of noise, including Gassian white noise, salt-and-pepper noise, Laplacian noise, signal dependent noise, impulse noise, and so forth. There are a number of trade-offs in reducing noise in an image. For example, whether sacrificing some image details is acceptable if we want to remove more noise in the image. In order to make better decision, the characteristics of the noise and the details in the images should also be taken into account.

In existing literature, the majority of denoising methods is dealing with Gaussian white noise, which can be modeled as:

$$B = A + \sigma_n Z, \tag{1}$$

where A is the noise-free image and B the image corrupted with Gaussian white noise, Z has a normal distribution N(0; 1) and σ_n is the noise standard deviation. There are a number of methods to deal with this kind of noise. Fathi and Naghsh-Nilchi [1] proposed an efficient image denoising method based on a new adaptive wavelet packet thresholding function. Chatterjee and Milanfar [2] studied patch-based

© Springer International Publishing Switzerland 2014
Z. Zeng et al. (Eds.): ISNN 2014, LNCS 8866, pp. 423–430, 2014.
DOI: 10.1007/978-3-319-12436-0_47

near-optimal image denoising. Rajwade et al. [3] worked on image denoising using the higher order singular value decomposition. Motta et al. [4] proposed the iDUDE framework for gray scale image denoising. Miller and Kingsburg [5] studied image denoising using derotated complex wavelet coefficients. Sendur and Selesnick [6] proposed a bivariate wavelet denoising technique for images. Dabov et al. [7] proposed a block matching 3D filtering (BM3D) technique for image denoising, which is the state-of-the-art in image denoising. Mäkitalo and Foi [8] developed a Poisson-Gaussian BM3D (PGBM3D) method for denoising. Chen and Kegl [9] proposed an Image denoising technique using complex ridgelets. Chen et al. [10] developed a wavelet-based image denoising method using three scales of dependency in wavelet coefficients. Chen et al. [11] invented an image denoising method using neighbouring wavelet coefficients. Chen et al. [12] developed an image denoising method with neighbour dependency and customized wavelet and threshold. Cho and Bui [13] proposed a multivariate statistical modeling technique for image denoising using wavelet transforms. Cho et al. [14] also studied Image denoising based on wavelet shrinkage using neighbour and level dependency.

Even though Gaussian white noise is well studied, there exist other kinds of noise in real-life images. For example, CMOS and CCD sensors are two special devices that suffer from noise. In CMOS sensors, there exists fixed pattern noise and a mixture of independent additive and multiplicative Gaussian noise. We formulate this kind of noisy image B as:

$$B = A + (k_0 + k_1 A)Z \tag{2}$$

where (k_0, k_1) are two parameters to determine the noise levels, A is the noise-free image, and Z is the Gaussian white noise with N(0,1) distribution. Only a few papers exist in the literature for reducing this kind of noise ([15], [16], [17]). Hirakawa and Parks [15] proposed an image denoising method for signal-dependent noise. Bosco et al. [16] studied signal-dependent raw image denoising using sensor noise characterization via multiple acquisitions. Goossens et al. [17] developed a wavelet domain image denoising technique for non-stationary noise and signal-dependent noise.

In this paper, we propose a new method for reducing this kind of noise. Our method is based on the block matching 3D filtering (BM3D) method [7], which is the state-of-the-art in image denoising. We perform the BM3D to the noisy image, and then conduct the BM3D to the noise residual. We merge these two denoised images and perform another BM3D to this merged image. Our new method is very simple, but it outperforms the standard BM3D [7], BivShrink [6] and Poisson-Gaussian BM3D (PGBM3D) [8] in term of peak signal to noise ratio (PSNR) for nearly all cases in our experiments for the mixture noise model discussed in this paper.

2   Proposed Method

In this paper, we propose a new method to reduce the noise in a noisy image. Our method deals with a mixture of the Gaussian white noise and the signal-dependent noise. Most denoising methods reduce the noise from the noisy images and only keep the denoised images. However, in our method, we denoise the noisy residual and merge this denoised residual into the denoised image. In this way, we can achieve

better denosing results because more fine features can be retained. We can still perform another denoising operation to this merged image if the noise-level is still higher
than a threshold. The noise variance σ_n can be approximated as [18]:

$$\sigma_n = \frac{median(|\ y_{1i}\ |)}{0.6745}, \quad y_{1i} \in \text{subband } HH_1. \tag{3}$$

where HH_1 is the finest scale of wavelet coefficient subband. We only need to perform the wavelet transform on the noisy image for one decomposition scale in order
to estimate σ_n.

In order to achieve better denoising results, we choose the BM3D algorithm [7] to
reduce noise in our proposed method. The BM3D algorithm is divided in two major
steps. The first step estimates the denoised image using hard thresholding during the
collaborative filtering. The second step is based on both the original noisy image and
the basic estimate obtained in the first step.

The collaborative filtering can be summarized as follows:

1. Locate the image patches similar to a given image patch and grouping them
   in a 3D block.
2. 3D linear transform of the 3D block.
3. Shrink the transform spectrum coefficients.
4. Inverse 3D transformation.

As a consequence, this 3D filtering can filter out all 2D image patches in the 3D
block simultaneously. By reducing the noise, the collaborative filtering retains the
finest details shared by grouped blocks and at the same time it preserves the essential
unique features of each individual block. The filtered blocks are then returned to their
original positions. Because there are overlapping in these blocks, we can obtain many
different estimates for each pixel. Aggregation is a particular averaging procedure,
which is exploited to take advantage of this redundancy in each 3D block.

In summary, we list the steps of our new method in this paper as follows:

1. Given the noisy image B, estimate the noise variance σ_n^1 from B according
   to equation (3).

2. Perform BM3D to B as $B_1 = BM3D(B, \sigma_n^1)$. Set $\tilde{B} = 255 \times B_1$ since BM3D
   scales the output image to the range of [0,1].

3. Get the residual image $B_2 = B - \tilde{B}$, and estimate the noise variance σ_n^2 from
   B_2 according to equation (3).

4. Perform BM3D as $C = BM3D(B_2, \sigma_n^2)$.

5. Normalize $C_1 = \tilde{B} + C \times \dfrac{mean(B_2)}{mean(C)}$. Estimate noise variance σ_n^3 from C_1
   according to equation (3).

6. If $\sigma_n^3 > T$ (T=1.0), then $D = BM3D(C_1, \sigma_n^3 \times \sigma_n^2 / 2)$. Here we use a bigger
   noise variance for BM3D because this can generate better denoising results.

7. Output $A = 255 \times D$ since BM3D scales the output image to the range of [0,1]. Stop.

8. If $\sigma_n^3 \leq T$, then output $A = \tilde{B}$. Stop.

The major contribution of this paper is the following. We have taken advantage of the BM3D method, which is the state-of-the-arts in image denoising, for a mixture of the Gaussian white noise and the signal-dependent noise. Our new method can retain more fine features in the denoised images than other existing denoising techniques for image denoising. Experimental results show that our proposed method is similar to the BM3D method for Gaussian white noise, and it is better than the BivShink [6], the BM3D [7], and the Poisson-Gaussian BM3D (PGBM3D) [8] for the mixture noise model for nearly all cases in our experiments.

The major limitation of our proposed method is that it is slower than the standard BM3D since it calls the BM3D for twice and at most three times. We are sacrificing some computation time in exchange for better image quality.

3   Experimental Results

We conducted a number of experiments in order to demonstrate the power of our proposed method in this paper. We tested our method with four grey-scale images: Fingerprint, House, Lena, and Pepper. These images are frequently used in other denoising papers in the literature. We compared our method with the BivShink, the BM3D, and the Poisson-Gaussian BM3D (PGBM3D). We considered both the Gaussian white noise and the signal-dependent noise in our experiments. Tables 1-4 tabulate the peak signal to noise ratio (PSNR) of the denoising methods mentioned above for the seven images, respectively. The PSNR is defined as

$$PSNR = 10 \log_{10} \left(\frac{M \times N \times 255^2}{\sum_{i,j} (B(i,j) - A(i,j))^2} \right) \tag{4}$$

where $M \times N$ is the number of pixels in the image, and A and B are the noise-free and denoised images. Fig. 1 shows the original noisy images, and the images generated by BivShrink, BM3D, PGBM3D, and our proposed method. It can be seen that our proposed method is comparable to BM3D for Gaussian white noise, and it is nearly always better than all other methods compared in this paper for signal-dependent noise. It should be pointed out that the standard BM3D is better than our new method in one case for the image Fingerprint. However, such cases are really rare in our experiments conducted in this paper. The PSNR improvement of our proposed method over standard BM3D sometimes can reach 5 dB. This indicates that our proposed denoising method in this paper is a good choice in enhancing real-life images.

In standard BM3D, the noise variance σ_n is a known parameter for the noisy image. We estimate it by using equation (2) in this paper. Since we only need to perform the wavelet transform for one decomposition scale, the time to estimate σ_n is fast.

Table 1. The peak signal to noise ratio (PSNR) of different denoising methods for image Fingerprint. The best results are highlighted in bold font.

Noise Type	Noise Level	Noisy	BivShrink	BM3D	PGBM3D	Proposed
Gaussian (σ_n)	20	20.10	28.56	**28.83**	26.48	**28.83**
	40	16.08	25.05	**25.51**	22.28	**25.51**
	60	12.56	23.17	**23.75**	21.64	**23.75**
	80	10.06	21.93	**22.54**	19.73	**22.54**
	100	8.12	21.01	**21.55**	17.74	**21.55**
Signal Dependant (k_0, k_1)	(10,0.1)	20.26	27.11	**27.39**	25.72	27.16
	(10,0.3)	13.35	21.66	**23.45**	19.56	**23.45**
	(10,0.5)	9.54	18.45	21.44	15.63	**21.84**
	(10,0.7)	6.90	16.20	19.76	12.95	**20.64**
	(10,0.9)	4.88	14.39	17.77	11.38	**18.97**

Table 2. The peak signal to noise ratio (PSNR) of different denoising methods for image House. The best results are highlighted in bold font

Noise Type	Noise Level	Noisy	BivShrink	BM3D	PGBM3D	Proposed
Gaussian (σ_n)	20	22.08	31.77	**33.78**	29.29	**33.78**
	40	16.06	28.62	**30.64**	27.32	**30.64**
	60	12.54	26.83	**28.76**	25.73	**28.76**
	80	10.04	25.58	**27.15**	24.37	**27.15**
	100	8.10	24.61	25.89	23.16	**26.17**
Signal Dependant (k_0, k_1)	(10,0.1)	20.42	30.31	**32.31**	28.97	**32.31**
	(10,0.3)	13.59	23.44	**27.06**	21.32	**27.06**
	(10,0.5)	9.80	19.39	23.97	17.15	**27.13**
	(10,0.7)	7.17	16.58	21.29	14.38	**25.30**
	(10,0.9)	5.16	14.49	18.49	12.37	**23.91**

Table 3. The peak signal to noise ratio (PSNR) of different denoising methods for image Lena. The best results are highlighted in bold font.

Noise Type	Noise Level	Noisy	BivShrink	BM3D	PGBM3D	Proposed
Gaussian (σ_n)	20	22.09	32.30	**33.03**	29.11	**33.03**
	40	16.08	29.20	**29.82**	27.03	**29.82**
	60	12.56	27.37	**28.15**	25.68	**28.15**
	80	10.06	26.05	**26.82**	24.74	**26.82**
	100	8.12	25.10	**25.76**	24.01	**25.76**
Signal Dependant (k_0, k_1)	(10,0.1)	20.92	30.05	**31.70**	28.86	**31.70**
	(10,0.3)	14.26	22.42	26.27	21.36	**28.87**
	(10,0.5)	10.52	18.22	23.43	17.34	**27.10**
	(10,0.7)	7.91	15.38	21.13	14.63	**25.75**
	(10,0.9)	5.91	13.26	18.87	12.57	**24.60**

Table 4. The peak signal to noise ratio (PSNR) of different denoising methods for image Peppers. The best results are highlighted in bold font.

Noise Type	Noise Level	Noisy	BivShrink	BM3D	PGBM3D	Proposed
Gaussian (σ_n)	20	22.08	29.93	**31.27**	28.09	**31.27**
	40	16.06	26.44	**27.64**	25.26	**27.64**
	60	12.54	24.54	**25.74**	23.67	**25.74**
	80	10.04	23.27	**24.37**	22.32	**24.37**
	100	8.10	22.32	**23.31**	20.93	**23.31**
Signal Dependant (k_0, k_1)	(10,0.1)	20.89	28.51	**29.80**	27.78	**29.80**
	(10,0.3)	14.19	21.83	24.35	20.65	**26.40**
	(10,0.5)	10.44	17.77	21.62	16.67	**24.38**
	(10,0.7)	7.83	14.98	19.51	13.92	**22.84**
	(10,0.9)	5.82	12.91	17.45	12.03	**21.66**

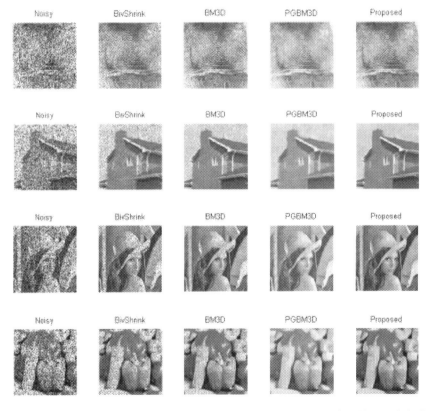

Fig. 1. The noisy images, the denoised images by BivShrink, BM3D, PGBM3D, and the Proposed method for Fingerprint, House, Lena, and Peppers, respectively

4   Conclusions

Reducing noise in digital images corrupted with additive, multiplicative, and mixed noise is a very important topic in image processing. In this paper, we have proposed a new method for reducing the noise in the noisy image. Our method reduces the noise in the residual image and merges this denoised residual image into the previously denoised main image. In this way, more fine features in the image will be retained. Our new denoising method in this paper works well for both the Gaussian white noise and signal-dependent noise. In addition, it nearly always outperforms the BM3D, Poisson-Gaussian BM3D (PGBM3D), and Bivariate shrinking (BivShrink) for signal dependent noise. It achieves similar results as the BM3D for Gaussian white noise.

Future research will be conducted in order to deal with other types of noise in the noisy 1D signals, 2D images, and 3D videos. We may replace the BM3D algorithm with our previous works ([9], [10], [11], [12], [13], [14]) for image denoising. We believe that our proposed method may be applied to multi-spectral or hyper-spectral satellite imagery as well. In addition, we will use other metrics to measure the image visual quality of the denoised images. For instance, we can use such metrics as MSSIM [20], VIF [21], MSE, etc.

Acknowledgments. We would like to thank the authors of [6], [7], [8] and [19] for posting their denoising software on their websites. This work was supported by the Natural Sciences and Engineering Research Council of Canada (NSERC).

References

1. Fathi, A., Naghsh-Nilchi, A.R.: Efficient image denoising method based on a new adaptive wavelet packet thresholding function. IEEE Transactions on Image Processing **21**, 3981–3990 (2012)
2. Chatterjee, P., Milanfar, P.: Patch-based near-optimal image denoising. IEEE Transactions on Image Processing **21**, 1635–1649 (2012)
3. Rajwade, A., Rangarajan, A., Banerje, A.: Image denoising using the higher order singular value decomposition. IEEE Transactions on Pattern Analysis and Machine Intelligence **35**, 849–862 (2013)
4. Motta, G., Ordentlich, E., Ramirez, I., Seroussi, G., Weinberger, M., J.: The iDUDE framework for grayscale image denoising. IEEE Transactions on Image Processing **20** (2011)
5. Miller, M., Kingsburg, N.: Image denoising using derotated complex wavelet coefficients. IEEE Transactions on Image Processing **17**, 1500–1511 (2008)
6. Sendur, L., Selesnick, J.W.: Bivariate shrinkage with local variance estimation. IEEE Signal Processing Letters **9**, 438–441 (2002)
7. Dabov, K., Foi, A., Katkovnik, V., Egiazarian, K.: Image denoising by sparse 3D transform-domain collaborative filtering. IEEE Transactions on Image Processing **16**, 2080–2095 (2007)
8. Mäkitalo, M., Foi, A.: Optimal inversion of the generalized Anscombe transformation for Poisson-Gaussian noise. IEEE Transactions on Image Processing **22**, 91–103 (2013)

 9. Chen, G.Y., Kegl, B.: Image denoising with complex ridgelets. Pattern Recognition **40,** 578–585 (2007)

10. Chen, G.Y., Zhu, W.P., Xie, W.F.: Wavelet-based image denoising using three scales of dependency. IET Image Processing **6**, 756–760 (2012)

11. Chen, G.Y., Bui, T.D., Krzyzak, A.: Image denoising using neighbouring wavelet coefficients. Integrated Computer-Aided Engineering **12**, 99–107 (2005)

12. Chen, G.Y., Bui, T.D., Krzyzak, A.: Image denoising with neighbour dependency and customized wavelet and threshold. Pattern Recognition **38**, 115–124 (2005)

13. Cho, D., Bui, T.D.: Multivariate statistical modeling for image denoising using wavelet transforms. Signal Processing: Image Communication **20**, 77–89 (2005)

14. Cho, D., Bui, T.D., Chen, G.Y.: Image denoising based on wavelet shrinkage using neighbour and level dependency. International Journal of Wavelets, Multiresolution and Information Processing **7**, 299–311 (2009)

15. Hirakawa, K., Parks, T.W.: Image Denoising For Signal-Dependent Noise. In: ICASSP 2005, pp. 29–32 (2005)

16. Bosco, A., Bruna, R.A., Giacalone, D., Battiato, S., Rizzo, R.: Signal-dependent raw image denoising using sensor noise characterization via multiple acquisitions, Digital Photography VI. In: Imai, F., Sampat, N., Xiao, F. (eds.) Proceedings of the SPIE, vol. 7537, article id. 753705 (2010)

17. Goossens, B., Pizurica, A., Philips, W.: Wavelet domain image denoising for non-stationary noise and signal-dependent noise. In: ICIP, pp. 1425–1428 (2006)

18. Donoho, D.L., Johnstone, I.M.: Ideal spatial adaptation by wavelet shrinkage. Biometrika **81**, 425–455 (1994)

19. Lebrun, M.: An Analysis and Implementation of the BM3D Image Denoising Method. Image Processing On Line (2012). http://dx.doi.org/10.5201/ipol.2012.l-bm3d

20. Wang, Z., Bovik, A.C., Sheikh, H.R., Simoncelli, E.P.: Image quality assessment: From error visibility to structural similarity. IEEE Transactions on Image Processing **13**, 600–612 (2004)

21. Sheikh, H.R., Bovik, A.C.: Image information and visual quality. IEEE Transactions on Image Processing 15, 430–444 (2006)

Different-Level Simultaneous Minimization with Aid of Ma Equivalence for Robotic Redundancy Resolution

Binbin Qiu[1,2,3], Dongsheng Guo[1,2,3], Hongzhou Tan[1,2], Zhi Yang[1,2],
and Yunong Zhang[1,2,3]([✉])

[1] School of Information Science and Technology, Sun Yat-sen University (SYSU),
Guangzhou 510006, China
zhynong@mail.sysu.edu.cn
[2] SYSU-CMU Shunde International Joint Research Institute, Shunde 528300, China
ynzhang@ieee.org
[3] Key Laboratory of Autonomous Systems and Networked Control,
Ministry of Education, Guangzhou 510640, China
gdongsh@ieee.org

Abstract. In this paper, with the aid of Ma equivalence (ME), a different-level simultaneous minimization (DLSM) scheme is proposed and investigated for robotic redundancy resolution. Such a DLSM scheme, combining the minimum kinetic energy (MKE) and minimum acceleration norm (MAN) solutions via a weighting factor, can prevent the occurrence of relatively high joint velocity/acceleration and can guarantee the final joint velocity of motion to be near zero. Simulation results based on PUMA560 robot manipulator further substantiate the efficacy and flexibility of the proposed DLSM scheme on robotic redundancy resolution.

Keywords: Different-Level Simultaneous Minimization (DLSM) · Minimum Kinetic Energy (MKE) · Minimum Acceleration Norm (MAN) · Robotic redundancy resolution · Ma Equivalence (ME)

1 Introduction

As for the research of robotics, being a fundamental issue, the redundancy-resolution problem (which closely relates to motion planning of redundant robot manipulators) is described as that, given the desired end-effector path $r_d(t) \in R^m$, the corresponding joint trajectory $\theta(t) \in R^n$ needs to be generated in real time [1,2]. The pseudoinverse-based approach is the conventional and analytic solution to the redundancy-resolution problem [3–5]; i.e., in the form of one minimum-norm particular solution plus a homogeneous solution. Evidently, such an approach can readily solve the redundancy-resolution problem (because it has an analytic-solution formulation). This characteristic has made the research of this approach popular in the past decades [3–7]. Being different from the

© Springer International Publishing Switzerland 2014
Z. Zeng et al. (Eds.): ISNN 2014, LNCS 8866, pp. 431–438, 2014.
DOI: 10.1007/978-3-319-12436-0_48

pseudoinverse-based approach, the quadratic-programming (QP) approach has also been reported on robotic redundancy resolution, in which various recurrent neural networks are involved [1,2]. In general, by combining the QP-based redundancy-resolution scheme and the corresponding neural-network solver, the purpose of motion planning of redundant robot manipulator is thus achieved [1,2,5]. Among these researches [1–8], an inspiring result was presented by Ma et al. [6], showing a relationship between the minimum velocity norm (MVN) scheme and its equivalent acceleration-level minimization scheme, i.e., the so-called Ma equivalence (ME) relationship. Based on the ME, Ma further developed and investigated a balancing scheme [7] with the pseudoinverse-based formulation for robotic redundancy resolution. Note that such a scheme but with the QP-based formulation has recently been presented and investigated in [8].

Being a study case of the pseudoinverse-based approach in this paper, the minimum acceleration norm (MAN) scheme [3,4] has been widely adopted for robotic redundancy resolution at the joint-acceleration level. Such an MAN scheme, minimizing the sum of squares of joint accelerations, is formulated as

$$\ddot{\theta} = J^{\dagger}(\ddot{r}_{\mathrm{d}} - \dot{J}\dot{\theta}), \tag{1}$$

where $\dot{\theta} \in R^n$ and $\ddot{\theta} \in R^n$ are the joint-velocity and joint-acceleration vectors, respectively. In addition, $J^{\dagger} \in R^{n \times m}$ is the pseudoinverse of the Jacobian matrix $J \in R^{m \times n}$, $\dot{J}$ is the time derivative of J, and $\ddot{r}_{\mathrm{d}} \in R^m$ is the second-order time derivative of the desired end-effector path r_{d}.

However, the joint-velocity and joint-acceleration solutions synthesized by the MAN scheme may be relatively large, and the final joint velocity of motion may be nonzero. These drawbacks are undesirable in engineering applications. To remedy the aforementioned phenomena, with the aid of the inspiring ME result [6–8], this paper develops and investigates a different-level simultaneous minimization (DLSM) scheme that combines the minimum kinetic energy (MKE) and MAN solutions via a weighting factor. Computer simulations based on PUMA560 robot manipulator are further performed to show the efficacy and flexibility of such a DLSM scheme. Before ending this section, it is worth pointing out the main contributions of this paper as follows.

1) This paper proposes and investigates the different-level simultaneous minimization (DLSM) scheme for redundancy resolution of robot manipulators, which is based on the weighted combination of MKE and MAN solutions. This is an important investigation for robotics.
2) In comparison with the single-criterion scheme, the DLSM scheme is more flexible in the sense that the latter can yield any suitable combination of MKE and MAN solutions if needed.
3) Simulation results are illustrated to substantiate that the proposed DLSM scheme is effective and flexible on robotic redundancy resolution.

2   Different-Level Simultaneous Minimization (DLSM)

In this section, the different-level simultaneous minimization (DLSM) scheme with the aid of ME is proposed, developed and investigated for robotic redundancy resolution. The resultant DLSM scheme is based on the weighted combination of MKE solution and MAN solution. Note that, since the MKE and MAN schemes are investigated originally at two different levels (i.e., the former corresponds to the joint-velocity level, while the latter corresponds to the joint-acceleration level), such a weighted-sum scheme is termed the DLSM scheme in this paper for robotic redundancy resolution.

To lay a basis for further discussion, the MKE scheme, which minimizes the weighted sum of squares of joint velocities, is formulated as $\dot{\theta} = J_H^\dagger \dot{r}_\mathrm{d} = H^{-1}J^\mathrm{T}(JH^{-1}J^\mathrm{T})^{-1}\dot{r}_\mathrm{d}$, where $J_H^\dagger \in R^{n \times m}$ denotes the weighted pseudoinverse matrix, $H \in R^{n \times n}$ is the positive definite inertia matrix, $\dot{r}_\mathrm{d}$ is the time derivative of the desired end-effector path r_d, and superscript $^\mathrm{T}$ denotes the matrix or vector transposition. By generalizing the inspiring result of MVN-type equivalence in Ma et al.'s work [6–8], the above velocity-level MKE scheme is mathematically equivalent to the following acceleration-level minimization scheme:

$$\ddot{\theta} = J_H^\dagger(\ddot{r}_\mathrm{d} - \dot{J}\dot{\theta}) + (I - J_H^\dagger J)H^{-1}(\dot{J}^\mathrm{T}(JH^{-1}J^\mathrm{T})^{-1} - \dot{H}J_H^\dagger)\dot{r}_\mathrm{d}, \qquad (2)$$

where $I \in R^{n \times n}$ is the identity matrix and $\dot{H}$ denotes the time derivative of H. Note that such an MKE-type equivalence (or termed, generalized Ma equivalence, GME) is the basis of the development of the DLSM scheme in this paper.

As mentioned previously, the solutions of joint velocity and joint acceleration synthesized by the MAN scheme (1) may be relatively large, and the final joint velocity of motion may be nonzero (which is not acceptable for engineering applications). To remedy the undesired phenomena encountered in the MAN scheme, with the aid of the presented GME result, the following DLSM scheme based on the weighted combination of MKE and MAN solutions is developed.

Definition. The DLSM scheme proposed in this paper is formulated as

$$\begin{aligned}\ddot{\theta}_\mathrm{(DLSM)} =&\,\alpha\ddot{\theta}_\mathrm{(MKE)} + (1 - \alpha)\ddot{\theta}_\mathrm{(MAN)} \\ =&\,(\alpha J_H^\dagger + (1 - \alpha)J^\dagger)(\ddot{r}_\mathrm{d} - \dot{J}\dot{\theta}) \qquad (3) \\ &+ \alpha(I - J_H^\dagger J)H^{-1}(\dot{J}^\mathrm{T}(JH^{-1}J^\mathrm{T})^{-1} - \dot{H}J_H^\dagger)\dot{r}_\mathrm{d},\end{aligned}$$

where weighting factor $\alpha \in (0, 1)$ is used to scale the combination of the MKE and MAN solutions. In addition, $\ddot{\theta}_\mathrm{(MKE)}$ and $\ddot{\theta}_\mathrm{(MAN)}$ correspond to the MKE solution computed by (2) and the MAN solution computed by (1).

Explanation. Let us consider the general formulation of an acceleration-level scheme for robotic redundancy resolution, i.e., $\ddot{\theta} = J^\dagger(\ddot{r}_\mathrm{d} - \dot{J}\dot{\theta}) + (I - J^\dagger J)z$, where $z \in R^n$ is an arbitrary vector usually selected by using some optimization criteria [3,4]. By introducing a weighting factor α and choosing z as $\ddot{\theta}_\mathrm{(MKE)}$, the following scheme can be developed:

$$\ddot{\theta} = J^\dagger(\ddot{r}_\mathrm{d} - \dot{J}\dot{\theta}) + \alpha(I - J^\dagger J)\ddot{\theta}_\mathrm{(MKE)}, \qquad (4)$$

where $\ddot{\theta}_\mathrm{(MKE)}$ is obtained by (2). It can be generalized from $JJ_H^\dagger = I \in R^{m \times m}$ that $(I - J^\dagger J)J_H^\dagger = J_H^\dagger - J^\dagger$ and $(I - J^\dagger J)(I - J_H^\dagger J) = I - J_H^\dagger J$, (4) is reformulated as

$$\ddot{\theta} = \alpha\ddot{\theta}_\mathrm{(MKE)} + (1 - \alpha)\ddot{\theta}_\mathrm{(MAN)}, \qquad (5)$$

where $\ddot{\theta}_{\text{(MAN)}}$ is obtained by (1). Thus, substituting (2) and (1) into (5) yields $\ddot{\theta}_{\text{(DLSM)}} = (\alpha J_H^{\dagger} + (1 - \alpha)J^{\dagger})(\ddot{r}_{\text{d}} - \dot{J}\dot{\theta}) + \alpha(I - J_H^{\dagger}J)H^{-1}(\dot{J}^{\text{T}}(JH^{-1}J^{\text{T}})^{-1} - \dot{H}J_H^{\dagger})\dot{r}_{\text{d}}$, which is the DLSM scheme (3). The explanation is thus complete. □

Evidently, such a scheme (3) reduces to the pure MAN scheme when $\alpha \to 0$ (i.e., only the MAN purpose is considered) or to the acceleration-level MKE scheme when $\alpha \to 1$ (i.e., only the MKE purpose is considered). In addition, the proposed DLSM scheme (3) is more flexible in the sense that it can yield various suitable combinations of MKE and MAN solutions (if needed). Such a scheme resolved at the joint-acceleration level can thus achieve the MKE and MAN purposes simultaneously. Note that, by using a suitable value of α, the proposed DLSM scheme (3) can prevent the occurrence of relatively high joint velocity and joint acceleration (which may be caused by the MAN scheme), and can guarantee the final joint velocity of motion to be near zero. This characteristic can be regarded as the actual significance of such a scheme (3).

Remark 1. It is worth mentioning here that, besides the aforementioned MAN scheme, various pseudoinverse-type schemes have been developed and studied for robotic redundancy resolution [3–5]. Thus, by means of the same way (i.e., the weighted combination of MKE and MAN schemes), various DLSM schemes can also be developed for further investigation of motion planning of redundant robot manipulators. In this paper, just the concept of different-level simultaneous optimization depicted in the pseudoinverse-type formulation is presented for redundant robot manipulators. With the aid of (MVN-type or MKE-type) ME, the presented weighted-sum approach can provide various choices on robotic redundancy resolution from a single level to different levels naturally.

3   Simulative Verifications and Comparisons

In this section, computer-simulation results (including comparisons and observations) are illustrated to show the efficacy and flexibility of the proposed DLSM scheme (3). Note that, in the simulations, the task duration is set as $T = 10$ s and the initial state is set as $\theta(0) = [0, -\pi/4, 0, \pi/2, -\pi/4, 0]^{\text{T}}$ rad. Specifically, such a scheme is simulated for the PUMA560 end-effector tracking a circular path with the radius being 0.25 m. Note that, for comparative purposes, the pure MAN scheme (1) is also simulated in this example. The corresponding simulation results are illustrated in Figs. 1 and 2, Table 1 and finally Fig. 3.

Figure 1 shows the simulation results, which are synthesized by the pure MAN scheme (1). As seen from Fig. 1(a), the end-effector trajectory is close to the desired circular path (with the maximal positioning error being less than 4.0×10^{-6} m), which illustrates that (1) is effective on robotic redundancy resolution. However, as illustrated in Figs. 1(b) through (d), the values of some joint velocities and joint accelerations appear to be relatively large for engineering applications, and the resultant $\dot{\theta}^{\text{T}}H\dot{\theta}$ (corresponding to the kinetic energy) and $\ddot{\theta}^{\text{T}}\ddot{\theta}$ (corresponding to the acceleration norm) are relatively large. Besides, Fig. 1(b) shows that some final velocities are not zero after motion and are too large for practical applications.

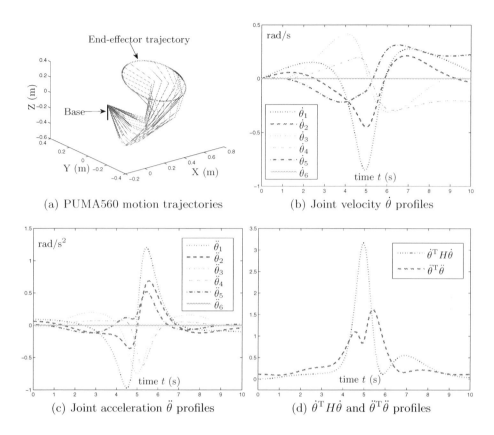

(a) PUMA560 motion trajectories

(b) Joint velocity $\dot{\theta}$ profiles

(c) Joint acceleration $\ddot{\theta}$ profiles

(d) $\dot{\theta}^{\mathrm{T}} H \dot{\theta}$ and $\ddot{\theta}^{\mathrm{T}} \ddot{\theta}$ profiles

Fig. 1. PUMA560 end-effector tracks a circular path synthesized by MAN scheme (1)

Figure 2(a) illustrates the motion trajectories of the PUMA560 robot manipulator, which is synthesized by the proposed DLSM scheme (3) with $\alpha = 0.5$. Figures 2(b) through (d) correspond to joint velocity, joint acceleration, and $\dot{\theta}^{\mathrm{T}} H \dot{\theta}$ and $\ddot{\theta}^{\mathrm{T}} \ddot{\theta}$ transients. As shown in Fig. 2(a), the simulated trajectory of the PUMA560 end-effector is close to the desired circular path (in which the maximal positioning error is less than 2.2778×10^{-6} m). In addition, Figs. 2(b) and (c) show that the corresponding values of joint velocity $\dot{\theta}$ and joint acceleration $\ddot{\theta}$ are relatively small, as compared with those shown in Fig. 1. Thus, the resultant $\dot{\theta}^{\mathrm{T}} H \dot{\theta}$ and $\ddot{\theta}^{\mathrm{T}} \ddot{\theta}$ given in Fig. 2(d) are smaller than those shown in Fig. 1(d), thereby implying that the MAN solution characteristic is still obtained by using the proposed DLSM scheme (3). More importantly, Fig. 2(b) shows that the joint velocities are near zero at the end of motion (with the absolute value being $|\dot{\theta}(T)| = [2.3186 \times 10^{-3}, 2.3513 \times 10^{-2}, 4.5981 \times 10^{-2}, 2.0115 \times 10^{-2}, 5.5819 \times 10^{-2}, 2.2245 \times 10^{-5}]^{\mathrm{T}}$ rad/s), which is acceptable for practical applications. These results substantiate the efficacy of the proposed DLSM scheme (3) for motion planning of redundant robot manipulators.

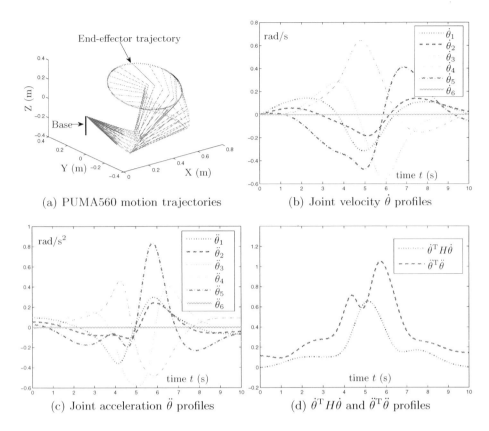

(a) PUMA560 motion trajectories

(b) Joint velocity $\dot{\theta}$ profiles

(c) Joint acceleration $\ddot{\theta}$ profiles

(d) $\dot{\theta}^{\mathrm{T}} H \dot{\theta}$ and $\ddot{\theta}^{\mathrm{T}} \ddot{\theta}$ profiles

Fig. 2. PUMA560 end-effector tracks the circular path synthesized by the proposed DLSM scheme (3) with $\alpha = 0.5$

For further investigation, we also simulate the proposed DLSM scheme (3) by using different values of α. Table 1 and Fig. 3 show the quantitative and qualitative evaluations of such a scheme with different α values, respectively. On the one hand, as seen from Table 1, the maximal end-effector positioning errors are small enough (i.e., of order 10^{-6} m). In addition, the final joint velocities of motion are near zero (shown in Table 1), which are suitable for practical applications. On the other hand, as shown in Fig. 3, the values of $\dot{\theta}^{\mathrm{T}} H \dot{\theta}$ and $\ddot{\theta}^{\mathrm{T}} \ddot{\theta}$ are small, which means that the MAN purpose is also achieved by using (3). Evidently, Table 1 and Fig. 3 provide an intuitive result about the performance of the proposed DLSM scheme (3) on robotic redundancy resolution. Besides, the related simulation results (omitted because of space limitation) show that the $\dot{\theta}$ and $\ddot{\theta}$ solutions via the proposed DLSM scheme (3) with different α values are small. These results show again the efficacy of the proposed DLSM scheme (3). By summarizing these results, the proposed DLSM scheme (3) with a suitable value of α can prevent the occurrence of relatively high joint-velocity/acceleration, and can guarantee the final joint-velocity of motion to be near zero.

Table 1. Maximal end-effector positioning errors (m) and final joint velocities (rad/s) when the PUMA560 end-effector tracks the circular path synthesized by the proposed DLSM scheme (3) with different values of α used

α	0.1	0.2	0.3	0.4
positioning error	1.4856×10^{-6}	1.6715×10^{-6}	2.0331×10^{-6}	3.5406×10^{-6}
$\|\dot{\theta}_1(T)\|$	6.3325×10^{-2}	3.3347×10^{-2}	9.5012×10^{-3}	3.1014×10^{-3}
$\|\dot{\theta}_2(T)\|$	4.0734×10^{-2}	3.4478×10^{-2}	1.6349×10^{-2}	3.7190×10^{-3}
$\|\dot{\theta}_3(T)\|$	3.4815×10^{-2}	5.2343×10^{-2}	2.5804×10^{-2}	1.2449×10^{-2}
$\|\dot{\theta}_4(T)\|$	1.6168×10^{-1}	7.2470×10^{-2}	3.4660×10^{-3}	$\mathbf{2.2886 \times 10^{-2}}$
$\|\dot{\theta}_5(T)\|$	1.8812×10^{-1}	1.1706×10^{-1}	4.5357×10^{-2}	1.3316×10^{-2}
$\|\dot{\theta}_6(T)\|$	5.6575×10^{-6}	1.3016×10^{-5}	1.9072×10^{-5}	2.1983×10^{-5}

α	0.6	0.7	0.8	0.9
positioning error	3.3771×10^{-6}	$\mathbf{1.4125 \times 10^{-6}}$	2.0554×10^{-6}	1.9342×10^{-6}
$\|\dot{\theta}_1(T)\|$	7.5610×10^{-4}	1.9167×10^{-3}	3.3026×10^{-3}	1.3442×10^{-3}
$\|\dot{\theta}_2(T)\|$	4.1493×10^{-2}	5.4359×10^{-2}	5.2967×10^{-2}	3.5061×10^{-2}
$\|\dot{\theta}_3(T)\|$	7.3764×10^{-2}	9.2616×10^{-2}	9.1463×10^{-2}	5.7169×10^{-2}
$\|\dot{\theta}_4(T)\|$	6.6480×10^{-3}	5.5930×10^{-3}	7.9130×10^{-3}	1.4460×10^{-3}
$\|\dot{\theta}_5(T)\|$	8.3348×10^{-2}	9.5893×10^{-2}	9.0577×10^{-2}	6.0597×10^{-2}
$\|\dot{\theta}_6(T)\|$	2.0598×10^{-5}	1.6269×10^{-5}	8.5640×10^{-6}	9.6910×10^{-7}

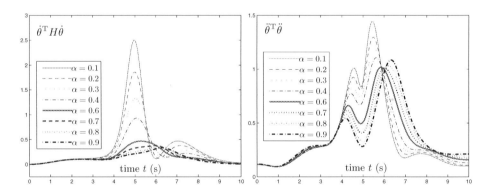

Fig. 3. Profiles of $\dot{\theta}^{\mathrm{T}} H \dot{\theta}$ and $\ddot{\theta}^{\mathrm{T}} \ddot{\theta}$ when the PUMA560 end-effector tracks the circular path synthesized by the proposed DLSM scheme (3) with different values of α used

In summary, Figs. 1 through 3 and Table 1 (as well as the related simulation results) have substantiated that the proposed DLSM scheme (3) is effective and flexible on redundancy resolution of robot manipulators.

Remark 2. The weighting factor α is used to scale the combination of velocity-level and acceleration-level solutions. Different α values can be chosen for different situations and/or requirements. For instance, to obtain a solution for a situation in which MKE has more effect compared with MAN, we can set α to have a larger value (e.g., 0.8 or 0.9). By choosing different values of α in the simulative tests (with results shown in Table 1 and Fig. 3), the value chosen for α is determined by the requirements needed to obtain an acceptable end-effector positioning error and solution realizability with a (near) zero final velocity.

4   Conclusions

By generalizing Ma et al.'s inspiring work (or termed, Ma equivalence) [6–8], this paper has proposed and investigated the different-level simultaneous minimization (DLSM) scheme (3), which combines the minimum kinetic energy (MKE) solution (2) and minimum acceleration norm (MAN) solution (1) via a weighting factor α, for robotic redundancy resolution. Computer-simulation results based on PUMA560 robot manipulator have further shown that the proposed DLSM scheme (3) with a suitable α value not only prevents the occurrence of relatively high joint-velocity/acceleration, but also guarantees the final joint-velocity of motion to be near zero. The efficacy and flexibility of the proposed DLSM scheme (3) for robotic redundancy resolution have thus been substantiated.

Acknowledgments. This work is supported by the 2012 Scholarship Award for Excellent Doctoral Student Granted by Ministry of Education of China (with number 3191004), and by the Foundation of Key Laboratory of Autonomous Systems and Networked Control, Ministry of Education, China (with number 2013A07). Besides, kindly note that all authors of the paper are jointly of the first authorship.

References

1. Zhang, Y., Zhang, Z.: Repetitive Motion Planning and Control of Redundant Robot Manipulators. Springer, New York (2013)
2. Cai, B., Zhang, Y.: Optimal and Efficient Motion Planning of Redundant Robot Manipulators. Lambert Academic Publishing, Germany (2013)
3. Siciliano, B., Khatib, O.: Springer Handbook of Robotics. Springer, Heidelberg (2008)
4. Siciliano, B., Sciavicco, L., Villani, L., Oriolo, G.: Robotics: Modelling, Planning and Control. Springer, London (2009)
5. Guo, D., Zhang, Y.: Li-Function Activated ZNN with Finite-Time Convergence Applied to Redundant-Manipulator Kinematic Control via Time-Varying Jacobian Matrix Pseudoinversion. Appl. Soft. Comput. **24**, 158–168 (2014)
6. Ma, S., Hirose, S., Yoshinada, H.: Dynamic Redundancy Resolution of Redundant Manipulators with Local Optimization of a Kinematic Criterion. Adv. Rob. **10**, 236–243 (1995)
7. Ma, S.: A Balancing Technique to Stabilize Local Torque Optimization Solution of Redundant Manipulators. J. Robot. Syst. **13**, 177–185 (1996)
8. Zhang, Y., Guo, D., Ma, S.: Different-Level Simultaneous Minimization of Joint-Velocity and Joint-Torque for Redundant Robot Manipulators. J. Intell. Robot. Syst. **72**, 301–323 (2013)

Content-Adaptive Rain and Snow Removal Algorithms for Single Image

Shujian Yu[1,3(✉)], Yixiao Zhao[1,2(✉)], Yi Mou[3], Jinghui Wu[2], Lu Han[2], Xiaopeng Yang[2], and Baojun Zhao[2]

[1] Department of Electrical and Computer Engineering, University of Florida, Gainesville, USA
{yusjlcy9011,beng0429}@ufl.edu
[2] School of Information and Electronics, Beijing Institute of Technology, Beijing, China
zbj@bit.edu.cn
[3] Department of Electronics and Information Engineering,
Huazhong University of Science and Technology, Wuhan, China

Abstract. In this paper, we present two content-adaptive rain and snow removal algorithms for single image based on filtering. The first algorithm treats rain and snow removal task as an issue of bilateral filtering, where a content-based saliency prior is introduced. While the other views the same task from the perspective of guided-image-filtering, and the guidance image is derived according to the statistical property of raindrops or snowflakes as well as image background content. A comparative study and quantitative evaluation with some main existing image assessment algorithms demonstrate better performance of our proposed algorithms. The main contributions of our works are twofold: firstly, to the best of our knowledge, our algorithms are among the first to introduce image content information for single-image-based rain and snow removal; and secondly, we are also among the first to introduce quantitative assessment for single-image-based rain and snow removal tasks.

Keywords: Rain removal · Snow removal · Bilateral filtering · Guided-image-filtering · Outdoor vision

1   Introduction

Computer vision of indoor situations has already been extensively studied, whereas vision algorithms that can handle complex and unpredictable behaviors caused by different weather conditions, such as rain, snow, fog, or haze, in outdoor situations still remain as challenging problems [1].

Garg and Nayar [2] classified weather effects into two types: steady weather such as fog and haze, and dynamic weather such as rain and snow, based on the size of weather particles. In [3], an novel dehaze algorithm with dark channel prior was proposed, it achieves pretty good performance for removing steady weather effects.

However, for larger particles such as raindrops and snowflakes, reducing or removing the weather effects while preserving scene information is a different and difficult

© Springer International Publishing Switzerland 2014
Z. Zeng et al. (Eds.): ISNN 2014, LNCS 8866, pp. 439–448, 2014.
DOI: 10.1007/978-3-319-12436-0_49

task due to two main reasons: firstly, the visual appearance of raindrops or snowflakes depends both on their backgrounds and lighting conditions, which makes it difficult to build a general appearance model; and secondly, unlike steady weather conditions, rain and snow effects vary significantly over spatial and temporal domain [8].

Previous works for reducing the visibility of dynamic weather effects are primarily based on video, where physical and photometric properties of raindrops or snowflakes can be well employed over the whole video sequence [4-7]. Nevertheless, when only a single image is available, such as an image taken by a camera or downloaded from internet, algorithms for single-image-based rain/snow removal are essential.

Fu [8] proposed a rain streak removal diagram with image decomposition and morphological component analysis. This method assumes that rain streaks distributed homogeneously over the image. However, if the raindrops distributed heterogeneous-ly and sparsely, it is difficult or impossible to learn a dictionary for rain streaks. Then, Xu [9] and Zheng [10] introduced guided-image-filtering [11] for rain/snow removal, where different refined guidance images are proposed separately. Both of the two algorithms ignored image content itself as well as statistical property of raindrops and snowflakes, and will inevitably introduce blurring artifacts to non-rain texture details.

In this paper, we propose two novel content-adaptive algorithms for single-image-based rain and snow removal. The first uses a content-based saliency a priori to seg-ment original image, then different parts in resulting image correspond to regions with different perception intensity. Thus an easy but effective strategy is to adjust filter parameters adaptively. The other employs a guided-image-filtering based algorithm to remove rain and snow for single image, where the guidance image is derived from the statistical chromatic and the photometric properties of raindrops or snowflakes.

A comprehensive analysis is performed and quantitative comparison with two fa-mous existing image assessment standards - visual information fidelity (VIF) metric [12] and feature-similarity (FSIM) index [13] are also carried out. Experimental re-sults demonstrate the effectiveness and efficiency of our proposed algorithms.

The remainder of this paper is organized as followings: in section II and III, the de-tails of our proposed two algorithms - bilateral filtering based algorithm with saliency a prior and guided-image-filtering based algorithm with statistical property are well explained; then in section IV, a comprehensive comparison analysis and quantitative evaluation is conducted; finally, section V concludes this paper.

2   Preliminary Knowledge

The intensity of rain and snow generally falls into four categories - light, moderate, heavy and violent (Fig. 1) - based on the rate of precipitation [14]. For images con-taminated by raindrops or snowflakes with light or moderate intensity, it is difficult or impossible to learn a construction model accurately due to lack of useful information provided by single image as well as their sparse distribution and random directions. If images are contaminated with weather effects of heavy or violent intensity, although

Fig. 1. The visual appearances of rain under different intensities. From left to right: (a) a beautiful girl in light rain; (b) a sitting man in moderate rain; (c) a building in heavy rain; and (d) crossroads in violent rain.

we can coarsely separate weather effects from background via dictionary learning and sparse coding [8], details in background (especially edges and corners resembling rain streaks or snowflakes) are often eliminated at the same time, only for the reason that raindrops or snowflakes are highly mixed with similar texture in almost each patch of the image. Therefore, unlike conventional image restoration problems, single-image-based rain/snow removal is not an easy and trivial task. However, algorithms based on edge-preserving filtering [11,15] provide a reliable solution.

2.1 Bilateral Filtering

As a simple, non-iterative scheme for edge-preserving smoothing, bilateral filtering is always the first step of computer vision based algorithms for different systems, such as vehicle tracking system, pedestrian detection and surveillance system [8], under rain or snow weather conditions. The basic idea of bilateral filtering is Gaussian distribution based averaging, which means that the intensity value at each pixel in an image is replaced by a weighted average of intensity values from nearby pixels [15]. However, the weights depend not only on Euclidean distance but also on the color intensity differences. This preserves sharp edges by systematically looping through each pixel and adjusting weights to the adjacent pixels accordingly [11].

2.2 Guided Image Filtering

In [11], He proposed a novel explicit image filter called guided filter. Derived from a local linear model, the guided filter computes the filtering output I_{guide} by considering the content of guidance image I, which can be the input image itself or another different image. In window ω_k, the output pixel q_i can be represented as:

$$q_i = a_k I_i + b_k, \forall i \epsilon \omega_k \tag{1}$$

where a_k and b_k are defined as:

$$a_k = ((\Sigma_{i \in \omega_k} I_i p_i)/|\omega| - u_k \overline{p_k})/(\sigma_k^2 + \varepsilon) \tag{2}$$

$$b_k = \overline{p_k} - a_k u_k \tag{3}$$

Here, p is the filter input, u_k and σ_k^2 are the mean and variance of I in ω_k, $\overline{p_k}$ is the mean of p in ω_k, $|\omega|$ is the pixel number in ω_k.

3   Bilateral Filtering Based Rain/Snow Removal with a Saliency Prior

According to aforementioned description in section II, as a widely used method for rain and snow removal for single image, conventional bilateral filtering has two major drawbacks: firstly, it will produce staircase effects at edges, especially at contours of un-degraded objects, and secondly, the filter parameters remain the same for all parts of the image without any emphasis, which will introduce similar blurring effects or flatting performance to both dominant objects and redundant backgrounds.

However, different parts of an image will cause different perception intensities to an observer: the salient parts, which always corresponding to dominant objects that are less influenced by raindrops or snowflakes, will lead to more perception intensity, whereas the less salient parts, which always corresponding to backgrounds or redundant image content, will result in low perception intensity. Therefore, it is essential to adaptively adjust filter parameters based on image content, i.e. content-aware saliency information in an image.

3.1   Content-Based Saliency Detection

Normally, raindrops or snowflakes result in low perception intensity compared with prominent objects in an image, due to the fact that such weather effects are always highly mixed with backgrounds and sparsely distributed.

In this section, we successfully introduce a saliency a priori for bilateral filtering. Different from the conventional bilateral filtering, our method can adaptively adjust filter parameters based on the intensity of saliency.

For the computation of saliency map, we use context-aware method [17] proposed by Goferman as our pre-processing step. The literatures on saliency detection contain nearly 65 vision attention models in the last 25 years [16], and we explain here why context-aware method is selected. Firstly, note that we have emphasized content-adaptive in this paper, and therefore methods designed for saliency detection only for dominant objects, regardless of surrounding context information, such as spectral residual approach [18] and global contrast based method [19], fall outside the scope of this paper. Secondly, of the existing methods for content-based saliency detection, methods based on symmetric surround or combined features introduced in [20] appear to be the closest in spirit to the context-aware saliency detection utilized here. However, only limited principles of human visual attention from psychology are utilized in [20], whereas [17] realized all of them mathematically. Fig. 2 demonstrates the saliency detection results.

| | | | |
| (a) | (b) | (c) | (d) |

Fig. 2. Ground-truth images with rain/snow and their corresponding saliency map

3.2   Ordered Sample Clustering for Histogram of Saliency Map

For the saliency map from section 3.1, larger pixel values represent high salient regions of the original image, while smaller ones correspond to low salient regions.

With ordered sample clustering algorithm, the histogram of saliency map can be partitioned into several segments, and each segment represents a specific saliency intensity level. In this paper, we used the well-known "optimal partition method (i.e. fisher method)" to separate the image histogram into several segments. The basic idea of "optimal partition" is minimizing the increment of sum of deviation squares of the ordered sample after segmentation.

Fig. 3(a) illustrates the clustering result for saliency map histogram of Fig. 2(b) (clustered into 3 categories). Fig. 3(b) illustrates the image segmentation of Fig. 2(a) according to clustering result, each segmented region corresponding to a specific segment of histogram of saliency map.

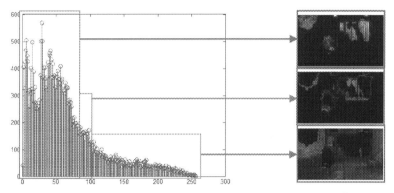

Fig. 3. Image segmentation according to ordered sample clustering to histogram of saliency map. (a) is the histogram of saliency map, the two breaking points are 75 and 101. (b) is image segmentation results.

3.3   Adaptive Parameter Adjustment

The performance of bilateral filtering depends on three parameters: filter width, standard deviations of geometric spread and photometric spread. The geometric spread σ_d controls the extent of low-pass filtering: a large value blurs more, and vice versa. Similarly, the photometric spread σ_r in the image range is set to achieve the desired amount of combination of pixel values. Therefore, it is reliable to allocate smaller filter parameters in the regions of high saliency, while in the low salient regions, larger parameters are preferable.

In our experiment, initial values for these three parameters, i.e. filter half-width, σ_d, and σ_r are set as 5, 3, and 0.1, respectively. For regions of lower saliency in the next level, we will increase each parameter by 2, 5, and 5 times, separately. The results are shown in Fig.4. Compared with conventional bilateral filtering method, our algorithm can preserve more image-related information (content) and remove raindrops and snowflakes with higher accuracy.

Fig. 4. Weather removal results for Fig. 3(a) and Fig.3(c) with refined bilateral filtering

4   Guided-Image-Filtering Based Rain/Snow Removal with Statistical Chromatic Property

Guided-image-filtering achieved good performance for dehaze. After that, feasible and practical extensions have been applied towards rain and snow removal [9,10]. Conventional revised guided-image-filtering algorithms are all based on the principle to preserve more useful details, especially edges. Here, we present a new method to extract guidance image based on statistical properties of raindrops/snowflakes.

4.1   Chromatic Property of Raindrops and Snowflakes

In [4], a chromatic model for spherical raindrop is presented (also applicable for snowflakes). It pointed out that raindrop refracts a wide range of light, therefore the projection of raindrop in the image is much brighter than its background. Because of the difference in wavelength, blue light has a larger index of refraction and a wider field of view than red light. Therefore, a raindrop should refract a little more blue light coming from the background. Followed with [2,4], we further investigated the subtle difference of refraction to the appearance of raindrops and snowflakes. According to our statistical observations[1], the intensity differences of R, G, and B channel caused by raindrops/snowflakes are roughly the same.

4.2   Photometric Property of Raindrops and Snowflakes

When a falling raindrop or a snowflake is captured by a camera, the intensity is a linear combination of irradiance of raindrops or snowflakes and the irradiance of background [9]. Their intensity values can be both expressed as:

$$I_{rs} = \int_0^{\tau} \overline{E_{rs}} \, dt + \iint_{\tau}^{T} \overline{E_b} dt \tag{4}$$

where I_{rs} is the intensity value of a pixel effected by raindrops or snowflakes, $\overline{E_{rs}}$ is the time-averaged irradiance of a stationary raindrop or snowflake, $\overline{E_b}$ is the time-averaged irradiance of background, T is the exposure time and during the time τ a raindrop or a snowflake is passing through the pixel.

---

[1] We verified these observations using two public videos from [21]. In each frame from the two videos, a fixed region of 50*50 is selected, and our observations are based on these 2500 pixel sequences.

If we define I_b as the background intensity, I_E as the intensity of a stationary raindrops or snowflakes at the time T. Eq.(4) can be simplified as:

$$I_{rs} = \alpha I_E + (1 - \alpha)I_b, \text{where } \alpha = \tau/T \tag{5}$$

Eq.(5) provides a photometric model of raindrops and snowflakes.

4.3   Refined Guidance Image Extraction

Here, we proposed a new method for guidance image extraction based on both chromatic and photometric properties of raindrops or snowflakes. Firstly, extraction of guidance image can be achieved through following procedures:

I. Smooth input image with bilateral filtering in R, G, B channels separately. The result images can be represented as I_{bf_R}, I_{bf_G} and I_{bf_B}, respectively.

II. Compute abstract differential images between three images from step I. Then we have three difference images: I_{R-G}, I_{G-B} and I_{B-R}.

III. Use Eq.(6) to compute the mean image of I_{R-G}, I_{G-B} and I_{B-R}, I_{mean} can be used as our first refined guided image.

$$I_{mean} = (I_{R-G} + I_{G-B} + I_{B-R})/3 \tag{6}$$

In addition, we note that Eq.(5) is established for R, G and B channels. If C indicates a coordinate of the RGB space and I_{rs-C} is the maximum value at RGB space of I_{rs}, I_{b-C} must be the maximum value at RGB space of I_b. This relation also holds for the minimum value of each vector at RGB space. Therefore, we have:

$$I_{rs-max} = \alpha I_{E-max} + (1 - \alpha)I_{b-max} \tag{7}$$

$$I_{rs-min} = \alpha I_{E-min} + (1 - \alpha)I_{b-min} \tag{8}$$

According to aforementioned description $I_{E-max} = I_{E-min}$. Subtract (8) from (7):

$$I_f = I_{rs-max} - I_{rs-min} = (1 - \alpha)(I_{b-max} - I_{b-min}) \tag{9}$$

Fig. 5. Raindrops/Snowflakes removal results with guided image filtering: first row shows removal reuslt for sitting man in rain weather; secod row shows removal result for mailbox in snow weather. From left to right: I_f, I_{mean}, $I_{guidance}$ and weather removal result.

Obviously, I_f is not affected by weather effects. Therefore, final guidance image $I_{guidance}$ can be represented as weighted combination of I_{mean} and I_f (See Fig.5):

$$I_{guidance} = \beta I_{mean} + (1 - \beta)I_f \tag{10}$$

5   Experiment and Result Analysis

We have conducted both qualitative and quantitative experiments to assess our proposed two algorithms with other state-of-art raindrop/snowflake removal algorithms. The goal of objective image quality assessment research is to provide computational models that can automatically predict perceptual image quality. In this paper, we will utilize VIF [12] and FSIM [13] to assess our raindrops/snowflakes removal results.

5.1   Qualitative Comparison

Fig.6 shows the weather effects removal results of several different filtering based methods[2]. The top row illustrates removal effects for raindrops, and the bottom row illustrates removal effects for snowflakes[3]. As can be seen, "GF" has good performance for rain/snow removal, but it introduces more blurring artificial effects. "BF" can keep more detail information, but it always preserve more weather effects. Compared with these two, "Xu" can keep more useful structure information, "Our I" can also keep more useful details and remove more weather effects. In addition, "Our II" outperforms "GF".

| (a)   (b)   (c)   (d)   (e) |

Fig. 6. Illustration of weather effects removal results with different algorithms: (a-e) removal results with "BF", "GF", "Xu", "Our I" and "Our II", respectively.

5.2   Quantitative Evaluation

In this section, VIF and FSIM are utilized to evaluate the raindrops and snowflakes removal effects quantitatively. Test images (See Fig.7) are downloaded from [21], where weather effects are added to the ground truth video frames with advanced rendering techniques [5]. Experimental results are presented in Table 1 and Table 2.

---

[2] More results are available from the author's homepage http://www.yushujian.com/index.html.

[3] We denote bilateral filtering in [15] as BF, guided-image-filtering in [11] as GF, Xu's method in [9] as Xu, our two proposed algorithm as Our I and Our II in section 5.1 and section 5.2.

| (a) | (b) | (c) | (d) |

Fig. 7. Two representative frames from two test videos: (a) and (c) are ground-truth images; (b) and (d) are ground-truth images with added weather effects

Table 1. Averaged VIF value for video frames with different algorithms

	BF	GF	Xu	Our I	Our II
Video I	0.1444	0.1172	0.1463	**0.1562**	0.1484
Video II	0.5378	0.5193	0.4463	**0.6477**	0.5745

Table 2. Averaged FSIM value for video frames with different algorithms

	BF	GF	Xu	Our I	Our II
Video I	0.3219	0.3039	0.3218	0.3302	**0.3360**
Video II	0.8180	0.7717	0.7553	**0.8395**	0.8119

6   Conclusion

In this paper, we have proposed two independent algorithms for raindrops/snowflakes removal in single image. Firstly, we successfully introduced a saliency-map-prior for bilateral filtering, the improved algorithm can automatically adjust filter parameters based on image content. In addition, we have also proposed a novel way for guidance image extraction based on properties of rain. The refined guided image filtering can achieve better performance than conventional version. Finally, we have conducted experiments to assess different rain streak removal methods both from subjective perspective and objective measurements. Experimental results demonstrate the effectiveness and efficiency of our proposed algorithms.

References

1. Zhang, X., Li, H., Qi, Y., Leow, W.K., Ng, T.K.: Rain Removal in Video by Combining Temporal and Chromatic Properties. In: Proceedings of the IEEE International Conference on Multimedia and Expo, pp. 461–464 (2006)
2. Garg, K., Nayar, S.K.: Detection and Removal of Rain from Videos. In: Proceedings of the IEEE Computer Society Conference on Computer Vision and Pattern Recognition, vol. 1, pp. I-528–I-535 (2004)
3. He, K., Sun, J., Tang, X.: Single Image Haze Removal Using Dark Channel Prior. IEEE Trans. Pattern Analysis and Machine Intelligence 33(12), 2341–2353 (2011)
4. Garg, K., Nayar, S.K.: Vision and Rain. International Journal of Computer Vision 75(1), 3–27 (2007)

5. Garg, K., Nayar, S.K.: Photorealistic rendering of rain streaks. ACM Transactions on Graphics (TOG) **25**(3), 996–1002 (2006)
6. Brewer, N., Liu, N.: Using the shape characteristics of rain to identify and remove rain from video. In: da Vitoria Lobo, N., Kasparis, T., Roli, F., Kwok, J.T., Georgiopoulos, M., Anagnostopoulos, G.C., Loog, M. (eds.) SSPR&SPR 2008. LNCS, vol. 5342, pp. 451–458. Springer, Heidelberg (2008)
7. Bossu, J., Hautière, N., Tarel, J.P.: Rain or snow detection in image sequences through use of a histogram of orientation of streaks. International Journal of Computer Vision **93**(3), 348–367 (2011)
8. Fu, Y.H., Kang, L.W., Lin, C.W., Hsu, C.T.: Single-frame-based rain removal via image decomposition. In: IEEE International Conference on Acoustics, Speech and Signal Processing (ICASSP), pp. 1453–1456 (2011)
9. Xu, J., Zhao, W., Liu, P., Tang, X.: An Improved Guidance Image Based Method to Remove Rain and Snow in a Single Image. Computer and Information Science **5**(3), 49 (2012)
10. Zheng, X., Liao, Y., Guo, W., Fu, X., Ding, X.: Single-Image-Based Rain and Snow Removal Using Multi-guided Filter. In: Lee, M., Hirose, A., Hou, Z.-G., Kil, R.M. (eds.) ICONIP 2013, Part III. LNCS, vol. 8228, pp. 258–265. Springer, Heidelberg (2013)
11. He, K., Sun, J., Tang, X.: Guided image filtering. In: Daniilidis, K., Maragos, P., Paragios, N. (eds.) ECCV 2010, Part I. LNCS, vol. 6311, pp. 1–14. Springer, Heidelberg (2010)
12. Sheikh, H.R., Bovik, A.C.: Image information and visual quality. IEEE Trans. Image Processing **15**(2), 430–444 (2006)
13. Zhang, L., Zhang, D., Mou, X.: FSIM: a feature similarity index for image quality assessment. IEEE Trans. Image Processing **20**(8), 2378–2386 (2011)
14. Glickman, T.S., Zenk, W.: Glossary of meteorology (2000)
15. Tomasi, C., Manduchi, R.: Bilateral filtering for gray and color images. In: International Conference on Computer Vision, pp. 839–846 (1998)
16. Borji, A., Itti, L.: State-of-the-art in visual attention modeling. IEEE Trans. Pattern Analysis and Machine Intelligence **35**(1), 185–207 (2013)
17. Goferman, S., Zelnik-Manor, L., Tal, A.: Context-aware saliency detection. IEEE Trans. Pattern Analysis and Machine Intelligence **34**(10), 1915–1926 (2012)
18. Hou, X., Zhang, L.: Saliency detection: A spectral residual approach. In: Proceedings of IEEE Conference on Computer Vision and Pattern Recognition (CVPR), pp. 1–8 (2007)
19. Cheng, M.M., Zhang, G.X., Mitra, N.J., Huang, X., Hu, S.M.: Global contrast based salient region detection. In: Proceedings of IEEE Conference on Computer Vision and Pattern Recognition (CVPR), pp. 409–416 (2011)
20. Liu, T., Yuan, Z., Sun, J., Wang, J., Zheng, N., Shum, H.: Y: Learning to detect a salient object. IEEE Trans. Pattern Analysis and Machine Intelligence **33**(2), 353–367 (2011)
21. Columbia University Computer Vision Laboratory Detection and Removal of Rain Project. http://www.cs.columbia.edu/CAVE/projects/rain_detection/

Data-Driven Bridge Detection in Compressed Domain from Panchromatic Satellite Imagery

Yixiao Zhao[1,2(✉)], Shujian Yu[1,3(✉)], Jinghui Wu[2], Lu Han[2], Zijing Chen[3], Xiaopeng Yang[2], and Baojun Zhao[2(✉)]

[1] Department of Electrical and Computer Engineering, University of Florida, Gainesville, USA
{beng0429,yusjlcy9011}@ufl.edu
[2] School of Information and Electronics, Beijing Institute of Technology, Beijing, China
zbj@bit.edu.cn
[3] Department of Electronics and Information Engineering,
Huazhong University of Science and Technology, Wuhan, China

Abstract. Bridge detection in panchromatic imagery is of great importance in civilian and military applications. Popular algorithms for bridge detection are often based on a priori knowledge to bridge structure or location features, where manually-introduced decision rules are incorporated into a complex algorithm in spatial domain. Instead of knowledge-based approach in spatial domain, in this paper, we proposed a fast data-driven algorithm in compressed domain for panchromatic satellite imagery. Our algorithm consists of two main steps: firstly, bridge region candidates detection with hierarchical saliency model in compressed domain; and secondly, bridge region candidates validation with Local Binary Patterns (LBP) and Extreme Learning Machine (ELM). Experiments are conduced, and detection results demonstrate the effectiveness and efficiency of our proposed algorithm. The main contributions of our work are twofold: 1) to the best of our knowledge, we are among the first to introduce the concept of compressed domain techniques for bridge detection; and 2) compared with other knowledge-based algorithms, no assumptions are made beforehand for our algorithm, which makes it applicable for bridges of various cases.

keywords: Bridge detection · Panchromatic satellite imagery · Remote sensing · Compressed domain · Data-driven

1 Introduction

Automatic bridge detection is of great importance in both civilian areas and military affairs. Information on bridge location, orientation is essential for geographical database maintaining, damage assessment caused by natural disasters, as well as battlefield monitoring and military reconnaissance [1]. The purpose of this study focuses on automatically detecting bridges over water in panchromatic satellite imagery, and a new data driven algorithm in compressed domain is proposed.

Literatures on bridge detection and extraction from satellite imagery is thin [6], and previous algorithms are typically based on knowledge-driven approach, where a priori knowledge from an observer or expert is introduced to support bridge region

© Springer International Publishing Switzerland 2014
Z. Zeng et al. (Eds.): ISNN 2014, LNCS 8866, pp. 449–458, 2014.
DOI: 10.1007/978-3-319-12436-0_50

(a)   (b)

Fig. 1. Bridge detection in panchromatic imagery with our proposed algorithm: (a) represents the original satellite data; (b) demonstrates bridge detection results

segmentation or bridge candidate validation. In [3], the priori information on relative location relationship between river canals and concrete bridges are utilized to guide the low-level pre-processing step. [4] proposed a seed bridge points based segmentation algorithm to extract bridge region, where the seed points are detected with three basic observations. Then, in [5], a geometric model was exploited for detection task, where the author depicted bridge as two parallel segments with minimum length and maximum accepted angle difference. Followed by that, the author further proposed six manually-produced decision rules to define and classify segments as bridge [7].

Almost for all knowledge-drive algorithms, the knowledge itself is not derived automatically, but introduced with human observation or common sense. Obviously, such algorithms require a highly generalized but accuracy model to describe bridges with various shapes, orientations, as well as backgrounds, in addition, a corresponding mathematical model is also essential. However, existing prior knowledge is not sufficient to generalize all cases, and it often causes misleading detection results to specific satellite imagery or bridges with certain structure [2].

Apart from aforementioned problem, another issue which is urgent to be solved comes from massive amounts of incoming data. Although higher resolution results in more distinct visual features to describe kinds of targets in satellite imagery, a challenging problem is how to improve efficiency of feature extraction and representation [8], since conventional algorithms are normally time consuming and computation demanding [9].

To overcome above two difficulties, in this paper, we proposed a novel data-driven algorithm in compressed domain for bridge detection from panchromatic satellite imagery (See Fig. 1). Different from knowledge based approach for bridge region segmentation, data-driven strategy is concerned in our algorithm, where bridge region candidates are extracted with a hieratical saliency detection method without any assumption. In addition, different bridge region candidates are further validated with a decision rule learned from Extreme Learning Machine (ELM) [10] based on well-known Local Binary Patterns (LBP) [11] feature.

The remainder of this paper is organized as follows: in section 2, we briefly describe the workflow for bridge detection algorithm from panchromatic satellite imagery; then, in 3, the details of our proposed algorithm, i.e. bridge region candidates extraction (including image decomposition, mean-shift smoothing and hieratical saliency detection), bridge region candidates validation (including LBP feature extraction and representation as well as classifier training with ELM) are well explained; in section 4, experiments are conducted, results analysis are also performed; finally, section 5 concludes this paper and also outlines future work.

2   Algorithm Overview

2.1   Image Data Source

The satellite images used are downloaded from "Google Earth". To verify the validity of our algorithm, we collected an image set of 1200 images. All of them are taken by "QuickBird" satellite along the shorelines of Florida, U.S., in addition, the bridges for our research are of different positions, orientations, as well as different kinds (including road over water, road over road, walkway over water, etc.). Among them, we randomly selected 400 images for training, and reserved the rest 800 images as test set. In this paper, we only selected two images (Fig. 1(a)) from our test set to demonstrate the superiority of our algorithm.

2.2   Review to Image Processing in Compressed Domain

The baseline block diagram of the JPEG2000 compression algorithm is shown in Fig. 2. According to [12], compressed domain is defined as anywhere in the compression or decompression procedure, after transform or before inverse transform. Therefore, any computer vision algorithms, including bridge detection can be conducted in compressed domain from points 1 to 6 in Fig. 2.

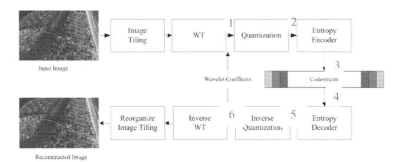

Fig. 2. Baseline block diagram of JPEG2000

In our proposed algorithm, point 1 is determined to be the ideal place for our compressed domain bridge detection algorithm based on three main reasons: firstly, entropy coder (points 3 and 4 in Fig. 2) will introduce serious destruction to spatial distribution of target features; secondly, since points 5, 6 are symmetry to points 1, 2, only points 1 and 2 are considered hereinafter; and thirdly, compared with implementation in point 1, implementation in point 2 is time consuming although it can preserve similar performance.

2.3   Overview to Our Proposed Algorithm

The flowchart of our proposed algorithm is illustrated in Fig. 3. It can be decomposed into two main steps: 1) bridge region candidates detection and extraction; 2) bridge region candidates validation. In the pre-processing, wavelet coefficients are extracted in the JPEG2000 codec. In order to have a fast implement of our proposed framework, only coefficients from low frequency subband (denoted as LL) is utilized for bridge region candidate extraction and ELM training. Experiment results and evaluation analysis demonstrate the effectiveness and efficiency of our proposed algorithm. Detailed description to our algorithm is provided in section 3.

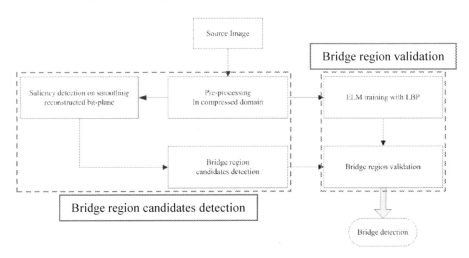

Fig. 3. Flow chart of proposed algorithm

3   Our Algorithm

3.1   Bridge Region Candidates Detection

Our bridge region candidates extraction is based on saliency detection in reconstructed bit-plane. Since bit-plane 6 and bit-plane 7 involve more details of targets information (See Fig. 4), also because a real time implement is preferred, a weighted combination of bit-plane 6 and bit-plane 7 is utilized for candidates extraction, regardless of other bit-planes.

Bit-Plane Reconstruction

Fig. 4 demonstrates the eight bit-plane images for Fig. 1(a) after discrete wavelet transform (DWT) in JPEG2000 codec. It can be observed that bit-plane 6 I_{Bit6} and bit-plane 7 I_{Bit7} reflect more information of different regions[1], thus a desired reconstructed bit-plane I_{Bit} can be represented as:

$$I_{Bit} = \alpha I_{Bit6} + (1 - \alpha)I_{Bit7}, \text{where } 0 \le \alpha \le 1 \tag{1}$$

In our framework, we set $\alpha = 0.9$ for all images in dataset. Fig. 5(a) illustrates our reconstructed bit-plane.

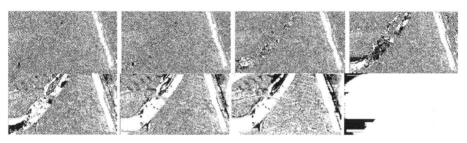

Fig. 4. Bit-plane images of Fig. 1(a) in gray level. First row from left to right: bit-plane 1 to bit-plane 4; second row from left to right: bit-plane 5 to bit-plane 8.

Mean-Shift Smoothing

Before bridge region candidates detection, the reconstructed bit-plane is smoothed with Mean-Shift which is extensively studied in [16]. Typically, an image can be represented as a $2D$ lattice of rD vectors. Therefore, each pixel x_i can be regarded as a dD vector ($d = r + 2$) in feature space $x_i = (x_i^2, x_i^r)$. Here x_i^2 records pixel spatial information, and x_i^r is the range (pixel value) part of feature vector.

Let x_i and z_i be the dD input and output of Mean-Shift filter. For each pixel:
1) Initialize $j = 1$ and $y_{i,1} = x_i$.
2) Update $y_{i,j+1}$ according to Eq.2 (g represents a kernel) until convergence, $y = y_{i,c}$.

$$y_{j+1} = \frac{\sum_{i=1}^{n} x_i g\left(\left\|\frac{y_i - x_i}{h}\right\|^2\right)}{\sum_{i=1}^{n} g\left(\left\|\frac{y_i - x_i}{h}\right\|^2\right)} \tag{2}$$

3) Assign $z_i = (x_i^2, y_{i,c}^r)$.

After mean shift filtering, each point converges into a point of convergence which represents the local mode of the density of the dD space. This process will achieve a high quality, discontinuity preserving effect (See Fig. 5(b)).

---

[1] We repeated the same experiment over 1200 images in our dataset, all of them share the same phenomenon.

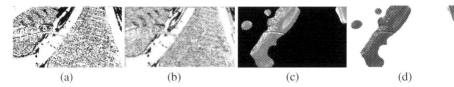

| (a) | (b) | (c) | (d) |

Fig. 5. Illustration of bridge region candidate extraction on smoothing reconstructed bit-plane. (a) is the reconstructed bit-plane image; (b) demonstrates smoothing effects with mean-shift to Fig. 5(a); (c) and (d) illustrate the saliency region with "spectral residual approach" in both gray level and color image of Fig. 1(a)

Saliency Detection with "Spectral Residual Approach"

The saliency detection method we used here is the well-known "spectral residual approach (SRA)" proposed by Hou [13]. According to Hou, information of the original image can be interpreted as the sum of the innovation and prior knowledge. The innovation stands for the interesting part with possible targets (like bridges in our paper), whereas the prior part stands for redundant or irrelevant information related to the background and invariant environment, which can be compressed. Also, Hou pointed out that image log spectrum provides an approximate description to image information.

Given an image $I(x)$, its log spectrum $L(f)$ can be represented as:

$$L(f) = \log\left(\tilde{F}(I(x))\right) \tag{3}$$

where $\tilde{F}$ represents Fourier Transform. Therefore, spectral residual $R(f)$ can be defined as:

$$R(f) = L(f) - A(f) \tag{4}$$

Here $A(f) = h(f) * L(f)$, where $h(f)$ denotes a local averaging filter. Using Inverse Fourier Transform, we can construct the saliency map in spatial domain.

The literatures on saliency detection contain nearly 65 vision attention models in the last 25 years [14], and we explain here why SRA is selected. Firstly, as we emphasized that our algorithm is data-driven, since SRA is independent of features and any other a priori knowledge of objects, it provides a reliable solution. In addition, SRA designed for saliency detection only for dominant objects, which also matches the scope of this paper. The final reason relates to the effectiveness and efficiency of SRA, since it can be implemented in 5 lines of Matlab codes [15]. Fig. 5(c),(d) demonstrate the saliency detection results with SRA. As it can be observed, a coarsely region involving bridge region and surrounding areas is detected and extracted thereby.

3.2   Bridge Region Validation

Obviously, regions extracted with saliency detection on smoothing reconstructed bit-plane are not guaranteed to be bridges. Especially when the source image including complex backgrounds and other objects which share similar appearance to bridges (See Fig. 6). Therefore, it is essential to validate the bridges region with supplementary mechanism. In this part, a novel approach based on ELM and LBP is presented.

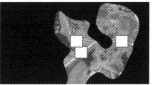

Fig. 6. Bridge region candidates detection results: (a) is source image; (b) is saliency detection result in smoothing reconstructed bit-plane; (c) is the corresponding saliency part in original panchromatic imagery. As it is shown, the three highlighted region share the similar appearance and both are detected in the last step. However, only red region is our desired.

LBP Feature Extraction

The feature descriptor we used in this step is Uniform LBP from LBP family [11]. LBP is considered here for three main reasons: firstly, as a data-driven system, object shape features, like ratio of long and wide axis, will not be considered; secondly, among the existing well-known statistical features, like co-occurrence matrix, Fourier descriptor, LBP is illumination invariant, which can effectively avoid the influence of weather effects which are commonly encountered in remote sensing imagery; and thirdly, LBP is convenient for computation, since only small neighborhood are involved for each pixel, which makes real-time application possible.

In our frameworks, instead of conventional LBP, Uniform LBP is used for feature extraction. Compared with conventional LBP, the main advantage of Uniform LBP relies on its property of generalization and dimensionality reduction, since we can use a vector of 59 dimension to generalize 90% of LBP patterns (See Fig. 7).

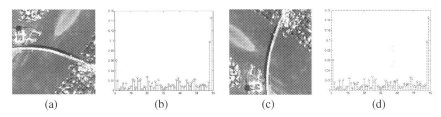

(a)   (b)   (c)   (d)

Fig. 7. Illustration of Uniform LBP descriptor: (a) is the bridge region from Fig. 1(a); (c) is the same region with 90 degree rotation; (b) and (d) demonstrate extracted Uniform LBP descriptor corresponding to (a) and (c), respectively. As can be seen, Uniform LBP is rotation invariant.

Verification of Bridge Candidates with ELM

We choose Extreme Learning Machine (ELM) based on single-hidden layer feed-forward networks (SLFN) to validate the bridge targets. ELM usually demonstrates outstanding performance in other remote sensing targets detection/classification tasks [17]. In addition, compared to traditional classifier, such as Nearest Neighbor (NN), Back Propagation (BP) algorithm, Support Vector Machine (SVM), ELM is fast and more accurate [10, 18].

In our frameworks, the bridge region candidates validation is regarded as a binary pattern classification issue, since we are not interest in what the false alarms are, and the samples for training various false alarms are not enough. As aforementioned, the training data were collected randomly from our dataset downloaded from "Google Earth" with total number of 400, which contains 512 typical false alarms and 493 bridge regions. After the training stage, the weighted vector is learned to discriminate two classes (bridge and non-bridge). The validation step (classification) consists of taking region of interests (ROIs) from coarsely bridge region in last step, computing the feature vectors and then applying the pre-computed ELM to verify the bridge targets.

4   Experiments and Results Analysis

In this section, extensively experiments are conducted. We tested our aforementioned algorithm on 800 images in our dataset. Among them, 1178 bridge targets were identified manually. With our proposed algorithm, we have successfully detected 1067 bridge targets with 132 false alarms. Some representative detection results are illustrated in Fig. 8. More results are available from http://www.yushujian.com/index.html.

In addition, we also compared our algorithm with other two state-of-art algorithms for bridge detection in satellite imagery with high resolution [6, 7]. The comparison results are shown in Table.1.

Table 1. Quantitative comparison results

Methods	Gedik [6]	Trias-Sanz [7]	Our algorithm
Accuracy	41.50%	82.35%	**90.58%**
Missing ratio	58.50%	17.65%	**9.42%**
False ratio	**10.53%**	17.68%	11.01%
Error ratio	69.03%	35.33%	**19.43%**

Above criteria are defined as [19]:

$$Accuracy = \frac{Number\ of\ correctly\ detected\ bridges}{Number\ of\ real\ bridges} * 100\% \tag{5}$$

$$Missing\ ratio = 100\% - Accuracy \tag{6}$$

$$False\ ratio = \frac{Number\ of\ false\ detected\ bridges}{Number\ of\ detected\ bridges} * 100\% \tag{7}$$

$$Error\ ratio = Missing\ ratio + False\ ratio \tag{8}$$

As we can see, our proposed algorithm outperform algorithms proposed in [6, 7] for three criteria. At the same time, we have also compared ELM with SVM for classification. Experimental results demonstrate that the training time with ELM is less than that with SVM by approximately 50%, and the detection accuracy with ELM is higher than that with SVM by three percents.

Fig. 8. More detection results illustration

5   Conclusion

In this paper, we proposed a novel algorithm for bridge detection on panchromatic satellite imagery based on data-driven methods in compressed domain. Compared with previous work, our algorithm can achieve better performance within less computation time. In addition, we are also among the first to introduce compressed domain processing techniques for bridge detection on satellite imagery. With saliency detection in smoothing reconstructed bit-plane, regions involving bridge can be coarsely extracted. After bridge candidates are extracted, ELM classification with LBP feature can provide a finer decision rule. Experiments demonstrate the superiority of our algorithm in terms of computation time and accuracy.

References

1. Chaudhuri, D., Samal, A.: An automatic bridge detection technique for multispectral images. IEEE Trans. Geoscience and Remote Sensing **46**(9), 2720–2727 (2008)
2. Han, Y., Zheng, H., Cao, Q., Wang, Y.: An effective method for bridge detection from satellite imagery. In: IEEE Conference on Industrial Electronics and Applications, pp. 2753–2757 (2007)

 3. Hou, B., Li, Y., Jiao, L.: Segmentation and Recognition of Bridges in High resolution SAR Images. In: IEEE CIE International Conference on Radar, pp. 479–482 (2001)
 4. Sithole, G., Vosselman, G.: Bridge detection in airborne laser scanner data. ISPRS Journal of Photogrammetry and Remote Sensing **61**(1), 33–46 (2006)
 5. Loménie, N., Barbeau, J., Trias-Sanz, R.: Integrating textural and geometric information for an automatic bridge detection system. In: Proceedings of IEEE International Geoscience and Remote Sensing Symposium, vol. 6, pp. 3952–3954 (2003)
 6. Gedik, E., Cinar, U., Karaman, E., Yardimci, Y., Halici, U., Pakin, K.: A new robust method for bridge detection from high resolution electro-optic satellite images. In: Proceedings International Conference on Geographic Object-Based Image Analysis, pp. 298–302 (2012)
 7. Trias-Sanz, R., Loménie, N.: Automatic bridge detection in high-resolution satellite images. In: Crowley, J.L., Piater, J.H., Vincze, M., Paletta, L. (eds.) ICVS 2003. LNCS, vol. 2626, pp. 172–181. Springer, Heidelberg (2003)
 8. Luo, J., Ming, D., Liu, W., Shen, Z., Wang, M., Sheng, H.: Extraction of bridges over water from IKONOS panchromatic data. International Journal of Remote Sensing **28**(16), 3633–3648 (2007)
 9. Chang, S.F.: Compressed-domain techniques for image/video indexing and manipulation. In: Proceedings IEEE International Conference on Image Processing, vol. 1, pp. 314–317 (1995)
10. Huang, G.B., Zhu, Q.Y., Siew, C.K.: Extreme learning machine: theory and applications. Neurocomputing **70**(1), 489–501 (2006)
11. Ojala, T., Pietikainen, M., Maenpaa, T.: Multiresolution gray-scale and rotation invariant texture classification with local binary patterns. IEEE Trans. Pattern Analysis and Machine Intelligence **24**(7), 971–987 (2002)
12. Delac, K., Grgic, M., Grgic, S.: Effects of JPEG and JPEG2000 compression on face recognition. In: Singh, S., Singh, M., Apte, C., Perner, P. (eds.) ICAPR 2005. LNCS, vol. 3687, pp. 136–145. Springer, Heidelberg (2005)
13. Hou, X., Zhang, L.: Saliency detection: A spectral residual approach. In: IEEE Conference on Computer Vision and Pattern Recognition, pp. 1–8 (2007)
14. Borji, A., Itti, L.: State-of-the-art in visual attention modeling. IEEE Trans. Pattern Analysis and Machine Intelligence **35**(1), 185–207 (2013)
15. Hou, X.: Spectral Residual. http://www.its.caltech.edu/~xhou/
16. Comaniciu, D., Meer, P.: Mean shift: A robust approach toward feature space analysis. IEEE Trans. Pattern Analysis and Machine Intelligence **24**(5), 603–619 (2002)
17. Pal, M., Maxwell, A.E., Warner, T.A.: Kernel-based extreme learning machine for remote-sensing image classification. Remote Sensing Letters **4**(9), 853–862 (2013)
18. Huang, G.B., Zhou, H., Ding, X., Zhang, R.: Extreme learning machine for regression and multiclass classification. IEEE Trans. Systems, Man, and Cybernetics, Part B: Cybernetics **42**(2), 513–529 (2012)
19. Zhu, C., Zhou, H., Wang, R., Guo, J.: A novel hierarchical method of ship detection from spaceborne optical image based on shape and texture features. IEEE Trans. Geoscience and Remote Sensing **48**(9), 3446–3456 (2010)

Fast Nonnegative Tensor Factorization by Using Accelerated Proximal Gradient

Guoxu Zhou[1,2](✉), Qibin Zhao[1], Yu Zhang[3], and Andrzej Cichocki[1]

[1] Laboratory for Advanced Brain Signal Processing RIKEN,
Brain Science Institute, Wako-shi, Saitama 3510198, Japan
{zhouguoxu,qbzhao,cia}@brain.riken.jp
[2] Faculty of Automation, Guangdong University of Technology,
Guangzhou, China
[3] School of Information Science and Engineering,
East China University of Science and Technology, Shanghai, China
yuzhang@ecust.edu.cn

Abstract. Nonnegative tensor factorization (NTF) has been widely applied in high-dimensional nonnegative tensor data analysis. However, existing algorithms suffer from slow convergence caused by the non-negativity constraint and hence their practical applications are severely limited. By combining accelerated proximal gradient and low-rank approximation, we propose a new NTF algorithm which is significantly faster than state-of-the-art NTF algorithms.

Keywords: CP (PARAFAC) decompositions · Nonnegative tensor factorization · Accelerated proximal gradient

1 Introduction

matrix factorization (NMF) is a problem of factorizing a given nonnegative matrix into two nonnegative, often lower-rank, matrices whose product optimally approximates the given matrix. NMF has been widely applied in nonnegative data analysis in order to provide more interpretable and meaningful representation of data [4,9]. Particularly, NMF has the ability of learning parts of objects as only addition operations are permitted, which makes it very attractive and an almost indispensable tool in many nonnegative data analysis tasks [4,9,17].

In contrast to matrices, high dimensional data, also referred to as tensors, are more and more common in modern scientific research and engineering applications. For example, a color image with RGB channels form a 3rd-tensor, and a clip of video forms a 4th-order tensor with additional dimension of frame. Similar to matrix factorization, tensor decomposition is one of the most fundamental problem in tensor analysis [4,7]. In the meanwhile, Canonical Polyadic (CP), also named as CANDECOMP/PARAFAC decomposition [3,6], has been extensively studied in the last four decades and found many applications [4]. In CPD a given tensor is represented as the sum of rank-1 tensors, which can be

© Springer International Publishing Switzerland 2014
Z. Zeng et al. (Eds.): ISNN 2014, LNCS 8866, pp. 459–468, 2014.
DOI: 10.1007/978-3-319-12436-0_51

viewed as an extension of singular value decomposition (SVD) in tensor field. One major advantage of CPD is that it is essentially unique under mild conditions [8,12], which makes it very useful in the case where only very limited *a priori* knowledge is available on factors.

Due to the essential uniqueness of CPD, it is generally unnecessary to impose additional constraints on the factors. However, uniqueness conditions are generally analyzed in noise-free. In practice the measured data are often corrupted by noise. Proper constraints reflecting some *a priori* knowledge on components can help us extract more interpretable components. Furthermore, the uniqueness of CPD relies on certain conditions. In the case where the uniqueness conditions are not satisfied, additional constraints help to extract specific components rather than arbitrary ones. Hence constrained CPD has also gained increasing importance. Nonnegative tensor factorization (NTF, or equivalently nonnegative CPD) is one of such an important topic, aiming to find compressed and parts-based representation of high-order tensors by imposing nonnegativity on factors. It is well known NMF algorithms often suffer from slow convergence due to the nonnegativity constraints. This issue is further aggravated in NTF. In fact, the efficiency has been a major bottleneck of NTF in practical applications. Motivated by recently major progresses in NMF/NTF, we proposed a new NTF algorithm in this paper, which can be significantly faster than existing NTF methods. While the basic idea of NTF based on a proceeding LRA has been briefly introduced in our recent overview paper [15], the detailed derivations are presented in this paper.

The following notations will be adopted. Bold capitals (e.g., $\mathbf{A}$) and bold lowercase letters (e.g., $\mathbf{y}$) denote matrices and vectors, respectively. $\mathbb{R}_+^{I \times J}$ denotes the set of $I \times J$ nonnegative matrices. Calligraphic bold capitals, e.g. $\mathcal{Y}$, denote tensors. Mode-n matricization (unfolding, flattening) of a tensor $\mathcal{Y} \in \mathbb{R}^{I_1 \times I_2 \times \cdots \times I_N}$ is denoted as $\mathbf{Y}_{(n)} \in \mathbb{R}^{I_n \times \Pi_{p \neq n} I_p}$, which consists of arranging all possible mode-n tubes (vectors) as the columns of it [7]. The Frobenius norm of a tensor is denoted by $\|\mathcal{Y}\|_F = (\sum_{i_1 i_2 \cdots i_N} y_{i_1 i_2 \cdots i_N}^2)^{\frac{1}{2}}$.

We use $\odot$ and $\circledast$ to denote the Khatri-Rao product (column-wise Kronecker product) and Hadamard product of matrices, respectively. We define $\bigodot_{k \neq n} \mathbf{A}^{(k)} = \mathbf{A}^{(N)} \odot \cdots \odot \mathbf{A}^{(n+1)} \odot \mathbf{A}^{(n-1)} \odot \cdots \odot \mathbf{A}^{(1)}$. Readers are referred to [4,7] for detailed tensor notations and operations.

2 NTF Using Accelerated Proximal Gradient

CP decomposition of a data tensor $\mathcal{Y} \in \mathbb{R}^{I_1 \times I_2 \cdots \times I_N}$ can be formulated as

$$\mathcal{Y} = \sum_{j=1}^{J} \lambda_j \, \mathbf{a}_j^{(1)} \circ \mathbf{a}_j^{(2)} \cdots \circ \mathbf{a}_j^{(N)} + \mathcal{E}, \tag{1}$$

where component (or factor, mode) matrices $\mathbf{A}^{(n)} = [\mathbf{a}_1^{(n)}, \mathbf{a}_2^{(n)}, \cdots, \mathbf{a}_J^{(n)}] \in \mathbb{R}^{I_n \times J}$, $n \in \mathcal{N} = \{1, 2, \cdots, N\}$, consist of unknown latent components $\mathbf{a}_j^{(n)}$

(e.g., latent source signals) that need to be estimated, ∘ denotes the outer product[1], and $\mathcal{E}$ denotes the tensor of error or residual terms. As the scalar factors λ_j can be absorbed into one factor matrix, e.g. $\mathbf{A}^{(N)}$ by letting $\mathbf{a}_j^{(N)} = \lambda_j \mathbf{a}_j^{(N)}, j \in \mathcal{J} = \{1, 2, \ldots, J\}$, we also use $\mathcal{Y} \approx [\![\mathbf{A}^{(1)}, \mathbf{A}^{(2)}, \cdots, \mathbf{A}^{(N)}]\!]$ as a shorthand notation of (1), where $\lambda_j = 1, \forall j$, has been implicitly assumed.

By using the CP model (1), the mode-n matricization of $\mathcal{Y}$ has the form of

$$\mathbf{Y}_{(n)} \approx \mathbf{A}^{(n)} \mathbf{B}^{(n)T}, \ (n \in \mathcal{N}), \tag{2}$$

where

$$\mathbf{B}^{(n)} = \bigodot_{p \neq n} \mathbf{A}^{(p)} \in \mathbb{R}^{(\Pi_{p \neq n} I_p) \times J}. \tag{3}$$

(2) is the foundation of standard alternating least squares methods (ALS) for CPD. With above preliminaries, we formulate the NTF problem as follows:

$$\min \ \left\| \mathcal{Y} - \sum_{j=1}^{J} \mathbf{a}_j^{(1)} \circ \mathbf{a}_j^{(2)} \cdots \circ \mathbf{a}_j^{(N)} \right\|_F^2 \tag{4}$$

$$s.t. \ \mathbf{a}_j^{(n)} \in \mathbb{R}_+^{I_n \times 1}, \ j \in \mathcal{J}, \ n \in \mathcal{N}.$$

To solve (4), we use the block coordinate descent method. That is, each time we update only one factor matrix while remaining the others unchanged. Then (4) is equivalent to

$$\min_{\mathbf{A}^{(n)}} \ f(\mathbf{A}^{(n)}) = \|\mathbf{Y}_{(n)} - \mathbf{A}^{(n)} \mathbf{B}^{(n)}\|_F^2,$$
$$s.t. \ \mathbf{A}^{(n)} \in \mathbb{R}_+^{I_n \times J}, \tag{5}$$

We update each $\mathbf{A}^{(n)}$ by solving (5) alternatively for $n \in \mathcal{N}$ till convergence. Note that (5) is also a basic sub-problem in NMF, which allows us to apply any existing NMF update rules to solve (5).

Here we consider the accelerated proximal gradient (APG) method to solve (5), which was originally proposed by Nesterov [10] for smooth optimization and has proven to be a very efficient method for NMF [5]. Follow the analysis in [5], f defined in (5) is convex and its gradient f' is Lipschitz continuous, that is, for any matrices $\mathbf{A}_1$ and $\mathbf{A}_2$ with proper sizes there holds that

$$\|f'(\mathbf{A}_1) - f'(\mathbf{A}_2)\|_F \leq L \|\mathbf{A}_1 - \mathbf{A}_2\|_F, \tag{6}$$

where $L = \|\mathbf{B}^{(n)T}\mathbf{B}^{(n)}\|_F$ is the Lipschitz constant. In the APG method two sequences, i.e. $\{\mathbf{A}_k^{(n)}\}$ and $\{\mathbf{Z}_k\}$ in our case, are alternatively updated in each iteration:

---

[1] The outer product of two vectors $\mathbf{a} \in \mathbb{R}^I$, $\mathbf{b} \in \mathbb{R}^T$ builds up a rank-one matrix $\mathbf{Y} = \mathbf{a} \circ \mathbf{b} = \mathbf{a}\mathbf{b}^T \in \mathbb{R}^{I \times T}$ and the outer product of three vectors: $\mathbf{a} \in \mathbb{R}^I$, $\mathbf{b} \in \mathbb{R}^T$, $\mathbf{c} \in \mathbb{R}^Q$ builds up a 3rd-order rank-one tensor: $\mathcal{Y} = \mathbf{a} \circ \mathbf{b} \circ \mathbf{c} \in \mathbb{R}^{I \times T \times Q}$, with entries defined as $y_{itq} = a_i b_t c_q$.

$$\mathbf{A}_k^{(n)} = \arg\min_{\mathbf{X} \geq 0} \phi(\mathbf{X})$$

$$= f(\mathbf{A}_{k-1}^{(n)}) + \langle f'(\mathbf{A}_{k-1}^{(n)}), \mathbf{X} - \mathbf{Z}_k \rangle + \frac{L}{2}\|\mathbf{X} - \mathbf{Z}_k\|_F^2 \tag{7}$$

and

$$\mathbf{Z}_{k+1} \leftarrow \mathbf{A}_k^{(n)} + \frac{\alpha_k - 1}{\alpha_{k+1}}(\mathbf{A}_k^{(n)} - \mathbf{A}_{k-1}^{(n)}), \tag{8}$$

where k is the iteration number, ϕ is the proximal function of f on $\mathbf{Y}_k$, and $\langle \cdot, \cdot \rangle$ is the inner product of two matrices. The update step size α_k is chosen as

$$\alpha_{k+1} = \frac{1 + \sqrt{4\alpha_k^2 + 1}}{2}. \tag{9}$$

By using the above update rules, an optimal convergence rate $\mathcal{O}(\frac{1}{k^2})$ can be achieved [5,10]. By using the Lagrange multiplier method to solve (7), optimal $\mathbf{A}_k^{(n)}$ is given as

$$\mathbf{A}_k^{(n)} \leftarrow \mathcal{P}_+ \left(\mathbf{Z}_k - \frac{1}{L}(\mathbf{Y}^{(n)}\mathbf{B}^{(n)} - \mathbf{A}_{k-1}^{(n)}\mathbf{B}^{(n)T}\mathbf{B}^{(n)}) \right), \tag{10}$$

where $\mathcal{P}_+(\mathbf{X})$ projects all negative entries of $\mathbf{X}$ to zeros, and $\mathbf{B}^{(n)}$ is given in (3). As the size of $\mathbf{B}^{(n)}$ is often huge, the computation of $\mathbf{B}^{(n)T}\mathbf{B}^{(n)}$ can be simplified as

$$\mathbf{B}^{(n)T}\mathbf{B}^{(n)} = \underset{p \neq n}{\circledast} \left(\mathbf{A}^{(p)T}\mathbf{A}^{(p)} \right). \tag{11}$$

Repeating (10) and (8) alternatively till convergence we obtain optimal $\mathbf{A}^{(n)}$ minimizing f in (5) with the fast convergence rate $\mathcal{O}(\frac{1}{k^2})$. Then repeat the above procedure alternatively for all $n \in \mathcal{N}$ till convergence, all nonnegative factors will be estimated. Based on the above analysis, we propose the NTF algorithm based on AGP updates which is presented in Algorithm 1. It can be seen that in the inner most iterations, the main computational load lies in the computation of $\mathbf{A}_k^{(n)}\mathbf{G}$, which is only $\mathcal{O}(I_n J^2)$. In the outer loops, however, the computation of $\mathbf{Y}_{(n)}\mathbf{B}^{(n)}$ is very expensive and has the complexity as high as $\mathcal{O}(J\prod_n I_n)$. It is therefore imperative to reduce the computational complexity of this part.

3   Fast NTF Algorithm Based on Low-Rank Approximation

To reduce the computational complexity of $\mathbf{Y}_{(n)}\mathbf{B}^{(n)}$ we consider replacing the huge matrix $\mathbf{Y}_{(n)}$ by its low-rank approximations. This idea has been applied to NMF and nonnegative tensor decompositions [11,16], where the low-rank approximation is used to reduce the computational complexity and filter out noise [16]. Here we consider similar idea. Suppose that $[\![\mathbf{U}^{(1)}, \mathbf{U}^{(2)}, \ldots, \mathbf{U}^{(N)}]\!]$ is

---

Algorithm 1. NTF Based on the APG Method (NTF_APG)

Require: $\mathcal{Y}$, J.

1. **while** Not converged **do**
2.   **for** $n = 1, 2, \ldots, N$ **do**
3.   Compute $\mathbf{B}^{(n)}$ and $\mathbf{G} = \mathbf{B}^{(n)T}\mathbf{B}^{(n)}$ using (3) and (11), respectively. $\mathbf{C} = \mathbf{Y}_{(n)}\mathbf{B}^{(n)}$, $L = \|\mathbf{G}\|_F$.
4. $\alpha_0 = 1$, $k = 1$, and $\mathbf{Z}_0 = \mathbf{A}_0^{(n)} = \mathbf{A}^{(n)}$.
5.   **repeat**
6. $\mathbf{A}_k^{(n)} = \mathcal{P}_+\left(\mathbf{Z}_k - \frac{1}{L}\left(\mathbf{C} - \mathbf{A}_{k-1}^{(n)}\mathbf{G}\right)\right)$,
7. $\alpha_k = \frac{1 + \sqrt{4\alpha_{k-1}^2 + 1}}{2}$,
8. $\mathbf{Z}_k = \mathbf{A}_k^{(n)} + \frac{\alpha_{k-1}-1}{\alpha_k}(\mathbf{A}_k^{(n)} - \mathbf{A}_{k-1}^{(n)})$.
9. $k \leftarrow k + 1$
10.   **until** a stopping criterion is satisfied
11.   **end for**
12. **end while**
13. **return** $\mathbf{A}^{(n)}, n = 1, 2, \ldots, N$.

---

the optimal rank-J approximation[2] to $\mathcal{Y}$. Then $\mathcal{Y}$ is updated as $\mathcal{Y} \leftarrow [\![\mathbf{U}^{(1)}, \mathbf{U}^{(2)}, \ldots, \mathbf{U}^{(N)}]\!]$ before iterations, which leads to

$$\mathbf{Y}_{(n)}\mathbf{B}^{(n)} = \mathbf{U}^{(n)}\left(\underset{p\neq n}{\circledast}(\mathbf{U}^{(p)T}\mathbf{A}^{(p)})\right), \tag{12}$$

from (2) and (3). Note that the time complexity of (12) is only about $\mathcal{O}(I_n J^2)$, which is significantly less than the case without low-rank approximation of $\mathcal{Y}$.

Due to the uniqueness of CPD, in the ideal noiseless case we should have $\mathbf{A}^{(n)} = \mathbf{U}^{(n)}\mathbf{P}_n\mathbf{D}_n$, where $\mathbf{P}_n$ and $\mathbf{D}_n$ are a permutation matrix and a nonsingular diagonal matrix, respectively. This relationship may not hold if noise exists. However, in this case $\mathbf{A}^{(n)}$ is often close to $\mathbf{U}^{(n)}$. So it is reasonable to use $\mathcal{P}_+(\mathbf{U}^{(n)})$ as the initialization to achieve fast convergence. However, a key factor to success of this initialization is that we need to remove the sign ambiguities caused by scale ambiguities of CPD. In other words, we should adjust the signs of factor matrices $\mathbf{U}^{(n)}$ such that as many as possible entries are nonnegative. To do this, we let

$$s_{nj} = \text{sign}(u_{i_0 j}^{(n)}),$$

where $u_{ij}^{(n)}$ is the ij-th entry of $\mathbf{U}^{(n)}$, $i_0 = \arg\max_i |u_{ij}^{(n)}|$, and the sign function returns the sign of a number. Let

$$\mathbf{u}_j^{(n)} \leftarrow \mathbf{u}_j^{(n)} s_{nj}, \quad \mathbf{u}_j^{(N)} \leftarrow \mathbf{u}_j^{(N)} s_{nj}, \quad \forall n \neq N. \tag{13}$$

---

[2] It can also be replaced by a rank-R approximation with $R \geq J$ which is obtained by solving the unconstrained CPD (1). The key idea is that unconstrained CPD is usually significantly faster than NTF algorithms.

Algorithm 2. The FastNTF_APG Algorithm

Require: $\mathcal{Y}$, J, and any efficient unconstrained CPD algorithm Ψ.
1. $[\![\mathbf{U}^{(1)}, \mathbf{U}^{(2)}, \cdots, \mathbf{U}^{(N)}]\!] = \Psi(\mathcal{Y}, J)$.
2. Adjust $\mathbf{U}^{(n)}$ using (13) and let $\mathbf{A}^{(n)} \leftarrow \mathcal{P}_+(\mathbf{U}^{(n)})$
3. **while** Not converged **do**
4.   **for** $n = 1, 2, \ldots, N$ **do**
5.   Compute $\mathbf{G} = \mathbf{B}^{(n)T}\mathbf{B}^{(n)}$ and $\mathbf{C} = \mathbf{Y}_{(n)}\mathbf{B}^{(n)}$ from (11) and (12), respectively. $L = \|\mathbf{G}\|_F$.
6. $\alpha_0 = 1$, $k = 1$, and $\mathbf{Z}_0 = \mathbf{A}_0^{(n)} = \mathbf{A}^{(n)}$.
7.   Repeat the procedure of line 5-10 in Algorithm 1
8.   **end for**
9. **end while**
10. **return** $\mathbf{A}^{(n)}, n = 1, 2, \ldots, N$.

After this adjustment, $\mathbf{U}^{(n)}$ are expected to be almost nonnegative and hence $\mathcal{P}_+(\mathbf{U}^{(n)})$ can be a very good initialization for NTF algorithms, provided that noise is mild (empirically, the SNR is higher 20dB) and the corresponding CPD is essentially unique.

With above analysis, the fast NTF algorithm based on APG (FastNTF_APG) is presented in Algorithm 2. The time complexity of FastNTF_APG is only about $\mathcal{O}(J^2 I_n)$, which is significantly lower than NTF_APG that is of $\mathcal{O}(J \prod_p I_p)$. Of course an additional unconstrained CPD is required in FastNTF_NTF. However, this step generally can be done very efficiently and it significantly reduces the time complexity of subsequent NTF procedure. Moreover, it is worth noticing that the tricks introduced in this section can be used to accelerate many existing NTF algorithms.

4   Simulations

In this section we present experimental results. All experiments were performed on a computer with Intel i7 CPU (3.33GHz) and 24GB memory running 64bit Windows 7. The MATLAB codes of FastNTF_APG can be downloaded from http://bsp.brain.riken.jp/TDALAB. The performance index Fit is defined as

$$\text{Fit}(\mathcal{Y}, \widehat{\mathcal{Y}}) = 1 - \|\mathcal{Y} - \widehat{\mathcal{Y}}\|_F / \|\mathcal{Y}\|_F,$$

where $\widehat{\mathcal{Y}}$ is an estimate of $\mathcal{Y}$, $\text{Fit}(\mathcal{Y}, \widehat{\mathcal{Y}}) = 1$ iff $\widehat{\mathcal{Y}} = \mathcal{Y}$.

Simulation 1: Application of NTF in tensor displays. Tensor displays introduced by Wetzstein *et al.* is a family of compressive light field displays comprising all architectures to develop glasses-free 3D displays (stereoscopic displays) [14]. The authors showed that any light field emitted by an N-layer, M-frame tensor can be represented by an Nth-order rank-M tensor. Hence NTF can be applied to tensor displays, which allows multilayer, multiframe decompositions and combines benefits of multiple layers and directional backlights. The major limitation of this innovative framework is the considerable computation resources required by the NTF algorithms [14]. In their implementation, the NTF algorithm based

Table 1. Performance comparison of the algorithms in the decomposition of a 5th-order tensor for multiplayer light field displays

Algorithms	NTF_MU	NTF_HALS	NTF_APG	FastNTF_APG
Fit	0.85	0.84	0.85	0.87
Time (s)	1199	427	437	245

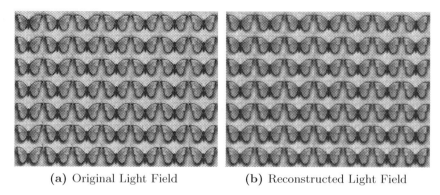

(a) Original Light Field   **(b)** Reconstructed Light Field

Fig. 1. Performance of FastNTF_APG in the decomposition of a light field tensor. We set $J = 50$ and the Fit was 0.87.

on multiplicative update rules were adopted, which was very slow, and a GPU-based solver was applied to approach satisfactory performance. In summary, efficient NTF algorithms play a very important role in their technique of tensor displays. Here we omitted the details about tensor displays but instead we simply decompose a light field tensor to evaluate the performance of the proposed algorithm. We used the code published in their website[3] to generate a light field tensor with the size of $7 \times 7 \times 384 \times 512 \times 3$. The proposed FastNTF_AGP algorithm was compared with the NTF based on multiplicative updates (NTF_MU) included in TensorToolbox [2], NTF using Hierarchical Alternating Least Squares Algorithm (NTF_HALS) [11], and the nPARAFAC algorithm in NWaytoolbox [1], all with the maximum iteration number 100 except FastNTF_APG. For all algorithms $J = 50$ was set. In FastNTF_APG, the maximum iteration number was 500, and we used the CP-ALS algorithm [2] with 20 iterations to perform unconstrained CPD. The result of a typical run of FastNTF_AGP is shown in Fig. 1. The performance averaged over 20 Monte-Carlo runs with random initial conditions of the compared algorithms are shown in TABLE 1 (The nPARAFAC algorithm ran *out of memory* in this experiment. To show the efficiency of pure

---

[3] http://web.media.mit.edu/~gordonw/courses/ComputationalDisplays/
TomographicLightFieldSynthesis-Code2.0.zip. More information is available at
http://web.media.mit.edu/~gordonw/TensorDisplays/

Fig. 2. Performance of FastNTF_AGP in the COIL-100 image clustering (we used the first 20 objects only). The two t-SNE components of $\mathbf{A}^{(4)}$ were used for visualization. The clustering accuracy was as high as 90.0%.

APG, the NTF_APG algorithm was also compared by setting the maximum iteration number 50). It can be seen that the FastNTF_APG algorithm not only achieved the best Fit but also was significantly faster than the others. Note that NTF_HALS also adopted low-rank approximation technique and theoretically it has the same time complexity as FastNTF_APG. However, from the experimental results FastNTF_APG converged faster than NTF_HALS.

Simulation 2: Application of NTF in image clustering. In this experiment we used NTF algorithms to extract features for image clustering. We used the database COIL-100 which consists of 7200 color images of 100 objects, each of which has 72 images taken from different directions. For simplicity we used images corresponding to the first 20 objects for clustering analysis. We generated a tensor $\mathcal{Y}$ with $128 \times 128 \times 3 \times 1440$ using these images. Then NTF algorithms were applied such that $\mathcal{Y} \approx [\![\mathbf{A}^{(1)}, \mathbf{A}^{(2)}, \mathbf{A}^{(3)}, \mathbf{A}^{(4)}]\!]$. Then two t-SNE [13] components of $\mathbf{A}^{(4)}$ were used for visualization and clustering. The K-means algorithm was adopted for clustering. As K-means is prone to be affected by initial cluster centers, in each run we repeated clustering 20 times, each with a new set of initial centers (see the help for K-means included in the MATLAB Statistics Toolbox). The performance averaged 20 Monte-Carlo runs are detailed in TABLE 2. A typical run of FastNTF_APG was visualized in Fig. 2 where the corresponding clustering accuracy was as high as 90%.

Table 2. Performance comparison of the algorithms in COIL-100 (first 20 objects) image clustering

Algorithms	NTF_MU	nPARAFAC	NTF_HALS	FastNTF_APG
Time (s)	872	635	421	180
Fit	0.68	0.71	0.69	0.69
Accuracy(%)	79.2	78.1	80.2	80.6

5   Conclusion

We proposed a new fast nonnegative tensor factorization (NTF) algorithm by combining accelerated proximate gradient (APG) and low-rank approximation techniques. Theoretically, the APG method offers the optimal convergence rate $\mathcal{O}(\frac{1}{k^2})$. Low-rank approximation pre-processing avoids manipulating huge matrices during iterations and suppresses noise. Simulations confirmed that the new algorithm is significantly faster than state-of-the-art NTF algorithms.

Acknowledgments. This work was partially supported by the JSPS KAKENHI (Grant 26730125), the Nation Nature Science Foundation of China under Grant (61305028, 61202155, and U1201253), and the Fundamental Research Funds for the Central Universities under Grant WH1314023.

References

1. Andersson, C.A., Bro, R.: The N-way toolbox for MATLAB (2000)
2. Bader, B.W., Kolda, T.G.: MATLAB tensor toolbox version 2.5 (February 2012)
3. Carroll, J., Chang, J.-J.: Analysis of individual differences in multidimensional scaling via an n-way generalization of Eckart-Young decomposition. Psychometrika **35**(3), 283–319 (1970)
4. Cichocki, A., Zdunek, R., Phan, A.-H., Amari, S.: Nonnegative Matrix and Tensor Factorizations: Applications to Exploratory Multi-way Data Analysis and Blind Source Separation. Wiley, Chichester (2009)
5. Guan, N., Tao, D., Luo, Z., Yuan, B.: NeNMF: An optimal gradient method for nonnegative matrix factorization. IEEE Transactions on Signal Processing **60**(6), 2882–2898 (2012)
6. Harshman, R.A.: Foundations of the PARAFAC procedure: Models and conditions for an 'explanatory' multi-modal factor analysis. UCLA Working Papers in Phonetics **16**(1) (1970)
7. Kolda, T.G., Bader, B.W.: Tensor decompositions and applications. SIAM Review **51**(3), 455–500 (2009)
8. Kruskal, J.B.: Three-way arrays: rank and uniqueness of trilinear decompositions, with application to arithmetic complexity and statistics. Linear Algebra and its Applications **18**(2), 95–138 (1977)
9. Lee, D.D., Sebastian Seung, H.: Algorithms for non-negative matrix factorization. In: Leen, T.K., Dietterich, T.G., Tresp, V. (eds.) Advances in Neural Information Processing Systems 13, pp. 556–562. MIT Press, Cambridge (2000)

10. Nesterov, Y.: A method of solving a convex programming problem with convergence rate $O(1/k^2)$. Soviet Mathematics Doklady **27**(2), 372–376 (1983)
11. Phan, A.-H., Cichocki, A.: Multi-way nonnegative tensor factorization using fast hierarchical alternating least squares algorithm (HALS). In: Proceedings of the 2008 International Symposium on Nonlinear Theory and its Applications, pp. 41–44 (2008)
12. Sidiropoulos, N.D., Bro, R.: On the uniqueness of multilinear decomposition of N-way arrays. Journal of Chemometrics **14**(3), 229–239 (2000)
13. Van Der Maaten, L., Detection, C.: Visualizing data using t-SNE. Journal of Machine Learning Research **9**(11), 2579–2605 (2008)
14. Wetzstein, G., Lanman, D., Hirsch, M., Raskar, R.: Tensor Displays: Compressive Light Field Synthesis using Multilayer Displays with Directional Backlighting. ACM Transactions on Graphics (Proc. SIGGRAPH) **31**(4), 1–11 (2012)
15. Zhou, G., Cichocki, A., Zhao, Q., Xie, S.: Nonnegative matrix and tensor factorizations: An algorithmic perspective. IEEE Signal Processing Magazine **31**(3), 54–65 (2014)
16. Zhou, G., Cichocki, A., Xie, S.: Fast nonnegative matrix/tensor factorization based on low-rank approximation. IEEE Transactions on Signal Processing **60**(6), 2928–2940 (2012)
17. Sun, Q., Wu, P., Wu, Y., Guo, M., Lu, J.: Unsupervised Multi-Level Non-Negative Matrix Factorization Model: Binary Data Case. International Journal of Information Security **3**(4), 245–250 (2012)

Multiclassifier System with Fuzzy Inference Method Applied to the Recognition of Biosignals in the Control of Bioprosthetic Hand

Marek Kurzynski[✉] and Andrzej Wolczowski

Department of Systems and Computer Networks, Wroclaw University of Technology,
Wyb. Wyspianskiego 27, 50-370 Wroclaw, Poland
marek.kurzynski@pwr.edu.pl

Abstract. The paper presents an original method of recognition of patient's intention to move of hand prosthesis during the grasping and manipulation of objects. The proposed method is based on a 2-level multiclassier system (MCS) with base classifiers dedicated to EMG and MMG signals, and with combining mechanism using a dynamic ensemble selection (DES) scheme and competence function. Competence function of base classifier is determined using validation set in the two step procedure. The first step consists in creating competence set using the methods based on relating the response of the classifier with the response obtained by a random guessing. In the second step, the competence set is generalized to the whole feature space using the learning procedure based on the Mamdani-type fuzzy inference system. The performance of MCS with proposed competence measure was experimentally compared against four benchmark classification methods using real data concerning the recognition of six types of grasping movements. The system developed achieved the highest classification accuracies demonstrating the potential of MC system for the control of bioprosthetic hand.

Keywords: Multiple classifier system · Fuzzy inference method · Biosignals · Prosthetic hand

1 Introduction

Loss of hand significantly reduces the activity of human life. The people who have lost their hands are doomed to permanent care. Restoring to these people even a hand substitute makes their life less onerous. The hand transplantations are still in a medical experiment, mainly due to the necessity of immune-suppression (permanent, to the end of patient's life). An alternative is to equip these people with cybernetic prostheses. Existing active prostheses of hand (the bioprostheses) are generally controlled on myoelectric way - they react to electrical signals that accompany the muscle activity (called electromyography signals - EMG signals). Nevertheless, reliable recognition of intended movement using only the EMG signals analysis is a hard problem. A recognition error increases along

© Springer International Publishing Switzerland 2014
Z. Zeng et al. (Eds.): ISNN 2014, LNCS 8866, pp. 469–478, 2014.
DOI: 10.1007/978-3-319-12436-0_52

with the cardinality of movement repertoire (i.e. with prosthesis dexterity). The natural solution to overcome this error is to improve the recognition method [12], [13], [14], [15]. Another approach consists in additional use of a different kind of modalities on recognition stage, i.e. to complement EMG signals with another type of biosignals. The authors studied the fusion of EMG signals and the mechanomyography signals (MMG signals), i.e. the mechanical vibrations propagating in the limb tissue as the muscle contracts [10].

According to the author's recent experience ([8], [9], [15]), increasing the efficiency of the recognition stage may be achieved through the following activities: (1) – by introducing the concept of simultaneous analysis of two different types of biosignals (EMG and MMG signals), which are the carrier of information about the performed hand movement; (2) – by the appropriate choice of feature extraction methods (biosignals parameterization) justified by the experimental results of comparative analysis; (3) – through the use of multiclassifier system with the heterogeneous base classifiers dedicated to particular registered biosignals; (4) – through development of the paradigm of dynamic ensemble classifier selection system using measures of competence in the selection and fusion procedures. Taking into account above observations and suggestions, the paper aims to solve the problem of recognition of the patient's intention to move the multiarticulated prosthetic hand during grasping and manipulating objects in a skillful manner, by measuring and analyzing multimodal signals coming from patient's body. The adopted solution takes into consideration the advantages given by the fusion of the EMG and MMG signals. The concept combines the recognition (of EMG and MMG signals) performed by multiclassifier system working in the dynamic ensemble selection (DES) fashions with measures of competence of base classifiers. The most DES schemes use the concept of classifier competence i.e. its capability to correct activity (correct classification) on a defined neighbourhood or local region of testing object.

In this study, the competence measure of base classifier is calculated in a two-step procedure. In the first step, the set of competences at all points of validation set (the competence set) is calculated. In the second step, this competence set is generalized into the whole feature space using Mamdani type fuzzy inference system. This paper is divided into five chapters and organized as follows. Chapter 2 provides an insight into biosignals acquisition procedure and method of feature extraction. Chapters 3 presents the key recognition algorithm based on the multiclassifier system with the dynamic ensemble classifier selection strategy. Chapter 4 presents experimental results confirming adopted solution and chapter 5 concludes the paper.

2   Biosignal Acquisition and Feature Extraction

The recognition of movement intention on the basis of the myopotentials comprises three stages: (1) – the acquisition of the EMG signals; (2) – extraction of the features differentiating the movements; (3) – the classification of the signals, that is assigning the signals to a particular classes (particular movements).

Biosignal acquisition and analysis processes influence essentially on the reliability of recognition of prosthesis motion control decisions. The acquisition process should take into account the nature of the measured signals and their measurement conditions. For the EMG signals the amplitude of voltages induced on the patient body as a result of the influence of external electric fields, may exceed more than 1000 times the value of useful signals. To overcome this difficulty a differential measurement system was applied. In the case of MMG signals the basic problem is to isolate the microphone sensor from the external sound sources along with the best acquisition of the sound propagating in the patient's tissue.

After the acquisition stage, the recorded signals have the form of strings of discrete samples. Their size is the product of measurement time and sampling frequency. For a typical motion action, that gives a record of size between 3 and 5 thousand of samples (time of the order of 3-5 s, and 1 kHz sampling). This primary representation of the signals hinders the effective classification and requires the reduction of dimensionality. This reduction leads to a representation in the form of a signal feature vector. To determine the algorithm of features extraction, the database records were divided into 256 ms frames and then analyzed in time and frequency using Short Time Fourier Transform (STFT).

The MMG histogram has two amplitude peaks: at the beginning and at the end of the movement, and relatively low amplitude in the middle while the EMG histogram shows a peak in the middle of the movement time span. The analyses of histograms for the tested movements allowed selecting the localization of the best signal features (the best points in time and frequency) securing the best differentiation of the movements.

The resulting algorithm which allows creating input vectors with an adjustable size, has the following form:

Step 1. Extract from the recorded EMG and MMG signals representing the specified movements the 256 sample segments. Each extracted segment has new time span ($t \in [0, T]$);

Step 2. Apply the STFT to each segment;

Step 3. Choose as signal features the values from the STFT product corresponding to the k (most representative) time slices;

Step 4. Repeat steps 2 and 3 for every channel;

Step 5. Use all the obtained (in steps 2 and 3) values as elements of the feature vector representing the analyzed signal segment.

3   Multiple Classifier System

3.1   Preliminaries

In the multiclassifier system (MCS) we assume that a set of trained classifiers $\Psi = \{\psi_1, \psi_2, \ldots, \psi_L\}$ called base classifiers is given. A classifier ψ_l is a function $\psi_l : \mathcal{X} \to \mathcal{M}$ from a feature space $\mathcal{X}$ to a set of class labels $\mathcal{M} = \{1, 2, \ldots, M\}$. Classification is made according to the maximum rule

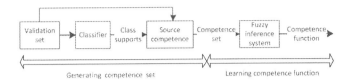

Fig. 1. Flowchart of the proposed method for calculating the competence function

$$\psi_l(x) = i \Leftrightarrow d_{li}(x) = \max_{j \in \mathcal{M}} d_{lj}(x), \tag{1}$$

where $[d_{l1}(x), d_{l2}(x), \ldots, d_{lM}(x)]$ is a vector of class supports (classifying function) produced by ψ_l. The value of $d_{lj}(x)$, $j \in \mathcal{M}$ represents a support given by the classifier ψ_l for the fact that the object x belongs to the j-th class. Without loss of generality we assume, that $d_{lj}(x) \geq 0$ and $\sum_j d_{lj}(x) = 1$.

The ensemble Ψ is used for classification through a combination function [5]. The proposed multiclassifier system uses dynamic ensemble selection (DES) strategy with trainable selection/fusion algorithm. The basis for dynamic selection of classifiers from the pool is a competence measure $C(\psi_l|x)$ of each base classifier ($l = 1, 2, \ldots, L$), which evaluates the competence of classifier ψ_l at a point $x \in \mathcal{X}$. For the training of competence it is assumed that a validation set

$$\mathcal{V} = \{(x_1, j_1), (x_2, j_2), \ldots, (x_N, j_N)\}; \quad x_k \in \mathcal{X}, \; j_k \in \mathcal{M} \tag{2}$$

containing pairs of feature vectors and their corresponding class labels is available.

The construction of the competence measure consists of the two steps. In the first step, a competence set $\mathcal{C}_l$ (set of competences for validation objects) for each classifier ψ_l in the ensemble is constructed:

$$\mathcal{C}_l = \{(x_1, C(\psi_l|x_1)), (x_2, C(\psi_l|x_2)), \ldots, (x_N, C(\psi_l|x_N))\}. \tag{3}$$

In the second step, the competence set (3) is used to construct the competence measure $c(\psi_l, x)$. The flowchart of the proposed method for calculating the competence measure of base classifier is shown in Fig. 1. The next two sections describe the steps of the method in detail.

3.2   The Competence Set

The competence $C(\psi_l|x_k)$ of the classifier ψ_l at a validation point $x_k \in \mathcal{X}$ from the set (2) is defined as [14]:

$$C(\psi_l|x_k) = 2 \cdot d_{j_k}(x_k)^{\frac{\log(2)}{\log(M)}} - 1. \tag{4}$$

The values of the function $C(\psi_l|x_k)$ lie within the interval $[-1, 1]$, where the interval limits -1 and 1 describe absolutely incompetent and absolutely competent classifier, respectively. The function $C(\psi_l|x_k)$ was defined in such a way because it satisfies the following criteria:

- it is strictly increasing, i.e. when the support $d_{j_k}(x_k)$ for the correct class increases the competence $C(\psi_l|x_k)$ also increases,
- it is equal to -1 (evaluates the classifier as absolutely incompetent) in the case of zero support for the correct class, i.e. $d_{j_k}(x_k) = 0 \Rightarrow C(\psi_l|x_k) = -1$,
- it is negative (evaluates the classifier as incompetent) in the case where the support for the correct class is lower than the probability of random guessing, i.e. $d_{j_k}(x_k) \in [0, \frac{1}{M}) \Rightarrow C(\psi_l|x_k) < 0$,
- it is equal to 0 (evaluates the classifier as neutral or random) in the case where the support for the correct class is equal to the probability of random guessing, i.e. $d_{j_k}(x_k) = \frac{1}{M} \Rightarrow C(\psi_l|x_k) = 0$,
- it is positive (evaluates the classifier as competent) in the case where the support for the correct class is greater than the probability of random guessing, i.e. $d_{j_k}(x_k) \in (\frac{1}{M}, 1] \Rightarrow C(\psi_l|x_k) > 0$,
- it is equal to 1 (evaluates the classifier as absolutely competent) in the case of maximum support for the correct class, i.e. $d_{j_k}(x_k) = 1 \Rightarrow C(\psi_l|x_k) = 1$.

3.3   Generalization Procedure

After the first step of the method the competence set (3) is given for each base classifier ψ_l ($l = 1, 2, ..., L$) of an ensemble Ψ.

In the second step, based on the competence set, the competence function $c(\psi_l, x)$ is determined. In other words, information contained in the set C_l, i.e. values of competence for validation points $x_k \in V$, is generalized to the whole feature space X which means supervised learning competence function of base classifier ψ_l.

In this study original generalizing system or method of supervised learning classifier competence is proposed. Method developed is based on Mamdani type fuzzy inference system which is a way of mapping an input space to an output space using fuzzy logic.

A fuzzy inference system applied to the supervised learning classifier competence is formed by four components [7]: (1) – The competence set C_l that describes the properties of the base classifier ψ_l for the validation set V; (2) – The point $x \in X$ which defines the problem and initiates the process of fuzzy reasoning; (3) – The set of fuzzy rules that determines the correspondence between the competence set and the competence at x; (4) – A fuzzy reasoning method which calculates a competence of classifier ψ_l at a point x using information given by the set C_l.

Mamdani fuzzy model is based on the collections of IF-THEN rules with both fuzzy antecedent and consequent propositions.

Taking into account the intuitive importance of competence of classifier ψ_l at validation point x_k and its distance from a point x to determine the competence of ψ_l at x, we propose these two quantities as antecedent variables of Mamdani fuzzy inference system. Thus, we have the following antecedent variables ($D(x, x_k)$ denotes Euclidean distance between x and x_k and D_{max} is a diameter of data set):

$$u_1 = D(x, x_k), \ u_1 \in [0, D_{max}], \ u_2 = C(\psi_l, x_k), \ u_2 \in [0, 1], \tag{5}$$

consequent variable:

$$y = c(\psi_l, x) \tag{6}$$

and the resulting set of fuzzy rules ($t = 1, 2, \ldots, T$ denotes number of rule):

$$IF\ u_1\ is\ A_1^{(t)}\ AND\ u_2\ is\ A_2^{(t)}\ THEN\ y\ is\ B^{(t)}. \tag{7}$$

From the perspective of computational complexity, we are interested in declaring fuzzy sets A_1, A_2 and B with membership functions that are efficient to evaluate. To this end, we use triangular and trapezoid fuzzy numbers for modeling linguistic values of antecedent and consequent variables. The membership functions of proposed fuzzy numbers are presented in Fig. 2.

Three fuzzy sets corresponding to values *Close*, *Medium* and *Far* of linguistic variable *Distance* cover uniformly (in the logarithmic scale) the space of antecedent variable u_1.

The value $1/M$ corresponding to probability of random guessing plays the key role in the fuzzy categorization of the space of antecedent variable u_2 and consequent variable y. The triangular fuzzy number corresponding to linguistic term of classifier competence *Neutral* is constructed around this value and next fuzzy sets are located to the left (*Incompetent* classifier) and to the right (*Weak*, *Competent* and *Very competent* classifier) of this set. Such a concept of fuzzy covering of input and output spaces leads to the set of 15 fuzzy IF-THEN rules, which structure is presented in Table 1.

Table 1. Fuzzy rules for the Mamdani system

	Close	Medium	Far
Incompetent	Incompetent	Neutral	Neutral
Neutral	Neutral	Neutral	Neutral
Weak	Weak	Neutral	Neutral
Competent	Competent	Competent	Weak
V. Competent	V. Competent	Competent	Competent

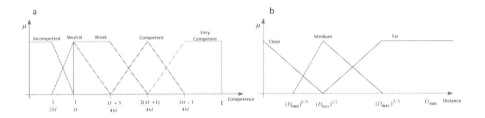

Fig. 2. Fuzzy partition with triangular and trapezoid membership functions for: (a) – the first antecedent variable/consequent variable (competence at validation point/competence at testing point), (b) – the second antecedent variable (distance between testing and validation points)

The Mamdani fuzzy inference system comprises five steps [2]: (1) fuzzification of the input variables, (2) calculating the degree of fulfilment of rules, (3) implication from the antecedent to the consequent, (4) aggregation of the consequents across the rules and (5) defuzzification.

3.4   Two-Level Multiclassifier System (2LMCS) with DES Scheme

DES based multiple classifier systems was constructed using proposed models of the base classifiers competences. In this system first a subset of the classifiers with the competences greater than the probability of random classification is selected from the ensemble for each x

$$\Psi_x = \{\psi_{l1}, \psi_{l2}, ..., \psi_{lK}\}, \quad c(\psi_{lk}, x) > 1/M. \tag{8}$$

This step eliminates inaccurate classifiers and keeps the ensemble relatively diverse. The selected classifiers are combined using a weighted vector of class supports, where the weights are equal to the competences, viz. $(j = 1, 2, \ldots, M)$:

$$d_j(x) = \sum_{t=1}^{K} c(\psi_{lt}, x) \, d_{lt,j}(x). \tag{9}$$

Since recognition of the patient's intent is made on the basis of analysis of two different biosignals (EMG and MMG), the multiple classifier system – according to the proposed concept of the recognition method – consisits of two submulticlassifiers: $\Psi^{(EMG)}$ and $\Psi^{(MMG)}$ – each of them dedicated to particular types of data. It leads to the two level structure of MC system presented in Fig. 3, in which the DES method is realized at the first level, whereas combining procedure at the second level is consistent with the continuous-valued dynamic fusion scheme.

At the second level of 2LMCS, supports (9) are combined by the weighted sum:

$$d_j^{(MC)}(x) = c^{(EMG)} d_j^{(EMG)}(x) + c^{(MMG)} d_j^{(MMG)}(x), \tag{10}$$

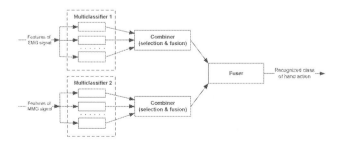

Fig. 3. Block diagram of the proposed multiclassifier system

where weight coefficients $c^{(EMG)}$ and $c^{(MMG)}$ denote mean competence of base classifiers from $\Psi^{(EMG)}$ and $\Psi^{(MMG)}$, respectively.

Finally, the 2LMCS system classifies $x = (x^{(EMG)}, x^{(MMG)})$ using the maximum rule:

$$\psi_{MC}(x) = i \;\Leftrightarrow\; d_i^{(MC)}(x) = \max_{j \in \mathcal{M}} d_j^{(MC)}(x). \tag{11}$$

4   Experiments

In order to study the performance of the proposed method of EMG and MMG signal recognition, some computer experiments were made. The experiments were conducted in MATLAB using PRTools package. In the recognition process of the grasping movements, 6 types of grips (tripoid, pinch, power, hook, column and mouse grip) were considered. Our choice is deliberate one and results from the fact that the control functions of simple bioprosthesis are hand closing/opening and wrist pronantion/supination, however for the dexterous hand these functions differ depending on grasped object [13].

Biosignals were registered using 3 EMG electrodes and 3 MMG microphones located on a forearm. The dataset set consisted of 400 measurements, i.e. pairs EMG and MMG signals segment/movement class. The values from STFT product corresponding to the k = 3, 4, 5 most representative time slices were considered as feature vector. The training and testing sets were extracted from each dataset using two-fold cross-validation. Half of objects from the training dataset were used as a validation dataset and the other half were used for the training of base classifiers. The experiments were carried out on healthy persons. Biosignals were registered using 3 EMG electrodes and 3 MMG microphones located on a forearm above the appropriate muscles. EMG and MMG signals were registered in specially designed 16-channel biosignals measuring circuit (Bagnoli Desktop EMG System made by DELSYS Inc.) with sampling frequency 1 kHz.

Three experiments were performed which differ in the biosignals used for classification (EMG signals, MMG signals, both EMG and MMG signals).

The experiments were conducted using the set of the following ten base classifiers [3] : (1-2) linear (quadratic) classifier based on normal distributions with

Tripoid grip   Pinch grip   Power grip   Hook grip   Column grip   Mouse grip

Fig. 4. Types of grips

Table 2. Classification accuracies of classifiers compared in the experiment. The best score for each dataset is highlighted. (k denotes the number of time slices per channel)

k	EMG signals			EMM signals			EMG and EMM signals		
	3	4	5	3	4	5	3	4	5
SB (1)	77.2	79.3	85.7	47.8	49.8	52.4	82.5	87.4	92.7
MV (2)	74.5	80.5	83.2	43.5	46.8	51.2	81.8	87.6	92.1
LA (3)	78.3	84.6	85.1	46.8	45.3	50.6	83.1	88.2	91.9
PF (4)	78.9	**85.2**	85.9	47.9	**50.3**	54.2	84.9	89.2	92.8
2LMCS	**79.8** 1,2	84.3 1,2	**86.2** 2	**48.4** 2,3	49.6 2,3	**55.6** 1,2,3	**85.5** 1,2,3	**90.2** 1,2	**95.3** 1,2,3,4

the same (different) covariance matrix for each class, (3) nearest mean classifier, (4-6) k-nearest neighbours classifiers with $k = 1, 5, 15$, (7) naive Bayes classifier (8) decision-tree classifier with Gini splitting criterion, (9-10) feed-forward back-propagation neural network with 1 hidden layer (with 2 hidden layers).

The performances of the DES system were compared against the following four multiple classifier systems: (SB) – The single best classifier in the ensemble [5]; (MV) – Majority voting (MV) of all classifiers in the ensemble [5]; (LA) – DES-local accuracy (LA) system [11]; (PF) – DES system with potential function as a generalization method (PF) [15].

5 Results and Conclusion

Classification accuracies (i.e. the percentage of correctly classified objects) for methods tested are listed in Table 2. The accuracies are average values obtained over 10 runs (5 replications of two-fold cross validation). Statistical differences between the performances of the DES-CD and the four MC systems were evaluated using Dietterich's 5x2cv test [1]. The level of $p < 0.05$ was considered statistically significant. In Table 2, statistically significant differences are given under the classification accuracies as indices of the method evaluated. These results imply the following conclusions:

1. The 2LMCS system produced statistically significant higher scores in 21 out of 36 cases (9 datasets $\times$ 4 classifiers compared);

2. The multiclassifier systems using both EMG and MMG signals achieved the highest classification accuracy for all datasets.

Experimental results indicate, that proposed methods of grasping movement recognition based on the dynamic ensemble selection with Mamdani-type fuzzy inference method for learning competence functions, produced accurate and reliable decisions, especially in the cases with features coming from the both EMG and MMG biosignals.

The problem of deliberate human impact on the mechanical device using natural biological signals generated in the body can be considered generally as a matter of "human – machine interface". The results presented in this paper significantly affect the development of this field and the overall discipline of biosignal recognition. But more importantly, these results will also find practical application in the design of dexterous prosthetic hand - in the synthesis of control algorithms for these devices, as well as development of computer systems for learning motor coordination, dedicated to individuals preparing for a prosthesis or waiting for a hand transplantation [6].

Acknowledgments. This work was financed from the National Science Center resources in 2012-2014 years as a research project No ST6/06168.

References

1. Alpaydin, E.: Combined 5x2cv F test for comparing supervised classification learning algorithms. Neural Computation **11**, 1885–1992 (1999)
2. Dubois, D., Prade, H.: Fuzzy sets and systems. Academic Press, New York (1988)
3. Duda, R., Hart, P., Stork, D.: Pattern Classification. Wiley-Interscience (2001)
4. Krysmann, M., Kurzynski, M.: Methods of learning classifier competence applied to the dynamic ensemble selection. Advances in Intelligent Systems and Computing **226**, 151–160 (2013)
5. Kuncheva, I.: Combining pattern classifiers: Methods and Algorithms. Wiley-Interscience (2004)
6. Kurzynski, M., Wolczowski, A.: Classification of EMG Signals in a System for Training of Bioprosthetic Hand Control in One Side Handless Human. In: Proc. International Conference on Electrical Engineering and Computer Science, EECS 2012, Szanghai, pp. 566–576 (2012)
7. Kurzynski, M., Krysmann, M.: Fuzzy Inference Methods Applied to the Learning Competence Measure in Dynamic Classifier Selection. In: Proc. XXVII Conf. on Graphics, Patterns and Images. IEEE Comp. Society Press (in press, 2014)
8. Kurzynski, M., Wolczowski, A.: Multiple Classifier System Applied to the Control of Bioprosthetic Hand Based on Recognition of Multimodal Biosignals. In: Goh, J. (ed.) The 15th International Conference on Biomedical Engineering. IFMBE Proceedings, vol. 43, pp. 577–580. Springer, Heidelberg (2013)
9. Kurzynski, M., Wolczowski, A.: Hetero- and Homogeneous Multiclassifier Systems Based on Competence Measure Applied to the Recognition of Hand Grasping Movements. In: Piętka, E., Kawa, J., Wieclawek, W. (eds.) Information Technologies in Biomedicine. AISC, vol. 284, pp. 163–174. Springer, Heidelberg (2014)
10. Orizio, C.: Muscle sound: basis for the introduction of a mechanomyographic signal in muscle studies. Critical Reviews in Biomedical Engineering **21**, 201–243 (1993)
11. Smits, P.: Multiple classifier systems for supervised remote sensing image classification based on dynamic classifier selection. IEEE Trans. on Geoscience and Remote Sensing **40**, 717–725 (2002)
12. Wolczowski, A., Kurzynski, M.: Control of dexterous hand via recognition of EMG signals using combination of decision-tree and sequential classifier. Advances in Soft Computing **45**, 687–694 (2007)
13. Wolczowski, A., Kurzynski, M.: Human-machine interface in bio-prosthesis control using EMG signal classification. Expert Systems **27**, 53–70 (2010)
14. Woloszynski, T., Kurzynski, M.: On a new measure of classifier competence applied to the design of multiclassifier systems. In: Foggia, P., Sansone, C., Vento, M. (eds.) ICIAP 2009. LNCS, vol. 5716, pp. 995–1004. Springer, Heidelberg (2009)
15. Woloszynski, T., Kurzynski, M.: A probabilistic model of classifier competence for dynamic ensemble selection. Pattern Recognition **44**, 2656–2668 (2011)

Target Detection Using Radar in Heavy Sea Clutter by Polarimetric Analysis and Neural Network

Ji Eun Kim[1(✉)], Sang Min Lee[1], Seung-Phil Lee[1], SooBum Kim[2], Young-Soo Kim[1], and Chan Hong Kim[3]

[1] POSTECH, Electrical Engineering, Pohang, South Korea
{jieun7,smlee926,feel86,ysk}@postech.ac.kr
[2] Digitron Co., Ltd., Seongnam, South Korea
ksbget@digitron.kr
[3] Agency for Defense Development (ADD), Deajeon, South Korea
chkim@add.re.kr

Abstract. To improve radar detection of targets in the presence of sea clutter, polarization decomposition analysis is used to output data to a neural network. To detect the target in heavy sea clutter, the received signal of each range bin is decomposed for polarimetric analysis. By applying the decomposed signal to the Self Organizing Map (SOM) network, the bin that contains the actual target range bin is identified. In simulations the algorithm located the target successfully in conditions of both mild and heavy clutter.

Keywords: Polarimetric decomposition · Self-organizing map · IPIX Radar dataset · Kohenen map

1   Introduction

The goal of marine surveillance is to detect targets, submarines, low-flying aircraft and small ships. Clutter is the term to denote unwanted echoes from the natural environments especially, in case the reflected signal form the sea surface is sea clutter. However, these targets are close to the ocean surface and the radar beam is nearly horizontal, so interference from sea clutter (e.g., waves) can distort the reflected beams and thereby reduce detection accuracy. To overcome this problem, methods to locate the target in the presence of sea clutter have been proposed. One idea is to use Constant False Alarm Rate (CFAR) detector [1]. To detect the target in the presence of clutter, the clutter signal is represented as statistical model, and based on this model, the signal is applied to a CFAR algorithm. Also, by exploiting the fact that a sea clutter signal has chaotic characteristic, a received signal can be analyzed using neural network (NN) [2]. By applying the received signal of each range bin to the NN, the real target position can be identified. Recently, use of time-frequency analysis to detect the target in sea clutter has been proposed [3].

© Springer International Publishing Switzerland 2014
Z. Zeng et al. (Eds.): ISNN 2014, LNCS 8866, pp. 479–488, 2014.
DOI: 10.1007/978-3-319-12436-0_53

However, each of these methods has a problem. Using CFAR algorithm is difficult when clutter is spiky because spike events can cause false alarms. NNs operate sell in the environments in which they were trained, but their detection accuracy degraded when used in other environments. Time-frequency analysis cannot detect slow target well.

In this paper we proposed an algorithm that combines polarization decomposition analysis and an NN to maintain good detection accuracy in the presence of heavy sea clutter. The algorithm is designed to detect the target in the presence of spiky clutter by using the polarimetrically decomposed signal as the input for the NN.

In section 2, we show the main idea of the polarimetric decomposition. In section 3 we introduce SOM network, an unsupervised learning network for clustering. In section 4 we explain how the proposed algorithm combines polarization decomposition and NN to detect the target. In section 5 we present simulation results of proposed algorithm. In section 6 we conclude the paper.

2   Polarimetric Decomposition

Polarimetric decomposition use the difference of the polarimetric characteristics of the received signal from the targets from that of the sea clutter, to detect the target in presence of sea clutter. A scattering matrix shows whether the mixed signal represents a target and clutter or clutter only. Therefore, the target can be distinguished from the clutter by connecting each component of scattering matrix and to the corresponding physical mechanism.

To identify the physical mechanism of the scattering matrix A, a scattering feature vector can be borrowed from vector signal estimation theory [4]. Pauli spin matrices

$$\Psi_P = \left\{ \sqrt{2}\begin{bmatrix} 1 & 0 \\ 0 & 1 \end{bmatrix}, \sqrt{2}\begin{bmatrix} 1 & 0 \\ 0 & -1 \end{bmatrix}, \sqrt{2}\begin{bmatrix} 0 & 1 \\ 1 & 0 \end{bmatrix}, \sqrt{2}\begin{bmatrix} 0 & -j \\ j & 0 \end{bmatrix} \right\} \tag{1}$$

were used to obtain the feature vector.

$$k = \frac{1}{2}Trace([A]\Psi_P) = \frac{1}{\sqrt{2}}\begin{Bmatrix} A_{HH} + A_{VV} \\ A_{VV} - A_{HH} \\ A_{HV} + A_{VH} \\ j(A_{VH} - A_{HV}) \end{Bmatrix} = \frac{1}{\sqrt{2}}\begin{Bmatrix} A_{HH} + A_{VV} \\ A_{VV} - A_{HH} \\ A_{HV} + A_{VH} \end{Bmatrix} \tag{2}$$

Where $A = \begin{bmatrix} A_{HH} & A_{HV} \\ A_{VH} & A_{VV} \end{bmatrix}$ is the scattering matrix.

The final term, $j(A_{VH} - A_{HV})$ disappear due to $A_{HV} = A_{VH}$ by reciprocal theorem.

Because the target is not a point target but a distributed target, the coherency matrix $\langle[C]\rangle$ must be written as a feature vector.

$$\langle[C]\rangle = \frac{1}{N}\sum_{i=1}^{N} k \bullet k^*$$
(3)

where N is the temporal or spatial averaging time or region and k is the feature vector and k^* is the conjugate of the k.

$\langle[C]\rangle$ is related to cloude and pottier model obtained by unitary transform [5].

$$<[C]>=[U][\Lambda][U]^H =[U]\begin{bmatrix} \lambda_1 & 0 & 0 \\ 0 & \lambda_2 & 0 \\ 0 & 0 & \lambda_3 \end{bmatrix}[U]^H$$
(4)

where U is unitary matrix, $\lambda_i , i = 1,2,3$ is the eigenvalues of $\langle[C]\rangle$ and H is the Hermition operator

$$[U] = \begin{bmatrix} \cos(\alpha_1) & \cos(\alpha_2) & \cos(\alpha_3) \\ \sin(\alpha_1)\cos(\beta_1)e^{j\delta_1} & \sin(\alpha_2)\cos(\beta_2)e^{j\delta_2} & \sin(\alpha_3)\cos(\beta_3)e^{j\delta_3} \\ \sin(\alpha_1)\cos(\beta_1)e^{j\gamma_1} & \sin(\alpha_2)\cos(\beta_2)e^{j\gamma_2} & \sin(\alpha_3)\cos(\beta_3)e^{j\gamma_3} \end{bmatrix}$$
$$\quad\quad e_1 \quad\quad\quad\quad\quad e_2 \quad\quad\quad\quad\quad e_3$$
(5)

We can calculate the eigenvalue $\lambda_1, \lambda_2, \lambda_3$ and matrix U and from the first row of the unitary matrix U the $\alpha_i, i = 1,2,3$ value can be calculated. The physical information can be obtained from the eigenvalue obtained by unitary transform and $\alpha_i, i = 1,2,3$ of U. The eigenvalue is related to the polarimetric entropy. For example, if $\lambda_1 = \lambda_2 = \lambda_3$, entropy becomes 1(maximum value), which means that there is no noticeable scattering occur and if $\lambda_2 = \lambda_3 = 0$, entropy becomes 0 (minimum value), which means that one scattering mechanism dominates the others.

The entropy is

$$H = \sum_{i=1}^{3} -P_i \log_3 P_i$$
(6)

Where $P_i = \lambda_i / \sum_{j=1}^{3} \lambda_j$

α value is the parameter showing the kind of the scatterer.

$$\alpha = P_1\alpha_1 + P_2\alpha_2 + P_3\alpha_3 \qquad (7)$$

Eq (6) and (7) yield the H - α plane (Fig. 1) of each range bin signal [6].

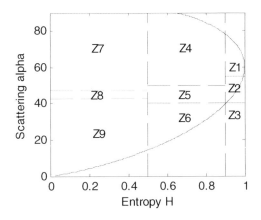

Fig. 1. H-α plane. Zone (Z) are described in the text

The H - α plane can be partitioned into zones that represent different forms of scattering, as follows. Zone 9(Z9): low entropy surface scattering. Z8: low entropy dipole scattering. Z7: low entropy multiple scattering events. Z6: medium entropy surface scatter. Z5: medium entropy vegetation scattering. Z4: medium entropy multiple scattering. Z3: high entropy surface scatter. Z2: high entropy vegetation scattering. Z1: high entropy multiple scattering [7]. The probability that a signal has a dominant scattering mechanism increases as signal entropy decreases.

3   Self Organizing Map

SOM is a form of unsupervised learning. In this paper, we use the Kohonen map which is one of the SOM network (Fig 2) [8]. By training the Kohonen map using the input as signal, high-dimension data can be projected to low-dimension data (output node). Also, input data can be clustered using the weight matrix. When the input data and output node are clustered, the topological characteristics of output node are kept the same as the number of output nodes, so by use the weight matrix, we cans identify the representative value of the input signal.

We use the 5 by 5 SOM in algorithm and SOM roles for clustering the H and α points, the 2 dimensional data. The size of the SOM is selected in heuristically. In 50000 H and α points, the weight matrix saves the 25 point as the representative value of the H and α points and we can use that point as the criteria of the detection algorithm. Details are showed in algorithm description section, section 4.

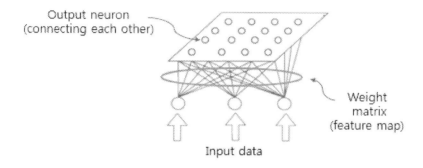

Fig. 2. Basic concept of the Kohonen map

4 Algorithm Description

4.1 Remove Spikes from Sea Clutter Data

Before applying the signal to the main algorithm, some spike events must be suppressed to improve the detection accuracy. The HH polarization signal is somewhat higher than the VV signal; this phenomenon can be used to remove the sea spike. If for observation i, copolarization ratio is [9]

$$r_i = 20\log_{10}\left(\left|A_{HH} / A_{VV}\right|_i\right), \quad i = 1, 2, ..., t \ . \tag{8}$$

Where r_i exceeds a threshold, the observation is removed.

4.2 Obtain the H - α Plane of Each Range Bin

To acquire the physical information from the signal polarimetric property, calculate Eq (6) and Eq (7) and plot the H - α plane of each range bin.

4.3   Apply the H - α Plane of Each Range Bin to 5 x 5 SOM

Kohonen map is trained applying 2D data (H, α) as the input. This training yields weight matrix that shows the clustering result.

4.4   Select and Average the Five Lowest Weights

H Value and α value are used to train the 5x5 SOM to obtain the 25 representative values of input data. To identify the dominant scattering mechanism, the weight point having the five lowest entropies are averaged. We can calculate this average value of each range bin, the range bin that has the lowest average to contain the target.

Because this method uses five lowest values in the multiple representative value, rather than just one representative value [6], the proposed algorithm can maintain good accuracy even in presence of heavy clutter.

5   Simulation Result

We used two sets of sea clutter data obtained using McMaster IPIX radar (Table 1) [10]. Each dataset has 14 range bins; the target is a Styrofoam sphere wrapped in wire mesh. Because the distance between range gates is 15 m and the range resolution is 30 m, except the real target position (primary target position), the target may be detected near the primary target range gate (secondary target position). In other word, The range bins where the targets are strongest detectable are called primary bins, while neigh-boring range bins where the targets may also be visible less than primary bins are called secondary bins[11].

Table 1. IPIX radar specifications and experiment environment

	Specifications		Environment
Frequency	X-band	Wave height	0.8-3.8 m
PRF	1 kHz	Wind speed	0-60 km/h
Polarization	Dual polarization	Range	500-8000 m
BW	25 MHz	Grazing angle	0.2-3.5°
Mode	Stare, surveillance mode	Target	Known floating objects

We use IPIX Radar data set files 54 and 310 (Table 2) for simulation to compare environments with heavy and moderate sea clutter, respectively.

Table 2. Specifications of the simulation environment

Datum	File	
	54	**310**
Primary target position	8th bin range	7th bin range
Secondary target position	7:10 bin range	6:9 bin range
Wave height (m)	0.7	0.91
Wind speed (km/h)	22	28

The proposed algorithm detected the target well for both file 54 and file 310. File 54 has a low sea wave height, so H - α plane of the range bin that contains only clutter is very different form the H - α plane of a range bin that contains both clutter and target (Fig. 3 a, b). File 310 has a moderate wave height, H - α plane is not much affected by the presence of the target (Fig. 4 a, b). Nevertheless, in the result of the SOM (Fig. 5) the average of the five lowest weights of the range bin that include the target is lower than the same average of the other range bins (Fig 5).

The averaged value of the 8th range bin that contains the target (8th range bin in file 54; 7th bin in file 310) was lower than that of the other range bins.

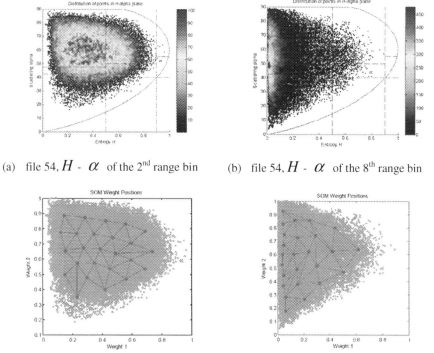

(a)   file 54, H - α of the 2nd range bin   (b)   file 54, H - α of the 8th range bin

(c)   file 54, SOM result of the 2nd range bin   (d)   file 54, SOM result of the 8th range bin

Fig. 3. H-alpha plane and SOM result of no.file54

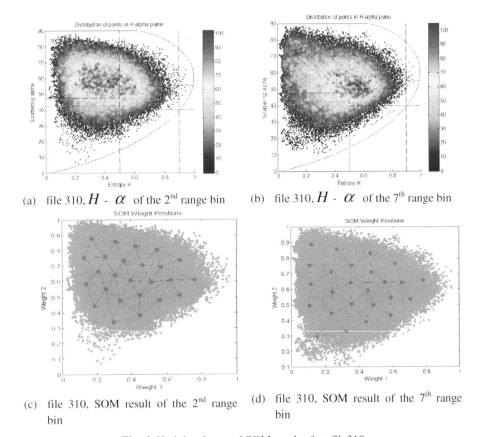

(a) file 310, H - α of the 2nd range bin

(b) file 310, H - α of the 7th range bin

(c) file 310, SOM result of the 2nd range bin

(d) file 310, SOM result of the 7th range bin

Fig. 4. H-alpha plane and SOM result of no.file310

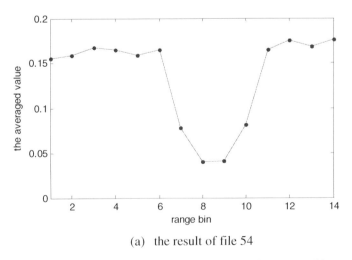

(a) the result of file 54

Fig. 5. Average of the five lowest weights of each range bin

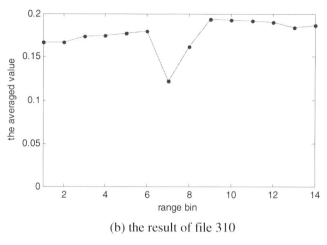

(b) the result of file 310

Fig. 5. (*Continued*)

6   Conclusion

We proposed an algorithm that combines polarization decomposition theorem and an NN to detect a target in presence of heavy sea clutter. In simulations using the IPIX radar data set, the proposed algorithm detected targets well in both conditions of low and heavy sea clutter.

Acknowledgements. This work was supported by the STRL (Sensor Target Recognition Laboratory) program of Defense Acquisition Program Administration and Agency for Defense Development

References

1. Siddiq, K., Irshad, M.: Analysis of the cell averaging CFAR in weibull background using a distribution approximation. In: Proceedings of the 2nd International Conference on Computer, Control and Communication (IC4 2009), pp. 1–5 (2009)
2. Li, Y., Wang, W., Sun, J.: Research of Small Target Detection within Sea Clutter Based On Chaos. In: 2009 International Conference on Environmental Science and Information Application Technology (ESIAT), vol. 2, pp. 469–472 (July 2009)
3. Lamont-Smith, T.: Azimuth dependence of Doppler spectra of sea clutter at low grazing angle. IET Radar, Sonar & Navigation **2**(2), 97–103 (2008)
4. Cloude, S.R., Pottier, E.: A review of Target Decomposition theorems in radar polarimetry. IEEE Transactions on Geoscience and Remote Sensing **34**(2), 498–518 (1996)

5. Cloude, S.R., Pottier, E.: An Entropy Based Classification Scheme for Land Applications of Polarimetric SAR. IEEE Transactions on Geoscience and Remote Sensing **35**(1), 68–78 (1997)
6. Wu, P., Wang, J., Wang, W.: A Novel Method of Small Target Detection in Sea Clutter. Hindawi Publishing Corporation International Scholarly Research Network ISRN Signal Processing 2011, Article ID 651790, 10 pages (2011). doi:10.5402/2011/651790
7. Boerner, W.-M., Yan, W.-L., Xi, A.-Q., Yamaguchi, Y.: Basic concepts in radar polarimetry. Springer, Netherlands (1992)
8. Kohonen, T.: Self-Organizing Maps. Springer, Heidelberg (2001)
9. Melief, H.W., Greidanus, H., van Genderen, P., Hoogeboom, P.: Analysis of Sea Spikes in Radar Sea Clutter Data. IEEE Transactions on Geoscience and Remote Sensing **44**(4), 985–993 (2006)
10. Drosopoulos, A.: Description of the OHGR database. Defence Research Establishment Ottawa (1995)
11. Guan, J., Liu, N., Zhang, J.: Low-Observable Target Detection in Sea Clutter Based on Fractal-Based Variable Step-Size Least Mean Square Algorithm. In: 2009 International Radar Conference - Surveillance for a Safer World (RADAR), pp. 1–5 (2009)

Human Action Recognition Based on Difference Silhouette and Static Reservoir

Danchen Zheng and Min Han[(✉)]

School of Control Science and Engineering,
Dalian University of Technology, Dalian 116023, Liaoning, China
{dcjeong,minhan}@dlut.edu.cn

Abstract. In this paper, the variation between features of frames for human action recognition is studied, and a new local descriptor extracted among the differences of human silhouettes is posed. This descriptor is represented by coarse histograms based on the distribution of sample points on the outlines of difference silhouettes. The static reservoir is employed as the classifier of human action. Two hyper-parameters, the scaling parameter γ and the regularization parameter C are taken to characterize a static reservoir, and the proper static reservoir for action recognition is identified on the $\gamma - C$ plane. We test our approach on two commonly used action datasets, and the experimental results show that the proposed method is effective.

Keywords: Image representation · Action classification · Difference silhouette · Static reservoir · $\gamma - C$ plane

1 Introduction

Human action recognition is one of the most important parts of human action analysis which has been receiving increasing attention from computer vision researchers. The recognition of human activities has a wide range of prospect and potential economic value such as automatic surveillance, interactive applications and efficient searching actions [1]. For clarity, the process of human action recognition can be divided into two stages: image representation and action classification [2]. The former course is aimed to extract certain areas in videos, and then to obtain the features following. Thereafter a classification method should be adopted for the next stage.

Shape and kinematics are two important cues in human movement analysis. Precise kinematics is difficult to be extracted from image sequences of videos, so humane actions are considered as a temporal process in which both the appearances and the locations of human silhouettes continuously change over time. However, it is not a good choice to directly adopt the features in each frame as an isolated human gesture. Obviously, the features should be extracted from

---

This research is supported by the project (61374154) of the National Nature Science Foundation of China.

Z. Zeng et al. (Eds.): ISNN 2014, LNCS 8866, pp. 489–498, 2014.
DOI: 10.1007/978-3-319-12436-0_54

silhouette sequences with the consideration of space-time characteristics, so that effective representation could be obtained.

Since the action classification problem is a practical pattern recognition problem, many classic methods such as supper vector machines (SVMs) [3,4] and extreme learning machines (ELMs) [5], are used to resolve the problem of the human action recognition. SVM has been widely used for researches, and is an effective method for nonlinear classification problems. Although kernel method can be applied for nonlinear problem, it is still difficult for SVM to adopt a proper nonlinear mapping $\phi(x)$. ELM is a large-scale feed-forward neural network with randomly generated hidden nodes, and it has been successful applied to classification problems [6]. However, the structure of ELM is difficult to determine.

In this paper, difference silhouette is proposed as a new local descriptor for image representation. Difference silhouettes extracted form silhouettes are global representation, and the features are represented by collections of patches of gestures. We further used histograms of polar coordinates which describe the distribution of all the sample points on contours relative to the shape center. Then we focus on the static reservoir for action classification, and the structure of static reservoir can be easily determined with two hyper-parameters, γ and C. A proper reservoir is identified on $\gamma - C$ plane for action classification.

2   Human Action Descriptor Based on Difference Silhouette

2.1   Difference Silhouette

In this section, without considering the problem of background subtraction, we assume that silhouettes have already been extracted. $F = \{f_1, \cdots, f_n\}$ is used to represent an silhouette sequence, where f_i is a binary silhouette of frame i and each binary pixel can be expressed by $f_i(x, y)$, $i = 1, \cdots, n$. We define forward difference silhouette (FDS) segment and backward difference silhouette (BDS) segment as follows:

$$d_j^F(x, y) = f_{i+\Delta}(x, y) - f_i(x, y) \cap f_{i+\Delta}(x, y) \tag{1}$$

$$d_j^B(x, y) = f_i(x, y) - f_i(x, y) \cap f_{i+\Delta}(x, y) \tag{2}$$

where $d_j^F(x, y)$ and $d_j^B(x, y)$ denote FDS segment and BDS segment, $j \leq n - \Delta$. Then, $D^F = \{d_1^F, \cdots, d_m^F\}$ and $D^B = \{d_1^B, \cdots, d_m^B\}$ are used to represent the FDS sequence and the BDS sequence, and difference silhouette (DS) is composed of FDS and BDS. In many cases, difference silhouettes contain some noises which are generated by the small differences between silhouettes. Therefore, filtering methods, such as opening operation and wavelet denoising, are adopted to eliminate noise effects, and opening operation is employed in this paper. Fig. 1 gives an example of generating FDS and BDS from two adjacent silhouettes.

(a)   (b)

Fig. 1. DS between two adjacent silhouettes. (a) The silhouette of different frames. (b) Forward difference silhouette and backward difference silhouette extracted form (a).

2.2   Histogram for Difference Silhouette

According to the approach of computing FDS and BDS, the shifts of postures between different frames are obtained by analyzing the differences of corresponding shapes. In our approach, DS is considered as a collection of samples on outlines, and we assume that the shape of DS can be essentially expressed by a finite subset of the points [7]. More practically, both of FDS and BDS are represented by a set of points sampled from contours which can be obtained by an edge detector. The numbers of points on the shapes of DS are quite different, η is introduced as the parameter of sample ratio which is set equal to 0.2. In the process of sampling, we prefer to sample the shape with roughly uniform spacing, and do not concern the key-points on contours such as maxima of curvature or inflection points. Given $p_{i,j}^F$ as the jth sample point in d_i^F, $p_{i,j}^B$ as the jth sample point in d_i^B, the corrsponding point set of DS can be represent as follows:

$$P_i = \{p_{i,1}^F, \cdots, p_{i,\alpha_i}^F, p_{i,1}^B, \cdots, p_{i,\beta_i}^B\} = \{p_{i,1}, \cdots, p_{i,N_i}\} \tag{3}$$

where N_i is the number of sample points in DS. Fig. 2 (c) gives an illustration of samples of DS form Fig. 2 (a) and Fig. 2 (b).

For all the sample points on the shape of DS, a coarse histogram is introduced to describe the distribution of them. In this case, the center of each DS o_i is defined as the coordinate origin, and this center is computed as follows:

$$o_i = \frac{c_j + c_{j+\Delta}}{2} \tag{4}$$

where c_j is the center of binary silhouette f_i, and $c_{j+\delta}$ is the center of binary silhouette $f_{j+\delta}$. For the center o_i of DS, a histogram h_i can be obtained as follows:

$$h_i(u) = \#\{p_{ij} : (p_{ij} - o_i) \in \text{bin}(k)\} \tag{5}$$

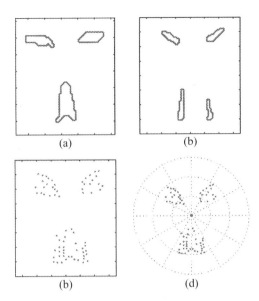

Fig. 2. Polar histogram and shape context computation. (a) and (b) Difference silhouette contours. (c) Contour sampling integration. (d) Polar histogram computation.

where p_{ij} is the jth sample point in P_i, k is the kth bin, and the bins are uniform in polar coordinates. Here the polar radius r is divided into 3 equal lengths, and the polar angle θ is divided into 12 equal parts. Fig. 2 (d) gives an example of dividing the polar space.

According to the proposed approach, the polar histogram is used to represent the human motion features in DS. With this descriptor lying in a low-dimensional space, different human actions are expressed as irregular sequences of vectors. A feature fusion computation $g_p = \sum\limits_{i=1}^{m} h_i$ is adopted for reducing the dimension of time space, and this classification problem of time series can be replaced by a general pattern recognition problem.

3   Static Reservoir for Human Action Classification

3.1   Static Reservoir

Classic multi-layer perception (MLP) models use adaptive basis functions with sigmoid nonlinearities, which can adapt the parameters so that the regions of input space over which the basis function vary corresponds to the data manifold. A MLP model with one hidden layer can be expressed as follows:

$$y = w^{\mathrm{T}} \mathrm{sig} \left(\sum_{j}^{p} \sum_{i}^{N} \gamma_{ij}^{w} \cdot W_{\mathrm{in}}^{ij} u_j + \sum_{i}^{N} \gamma_i^{b} \cdot b^i \right) + b_{\mathrm{o}} \tag{6}$$

where p is the dimension of input, u_j is the jth input, and $\boldsymbol{y}$ is the output matrix, N is the number of hidden layer neurons, the input weight connections and bias are initialized to W_{in}^{ij} and b^i, $i = 1, \cdots, N$, $j = 1, \cdots, p$, respectively. γ_{ij}^w is the local scaling parameter for the input weight W_{in}^{ij}, and γ_i^b is the local scaling parameter for the bias b^i. The standard form of static reservoir can be expressed as follows:

$$y = \boldsymbol{w}^{\text{T}} \text{sig}\left(\gamma W_{\text{in}} \boldsymbol{u}\right) + b_{\text{o}} \tag{7}$$

where $\boldsymbol{u}$ is the input of the reservoir, bias vector $\boldsymbol{b}$ has been included in the input matrix W_{in} for convince, and the initial input $\boldsymbol{u}$ is converted into $\boldsymbol{u} = \begin{bmatrix} \boldsymbol{u}^{\text{T}} & 1 \end{bmatrix}$. The input matrix W_{in} is generated randomly, γ is the global input scaling parameter as described before.

The activation matrix A and target output Y can be defined as follows:

$$A = \begin{bmatrix} \boldsymbol{a}_1, \cdots, \boldsymbol{a}_l \end{bmatrix}^{\text{T}} \tag{8}$$

$$Y = \begin{bmatrix} \boldsymbol{y}_1, \cdots, \boldsymbol{y}_l \end{bmatrix}^{\text{T}} \tag{9}$$

where $\boldsymbol{a}_i = \text{sig}\left(\gamma W_{\text{in}} \boldsymbol{u}_i\right)$, $\boldsymbol{u}_i$ is the input vector of training data, $\boldsymbol{y}_i$ is the classification result, and i is the index of training examples, $1 \leq i \leq l$.

The Tikhonov-type regularization is used here, with a Cholesky decomposition of matrix $A^{\text{T}}A + I/C \in R^{N \times N}$, which is not related to the number of training samples but the scale of the reservoir. The solution can be expressed as follows:

$$A^{\text{T}}A + I/C = L^{\text{T}}L, L\boldsymbol{z} = A^{\text{T}}Y, L^{\text{T}}\boldsymbol{w} = \boldsymbol{z} \tag{10}$$

where $\boldsymbol{z}$ is the intermediate variable, L is the lower triangular matrix in Cholesky decomposition.

Therefore, we select two hyper-parameters for static reservoir: the global scaling parameter γ and the regularization parameter C. With γ and C, static reservoir construction will be discussed in the next subsection.

3.2 Static Reservoir with $\gamma - C$ Plane for Classification

As discussed in the previous section, we prefer using the scaling parameter γ and the regularization parameter C to manage the magnitudes of the weights. The global scaling parameter γ is adopted to restrict the original input weight matrix W_{in} which is generated at random. Therefore, the new input weight connection γW_{in} fully depends on the scaling parameter γ. The magnitudes of output weights are determined by the regulation parameter C given the activation matrix A and target output Y. Therefore, we take the $\gamma - C$ plane as an evaluation of the static performance, and a reservoir can be constructed through the computation of the classification accuracy with different parameters. The static reservoir constructed with small γ and small C performances weak nonlinearity and small capacity, which lead to underfitting. On the contrary, large γ

and large C denote strong complexity and overfitting. The two parameters are greater than zero, and the plane should be established in logarithmic space [8].

In order to obtain effective working area of $\gamma - C$ plane for classification problem, the procedure of searching good $\gamma - C$ pair should be settled. The error measure can be defined as the error rate of the classification. A good pair of γ and C can be obtained by many existing methods, such as the grid search method or Nelder and the Meads downhill simplex method. For the experiments in this paper, we use the gird search to solve the problem.

4   Experimental Results

Human action recognition experiments are carried out based on two datasets, Weizmann dataset and KTH dataset. Weizmann dataset is first reported in [9], and Fig. 3 gives several examples of this dataset. It contains 93 low-resolution videos of 9 different persons, each performing 10 actions (one person perform each of 3 different actions for twice). These actions include run, walk, skip, jack, bend, jump, pjump, side, wave2, and wave1. Both the backgrounds and the view point of this dataset are static. As the backgrounds of these videos are included in the dataset, the silhouettes can be easily extracted from each frames. In order to make fair comparisons with the results of other researches, we directly use the foreground silhouettes included in the given dataset.

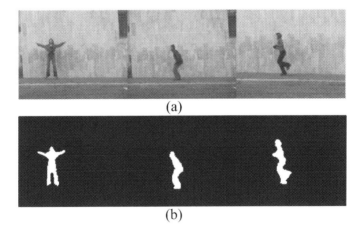

(a)

(b)

Fig. 3. Human action examples from Weizmann dataset. (a) Original image. (b) Silhouette image.

KTH human motion dataset has been introduced by Schuldt et al. in [3]. It contains 600 AVI files of 6 actions exhibited by 25 persons in 4 different conditions: static background, scale variations, different clothes, and lighting variations. The actions include walking, jogging, running, boxing, handwaving, and handclapping. Since the scenes in the KTH dataset are not fixed, and the videos are grayscale, direct subtractions in these videos are not feasible. Therefore we firstly use human detection method to localize the human figure, and then applied background subtractions for the localized results. However, the extracted silhouettes is still imperfect but available.

For different datasets, we use a fixed reservoir size ($N = 100$), and the input of weight connections are now dependent on the hyper-parameter γ. The $\gamma - C$ plane is demonstrated by calculating the classification error. The two hyper-parameters are combinations of exponentially growing sequences of γ and C. Fig. 4 and Fig. 5 are the $\gamma - C$ planes for the Weizmann and KTH datasets.

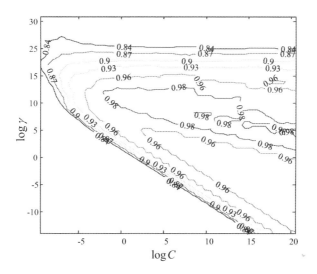

Fig. 4. The process of static reservoir construction with $\gamma - C$ plane for Weizmann dataset

From the $\gamma - C$ plane we can get the range of $\gamma - C$ selection. In Weizmann Dataset, we select $\gamma = 2^{1}0$ and $C = 2^{5}$, the confusion matrices for Weizmann dataset is shown in Fig. 6, and the classification accuracy is 98.75%. And in KTH dataset, we select $\gamma = 2^{3}$ and $C = 2^{1}0$, the confusion matrices for KTH dataset is shown in Fig. 7, and the classification accuracy is 95.29%.

The recognition accuracy of Weizmann dataset and KTH dataset obtained via the proposed approach is compared with works of other researchers as reported in Table 1 and Table 2. The Weizmann is widely used by many researchers, while only experimental results achieved under similar condition are included in the comparison.

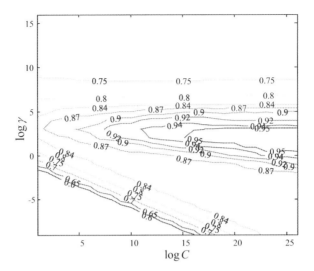

Fig. 5. The process of static reservoir construction with $\gamma - C$ plane for KTH dataset

	Bend	Jack	Jump	Pjump	Run	Side	Skip	Walk	Wave1	Wave2
Bend	1.00	0	0	0	0	0	0	0	0	0
Jack	0	1.00	0	0	0	0	0	0	0	0
Jump	0	0	0.98	0	0	0	0.02	0	0	0
Pjump	0	0	0	1.00	0	0	0	0	0	0
Run	0	0	0	0	0.97	0	0	0.03	0	0
Side	0	0	0	0	0	1.00	0	0	0	0
Skip	0	0	0	0	0	0	1.00	0	0	0
Walk	0	0	0	0	0	0	0	1.00	0	0
Wave1	0	0	0	0	0	0	0	0	1.00	0
Wave2	0	0	0	0	0	0	0	0	0	1.00

Fig. 6. Confusion matrix for classification results of Weizmann dataset

	Box	Wave	Clap	Jog	Run	Walk
Box	1.00	0	0	0	0	0
Wave	0	0.93	0.07	0	0	0
Clap	0	0.02	0.98	0	0	0
Jog	0	0	0	0.91	0.06	0.03
Run	0	0	0	0.07	0.91	0.02
Walk	0	0	0	0.01	0.01	0.98

Fig. 7. Confusion matrix for classification results of KTH dataset

Table 1. Recognition rates over Weizmann dataset

Method	Accurancy
Our method	98.75%
Gorelick et al.[7]	97.50%
Gkalelis et al.[10]	96.00%
Niebles et al[11]	72.80%

Table 2. Recognition rates over KTH dataset

Method	Accurancy
Our method	95.29%
Minhas et al.[5]	94.83%
Liu et al.[12]	94.00%
Ikizler et al.[4]	89.40%
Schuldt et al.[3]	71.70%

5 Summary and Future Work

In this paper, we proposed a new descriptor: difference silhouette for image representation. DS obtained by the adjacent binary silhouettes is a coarse histogram of the distribution with sample points on contours of DS. With the fusion of DS as the features of actions, we use static reservoir for action classification. On $\gamma - C$ plane, the proper static reservoir is constructed. We demonstrate the effectiveness of our method over two datasets, Weizmann dataset and KTH dataset. The results of our method are comparable and even superior to the results over these datasets.

One shortcoming of our approach is its dependence on silhouette extraction. We observed that most of the confusion, especially in the KTH dataset, occurs because of the imperfect silhouettes. However we should also note that, even with imperfect silhouettes, our method achieves high recognition rates which shows our method robustness to noise.

Future work includes application of view-invariance case, by means of orthographic projections of rectangular regions.

References

1. Weinland, D., Ronfard, R., Boyer, E.: A survey of vision-based methods for action representation, segmentation and recognition. Computer Vision and Image Understanding **115**(2), 224–241 (2011)
2. Poppe, R.: A survey on vision-based human action recognition. Image and Vision Computing **28**(6), 976–990 (2010)
3. Schuldt, C., Laptev, I., Caputo, B.: Recognizing human actions: A local svm approach. In: Proc. of 17th International Conference on Pattern Recognition, pp. 32–36. IEEE Press, Cambridge (2004)

4. Ikizler, N., Duygulu, P.: Histogram of oriented rectangles: A new pose descriptor for human action recognition. Image and Vision Computing **27**(10), 1515–1526 (2009)
5. Minhas, R., Baradarani, A., Seifzadeh, S., Jonathan Wu, Q.M.: Human action recognition using extreme learning machine based on visual vocabularies. Neurocomputing **73**(10–12), 1906–1917 (2010)
6. Huang, G.-B., Zhu, Q.-Y., Siew, C.-K.: Extreme learning machine: Theory and applications. Neurocomputing **70**(1–3), 489–501 (2006)
7. Belongie, S., Malik, J., Puzicha, J.: Shape matching and object recognition using shape contexts. IEEE Transactions on Pattern Analysis and Machine Intelligence **24**(4), 509–522 (2002)
8. Shi, Z., Han, M.: $\gamma - C$ plane and robustness in static reservoir for nonlinear regression estimation. Neurocomputing **72**(7–9), 1732–1743 (2009)
9. Veeraraghavan, A., Chowdhury, A.R., Chellappa, R.: Role of shape and kinematics in human movement analysis. In: IEEE Computer Society Conference on Computer Vision and Pattern Recognition 2004, Washington DC, pp. I-730–I-737 (2004)
10. Gkalelis, N., Tefas, A., Pitas, I.: Combining fuzzy vector quantization with linear discriminant analysis for continuous human movement recognition. IEEE Transactions on Circuits and Systems for Video Technology **18**(11), 1511–1521 (2008)
11. Niebles, J.C., Li, F.-F.: A hierarchical model of shape and appearance for human action classification. In: IEEE Conference on Computer Vision and Pattern Recognition 2007, Minneapolis, pp. 1–8 (2007)
12. Jingen, L., Mubarak, S.: Learning human actions via information maximization. In: IEEE Conference on Computer Vision and Pattern Recognition 2008, Anchorage, pp. 1–8 (2008)

Metric-Based Multi-Task Grouping Neural Network for Traffic Flow Forecasting

Haikun Hong[✉], Wenhao Huang, Guojie Song, and Kunqing Xie

Key Laboratory of Machine Perception, Ministry of Education, School of Electronics
Engineering and Computer Science, Peking University, Beijing 100871, China
honghaikun@hotmail.com
rubio8741@gmail.com

Abstract. Traffic flow forecasting is a fundamental problem in transportation modeling and management. Among various methods multi-
task neural network has been demonstrated to be a promising and effective model for traffic flow forecasting, while there are still two issues
unconsidered: 1) learning unrelated tasks together tends to reduce the
model's performance; 2) how to define or learn the distance metric for
distinguishing related tasks and unrelated tasks. In this paper, a metric learning based K-means method is proposed to group related tasks
together which effectively reduces the semantic gap between domain
knowledge and handcrafted feature engineering. Then for each group of
tasks, a deep neural network is built for traffic flow forecasting. Experimental results show the metric-based grouping method clusters tasks
more reasonably with a better metric than classic Euclidean-based K-
means. The final results of traffic flow forecasting on real dataset show
the metric-based multi-task neural network outperforms the Euclidean-
based multi-task neural network.

Keywords: Metric learning · Traffic flow forecasting · Multi-task learning · Deep neural network · Metric-based MTGNN

1 Introduction

Intelligent Transportation Systems (ITS) have been the main solution to improve
transportation performance and relief traffic congestion. As one of the fundamental components in ITS, traffic flow forecasting (TFF) has been a hot spot
in transportation community for decades of years. Without accurate and efficient traffic flow forecasting, none of the ITS can work well. So far, researchers
have proposed a variety of methods for traffic flow forecasting, such as time-
series model [1,14,21], Kalman filtering [15], simulation-based model [5], non-
parametric regression [17], support vector machine [4,9], bayesian network [19]
and neural network [6,7,11,12,16,18,20,23]. Given the good ability of nonlinear representation for complicated systems, many researchers have developed
plenty of approaches based on neural network, such as fuzzy neural network
[23], wavelet-based neural network [11], graphical-lasso neural network [7], EMD-
based neural network [20], multi-task neural network [6,12], multi-task ensemble

© Springer International Publishing Switzerland 2014
Z. Zeng et al. (Eds.): ISNN 2014, LNCS 8866, pp. 499–507, 2014.
DOI: 10.1007/978-3-319-12436-0_55

of neural networks [18]. Among these works, multi-task neural network seems to be a popular trend in recent works. Jin et al. [12] proposes a multi-task neural network for traffic flow forecasting, which incorporates traffic flows at continuous multi-time into an output layer in neural network. Gao et al. [6] extends multi-time neural network to multi-neighbor-link multi-time output layer.

There are still two issues unsolved in the real-life situation: first, aforementioned works do not consider the other inductive bias that learning unrelated tasks together may reduce the model performance. So it should be conscious in learning multiple tasks together. Second, the traffic networks in real-life are always too complex to define the similarities between the "related links" or "unrelated links" simply by network topological structure. In this paper, a metric-based approach is proposed to group tasks into several clusters and in each cluster one multi-task neural network is trained for traffic flow forecasting.

The rest of the paper is organized as follows. Section 2 introduces Multi-Task Neural Network (MTNN) and discusses effects of "related tasks" and "unrelated tasks" in multi-task learning. In Section 3, we propose a multi-task grouping neural network for traffic flow forecasting based on metric learning algorithm. Section 4 gives the experimental results. Conclusion is presented in Section 5.

2   Multi-Task Neural Network

Multi-Task learning [3] is an inductive transfer mechanism whose principle goal is to improve generalization performance by leveraging the domain-specific information contained in the training signals of related tasks. It does this by training tasks in parallel while using a shared representation. With the representation of a hidden layer shared by all output tasks, neural network is naturally used for multi-task learning problems.

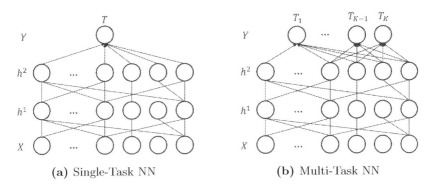

(a) Single-Task NN   (b) Multi-Task NN

Fig. 1. Neural Network Architectures

As Fig. 1 shows, Fig. 1a is a single-task neural network. X and Y represent input layer and output layer, respectively. h^1 and h^2 represent the hidden layers. Fig. 1b is a multi-task neural network whose output layer Y consists of K tasks as shown. Besides multi-task learning on traditional neural network,

Huang et al. [10] propose a deep neural network (DNN) architecture for traffic flow prediction with deep belief network which outperforms most classic traffic flow forecasting models. Experiments in Section 4 are constructed on both classic neural network and deep neural network.

In Section 1, we have introduced the related works in multi-task neural network. The inductive bias of "related" tasks that learning multiple related tasks together tends to improve the performance by leveraging the domain knowledge contained in training data of related tasks has been explored and demonstrated by previous researches. On the other hand, the inductive bias of "unrelated" tasks is unconsidered that noisy information contained in unrelated tasks tends to reduce model's performance. As a result, tasks need to be grouped into different groups by their relationships before put into multi-task learning model to ensure the tasks in the same group are related.

3 Metric-Based Multi-Task Grouping Neural Network

In this section, a novel metric-based multi-task grouping neural network algorithm is proposed to deal with the issues of task grouping in multi-task learning and traffic flow forecasting.

3.1 Metric Learning

The metric learning problem is concerned with learning a distance function tuned to a particular task, and has been shown to be useful when cooperates with nearest-neighbor methods and other techniques that rely on distances or similarities. Metric learning emerged in 2002 with the pioneering work of Xing et al. [22] which formulates it as a convex optimization problem.

Suppose we have some training set of points $\{x_i\}_{i=1}^n \subseteq \mathbb{R}^d$, and are given information that certain pairs of them are "similar" or "dissimilar":

$$
\begin{aligned}
S &: (x_i, x_j) \in \mathcal{S} \quad \text{if } x_i \text{ and } x_j \text{ are similar} \\
D &: (x_i, x_j) \in \mathcal{D} \quad \text{if } x_i \text{ and } x_j \text{ are dissimilar}
\end{aligned}
\tag{1}
$$

Thus, the goal of metric learning is to learn a Mahalanobis distance

$$
d(x, x') = d_M(x, x') = \|x - x'\|_M = \sqrt{(x - x')^T M (x - x')}
\tag{2}
$$

which make similar points close to each other and dissimilar points far away from each other. A simple way to define a criterion for the desired metric is to constrain that pairs of points (x_i, x_j) in $\mathcal{S}$ have small squared distance between them while pairs of points in $\mathcal{D}$ have large distance. This gives the basic optimization problem:

$$
\begin{aligned}
M = \arg\min_M &\sum_{(x_i, x_j) \in \mathcal{S}} \|x_i - x_j\|_M^2 \\
s.t. \sum_{(x_i, x_j) \in \mathcal{D}} &\|x_i - x_j\|_M \geq 1, \\
M &\succeq 0.
\end{aligned}
\tag{3}
$$

The optimization problem is convex which can be solved by gradient descent techniques. More details can be found in works [2, 13, 22].

3.2   Metric-Based MTG Neural Network

To handle the issue of distinguishing between related tasks and unrelated tasks in network-level traffic flow forecasting, a metric learning based multi-task grouping neural network algorithm for traffic flow forecasting is proposed as shown in Fig. 2.

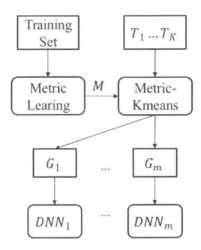

Fig. 2. Metric-based Multi-Task Grouping Neural Network

- Given the training set containing pairwise samples labeled by domain experts, a better metric M is learned through metric learning method.
- Using the learned metric instead of traditional Euclidean metric, a metric-based K-means clustering algorithm groups the T tasks into m groups. Tasks in each group G_i are more "similar" or "related" than those in other groups.
- For each group, we build a multi-task deep neural network as its own traffic flow forecasting model.

The multi-task neural network is a deep belief network with a multi-task regression layer on top. Deep Belief Network (DBN) is the most common and effective approach among all deep learning models. It is a stack of Restricted Boltzmann Machines (RBM) having only one hidden layer for each RBM. The learned unit activations of one RBM are input as the "data" for the next RBM in the stack. Hinton et al. proposed a way to perform fast greedy layer-wise trainning of DBN [8]. The RBM defines a probability distribution over hidden variables (h) and visible variables (v) via an energy function:

$$-\log P(v, h) \propto E(v, h; \theta)$$
$$= -\sum_{i=1}^{|V|} \sum_{j=1}^{|H|} w_{ij} v_i j_j - \sum_{i=1}^{|V|} b_i v_i - \sum_{j=1}^{|H|} a_j h_j \tag{4}$$

Then the conditional probability distributions of h and v can be computed as:

$$p(h_j\,|v;\theta) = sigm(\sum_{i=1}^{|V|} w_{ij}v_i + a_j)$$
$$p(v_i\,|h;\theta) = sigm(\sum_{j=1}^{|H|} w_{ij}h_j + b_i)$$

(5)

More details about the prediction model can be found in our previous work [10] which is beyond the discussion of this work.

4   Experiment

4.1   Experiment Setting

In this paper, we use the highway traffic toll collection data of Anhui province in China from Sep.2, 2010 to Sep.30, 2011. Input flows and output flows of all toll stations are detected and uploaded every 15 minutes. The flows of all road sections can be gained through a micro traffic simulation model. The task of traffic flow forecasting is to forecast the flows in next time interval based on historical time series data. For the task grouping, we construct the feature space using the traffic flows of 96 time interval. As the Fig. 3 shows, we choose top 10 sections and 5 other sections according to their average daily flow as 15 tasks. Their IDs are sorted by total traffic flow. Thus $section_1$ has the highest traffic flow and $section_{15}$ has the lowest traffic flow. The neural network architecture adopted in this section includes both traditional multi-task neural network and multi-task deep neural network with deep belief network for traffic flow forecasting.

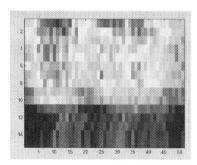

Fig. 3. Example: traffic flows of 15 sections

4.2   Experimental Results

In this subsection, we present mainly two parts of experimental results: task grouping visualization and error rates of traffic flow forecasting.

Table. 1 shows the grouping results of traditional Euclidean-based K-means and metric-based K-means. The result of the metric-based method is similar with the former one except for G_1 and G_2. The Eu-method groups $Task_{1,2}$ together and $Task_{3-8}$ together. While the metric-method groups $Task_{1,4,5,7}$ together and $Task_{2,3,6,8}$ into another $Group$. As the Euclidean metric treats each feature equally, when used in high-dimension feature space the influence of each feature's weight is reduced. So the Eu-method tends to group sections with similar total flow together.

Table 1. Results of Tasks Grouping

	G_1	G_2	G_3	G_4
Eu-Group	1, 2	3-8	9, 10	11-15
Met-Group	1, 4, 5, 7	2, 3, 6, 8	9, 10	11-15

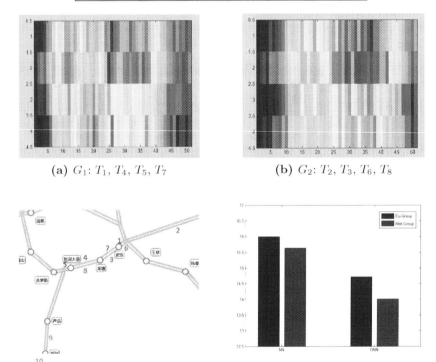

(a) G_1: T_1, T_4, T_5, T_7

(b) G_2: T_2, T_3, T_6, T_8

(c) Tasks on Real-life Traffic Network

(d) Forecasting Error Rates

Fig. 4. Results of Metric-based Multi-Task Neural Network

On the other hand, the metric-based method seems to group the tasks according to other traffic patterns. Fig. 4a and Fig. 4b show the tasks of G_1 and G_2 using metric-based K-means. The traffic peaks of sections in G_1 commonly come earlier than those in G_2. Fig. 4c may explain this interesting phenomenon that road sections grouped in the same G (for example, road section 1, 4, 5, 7) are located closely on geographical space and are on the same travel direction, that is to say they are upstream and downstream sections. As in traffic situation the vehicles travel from upstream sections to downstream sections, the traffic flows of sections on the same direction present a propagation trend on the time-space dimension like Fig. 4a and Fig. 4b show. In addition, the two grouping methods are applied with both traditional neural networks and deep neural networks. As Fig. 4d shows, the metric-based grouping method works better than classic Euclidean-based method both on both traditional neural network and deep neural network. Besides, the deep neural network achieves better performance than classic neural network in this experimental setting as we have demonstrated in precious work. According to the discussion above, metric-based grouping approach is effective to represent potential traffic domain patterns and fit to the complicated traffic situation. With a more reasonable grouping scheme, both traditional neural network and deep neural network achieve better performance.

5   Conclusions

In this study, we propose a metric-based multi-task grouping neural network algorithm for traffic flow forecasting. Firstly, a "better" metric is learned from pairwise training set labeled by domain experts using metric learning method which reduces the semantic gap between domain knowledge and handmade feature engineering. Secondly, a metric-based K-means algorithm is adopted to group "related" tasks into groups. Thirdly, for each group G we build a multi-task deep neural network for traffic flow forecasting.

As far as we know, this is the first work to consider the influence of "unrelated" tasks in multi-task learning neural network. We propose a novel metric-based algorithm to deal with the distance metric between related tasks and unrelated tasks. The task grouping method with learned metric seems to get a reasonable grouping result and explores the potential patterns under the complex traffic conditions. Results of our algorithm for traffic flow forecasting outperform the traditional Euclidean-based algorithm with near 1% improvement which is impressive for traffic flow forecasting problem.

As mentioned above, the metric-based multi-task neural network is a promising and effective algorithm and can be widely used in other applications. The metric learning paradigm can be further adopted to other nearest-neighbor or distance-based methods.

Acknowledgments. This work is supported by the National Natural Science Foundation of China under Grant No.61103025.

References

1. Ahmed, M.S., Cook, A.R.: Analysis of freeway traffic time-series data by using box-jenkins techniques. Transportation Research Record (722) (1979)
2. Bellet, A., Habrard, A., Sebban, M.: A survey on metric learning for feature vectors and structured data. CoRR abs/1306.6709 (2013)
3. Caruana, R.: Multitask learning. Machine Learning **28**(1), 41–75 (1997)

4. Castro-Neto, M., Jeong, Y.S., Jeong, M.K., Han, L.D.: Online-svr for short-term traffic flow prediction under typical and atypical traffic conditions. Expert Systems with Applications **36**(3), 6164–6173 (2009)
5. Chrobok, R., Wahle, J., Schreckenberg, M.: Traffic forecast using simulations of large scale networks. In: 2001 IEEE Proceedings of the Intelligent Transportation Systems, pp. 434–439. IEEE (2001)
6. Gao, Y., Sun, S.: Multi-link traffic flow forecasting using neural networks. In: 2010 Sixth International Conference on Natural Computation (ICNC), vol. 1, pp. 398–401. IEEE (2010)
7. Gao, Y., Sun, S., Shi, D.: Network-scale traffic modeling and forecasting with graphical lasso. In: Liu, D., Zhang, H., Polycarpou, M., Alippi, C., He, H. (eds.) ISNN 2011, Part II. LNCS, vol. 6676, pp. 151–158. Springer, Heidelberg (2011)
8. Hinton, G.E., Osindero, S., Teh, Y.W.: A fast learning algorithm for deep belief nets. Neural Comput. **18**(7), 1527–1554 (2006). http://dx.doi.org/10.1162/neco.2006.18.7.1527
9. Hong, W.C., Dong, Y., Zheng, F., Lai, C.Y.: Forecasting urban traffic flow by svr with continuous aco. Applied Mathematical Modelling **35**(3), 1282–1291 (2011)
10. Huang, W., Hong, H., Li, M., Hu, W., Song, G., Xie, K.: Deep architecture for traffic flow prediction. In: Motoda, H., Wu, Z., Cao, L., Zaiane, O., Yao, M., Wang, W. (eds.) ADMA 2013, Part II. LNCS, vol. 8347, pp. 165–176. Springer, Heidelberg (2013)
11. Jiang, X., Adeli, H.: Dynamic wavelet neural network model for traffic flow forecasting. Journal of Transportation Engineering **131**(10), 771–779 (2005)
12. Jin, F., Sun, S.: Neural network multitask learning for traffic flow forecasting. In: IEEE International Joint Conference on Neural Networks, IJCNN 2008. (IEEE World Congress on Computational Intelligence), pp. 1897–1901. IEEE (2008)
13. Kulis, B.: Metric learning: A survey. Foundations & Trends in Machine Learning **5**(4), 287–364 (2012)
14. Min, W., Wynter, L.: Real-time road traffic prediction with spatio-temporal correlations. Transportation Research Part C: Emerging Technologies **19**(4), 606–616 (2011)
15. Okutani, I., Stephanedes, Y.J.: Dynamic prediction of traffic volume through kalman filtering theory. Transportation Research Part B: Methodological **18**(1), 1–11 (1984)
16. Park, B., Messer, C.J., Urbanik II, T.: Short-term freeway traffic volume forecasting using radial basis function neural network. Transportation Research Record: Journal of the Transportation Research Board **1651**(1), 39–47 (1998)
17. Smith, B.L., Williams, B.M.: Keith Oswald, R.: Comparison of parametric and nonparametric models for traffic flow forecasting. Transportation Research Part C: Emerging Technologies **10**(4), 303–321 (2002)
18. Sun, S.: Traffic flow forecasting based on multitask ensemble learning. In: Proceedings of the First ACM/SIGEVO Summit on Genetic and Evolutionary Computation, pp. 961–964. ACM (2009)
19. Sun, S., Zhang, C., Yu, G.: A bayesian network approach to traffic flow forecasting. IEEE Transactions on Intelligent Transportation Systems **7**(1), 124–132 (2006)
20. Wei, Y., Chen, M.C.: Forecasting the short-term metro passenger flow with empirical mode decomposition and neural networks. Transportation Research Part C: Emerging Technologies **21**(1), 148–162 (2012)
21. Williams, B.M., Hoel, L.A.: Modeling and forecasting vehicular traffic flow as a seasonal arima process: Theoretical basis and empirical results. Journal of Transportation Engineering **129**(6), 664–672 (2003)

22. Xing, E.P., Jordan, M.I., Russell, S., Ng, A.Y.: Distance metric learning with application to clustering with side-information. In: Advances in Neural Information Processing Systems, pp. 505–512 (2002)
23. Yin, H., Wong, S., Xu, J., Wong, C.: Urban traffic flow prediction using a fuzzy-neural approach. Transportation Research Part C: Emerging Technologies **10**(2), 85–98 (2002)

Object Tracking with a Novel Method
Based on FS-CBWH within Mean-Shift Framework

Dejun Wang[1], Yongtao Shi[2], Weiping Sun[1(✉)], and Shengsheng Yu[1]

[1] School of Computer Science and Technology,
Huazhong University of Science and Technology, Wuhan, Hubei, China
dejunw123@gmail.com, {wpsun,ssyu}@hust.edu.cn
[2] Department of Computer Science, Three Gorge University, Hubei, China
shi.yungtao@gmail.com

Abstract. Effective appearance models are one critical factor for robust object tracking. In this paper, we introduce foreground feature salience concept into the background modelling, and put forward a novel foreground salience-based corrected background weighted-histogram (FS-CBWH) scheme for object representation and tracking, which exploits salient features of both foreground and background. We think that background and foreground salient features are both crucial for object representation and tracking. Experimental results show that the proposed FS-CBWH scheme can improve the robustness and performance of mean-shift tracker significantly especially in heavy occlusions and large background variation scenes.

Keywords: Target tracking · Weighted histogram · Foreground feature saliency

1   Introduction

Real-time object tracking has been extensively studied over the years, since it is an important step in many computer vision tasks such as human-computer interaction [1], robotics [2], video surveillance [3]. But there are still considerable difficulties, such as pose changes, illumination changes, occlusions. Many tracking algorithms have been proposed to overcome these challenges (e.g., GOA tracker [4], Kalman filter [5], particle filter [6] and mean-shift [8]). In recent years, because of the pretty good robustness to illumination, rotation, partial occlusion and low complexity, mean shift algorithms are used in object tracking by many researchers [8, 9, 10].

Effective appearance models are one critical factor for robust object tracking. In the conventional mean-shift tracking algorithm, object is represented by kernel-weighted color histogram. However, it is not always discriminative enough especially when object has similar feature with its background. To address this challenging problem, the background-weighted histogram (BWH) algorithm has been adopted to improve color histogram by Comaniciu *et al.* [7]. It decreases background interference from salient background features in the target model and candidate model. However in [9], Ning *et al.* proved the BWH transformation formula is equivalent to the standard mean-shift tracking. They proposed a corrected BWH (CBWH) algorithm to transform

© Springer International Publishing Switzerland 2014
Z. Zeng et al. (Eds.): ISNN 2014, LNCS 8866, pp. 508–515, 2014.
DOI: 10.1007/978-3-319-12436-0_56

only the target model but not the target candidate model. In [10] Wang *et al.* proposed to a novel weight calculation approach which incorporates the local background information. While these algorithms achieve relatively good effects, they often fail to track objects in challenging image sequences with drastic background appearance change and occlusion due to discriminative enough.

Motivated by Comaniciu *et al.* [7]'s work, we propose foreground salience-based corrected background-weighted histogram (FS-CBWH) scheme. Compared to CBWH, FS-CBWH takes background and foreground salient features into account. Since foreground salience and background salience are both employed as weighted-histogram criterion, FS-CBWH scheme further improves the object representation and tracking. Experimental results show that FS-CBWH outperform CBWH method in heavy occlusions and large background variation, meanwhile FS-CBWH has more advantage than CBWH in the time cost. It is more robust and efficient.

2   Related Works

2.1   Background-Weighted Histogram

Assume that we have an original background model $b = \{b_u\}_{u=1,...,m}$ (with $\sum_{u=1}^{m} b_u = 1$) and its minimal non-zero entry b^* of the background model in an image. The background window of the target is surrounded around it as a rectangular ring with a fixed three times of the target area. To get the model discriminative enough against the background is accomplished via a background-weighted histogram (BWH) procedure [8] as below:

$$\left\{ \tau_u = \min\left(\frac{b^*}{b_u}, 1 \right) \right\}_{u=1...m} . \tag{1}$$

The τ_u is the weight coefficient within histogram bin u, lower τ_u are more prominent in the background and less important for the target representation. So it is used to transform the representations of both target model and target candidate model. Then the target model can be obtained as:

$$q' = \{q'_u\}_{u=1,...,m} ; q'_u = c'_1 \tau_u \sum_{i=1}^{n} k\left(\left\| \frac{x_i}{h} \right\| \right)^2 \delta\left[b(x_i) - u \right] , \tag{2}$$

where q'_u represents the density of feature u in target model q', $k(x)$ is an iso-tropic kernel profile, m is the number of feature bins, x_i $(i=1,...n)$ is pixel position in target region centered at original position, δ is the Kronecker delta function,

$b(x_i)$ maps the pixel x_i to the histogram bin index, h is the bandwidth, c_1' is a normalization constant. Similarly, the target candidate model p' can be got as:

$$p'=\{p_u'\}_{u=1,...,m} ; p_u'=c_2'\tau_u\sum_{i=1}^{n_h}k\left(\left\|\frac{y-x_i}{h}\right\|\right)^2 \delta\left[b(x_i)-u\right] ,$$ (3)

where p_u' represents the density of feature u in target candidate model p', x_i ($i=1,...n_h$) is the pixel position in the target candidate region centered at y, c_2' is a normalization constant.

2.2   Corrected Background-Weighted Histogram

A corrected BWH (CBWH) algorithm is proposed by Ning *et al.* [9]. Rather than both transforming the target model and the target candidate model, it just transforms the target model. Ning *et al.* proved above BWH transformation result is identical to usual target representation under the mean shift tracking framework. Therefore, in CBWH algorithm the target candidate model still uses the original model as follows:

$$p=\{p_u\}_{u=1,...,m} ; p_u=c_2\sum_{i=1}^{n_h}k\left(\left\|\frac{y-x_i}{h}\right\|\right)^2 \delta\left[b(x_i)-u\right] ,$$ (4)

where p_u represents the density of feature *u* in original target candidate model p, c_2 is a normalization constant.

Notwithstanding the demonstrated success of CBWH, no attempts have been made to directly exploit the foreground salience information. If background salient features are also foreground salient, foreground salience information is also critical for object tracking.

3   The Proposed Scheme

3.1   Foreground Salience Modelling

Assume that we have a foreground model $f=\{f_u\}_{u=1,...,m}$ (with $\sum_{u=1}^{m}f_u=1$) and its maximal non-zero entry f^* and above original background model. We propose a foreground salience model in which the salience of feature in the target model as below:

$$S=\{S_u\}_{u=1,...,m} ; \left\{S_u=\max\left(\frac{f_u}{f^*},1\right)\right\}_{u=1...m} ,$$ (5)

original image   *foreground saliency image*

Fig. 1. A sample of foreground salience image

where S represents foreground salience model in target model and the S_u is the salience coefficient. Those features with larger S_u are more prominent in the foreground and more important for the target representation. An example of foreground feature salience image is illustrated in Fig. 1.

3.2   FS-CBWH

We adopt foreground salience to transform the representation of original background model in order to select the salient foreground feature components in the background model. Thus we get a more complex representation of the background features, namely foreground salience-based background model (FSB):

$$b^{fg} = \left\{b_u^{fg}\right\}_{u=1,\dots,m}; b_u^{fg} = C_3 S_u b_u, \qquad (6)$$

where b_u^{fg} represents the justified density of feature u in new background model b^{fg}, C_3 is a normalization constant.

Different from Comaniciu *et al.* ' [7] background model, our FSB implicitly includes some feature salience information from the foreground. Assume that we have the minimal non-zero entry b_{fg}^* of the background model FSB in an image; we define a transformation for the representation of target model as follows:

$$\left\{\tau_u^{new} = \min\left(\frac{b_{fg}^*}{b_u^{fg}}, 1\right)\right\}_{u=1\dots m}. \qquad (7)$$

The τ_u^{new} is the weight coefficient, lower τ_u^{new} are more prominent in the background and less prominent in the foreground and less important for the target representation. Similarly in [9], τ_u^{new} is only used to transform the target model.

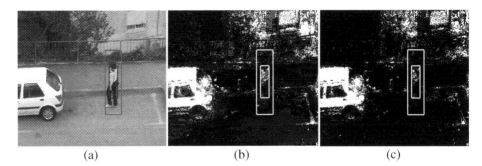

Fig. 2. An example of the weight image example

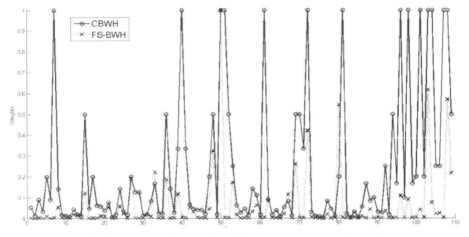

Fig. 3. Corresponding weights of the non-zero features in Fig. 2

Then the new target model can be obtained as:

$$q^{new} = \left\{ q_u^{new} \right\}_{u=1,...,m}; q_u^{new} = C_4 \tau_u^{new} \sum_{i=1}^{n} k \left(\left\| \frac{x_i}{h} \right\| \right)^2 \delta \left[b(x_i) - u \right] , \qquad (8)$$

where q_u^{new} represents the density of feature u in target model q_u^{new}, C_4 is a normalization constant.

An example of the weight image of CBWH and FS-CBWH is illustrated in Fig. 2. The corresponding weights of the non-zero features therein are shown in Fig. 3. We compute the Bhattacharyya similarities between the tuned target model and its surrounding background region by CBWH and FS-CBWH for original image in Fig. 2. CBWH and FS-CBWH have Bhattacharyya similarity of 0.04 and 0.02 respectively, which implies that FS-CBWH can better distinguish the target from background. Since the weight of each feature in the target model for FS-CBWH is determined by the feature distinctiveness of background region and foreground region, the FS-CBWH scheme is more robust and discriminant against the background.

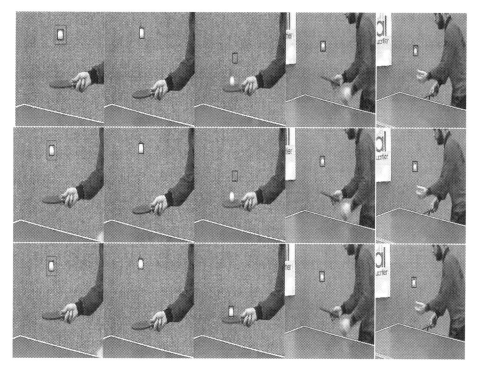

Fig. 4. Tracking results from Ping-Pong ball sequence. Rows 1, 2, and 3 are for standard Mean-Shift tracker, CBWH tracker, and our proposed tracker, respectively. The frame indexes are 1, 22, 26, 45 and 51.

4   Experimental Results

Our algorithm has been tested on two kinds of video datasets using standard mean-shift tracking algorithm, the CBWH algorithm and our FS-CBWH algorithm, One is the standard ping-pang ball test sequence used in [9], the other is the sequence obtained in [11]on WalkingWoman sequence.

In the first video, the target moves quickly and undergoes sudden background change. The experimental results are shown in Fig. 4 and Table 1. For FS-CBWH, since it incorporates feature saliency to enhance the approximation of the target model, it is more discriminative and robust to sudden background change than others. Thus when the target touches the bat at frame 26, our method can successfully capture the target while the other two algorithms fail. Compared with other two algorithms, it shows our method is less sensitive to sudden background change.

In the second experiment, the video is complex and challenging since the first car has the similar color to the women and partly occludes her. As far as target localization accuracy, the FS-CBWH performs best since the foreground salient features and background salient features are well exploited. The experimental results are shown in Fig. 5 and Table 1. Our algorithm can still track the object steadily even in such a

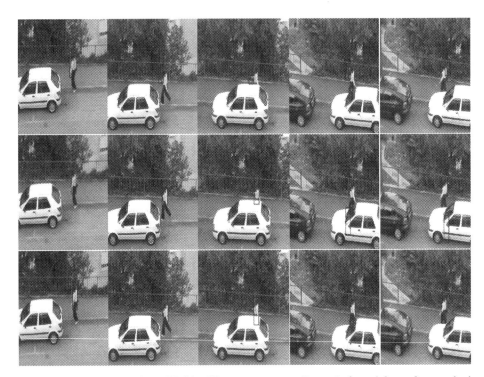

Fig. 5. Tracking results from WalkingWoman sequence. Rows 1, 2, and 3 are for standard Mean-Shift tracker, CBWH tracker, and our proposed tracker, respectively. The frame indexes are 22, 32, 54, 79 and 86.

Table 1. Average error and average number of iterations

Video sequences	standard mean-shift		CBWH		FS-CBWH	
	Average error	Average number of itera-tions	Average error	Average number of itera-tions	Average error	Average number of itera-tions
ping-pang ball	3.5619	3.9423	3.3872	3.2308	2.6358	3.0192
walking woman	33.8516	6.0	35.6829	4.9333	11.4839	4.123

challenging scene while the other two algorithm loss the target in the end. It shows our method increases the discrimination between object and background, and is more robust to occlusion and similar background.

From Table 1, our method need lest number of iterations. It means the proposed method converges more quickly and requires less computation.

5   Conclusion

In this paper, we proposed foreground feature saliency concept. Based on this novel concept, we derived a more complex background model, foreground salience-based background model (FSB), and used it to propose a novel weighted-histogram scheme, foreground saliency-based background-weighted histogram (FS-CBWH) scheme. Then mean-shift tracking was performed. The major advantage of this scheme lies in that it encodes feature distinctiveness information from both background and fore-ground. Our experiments demonstrated the proposed FS-CBWH scheme improves the efficiency and robustness of mean-shift tracker significantly.

Acknowledgement. This work is supported by the National Natural Science Foundation of China (NSFC) under Grant 61300140.

References

1. Bradski, G.R.: Real time face and object tracking as a component of a perceptual user in-terface. In: Fourth IEEE Workshop on Applications of Computer Vision, pp. 214–219. IEEE Press, New York (1998)
2. Papanikolopoulos, N.P., Khosla, P.K., Kanade, T.: Visual tracking of a moving target by a camera mounted on a robot: a combination of control and vision. IEEE Trans. Robotics and Automation **9**(1), 14–35 (1993)
3. Stauffer, C., Grimson, W.E.L.: Learning Patterns of Activity Using Real-Time Tracking. IEEE Trans. Pattern Anal. and Mach. Intell. **22**(8), 747–757 (2000)
4. Devi, M.S., Bajaj, P.R.: Active Facial Tracking. In: 3rd International Conference on Emerging Trends in Engin. and Tech., pp. 91–95. IEEE Press, New York (2010)
5. Isard, M., Blake, A.: CONDENSATION—Conditional Density Propagation for Visual Tracking. Int. J. Comput. Vis. **29**(1), 5–28 (1998)
6. Pérez, P., Hue, C., Vermaak, J., Gangnet, M.: Color-Based Probabilistic Tracking. In: Heyden, A., Sparr, G., Nielsen, M., Johansen, P. (eds.) ECCV 2002, Part I. LNCS, vol. 2350, pp. 661–675. Springer, Heidelberg (2002)
7. Comaniciu, D., Ramesh, V., Meer, P.: Kernel-based object tracking. IEEE Trans. Pattern Anal. Mach. Intell. **25**(5), 564–577 (2003)
8. Vojir, T., Noskova, J., Matas, J.: Robust Scale-Adaptive Mean-Shift for Tracking. In: Kämäräinen, J.-K., Koskela, M. (eds.) SCIA 2013. LNCS, vol. 7944, pp. 652–663. Springer, Heidelberg (2013)
9. Ning, J.F., Zhang, L., Zhang, D., Wu, C.: Robust mean-shift tracking with corrected back-ground-weighted histogram. IET Comput. Vision **6**(1), 62–69 (2010)
10. Wang, L.F., Pan, C.H., Xiang, S.M.: Mean-shift tracking algorithm with weight fusion strategy. In: 18th Inter. Conf. on Image Proc., pp. 473–476. IEEE Press, New York (2011)
11. Adam, A., Rivlin, E., Shimshoni, I.: Robust fragments-based tracking using the integral histogram. In: IEEE Computer Society Conference on Computer Vision and Pattern Recognition, pp. 798–805. IEEE Press, New York (2006)

Bayesian Covariance Tracking
with Adaptive Feature Selection

Dejun Wang[1], Lin Li[1,2], Wei Liu[1,3], Weiping Sun[1(✉)], and Shengsheng Yu[1]

[1] School of Computer Science and Technology,
Huazhong University of Science and Technology, Wuhan, China
dejunw123@gmail.com, {wpsun,ssyu}@hust.edu.cn
[2] School of Computer Science and Technology,
Wuhan University of Technology, Wuhan, China
lilyfour@163.com
[3] School of Computer Science, Central China Normal University, Wuhan, China
liuwei@mail.ccnu.edu.cn

Abstract. Effective appearance models are one important factor for robust object tracking. In this paper, a more elaborate object representation model via a simultaneous online feature selection and feature fusion algorithm is proposed, in which extended variance ratio is used to select the most discriminative power features, and thereby account for appearance model using region covariance descriptor which takes into account feature correlation information during tracking. Fusing all selected features, we get a more discriminative appearance model. Furthermore, our simultaneous online feature selection and feature fusion method is integrated into particle filter framework for robust tracking. Experimental results show that this proposed method is robust in heavy occlusions scenes and is able to handle variations in illumination and scale.

Keywords: Target tracking · Particle filter · Feature selection · Feature fusion

1   Introduction

Appearance models are one important factor for object tracking. Extensive studies have been presented (e.g. color histogram [1], subspace method [2], and sparse representation [3]). A good object representation should be robust to cope with pose change, scale variation, partial occlusion, illumination change etc. In order to successfully handle the object and background appearance variation in visual tracking, the object appearance model must be adapted over time.

Online feature selection has become an important technique to model the appearance in order to adapt to object and background appearance changes for visual tracking. Recently many adaptive feature selection techniques have been developed. Collins *et al.* [4] proposed to online select discriminative features from color feature spaces. Features are ranked according to variance ratio and the top N features are selected with each producing a likelihood map. Then Mean-shift is separately performed on each likelihood map to locate target. Finally, target is located by selecting the median

© Springer International Publishing Switzerland 2014
Z. Zeng et al. (Eds.): ISNN 2014, LNCS 8866, pp. 516–523, 2014.
DOI: 10.1007/978-3-319-12436-0_57

location. However, their method treats selected features independently within the Mean-shift framework; this can fail to track the target. Chen *et al.*[5] applied particle filtering for feature selection. Feature effectiveness depends on the Kullback-Leibler between foreground and background. By weighting all feature particles by the Kullback-Leibler observation model, a compound likelihood map is generated. Finally another particle filter is performed on previous compound likelihood image. Chen [5] *et al.* extended the work of Collins [4] *et al.* by feature weighting sum scheme, but their method   extracts single feature and ignores features covariance.

Recently the success of region covariance descriptor[6] addressing appearance modelling inspires us to base our tracking method on it. We employ it to capture the correlations among selected features inside a region, instead of feature weighting sum scheme. It affords significant insensitivity to changes in illumination and scale.

We also employ particle filters to track the target object. In summary, we proposed covariance-based particle filter with adaptive feature selection (CVPF-FS) tracking algorithm. It includes two stages. Firstly, the most discriminative power features is obtained by ranking the candidate features according to extended variance ratio score. Secondly, region covariance descriptor is constructed online from selected features forming covariance-based observation model, then a particle filter is applied to propagate sample distributions over time. Experimental results show that the CVPF-FS method is robust in illumination and scale variations and heavy occlusions scenes.

2   Particle Filter Algorithm

The standard particle filter algorithm consists of the well-known two models: the observation model and the state transition model. Let s_t and z_t denote the state of a target and its observation at time t respectively. Provided the state transition model $p(s_t \mid s_{t-1})$ and the observation model $p(z_t \mid s_t)$ and previous observations $z_{0:t} = \{z_0, ..., z_{t-1}\}$, our goal is to estimate posterior distribution of unknown state s_t : $p(s \mid z_{0:t})$. By Bayes rule we can derive its recursive form as follows:

$$p(s_t \mid z_{0:t}) \propto p(z_t \mid s_t) \int p(s_t \mid s_{t-1}) p(s_{t-1} \mid z_{0:t-1}) ds_{t-1} \ , \qquad (1)$$

where $p(z_t \mid s_t)$ is observation density, and $p(s_t \mid s_{t-1})$ is state transition density.

Observation density is generally multimodal so $p(s \mid z_{0:t})$ cannot be in closed form. By sequential Monte Carlo simulations, $p(s \mid z_{0:t})$ is approximated by normalized weighted sample set $\left\{ s_t^i, w_t^i \right\}_{i=1}^N$. All sample particles are sampled from a proposal importance distribution $q(s_t^i \mid s_{t-1}^i, z_t)$, the weight associated with each particle is formulated as follows:

$$w_t^i \propto \frac{p(z_t \mid s_t^i) p(s_t^i \mid s_{t-1}^i)}{q(s_t^i \mid s_{t-1}^i, z_t)} w_{t-1}^i \quad . \tag{2}$$

To avoid weight degeneracy, the particles are resampled to obtain the equal weight particle set $\left\{ s_t^i, \dfrac{1}{N} \right\}_{i=1}^{N}$.

The proposal importance density is set as $q(s_t \mid s_{t-1}, z_t) = p(s_t \mid s_{t-1})$ for simplicity, namely condensation algorithms [7]. The Monte Carlo approximation of expectation, $\hat{x}_t = \dfrac{1}{N} \sum_{i=1}^{N} x_t^i \approx E(x_t \mid z_{0:t})$, is used for state estimation at time t.

3   Adaptive Feature Selection

3.1   Feature Likelihood Ratio

Given a feature f, we first determine its direct background window: its surrounding neighborhood of three times the target area. Given normalized histograms h_{fg} and h_{bg} of specified feature space, we yield a set of tuned likelihood values [4] $L(i)$ that are defined as Eq. 3. A typical feature likelihood image is shown in Fig. 1.

$$L(i) = \log \frac{\max\left(h_{fg}(i), \delta\right)}{\max\left(h_{bg}(i), \delta\right)} \quad , \tag{3}$$

where δ is a small positive constant that avoid division by zero, and i is the feature bin. $L(i)$ of the feature actually forms a new likelihood feature space.

There are so many features can be as candidate feature, such as color, texture, shape contexts. We adopt color feature set proposed in [4]. It consists of linear combinations of RGB values. The candidate feature set be defined as:

$$F = \{c_1 R + c_2 G + c_3 B \mid c_1, c_2, c_3 \in \{-2, -1, 0, 1, 2\}\} \quad . \tag{4}$$

There are totally 49 valid features in the feature set by discarding invalid features.

3.2   Feature Discriminability Evaluation

There are many potential feature selection criterions, such as variance ratio (VR) [4], KL divergence [8]. VR is the ratio of the between-class variance of the feature to the within-class variance of feature; it is according to the fact that the best feature is the one that best distinguishes the object from the background.

Original image   Likelihood image

Fig. 1. A typical feature likelihood image

Different from [4], we do not reuse the initial color distributions. This is thought of as the extended variance ratio (EVR). EVR of each under the discrete probability h'_{fg} for foreground and h'_{bg} for background in likelihood feature space is as follows:

$$EVR\left(L, h'_{fg}, h'_{bg}\right) = \frac{\mathrm{var}\left(L;\left(h'_{fg} + h'_{bg}\right)/2\right)}{\mathrm{var}\left(L; h'_{fg}\right) + \mathrm{var}\left(L; h'_{bg}\right)} \quad . \tag{5}$$

The denominator of the EVR is the sum of the variances within foreground and background, it should be small so that the object and background are both tightly clustered, while the numerator is the total variance over both foreground and background, it should be big so that the object and background are widely spread apart. We use Eq. 5 to evaluate and rank all the features' discriminative power according to extended variance ratio.

3.3   Feature Selection Tracker

In Collins' [4] tracker top-ranked K features likelihood maps are embedded in the Mean-shift [9] tracking algorithm. Then K estimates of object location are obtained. By naive median estimator, the median is chosen as final estimate of object location. We implement a Collins-like' (CL) tracker. Different from naive median estimator [4], our tracker uses a weighted sum approach. The K estimates of object location are weighted by corresponding normalized feature EVR scores, so that in our compound final estimate, we increase the contribution from the more discriminative features.

4   Feature Adaptive Covariance-Based Bayesian Tracking

Our Collins-like' tracker does not preserve the correlation information of the feature. To overcome this, we adopt region covariance to model the target for tracking.

4.1   Region Covariance Descriptor

Here we adopt above top-ranked K features. Region covariance descriptor involves evaluation of correlation of the selected region. Consider a rectangular region likelihood maps features. Region covariance descriptor is recently proposed and commonly used in the object detection, R of size $W \times H$, we can extract $W \times H \times k$ dimensional feature image as follows:

$$F(x, y) = \varphi(R, x, y), \qquad (6)$$

where φ is feature mapping such as color, gradients and texture. (x, y) denotes the pixel location, $F(x, y)$ is the k dimensional feature vector. let N is the number of pixels in the region R, hence for rectangular region R containing N pixels, the all k dimensional feature vector points be $\{F_i\}_{i=1...N}$, and the region covariance descriptor is given by:

$$C_R = \frac{1}{N-1} \sum_{i=1}^{N} (F_i - \overline{F})(F_i - \overline{F})^T, \qquad (7)$$

where $\overline{F}$ is mean value of F_i.

4.2   Bayesian Tracking

Given the region covariance, we aim to compute $p(z_t \mid s_t)$. Notice that the covariance matrices are in general symmetric positive definite(SPD), thus the nonsingular covariance matrix can be formulated as a connected Riemannian manifold that has the Lie group structure. We adopt the Log-Euclidean Riemannian metric [10] for measuring distance between covariance matrices, it is defined as:

$$\rho_{\log e}(C_1, C_2) = \left\| \log(C_1) - \log(C_2) \right\|, \qquad (8)$$

where $\|\cdot\|$ is the Euclidean norm in the vector space. This metric maps the Riemannian manifold to the Euclidean space, so complex matrix operations are avoided and it enables computational cost low. Hence, the measurement likelihood is formulated as:

$$p(z_t \mid s_t^{(i)}) \propto \exp\left(-\frac{\left\| \log(C_i) - \log(C_T) \right\|^2}{\sigma_{\mathrm{cov}}^2}\right), \qquad (9)$$

where C_T and C_i are respectively the object covariance and particle image covariance corresponding to each . The computation of the covariance descriptor is intensive, but it can be computed efficiently using integral images [6].

5   Experimental Results

We compare the tracking results of our proposed CVPF-FS tracker with CL tracker in Section 3.3 and covariance-based particle filter (CVPF) tracker. The first sequence is WalkingWoman sequence [11]. In this video the woman undergoes partial occlusion. Sample frames are shown in Fig.2; it can be observed that the three trackers keep tracking the woman when the woman is partially occluded after frame 56 except CL tracker. Partial occlusion can make similar "noise" in the target region; therefore covariance could be used for further improvement of appearance models. Compared with CL tracker, covariance-based trackers are more robust to the occlusion.

Fig. 2. Tracking results of WalkingWoman sequence. Rows 1, 2, and 3 are for CL tracker, CVPF tracker, and CVPF-FS tracker, respectively. Frame indexes are 5, 36, 56, 69, and 82.

The second sequence is LeftBag sequence [12]. This video has obvious illumination and scale change. To represent the variability of appearance, adaptive feature selection will be useful. As shown in Fig. 3, although the CVPF tracker combines the advantage of covariance and particle filter, it begins to drift in frame 8. Whereas CVPF- FS tracker and CL tracker can lock target by robustly selecting features and updating target model. It is obvious that adaptive feature selection methods are more discriminative than no feature selection method. Moreover, our CVPF-FS tracker in frame 38 also captures this significant scale change.

The third sequence is more complex ThreePastShop2cor sequence [12]; this video has illumination variation and heavy occlusions and similar background, these are the main challenges. Sample frames are shown in Fig.4. Our CVPF tracker can track the target well even with consecutive occlusions and other challenges. Since CVPF tracker ignores adaptive feature selection, thus it drifts off the target in frame 181. Since CL tracker ignores feature correlation information, thus it fails to track the target in frame 153 due to consecutive occlusions. Whereas CVPF-FS tracker combines the advantage of covariance and adaptive feature selection, it makes the appearance model is more robust to the interference of similar background and partial occlusion.

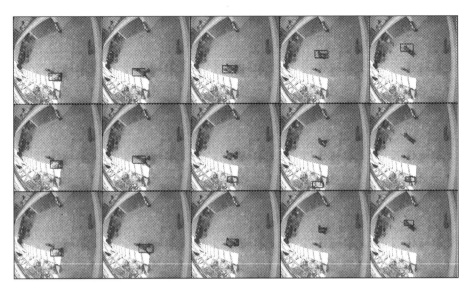

Fig. 3. Tracking results of LeftBag sequence. Rows 1, 2, and 3 are for CL tracker, CVPF tracker, and CVPF-FS tracker, respectively. Frame indexes are 3, 8, 13, 38, and 91.

Fig. 4. Tracking results of ThreePastShop2cor sequence. Rows 1, 2, and 3 are for CL tracker, CVPF tracker, and CVPF-FS tracker, respectively. Frame indexes are 5, 32, 89, 153, and 181.

6   Conclusion

We considered an appearance model based on the simultaneous feature selection and fusion. The extended variance ratio was used to select the most discriminative power features. Moreover we extended online feature selection process by computing the selected features region covariance, which can capture correlation information and is one efficient approach for feature fusion. In contrast to the existing feature selection tracking method and covariance tracking method, our method faithfully model the appearance of the target. Finally we embedded the new target model into particle filters to track the target.

Acknowledgement. This work is supported by the National Natural Science Foundation of China (NSFC) under Grant 61300140.

References

1. Comaniciu, D., Ramesh, V., Meer, P.: Kernel-based object tracking. IEEE Trans. Pattern Anal. Mach. Intell. **25**(5), 564–577 (2003)
2. Ross, D.A., Lim, J., Lin, R.S., Yang, M.H.: Incremental Learning for Robust Visual Tracking. Int. J. Comput. Vis. **77**(1–3), 125–141 (2007)
3. Mei, X., Ling, H.: Robust Visual Tracking using l1 Minimization. In: IEEE Int. Conf. Comput. Vision, pp. 1436–1443. IEEE Press, New York (2009)
4. Collins, R., Liu, Y., Leordeanu, M.: Online selection of discriminative tracking features. IEEE Trans. Pattern Anal. Mach. Intell. **27**(10), 1631–1643 (2005)
5. Chen, H.T., Liu, T.L., Fuh, C.S.: Probabilistic Tracking with Adaptive Feature Selection. In: 17th Inter. Confe. on Pattern Recog., pp. 736–739. IEEE Press, New York (2004)
6. Tuzel, O., Porikli, F., Meer, P.: Region Covariance: A Fast Descriptor for Detection and Classification. In: Leonardis, A., Bischof, H., Pinz, A. (eds.) ECCV 2006. LNCS, vol. 3952, pp. 589–600. Springer, Heidelberg (2006)
7. Isard, M., Blake, A.: CONDENSATION—Conditional Density Propagation for Visual Tracking. Int. J. Comput. Vis. **29**(1), 5–28 (1998)
8. Collins, R.: Spatial Divide and Conquer with Motion Cues for Tracking through Clutter. In: IEEE Comp. Soc. Confer. on Comp. Vis. and Patt. Rec., pp. 570–577. IEEE Press, New York (2006)
9. Comaniciu, D., Ramesh, V., Meer, P.: Real-time tracking of non-rigid objects using mean shift. In: IEEE Comp. Society Confer. on Comp. Vis. and Patt. Rec., pp. 142–149. IEEE Press, New York (2000)
10. Arsigny, N., Fillard, V., Pennec, P., Ayache, X.: Geometric means in a novel vector space structure on symmetric positive-definite matrices. SIAM J. Matrix Ana. App. **29**(1), 328–347 (2006)
11. Adam, A., Rivlin, E., Shimshoni, I.: Robust fragments-based tracking using the integral histogram. In: IEEE Comp. Soc. Conf. on Comp. Vis. and Patt. Rec., pp. 798–805. IEEE Press, New York (2006)
12. CAVIAR Test Case Scenarios. http://www.homepages.inf.ed.ac.uk/rbf/CAVIAR

Single-Trial Detection of Error-Related Potential by One-Unit SOBI-R in SSVEP-Based BCI

Janir Nuno da Cruz, Ze Wang, Chi Man Wong, and Feng Wan[✉]

Department of Electrical and Computer Engineering,
Faculty of Science and Technology, University of Macau, Macau, China
fwan@umac.mo

Abstract. Error-related potential (ErrP) is a form of event-related potential (ERP) that is triggered in the brain when a user either makes a mistake or the application behaves differently from his/her intend. Unfortunately, due to its short-time duration, low signal-to-noise ratio, non-stationarity and transient characteristic, a single-trial extraction of ErrP remains a difficult task. In this study, we propose the use of one-unit second order blind identification with reference (SOBI-R) for extraction of ErrP in the context of steady-state visual evoked potentials based brain-computer interface (SSVEP-based BCI). Fractal features are extracted from the one-unit SOBI-R data by means of Katz fractal dimensional. At last, the ErrP classification is obtained using a regularized version of the linear discriminant analysis (LDA). The proposed method was tested on 6 subjects data and achieved an average recognition rate of correct and erroneous single-trials of 87.03% and 80.7%, respectively. These results show that single-trial detection of ErrP is feasible for SSVEP-based BCI.

Keywords: Brain-computer interface (BCI) · One-unit second-order blind identification with reference (SOBI-R) · Error-related potentials (ErrP) · Steady-state visual evoked potentials (SSVEP)

1 Introduction

People with disabilities are confronted daily with obstacles to independent living. Brain-computer interfaces (BCIs) can, to a certain extent, reduce these limitations by providing a direct pathway between the human brain and controllable devices. Among all kinds of BCIs, the steady-state visual evoked potential (SSVEP) based BCI is the most common due to its relatively high operation speed and relatively little user training [1]. Nevertheless, SSVEP-based BCIs are prone to errors in translating users intention into control signals. Yet, humans have a strong ability to identify errors in a precise way. Several studies have found the existence of error-related potentials (ErrP) in the EEG signal over the fronto-central region of the scalp of subjects while perceiving errors and these potentials can be used to correct the errors [2].

Researchers in BCI field are aware that errors can slow the interaction with the system down and be frustrating. Thus having an accurate method to detect

© Springer International Publishing Switzerland 2014
Z. Zeng et al. (Eds.): ISNN 2014, LNCS 8866, pp. 524–532, 2014.
DOI: 10.1007/978-3-319-12436-0_58

the error can improve the system in speed and become less frustrating. However, only few studies have been done regarding the use of ErrP in BCI. Ferrez et al. studied ErrPs in motor-imaginary BCI and trained a Gaussian classifier to recognize ErrPs, achieving an accuracy of 80% [3]. Dal Seno et al. investigated the single-sweep detection of ErrP in a P300-based speller, achieving an accuracy of around 60% [4]. Combaz et al. studied the amount of training data necessary for accurate classification of ErrPs for P300-based BCI [5].

The low classification accuracies of the ErrP can be explained by its high variation in the shape, size and inter-subject latency variability, and the nature of this is still not certain [6]. In addition, a single-trial ErrP extraction remains a difficult task because its short-time duration, low signal-to-noise ratio, non-stationarity and transient characteristic. One approach to deal with this kind of signals is the blind source separation (BSS). The BSS can ideally extract the desired sources from the observed mixtures while discarding sources that are of no interest. Moreover, making full use of the prior knowledge about ErrP, constrained BSS algorithm with reference can be used. Therefore, in this paper we propose the use of one-unit second-order blind identification (SOBI-R), a robust BSS method to extract small-length signal, non-periodic, transient signals [7], to obtain the ErrP from EEG data.

To the best knowledge of the authors, no report has been made of the use of ErrP in SSVEP-based BCIs. Thus, the goal of this preliminary study is to evaluate the detection and classification accuracy of ErrP in the context of SSVEP-based BCI and its possible application as an error detection mechanism for assessing the real-time online performance of SSVEP-based BCI. In essence, to evaluate its suitability to substitute a feedback from the user, proposed in our previous work, as a faster and more elegant way of notifying the SSVEP-based BCI system of errors in order to correct the output and\or modify the systems parameters [8].

2   Materials and Methods

2.1   Data Recording

The EEG signals were recorded with an amplifier (g.USBamp, Guger Technologies, Graz, Austria) with 8 electrodes distributed according to an extended 10/20 international system (FCz, Cz, C1, C2, FC1, FC2, Fz, and CPz), with the ground on the forehead and the reference on the left earlobe. The EEG signals were digitized with a sampling frequency of 256Hz, powerline notch-filtered, and band-pass filtered between 0.5 and 10Hz.

2.2   Experimental Design

In order to study the existence of ErrP in an SSVEP-based BCI context a Speller was used. The proposed speller allows to the input 16 letters, 'A to P', as shown in Fig. 1. The frequencies of these 16 targets ranged from 8 to 15.5Hz with increase steps of 0.5Hz. An LCD monitor (ViewSonic 22", refresh rate 120Hz, 1680×1050 pixel resolution) was used as the visual stimulator, which was programmed in Microsoft Visual C++ 2010 and DirectX SDK 2010.

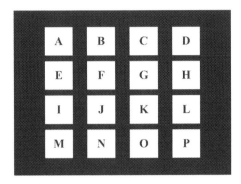

Fig. 1. Graphical User Interface for the study of ErrP in SSVEP-based BCI Speller

The speller was programmed to automatically output a letter every 3 seconds sequentially from A to P. The probability that the speller would output the correct letter was set to 70% and the probability to output erroneously was set 30%. In case the speller does not output the correct letter it will continue to output letters until it can output the correct one. After that, it will move to the next letter in the sequence. This is done in order to elicit the subject's error-related potentials. In this situation there are always 16 correct outputs; however, the number of errors depends on chance.

Six healthy subjects (aged from 22 to 29 years old) with normal or corrected-to-normal vision participated in this study. The subjects were seated on a confortable chair in front of the visual stimulator with a distance of around 60cm.

The subjects were asked to input the 16 characters (from A to P) by gazing at the target frequency one at a time until the system output the desired character. The subjects were not told that the system outputs the characters automatically, in order to make them think that they were controlling the system. Each subject did the experiment for 6 sessions.

2.3   One-Unit SOBI-R

The SOBI algorithm with reference is derived from the traditional BSS algorithm [9]. Given N observed mixtures $\boldsymbol{x}(t) = [x_1(t)\ x_2(t) \cdots x_n(t)]^T$, it can be modeled as

$$\boldsymbol{x}(t) = \boldsymbol{A}\boldsymbol{s}(t) \tag{1}$$

where $\boldsymbol{A}$ is an $N \times M$ unknown full column-rank mixing matrix and $\boldsymbol{s}(t) = [s_1(t)\ s_2(t) \cdots s_n(t)]^T$ are assumed to be independent source signals. BSS is the problem of estimating $\boldsymbol{A}$ and $\boldsymbol{s}(t)$. The BSS determines an $M \times N$ demixing matrix $\boldsymbol{W}$, with M output signals:

$$y(t) = \boldsymbol{W}\boldsymbol{x}(t) = \boldsymbol{W}\boldsymbol{A}\boldsymbol{s}(t) = \boldsymbol{s}(t) \tag{2}$$

Given a set of *references*, the SOBI with reference algorithm extracts only the independent components (ICs) that are most relevant to the *references*. Moreover, for signals as the ErrP, only a single source needs to be extracted. In this case the computation cost can be decreased greatly [7]. Hence, the objective of the one-unit SOBI-R algorithm is then to identify a demixing vector, $\boldsymbol{w}$ (one column of the demixing matrix $\boldsymbol{W}$), which minimizes the contrast function defined as

$$J(y) = -\sum_{\tau=\tau_1}^{\tau_p} \left(\boldsymbol{w}E\left(\boldsymbol{x}\boldsymbol{x}^T\right)\boldsymbol{w}^T\right)^2 = -\sum_{\tau=\tau_1}^{\tau_p} E\left(\boldsymbol{y}(\boldsymbol{t})y(\boldsymbol{t}-\boldsymbol{\tau})\right)^2 \tag{3}$$

where $E(\cdot)$ is the deterministic averaging operation and τ is time delay.

The one-unit SOBI-R algorithm can be summarized as follows:

Algorithm 1. One-unit SOBI-R

Input: μ and λ: initial values of Lagrange multipliers; η: update rate; ϵ: error limit (in this project, $\epsilon = 0.01$)

1: Whiten and decentralize all the observations, normalize the reference to zero mean and unit variance;
2: Choose an initial vector, $\boldsymbol{w}_0$, where $\boldsymbol{w}_0 \neq 0$;
3: **repeat**
4:    Update μ and λ by $\mu_{i+1} = \mu_i + \Delta\mu$ and $\lambda_{i+1} = \lambda_i + \Delta\lambda$;
5:    Calculate the first and second derivatives of the contrast function (3);
6:    Update vector $\boldsymbol{w}$ to $\boldsymbol{w}_{i+1} = \boldsymbol{w}_i + \Delta\boldsymbol{w}$ and normalize $\boldsymbol{w}$ as $\boldsymbol{w} = \boldsymbol{w}/\|\boldsymbol{w}\|$;
7:    Minimize $|J_{i+1}(y) - J_i(y)|$
8: **until** $\|\Delta\boldsymbol{w}\| \geq \epsilon$
Output: $\boldsymbol{w}$: demixing vector

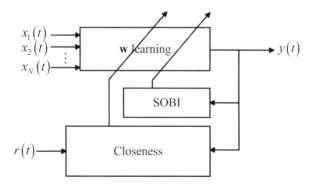

Fig. 2. Block Diagram of the one-unit SOBI-R algorithm with reference

2.4   Feature Extraction

Feature extraction is performed on the 1s long active segment after the system output of a character rather than directly classifying the native EEG data without feature extraction. It greatly affects the performance of the classification, *i.e.* the better the extracted features, the higher the classification accuracy. In this study, the active segments are first passed through the one-unit SOBI-R to isolate the brain signal of interest, *i.e.* the ErrP. Hjorth parameters can be seen as morphological characteristics of the signal. While Fractal dimension (FD) describes the signal in terms of entropy.

Hjorth Parameters. Hjorth parameters [10] describe the signal characteristics in terms of activity (variance of the signal), mobility (a measure of the mean frequency) and complexity (a measure of the deviation from the sine shape). The three parameters can be briefly described as follows:

$$\text{Activity} = \text{var}\,(x(t)) \tag{4}$$

$$\text{Mobility} = \sqrt{\frac{\text{Activity}\,(x')}{\text{Activity}\,(x)}} \tag{5}$$

$$\text{Complexity} = \frac{\text{Mobility}\,(x')}{\text{Mobility}\,(x)} \tag{6}$$

where x is the signal and x' is the derivative of the signal.

Fractal Dimension. FD has a relation with the entropy, which in turn has a direct relation with the amount of information inside a signal. FD can be seen as the degree of irregularity or roughness of a signal. There are several methods for calculating FD. In this study we used Katz method since it has been reported to be more robust than others [11]. The Katz FD is derived directly from the waveform and it can be defined as

$$\text{FD} = \frac{\log_{10} L}{\log_{10} d} \tag{7}$$

where L is the total length of the curve and d is the diameter estimated as the distance between the first point of the sequence and the point of the sequence that provides the farthest distance.

2.5   Single-Trial Classification

In this study, a regularized version of the linear discriminant analysis (LDA) was used for classification [12]. The hyperplane LDA discriminant function $D(\boldsymbol{f})$ maximally separates the feature distributions corresponding to two classes: $D(\boldsymbol{f}) = \boldsymbol{w}^T \boldsymbol{f} + b$, where $\boldsymbol{f}$ is the feature vector to be classified, and $\boldsymbol{w}$ and b the normal vector to the hyperplane and the corresponding bias, respectively. The two are computed by $\hat{\boldsymbol{\mu}} = \frac{1}{2}(\hat{\boldsymbol{\mu}}_1 + \hat{\boldsymbol{\mu}}_2)$, $\boldsymbol{w} = \widetilde{\Sigma}^{-1}(\hat{\boldsymbol{\mu}}_2 - \hat{\boldsymbol{\mu}}_1)$, $b = -w^T\hat{\boldsymbol{\mu}}$, where $\hat{\boldsymbol{\mu}}_j$ is the sample mean of class j, $\hat{\boldsymbol{\mu}}$ is the sample global mean and $\widetilde{\Sigma}$ is the regularized sample covariance matrix, which is divided by the two classes. The regularization aims to minimize the covariance estimation error $E = \left|\Sigma - \widetilde{\Sigma}\right|$, with Σ being the real covariance matrix, by penalizing very large and very small eigenvalues.

3   Results

The recorded EEG signals were analyzed using a time window of 1s, following the output of character by the SSVEP-based BCI Speller. Since the correctness of the output is known, it is easy to separate the EEG signals in erroneous and correct responses. As described before, a band-pass filtering stage is applied to the raw data before being fed to the one-unit SOBI-R algorithm. Fig. 3 shows the signals from the electrodes after band-pass filtering, the reference signal and the extracted signal with the one-unit SOBI-R algorithm, for an erroneous and a correct response, for the same subject in a single-trial.

From the Fig. 3, we can see that the ErrP extracted from the erroneous EEG is noticeably different from the correct response. The ErrP shows two outstanding peaks: a negative one at around 200ms and a positive one at around 300 ms. These peaks correspond to the N2 and P3, respectively. While, the correct output only has the P3 at around 300ms, which is the event-related potential (ERP) that is present after the presence of a stimulus. These results are consistent with the literature [2,3].

Ten-fold cross validation was used for classification tests of the ErrP data. In essence, the dataset of all the experimental sessions for each subject was divided randomly into ten subsets. The following procedure was repeated ten times. For each subject, each time, one of the ten subsets is used as the testing data and the remaining are used as training data. The average classification rate is evaluated readily across all ten folds.

First, to verify the effectiveness of the proposed one-unit SOBI-R, we compare the classification accuracy for FD without and with the proposed one-unit SOBI-R. The results were found to be as shown in Table 1. The classification accuracy without the proposed method was found to be 67.0%±3.69%, while the classification accuracy with the one-unit SOBI-R increased to 84.9% ± 5.07%. This indicates that the proposed method can improve the overall performance in ErrP detection on average around 17.9%. Moreover, two-way ANOVA was performed to verify if the results are significantly different. The p-value was found to be 0.0001, which indicates that performance improvement is significant.

In addition, higher accuracy classification is expected if the extracted features are better. Since the Hjorth parameters are popular among different feature extraction approaches, it was done a comparison with the FD. The same experiment as before was performed to evaluate the performance of the features. Table 2 shows the comparisons of the classification accuracy as well as the sensitivity (true classification of ErrP) and specificity (true classification on non-ErrP) for the two different features using LDA as a classifier.

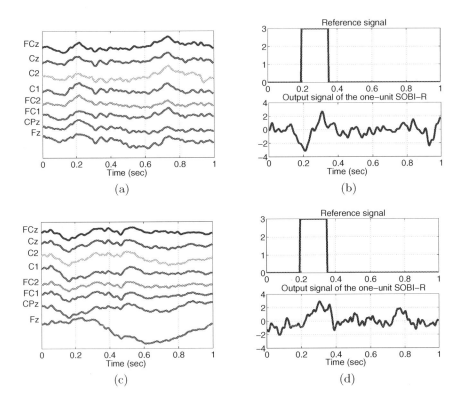

Fig. 3. EEG signals from different electrodes of one subject after output of a character fed to the one-unit SOBI-R: a) erroneous output, c) correct output. The reference signal and the extracted signal using the one-unit SOBI-R: b) erroneous output, d) correct output.

Table 1. Comparison of ErrP classification accuracy between with and without one-unit SOBI-R

Subject	With one-unit SOBI-R Accuracy (%)	Without one-unit SOBI-R Accuracy (%)
S1	91.5	70.0
S2	82.8	68.6
S3	84.7	64.9
S4	90.3	71.3
S5	78.4	65.8
S6	81.9	61.3
Mean	84.9	67.0
S. D.	5.07	3.69
p-value	0.0001	

Table 2. Comparison of classification of ErrP between Hjorth Parameters and Fractal Dimension under LDA

Subject	Fractal Dimension			Hjorth Parameters		
	Accuracy	Sensitivity	Specificity	Accuracy	Sensitivity	Specificity
	(%)	(%)	(%)	(%)	(%)	(%)
S1	91.5	86.8	93.2	83.0	81.9	84.1
S2	82.8	78.2	84.2	74.3	73.1	75.6
S3	84.7	79.1	87.2	81.7	81.0	83.1
S4	90.3	88.3	92.3	83.0	81.9	83.4
S5	78.4	74.1	80.1	75.1	72.3	75.5
S6	81.9	77.5	85.2	79.1	76.9	80.6
Mean	84.9	80.7	87.03	79.4	77.9	80.4
S. D.	5.07	5.61	5.01	3.89	4.40	3.93

It can be seen from the Table 2, the Fractal Dimension outperforms the Hjorth Parameters for all the studied performance indices (accuracy, sensitivity and specificity). Moreover, we can see from the table that it is easier for the classifier to correctly classify to non-ErrP than ErrP (for both feature extraction methods). This seems reasonable since far more examples of non-ErrP than ErrP for training. Nevertheless, we can see that the average true classification of ErrP and non-ErrP, sensitivity and specificity, respectively under Hjorth Parameters is more uniform than Fractal Dimension. In addition, the average classification accuracy of ErrP signal under Hjorth parameters is on average 9.4% higher than without using the one-unit SOBI-R. Moreover, using two-away ANOVA, the p-value was found to be 0.0012, which still indicates that the performance improvement is significant by using the one-unit SOBI-R for ErrP extraction. Accordingly, the one-unit SOBI-R is a potential method for single-trial detection of short-time, transient biomedical signals such as the ErrP.

Since the setup of the experiment was similar to a real SSVEP-based BCI experiment and the subjects believed that the system was translating their intend, we can claim that the results show the feasibility of single-trial ErrP extraction in the context of SSVEP-based BCI. However, in order to support the claim the proposed method needs to be applied in real-time condition and during actual SSVEP-based BCI control.

4   Conclusions

In this study, an one-unit SOBI-R is proposed for single-trial extraction of ErrP for SSVEP-based BCI. The one-unit SOBI-R is a kind of constrained BSS method that makes full use of a reference signal. The method is optimized for extraction of short-time duration, low SNR, non-stationarity and transient characteristic, which the case of several biomedical signals, such as the ErrP.

Experiment results on 6 subjects demonstrated that the proposed one-unit SOBI-R can greatly improve the detection accuracy of single-trial ErrP.

Moreover, the same results showed that FD provides better features than Hjorth Parameters for ErrP classification.

In future works, the single-trial of ErrP detection will be incorporated in an online error detection and correction mechanism for SSVEP-based BCI. Moreover, other classifiers instead of the regularized LDA will be employed to further improve the performance.

Acknowledgments. This work is supported in part by the Macau Science and Technology Development Fund under grant FDCT 036/2009/A and the University of Macau Research Committee under grants MYRG139(Y1-L2)-FST11-WF, MYRG079(Y1-L2)-FST12-VMI and MYRG069(Y1-L2)-FST13-WF and MYRG2014-00174-FST.

References

1. Volosyak, I.: SSVEP-based Bremen-BCI Interface-Boosting Information Transfer Rates. J. Neural Eng. **8**, 036020 (2011)
2. Gehring, W.J., Goss, B., Coles, M.G., Meyer, D.E., Donchin, E.: a Neural System for Error Detection and Compensation. Psychol. Sci. **4**, 385–390 (1993)
3. Ferrez, P.W., Millan, J.D.: Simultaneous Real-time Detection of Motor Imaginary and Error-related Potentials for Improved BCI Accuracy. In: 4th International Brain-Computer Interface Workshop and Training Course, Graz, Austria, pp. 197–202 (2008)
4. Dal Seno, B., Matteucci, M., Mainardi, L.: Online Detection of P300 and Error Potentials in a BCI Speller. Comput. Intell. Neurosci. **11**, 1–5 (2010)
5. Combaz, A., Chumerin, N., Manyakov, N.V., Robben, A., Suykens, J.A., Van Hulle, M.M.: Towards the Detection of Error-related Potentials and Its Integration in the Context of a P300 Speller Brain-computer Interface. Neurocomputing **80**, 73–82 (2012)
6. Falkenstein, M.: ErrP Correlates of Erroneous Performance. In: Errors, Conflicts, and the Brain. Current Opinions on Performance Monitoring, 5–14 (2004)
7. Liu, H., Xie, X., Xu, S., Wan, F., Hu, Y.: One-unit Second-order Blind Identification with Reference for Short Transient Signals. Inform. Sciences **227**, 90–101 (2013)
8. da Cruz, J.N., Wong, C.M., Wan, F.: An SSVEP-Based BCI with Adaptive Time-Window Length. In: Guo, C., Hou, Z.-G., Zeng, Z. (eds.) ISNN 2013, Part II. LNCS, vol. 7952, pp. 305–314. Springer, Heidelberg (2013)
9. Lin, Q.H., Zheng, Y.R., Yin, F.L., Liang, H., Calhoun, V.D.: a Fast Algorithm for One-unit ICA-R. Inform. Sciences **5**, 1265–1275 (2007)
10. Hjorth, B.: EEG Analysis Based on Time Domain Properties. Electroencephalogr. Clin. Neurophysiol. **29**, 306–310 (1970)
11. Esteller, R., Vachtsevanos, G., Echauz, J., Lilt, B.: a Comparison of Fractal Dimension Algorithms using Synthetic and Experimental Data. In: IEEE International Symposium on Circuits and Systems (ISCAS 1999), pp. 199–202. IEEE (1999)
12. Blankertz, B., Lemm, S., Treder, M., Haufe, S., Muller, K.R.: Single-Trial Analysis and Classification of ERP Components-a Tutorial. Neuroimage **56**, 814–825 (2011)

Generalized Regression Neural Networks with K-Fold Cross-Validation for Displacement of Landslide Forecasting

Ping Jiang[1,2,3], Zhigang Zeng[1,3(✉)], Jiejie Chen[1,3],
and Tingwen Huang[4]

[1] School of Automation, Huazhong University of Science and Technology,
Wuhan 430074, China
zgzeng@mail.hust.edu.cn
[2] School of Computer Science and Technology, Hubei PolyTechnic University,
Huangshi 435002, China
[3] Key Laboratory of Image Processing and Intelligent Control of Education Ministry
of China, Wuhan 430074, China
[4] Texas A&M University at Qatar, Doha 5825, Qatar

Abstract. This paper proposes a generalized regression neural networks (GRNNS) with K-fold cross-validation (GRNNSK) for predicting the displacement of landslide. Furthermore, correlation analysis is a fundamental analysis to find the potential input variables for a forecast model. Pearson cross-correlation coefficients (PCC) and mutual information (MI) are applied in the paper. Test on the case study of Liangshuijing (LSJ) landslide in the Three Gorges reservoir in China demonstrate the effectiveness of the proposed approach.

Keywords: Generalized regression neural networks · Pearson cross-correlation coefficients · Mutual information · Landslide

1 Introduction

A landslide own is a geological phenomenon [see 1-4], which causes by the large number of interacting factors. In this process, the stability of the slope changes from a stable to an unstable condition [see 3-4]. Frequent landslides constitute significant risk in the Three Gorges Reservoir area [see 5], which is located at the upper reaches of the Yangtze River in China, causing damage that affects people and property almost every year. Hence, there have been considerable

This work was supported by the Natural Science Foundation of China under Grant 61125303, National Basic Research Program of China (973 Program) under Grant 2011CB710606, the Program for Science and Technology in Wuhan of China under Grant 2014010101010004, the Program for Changjiang Scholars and Innovative Research Team in University of China under Grant IRT1245. This publication was made possible by NPRP grant # NPRP 4-1162-1-181 from the Qatar National Research Fund (a member of Qatar Foundation).

Z. Zeng et al. (Eds.): ISNN 2014, LNCS 8866, pp. 533–541, 2014.
DOI: 10.1007/978-3-319-12436-0_59

researches in developing algorithms and technology for landslide [see 6] prediction and warning. The evolution process of landslides [see 1-2] can be taken into account an open nonlinear [see 7] dynamic system with the complexity and uncertainty.

In recent years, a large number of approaches have been used in the problem of displacement of landslide predicting [see 8-11], such as statistical, artificial intelligent methods, linear or multiple regression and so on. A linear combination model with optimal weight which is applied in landslide displacement prediction [see 8]. Artificial neural networks (ANNS) are computational models and benefit for the analysis and prediction of landslide hazards [see 9-11].

According to above researches, ANNS are qualified to predict displacement of landslide. However, there exists some drawbacks, such as slow convergence rate and local minimum traps. The general regression neural networks (GRNNS) is a feed forward ANNS and originally proposed by [see 12]. GRNNS has ability in approximating continuous functions. Then, It has been shown to be a competent tool for predicting many engineering problems [see 13-14]. GRNNSK is applied to two typical colluvial landslides in Three Gorges Reservoir in China, the LSH and BSH landslides in this paper. Moreover, a comparative study is conducted between the results obtained through GRNNSK and two ANNS models.

2   Forecast Data Analysis

2.1   PCC

Input variable selection is an initial, necessary step of the process of modeling neural network and a proper selection can convince model accuracy. In this study, we choose PCC [see 15] and MI [see 16] for selecting the input variables.

PCC can measure the strength of a linear association between two variables X and Y, where the value $\rho = 1$ signifies a perfect positive correlation and the value $\rho = -1$ signifies a perfect negative correlation. It is easily to find out whether variable X and variable Y are correlated:

$$\rho = Cor(X,Y) = \frac{Cov(X,Y)}{\sqrt{Var(X)Var(Y)}} \tag{1}$$

2.2   MI

MI [see 16] can be used to identify the linear and nonlinear statistical dependence between a set of candidate input and output variables. MI between random variables X and Y can be designed as:

$$MI = \int \int \varphi_{X,Y}(x,y) \log[\frac{\varphi_{X,Y}(x,y)}{\varphi_X(x)\varphi_Y(y)}]dxdy \tag{2}$$

Where $\varphi_X(x)$ and $\varphi_Y(y)$ are the marginal probability density functions of X and Y, respectively. $\varphi_{X,Y}(x,y)$ is the joint probability density function of X

and Y. On account of the data used in this paper is not much, the kth nearest neighbor approach is applied to estimate MI. This estimator is fit for small data sets. And it is suggested to set $k = 2 \sim 4$. Considering the size of the data sample is small, we set $k = 3$ in this study.

2.3   Cross-Validation

Cross-validation [see 17] is a measure of assessing the performance of a predictive model, and statistical analysis will generalize to an independent data set.

The K-fold cross-validation is a technique of dividing the original sample randomly into K sub-samples. Then, a single sub-sample is regarded as the validation data for testing the model, and the remaining K-1 sub-samples are used as training data. An example of estimating a turning parameter γ with K-fold cross-validation as follow:

step 1: Divide the data into K roughly equal parts;

step 2: For each $i = 1, 2, 3, ..., K$, fit the model with parameter γ to other K-1 parts, giving $\hat{\alpha}^{-k}(\gamma)$ and compute its error in predicting the kth part;

$$E_k(\gamma) = \sum_{i \in kthpart} (y_i - x_i \hat{\alpha}^{-k}(\gamma))^2 \tag{3}$$

This gives the cross-validation error

$$CV(\gamma) = \frac{1}{K} \sum_{i=1}^{K} E_i(\gamma) \tag{4}$$

step 3: Do this for many values of γ and choose the value of γ that makes $CV(\gamma)$ smallest. In this paper, we set $K = 10$.

3   GRNNS

GRNNS is a variation of the radial basis neural networks ,which is introduced to perform general (linear or nonlinear) regressions [see 13]. GRNNS figures out the joint probability density function (PDF) of x and y with a training set. The system is perfectly general, on account of the pdf is derived from the data with no preconceptions about its form. If $f(x, y)$ stands for the known joint continuous probability density function of a vector random variable x and a scalar random variable y, the conditional mean of y given X (also called the regression of y on X) is given by

$$E[y|X] = \frac{\int_{-\infty}^{\infty} y f(X, y) dy}{\int_{-\infty}^{\infty} f(X, y) dy} \tag{5}$$

When the density $f(X, Y)$ is unknown, it must usually be calculated from a sample of observations of x and y. The probability estimator $f(X, Y)$ is based on sample values X_i and Y_i of the random variables x and y. The number of sample of observations is n, and the dimension of the vector variable x is p:

$$f(X,Y) = \frac{1}{(2\pi)^{(p+1)/2}\sigma^{(p+1)}}\frac{1}{n}$$
$$\times \sum_{i=1}^{n}\exp[-\frac{(X-X_i)^T(X-X_i)}{2\sigma^2}]$$
$$\times \exp[-\frac{(Y-Y_i)^2}{2\sigma^2}]. \tag{6}$$

The probability estimate $f(X,Y)$ is designed in a physical way, which allocates sample probability of width σ for each sample X_i and Y_i. The probability estimate is the sum total of those sample probabilities. The scalar function D_i^2 is defined,

$$D_i^2 = (X-X_i)^T(X-X_i) \tag{7}$$

and $Y(X)_i$ is:

$$Y(X)_i = \frac{\sum_{i=1}^{n}Y_i\exp(-\frac{D_i^2}{2\sigma^2})}{\sum_{i=1}^{n}\exp(-\frac{D_i^2}{2\sigma^2})} \tag{8}$$

The resulting regression is straight suited to problems concerning numerical data. Since the smoothing σ sets large, the estimated density is obliged to be smooth and becomes a multivariate Gaussian with covariance σI in the limit. On the contrary, a smaller value of σ lets the estimated density to assume non-Gaussian shapes, so the wild points may play an important effect on the estimate. An architecture of the GRNNS that consists of four layers: input layer, pattern layer, summation layer, and output layer, which is shown in Fig. 1.

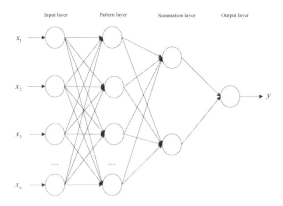

Fig. 1. Schematic diagram of a GRNNS architecture

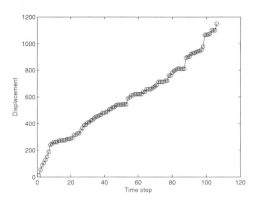

Fig. 2. LSJ displacement

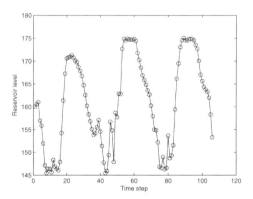

Fig. 3. LSJ reservoir level

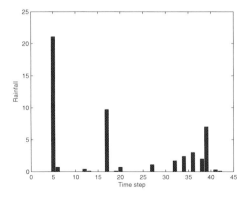

Fig. 4. LSJ rainfall

4   Case Study

4.1   Dataset

The formation of the landslides is very complicated, with the easy sliding loess material, complex geologic structure, precipitation and human engineering activities. We choose the LSJ landslide as our first case study which is located in the town of Yunyang, in the northeast of Chongqing city in China, and in the centre of the Three Gorges Dam of China. There are 24 GPS monitoring points located on the landslide surface. The monitoring data at ZJG24 point is selected to establish the prediction model.

Fig. 2 and Fig. 3 describe the temporal curves of the displacement, reservoir level of LSJ landslide, spanning from April 6 2009 to May 25 2011. The data of the sixth day, the sixteenth day, the twenty-sixth day of each month are chosen. And total number of the displacement and reservoir level are 106. Fig. 4 depicts the temporal curves of rainfall, from April 6 2009 to June 16 2010. The total number of rainfall data is 43, for the reason that the rest data record 0. PCC, which are incline to solve up a linear relationship between two variables.

Then, we should handle relation among displacement, reservoir level and rainfall. So, we do the statistical analysis and compare the test results with those mentioned in Section II. Then, data sets should be normalized into the intervals $[-1, 1]$. In Table. 1, A presents relation between displacement and reservoir level, B presents relation between displacement and rainfall, C presents relation between reservoir level and rainfall.

Table 1. PCC and MI between and among displacement, reservoir level and rainfall

Parameter	A	B	C
PCC	0.9338	0.1100	0.0091
MI	0.7236	-0.0063	0.1367

The relative high coefficients in Table. 1 confirms the close relationship between the variables of displacement and reservoir level. Then, the variables of displacement and reservoir level are as inputs for a forecast model.

4.2   Analysis and Results

The total number of data points in Fig. 2 and Fig. 3 are 106, and all of them are divided into two groups. The first group including data 71 is taken for the training process to build the forecasting model. The rest data is used for testing data. All data sets should be normalized into the intervals $[-1, 1]$. The initial parametric of neural network is very fundamental for results. GRNNS is a kind of radial basis network that is often used for function approximation. GRNNS can be designed very quick and return a new generalized regression neural network. GRNNS has one remarkable advantage that is only one variable required in the initial stage of building the structure of model.

In this paper, GRNNS creates a two-layer network. The first layer has rad-bas neurons, and calculates weighted inputs with dist and net input. The second layer has purelin neurons. The cross-validation methodology is used to find fittest value for parameter 'spread'. The searching ranges for spread is as : spread $\in [0.1, 20]$. And it selects $10 - fold$ cross-validation. Then, 9 folds are used for training and the last fold is used for evaluation. This process is repeated 10 times, leaving one different fold for evaluation each time. After training, the best spread is equal to 0.2. In addition, two ANNS models, Back-Propagation Neural Networks (BPNN), Radial Basis Function (RBF) are also used to forecast landslide displacement for comparison purposes. In order to measure the prediction performance, the root mean square error (RMSE) are used as the criteria to evaluate the proposed models.

$$RMSE = \sqrt{\frac{1}{N} \sum_{i=1}^{N} (\hat{Y}_i - Y_i)^2} \tag{9}$$

where $\hat{Y}_i$ is the predicted value for the time period i, Y_i is the actual value for the time for the same period, and N is the number of predictions. The comparison between monitoring data and prediction results is shown in Fig. 5. Some points of predicted displacement of GRNNSK, RBF and BPNN are similar. Form the fifteenth point to the eighteenth point, the SVM and RBF are different from GRNNSK. On the whole, the curve of predicted displacement of GRNNSK is most close to the real data. The experience results show that, GRNNSK obtains a lowest RMSE 51.7413 among three methods:RBF 78.9520 and BPNN 118.9718. So GRNNSK gets best results for predicting displacement of landslide.

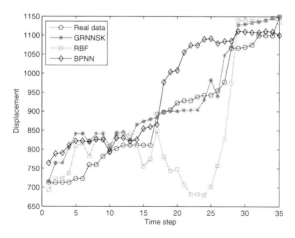

Fig. 5. The comparison between real data and prediction results

5   Concluding Remarks

A novel short-term landslide displacement forecast approach has been proposed in this paper. The capabilities of the GRNNSK is revealed by application to a practical problem: LSJ landslide. Compared with RBF and BPNN, the test result illustrated the effectiveness of the GRNNSK. Moreover, GRNNSK has a better ability to forecast most of data except some points than RBF and BPNN. As we known, landslides considered as an open nonlinear dynamic system with complexity and uncertainty. It is difficult to set up prediction systems for landslide, because it still remains important but largely unsolved problems. In this paper, the displacement of landslide is regarded as the final results of all factors. In future studies, we should pay some attention to some points, such as more geologic factors information, GRNNS-based model and multi-step head predict.

References

1. Qin, S.Q., Jiao, J.J., Wang, S.J.: The predictable time scale of landslides. Bull Eng Geol Environ **59**, 307–312 (2001)
2. Qin, S.Q., Jiao, J.J., Wang, S.J.: A nonlinear dynamical model of landslide evolution. Geomorphology **43**, 77–85 (2002)
3. Chen, C.T., Lin, M.L., Wang, K.L.: Landslide seismic signal recognition and mobility for an earthquake-induced rockslide in Tsaoling. Taiwan. Engineering Geology **171**, 31–44 (2014)
4. Sorbino, G., Sica, C., Cascini, L.: Susceptibility analysis of shallow landslides source areas using physically based models. NAT HAZARDS **53**, 313–332 (2010)
5. Miao, H.B., Wang, G.H., Yin, K.L., Kamai, T., Lin, Y.Y.: Mechanism of the slow-moving landslides in Jurassic red-strata in the Three Gorges Reservoir. China. Engineering Geology **171**, 59–69 (2014)
6. Zhang, Y.B., Chen, G.Q., Zheng, L., Li, Y., Wu, J.: Effects of near-fault seismic loadings on run-out of large-scale landslide: A case study. Engineering Geology **166**, 216–236 (2013)
7. Inoussa, G., Peng, H., Wu, J.: Nonlinear time series modeling and prediction using functional weights wavelet neural network-based state-dependent AR model. Neurocomputing **86**, 59–74 (2012)
8. Li, X.Z., Kong, J.M., Wang, Z.Y.: Landslide displacement prediction based on combining method with optimal weight. NAT HAZARDS **61**, 635–646 (2012)
9. Chen, H.Q., Zeng, Z.G.: Deformation prediction of landslide based on improved back-propagation neural network. Cognitive Computation **5**, 56–62 (2013)
10. Melchiorre, C., Matteucci, M., Azzoni, A., Zanchi, A.: Artificial Neural Networks and Cluster Analysis in Landslide Susceptibility Zonation. Geomorphol **94**, 379–400 (2008)
11. Kanungo, D.P., Sarkar, S., Sharma, S.: Combining neural network with fuzzy, certainty factor and likelihood ratio concepts for spatial prediction of landslides. Nat Hazards Rev **59**, 1491–1512 (2011)
12. Specht, D.: A general regression neural network. IEEE Transactions on Neural Networks **2**, 568–76 (1991)

13. Firat, M., Gungor, M.: Generalized Regression Neural Networks and Feed Forward Neural Networks for prediction of scour depth around bridge piers. Advances in Engineering Software **40**, 731–737 (2009)
14. Bowden, G.J., Nixon, J.B., Dandy, G.C., Maier, H.R., Holmes, M.: Forecasting chlorine residuals in a water distribution system using a general regression neural network. Mathematical and Computer Modelling **44**, 469–484 (2006)
15. Wang, G.J., Xie, C., Chen, S., Yang, J.J., Yang, M.Y.: Random matrix theory analysis of cross-correlations in the US stock market: Evidence from Pearson's correlation coefficient and detrended cross-correlation coefficient. Physica A: Statistical Mechanics and its Applications **392**, 3715–3730 (2013)
16. Frenzel, S., Pompe, B.: Partial mutual information for coupling analysis of multivariate time series. Physical Review Letters **99**, 1–4 (2007)
17. Shao, C., Paynabar, K., Kim, T.H., Jin, J.H., Hu, S.J., Spicer, J.P., Wang, H., Abelld, J.A.: Feature selection for manufacturing process monitoring using cross-validation. Journal of Manufacturing Systems **32**, 550–555 (2013)

One-Class Classification Ensemble with Dynamic Classifier Selection

Bartosz Krawczyk[✉] and Michał Woźniak

Department of Systems and Computer Networks, Wrocław University
of Technology, Wybrzeże Wyspiańskiego 27, 50-370 Wrocław, Poland
{bartosz.krawczyk,michal.wozniak}@pwr.edu.pl

Abstract. The main problem of one-class classification lies in selecting
the model for the data, as we do not have any access to counterexamples, and cannot use standard methods for estimating the quality of the
classifier. Therefore ensemble methods that can utilize more than one
model, are a highly attractive solution which prevents the situation of
choosing the weakest model and improves the robustness of our recognition system. However, one cannot assume that all classifiers available
in the pool are in general accurate - they may have some local areas
of competence in which they should be utilized. We present a dynamic
classifier selection method for constructing efficient one-class ensembles.
We propose to calculate the individual classifier competence in a given
validation point and use them to estimate competence of each classifier
over the entire decision space with a Gaussian potential function. Experimental analysis, carried on a number of benchmark data and backed-up
with a thorough statistical analysis prove its usefulness.

Keywords: One-class classification · Classifier selection · Competence
measure

1 Introduction

One-class classification is among the most difficult areas of the contemporary
machine learning. It works with the assumption that during the training phase,
we have only objects originating from a single class at our disposal. As we have
no access to any counterexamples during the training phase, constructing an
efficient model and selecting optimal parameters for it becomes a very demanding
task. Therefore, methods that can improve the accuracy and robustness of one-
class classifiers are highly demanded. Among them a combined classification
seems as a promising direction [5].

In the last decade, we have seen a significant development of algorithms
known as multiple classifier systems. Their success lies in ability to tackle complex tasks by decomposition, utilizing different properties of each model and
taking advantage of collective classification abilities. For ensemble to work properly two assumptions must be satisfied: base classifiers should be characterized

© Springer International Publishing Switzerland 2014
Z. Zeng et al. (Eds.): ISNN 2014, LNCS 8866, pp. 542–549, 2014.
DOI: 10.1007/978-3-319-12436-0_60

by high individual quality and be at the same time mutually complementary. This is the main reasoning behind the ensemble idea, but it can be realized in a plethora of ways.

The most common approach is known as static selection. It concentrates on creating a pool of base learners and then establishing a combination method for efficient exploitation of the pool members. Firstly, one needs to prepare a set of classifiers for a given task - they can be supplied with the problem (e.g., coming from different sensors in network) or must be carefully designed from the dataset. Several different proposals exist on how to produce different classifiers for one problem - obtaining heterogeneous ensemble is easier (as one need to train several different models), while homogeneous requires manipulating input information to obtain initial diversity among members. Then, one need to choose appropriate combination method.

Second group is known as dynamic selection [4], where we assume that the structure of the ensemble varies for each new incoming object. This is based on the assumption, that each classifier has its own local area of competence [6]. Therefore, for classifying new object the most competent model(s) should be delegated. To establish the competence, a dedicated measure based on correctness of classification is needed. As it can only be calculated locally for objects provided during the training phase, one must extend it over the entire decision space. Dynamic ensembles are divided into two categories: dynamic classifier selection (DCS) [2] and dynamic ensemble selection (DES) [7]. First model assumes, that for each new object we select the single classifier with highest competence and the decision of the ensemble is based on the output of this individual classifier, while in DES systems, one select l most competent classifiers and construct a local sub-ensemble.

In this paper, we propose a DCS system dedicated to the specific nature of one-class problems. Up to the best of authors knowledge, this is the first work on dynamic ensembles for single-class task.

2 One-Class Classification

One-class classification (OCC) aims at distinguishing a given single-class from a more broad set of classes. This class is known as the target concept and denoted as ω_T. All other objects, that do not satisfy the conditions of ω_T are labeled as outliers ω_O. This may seem as a binary classification problem, but the biggest difference lies in the learning procedure [9]. An OCC model needs to estimate the classification rules without an access to counterexamples. At the same time, it must display good generalization properties as during the exploitation phase both objects from the target concept and unseen outliers may appear. OCC aims at finding a trade-off between capturing the properties of the target class (too fitted or too lose boundary may lead to high false rejection / false acceptance rates) and maintaining good generalization (as it is easy to overfit model when having only objects from a single class for training). There is a number of different methods for OCC, that can be categorized in three groups: based on density estimation, based on clustering and based on optimizing the boundary volume. Each of these methods can output significantly different shape of a decision hyperplane (see Fig. 1).

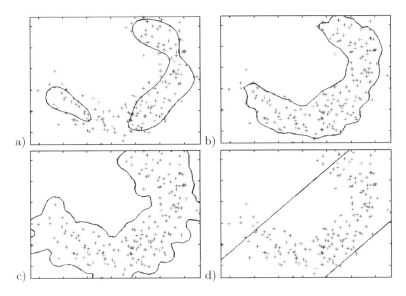

Fig. 1. Exemplary differences between decision boundaries created by different one-class classifiers: (*a*) Support Vector Data Description, (*b*) Parzen Density Data Description, (*c*) Minimum Spanning Tree Data Description, and (*d*) Principal Component Analysis Data Description.

As we do not have any counterexamples at our disposal during the training phase, selecting the best model and fitting its parameters can be hard and time-consuming task [8]. At the same time more than one model may have desirable properties for a given problem, especially in case of complex datasets. Therefore, utilizing more than single model in an ensemble may improve the robustness of the constructed system and prevent us from choosing a weaker model.

3   One-Class Ensemble with Dynamic Classifier Selection

Selecting the best model for an one-class classification task is often accompanied by observations, that more than one type of classifier can be useful for analyzing the given dataset. This assumptions are the basis of our proposal of one-class dynamic classifier selection (OCDCS) system, in which we prepare a pool of heterogeneous models and delegate one of them dynamically to the decision area in which it is the most competent one.

3.1   Preliminaries

The model deals with a one-class problem described by a set of class labels $\mathcal{M}_{occ} = \{\omega_T, \omega_O\}$. A one-class classifier $\Psi_{occ} : \mathcal{X} \mapsto \mathcal{M}_{occ}$ produces the pair of discriminant functions $(d_{\omega_T}(x), d_{\omega_O}(x))$ for a given object $x \in \mathcal{X} \subseteq \mathcal{R}^n$. Discriminants are characterized by following properties: $d_i(x) \geq 0$ and $\sum_i d_i(x) = 1$. Classification is made with the usage of maximum rule. During the ensemble training stage, we should have at our disposal a training set $\mathcal{TS}$ and a validation set $\mathcal{VS}$, both with objects described by feature vector x and with known true class labels. In case of one-class classification both training and validation sets consists only of objects belonging to the target concept ω_T.

3.2   Measuring Classifier Competence

Using the provided $\mathcal{VS}$, one may calculate the competence measure $C(\Psi_{occ}|x)$. It reflects the competence given classifier to correctly classify a given point $x \in \mathcal{X}$. However, this only gives us an outlook on the performance of the classifier for given validation points. We would like to extend this to the entire decision space, so one may select the most competent classifier for a given object. Following the suggestion for multi-class problems, we use a two-step procedure for estimating the competence of a given classifier for the entire decision space: (i) calculate a source competence $C_{SRC}(\Psi_{occ}|x_k)$ for each $x_k \in \mathcal{VS}$, (ii) extend these source competencies for the entire decision space according to normalized Gaussian potential function [10].

The normalized Gaussian potential function allows us to estimate the competence of l-th one-class classifier over the entire space (and not only it part, described by validation objects). This can be formulated as:

$$C(\Psi_{occ}|x_n) = \frac{\sum_{x_k \in \mathcal{VS}} C_{SRC}(\Psi_{occ}|x_k) exp(-dist(x_n, x_k))^2}{\sum_{x_k \in \mathcal{VS}} exp(-dist(x_n, x_k))^2}, \qquad (1)$$

where x_n is the new, incoming object to be classified and $dist(x_n, x_k)$ is the Euclidean distance between the new object and object from $\mathcal{VS}$ with already known source competence.

Let's define the method for calculating the source competence $C_{SRC}(\Psi_{occ}|x_k)$ based on the entropy criterion. The competence measure is a combination of the absolute value of the competence and its sign. The value is inverse proportional to the normalized entropy value of the discriminant values of given one-class classifier, while the sign of the competence is determined by the correctness of the classification of validation point x_k. With this, one may calculate the entropy-based competence function as follows:

$$C_{SRC}(\Psi_{occ}|x_k) = (-1)^{I_{\{\Psi_{occ}(x_k) \neq j_k\}}} \left[1 - \frac{1}{\log n} \sum_{i=1}^{2} d_i(x_k) \log d_i(x_k)\right], \qquad (2)$$

where $j \in \mathcal{M}$ and $I_{\{A\}}$ is the indicator of the set A.

This competence measure satisfies the properties required for Gaussian-based estimation of competence values over the decision space,i.e., $-1 \leq C_{SRC}(\Psi_{occ}|x_k)$ ≤ 1 and $C_{SRC}(\Psi_{occ}|x_k)$ is a strictly increasing function of $d_i(x_k)$. If C_{SRC} $(\Psi_{occ}|x_k) < 0$ then the considered classifier is deemed as incompetent, if its value is grater than 0 that the considered classifier is deemed as competent, but in the case that the value is equal 0 then the considered classifier is recognized as neutral.

3.3   Dynamic Classifier Selection

We can easily extend the presented competence functions for classifier ensemble system. Let us assume, that for a given pattern classification problem we have a pool of L one-class classifiers at our disposal $\mathcal{L} = \{\Psi_1, \Psi_2, \ldots \Psi_L\}$. The robustness and diversity of the one-class classification system could be improved by utilizing a pool of different one-class models and dynamically selecting for each incoming object the single most-competent classifier, i.e.,

$$C(\Psi_{occ_l}|x_n) > 0 \wedge C(\Psi_{occ_l}|x_n) = \max_{g=1,2,\ldots,L} C(\Psi_{occ_g}|x_n) \tag{3}$$

4   Experimental Analysis

The aim of the experimental analysis was to investigate the usefulness of applying DCS system in one-class classification problems and checking if it can outperform single-model and static ensemble approaches.

4.1   Datasets

We have chosen 10 binary datasets described in Tab. 1. Due to the lack of one-class benchmarks we use the canonical multi-class ones. The training set was composed from the part of objects from the target class (according to cross-validation rules), while the testing set consisted of the remaining objects from the target class and outliers (to check both the false acceptance and false rejection rates). Majority class was used as the target concept.

Table 1. Details of datasets used in the experimental investigation. Numbers in parentheses indicates the number of objects in the minor class in case of binary problems.

Name	Objects	Features	Name	Objects	Features
Breast-cancer	286 (85)	9	Hepatitis	155 (32)	19
Breast-Wisconsin	699 (241)	9	Ionosphere	351(124)	34
Colic	368 (191)	22	Sonar	208 (97)	60
Diabetes	768 (268)	8	Voting records	435 (168)	16
Heart-statlog	270 (120)	13	CYP2C19 isoform	837 (181)	242

4.2   Set-up

To apply the proposed OCDCS system, we need to have a pool of base classifiers. We propose to construct them our of 5 heterogeneous models, that are presented together with their parameters in Table 2.

Table 2. Details of one-class classification models and their parameters, that were used as a base for the dynamic ensemble

Classifier	Abbreviation	Parameters
Nearest Neighbor Data Description	NNdd	frac. rejected = 0.05
Parzen Density Data Description	PDdd	frac. rejected = 0.05
Auto-Encoder Neural Network	AENN	hidden units = 10, frac. rejected = 0.1
Minimum Spanning Tree	MST	max. path = 20, frac. rejected = 0.1
Support Vector Data Description	SVDD	kernel = RBF, $\sigma = 0.3$, frac. rejected = 0.05

To put the obtained results into context, we compared our proposed method (**OCDCS**) with reference approaches: Single Best (**SB**, using single model with highest accuracy), Majority Voting (**MV**, standard voting based on discrete outputs), Maximum (**MAX**, static selection of maximum support value) and Average (**AVG**, using averaged values of support functions).

In order to present a detailed comparison among a group of machine learning algorithms, one must use statistical tests to prove, that the reported differences among classifiers are significant. For training/testing and a pairwise comparison a 5x2 combined CV F-test is used, while Friedman ranking test [1] and Shaffer post-hoc test [3] are used for comparing classifiers over multiple datasets. We fix the significance level $\alpha = 0.05$ for all comparisons.

To calculate the competence for OCDCS system, we need to have a validation set $\mathcal{VS}$. Therefore, for each iteration of 5x2 CV, we separate 20% of the training data for validation purposes. The results are presented in Table 3, while results of the Shaffer post-hoc test are depicted in Table 4.

4.3   Discussion

FAs we can see the proposed OCDCS outperformed all other reference methods for 6 out of 10 benchmark datasets. Let us have a closer look on these experimental findings. In 2 cases (diabetes and hepatitis datasets) was unable to outperform a single-best classifier from its pool. This can be explained by a situation, in which we have a single dominant model (strong classifier). In such cases this model outputs the best performance over the entire decision space and applying dynamic selection cannot improve the ensemble performance. But this is a quite rare situation in one-class classification. Usually the structure of the target class is too complex to be handled by only one specific type of classifier's model. In 4 cases (breast-wisconsin, diabetes, hepatitis and voting records datasets) the OCDCS ensemble was similar or slightly worse than the

Table 3. Results of the experimental results with the respect to the accuracy [%] and statistical significance. Small numbers under each method stands for the indexes of models from which the considered one is statistically better.

Dataset	SB[1]	MV[2]	MAX[3]	AVG[4]	OCDCS[5]
Breast-cancer	62.87	60.04	63.72	61.15	**64.28**
	2	–	2,4	2	*ALL*
Breast-Wisconsin	83.29	83.29	**85.10**	80.94	84.73
	4	4	1,2,4	–	1,2,4
Colic	73.28	72.15	75.37	72.48	**77.13**
	2,4	–	1,2,4	–	*ALL*
Diabetes	62.04	58.38	**62.27**	60.02	**62.27**
	2,4	–	2,4	2	2,4
Heart-statlog	82.19	84.72	85.07	83.23	**87.82**
	–	1,4	1,4	1	*ALL*
Hepatitis	55.63	54.29	**55.91**	52.28	55.28
	2,4	4	2,4	–	2,4
Ionosphere	75.26	72.81	77.82	73.79	**80.01**
	2,4	–	1,2,4	2	*ALL*
Sonar	85.27	86.17	86.72	86.48	**88.58**
	–	1	1	1	*ALL*
Voting records	91.36	88.47	**93.90**	90.18	92.45
	2,4	–	*ALL*	2	1,2,4
CYP2C19 isoform	80.16	78.05	84.78	81.79	**89.58**
	2	–	1,2,4	1	*ALL*
Rank	3.70	4.50	1.70	4.10	1.45

Table 4. Shaffer test for comparison between the proposed OCDCS and reference methods. Symbol '+' stands for for situation in which the method on the left is superior.

hypothesis	p-value	hypothesis	p-value
OCDCS vs SB	+ (0.0112)	OCDCS vs MAX	+ (0.0319)
OCDCS vs MV	+ (0.0163)	OCDCS vs AVG	+ (0.0103)

MAX operator. This can be explained by a situation, in which competence measure estimation becomes very sparse (for objects located far from the validation points). This shows us, that there is a need for developing different methods of competence estimation that can handle such situations. OCDCS was superior to majority voting and average operator in all examined cases. OCDCS displays statistically significant improvement over reference methods, when considering its performance over multiple datasets.

5   Conclusions

We have presented a novel approach for constructing ensembles for one-class classification problems based on dynamic classifier selection. We described the complete dynamic selection system designed for purpose of learning in the absence of counterexamples. To properly calculate the competence of one-class methods, a competence measure based on evaluating the entropy of discriminant functions was introduced. To estimate the competence of each classifier from the pool, we proposed to use a Gaussian potential function. These steps allowed to create an efficient ensemble system for one-class classification, that is able to exploit the local competencies of classifiers from the pool, and delegate them to local decision areas. We showed that our ensemble works very well with pool of heterogeneous classifiers, but there is no restrictions that prohibit for using OCDCS with homogeneous classifiers.

Acknowledgments. This work was supported by the Polish National Science Centre under the grant PRELUDIUM number DEC-2013/09/N/ST6/03504 realized in years 2014-2016.

References

1. Demsar, J.: Statistical comparisons of classifiers over multiple data sets. Journal of Machine Learning Research **7**, 1–30 (2006)
2. Galar, M., Fernández, A., Barrenechea Tartas, E., Bustine Sola, H., Herrera, F.: Dynamic classifier selection for one-vs-one strategy: Avoiding non-competent classifiers. Pattern Recognition **46**(12), 3412–3424 (2013)
3. García, S., Fernández, A., Luengo, J., Herrera, F.: Advanced nonparametric tests for multiple comparisons in the design of experiments in computational intelligence and data mining: Experimental analysis of power. Inf. Sci. **180**(10), 2044–2064 (2010)
4. Hung-Ren Ko, A., Sabourin, R., de Souza Britto Jr., A.: From dynamic classifier selection to dynamic ensemble selection. Pattern Recognition 41(5), 1718–1731 (2008)
5. Krawczyk, B., Woźniak, M.: Diversity measures for one-class classifier ensembles. Neurocomputing **126**, 36–44 (2014)
6. Lin, C., Chen, W., Qiu, C., Wu, Y., Krishnan, S., Zou, Q.: Libd3c: Ensemble classifiers with a clustering and dynamic selection strategy. Neurocomputing **123**, 424–435 (2014)
7. Łysiak, R., Kurzyński, M., Wołoszynski, T.: Optimal selection of ensemble classifiers using measures of competence and diversity of base classifiers. Neurocomputing **126**, 29–35 (2014)
8. Tax, D.M.J., Müller, K.: A consistency-based model selection for one-class classification. In: Proceedings of the International Conference on Pattern Recognition, vol. 3, pp. 363–366 (2004) (cited by since 1996:12)
9. Tax, D.M.J., Duin, R. P. W.: Characterizing one-class datasets. In: Proceedings of the Sixteenth Annual Symposium of the Pattern Recognition Association of South Africa, pp. 21–26 (2005)
10. Woloszynski, T., Kurzynski, M.: On a New Measure of Classifier Competence Applied to the Design of Multiclassifier Systems. In: Foggia, P., Sansone, C., Vento, M. (eds.) ICIAP 2009. LNCS, vol. 5716, pp. 995–1004. Springer, Heidelberg (2009)

A Massive Sensor Data Streams Multi-dimensional Analysis Strategy Using Progressive Logarithmic Tilted Time Frame for Cloud-Based Monitoring Application

Xin Song[✉], Cuirong Wang, Yanjun Chen, and Jing Gao

Northeastern University at Qinhuangdao, Qinhuangdao 066004, China
bravesong@163.com

Abstract. The massive sensor data streams multi-dimensional analysis in the monitoring application of internet of things is very important, especially in the environments where supporting such kind of real time streaming data storage and management. Cloud computing can provide a powerful, scalable storage and the massive data processing infrastructure to perform both online and offline analysis and mining of the heterogeneous sensor data streams. In order to support high-volume and real-time sensor data streams processing, in this paper, we propose a massive sensor data streams multi-dimensional analysis strategy using progressive logarithmic tilted time frame for cloud based monitoring application. The proposed strategy is sufficient for many high-dimensional streams analysis tasks using map-reduce platform of cloud computing. Finally, the simulation results show that proposed strategy achieves the enhancing storage performance and also can ensures that the total amount of data to retain in memory or to be stored on disk is small for achieving the performance improvement of the massive sensor data streams analysis.

Keywords: Massive data streams analysis · Multi-dimensional streams data processing · Progressive logarithmic tilted time frame · Cloud computing

1   Introduction

In recent years, an increasing number of emerging applications deal with a large number of heterogeneous sensor data objects in Internet of Things (IoT) because of a wide variety of sensor devices on sensing layer. Unlike traditional data sets, the sensor data streams flow in and out of a computer monitoring system continuously and with varying update rates. They are temporally ordered, fast changing, massive, and potentially infinite. In order to process the continuously changing sensor data streams, the IoT application system terminal equipment must implement the massive sensor data storage and the powerful computing ability for real-time collection, dissemination and extracting of sensor data to users and administrators anytime and from anywhere. However, due to the continuity and infiniteness of the sensor data streams, the traditional sensor data processing model focuses on a relatively small scale historical data and is not able to meet the increasing data processing requirement of IoT system

© Springer International Publishing Switzerland 2014
Z. Zeng et al. (Eds.): ISNN 2014, LNCS 8866, pp. 550–557, 2014.
DOI: 10.1007/978-3-319-12436-0_61

because of some restricted factors such as memory capacity, data collection speed, transmission bandwidth and so on. As a result, the monitoring system was deployed based on Cloud computing platform that is a model for enabling convenient, on-demand network access to a shared pool of configurable computing resources [1]. It may be impossible to store an entire data stream or to scan through it multiple times due to the tremendous volume of the sensor data streams on the Cloud-based monitoring application system. Moreover, the sensor data streams tend to be of a rather low level of abstraction, whereas most analysts are interested in relatively high-level dynamic changes, such as trends and deviations. To discover knowledge or patterns form data streams, it is necessary to develop multi-dimensional stream processing and analysis methods. In this paper, we proposed a massive sensor data streams multi-dimensional analysis strategy using progressive logarithmic tilted time frame for cloud based monitoring application. The proposed strategy can realize the approximate representation of original sensor data streams by compressing the time dimension to reduce the storage space and the communication energy requirements. Using progressive logarithmic tilted time frame model, the simulation results show that proposed strategy achieves the enhancing storage performance and also can ensure that the total amount of data to retain in memory or to be stored on disk is small.

The rest of this paper is organized as follows: in section 2, we briefly review some closely related works. The proposed massive sensor data streams multi-dimensional analysis strategy is derived and discussed in section 3. The validity analysis and performance evaluation are presented in section 4. Finally, the conclusions and future work directions are described in section 5.

2   Related Works

There have been a few of studies on the management of the sensor data streams using cloud computing. An increasing number of data-intensive application deals with continuously changing data streams from sensors. One requires the data processing system that can store, update, and retrieve large sets of multidimensional sensor data. The conventional information system technology cannot manage the continuously changing properties of the sensor data. Therefore, it is very necessary for managing massive and heterogeneous sensor data via combining the cloud computing and wireless sensor networks technology. Ref. [2] formally presented a comprehensive framework for managing the continuously changing data objects with insights into the spatiotemporal uncertainty problem and presented an original parallel-processing solution for efficiently managing the uncertainty using the map-reduce platform of cloud computing. The proposed quantitative computation and cloud computing paradigm provide sound guidelines for any relevant application design where a scalable data management solution for a sensor and sensing system is required. However, the proposed framework provided only a theoretical and practical basis for parallel quantification and application of the spatiotemporal uncertainty in a highly scalable manner. Many run-time data analysis tools can give the most up-to-date knowledge of the system to administrators. However, when troubleshooting a problem in depth, the offline data analysis functionality is necessarily required to get the complete knowledge for system diagnosis.

Ref. [3] proposed a high volumes of event stream indexing and efficient multi-keyword searching framework for cloud monitoring. By integrating the composite tree index structure to the run-time correlation engine framework, the analysts are able to further enhance the tool to perform offline data analysis tasks to provide a more sophisticated monitoring service in the cloud computing environment. The next generation of cloud-oriented systems will require a novel approach to monitoring that crosses boundaries, federating millions of metrics from heterogeneous sensor sources. Michael Smit, Bradley Simmons and Marin Litoiu presented and implemented an architecture using stream processing to provide near real-time, cross-boundary, distributed, scalable, fault-tolerant monitoring. The pluggable, extensible architecture allowed for metrics from existing sensors to be published as streams. These streams were made available to subscribers by push notifications, by pull polling, and by a pluggable architecture allowing the subscriber to provide their own component to manage streams. Architecturally, streams can be aggregated, and the entire monitoring infrastructure can be managed adaptively [4]. For analyzing the data streams from city sensing infrastructures, Ref. [5] introduced an algorithmic architecture for kernel-based modeling of data streams. The approach was focused on a kernel dictionary implementing a general hypothesis space which is update incrementally, accounting for memory and processing capacity limitations. The presented implementation builds on top of the MapReduce framework designed for robust distributed computation. The most research achievements focus on data streams analysis from the healthcare applications and the electric power grids. Zhang fan et al. proposed a task-level adaptive MapReduce framework for real-time streaming data in healthcare scientific applications. The framework extended the traditional Hadoop MapReduce and specifically addressed the varied arrival rate of big-data splits. The designer applied four scaling theorems and scaling corollaries to implement heterogeneous cloud platform for real-life healthcare scientific applications [6]. Sudip Misra and Subarna Chatterjee proposed social choice considerations in cloud-assisted WBAN (Wireless Body Area Network) architecture for post-disaster healthcare: data aggregation and channelization. Their work focuses on two fundamental research issues in this context- aggregation of health data transmitted by the local data processing units within the mobile monitoring nodes, and channelization of the aggregated data by dynamic selection of the cloud gateways [7]. In addition to these, there are more aspects for the healthcare services system [8], including the rich media [9], the secure monitoring and sharing of generic data [10, 11]. For the data streams analysis of the power supply system, some researcher discussed how cloud computing model can be used for developing smart grid solutions. Flexible resources and services shared in network, parallel processing and omnipresent access are some features of Cloud Computing that are desirable for Smart Grid applications. Even though the Cloud Computing model is considered efficient for Smart Grids, it has some constraints such as security and reliability [12, 13]. A fundamental difference in the analysis of stream data from that of relational and warehouse data is that the stream data are generated in huge volume, flowing in and out dynamically and changing rapidly. Due to limited memory, disk space, and processing power, it is impossible to register completely the detailed level of data stream. Therefore, it is necessary for managing massive sensor data streams via combining the cloud computing, stream data compressed and mining technology. In this paper, we built a large-scale heterogeneous sensing data stream processing system based on cloud computing. The managing platform has both the distributed storage technology of massive sensor data streams and the data streams compression algorithm based progressive logarithmic tilted time frame.

3   Implementation of Massive Sensor Data Streams Multi-dimensional Analysis Strategy

Stream data are generated continuously in a dynamic environment, with huge volume, infinite flow, and fast-changing behavior. To find interesting or unusual patterns, it is essential to perform multi-dimensional analysis with the faster response time. The proposed massive sensor data streams multi-dimensional analysis strategy includes the distributed compressed storage solution of data streams using progressive logarithmic tilted time frame in cloud computing and the multi-dimensional analysis strategy based on map-reduce.

3.1   The Distributed Compressed Storage Solution of Data Stream Using Progressive Logarithmic Tilted Time Frame

Due to limited memory, disk space, and processing power, it is impossible to register completely the detailed level of data. A realistic design is to explore several data compression technique. The stream data so constructed are much smaller than those constructed from the raw stream data but will still be effective for multi-dimensional stream data analysis. The proposed strategy applied the progressive logarithmic tilted time frame to achieve the distributed compressed storage solution of data stream in cloud computing.

In stream data analysis, people are usually interested in recent changes at a fine scale but in long-term changes at a coarse scale. Naturally, we can register time at different levels of granularity. The most recent time is registered at the finest granularity; the more distant time is registered at a coarser granularity; and the level of coarseness depends on the application requirements and on how old the time point is (from the current time). Such a time dimension model is called a tilted time frame. This model is sufficient for many analysis tasks and also ensures that the total amount of data to retain in memory or to be stored on disk is small.

The proposed compressed data streams method is the progressive logarithmic tilted time frame model, where snapshots are stored at differing levels of granularity depending on the recency. Let Eps_time be the clock time elapsed since the beginning of the stream. Snapshots are classified into different frame numbers ($Frame\_number = i$), which can vary from 0 to $Frame\_max$ where $\log_2(Eps\_time) - Capacity\_max \leq Frame\_max \leq \log_2(Eps\_time)$, $Capacity\_max$ is the maximal number of snapshots held in each frame. Each snapshot is represented by its timestamp. The rules for insertion of a snapshot Spt_time (at time Spt_time) into the snapshot frame table are defined as follows: (1) If (Spt_time mod 2^i)=0 but (Spt_time mod 2^{i+1}) $\neq 0$, Spt_time is inserted into $Frame\_number = i$ if

$i \le Frame\_max$; otherwise (i.e., $i \ge Frame\_max$), Spt_time is inserted into $Frame\_max$. (2) Each slot has a $Capacity\_max$. At the insertion of Spt_time into $Frame\_number = i$, if the slot already reaches its $Capacity\_max$, the oldest snapshot in this frame is removed and the new snapshot inserted.

Consider the snapshot frame table of Fig. 1, where $Frame\_max$ is 5 and $Capacity\_max$ is 3. A example was explained how timestamp 64 was inserted into the table. We know (64 mod 2^6)=0 but (64 mod 2^7) $\ne$ 0, that is, $Frame\_number = i = 6$. However, since this value of $Frame\_number = i$ exceeds $Frame\_max$, 64 was inserted into frame 5 instead of frame 6. Suppose we now need to insert a timestamp of 70. At time 70, since (70 mod 2^1)=0 but (70 mod 2^2) $\ne$ 0, we would insert 70 into $Frame\_number = i = 1$. This would knock out the oldest snapshot of 58, given the slot capacity of 3. From the table, we see that the closer a timestamp is to the current time, the denser are the snapshots stored.

frame number	Snapshots(by clock time)
0	69 67 65
1	66 62 58
2	68 60 52
3	56 40 24
4	48 16
5	64 32

Fig. 1. A progressive logarithmic tilted time frame table

3.2   The Distributed Storage Architecture and Multi-dimensional Analysis Strategy Based on Map-reduce in Cloud Computing

The Distributed storage architecture of the polymorphism sensor data stream in cloud computing environment is three-level storage architecture, as shown in figure 2.

The operation support data layer is responsible for the storage and dynamic update of the sensor data streams (or the intermediate results). The operation result data layer is responsible for the storage and dynamic update of the final processing results. The historical data layer is responsible for the storage and additional update of the historical sensor data. The candidate historical data in the operation result data layer were stripped out and appended to the historical data layer after each time data processing. The central storage scheduling module controls respectively the three layers data collection according to the relevant instruction and keeps the data consistency between the operation support data layer and the operation result data layer in the system operation process. The basic storage management layer provides the data acquisition and update services using the cluster distributed file system.

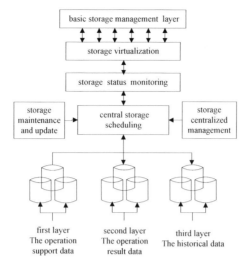

Fig. 2. The three-level storage architecture

To ensure real-time processing of a large-scale heterogeneous sensor data streams in cloud computing, the intermediate results of the preprocessing historical data were distributed at each cache nodes for reducing the duplication processing of the historical sensor data and avoiding the frequent transmission between nodes. Each node redundantly received data stream, so the pending processing data of the node were filtered by Map stage and operated Reduce calculation in the node local cache. When the existing node's local computing and storage resources cannot meet the real needs, the new increased node will be utilized for mobile cache data extension using the re-division technology. Finally, the local calculation results were synchronized to the distributed storage area. The Map-Reduce process is shown in Fig. 3.

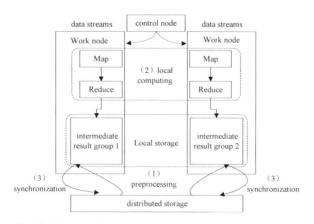

Fig. 3. The workflow of the sensor data streams processing

4   Performance Evaluation

In this section, the intermediate result local storage performance of proposed strategy was evaluated. The data flow velocity is 1MB/s (that is, the data is sent by 200B each, and 5000 sequence /s), the scale of the intermediate result is 50GB, Each test for 10 times, each time 10 min, the experimental results is the average value. The Fig. 4 shows the performance contrast for the LRU algorithm (Least Recently Used), the recent research RTMR algorithm [14], Naive algorithm and the proposed strategy (PLTTF). The memory read/write performance is improved 19.5% by the proposed strategy. The external storage read/write and memory hit rate performance are improved 26.9% and 14.1% respectively. Entirety read/write performance is improved 23.5%.

Performance indicators	Test method	Experimental results	Enhancing effect
Memory read/write	Read/write synchronization	75385.2t/s	19.5%
	synchronization elimination(RTMR)	84506.8t/s	
	synchronization elimination(PLTTF)	90083.7t/s	
The external storage read/ write	Big Table	4425.5t/s	26.9%
	RTMR	5111.4t/s	
	PLTTF	5620.2t/s	
Memory hit rate	LRU	66.7 %	14.1%
	RTMR	72.9 %	
	PLTTF	76.1%	
Entirety read/write performance	Naive algorithm	73901.4t/s	23.5%
	RTMR	88608.5t/s	
	PLTTF	91254.8t/s	

Fig. 4. The performance optimization of the intermediate result storage

5   Conclusion

In this paper, we have proposed and described a massive sensor data streams multi-dimensional analysis strategy using progressive logarithmic tilted time frame for cloud-based monitoring application. The progressive logarithmic tilted time frame model is sufficient for typical time-related queries, and at the same time, ensures that the total amount of data to retain in memory and to be computed is small. Depending on the given application, the proposed strategy can provide a fading factor, such as by placing more weight on the more recent time frames. Furthermore, the Map-Reduce Processing Strategy of Massive Sensor Data Streams was presented for improving the real-time processing performance.

Acknowledgment. The research work was supported by the Fundamental Research Funds for the Central University under Grant no. N120323009, the Doctoral Fund of Northeastern University at Qinhuangdao under Grant No. XNB201301, the Natural Science Foundation of Hebei Province under Grant No. F2014501055, the Program of Science and Technology Research of Hebei University No. ZD20132003 , and the National Natural Science Foundation of China under Grant No. 61403069 and 61473066.

References

1. McDaniel, P., Smith, S.W.: Outlook: Cloud Computing With a Chance of Security Challenges and Improvements. In: Proceeding of the IEEE Computer and Reliability Societies, pp. 77–80 (2010)
2. Yu, B., Sen, R., Jeong, D.H.: An Integrated Framework for Managing Sensor Data Uncertainty Using Cloud Computing. Information Systems **38**(8), 1252–1268 (2013)
3. Wang, M., Holub, V., Murphy, J., Sullivan, P.O.: High Volumes of Event Stream Indexing and Efficient Multi-keyword Searching for Cloud Monitoring. Future Generation Computer Systems **29**(8), 1943–1962 (2013)
4. Smit, M., Simmons, B., Litoiu, M.: Distributed, Application-level Monitoring for Heterogeneous Clouds Using Stream Processing. Future Generation Computer Systems **29**(8), 2103–2114 (2013)
5. Kaiser, C., Pozdnoukhov, A.: Enabling Real-time City Sensing with Kernel Stream Oracles and MapReduce. Pervasive and Mobile Computing **9**, 708–721 (2013)
6. Zhang, F., Cao, J.W., Khan, S.U., Li, K.Q., Hwang, K.: A Task-level Adaptive MapReduce Framework for Real-time Streaming Data in Healthcare Applications. Future Generation Computer Systems (in press, July 5, 2014)
7. Misra, S., Chatterjee, S.: Social Choice Considerations in Cloud-assisted WBAN Architect- ure for Post-disaster Healthcare: Data Aggregation and Channelization. Information Sciences **284**, 95–117 (2014)
8. Sultan, N.: Making Use of Cloud Computing for Healthcare Provision: Opportunities and Challenges. International Journal of Information Management **34**(2), 177–184 (2014)
9. Chen, M.: NDNC-BAN: Supporting Rich Media Healthcare Services via Named Data Networking in Cloud-assisted Wireless Body Area Networks. Information Sciences **284**, 142–156 (2014)
10. Thilakanathan, D., Chen, S., Nepal, S., Calvo, R., Alem, L.: A Platform for Secure Monitoring and Sharing of Generic Health Data in the Cloud. Future Generation Computer Systems **35**, 102–113 (2014)
11. Castiglione, A., Pizzolante, R., Santis, A.D., et al.: Cloud-based Adaptive Compression and Secure Management Services for 3D Healthcare Data. Future Generation Computer Systems (in press, July 16, 2014)
12. Markovicn, D.S., Zivkovic, D., Branovic, I., et al.: Smart Power Grid and Cloud Computing. Renewable and Sustainable Energy Reviews **24**, 566–577 (2013)
13. Yigit, M., Gungor, V.C., Baktir, S.: Cloud Computing for Smart Grid applications. Computer Networks **70**, 312–329 (2014)
14. Qi, K.Y., Zhao, Z.F., Fang, J., Ma, Q.: Real-Time Processing for High Speed Data Stream over Large Scale Data. Chinese Journal of Computers **35**(3), 477–490 (2012)

Evolutionary Clustering Detection of Similarity in Neuronal Spike Patterns

Hu Lu[1,2](✉), Zhe Liu[2], and Yuqing Song[2]

[1] School of Computer Science, Fudan University, Shanghai, China
myluhu@126.com
[2] School of Computer Science and Communication Engineering,
Jiangsu University, Jiangsu, China

Abstract. The key to interpreting multi-electrode recorded neuronal spike trains are the firing patterns hidden in a population of neurons. Here, we present a new firing pattern detection method based on community structure partitioning method, in which we apply the genetic evolutionary algorithm to maximize modularity function Q. We propose a new genotype encoding method to represent the functional connections between neurons. Independent of prior 'knowledge,' this method automatically finds the number and type of firing patterns in neuronal populations, an advantage over current leading methods.

Keywords: Optimization algorithm · Modularity function · Spike trains · Firing patterns

1   Introduction

Perceiving the external world and relating information, brain processes are detectable by the biochemical spikes generated by myriad synaptic connections between neurons. The ordered firing of spikes forms spatio-temporal firing patterns. Analysis of these firing patterns is a core computational problem in neural information processing [1]. Since we cannot know in advance whether given firing patterns are meaningful, nor the number and type of patterns in the original spike trains, neural information processing relies on the performance of new pattern analysis methods.

Clustering algorithm as an unsupervised machine learning method is undoubtedly an effective tool for detecting spike trains and uncovering their patterns. In recent years, some clustering algorithms have been proposed to treat data on firing patterns of spike trains [2,3]. A k-means algorithm to cluster multiple trials of recorded spike trains has been proposed [4]. A spectral clustering method has been applied to detect synchronization in spike trains [5]. Both these clustering algorithms require specifying the number k of clusters. From these the k-means, FCM or spectral clustering algorithm are applied to cluster the spike trains in k classes. Dealing with the original data sets, we cannot know a priori the number of clusters. Artificial determination of cluster numbers tends to be highly subjective. Then a new method has been proposed, derived from the problem of community detection, which can self-determine the number of groups by maximizing the modularity function Q, and thus group the

© Springer International Publishing Switzerland 2014
Z. Zeng et al. (Eds.): ISNN 2014, LNCS 8866, pp. 558–567, 2014.
DOI: 10.1007/978-3-319-12436-0_62

corresponding firing patterns [6]. However, this method requires we calculate the value of Q when the total number of groups varies from 2 to N (N is the number of neurons), then select the maximum value of Q. Since community partitioning is an NP-hard problem, many traditional community partitioning methods can only find the local optimal value of Q [7].

In community structure detection, researchers have proposed many community partitioning methods based on optimized network modularity, using evolutionary algorithms, such as extremal optimization and genetic algorithm [8-10]. By encoding the network structures into chromosomes to calculate the global optimization value of Q, this approach avoids repeated calculation of the Q value that corresponds to varying numbers of modules. Community structure detection automatically recognizes the optimal number of modules.

In this study, we applied the community partitioning method to the analysis of neuronal spike trains, based on an evolution algorithm. We propose a new genotype encoding method that encodes the functional connections between neurons and possible divisions into various genotypes, a distinction from the existing community structure detection method based on genetic algorithms. Genetic algorithm approach uses the adjacency matrix graph. Since the correlation matrix between neurons is a weighted matrix, we implement genetic community partitioning based on the weighted matrix. Our tests on different spike train data sets show that this method can self-determine the number of patterns and detect the type of patterns. We then compared our results with two other community partitioning methods, one of which is fast Newman community detection method [7]. The other is also a genetic-based community partitioning method, but with a genotype encoding method which adopts the random allocation principle (GA-random), different from the method proposed in this paper. We feel our method yields superior results.

2   Method and Realization

Before using the clustering algorithm to analyze the spike trains, pairwise relations between neurons must be computed, then an N*N similarity matrix can be inferred from pairwise relations [11,12]. N is the number of neurons (figure 1). Figure 1A is a raster plot of spike trains in the 10 seconds for 10 neurons. We used non-overlapping, short time windows (known as bins) to bin the spike trains (Fig. 1B). Each line represents neuronal spike trains. Each element represents the number of spikes in each bin. Then we use the vector similarity distance to calculate spike train similarity. Figure 1C represents the similarity relations between neurons. The thickness of lines represents the strength of similarity between two neuronal spike trains. Our method applies the following formula:

$$s_{xy} = 1 - \frac{\sum_{i=1}^{n}(x_i - y_i)^2}{\sum_{i=1}^{n}x_i^2 + \sum_{i=1}^{n}y_i^2} \tag{1}$$

X_i is the number of spikes of xth neuron in the ith bin. The value of s_{xy} varies from 0 to 1. The greater the value of s_{xy}, the more similarity found between two given neurons.

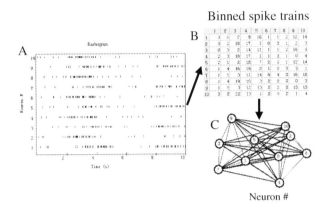

Fig. 1. Overview of construction of neuronal functional network. (A) Raster plot of ten neuronal spike trains. (B) Binned spike trains matrix obtained from raster plot in A, bin size=1sec. (C) Neuronal functional network constructed according to similarity between spike trains. The location of neurons is arbitrary.

Figure 1C shows an undirected graph composed of nodes and links. If this graph is divided into sub-graphs, we can then partition it with the graph clustering algorithm. In social network analysis, community structure partition method is a prominent graph clustering method. Researchers have carried out a great deal of research on this subject [13,14]. Many algorithms have been proposed to detect community structure, such as the earliest GN algorithm, fast Newman algorithm and hierarchical tree partitioning method. But before the division of community, a key problem to be solved is: How to determine the number of networks in the community? Newman et al. proposed a concept of modularity where the modularity Q is defined as:

$$Q = \frac{1}{2m} \sum_{ij} \left[a_{ij} - \frac{k_i . k_j}{2m} \right] \delta(c_i, c_j) \qquad (2)$$

The number of communities as pre-specified varies from 1 to N. The partitioning algorithm is used to divide the graph into several modules and calculate the corresponding values of Q. Last, we discover the maximum value of Q to determine the number of communities. However, the calculation Q's value is an NP-hard problem. Many traditional algorithms can only find the local optimal value of Q. Therefore, various community detection algorithms based on evolutionary algorithm have been proposed to optimize Q. These methods do not require specified number of modules and have performed well. These algorithms are used in the binary network, represented by an adjacency matrix, the value of connections only taken as 0 or 1. Our similarity matrix is a weighted graph. If we convert the weighted graph to the un-weighted graph, the parameter choices are difficult. Researchers are familiar with the problematic challenge of parameter choice. We carried out this study by dividing the network directly in a weighted graph.

2.1   Genetic Algorithm

Encoding and Decoding

Genetic algorithm is a widely applied global searching optimization algorithm, as proposed by [15]. To use a genetic algorithm on dividing the network community, first we need to encode the structures of Figure 1C into different genotypes of genetic algorithm.

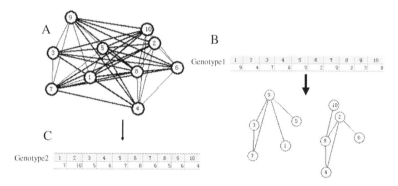

Fig. 2. Genetic algorithm operation. (A) Neuronal network, as shown in Figure 1C. (B) Generated genotype1 based on the encoding strategy proposed in this paper. The graph-based structure of the genotype is created. Ten neurons are automatically divided into two groups and the number of patterns is determined to be two. (C) Another genotype.

Since neuronal similarity matrix is a weighted matrix, we encode the genotype according to the node minimum similarity principle. Here, we prohibited node back-searching. For example, if nodes 1 and 9 meet minimum similarity, then there is a link between nodes 1 and 9. At the same time, if the minimum similarity node of node 9 is 1, then node 1 is the starting point of node 9, and it thus cannot be the ending point. It must to find another minimum similarity node, such as node 3, avoiding a back search. After all nodes have been searched, we get genotype1. It consists of ten nodes numbered in sequence. In the decoding step, we construct the connections between nodes according to genotype 1. If the value in the 1st position is 9, then we add a link between nodes 1 and 9. Ultimately, links form a connection structure shown in Figure 2B. It can be inferred that Figure 2B contains two modules. Identification of the number of modules in the community structure, without need to specify in advance, is the greatest advantage of this approach.

Crossover

The crossover operation produces new generation of genotypes, which can then generate new network community structures. Here, based on the two genotypes, we randomly select two positions and change their values. Two new chromosomes are then produced (Fig. 3A). New network structures can be generated according to new genotypes. We can compare the differences with the network shown in Figure 2B.

Mutation

In the genetic algorithm, the goal of mutation operation is to prevent genomic prematurity and early search stops. Here, we use two different mutation strategies. First, we choose a random position 'i,' to be mutated. The minimum similarity node j for node i is then searched. The value of position i is changed to j. Because $s_{xy}(i,j)$ is the minimum, this means that in the clustering solution nodes i and j will be in the same module. In the second approach, we choose a random position "i,' and change its value to k, ($1=<k<=N$), and can be sure new genotypes will result.

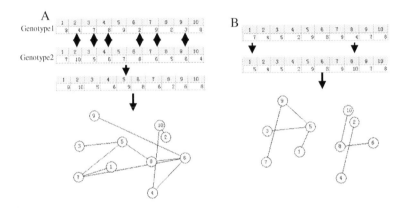

Fig. 3. (A) Crossover of two individuals and their graph-based representation. (B) Mutation operation and a new genotype generated.

In the iteration process of genetic algorithms, there must be a fitness function to evaluate genotypes. Here, we also used the modularity function to restrain the genetic algorithm. At the decoding process, a different number k of components {s1, s2, s3... ., sk} is provided. When calculating the value of Q, the larger Q's value is, the better the community structure obtained. The number of modules corresponding to the maximum value of Q is the optimum value in the communities. In this paper, we use the global optimization ability of genetic algorithms to find the maximum value of Q.

3   Experimental Results

In this section we study the effectiveness of our approach on several spike trains data sets.

3.1   Synthetic Spike Train Data

Firstly, we tested this method on a series of surrogate data sets. These data sets were artificially generated. The firing patterns were predetermined. The total comprised 4 data sets. We created the first data set. The others are created by Fellous et al. (2004) (available from http://cnl.salk.edu/fellous/data/JN2004data/data.html).

3.2   Multi-Electrode Recording Spike Train Data

Subsequently, we applied this method to the multi-electrode recorded neuronal spike trains. The spike trains were recorded from the hippocampus in a rat which performed the U-maze behavioral task. The recording process was executed by the Institute of Brain Functional Genomics of East China Normal University. This behavioral experiment and spike signal recording processes have been described previously [16,17]. The set of data set we used contained 24 neurons and a total of 75 trials. We selected the initial 10 trials to analyze. Each trial contains all 24 neurons.

3.3   Results

Table 1. The modularity function Q and iterative number of the four data sets

	This paper		fastnewman		GA-random	
	Q value	Iter num	Q value	Iter num	Q value	Iter num
Data set 1	0.6135	1	0.6087	N/A	0.3482	28
Data set 2	0.1810	1	0.1810	N/A	0.1461	5
Data set 3	0.2172	5	0.1196	N/A	0.1775	20
Data set 4	0.2477	2	0.2179	N/A	0.1712	24

It can be seen from table 1, the modularity function Q obtained in this paper is generally larger than that from the other two methods. Results show that the partitioning obtained in this paper indicates strong community structure and the corresponding firing patterns are more accurate (Fig. 4). Compared with the genotypes based on random encoding strategy, using the encoding proposed in this paper needs few iterations to get the larger value of modularity function Q. 'N/A' represents that this method does not need iterative computation. We illustrate the raster plot of firing patterns for the first two data sets. As we can see from Figure 4, the method proposed in this paper does depend on prior knowledge of pattern numbers. According to the self-similarity between the spike trains, spike trains can be divided into two or three different types of firing patterns, which the human eye cannot distinguish.

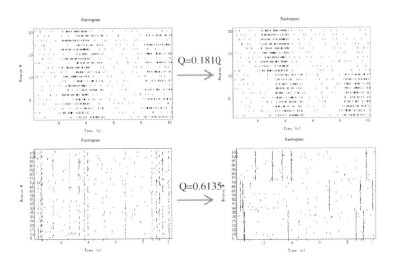

Fig. 4. On the left spike trains are mixed. On the right the raster plot sorted by group membership, showing a meaningful firing patterns.

In addition, we applied this method to the multi-electrode recorded spike trains. Since these data sets are original data, we cannot know how many firing patterns or what kinds of patterns are contained in the data sets. We can only compare with values of Q of different methods to analyze the performance of the methods.

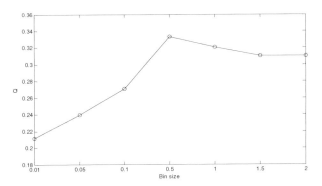

Fig. 5. Comparison of different value of Q of ten trials

Figure 5 shows the values of Q from ten trials. The method proposed in this paper obtained the largest values of Q in several trials. Yet, the values of Q from random genotypes encoding may be larger than those in this paper for some trials in which community structures is not obvious, This is due to neuronal functional networks constructed from spike trains in which strong community structure may not exist, and corresponding value of Q is small.

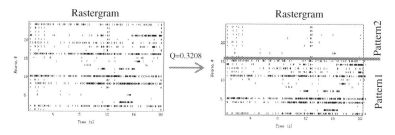

Fig. 6. Detection results of first trial from multi-electrode recording spike trains

Here, we illustrate the raster plot of first trial process. According to the method proposed in this paper, the maximum value of modularity function Q equal 0.3208 and corresponding number of modules equals two. Then we obtain two groups of firing patterns, shown in Figure 6. Because we cannot know prior patterns in the original data set, evaluating the pattern search method is a problem.

We use short time windows to bin the spike trains into vectors. Different bin sizes may affect the number of spikes in each bin, resulting in different values for modularity function Q. For our first trial on multi-electrode recorded spike trains, we set different bin sizes respectively and compared the corresponding value of Q.

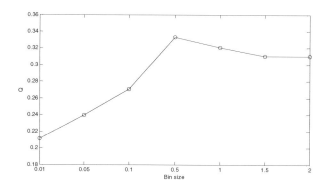

Fig. 7. Modularity function Q of different bin size

Figure 7 shows that the selection of time windows has an influence on calculating modularity function Q, resulting in different values of Q. Without a standard method, selection of Q's value is difficult.

4   Conclusions

We present a new search method for firing patterns in neuronal spike trains, based on genetic algorithms for detecting communities. To our knowledge, this is the first study using genetic optimal partitioning method to find spike train modules. Compared to previous detection methods described in this paper, the values of Q obtained by our approach are larger, indicating that partitioning results are more reasonable. However, the modularity function Q itself cannot identify some modules smaller than a certain scale. Some researchers have proposed new evaluation functions, such as density, community coefficient and so on [18]. In future research, we will target application of evolutionary algorithms to optimize these evaluation functions and compare them against the results of modularity function Q.

Acknowledgments. We are grateful to Pro. Longnian Ling (East China Normal University) for providing us with multiple electrodes neural spike trains data. This study was supported by the National Natural Science Foundation of China (Project No. 61375122), China Postdoctoral Science Foundation (Project No. 2014M551324), Scientific Research Foundation for Advanced Talents of Jiangsu University (Project No. 14JDG040), The natural science foundation of Jiangsu Province(BK20130529), Research Fund for the Doctoral Program of Higher Education of China (20113227110010), Science and technology project of Zhenjiang City(SH20140110).

References

1. Brown, E.N., Kass, R.E., Mitra, P.P.: Multiple Neural Spike Train Data Analysis: State-of-the-art and Future Challenges. Nature Neuroscience **7**(5), 456–461 (2004)
2. Lopes-dos-Santos, V., Conde-Ocazionez, S., Nicolelis, M.A.L., et al.: Neuronal Assembly Detection and Cell Membership Specification by Principal Component Analysis. PLoS One **6**(6), e20996 (2011)
3. Hu, L., Hui, W.: Discovering the Multi-Neuronal Firing Patterns Based on a New Binless Spike Trains Measure. In: International Symposium on Neural Networks, July 4-6 (2013)
4. Fellous, J.M., Tiesinga, P.H.E., Thomas, P.J., et al.: Discovering Spike Patterns in Neuronal Responses. The Journal of Neuroscience **24**(12), 2989–3001 (2004)
5. Paiva, A.R.C., Rao, S., Park, I., et al.: Spectral Clustering of Synchronous Spike Trains. In: 2007 International Joint Conference on Neural Networks, pp. 1831–1835. IEEE Press, New York (2007)
6. Humphries, M.D.: Spike-train Communities: Finding Groups of Similar Spike Trains. The Journal of Neuroscience **31**(6), 2321–2336 (2011)
7. Newman, M.E.J.: Fast Algorithm for Detecting Community Structure in Networks. Physical Review E **69**(6), 066133 (2004)
8. Pizzuti, C.: GA-Net: A Genetic Algorithm for Community Detection in Social Networks. In: Rudolph, G., Jansen, T., Lucas, S., Poloni, C., Beume, N. (eds.) PPSN 2008. LNCS, vol. 5199, pp. 1081–1090. Springer, Heidelberg (2008)
9. Tasgin, M., Herdagdelen, A., Bingol, H.: Community Detection in Complex Networks using Genetic Algorithms. arXiv preprint arXiv:0711.0491 (2007)
10. Duch, J., Arenas, A.: Community Detection in Complex Networks using Extremal Optimization. Physical Review E **72**(2), 027104 (2005)
11. Mehboob, Z., Panzeri, S., Diamond, M.E., et al.: Topological Clustering of Synchronous Spike Trains. In: 2008 IEEE International Joint Conference on Neural Networks, pp. 3889–3894. IEEE Press, New York (2008)
12. Dauwels, J., Vialatte, F., Weber, T., Cichocki, A.: On Similarity Measures for Spike Trains. In: Köppen, M., Kasabov, N., Coghill, G. (eds.) ICONIP 2008, Part I. LNCS, vol. 5506, pp. 177–185. Springer, Heidelberg (2009)

13. Newman, M.E.J.: Detecting Community Structure in Networks. The European Physical Journal B-Condensed Matter and Complex Systems **38**(2), 321–330 (2004)
14. Girvan, M., Newman, M.E.J.: Community Structure in Social and Biological Networks. Proceedings of the National Academy of Sciences **99**(12), 7821–7826 (2002)
15. Holland, J.H.: Adaptation in Natural and Artificial Systems. MIT Press, Cambridge (1992). ISBN 0262581116
16. Kuang, H., Lin, L., Tsien, J.Z.: Temporal Dynamics of Distinct CA1 Cell Populations During Unconscious State Induced by Ketamine. PloS One **5**(12), e15209 (2010)
17. Wang, D.V., Wang, F., Liu, J., et al.: Neurons in the Amygdala with Response-Selectivity for Anxiety in Two Ethologically Based Tests. PloS One **6**(4), e18739 (2011)
18. Hu, L., Hui, W.: Detecting Community Structure in Networks Based on Community Coefficients. Physica A: Statistical Mechanics and its Applications **391**, 6156–6164 (2012)

Fast and Effective Image Segmentation via Superpixels and Adaptive Thresholding

Yunsheng Jiang and Jinwen Ma[✉]

Department of Information Science, School of Mathematical Sciences and LMAM,
Peking University, Beijing 100871, China
jwma@math.pku.edu.cn

Abstract. This paper proposes a fast and effective image segmentation algorithm by firstly clustering image pixels into a small number of *superpixels* and then merging these superpixels whose distances are below an *adaptive threshold* together to get the final segmented fields. The adoption of superpixels dramatically decreases the computation cost, while the adaptive thresholding aims to select a reasonable segmentation from a set of possible segmentations with hierarchical scales. The adaptive threshold can be calculated with a fast sequential procedure. Experiments on Berkeley Segmentation Data Set (BSDS500) demonstrate that our proposed algorithm is competitive to other state-of-the-art segmentation methods. Moreover, this segmentation framework can be improved to excellent performance by using more elaborate superpixel algorithms.

Keywords: Image segmentation · Superpixels · Adaptive thresholding

1 Introduction

Image segmentation is a fundamental issue in computer vision, and it often serves as the first step of some visual tasks (*e.g.*, object recognition). Image segmentation tries to partition an image into several meaningful and homogeneous regions, with each region corresponding to a semantic object or background.

Clustering analysis is a popular method for image segmentation. The simplest way is extracting a feature vector for each pixel and using the classic k-means algorithm [1] to cluster these pixels into image segments (in the feature space). Mean Shift (MS) algorithm [2] was originally proposed for the mode detection of density functions, and it has successfully applied to clustering analysis, image segmentation and visual detection/tracking since [3].

On the other hand, Normalized Cuts (N-Cuts) algorithm [4] constructs a graph for each image, and converts the segmentation task to a graph partitioning problem, which can further be reduced to a generalized eigenvalue problem (*i.e.*, spectral clustering). Besides, Felz-Hutt algorithm [5] also measures the evidence of objects' boundaries by a graph-based representation of the image, and then obtains the segmented objects by greedy decisions.

Recently, Arbelaez et al. [6] proposed the powerful "gpb-owt-ucm" algorithm, which got the art-of-the-state performance on BSDS500. This algorithm utilizes

© Springer International Publishing Switzerland 2014
Z. Zeng et al. (Eds.): ISNN 2014, LNCS 8866, pp. 568–575, 2014.
DOI: 10.1007/978-3-319-12436-0_63

the Oriented Watershed Transform (owt) to construct a set of initial regions from the Global Probability of Boundary (gpb) signals. The output of gpb-owt-ucm is a set of segmentations with hierarchical scales, which are represented by the Ultrametric Contour Map (ucm).

In this paper we propose a simple but effective image segmentation algorithm. It clusters image pixels into superpixlels, computes the distances between different superpixels, and merges the superpixels whose distances are smaller than an adaptive threshold. Its competitive performance is demonstrated by the experiments on BSDS500. Beyond that, we also modify this framework by applying a more elaborate superpixel algorithm, which gains more excellent segmentation performance at the cost of relatively larger computation time.

The rest of this paper is organized as follows. Section 2 presents the overall framework of our algorithm. The experimental results on BSDS500 are contained in Section 3 for both the original *fast* algorithm and the modified *fine* version. Finally, a brief conclusion is made in Section 4.

2   Algorithm Framework

The overall framework of our image segmentation algorithm is shown in Figure 1. For an input image, we firstly cluster the image pixels into **superpixels**, then define a **distance function** to measure the dissimilarities between different superpixels, and finally merge the superpixels whose distances are smaller than an **adaptive threshold** to get the image segments.

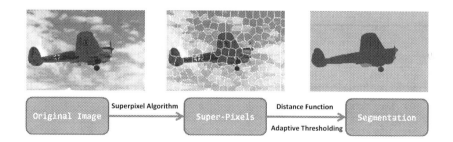

Fig. 1. The framework of our image segmentation algorithm

2.1   Superpixels Generation

The huge number of pixels in an image make it time-consuming for the application of some elaborate pixel-level features. The term *superpixel* [7] represents a set of pixels which have high similar properties (*e.g.*, position, color, texture, etc). If we cluster the image pixels into superpixels, and extract features on superpixel-level instead of pixel-level, then the number of features to be extracted can be dramatically reduced, which will save a lot of computation time.

Superpixels can be obtained by some typical clustering algorithms, like the k-means or Means-Shift algorithm. However, we choose to generate superpixels with the popular Simple Linear Iterative Clustering (SLIC) algorithm [8], due to its linear computation complexity and competitive performance. The parameters in the SLIC algorithm are set as $S = 30 \times 30$ (the normal size of superpixels) and $\lambda = 15$ (the spatial regularizer), which are selected by balancing the number of superpixels, clustering appearance and spatial regularization.

It should be noted that, for a given segmentation algorithm, if we "tune" some parameters to make it output over-segmentation (*e.g.*, reduce the threshold for boundary/contour map), the resultant *over-segments* can also be seen as superpixels. Usually, this kind of superpixels is more effective than the simple SLIC superpixels, but also more time-consuming, which will be demonstrated by our following experiments.

2.2   Distance Function

Generally, the distance between two superpixels depends on their positions (*e.g.*, center coordinates), region features (*e.g.*, mean colors) and boundary intensity (*e.g.*, gradient). As we only merge adjacent superpixels in the next mergence step, we will not bring the position information to the calculation of distance directly, but only *set the distance to infinity* if two superpixels are not adjacent.

When two superpixels A and B are **adjacent**, their distance is composed by the difference of region features and the intensity of their boundary. For computational simplicity, we use the *mean LUV colors* in the A and B as the region features, denoted as $L(A)$ and $L(B)$, and use the *mean gradient amplitude* on the border line of A and B as the boundary feature, denoted as $G(A, B)$. Therefore, the distance between A and B can be calculated as follows:

$$D(A, B) = ||L(A) - L(B)|| + \alpha \cdot |G(A, B)| , \qquad (1)$$

where α is the regularizer for boundary intensity. Clearly, the features $L(A), L(B)$ and $G(A, B)$ should be normalized to interval $[0, 1]$ at first.

2.3   Adaptive Thresholding

After we compute the distances between different superpixels, we can obtain the final image segmentation by merging similar superpixels whose distances are below a particular threshold. The determination of this threshold is an important issue. Some algorithms (like gpb-owt-ucm) avoid this problem by using a set of thresholds and thus output a set of hierarchical segmentations. However, this is not convenient in practical application. Here, we give a fast algorithm to select a proper threshold adaptively for different images as follows.

Firstly, we need a criterion to *estimate* (not evaluate) the performance for a given segmentation. For the simplicity of constructing a fast algorithm, we use the mean WGSSEs (Within-Group-Sum-of-Squared-Error) of LUV color values plus a regularization item of #*groups* to estimate the segmentation performance:

Input   : The distance matrix of superpixels: $D = [D(i,j)]_{K_0 \times K_0}$,
                The total number of pixels in the image: N_{total},
                The regularizer for segment nunmber: β
Output: The adaptive threshold for segmentation: T_{best}

1 Count $\{S_1(i), S_2(i), N(i)\}_{i=1}^{K_0}$ for the current superpixel segmentation;

2 Compute the sum of WGSSEs: $S_w = \sum_{i=1}^{K_0} [S_2(i) - \frac{S_1^2(i)}{N(i)}]$;

3 Set $J_{best} = \frac{1}{N_{total}} S_w + \beta \cdot K_0$ and $T_{best} = 0$;

4 **for** $K \leftarrow (K_0 - 1)$ **to** 1 **do**

5   Find out (p, q) such that $p \neq q$, and $D(p, q)$ is the minimum value of D;

6   Let $S_w = S_w - [S_2(p) - \frac{S_1^2(p)}{N(p)}] - [S_2(q) - \frac{S_1^2(q)}{N(q)}]$;

7      $S_1(p) \leftarrow S_1(p) + S_1(q), \quad S_1(q) = 0$;   `// merge superpixel q to p`

8      $S_2(p) \leftarrow S_2(p) + S_2(q), \quad S_2(q) = 0$;

9      $N(p) \leftarrow N(p) + N(q), \quad N(q) = 0$;

10   Update $S_w = S_w + [S_2(p) - \frac{S_1^2(p)}{N(p)}]$;

11   Compute $J_{tmp} = \frac{1}{N_{total}} S_w + \beta \cdot K$;

12   **if** $J_{tmp} < J_{best}$ **then**   `// record the optimal threshold`

13         $J_{best} = J_{tmp}$;

14         $T_{best} = D(p, q)$;

15   **end**

16   **for** $i \leftarrow 1$ **to** K_0 **do**   `// update the distance matrix`

17   Set $D(p, i) = D(i, p) = \min \{D(p, i), D(q, i)\}$;

18   Set $D(q, i) = D(i, q) = \infty$;

19   **end**

20 **end**

21 **return** *The final* T_{best}

Algorithm 1. The fast sequential procedure for adaptive threshold

$$J(seg) = \frac{1}{N_{total}} \sum_{A \in seg} S_w(A) + \beta \cdot K, \tag{2}$$

where seg denotes a image segmentation, N_{total} is the total number of pixels in the image, A is a segment in seg, K is the number of segments in seg, $S_w(A)$ is the WGSSE of segment A, and β is a regularizer. Note that small $J(seg)$ corresponds to good segmentation performance. If we denote $S_1(A)$ as the sum of pixels' color values in superpixel A, $S_2(A)$ as the sum of *squared* color values, and $N(A)$ as the number of pixels in A, then $S_w(A)$ can be obtained by

$$S_w(A) = \sum_{x_i \in A} (x_i - \bar{x})^2 = \left(\sum x_i^2\right) - \frac{(\sum x_i)^2}{N(A)} = S_2(A) - \frac{S_1^2(A)}{N(A)}. \tag{3}$$

When two superpixels A and B are merged, we update A with

$$\phi(A) \leftarrow \phi(A) + \phi(B), \qquad \phi \in \{S_1, S_2, N\} \tag{4}$$

and delete B to maintain S_1, S_2, N.

Therefore, if we merge superpixels sequentially, for example, form small distances to large distances, then we can calculate $J(seg)$ directly by only storing S_1, S_2, N and updating them sequentially. The detailed procedure of this sequential algorithm is shown in Algorithm 1. Note that, due to the definition of distances and the updating of distance matrix D on *line-18*, most elements of D are infinities. Therefore, the minimization step on *line-5* can be effectively solved by a sparse representation of D, and this sequential procedure is very fast and efficient.

3   Experimental Results

We test our proposed algorithm on the popular **Berkeley Segmentation Data Set (BSDS500)** [6], which covers a variety of images with complex scenarios and has manually produced ground-truth segmentation for each image. As for the evaluation of segmentation performance, we use three different metrics, *i.e.*, the Segmentation Covering (SC), Probability Rand Index (PRI) and Variation of Information (VI) [6]. These three metrics have different characters, for example, the PRI tends to over-segmentation, while the SC and VI encourage under-segmentation. Therefore, an overall consideration of these three metrics is necessary and reasonable. Note that, good segmentation corresponds to high SC value, high PRI value, but *low* VI value.

In this section we firstly show the results of our *fast* segmentation algorithm based on slic-superpixels, and then give the experiments for the modified *fine* version based on ucm-superpixels.

3.1   Results for the Fast SLIC-Based Algorithm

In this subsection we use the simple and fast SLIC algorithm to generate the superpixels. The hyper-parameters α and β are optimized on the BSDS500 trainset. For example, α is set as $\alpha = 1$. As for β, it is set as $\beta = 1.8$ for SC metric, $\beta = 1.0$ for PRI metric and $\beta = 5.1$ for VI metric. These hyper-parameters are fixed on the whole testset once obtained.

Some segmentation examples from BSDS500 testset are shown in Figure 2. From this figure we can see that, the slic-based superpixels are just small image patches with approximatively fixed size and high within-similarities. The final segmentation images show that most superpixels are correctly merged, *i.e.*, the superpixels from the same semantic object are merged together, while the superpixels from different objects are keep intact. This indicates the effectiveness of our distance function and adaptive thresholding.

Table 1 lists the mean scores of SC, PRI and VI on the whole BSDS500 testset, as well as comparisons with some other competitive segmentation algorithms. The row "*Human*" denotes the segmentation scores gained by human beings. From this table we can see that, the performance of our fast algorithm (denoted as "SP-AdaThresh (slic-based)") is very competitive. In fact, only the "gpb-owt-ucm" algorithm gains better performance than our slic-based algorithm,

(a) $SC = 0.81002$, $PRI = 0.96086$, $VI = 0.86197$

(b) $SC = 0.70322$, $PRI = 0.97182$, $VI = 1.4959$

Fig. 2. Segmentation examples for our **fast slic-based algorithm**. Left column: original images. Middle column: superpixel images. Right column: segmentation images.

but its segmentation time is much longer than ours (about 4 **minutes** V.S. our 1 **second**). Besides, "gpb-owt-ucm" algorithm needs much more memory space (about 8GB). Therefore, by comprehensive consideration of performance, time and hardware, our "SP-AdaThresh(slic-based)" algorithm is competitive, and it suitable to serve as the first step of some visual recognition tasks.

Note that, the segmentation time is obtained in our personal computer, with 3.2GHz 4-core Intel Core CPU, 8G Memory, Linux-3.5 OS, Matlab R2012b.

3.2   Results for the Fine UCM-Based Algorithm

By analyzing the results of the powerful "gpb-owt-ucm" algorithm, we find that the Ultrametric Contour Map (UCM) is very effective though sophisticated. These contour profiles can indicate the true boundaries between different objects while filter out the noises and interference edges. If we threshold the "ucm" with a small value, the resultant over-segmentation can serve as superpixels. These ucm-based superpixels are much better than the simple slic-based superpixels. Therefore, when the requirements for segmentation time or memory space are loose, we can replace the slic-based superpixels with these powful ucm-based superpixels, and follow the rest of the framework discussed above, then we can obtain the fine ucm-based algorithm. In this paper the low threshold for ucm is set as $T_{ucm} = 0.1$, which is also optimized on the BSDS500 trainset.

Some segmentation examples of this ucm-based algorithm are shown in Figure 3. From this figure we can see that, compared to slic-based superpixels,

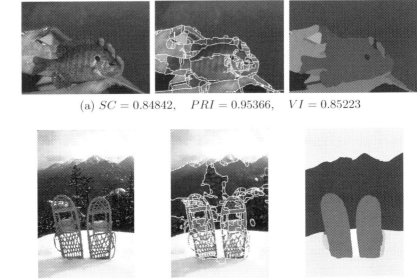

(a) $SC = 0.84842,$ $PRI = 0.95366,$ $VI = 0.85223$

(b) $SC = 0.77393,$ $PRI = 0.92299,$ $VI = 0.85552$

Fig. 3. Segmentation examples for our **fine UCM-based** algorithm. Left column: original images. Middle column: superpixel images. Right column: segmentation images.

Table 1. Segmentation Results on BSDS500 testset

Method	SC	PRI	VI
Human	*0.72*	*0.88*	*1.17*
Quad-Tree	0.32	0.73	2.46
N-Cuts [9]	0.45	0.78	2.23
Canny-owt-ucm [6]	0.49	0.79	2.19
Felz-Hutt [5]	0.52	0.80	2.21
Mean Shift [3]	0.54	0.79	1.85
gPb-owt-ucm [6]	0.59	0.83	1.69
SP-AdaThresh (slic-based)	0.55	0.80	1.90
SP-AdaThresh (ucm-based)	0.61	0.83	1.65

the ucm-based superpixels are more "close" to the truth: They are very coarse on smooth regions, but very fine-grained on regions with abrupt changes. In other words, these ucm-based superpixels focus on the regions with abundant contours and can capture more boundary information. The good segmentation results also indicate the effectiveness of these ucm-based superpixels.

The mean scores of this fine ucm-based algorithm are also listed in Table 1 as "SP-AdaThresh(ucm-based)". This ucm-based algorithm is even slightly better than the "gpb-owt-ucm", and its segmentation time is almost the same with "gpb-owt-ucm" due to the fast sequential procedure for adaptive threshold. Note that our algorithm outputs only one segmentation for each image, while "gpb-owt-ucm" outputs a set of hierarchical segmentations. Hence, our algorithm is more competitive and more suitable for practical application.

4   Conclusions

We have established an effective framework for fast image segmentation. It consists of clustering pixels into superpixels and merging superpixels whose distances are below an adaptive threshold. The experiments on BSDS500 demonstrate its competitive performance for both the fast slic-based superpixels and the fine ucm-based superpixels. Currently, the ucm-based algorithm is still time-consuming and our future work will focus on speeding up this algorithm without decreasing the segmentation performance.

Acknowledgments. This work was supported by the Natural Science Foundation of China for Grant 61171138.

References

1. MacQueen, J.: Some Methods for Classification and Analysis of Multivariate Observations. In: Berkeley Symposium on Mathematical Statistics and Probability, California, USA, vol. 1, pp. 281–297 (1967)
2. Fukunaga, K., Hostetler, L.: The Estimation of the Gradient of a Density Function, with Applications in Pattern Recognition. IEEE Transactions on Information Theory **21**(1), 32–40 (1975)
3. Comaniciu, D., Meer, P.: Mean Shift: A Robust Approach toward Feature Space Analysis. Pattern Analysis and Machine Intelligence **24**(5), 603–619 (2002)
4. Shi, J., Malik, J.: Normalized Cuts and Image Segmentation. Pattern Analysis and Machine Intelligence **22**(8), 888–905 (2000)
5. Felzenszwalb, P., Huttenlocher, D.: Efficient Graph-based Image Segmentation. International Journal of Computer Vision **59**(2), 167–181 (2004)
6. Arbelaez, P., Maire, M., Fowlkes, C., Malik, J.: Contour Detection and Hierarchical Image Segmentation. Pattern Analysis and Machine Intelligence **33**(5), 898–916 (2011)
7. Ren, X., Malik, J.: Learning A Classification Model for Segmentation. In: International Conference on Computer Vision, vol. 1, 10–17 (2003)
8. Achanta, R., Shaji, A., Smith, K., Lucchi, A., Fua, P., Susstrunk, S.: Slic Superpixels Compared to State-of-the-art Superpixel Methods. Pattern Analysis and Machine Intelligence **34**(11), 2274–2282 (2012)
9. Cour, T., Benezit, F., Shi, J.: Spectral Segmentation with Multiscale Graph Decomposition. In: Computer Vision and Pattern Recognition, vol. 2, pp. 1124–1131 (2005)

Diagonal Log-Normal Generalized RBF Neural Network for Stock Price Prediction

Wenli Zheng and Jinwen Ma[✉]

Department of Information Science, School of Mathematical Sciences and LMAM,
Peking University, Beijing 100871, China
jwma@math.pku.edu.cn

Abstract. Stock price prediction is one of the most important topics in financial engineering. In this paper, for stock closing price prediction, we propose a diagonal log-normal generalized RBF neural network in which the diagonal log-normal density functions serve as the RBFs. Specifically, it utilizes the dynamic split-and-merge EM algorithm to select the number of hidden units (or RBFs) as well as the initial values of the parameters, and implements a synchronous LMS learning algorithm for parameter learning. It is demonstrated by the experiments that the diagonal log-normal generalized RBF neural network has a competitive performance on stock closing price prediction.

Keywords: RBF neural network · Log-normal distributions · EM algorithm · Stock price prediction

1 Introduction

Stock is one major security for the companies to raise money. The fluctuation of stock prices influences investors' profits and thus successful stock price prediction or forecasting has been a hot topic in financial engineering. Actually, stock price prediction is the act of trying to precisely forecast future prices based on previous stock price data. However the stock price system is a complicated nonlinear dynamic system [1], and it is affected by internal and external factors, such as bank interest rates, national polices, the price index, performance of companies and investors' psychological reaction [2,3]. This uncertainty in the stock price system imposes great challenges on the stock price prediction task.

The stock price data have a natural temporal ordering, thus time series analysis is a conventional method to predict stock prices. The classical techniques for time series forecasting include moving average, exponential smoothing and decomposition methods. However, those traditional methods cannot always have good performance on time series forecasting, therefore there emerge many new methods for time series forecasting, including generalized autoregressive conditional heteroskedasticity (GARCH, Bollerslev(1986)) and artificial neural networks (ANNs) [2,4]. Due to their nonlinearity, adaptivity and generalization, ANNs are widely applied on time series forecasting.

© Springer International Publishing Switzerland 2014
Z. Zeng et al. (Eds.): ISNN 2014, LNCS 8866, pp. 576–583, 2014.
DOI: 10.1007/978-3-319-12436-0_64

Particularly, RBF neural networks are majorly used for the stock price prediction [5–7]. In fact, Radial basis function (RBF) neural network is a three-layer feed forward neural network, firstly proposed by Broomhead and Lowe [8]. Radial basis functions are a kind of special functions, with their responses monotonously decreasing with the distance from a center. In comparison with the other ANNs, the structure of RBF neural network is simple, and the speed of learning is fast [10]. Due to its special structure, RBF neural network has a good ability of local approximation. But certain disadvantages and limitations of RBF neural network were also found from its application on time series prediction. In order to improve the performance of RBF neural network, the generalized RBF neural network were proposed in [9,10]. It uses Gaussian distributions instead of Gaussian kernel functions. Furthermore, the diagonal generalized RBF neural network was proposed to improve the learning and generalization ability via the constraint of covariance matrices in a subspace [3,10]. Actually, the diagonal generalized RBF neural network is shown to have a better performance on nonlinear time series prediction [10].

In finance, log-normal distributions are often used to describe the stock prices. In fact, the Black-Scholes-Merton option pricing model assumes that stock prices subject to log-normal distributions. In addition, due to their skewness and long tails, log-normal distributions have the advantages to analyze stock prices [11,12]. So, if the radial basis functions become log-normal probability densities or distributions, the generalized RBF neural network can be more powerful for stock price prediction. Therefore, in this paper, we propose the diagonal log-normal generalized RBF neural network to predict stock prices, in which the radial basis functions are log-normal distributions with the covariance matrixes being diagonal.

The rest of this paper is organized as follow. In Section 2, we introduce log-normal distribution and stock price data. We propose the diagonal log-normal generalized RBF neural network and the LMS learning algorithm in Section 3. Experimental results are contained in Section 4, and Section 5 gives a brief conclusion.

2   Log-Normal Distribution and Stock Price Time Series

We begin with a brief introduction of log-normal distribution. Log-normal distributions are obtained by the derivation of gaussian probability density functions, but log-normal distributions are different from normal distributions. Firstly, random variables of log-normal distributions are positive, but for normal distributions it takes both negative and positive values. Secondly, log-normal distributions are skewed to right and have a long right tail, in contrast normal distributions are symmetrical at its maximum point.

For stock price time series, we draw the histograms of Shangzheng and Huaxia stock closing prices as examples to illustrate stock price time series characteristics, shown in Figure 1 (a) & (b), respectively. Shangzheng stock closing price data range from 2000 to 3000, and Huaxia stock closing price data are from 6 to 14.

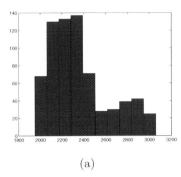 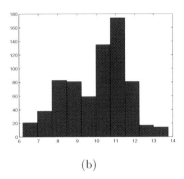

(a)   (b)

Fig. 1. The histograms of Shangzheng and Huaxia stock closing price data. (a). The histogram of Shangzheng stock closing price data. (b). The histogram of Huaxia stock closing price data.

They are both positive. The Shangzheng and Huaxia stock closing price data are asymmetric, and they are both skew to right. Regarding to these features, we assume that stock price data sets subject to the log-normal distributions.

3   Diagonal Log-Normal Generalized RBF Neural Network and LMS Learning Algorithm

3.1   Diagonal Log-Normal Generalized RBF Neural Network

The diagonal log-normal generalized RBF neural networks have three layers including a input layer, a hidden layer and a output layer. The hidden unit activation functions are log-normal distributions instead of gaussian distributions, and the mapping relationship between input and output takes the following form:

$$y_l = \sum_{j=1}^{m} w_{jl} R_j(x),$$ (1)

where input variables $x \in R^n$, outputs $y \in R^p$, the number of hidden units m, and w_{jl} is the weight between the jth hidden unit and the lth output unit.

$$R_j(x) = \frac{1}{(2\pi)^{\frac{n}{2}} |\Sigma_j|^{\frac{1}{2}} \prod_{i=1}^{n} x_i} \exp\{-\frac{1}{2}(\ln x - m_j)^T \Sigma^{-1}(\ln x - m_j)\},$$ (2)

where m_j and Σ_j are the mean vector and covariance matrix of jth radial function.

3.2   LMS Learning Algorithm

In order to learn parameters of the network, we use a synchronous least mean square error (LMS) learning algorithm to find parameters on training sets. Firstly the mean square error function of sample data takes the following form:

$$E = \frac{1}{2} \sum_{t=1}^{N} \sum_{l=1}^{p} [y_{tl} - \hat{y}_{tl}]^2 = \frac{1}{2} \sum_{t=1}^{N} \sum_{l=1}^{p} [y_{tl} - \sum_{j=1}^{m} w_{jl} R_j(x)]^2. \qquad (3)$$

Then we can get the partial derivatives of the error function E. The LMS learning criteria is that:

$$\Delta w_{jl} = \eta \sum_{t=1}^{N} (y_{tl} - \hat{y}_{tl}) R_j(x_t), \qquad (4)$$

$$\Delta m_j = \eta \sum_{t=1}^{N} \sum_{l=1}^{p} (y_{tl} - \hat{y}_{tl}) w_{jl} R_j(x_t) \Sigma^{-1} (\ln x_t - m_j). \qquad (5)$$

In order to improve the ability of generalization, we let the covariance matrix be a diagonal matrix, i.e $\Sigma_j = diag(\sigma_{j1}, ..., \sigma_{jn})$ where $\sigma_{ji} > 0$. The iterative formula is as following forms:

$$\Delta \sigma_{ji} = \frac{\eta}{2} \sum_{t=1}^{N} \sum_{l=1}^{p} (y_{tl} - \hat{y}_{tl}) w_{jl} R_j(x_t) [\frac{1}{\sigma_{ji}} - \frac{(\ln x_{ti} - m_{ji})^2}{\sigma_{ji}^2}]. \qquad (6)$$

We further discuss how to implement our proposed model. The number of hidden units has big effects on results of learning, and specifying the number of hidden units is not a easy work. In this paper we utilize the dynamic split-and-merge EM algorithm [13,14] to initialize the number of hidden units and mean vectors of radial basis functions. The initial covariance matrix Σ_j is set to be a unit matrix. For the other parameters in our model, a better initialization is necessary. We set the learning rate as following forms [3]:

$$\eta = \begin{cases} 1.05 * \eta & E^k < E^{k-1} \\ \frac{\eta}{10} & E^k > E^{k-1}, \end{cases} \qquad (7)$$

where the initial learning rate is set to be 10^{-3}. For the stop criterion: $|\triangle E| = |E^k - E^{k-1}| \leq \epsilon$, where the value of ϵ is set to be 10^{-8}. After the number of hidden units and parameters of the network are initialized, we implement the LMS algorithm to learn the parameters of radial basis functions and the weights between the hidden layer and the output layer.

4   Experimental Results

In this section, we utilize the diagonal log-normal generalized RBF neural networks to predict Shangzheng, Huaxia, Shangzheng380, and Shangye stocks closing prices (contained in Dazhijui software), with being compared with the diagonal generalized RBF neural networks [3]. The radial basis functions are log-normal distributions in our proposed model, but they are Gaussian distributions in the diagonal generalized RBF neural networks [3]. For convenience, our proposed model is referred to as DLN-RBFNN, while the diagonal generalized RBF neural network [3] is referred to as DG-RBFNN.

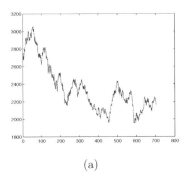
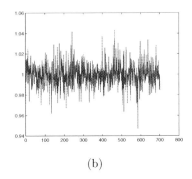

(a)   (b)

Fig. 2. The skecthes of Shangzheng stock closing price data before and after the preprocessing. (a). Shangzheng stock closing price data before the preprocessing (b). Shangzheng stock closing price data after the preprocessing.

4.1   The Prediction Model

Let stock price time series be $X = x_1, x_2, ...x_i, ...$, where $x_i > 0$ and $i = 1, 2, 3...L$. The stock price of future M days can be forecasted by previous D days' stock price. D is embedding dimensions. Generally, there are three ways of prediction, including one-step prediction, i.e $M = 1$, multi-step prediction, i.e $M > 1$ and roll prediction. Here, we just concentrate on one-step prediction.

We firstly generate 700 samples, and each sample is a $D + 1$ dimensions vector. The input layer has D units and the output layer has one unit. We then divide the 700 samples into two parts which contain 600 training samples and 100 test samples, respectively. Finally, we train the DLN-RBFNN on the training samples, and verify its prediction performance on the test samples.

4.2   Stock Price Data Preprocessing

The stock price time series are nonlinear and non-stationary data, and such non-stationary characteristics influence the prediction results. So, we can use the difference method to smooth the stock price data before training the network. The difference method is the following forms:

$$d_t = \exp\left(\ln x_{t+1} - \ln x_t\right), \tag{8}$$

In order to show the performance of stock price data preprocessing, we plot the sketches of Shangzheng stock closing price data before and after the data preprocessing, shown in Figure 2 (a) & (b), respectively. It can be observed that the preprocessed stock price data are more stationary.

4.3   Prediction Results and Analysis

Here, we use Shangzheng, Huaxia, Shangzheng380 and Shangye stock closing price data from 2011 to 2013 to generate training and test sets as described

Table 1. The RMSEs of the diagonal log-normal generalized RBF neural network (*refer to DLG-RBFNN*) on Shangzheng, Huaxia, ShangZheng380, Shangye stock closing price test data

Embedding Dimensions	Shangzheng	Huaxia	ShangZheng380	Shangye
2	21.924	0.3092	43.3000	42.4223
	±0.0205	±0.0083	±0.0149	±0.0576
3	21.9930	0.3080	43.4687	42.6446
	±0.0299	±0.0019	±0.0292	±0.1652
4	21.8795	0.3396	44.0850	43.3931
	±0.0144	±0.0069	±0.0579	±0.1564
5	21.8838	0.3708	44.1925	43.4333
	±0.0078	±0.0118	±0.0390	±0.0867

in Section 4.1. For each pair of training and test sets, we train the the DLN-RBFNN by using the LMS learning algorithm on the training data and then verify it for the prediction on the test data, and finally use the root mean square error (RMSE) and the correct rate (CR) of directional predictions to measure the predicted results of raw stock prices. The RMSE takes the following form:

$$RMSE = \sqrt{\frac{1}{N} \sum_{t=1}^{N} (\hat{x}_t - x_t)^2}. \tag{9}$$

Moreover, we adopt $CR = \frac{1}{N} \sum_{t=1}^{N} 1(\hat{dir}_t = dir_t)$ as defined in [15], where the dir takes the following form:

$$dir = \begin{cases} -1 & \Delta x_t < c_0 \\ 0 & c_0 < \Delta x_t < c_1 \\ 1 & \Delta x_t > c_1. \end{cases} \tag{10}$$

and $\Delta x_t = x_{t+1} - x_t$ is the difference between two sequent stock prices.

Table 1 and Table 3 list the RMSEs and CRs of the DLN-RBFNN on four stock closing price datasets, respectively. For comparison, Table 2 & Table 4 list the RMSEs and CRs of the DG-RBFNN on four stock closing price datasets, respectively. It can be seen from Tables 1 & 2 that the RMSEs of the DLN-RBFNN are all less and more stable than those of the DG-RBFNN on the four datasets. Moreover, The CRs of the LDG-RBFNN are all higher than those of the DG-RBFNN on the four datasets. Therefore, these experimental results demonstrate that our DLN-RBFNN as well as the LMS learning algorithm can be successfully applied to stock price prediction, and have a competitive performance on stock closing price prediction.

Table 2. The RMSEs of the diagonal generalized RBF neural network (*refer to DG-RBFNN*)on Shangzheng, Huaxia, ShangZheng380, Shangye stock closing price test data

Embedding Dimensions	Shangzheng	Huaxia	ShangZheng380	Shangye
2	23.2131	0.3371	46.2443	45.7020
	±0.2280	±0.0269	±3.0463	±1.4642
3	23.6499	0.3341	49.5989	48.6047
	±0.8178	±0.0020	±5.4353	±3.6431
4	24.1317	0.3582	51.8794	53.4201
	±0.9475	±0.0152	±5.3981	±2.9311
5	28.7286	0.7187	49.6149	60.3121
	±2.2037	±0.2224	±3.9069	±5.7065

Table 3. The CRs of the diagonal log-normal generalized RBF neural network (*refer to DLG-RBFNN*)on Shangzheng, Huaxia, ShangZheng380, Shangye stock closing price test data

Embedding Dimensions	Shangzheng	Huaxia	ShangZheng380	Shangye
2	0.720	0.740	0.600	0.530
3	0.720	0.750	0.600	0.530
4	0.720	0.740	0.600	0.530
5	0.720	0.730	0.600	0.530

Table 4. The CRs of the diagonal generalized RBF neural network (*refer to DG-RBFNN*)on Shangzheng, Huaxia, ShangZheng380, Shangye stock closing price test data

Embedding Dimensions	Shangzheng	Huaxia	ShangZheng380	Shangye
2	0.718	0.560	0.588	0.520
3	0.704	0.610	0.532	0.472
4	0.700	0.392	0.490	0.402
5	0.564	0.154	0.536	0.354

5   Conclusions

We have applied the diagonal log-normal generalized RBF neural network to stock price prediction, and compared it with the diagonal generalized RBF neural network which has gaussian distributions radial basis functions. It is demonstrated by the experiments on four datasets that the diagonal log-normal generalized RBF neural network has a competitive performance on stock closing price prediction.

Acknowledgments. This work was supported by the Natural Science Foundation of China for Grant 61171138.

References

1. Jiang, Y., Lin, Y.: The Application of RBF Neural Networks to Stock Price Forecasting. Mind and Computation 1(4), 413–419 (2007) (in Chinese)
2. Liu, H., Bai, Y.: Analysis of AR Model and Neural Network for Forecasting Stock Price. Mathmatics in Practice and Theory 41(4), 14–19 (2011) (in Chinese)
3. Liu, S., Ma, J.: The Application of Diagonal Generalized RBF Neural Network to Stock Price Prediction. China Sciencepaper Online (2014) (in Chinese)
4. Tsang, P.M., Kwok, P., Choy, S.O., et al.: Design and Implementation of NN5 for Hong Kong Stock Price Forcasting. Enginerring Applications of Artificial Intelligence 20(4), 453–461 (2007)
5. Fu, C., Fu, M., Que, J.: Prediction of Stock Price Base on Radial Basic Function Neural Networks. Technological Development of Enterprise 23(4), 14–15 (2004) (in Chinese)
6. Zheng, P., Ma, Y.: RBF Neural Network Based Sock Market Modeling and Forecasting. Journal of Tianjin University 33(4), 483–486 (2000) (in Chinese)
7. Lee, R.S.: iJADE Stock Advisor: An Intelligent Agent Based Stock Prediction System Using Hybrid RBF Recurrent Network. IEEE Transctions on Systems, Man, and Cybernetics, Part A: Systems and Humans **34**(3), 421–428 (2004)
8. Broomhead, D.S., Lowe, D.: Mutivariable Functional Interpolation and Adaptive Networks. Complex Systems **2**, 321–355 (1988)
9. Huang, K., Wang, L., Ma, J.: Efficient Training of RBF Networks Via the BYY Automated Model Selection Learning Algorithms. In: Liu, D., Fei, S., Hou, Z.-G., Zhang, H., Sun, C. (eds.) ISNN 2007, Part I. LNCS, vol. 4491, pp. 1183–1192. Springer, Heidelberg (2007)
10. Ma, J., Qing, C.: Diagonal Generalized RBF Neural Network and Nonelinear Time Series Prediction. Journal of Signal Processing 29(12), 1609–1614 (2013) (in Chinese)
11. Yu, Y.: The Application of the Lognormal Distribution in the Stock Price Models. Journal of Langfang Teachers College (Natural Science Edition) 12(5), 69–72 (2012) (in Chinese)
12. Long, S., Xiang, L.: Emirical Analysis of Stock Price's Lognormal Distribution. Journal of Huanggang Normal University 33(3), 9–11 (2013) (in Chinese)
13. Wang, L., Ma, J.: A Kurtosis and Skewness Based Criterion for Model Selection on Gaussian Mixture. In: The 2nd International Conference on Biomedical Engineering and Information, pp. 1–5 (2009)
14. Wang, L., Ma, J.: Efficient Training of RBF Networks via the Kurtosis and Skewness Minimization Learning Alogrithm. Journal of Theoretical and Applied Information Technology **48**(1), 496–504 (2013)
15. Kim, T.H., Mizen, P., Thanaset, A.: Forecasting Changes in UK Interest Rates. Journal of Forecasting **27**(1), 53–74 (2008)

Design of a Greedy V2G Coordinator Achieving Microgrid-Level Load Shift

Junghoon Lee and Gyung-Leen Park[✉]

Department of Computer Science and Statistics, Jeju National University,
Jeju-si, Republic of Korea
{jhlee,glpark}@jejunu.ac.kr

Abstract. This paper designs a microgrid-level V2G (Vehicle-to-Grid) coordinator capable of controlling the electricity flow from EV (Electric Vehicle) batteries to a grid, aiming at achieving temporal and spatial load shift in energy consumption. A bidding request specifies earliest and latest arrival times, amount to sell, and plug-in duration, while the controller creates an on-off schedule by which EVs are connected or disconnected to the grid on each time slot. After defining the search space made up of all feasible solutions, a greedy scheduler traverses the space to find an optimal slot allocation, which can enhance the demand-supply balance by means of minimizing the amount of surplus and lacking energy. The schedule also tells EV owners when to come and be connected to the microgrid. The performance evaluation result obtained from a prototype implementation shows that the proposed scheme reduces the energy lack by 49.0 % and the energy waste by 63.7 %, compared with the uncoordinated scheduling strategy.

Keywords: Vehicle-to-grid · Battery discharge control · Surplus and lacking energy · Exhaustive search · Demand-supply balance

1 Introduction

The smart grid technology makes the power network much intelligent and reliable, integrating sophisticated information and communication technologies [1]. It embraces a variety of grid entities chained from power generation to consumption. In the meantime, EVs (Electric Vehicles) allow even the transportation system to be a part of the smart grid, as they are powered by electricity stored in rechargeable batteries [2]. From the environmental aspect, they can avoid burning fossil fuels and take advantage of diverse energy sources including nuclear power and even renewable energies such as wind, sunlight, and the like, to charge their batteries [3]. It is true that there are inherent drawbacks of short driving range and long charging time [4]. However, low-cost overnight charging

This research was financially supported by the Ministry of Knowledge Economy (MKE), Korea Institute for Advancement of Technology (KIAT) through the Inter-ER Cooperation Projects.

© Springer International Publishing Switzerland 2014
Z. Zeng et al. (Eds.): ISNN 2014, LNCS 8866, pp. 584–593, 2014.
DOI: 10.1007/978-3-319-12436-0_65

with sufficient plug-in duration can give enough energy economically to most EVs for their daytime driving [5].

Not just restricted to their primary goal, namely, driving, EVs can be considered large-capacity batteries. EV batteries can send power back to the grid when they are connected to the grid, possibly through the outlet in the parking area [6]. Bidirectional chargers enable such V2G (Vehicle-to-Grid) technologies, while large-scale deployment of EVs in the transportation opens a possibility of integrating many intelligent control strategies for their charging and discharging [7]. With an appropriate control mechanism mainly implemented in EV aggregators, the grid-connected EVs can help maintain grid reliability, balance supply-and-demand, and bring many other benefits [10]. Moreover, EV batteries can be used to integrate renewable energy sources into the grid, managing their fluctuation problems, which usually lead to unpredictable excess and insufficient production [8].

Practically, electricity flow from EVs to the main grid is not easily allowed due to several reasons such as unverified reliability, price plan complexity, and security issues. However, microgrids can autonomously implement their own V2G strategies, charging and discharging EVs via regulation signals from grid-specific control logic. In addition, it is important that EVs can move. That is, they can achieve not just temporal but also spatial shift in power consumption. Even EVs are parked most of the time during a day, they are highly likely to move at least once for commute, shopping, and so on. If a microgrid, such as buildings, shopping malls, and factories, consuming a significant amount of energy, can consistently gather EVs during its peak time and get home-charged electricity from them, its power provisioning cost will be correspondingly cut down. The mobility of EVs brings better efficiency in consuming electricity cheaply charged overnight.

EVs are usually charged to their full capacity during the night. A fully charged EV can drive up to about 130 km in practice. However, the daily driving distance for most vehicles hardly exceeds this limit. Hence, the rest of the stored energy can be provided to other appliances or sold to a microgrid, which can be different from the one the EV has been charged at. During the peak time, the price rate gets higher, so the EV-stored energy can reduce the cost quite much for each microgrid. Here, if multiple EVs gather at a single grid during a specific time interval, not all electricity can be sold. On the contrary, if no EV is available, the microgrid must use the expensive grid-supplying electricity. So as to minimize the surplus and lacking energy, it is necessary to distribute EVs over the time interval according to the power demand on the microgrid, taking advantage of intelligent computer algorithms running on a robust reservation mechanism.

In this regard, this paper designs an EV-to-microgrid electricity flow control scheme which determines when EVs will be plugged-in to the grid and which EVs will send electricity to the grid on each time slot. Here, each EV which wants to sell its excessive energy to a microgrid is characterized by the earliest and latest times it can be plugged-in, the amount of energy to sell, and how long it can stay once plugged-in. Here, the modern communication mechanism essentially allows real-time two-way interaction between EVs and a microgrid. In addition, current computing capability can achieve fast calculation even if an EV changes its requirement. For the given parameter set, the potentially complex search space, consisting of all feasible solutions, is clarified, and an exhaustive search is designed to find an optimal schedule.

The rest of this paper is organized as follows: Section 2 reviews some related work. Section 3 describes the proposed scheme in detail. After Section 4 demonstrates and discusses the performance measurement results, Section 5 concludes this paper with a brief introduction of future work.

2   Related Work

With the capability of bidirectional energy exchange, EVs can participate in a frequency regulation service, when they are connected to the grid, be it a global grid or an autonomous microgrid such as buildings, shopping malls, and the like. [9] addresses a distributed V2G control scheme, mainly focusing on frequency regulation, aiming at efficiently suppressing system frequency fluctuation. Detecting a system-level frequency drop, the control mechanism reduces the load by shifting EV charging or even makes EVs inject power back to the grid. The authors consider two policies. The first part, called *Battery SoC (State of Charge) Holder* is designed for EVs having enough SoC to potentially give to the grid. The second one, called *Charging with Frequency Regulation*, works for EVs currently participating in scheduled charging.

Next, [10] categorizes GIV (Grid Interconnected Vehicle) mechanisms into centralized and decentralized ones. Decentralized schemes enforce each EV locally to decide the energy amount for V2G services. On the contrary, in centralized schemes, with all information collected from EVs, the central server controls the whole V2G-related actions in the system. In their scenario, as one of the most important elements in ancillary services, power regulation is used to keep the grid frequency and voltage within acceptable limits. Here, to participate in the regulation service, EVs submit bids, specifying the amount of its regulation power and price. Accepted EVs follow an on-line regulation signal from the grid. The performance evaluation reveals that while the centralized optimization achieves better regulation performance, it may suffer from long decision-making time for a large number of EVs.

V2G also makes it possible for the connected EVs to provide potential storage for the intermittent renewable energies, efficiently matching the difference between time-of-generation and time-of-load [5]. During the plug-in time, EVs supply a distributed spinning reserve according to the frequency deviation resulted from temporal demand-and-supply imbalance. In addition, [8] analyzes the effect of intelligent control, specifically, CHP (Combined Heat and Power). In this design, the grid-connected EV batteries are charged during low-demand hours and discharges guided by the real-time signal issued by the control system.

The analysis result shows that EVs can absorb excess power from renewable energies and return the power when necessary, while night time charging can benefit from abundant wind.

3   Electricity Flow Control

3.1   System Model

Figure 1 illustrates our system model. We mainly consider shopping malls, as they consume much energy during their operation hours, and EVs visit flexibly without tight restriction. For example, a shopper wants to visit a shopping mall between 1 PM to 5 PM and stay for about 2 hours. In addition, the shopping mall is assumed to be able to estimate its power demand using an appropriate forecast mechanism and tries to buy as much electricity as possible from EVs which will visit it. For the purchased energy, the shopping mall gives cash reward or gift cards. Moreover, the reward level can be different according to the time-of-purchase, that is, how much the mall saves energy cost by avoiding the expensive peak rate. If there are no EVs available, the mall must use just the electricity from its main power line. If we can arrange the visiting time of each EV, the mall can better benefit from EV-stored energy, saving more money.

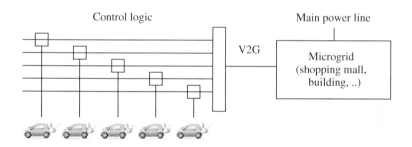

Fig. 1. System architecture

Shoppers, who want to sell electricity, submit their requests specifying their earliest and latest arrival times as well as the amount to sell. In addition, the plug-in duration, which corresponds to the stay time in the mall, is also given. The control switch can connect or disconnect each EV according to control logic. For each request, the server makes an electricity purchase plan and calculates its efficiency in terms of surplus and unused amount of energy in EV batteries. Here, like other scheduling policies [11], the time axis is divided into fixed-size time slots. This makes the scheduling problem more manageable and its execution time predictable. The EVs plugged-in to the grid are connected or disconnected to the grid just on the time slot boundary. For each time slot, whose demand is forecasted in advance, the scheduler selects the EVs to discharge to the microgrid out of available ones.

3.2   Scheduling Mechanism

Our task model can be explained with an example shown in Figure 2. In Figure 2(a), a bid request, R_i, consists of (E_i, L_i, A_i, D_i), where each element denotes earliest arrival time, latest arrival time, amount to sell, and plug-in duration, sequentially. With modern communication technologies, the interaction between servers and clients, even in the case they are running on mobile devices, is very common. In the example, the submitter of R_0 can arrive between time 0 to time 2 according to the decision of the scheduler. The time scale is aligned with the length of a time slot, for example, 0.5 hours. Once arrived, the driver stays in the mall for 4 units for shopping and other activities. Here, we assume that the amount of electricity flow from the EV to the grid is fixed and thus linear to the time length during which the EV is connected to the grid. After all, the amount to sell can be also represented by the number of time slots. Additionally, the amount to sell and the plug-in duration of R_2 is same. It implies that once the R_2 issuer is connected, the connection cannot be suspended.

	Earliest	Latest	Amount	Plug–in duration
R_0	0	2	3	4
R_1	1	5	3	6
R_2	2	4	4	4
R_3	0	3	3	8

(a) sell record

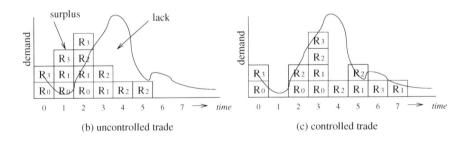

(b) uncontrolled trade   (c) controlled trade

Fig. 2. Task model and electricity flow coordination

Without any coordination, EVs will arrive anytime they want. Figure 2(b) illustrates the case each EV arrives at its earliest time and is connected to the grid as many slots as the amount to sell without being disconnected. In the figure, actual demand is plotted by a curve. For each slot, the number of connected EVs corresponds to the amount of available EV-stored electricity on the grid. If it is above the demand curve, the energy cannot be used but unavoidably wasted (surplus amount). On the contrary, if it is below the demand curve, the lacking part (lacking amount) is complemented only by the main power line. In Figure

2(b), we can find not a little surplus and lacking energy for the given demand curve. On the contrary, EVs arrive and are connected under the coordination of a controller, the discrepancy between the demand curve and the amount of available EV-stored energy will be much cut down as shown in Figure 2(c).

According to the schedule in Figure 2(c), R_1 can be plugged-in at time slot 3, not its earliest arrival time 1. In addition, it is connected to the grid at time slot 3 and disconnected at 4. Here, the R_1 issuer can arrive at slot 3 and leave at slot 7. Formally, speaking, for R_i, the EV can arrive during the interval from E_i to L_i, so the number of feasible arrival times is $L_i - E_i + 1$. It is plugged-in during D_i slots and connected to the grid for A_i slots. We assume that D_i is fixed irrespective of when the EV arrives. It cannot be always valid, but its dependency relation can be combined in the task model. After all, the number of feasible connection/disconnection schedules for R_i, namely, F_i is calculated as in Eq. (1).

$$F_i = (L_i - E_i + 1) \times \binom{D_i}{A_i} \tag{1}$$

As a consequence, the size of the whole search space is shown in Eq. (2).

$$\prod F_i \tag{2}$$

The scheduling problem is analogous to assigning resources to a task. As there is no restriction on the maximum number of allocated resources, the allocation of each resource is mutually independent. Our scheduler recursively builds an allocation table with the precomputation of all feasible combinations of $\binom{D_i}{A_i}$. Reaching a leaf, a complete allocation is evaluated primarily in terms of lacking amount. Actually, as the whole available amount of EV-stored energy is fixed for the given bid set, a schedule lacking smallest has the smallest surplus. Here, the price effect is not considered in this paper. However, many other constraints can be integrated into our cost function. For each evaluation, if a solution is better than the current best, the solution will replace it. Finally, after the whole search space traversal, the best solution remains in the current best.

4   Experiment Result

This section measures the performance of the proposed scheme through a prototype implementation. Main performance metrics include lacking amount, surplus amount, meet ratio, and consumption ratio. Lacking amount and surplus amount are already explained in the previous section and they give an insight on energy efficiency, indicating how close a schedule can make the amount of purchased energy to the demand curve. They are measured on request basis. In addition, the meet ratio denotes how much the demand is satisfied from the side of a shopping mall. Next, the consumption ratio measures how much EV-stored energy is consumed on the grid. Those metrics are measured according to not only the number of EVs which are willing to sell energy but also power demand from the microgrid side.

In our experiment setting, the time slot is set to 0.5 hours, while the scheduling window size is set to 14 slots. Actually, if the number of slots in the scheduling window increases, the computation time can explode, especially when the number of EVs increases. For less than or equal to 14 slots, the execution time is maintained within a reasonable range. R_i, E_i, L_i, A_i, and D_i are selected randomly within 14 slots, with the restriction that L_i is larger than E_i and D_i is larger than A_i. In addition, power demand for each slot distributes exponentially with the given average. For each parameter setting, 20 sets are generated and the results are averaged. We compare the performance with the uncoordinated scheme. It employs no intelligent mechanism, but gives us a good reference to compare and estimate how well our scheme works.

The first experiment measures the effect of the number of EVs to energy efficiency, its result being plotted in Figure 3. In the y-axis, no scale unit is explicitly given, but a single unit matches the amount of electricity which flows from an EV to the grid during a single time slot. The experiment changes the number of EVs, or interchangeably, tasks from the viewpoint of the scheduler, from 2 to 6. The average demand per slot is 1.5. When there are just 2 tasks, the difference between the two schemes remains at just 1.6, which corresponds to 12.4 % improvement over the uncoordinated case. However, the performance gap gets larger as the number of tasks increases, when the search space size gets extended and the greedy search can find an efficient schedule. The proposed scheme outperforms by 49.0 %. From the perspective of surplus amount, the proposed scheme can reduce the amount of energy not used on the grid by up to 63.7 % in the case of 6 tasks.

Next experiment measures the meet ratio and the consumption ratio also according to the number of tasks, and its result is plotted in Figure 4. More energy demand from the grid can be met when more EV-stored electricity is available. Hence, with more tasks, the meet ratio increases for both cases. Our scheme starts from 25 % for 2 tasks and reaches 65 % on 6 tasks, as shown in Figure 4(a). On the contrary, the uncoordinated scheme hardly benefits from

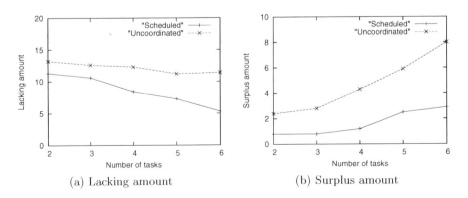

(a) Lacking amount   (b) Surplus amount

Fig. 3. Effect of the number of tasks to coordination efficiency

the increase in the available EV-stored energy, obtaining just 17 % increase in the meet ratio. Figure 4(b) plots the consumption ratio. For the given parameter range, both schemes are not significantly affected, but our scheme shows better performance by from 33 to 40 %. This result indicates that the amount of consumed electricity more depends on power demand from the grid.

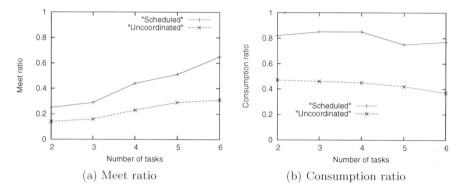

(a) Meet ratio   (b) Consumption ratio

Fig. 4. Effect of the number of tasks to system-level efficiency

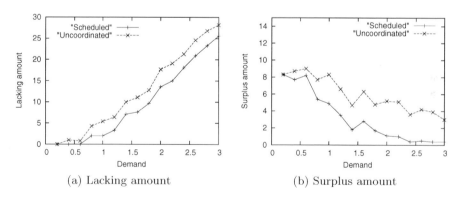

(a) Lacking amount   (b) Surplus amount

Fig. 5. Effect of power demand to coordination efficiency

Figure 5 measures the effect of the power demand to the lacking amount and the surplus amount. The experiment makes the per-slot power demand range from 0.2 to 3.0, while the number of tasks is set to 5. Necessarily, according to the increase in the energy demand from a microgrid, the lacking amount also increases while the surplus amount decreases. Figure 5(a) shows that the lacking amount is 0 until the demand is 0.6 for the proposed scheme and 0.2 for the uncoordinated scheme. From those points, both schemes show almost the same slope in their lacking amount. The lacking amount of the proposed scheme is

37.0 % of that of the uncoordinated case when the demand is 1.0 and 90.7 % when the demand is 3.0, while the absolute difference remains in the range of 2.6 to 4.1. In addition, coordinated scheduling achieves significant efficiency in reducing the surplus amount as shown in Figure 5(b). The performance gap gets larger according to the increase in the power demand, reaching 86.6 % when the demand is 3.0.

5   Conclusions

Besides their original role of driving, EVs can be used for power regulation by giving remaining electricity cheaply charged overnight back to the grid during the daytime peak hours. The microgrid, such as shopping malls, can predict its power demand with an appropriate forecast model and make an energy purchase plan to avoid electricity consumption from the power vendor during the expensive peak-rate interval. If many EVs are concentrated on a specific interval, not every battery-stored electricity can be bought by the grid. If there are not sufficiently available EVs, the grid cannot but use expensive grid-supplying electricity. This problem can be efficiently alleviated based on sophisticated computational intelligence and modern ubiquitous connectivity.

With an efficient coordination mechanism, it is possible to reduce such surplus and lacking energy. In our service model, each bid request specifies earliest and latest arrival times, amount to sell, and plug-in duration. The coordinator builds and traverses the search space to find an optimal schedule, by which EVs are connected to or disconnected from the grid, achieving an efficient temporal and spatial load shift in power consumption. Importantly, the bidder is informed of when it will arrive at the purchaser's site. The experiment result shows that the proposed scheme reduces the lacking amount by 49.0 % and the surplus amount by 63.7 %, compared with the uncoordinated scheduling strategy. In addition, on the high-demand interval, the surplus amount can be reduced by up to 86.6 % with the coordinated V2G.

As future work, we are planning to design an integrative tour scheduler for EV rent-a-cars which will visit multiple destinations. Here, some places are equipped with charging facilities allowing EVs to be charged while the drivers are taking a tour. Other places want to buy electricity from visiting EVs if available. In this V2G-combined tour, an efficient visiting order can improve user-side profits, giving more convenience to EV rent-a-car tourists. As the rate difference will get larger between low-load and high-load hours due to demand response policy, such an application will be more common in our daily lives.

References

1. Ipakchi, A., Albuyeh, F.: Grid of the Future. IEEE Power & Energy Magazine, 52–62 (2009)
2. Goebel, C., Callaway, D.: Using ICT-Controlled Plug-in Electric Vehicles to Supply Grid Regulation in California at Different Renewable Integration Levels. IEEE Transactions on Smart Grid 4(2), 729–740 (2013)

3. Budischak, C., Sewell, D., Thomson, H., Mach, L., Veron, D.: Cost-Minimized Combinations of Wind Power, Solar Power and Electrochemical Storage, Powering the Grid up to 99.9 % of the Time. Journal of Power Sources **225**, 60–74 (2013)

4. Tarantilis, C., Zachariadis, E., Kiranoudis, C.: A Hybrid Metaheuristic Algorithm for the Integrated Vehicle Routing and Three-Dimensional Container-Loading Problem. IEEE Transactions on Intelligent Transportation Systems **10**(2), 255–271 (2009)

5. Ota, Y., Taniguchi, J., Nakajima, T., Liyanage, K., Baba, J., Yokoyama, A.: Autonomous Distributed V2G (Vehicle-to-Grid) Satisfying Scheduled Charging. IEEE Transactions on Smart Grid **3**(1), 559–564 (2012)

6. Tomic, J., Kempton, W.: Using Fleets of Electric-Drive Vehicles for Grid Support. Journal of Power Sources **168**, 459–468 (2007)

7. Kisacikoglu, M., Ozpineci, B., Tolbert, L.: EV/PHEV Bidirectional Charger Assessment for V2G Reactive Power Operation. IEEE Transactions on Power Electronics **28**(12), 5717–5727 (2013)

8. Lund, H., Kempton, W.: Integration of Renewable Energy into the Transport and Electricity Sectors through V2G. Energy Policy **36**, 3578–3587 (2008)

9. Liu, H., Hu, Z., Song, Y., Lin, J.: Decentralized Vehicle-to-Grid Control for Primary Frequency Regulation Considering Charging Demands. IEEE Transactions on Power Systems **28**(3), 3480–3489 (2013)

10. Vandael, S., Holvoet, T., Deconinck, G., Kamboj, S., Kempton, W.: A Comparison of Two GIV Mechanisms for Providing Ancillary Services at the University of Delaware. In: IEEE International Conference on Smart Grid Communications (2013)

11. Vedova, M., Palma, E., Facchinetti, T.: Electric Load as Real-time Tasks: An Application of Real-time Physical Systems. In: International Wireless Communications and Mobile Computing Conference, pp. 1117–1123 (2011)

PolSAR Image Segmentation Based on the Modified Non-negative Matrix Factorization and Support Vector Machine

Jianchao Fan[1,2(✉)], Jun Wang[1,3], and Dongzhi Zhao[2]

[1] School of Control Science and Engineering, Dalian University of Technology,
Dalian 116023, Liaoning, China
jcfan@nmemc.gov.cn, jwang@mae.cuhk.edu.hk
[2] Department of Ocean Remote Sensing, National Marine Environment Monitoring
Center, Dalian 116023, Liaoning, China
dzzhao@nmemc.gov.cn
[3] Department of Mechanical and Automation Engineering,
The Chinese University of Hong Kong, Shatin, Hong Kong

Abstract. To improve polarimetric synthetic aperture radar (PolSAR) imagery segmentation accuracy, a modified non-negative matrix factorization algorithm based on the support vector machine is proposed. Focusing on PolSAR remote sensing images, the modified non-negative matrix factorization algorithm with the neurodynamic optimization achieves the image feature extraction. Compared with basic features, such as the basic backscatter coefficient, structuring more targeted localization non-negative character fits better for the physical significance of remote sensing images. Furthermore, based on the new constructive features, a support vector machine is employed for remote sensing image classification, which remedies the small sample training problem. Simulation results on PolSAR image classification substantiate the effectiveness of the proposed approach.

Keywords: PolSAR · Non-negative matrix factorization · Image segmentation · Support vector machine

1 Introduction

Image segmentation is regarded as a process of partitioning digital images into multiple regions or objects. These objects provide more information than individual pixels since the interpretation of images based on objects is more meaningful

This research was supported by the project (61273307) of the National Nature Science Foundation of China, and supported by the Foundation (201313) of Key Laboratory of Marine Spill Oil Identification and Damage Assessment Technology and also supported by China Postdoctoral Science Foundation (2014M551082). The work described in the chapter was also supported by the Research Grants Council of the Hong Kong Special Administrative Region, China, under Grants CUHK416812E.

© Springer International Publishing Switzerland 2014
Z. Zeng et al. (Eds.): ISNN 2014, LNCS 8866, pp. 594–601, 2014.
DOI: 10.1007/978-3-319-12436-0_66

than the interpretation based on individual pixel only [1]. The modern generation of SAR sensors offer possibilities for global environmental monitoring and
land-cover mapping at high-resolution scale using multi-polarization data. SAR
sensors could work with day and night ability, and is independent on the weather
condition [2]. Thus it has a unique advantage over other passive satellites.

There is now an increasing volume of fully polarimetric data due to the
launch of sensors capable of fully polarimetric imaging. Thus, automated image
segmentation and classification methods are desired to replace manual interpretation, which is subjective and labor intensive. Cao [3] employed the $H/A/\alpha$ and
total backscattering power (SPAN) together to the complex Wishart clustering.
A region-based unsupervised segmentation algorithm for PolSAR imagery that
incorporate region growing and a Markov random field edge strength model was
designed and implemented [4]. Neural networks are also adopted in the PolSAR
processing. An Evolutionary robust radial basis function network based classifier
is presented by Ince *et al.* [5]. Because of its superiority for small sample learning problems, support vector machine (SVM) has great advantages for remote
sensing image processing. This is because obtaining a lot of ground true training samples in a remote sensing image will take a lot of manpower and other
resources. Furthermore, most SVM models are employed in the visible spectrum
[6]. Thus, that how to use the advantage of SVM for PolSAR image segmentation
is still needed to be explored.

In addition, more and more PolSAR image information, such as texture,
decomposition matrix etc. [7], are needed in the segmentation process to improve
the accuracy. Most algorithms already reported in the literatures only could deal
with the basic data type. How to find the best feature for image segmentation
is a new challenge. In this study, the modified non-negative matrix factorization
algorithm is proposed to construct the feature vectors for the classifier. Then the
support vector machine is adopted to complete the classification process.

The remainder of this paper is organized as follows. Section II presents some
preliminaries on Polarimetric SAR data processing. Section III describes the
modified non-negative matrix factorization algorithm. The PolSAR classification algorithm based on the support vector machine is presented in section IV.
Section V highlights the potential of the proposed approach through experimental examples. Concluding remarks are presented in Section VI.

2   Preliminaries of PolSAR Data Processing

Basic PolSAR images are often constructed by the complex scattering matrix
$[S]$ produced by a target under study with the objective to infer its physical
proprieties, which could be expressed as

$$S = \begin{bmatrix} S_{hh} & S_{hv} \\ S_{vh} & S_{vv} \end{bmatrix} \tag{1}$$

The measured scattering matrix $[S]$ is transformed by the Pauli decomposition [8]. If the conventional orthogonal linear (h,v) basis considers as the Pauli

basis $|S|_a = \frac{1}{\sqrt{2}} \begin{bmatrix} 1 & 0 \\ 0 & 1 \end{bmatrix}, |S|_b = \frac{1}{\sqrt{2}} \begin{bmatrix} 1 & 0 \\ 0 & -1 \end{bmatrix}, |S|_c = \frac{1}{\sqrt{2}} \begin{bmatrix} 0 & 1 \\ 1 & 0 \end{bmatrix}, |S|_d = \frac{1}{\sqrt{2}} \begin{bmatrix} 0 & -1 \\ 1 & 0 \end{bmatrix},$

$[S]$ can be expressed as,

$$S = \begin{bmatrix} S_{hh} & S_{hv} \\ S_{vh} & S_{vv} \end{bmatrix} = \alpha[S]_a + \beta[S]_b + \gamma[S]_c \tag{2}$$

where $\alpha = (S_{hh} + S_{vv})/\sqrt{2}$, $\beta = (S_{hh} - S_{vv})/\sqrt{2}$, $\gamma = \sqrt{2}S_{hv}$. Hence by means of the Pauli decomposition, all polarimetric information in S could be represented in a single RGB image by combining the intensities $|\alpha|^2 \rightarrow$Red, $|\beta|^2 \rightarrow$Blue, and $|\gamma|^2 \rightarrow$Green, which determines the power scattered by different type of scatterers such as single- or odd-bounce scattering, double- or even-bounce scattering, and orthogonal polarization returns by the scattering.

In order to simplify the analysis of the physical information provided by eigenvectors decomposition, three secondary parameters are defined as a function for second-order polarimetric descriptors $\langle[T_3]\rangle$: entropy H, anisotropy A and mean alpha angle α, which are expressed as follows:

$$H = -\sum_{i=1}^{3} p_i \log_3 (p_i), \tag{3}$$

$$A = \frac{\lambda_2 - \lambda_3}{\lambda_2 + \lambda_3}, \tag{4}$$

$$\alpha = \sum_{i=1}^{3} p_i \alpha_i, \tag{5}$$

where p_i is the eigenvector, and also called the probability of the eigenvalue λ_i. It represents the relative importance of this eigenvalue with respect to the total scattered power, since

$$SPAN = |S_{hh}|^2 + |S_{vv}|^2 + 2|S_{hv}|^2 = \sum_{k=1}^{3} \lambda_k . \tag{6}$$

3　Modified Non-negative Matrix Factorization

Analysis on separate matrices or slices extracted from a data block often faces the risk of losing the covariance information among various modes. The $H/A/\alpha$ and SPAN above are stationary matrix decomposition approach, which could not adjust adaptively according to each special PolSAR data. Thus, in order to discover hidden components within the data, the analysis tools is adopted to explore the multidimensional structure of image data. Non-negative matrix factorization (NMF) algorithm is first proposed by Lee and Seung [9]. NMF is distinguished from the other methods, such as principal component analysis

(PCA), by its use of non-negativity constraints, which has more realistic application characteristics with real physical significance. Furthermore, more newly constructive images are localized features that correspond to better intuitive notions in PolSAR images.

Let $V_{n \times m}$ denote each single polarimetric SAR image data S_{hh}, S_{hv}, S_{vh} and S_{vv}, where n is the number of pixels, m is the number of basic feature in the original image. The matrix factorization is expressed as follows,

$$V \approx WH \tag{7}$$

where $W_{n \times r}$ is the basis matrix, $H_{r \times m}$ is the corresponding coefficient matrix, r is the feature dimension after factorization. Each column vector of V is approximately equals to the linear combination of the matrix W, and H is the combination coefficient.

In order to perform non-negative matrix factorization, Eqn. (7) can be defined as a constrained optimization problem as follows:

$$\begin{aligned} \min \quad & f(W, H) \\ \text{s.t. } & W \geq 0, H \geq 0 \end{aligned} \tag{8}$$

where $f(W, H)$ is an objective function, which characterizes the degree after decomposition similar to the original remote sensing data matrix. The constraints ensure the nonnegativity. Considering the noises of PolSAR remote sensing images, an alternative objective function is built as,

$$f(W, H) = \sum_{i=1}^{n} \sum_{j=1}^{m} (V_{ij} - (WH)_{ij})^2. \tag{9}$$

Considering (9) objective function optimization problems with bound constraints, these could all change to the general form

$$\begin{aligned} \min \quad & f(x) \\ \text{subject to } & x \in \Omega \end{aligned} \tag{10}$$

where $x = (x_1, x_2, \cdots, x_{(n+m)*r}) \in \Re^{(n+m)*r}$ is the independent variable, f is an objective function, Ω is a nonempty and closed convex set in $\Re^{(n+m)*r}$. A one-layer projection neural network is presented for this problem. It has the lowest model complexity among neurodynamic optimization models. On the basis of that, the recurrent neural network model for solving (10) optimization problem is presented by the following dynamical equation:

$$\dot{x}(t) = -x(t) + P_\Omega(x(t) - \nabla f(x(t))) \tag{11}$$

where x is the state of recurrent neural network, which are corresponding to the independent variable of NMF algorithm, ∇f denotes the gradient of f, and P_Ω is a projection operator. Each feature extraction method for matrix factorization factor W and H imposes different constraints limiting conditions. In the PCA algorithm, the matrix W is imposed column vectors on the orthogonal constraint.

Although the direction of the extracted characteristic has a large variance of statistical significance, that wherein the linear combination of positive and negative is not visually apparent on the understanding. MNMF algorithm matrix decomposition process, due to the non-negative constraints, there is no redundancy negative generated features .

4   The PolSAR Classification Algorithm Based on Support Vector Machine

After the feature extraction process, a Support Vector Machine (SVM) is employed to classify the PolSAR images. SVM is a typical representative of kernel methods. In the regression prediction and pattern classification fields, it has a wide range of applications. Because SVM is built on the basis of statistical learning theory, it has good generalization ability under limited training samples [10]. First, from two-class problem, analysis the SVM classification model. Adopt the form of a linear model:

$$y(\mathbf{x}) = \mathbf{w}^{\mathrm{T}} \phi(\mathbf{x}) + b \tag{12}$$

where $\phi(\mathbf{x})$ denotes the feature spatial transformation, b is the bias. The dataset of training include the Pauli decomposition and MNMF features, which are $\mathbf{x}_1, \mathbf{x}_2, \cdots, \mathbf{x}_N$, the objective value is $t_1, t_2, \cdots, t_N$, here $t_n \in \{-1, 1\}$. , solve (13),

$$\arg \max_{\mathbf{w}, b} \left\{ \frac{1}{\|\mathbf{w}\|} \min[t_n(\mathbf{w}^{\mathrm{T}} \phi(\mathbf{x}_n) + b)] \right\} \tag{13}$$

Under the constraint, the best $\{\mathbf{w}, b\}$ point can be obtained,

$$t_n(\mathbf{w}^{\mathrm{T}} \phi(\mathbf{x}_n) + b) \geq 1, \qquad n = 1, \cdots, N \tag{14}$$

SVM is basically two-class classifiers. However, in PolSAR segmentation problems, the majority have $K > 2$. Therefore, multi-class classification SVMs are employed to build different multi-class classifiers. Then, sample x_i is grouped into a class according to the largest number of votes. In remote sensing image classification process, there usually exist too few training samples. Thus, SVM algorithm has a unique advantage in the small sample study and have a better applicability.

5   Simulation Results

In order to substantiate the effectiveness and efficiency of the proposed MNMF+ SVM model, the following simulation of PolSAR image classification was performed. A quad polarimetric Radarsat-2 SAR image in Quebec City, CA, was adopted in this experiment. The satellite carries a C-band active phased array SAR. This PolSAR image has 2055 rows × 1720 columns with HH, HV, VH and VV polarimetric types. The false color composite image is produced using

Pauli matrix factorization, as shown in Fig. 1. There are four land covers, which are river, green land, wetland, and urban land. For this geographic area, 37322 pixels are carefully recorded and registered ground cover information as learning samples by human observers based on ground reference and land-use map, shown in Fig. 1. 80% samples are used as the training data, the others are the test data. The segmentation results based on MNMF+SVM are shown in Fig. 2. It is seen that the proposed approach can obtain the good segmentation results.

Fig. 1. The false color composite image with learning samples based on Pauli matrix factorization

In order to compared with other algorithms, introduce the index of overall accuracy and Kappa coefficient to evaluate. The overall accuracy is defined as follows:

$$O = \frac{\text{the number of correctly segmentation samples}}{\text{the total number of samples}} \qquad (15)$$

In addition, a discrete multivariate analysis technique is used to test whether the overall agreement in the different separate error matrices is significantly different. The measure of agreement called Kappa coefficient [11] is adopted to assess the significant difference, which is defined as follows:

$$\kappa = \frac{N \sum_{k=1}^{C} N_{kk} - \sum_{k=1}^{C} N_{k+} N_{+k}}{N^2 - \sum_{k=1}^{C} N_{k+} N_{+k}} \qquad (16)$$

where N is the number of samples, C is the number of clusters, N_{kk} denotes the number of correctly classified. N_{k+} and N_{+k} indicate, respectively, the number for class i and the number of clustering to class i.

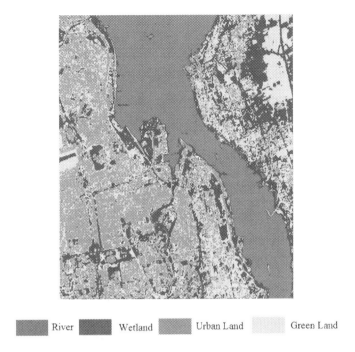

River   Wetland   Urban Land   Green Land

Fig. 2. The PolSAR image segmentation of MNMF+SVM

The specifical overall accuracy and Kappa coefficient are tabulated in Table 1. Classic PCA with SVM model and $H/A/\alpha$[3] are added in the simulation. According to the results, MNMF+SVM algorithm outperforms other ones. The MNMF method is superior to the basic PCA and $H/A/\alpha$ method, which obtains the best image features.

Table 1. Clustering Performance of Three Algorithms for PolSAR image segmentation

	Quebec City	
	Overall Accuracy	κ coefficient
PCA+SVM	80.24%	0.7392
$H/A/\alpha$[3]	85.24%	0.8192
MNMF+SVM	89.23%	0.8424

6   Conclusions

A method of modified non-negative matrix factorization and support vector machine is developed for PolSAR image segmentation, which can effectively extract target feature information. Through the neurodynamic optimization approach, the nonconvex constraint optimization of NMF is solved. Based on these extracted features, the use of support vector machine image classification model can obtain the better classification results. Compared with other methods, the simulation results demonstrates the validity of MNMF+SVM for PolSAR image segmentation.

References

1. Fan, J.C., Han, M., Wang, J.: Single point iterative weighted fuzzy c-means clustering algorithm for remote sensing image segmentation. Pattern Recognition **42**, 2527–2540 (2009)
2. Suwa, K., Iwamoto, M., Wakayama, T.: Analysis on the resolution of polarimetric radar and performance evaluation of the polarimetric bandwidth extrapolation method. IEEE Transactions on Geoscience and Remote Sensing **51**, 4260–4278 (2013)
3. Cao, F., Hong, W., Wu, Y., Pottier, E.: An unsupervised segmentation with an adaptive number of clusters using the Span/H/alpha/A space and the complex wishart clustering for fully polarimetric SAR data analysis. IEEE Transactions on Geosciences and Remote Sensing **45**, 3454–3466 (2007)
4. Yu, P., Qin, A.K., Clausi, D.A.: Unsupervised polarimetric SAR image segmentation and classification using region growing with edge penalty. IEEE Transactions on Geosciences and Remote Sensing **50**, 1302–1317 (2012)
5. Ince, T., Kiranyaz, S., Gabbouj, M.: Evolutionary RBF classifier for polarimetric SAR images. Expert Systems with Applications **39**, 4710–4717 (2012)
6. Shao, Y., Lunetta, R.S.: Comparison of support vector machine, neural network, and CART algorithms for the land-cover classification using limited training data points. ISPRS Journal of Photogrammetry and Remote Sensing **70**, 78–87 (2012)
7. Akbari, V., Doulgeris, A.P., Moser, G., Eltoft, T., Anfinsen, S.N., Serpico, S.B.: A textural-contextual model for unsupervised segmentation of multipolarization synthetic aperture radar images. IEEE Transactions on Geoscience and Remote Sensing **51**, 2442–2453 (2013)
8. Ince, T.: Unsupervised classification of polarimetric SAR image with dynamic clustering: An image processing approach. Advances in Engineering Software **41**, 636–646 (2010)
9. Lee, D.D., Seung, H.: Learning the parts of objects by non-negative matrix factorization. Nature **401**, 840–850 (1999)
10. Hsu, C.W., Lin, C.J.: A comparison of methods for multiclass support vector machines. IEEE Transactions on Geoscience and Remote Sensing **13**, 415–425 (2002)
11. Tao, W.B., Jin, H., Zhang, Y.M.: Color image segmentation based on mean shift and normalized cuts. IEEE Transactions on Systems, Man and Cybernetics, Part B: Cybernetics **37**, 1382–1389 (2007)

Solving Path Planning of UAV Based on Modified Multi-Population Differential Evolution Algorithm

Zhengxue Li$^{(\boxtimes)}$, Jie Jia, Mingsong Cheng, and Zhiwei Cui

School of Mathematical Sciences, Dalian University of Technology,
Dalian 116024, People's Republic of China
lizx@dlut.edu.cn

Abstract. In this paper we solve the path planning of Unmanned Aerial Vehicle (UAV) using differential evolution algorithm (DE). Based on traditional DE, we proposed a modified multi-population differential evolution algorithm (MMPDE) which adopts the multi-population framework and two new operators: chemical adsorption mutation operator and selection mutation operator. The simulation experiments show that the new algorithm has good performance.

Keywords: Differential evolution · Chemical adsorption mutation operator · Selection mutation operator · Multi-population evolution · Path planning of UAV

1 Introduction

Unmanned Aerial Vehicle (UAV) has been widely used. In the military, it can be used on battlefield detection and monitoring, location emendation, damage assessment, electronic warfare, and so on. In the civil field, it can be used on aerial photography, disaster condition monitoring, traffic patrolling, public security control, and so on. Comparing with the general airplanes, the UAV has the advantage of low-cost, high security, strong survival ability, good maneuvering performance. Furthermore, there is no need to worry about the casualty problems. Therefore, the UAV can be convenient to use widely.

The aim of path planning of UAV is to find an optimal route subject to some certain performance indexes based on the task objective. One must consider topography, data, threat information, fuel oil, constrained time in path planning and build mathematical model based on these constraints. Now there are some algorithms in path planning such as genetic algorithm, particle swarm optimization algorithm, neural network, simulated annealing algorithm, and so on [1-3]. In this paper we solve the path planning of UAV based on the MMPDE and compare the results of the new and old algorithms.

This work was supported by the National Natural Science Foundation of China (No. 11171367, 91230103), the Fundamental Research Funds for the Central Universities (DUT13LK04).

© Springer International Publishing Switzerland 2014
Z. Zeng et al. (Eds.): ISNN 2014, LNCS 8866, pp. 602–610, 2014.
DOI: 10.1007/978-3-319-12436-0_67

2   Path Planning of UAV and its Mathematical Model

2.1   Description of Problem

In this section we first translate the path planning into an optimization problem with dimension n (see figure 1) [4].

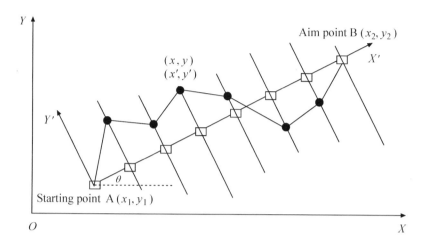

Fig. 1. Graph of Coordinate Transformation

Build a new coordinate system, where the new X' axis is from the starting point $A(x_1, y_1)$ to the aim point $B(x_2, y_2)$, and the coordinate transformation formula is as follows:

$$\theta = \arcsin \frac{y_2 - y_1}{|\overrightarrow{AB}|}, \tag{1}$$

$$\begin{pmatrix} x \\ y \end{pmatrix} = \begin{pmatrix} \cos\theta & \sin\theta \\ -\sin\theta & \cos\theta \end{pmatrix} \begin{pmatrix} x' \\ y' \end{pmatrix} + \begin{pmatrix} x_1 \\ y_1 \end{pmatrix}. \tag{2}$$

Next we equally divide AB into n line segments whose partition points are a_i $(i = 1, 2, \cdots, n)$, then we can get the corresponding b_i $(i = 1, 2, \cdots, n)$ via optimization, and so form the vector $(b_1, b_2, \cdots, b_n)^T$. Link these points in turn, we will get a polyline path connected the starting point and the aim point, hence we transform path planning of UAV into an optimization problem with dimension n.

2.2   Threat Cost

On the path $L_{i,j}$, the total threat cost originated from N_t threat source is

$$w_{t,L_{ij}} = \int_0^{L_{ij}} \sum_{k=1}^{N_t} \frac{t_k}{[(x - x_k)^2 + (y - y_k)^2]^2} \, dl. \tag{3}$$

For simplicity (see figure 2), every edge is divided into 5 segments. We calculate the threat cost of this edge only taking 5 threat points. If the distance between the threat point and the edge is less than the threat radius, then calculate the threat cost according to the following equation:

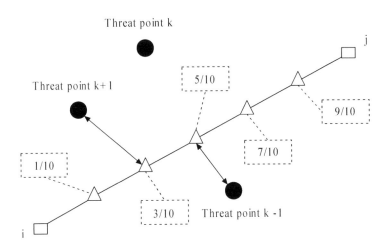

Fig. 2. Calculation of Threat Cost

$$w_{t,L_{ij}} = \frac{L_{ij}}{5} \sum_{k=1}^{N_t} t_k \left(\frac{1}{d_{0.1,k}^4} + \frac{1}{d_{0.3,k}^4} + \frac{1}{d_{0.5,k}^4} + \frac{1}{d_{0.7,k}^4} + \frac{1}{d_{0.9,k}^4} \right), \qquad (4)$$

where L_{ij} denotes the edge length connected the nodes i and j, $d_{0.1,k}$ denotes the distance between the $1/10$ partition point on edge $L_{i,j}$ and the kth threat center, and t_k denotes the threat level of the threat source.

Besides, since the fuel cost is related to the length, we denote $w_f = L$ for simplicity and the fuel cost for every edge is $w_{f,L_{ij}} = L_{ij}$.

2.3   Performance Indexes

The performance indexes of path planning of UAV mainly include safe performance index and fuel performance index completing the given task, that is, the minimum performance indexes of threat cost and fuel cost.

Denote L the length of path, then the minimum performance index of threat cost is

$$\min J_t = \int_0^L w_t \, dl, \qquad (5)$$

and the minimum performance index of fuel cost

$$\min J_f = \int_0^L w_f \, dl, \qquad (6)$$

then the total performance index of path of UAV is

$$\min J = kJ_t + (1 - k)J_f. \tag{7}$$

In the above equations, w_t denotes the threat cost of every point, w_f is the fuel cost of every point which is a function of length (w_f is a constant 1 in simulation), $k \in [0, 1]$, denotes the weight coefficient between the safe and fuel performance indexes which can be determined by the task of UAV.

3   Differential Evolution Algorithm

3.1   Basic Differential Evolution

In 1995, R. Storn and K. V. Price broke the shackles of traditional algorithm frameworks, put forward a simple and effective method (called differential evolution, DE) to solve global optimization problems with continuous variables. From then on, DE has been attracting the attention of the researchers from diverse domains of knowledge [5-8].

Based on real-coding and multi-parent crossover, DE replaces a current individual by a new one only if the new generation individual is excellent. Next we introduce the basic operation via solving the minimum problem [9, 10].

(1) Mutation Operator

For the individual $X_i(t)$ in the population, choose randomly three integers r_1, r_2, $r_3 \in \{1, 2, \cdots, N\}$, $r_1 \neq r_2 \neq r_3 \neq i$, and create the mutation individual:

$$V_i(t) = X_{r_1}(t) + F \cdot [X_{r_2}(t) - X_{r_3}(t)] \triangleq (v_{i1}(t), v_{i2}(t), \cdots, v_{in}(t))^T, \tag{8}$$

where F is the magnification factor of finite difference vector, usually $F \in (0, 1)$.

(2) Crossover Operator

Crossover the mutation individual $V_i(t)$ and the current one $X_i(t)$ to get the competitive individual:

$$U_i(t) \triangleq (u_{i1}(t), u_{i2}(t), \cdots, u_{in}(t))^T.$$

Precisely, generate randomly integers $rand_j \in \{1, 2, \cdots, n\}$ for $j = 1, 2, \cdots, n$, and we have

$$u_{ij}(t) = \begin{cases} v_{ij}(t), & \text{if } \text{rand}\,[0, 1) \leq CR \text{ or } j = rand_j, \\ x_{ij}(t), & \text{otherwise}, \end{cases} \tag{9}$$

where CR is the crossover control parameter, usually $CR \in [0, 1]$.

(3) Selection Operator

$$X_i(t + 1) = \begin{cases} U_i(t), & \text{if } f(U_i(t)) < f(X_i(t)), \\ X_i(t), & \text{otherwise}. \end{cases} \tag{10}$$

3.2   Modification of Evolution Operators

3.2.1 Chemical Adsorption Mutation Operator

Adsorption is an phenomenon or a process of a particle adhesion to some body because of the object's attraction, such as chemical adsorption and physical one. If the property doesn't changed during the process, it can be called a physical adsorption. On the other hand, it can be called a chemical adsorption, if chemical bond(s) is(are) changed during the process besides gravitation. In mathematical words, if the physical adsorption is equivalent to linear operator, then nonlinear operator is the chemical adsorption.

Notice that in the generation process of mutation (see (8)), each individual can be regarded as a physical adsorption merely with position shifting, that is, the individual $X_{r_1}(t)$ is perturbed by the gravitation of $X_{r_2}(t)$ and $X_{r_3}(t)$. It is easy to know that the operator (8) is a linear operator. Since the mutation individual $V_i(t)$ is restricted on the hyperplane defined by the parent individuals $X_{r_1}(t)$, $X_{r_2}(t)$ and $X_{r_3}(t)$, the algorithm may fall into local optimization. Hence, we can regard the components of the individual $X_{r_1}(t)$ as chemical bonds, and make some components adhere to a certain individual firmly under these chemical bonds. This is a chemical adsorption corresponding to a nonlinear operator in mathematics. If we take the geometric center of the parent individuals $X_{r_1}(t)$, $X_{r_2}(t)$ and $X_{r_3}(t)$ as the certain adsorbed substance, the chemical adsorption mutation operator is as follows:

$$v_{ij}(t) = \begin{cases} (x_{r_1j}(t) + x_{r_2j}(t) + x_{r_3j}(t))/3, & \text{if } \text{rand}\,[0,1) < P_\alpha, \\ x_{r_1j}(t) + F \cdot (x_{r_2j}(t) - x_{r_3j}(t)), & \text{otherwise}, \end{cases} \tag{11}$$

where $P_\alpha \in [0,1)$ is the adsorption strength, we can choose $P_\alpha = 1 - \frac{1}{n}$.

3.2.2 Selection Mutation Operator

In DE, the current individual $X_i(t)$ is replaced by the competitive individual $U_i(t)$ if and only if $U_i(t)$ is better than $X_i(t)$, and this is a determined substitution rule. Though this selection is beneficial to the rapid convergence for the population, usually it goes against the diversification of the population. In view of this, we give the weak individual a more mutation chance when selecting the competitive individual. That is, if the competitive individual doesn't replace the current one and it is a weak individual, then mutate the current individual directly according to a certain probability. This operation attends the weak group and may increase the diversification of the population. In this paper, the weak individual is defined as the one who doesn't reach average level. For example, in the minimum problem, if $X_i(t)$ satisfies

$$f(X_i(t)) \geq \sum_{l=1}^{N} f(X_l(t))/N, \tag{12}$$

then $X_i(t)$ is called a weak individual. The mutation operator can be chosen as the Cauchy-Lorentz mutation for the optimal individual center of the current population:

$$X_i(t+1) = X_{best}(t) + Cauchy(0, \gamma), \tag{13}$$

where $\gamma \in (0, 1)$.

3.3 Multi-Population Evolutionary

The so-called multi-population evolutionary means partitioning all the individuals in the group into some small subgroups. Under this significance, every subgroup denotes a subspace or sub-domain in the solution space, and every individual denotes a solution. All the subgroups run parallel local searching and interconnect each other via emigration operator. Emigration operator termly (for some certain generations) emigrates the optimal individual in the subgroup into other subgroup during the evolution process, in order to exchange information between the subgroups. That is, the weakest individual in the objective subgroup is replaced by the optimal individual in the source subgroup. It is very important for the evolution. Without emigration operator, all subgroups would not contact each other, and the multi-population differential evolution would be only a repeated DE with different parameters, which would lose its characteristics [11].

3.4 Modified Multi-Population Differential Evolution Algorithm (MMPDE)

In order to increase the diversity of the population, and improve its ability of jumping out of local minimum, next we propose MMPDE which, uses the multi-population framework in the DE algorithm, replaces the traditional mutation operator (8) by the chemical adsorption mutation one (11), and introduces (12) and (13) into the selection operator (10). The steps are as follows for the minimum optimal problem:

Step 1 (Initialize): Input the evolution parameters, the number of subgroups MP, the size of subgroup N, the magnification factor F of difference vector, the crossover control parameter $CR \in [0,1]$, the adsorption strength $P_\alpha \in [0,1)$ and the maximal iteration number $N_{c_{max}}$. Randomly initialize each subgroup $X^k(0) = (X_1^k(0), X_2^k(0), \cdots, X_N^k(0))$, where $X_i^k(0) \in \mathbb{R}^n$, $k = 1, 2, \cdots, MP$. Note that the vector $X_{best}^k(0)$ is the optimal individual in $X^k(0)$, and $E_0 = \{X_{best}^1(0), X_{best}^2(0), \cdots, X_{best}^{MP}(0)\}$ be the set of elite individuals. Let $t = 0$.

Step 2 (Population Evolution): For $k = 1, 2, \cdots, MP$, for each individual $X_i^k(t)$ in subgroup $X^k(t)$, do the following operations:
    (1) mutation

$$v_{ij}(t) = \begin{cases} (x_{r_1 j}(t) + x_{r_2 j}(t) + x_{r_3 j}(t))/3, & \text{if } rand\,[0,1) < P_\alpha, \\ x_{r_1 j}(t) + F \cdot (x_{r_2 j}(t) - x_{r_3 j}(t)), & \text{otherwise.} \end{cases}$$

(2) crossover

$$u_{ij}(t) = \begin{cases} v_{ij}(t), & \text{if } \text{rand}\,[0,1) \leq CR \text{ or } j = rand_j, \\ x_{ij}(t), & \text{otherwise.} \end{cases}$$

(3) selection

$$X_i^k(t+1) = \begin{cases} U_i^k(t), & \text{if } f(U_i^k(t)) < f(X_i^k(t)), \\ X_{best}^k(t) + Cauchy(0, \gamma), & \text{else if } f(X_i(t)) \geq \sum_{l=1}^{N} f(X_l(t))/N, \\ X_i^k(t), & \text{otherwise.} \end{cases}$$

(14)

Step 3 (Select Elite Individuals)**:** Select the optimal MP individuals from the set

$$\{X_{best}^1(t), X_{best}^2(t), \cdots, X_{best}^{MP}(t), X_{best}^1(t+1), X_{best}^2(t+2), \cdots, X_{best}^{MP}(t+1)\}$$

to get the set of elite individuals $E(t+1)$.

Step 4 (Emigration)**:** Replace the weakest individual in subgroup $i+1$ by the optimal individual in subgroup i, and replace the weakest individual in the first subgroup by the optimal individual in the last subgroup.

Step 5 (Replacement)**:** Replace the weakest MP individuals in the first subgroup by the individuals in the elite population $E(t+1)$.

Step 6 (Criteria)**:** If the optimal individual $E_{best}(t+1)$ in the elite population $E(t+1)$ satisfies the requirement or it is the evolution deadline, then stop and output $E_{best}(t+1)$ as the approximate solution; else let $t = t+1$, goto step 2.

4   Roughly Steps of MMPDE for Path Planning of UAV

Step 1: Build a new coordinate system, and transform the threat information into the new coordinate system. Divide the segment AB into n segments, then every feasible solution is a real vector of dimension n.

Step 2: Randomly generate $MP \times N$ initial paths within permissible range on the battle field, and calculate the cost of every feasible path based on every threat source information.

Step 3: Evolution computation.

Step 4: For every subgroup formed by the MP feasible solutions, execute mutation, crossover and selection in turn.

Step 5: Emigrate and update the set of elite individuals.

Step 6: If iteration number is greater than the permissive maximal number, then break the circulation; else goto step 3.

Step 7: Transform the final optimal path inversely and output.

5   Simulation Experiments

The battlefield environment of UAV task is as shown in Table 1.

Table 1. Settings of task

starting point	[10,10]	aim point	[55,100]	cost weight	0.5
threat center	[45,50]	[12,40]	[32,68]	[36,26]	[55,80]
threat radius	10	10	8	12	9
threat level	2	10	1	2	3

The number of subgroups $MP = 5$, the size of every subgroup $N = 30$, the dimension of the solution space $n = 20$, the maximal iteration number $N_{c_{max}} = 400$, the mutation factor $F = 0.5$, the crossover factor $CR = 0.9$. Take four experiments randomly and the results are shown in the following figures.

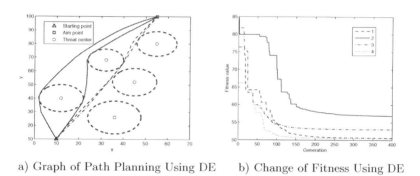

a) Graph of Path Planning Using DE   b) Change of Fitness Using DE

Fig. 3. Graph of the Results After 4 Times DE Algorighm

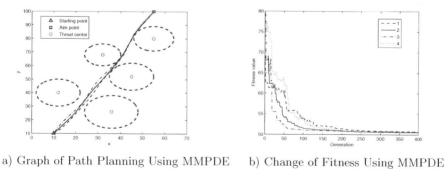

a) Graph of Path Planning Using MMPDE   b) Change of Fitness Using MMPDE

Fig. 4. Graph of the Results After 4 Times MMPDE Algorighm

From the experiments we can see that all the UAV paths generated by DE and MMPDE successfully bypass every threat within the battlefield and reach the target. However, from figure 3 we can see that two of the four paths generated by DE goes peripherally from the threat center, so the paths are longer and the effect is worse, hence the DE algorithm isn't stable. Figure 4 show that the four results generated by MMPDE differ little, and all the four paths safely goes through the intermediate threat, so the paths are short and the effect is better. These show that the new algorithm MMPDE has faster convergence rate, more stable effect, and better universality.

6   Conclusion

Based on the differential evolution, we introduce two new operators: chemical adsorption mutation operator and selection mutation operator, in order to strength local searching ability and global optimizing ability of the DE. Meanwhile, based on the emigration and sharing of information, a multi-population differential evolution algorithm is proposed which improves the global convergence effect. The path planning of UAV is solved by MMPDE which shows that the new algorithm has faster convergence property and good performance, so it can solve sophisticated optimization problems.

References

1. Wang, Q., Ma, L., Deng, H.H.: Adaptive Path Planning of the UAV Based on Genetic Algorithm. Computer Systems & Applications **22**(1), 200–203 (2013)
2. Zhu, H.G., Zheng, C.W.: Multi- UAV Route Planning Based on PSO. Computer Engineering & Science **32**(10), 142–144 (2010)
3. Zhang, Y.P., Zhu, L.C., Sun, T.: UAV Route Planning Using Hopfield Neural Network and Simulated Annealing. Journal of Naval Aeronautical Engineering Institute **22**(4), 451–453 (2007)
4. Duan, H.B., Zhang, X.Y., Xu, C.F.: Bio-inspired Computing, Science Press (2011)
5. Storn, R., Price, K.: Differential Evolution - A Simple and Efficient Adaptive Scheme for Global Optimization over Continuous Spaces. International Computer Science Institute (1995)
6. Price, K.: Differential Evolution: A Fast and Simple Numerical Optimizer. In: Proceedings of Biennial Conference of the North American Fuzzy Information Processing Society, pp. 524–527 (1996)
7. Storn, R., Price, K.: Differential Evolution - A Simply and Efficient Heuristic for Global Optimization over Continuous Spaces. Journal of Global Optimiaativn **11**(4), 341–359 (1997)
8. Yang, M.S., Luo, C.T.: Theory. Science Press, Method and Software of Optimization (2006)
9. Su, H.J., Yang, Y.P., Wang, Y.J.: Research on Differential Evolution Algorithm: A Survey. Systems Engineering and Electronics **30**(9), 1793–1796 (2008)
10. Xu, Z.B.: Computational Intelligence 1 - Simulated Evolutionary Computation, China Higher Education Press (2004)
11. Shi, F., Wang, H., Yu, L., Hu, F.: 30 Cases Analysis for MATLAB Intelligence Computation, Beihang University Press (2011)

Consensus for Higher-Order Multi-agent Networks with External Disturbances

Deqiang Ouyang, Haijun Jiang[(⊠)], Cheng Hu, and Yingying Liu

College of Mathematics and System Sciences, Xinjiang University, Urumqi 830046,
Xinjiang, People's Republic of China
jhj@xju.edu.cn

Abstract. In this paper, a class of consensus protocol for detail-balanced networks of agents with higher-order Lipschitz-type nonlinear dynamics and external disturbances is investigated. To guarantee asymptotic consensus in such a multi-agent system, several distributed controllers are constructed based only on the relative state information of neighboring agents. By appropriately constructing Lyapunov function and using tools from M-matrix theory, some sufficient conditions for achieving distributed consensus are provided. Finally, a example and simulation result is given to illustrate the effectiveness of the obtained theoretical result.

Keywords: Higher-order multi-agent · Consensus · Lipschitz nonlinear · External disturbances

1 Introduction

In recent years, there has been increasing interest in the collective behavior of multi-agent networks. Multi-agent networks, which consist of lots of interacting autonomous agents, have broad applications in various fields, such as cooperative control of unmanned air vehicles or unmanned underwater vehicles [1], distributed sensor networks [2–4], formation control for multi-robot systems [5–7], attitude alignment of clusters of satellites , synchronization of complex networks [8,9].

In this paper, we shall study a class of multi-agent systems with higher-order Lipschitz-type nonlinear dynamics with external disturbance. The main contribution of this paper will be presented in the following aspects. Firstly, the protocol to solve the consensus problems is designed. Secondly, some useful criteria have been derived analytically which can guarantee multi-agent systems with external disturbance to achieve consensus. To our knowledge there are few research papers dealing with this issue. Thirdly, by appropriately constructing Lyapunov function and using tools from M-matrix theory, we prove that the consensus can be reached.

The rest of this paper is organized as follows. Some concepts and consensus of multi-agent systems are briefly reviewed in Section 2. The dynamic gains based

© Springer International Publishing Switzerland 2014
Z. Zeng et al. (Eds.): ISNN 2014, LNCS 8866, pp. 611–620, 2014.
DOI: 10.1007/978-3-319-12436-0_68

disturbance inputs for consensus of leader-following multi-agent systems with external disturbances are presented in Section 3. A numerical example is given in Section 4 to show the effectiveness of the proposed protocol. Finally, some useful conclusions are drawn in Section 5.

Notion. Let R and N be the set of real and natural numbers, respectively, and $R^{N \times N}$ be the set of $N \times N$ real matrices. Let I_N ($\mathbf{0}_N$) be the N-dimensional identity (zero) matrix, and $1_N \in R^N$ ($0_N \in R^N$) be the N-dimensional column vector with all entries equal to one (zero). $diag\{d_1, \cdots, d_N\}$ indicates the diagonal matrix with diagonal entries d_1 to d_N and A^T represent the transpose of A. Let $||A||_1$, $||A||$ and $||A||_\infty$ denote 1-norm, 2-norm and ∞-norm of a matrix A, respectively. Let $\lambda_{min}(A)$ and $\lambda_{max}(A)$ denote respectively the smallest and largest eigenvalues of matrix A. The matrix inequality $A > 0$ means that A is positive definite. Let $\otimes$ denote the Kronecker product.

2   Preliminaries

2.1   Graph Theory

Let $G = \{V, \mathcal{E}, A\}$ be a weighted directed graph with the vertex set $v = \{v_1, v_2, \cdots, v_N\}$, where node v_i, $i \in I_N$, represents the ith agent, and the finite index set $I_N = \{1, 2, \cdots, N\}$ is the node indexes of G. $\mathcal{E} \subseteq V \times V$ is the set of edges, whose elements denote the communication links between the agents. $A = [a_{ij}]_{N \times N}$ is the weighted adjacency matrix of the weighted directed graph G with nonnegative adjacency elements $a_{ij} > 0$, $\forall i, j \in I_N$. An edge e_{ij} in G is denoted by the ordered pair of nodes (v_j, v_i), where v_j and v_i are called the parent and child nodes, respectively, and $e_{ij} \in \mathcal{E}$ if and only if $a_{ij} > 0$, otherwise $a_{ij} = 0$. Furthermore, self-loops are not allowed, i.e., $a_{ii} = 0$, $\forall i \in I_N$. The Laplacian matrix $L = [l_{ij}]_{N \times N}$ is defined as $l_{ii} = \sum_{j=1, j \neq i}^{N} a_{ij}$ and $l_{ij} = -a_{ij}$, $i \neq j$, for all $i, j \in I_N$. For undirected graphs, both A and L are symmetric. A directed path form node v_i to node v_j is a finite ordered sequence of edges, $(v_i, v_{k_1}), (v_{k_1}, v_{k_2}), \cdots, (v_{k_l}, v_j)$, with distinct nodes v_{k_m}, $m = 1, 2, \cdots, l$. A directed graph is called strongly connected if and only if there is a directed path between any pair of distinct nodes. Moreover, a directed tree is a directed graph where every vertex v, except one special vertex r without any parent, which is called the root vertex, has exactly one parent, and there exists a unique directed path from r to v. A directed spanning tree of a network G is a directed tree, which contains all the vertices and some edges of G. The graph is said to contain a directed-balanced if there exist some scalars $w_i > 0$, $i \in I_N$, such that $w_i a_{ij} = w_j a_{ji}$ for all $i, j \in I_N$. Moreover, the interaction topology may be dynamically changing. For a fixed directed graph G, its Laplacian matrix L has the following property.

Lemmas 1. ([8]) The Laplacian matrix L of G has a simple zero eigenvalue and all the other eigenvalues have positive real parts if and only if G contains a directed spanning tree.

Before moving on, the following notion and lemmas are introduced.

Definition 1. ([10]) A nonsingular matrix A is called an M-matrix if $A \in Z_N$ and all the eigenvalues of A have positive real parts.

Lemmas 2. ([10]) If $A \in Z_N$, the following statements are equivalent.
1) A is an M-matrix.
2) Matrix A can be expressed by $A = \gamma I_N - B$, where $B \geq 0$ and $\gamma > \rho(B)$.
3) A^{-1} exists, and $A^{-1} \geq 0$.
4) There exists a positive definite diagonal matrix $\Xi = diag(\xi_1, \cdots, \xi_N)$ such that $\Xi A + A^T \Xi > 0$.
5) There is a positive vector $x \in R^N$ such that $Ax > 0$.

Lemmas 3. ([9]) Let $Q \in R^{N \times N}$ be any symmetric matrix and $W \in R^{n \times n}$ be any symmetric positive-semidefinite matrix. Then, for any column vector $x \in R^{nN}$, there holds

$$\lambda_{min}(Q)x^T(I_N \otimes W)x \leq x^T(Q \otimes W)x \leq \lambda_{max}(Q)x^T(I_N \otimes W)x.$$

2.2   Problem Formulation

Consider a group of N agents, where an agent indexed by 1 is assigned as the leader and the agents indexed by $2, 3, \cdots, N$, are referred to as followers. In multi-agent networks, the agent i is with the following dynamics

$$\dot{x}_i(t) = Ax_i(t) + Cf(x_i(t), t) + Bu_i(t) + Bd_i(t), \quad i = 2, 3, \cdots, N, \quad (1)$$

where $x_i(t) \in R^n$ is the state of the ith follower, $f : R^n \times [0, +\infty) \to R^m$ is a continuously differentiable vector-valued function representing the intrinsic non-linear dynamics of the ith follower, and $u_i(t) \in R^p$ is the control input to be designed, and $d_i(t) \in R^p$ is the external disturbance of the ith follower, and A, B, and C are constant real matrices. It is assumed that matrix pair (A, B) is stabilizable. For notational convenience, let $f(x_i(t), t) = (f_1(x_i(t), t), f_2(x_i(t), t), \cdots, f_m(x_i(t), t))^T$, $i = 2, 3, \cdots, N$.

Assumption 1. There exists a nonnegative constant γ, such that

$$\|f(y, t) - f(z, t)\| \leq \gamma \|y - z\|, \quad \forall \, y, \, z \in R^n, \, t \geq 0. \quad (2)$$

Assumption 2. There exists a constant $l > 0$, such that $\|d_i(t)\|_\infty \leq l$, $i = 2, 3, \cdots, N$.

In many practical cases, the leader plays the role of a command generator providing a reference state and the followers are close to the leader. Therefore, it is sensible to assume that the state of the leader evolves without being affected by those of the followers. The dynamics of the leader is described by

$$\dot{x}_1(t) = Ax_1(t) + Cf(x_1(t), t), \quad (3)$$

where $x_1(t) \in R^n$ is the state of the leader and $f : R^n \times [0, +\infty) \rightarrow R^m$ is a continuously differentiable vector-valued function representing the intrinsic nonlinear
dynamics of the leader, and A and C are constant real matrices. For notational
convenience, let $f(x_1(t), t) = (f_1(x_1(t), t), f_2(x_1(t), t), \cdots, f_m(x_1(t), t))^T$.

Definition 2. Consensus of the leader-following multi-agent systems (1) and
(3) can be achieved, if for any initial states

$$\lim_{t \to \infty} ||x_i(t) - x_1(t)|| = 0, \quad \forall \, i = 2, 3, \cdots, N.$$

Let us consider the following example for integrator agents.

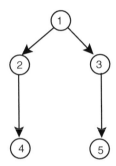

Fig. 1. Network topology with four followers and one leader

Example 1. Fig. 1 provides a topology for group of 5 single-integrator with
$V = \{1, 2, 3, 4, 5\}$, where a is the strength of the coupling for agents. In [13], some
sufficient for achieving distributed consensus are provided which are shown in
fig. 2. However, consensus of agents within the same interaction topology cannot
be achieved when one of agents occurs external disturbance which are shown in
fig. 2.

Example 1 implies that there are some interaction topologies which cannot
guarantee the hold of such distributed protocol. In this paper, we will be focusing
on further explore under what kind of interaction topologies with distributed
protocol can the consensus be achieved while regardless how weak or strong
the coupling among the agents are. In order to achieve consensus tracking, a
distributed protocol based on relative information between neighbouring agents
for (1) is proposed as

$$u_i(t) = \alpha F \left(\sum_{j=1, j \neq i}^{N} a_{ij}(x_j - x_i) \right)$$

$$+ \beta sgn \left(F \sum_{j=1, j \neq i}^{N} a_{ij}(x_j - x_i) \right), \quad \forall \, i = 2, 3, \cdots, N,$$

(4)

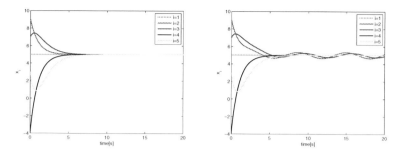

Fig. 2. Trajectories of x_i

where $\alpha > 0$ represents the coupling strength, $\beta > 0$ represents the gain constant, $F \in R^{p \times n}$ is the feedback gain matrix to be designed, and $\text{sgn}(\cdot)$ is the signum function.

3   Main Results

In this section, the main theoretical results of this paper are presented and proved. For further analysis, the following assumptions and lemmas are need.

Assumption 3. The graph G contains a directed spanning tree with the leader being the root. So there exists a directed path from the leader to each follower.

Since the leader has no neighbors, the laplacian matrix L in connection with the communication topology G can be divided into

$$L = \begin{pmatrix} 0 & 0_{N-1}^T \\ q & \hat{L} \end{pmatrix}, \tag{5}$$

where $q \in R^{N-1}$, and $\hat{L} \in R^{(N-1) \times (N-1)}$. Under Assumption 3, it follows from Lemma 1 that the Laplacian matrix L of G has a simple zero eigenvalue and all the other eigenvalues has positive real parts. Obviously, all the eigenvalues of $\hat{L}$ defined in (5) have positive real parts.

Define $W = diag\{\omega_1, \omega_2, \cdots, \omega_{N-1}\}$ where $w_i a_{ij} = w_j a_{ji}$ for all $i, j \in I_N$. Let ω_{max} and ω_{min} be the largest and smallest one of ω_i, $i = 1, 2, \cdots, N$, respectively. Then one has the following result.

Lemma 4. ([11]) The matrix $\hat{L}W$ is positive definite.

Let $e_i(t) = x_{i+1} - x_1(t)$, $i = 1, 2, \cdots, N-1$, and $e(t) = (e_1^T(t), e_2^T(t), \cdots, e_{N-1}^T(t))^T$. Obviously, $e(t) = 0_{(N-1)n}$ if and only if $x_1(t) = x_2(t) = \cdots = x_N(t)$, for all $t > 0$. Using (4) for (1), we obtain a compact form, one has

$$\dot{e}(t) = (I_{N-1} \otimes A)e(t) + (I_{N-1} \otimes C)(F(x) - F(x_1)) - \alpha(\hat{L} \otimes BF)e(t)$$
$$- \beta(I_{N-1} \otimes B)sgn((\hat{L} \otimes F)e(t)) + (I_{N-1} \otimes B)d(t) \tag{6}$$

where $d(t) = (d_2^T(t), d_3^T(t), \cdots, d_N^T(t))^T$, $F(x) = (f^T(x_2(t), t), \cdots, f^T(x_N(t), t))^T$ and $F(x_1) = (f^T(x_1(t), t), \cdots, f^T(x_1(t), t))^T$.

Before moving on, a multi-step design procedure is given for selecting the control parameters of protocol (4) under a fixed topology G.

Algorithm 1. Suppose that the matrix pair (A, B) is stabilizable. Under Assumptions 1, 2 and 3 the consensus protocol (4) can be designed as follows:

Solve the following linear matrix inequality (LMI):

$$\begin{pmatrix} AP + PA^T - 2\alpha\omega_{max}^{-1}\lambda_{min}BB^T + \sigma CC^T + \beta P & P \\ P & -\frac{\sigma}{\gamma^2}I \end{pmatrix} < 0 \qquad (7)$$

to get a matrix $P > 0$, and scalars $\sigma > 0$ and $\beta > 0$. Where $\lambda_{min} = \min\{\lambda_i,\ i = 1, 2, \cdots, N-1\}$ and λ_i is the eigenvalue of matrix $\hat{L}W$. Then, take $F = B^T P^{-1}$.

Then, one can establish the following theorem.

Theorem 1. Suppose the LMI (7) has a feasible solution. Under Assumptions 1, 2, and 3, the distributed control protocol (4) solves the consensus tracking problem for the leader-follower networks (1) and (3), if choose $\beta\omega_{min} \geq l\omega_{max}$, and $F = B^T P^{-1}$.

proof. Consider the following Lyapunov function candidate for the error dynamical system (6):

$$V(t) = e^T(t)(\hat{L}W \otimes P^{-1})e(t), \qquad (8)$$

where $\hat{L}$ is defined in Assumption 4, $W = diag\{\omega_1, \omega_2, \cdots, \omega_{N-1}\}$ and the positive definite matrix P is a solution of (7). The time derivative of $V(t)$ along the trajectory of (6) gives

$$\begin{aligned} \dot{V}(t) = {} & e^T(t)[\hat{L}W \otimes (A^T P^{-1} + P^{-1}A) - \alpha(\hat{L}^T\hat{L}W \otimes F^T B^T P^{-1}) \\ & - \alpha(\hat{L}W\hat{L} \otimes P^{-1}BF)]e(t) + 2e^T(t)(\hat{L}W \otimes P^{-1}C)(F(x) - F(x_1)) \\ & - 2\beta e^T(t)(\hat{L}W \otimes P^{-1}B)sgn((\hat{L} \otimes F)e(t)) \\ & + 2e^T(t)(\hat{L}W \otimes P^{-1}B)d(t). \end{aligned} \qquad (9)$$

By substituting $F = B^T P^{-1}$ into (9), one gets

$$\begin{aligned} \dot{V}(t) \leq {} & e^T(t)[\hat{L}W \otimes (A^T P^{-1} + P^{-1}A) \\ & - \alpha((\hat{L}^T\hat{L}W + \hat{L}W\hat{L}) \otimes P^{-1}BB^T P^{-1})]e(t) \\ & + 2e^T(t)(\hat{L}W \otimes P^{-1}C)(F(x) - F(x_1)) \\ & - 2\beta\omega_{min}\|(\hat{L} \otimes B^T P^{-1})e(t)\|_1 \\ & + 2\omega_{max}\|d(t)\|_\infty\|(\hat{L} \otimes B^T P^{-1})e(t)\|_1 \\ \leq {} & e^T(t)[\hat{L}W \otimes (A^T P^{-1} + P^{-1}A) \\ & - \alpha((\hat{L}^T\hat{L}W + \hat{L}W\hat{L}) \otimes P^{-1}BB^T P^{-1})]e(t) \\ & + 2e^T(t)(\hat{L}W \otimes P^{-1}C)(F(x) - F(x_1)) \\ & - 2(\beta\omega_{min} - l\omega_{max})\|(\hat{L} \otimes B^T P^{-1})e(t)\|_1. \end{aligned} \qquad (10)$$

Since $\beta\omega_{min} \geq l\omega_{max}$, one has

$$
\begin{aligned}
\dot{V}(t) \leq\ & e^T(t)[\hat{L}W \otimes (A^T P^{-1} + P^{-1}A) \\
& - \alpha((\hat{L}^T\hat{L}W + \hat{L}W\hat{L}) \otimes P^{-1}BB^T P^{-1})]e(t) \\
& + 2e^T(t)(\hat{L}W \otimes P^{-1}C)(F(x) - F(x_1)) \\
\leq\ & e^T(t)[\hat{L}W \otimes (A^T P^{-1} + P^{-1}A) \\
& - \alpha\omega_{max}^{-1}(((\hat{L}W)^2 + (W\hat{L})^2) \otimes P^{-1}BB^T P^{-1})]e(t) \\
& + 2e^T(t)(\hat{L}W \otimes P^{-1}C)(F(x) - F(x_1)).
\end{aligned}
\tag{11}
$$

By choosing a orthogonal matrix $Q \in R^{(N-1)\times(N-1)}$, such that $Q^T(\hat{L}W)Q = \Lambda$, where $\Lambda = diag(\lambda_1, \lambda_2, \cdots, \lambda_{N-1})$ and by letting $e(t) = (Q \otimes I)\tilde{e}(t)$, one has

$$
\begin{aligned}
\dot{V}(t) \leq\ & \tilde{e}^T(t)[\Lambda \otimes (A^T P^{-1} + P^{-1}A) \\
& - 2\alpha\omega_{max}^{-1}(\Lambda^2 \otimes P^{-1}BB^T P^{-1})]\tilde{e}(t) \\
& + 2\tilde{e}^T(t)(\Lambda \otimes P^{-1}C)(Q^T \otimes I)(F(x) - F(x_1)).
\end{aligned}
\tag{12}
$$

Using the Lipschitz condition (2) gives

$$
\begin{aligned}
& 2\tilde{e}^T(t)(\Lambda \otimes P^{-1}C)(Q^T \otimes I)(F(x) - F(x_1)) \\
& \leq \sigma\tilde{e}(t)(\Lambda\Theta\Lambda \otimes P^{-1}CC^T P^{-1})\tilde{e}(t) + \frac{\gamma^2}{\sigma}\tilde{e}^T(t)(\Theta^{-1} \otimes I)\tilde{e}(t),
\end{aligned}
\tag{13}
$$

where diagonal matrix $\Theta > 0$ and scalar $\sigma > 0$. By choosing $\Theta = \Lambda^{-1}$, it followers from (12) and (13) that

$$
\begin{aligned}
\dot{V}(t) \leq\ & \tilde{e}^T(t)[\Lambda \otimes (A^T P^{-1} + P^{-1}A) - 2\alpha\omega_{max}^{-1}(\Lambda^2 \otimes P^{-1}BB^T P^{-1}) \\
& + \sigma(\Lambda \otimes P^{-1}CC^T P^{-1}) + \frac{\gamma^2}{\sigma}(\Lambda \otimes I)]\tilde{e}(t).
\end{aligned}
\tag{14}
$$

Let $\varepsilon(t) = (\varepsilon_1^T(t), \varepsilon_2^T(t), \cdots, \varepsilon_{N-1}^T(t))^T$, where $\varepsilon_i(t) = P^{-1}\tilde{e}_i(t)$, $i = 1, 2, \cdots, N-1$. Obviously, $\tilde{e}(t) = (I_{N-1} \otimes P)\varepsilon(t)$. It thus follows from (14) that

$$
\begin{aligned}
\dot{V}(t) \leq\ & \varepsilon^T(t)[\Lambda \otimes (AP + PA^T) - 2\alpha\omega_{max}^{-1}(\Lambda^2 \otimes BB^T) \\
& + \sigma(\Lambda \otimes CC^T) + \frac{\gamma^2}{\sigma}(\Lambda \otimes P^T P)]\varepsilon(t) \\
\leq\ & \varepsilon^T(t)[\Lambda \otimes (AP + PA^T - 2\alpha\omega_{max}^{-1}\lambda_{min}BB^T \\
& + \sigma CC^T + \frac{\gamma^2}{\sigma}P^T P]\varepsilon(t),
\end{aligned}
\tag{15}
$$

where $\lambda_{min} = \min_i \lambda_i$, $i \in \{1, 2, \cdots, N-1\}$. Using (7), it follows from (15) that

$$
\begin{aligned}
\dot{V}(t) &\leq -\beta\varepsilon^T(t)(\Lambda \otimes P)\varepsilon(t) \\
&= -\beta e^T(t)(\hat{L}W \otimes P^{-1})e(t).
\end{aligned}
$$

Thus, one gets

$$
V(t) \leq e^{-\beta t}V(0)
$$

for all $t > 0$. Then, one concludes that $e(t) \to 0$ as $t \to \infty$. Thus, the consensus tracking problem in multi-agent system (1) is solved by the distributed consensus tracking protocol (4), constructed in Algorithm 1. This completes the proof.

Remark 1. Theorem 1 shows that the interaction topologies with external disturbances can the consensus be achieved while regardless how weak or strong the coupling among the agents are.

Corollary 1. Suppose the LMI (7) has a feasible solution. Under Assumptions 1 and 3, the distributed control protocol (4) solves the consensus tracking problem for the leader-follower networks (1) and (3) without external disturbances, if choose $\beta \geq 0$, and $F = B^T P^{-1}$.

4   Numerical Simulations

In this section, a numerical example is provided to verify the theoretical analysis.

Example 2. Consider a multi-agent system with topology G visualised by Figure 3, where the weights are indicated on the edges. Agents 2-5 are followers and agent 1 is the single leader.

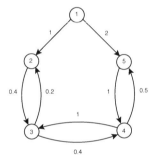

Fig. 3. Network topology with four followers and one leader

Define $W = diag\{\ 1,\ 2,\ 0.8,\ 0.4\ \}$. Obviously, $\hat{L}W$ is positive definite. The external disturbances are given as: $d_2(t) = 1.5sin(t)$, $d_3(t) = -sin(t)$, $d_4(t) = 0.5sin(t^2)$, $d_5(t) = -0.8cos(t^2)$, which implies that l can be selected as $l = 1.5$. The control parameters in (4) are chosen as $\alpha = 17$ and $\beta = 8$. Clearly, $f(x_i(t), t)$ is Lipschitz nonlinear function with a Lipschitz constant $\gamma = 0.8$. According to Algorithm 1, solving LMI (7) gives

$$F = (\ 7.9210\ ,\ -19.3687\ ,\ -0.9659\ ,\ 0.0512\ ). \tag{16}$$

Therefore, The protocol (4) with feedback matrix F solves the distributed consensus tracking problem of given nonlinear network. The simulation results are depicted in Figures 4.

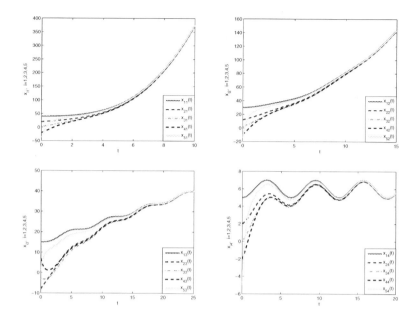

Fig. 4. Consensus of trajectories of $x_i, i = 1, 2, 3, 4, 5$

5   Conclusion

By using tools from Lyapunov stability analysis and M-matrix theory, the distributed consensus problem has been studied for a class of multi-agent systems with higher-order Lipschitz-type nonlinear dynamics and external disturbances in this paper. To achieve consensus, a new class of distributed controllers has been constructed. Some sufficient conditions have been further provided for guiding the states of the followers to asymptotically track those of the leader. Future works will focus on the distributed consensus tracking problems under finite time.

Acknowledgments. This work was supported by National Natural Science Foundation of Peoples Republic of China (Grants No. 61164004, No. 61473244, No. 11402223), Natural Science Foundation of Xinjiang (Grant No. 2013211B06), Project funded by China Postdoctoral Science Foundation(Grant No. 2013M540-782), Natural Science Foundation of Xinjiang University (Grant No. BS120101), Specialized Research Fund for the Doctoral Program of Higher Education (Grant No. 20136501120001).

References

1. Ren, W., Moore, K., Chen, Y.Q.: High-order consensus algorithms in cooperative vehicle systems. In: Proceedings of the 2006 IEEE International Conference on Networking, Sensing and Control, ICNSC 2006, pp. 457–462. IEEE (2006)

2. Kar, S., Moura, J.M.F.: Distributed consensus algorithms in sensor networks: Quantized data and random link failures. IEEE Transactions on Signal Processing **58**(3), 1383–1400 (2010)

3. Yu, W., Chen, G., Wang, Z.: Distributed consensus filtering in sensor networks. IEEE Transactions on Systems, Man, and Cybernetics, Part B: Cybernetics **39**(6), 1568–1577 (2009)

4. Xiao, F., Wang, L., Chen, J., et al.: Finite-time formation control for multi-agent systems. Automatica **45**(11), 2605–2611 (2009)

5. Abdollahi, F., Khorasani, K.: A decentralized Markovian jump control routing strategy for mobile multi-agent networked systems. IEEE Transactions on Control Systems Technology **19**(2), 269–283 (2011)

6. Hu, C., Yu, J., Jiang, H., et al.: Synchronization of complex community networks with nonidentical nodes and adaptive coupling strength. Physics Letters A **375**(5), 873–879 (2011)

7. Hu, C., Yu, J., Jiang, H., et al.: Exponential synchronization of complex networks with finite distributed delays coupling. IEEE Transactions on Neural Networks **22**(12), 1999–2010 (2011)

8. Godsil, C.D., Royle, G., Godsil, C.D.: Algebraic graph theory. Springer, New York (2001)

9. Qin, J., Yu, C.: Cluster consensus control of generic linear multi-agent systems under directed topology with acyclic partition. Automatica **49**(9), 2898–2905 (2013)

10. Horn, R.A., Johnson, C.R.: Topics in matrix analysis. Cambridge University Press (1991)

11. Zhang, Y., Yang, Y., Zhao, Y., et al.: Distributed finite-time tracking control for nonlinear multi-agent systems subject to external disturbances. International Journal of Control **86**(1), 29–40 (2013)

12. Wen, G., Duan, Z., Ren, W., et al.: Distributed consensus of multi-agent systems with general linear node dynamics and intermittent communications. International Journal of Robust and Nonlinear Control (2013)

13. Wen, G., Duan, Z., Chen, G., et al.: Consensus tracking of multi-agent systems with lipschitz-type node dynamics and switching topologies. IEEE Transactions on Circuits and Systems I: Regular Papers **61**(2), 499–511 (2013)

14. Wen, G., Hu, G., Yu, W., et al.: Consensus tracking for higher-order multi-agent systems with switching directed topologies and occasionally missing control inputs. Systems Control Letters **62**(12), 1151–1158 (2013)

The Research of Document Clustering Topical Concept Based on Neural Networks

Xian Fu$^{(\boxtimes)}$ and Yi Ding

Hubei Normal University, Huangshi 435002, China
{teacher.fu,teacher.dingyi}@live.com

Abstract. Nowadays, document clustering technology has been extensively used in text mining, information retrieval systems and etc. The input of network is the key problem for topical concept utilizing the Neural Network. This paper presents an input model of Neural Network that calculates the Mutual Information between contextual words and ambiguous word by using statistical method and taking the contextual words to certain number beside the topical concept according to (-M, +N). In this paper, we introduce a novel topical document clustering method called Document Characters Indexing Clustering (DCIC), which can identify topics accurately and cluster documents according to these topics. In DCIC, "topic elements" are defined and extracted for indexing base clusters. Additionally, document characters are investigated and exploited. Experimental results show that DCIC based on BP Neural Networks models can gain a higher precision (92.76%) than some widely used traditional clustering methods.

Keywords: Document clustering · Clusters indexing · Topical concept

1 Introduction

With the abundance of text documents available through the Web or corporate document management systems, the dynamic partitioning of document sets into previously unseen categories ranks high on the priority list for many applications like business intelligence systems. However, current text clustering approaches tend to neglect several major aspects that greatly limit their practical applicability. In this paper, a new clustering method is presented, which is named Document Characters Indexing Clustering (DCIC) method [1].

All clustering approaches based on frequencies of terms/concepts and similarities of data points suffer from the same mathematical properties of the underlying spaces. These properties imply that even when "good" clusters with relatively small mean squared errors can be built, these clusters do not exhibit significant structural information as their data points are not really more similar to each other than to many other data points. Therefore, we derive the high level requirement for text clustering

The work is supported by the S&T plan projects of Hubei Provincial Education Department of China (No.Q20122207).

Z. Zeng et al. (Eds.): ISNN 2014, LNCS 8866, pp. 621–628, 2014.
DOI: 10.1007/978-3-319-12436-0_69

approaches that they either rely on much more background knowledge or that they cluster in subspaces of the input space. In order to improving this problem, the most relevant works are that of Zamir et al. [2][3]. They proposed a phrase-based document clustering approach based on Suffix Tree Clustering (STC) [4][5]. The method basically involves the use of a tree structure to represent share suffix, which generate base clusters of documents and then combined into final clusters based on connect-component graph algorithm.

2   Document Characters Indexing Clustering

Formal Concept Analysis (FCA) is a method mainly used for the analysis of data, i.e. for investigating and processing explicitly given information [7]. Such data are structured into units which are formal abstractions of concepts of human thought allowing meaningful comprehensible interpretation. The reader is referred to [8] for the definitions of a formal context, formal concept and the sub-concept-sup-concept relation between formal concepts. In the approach presented in this paper we use Formal Concept Analysis as a conceptual clustering technique to automatically derive a partial order or concept hierarchy between terms on the basis of syntactic dependencies as features [9].

2.1   BP Neural Network

At the moment, there are about more than 30 kinds of artificial neural network (ANN) in the domain of research and application. Especially, BP neural network is a most popular model of ANN nowadays.

The BP model provides a simple method to calculate the variation of network performance caused by variation of single weight. This model contains not only input nodes and output nodes, but also multi-layer or mono-layer hidden nodes. Fig. 1. is a construction chart of triple-layer BP neural network. As it is including the weights modifying process from the output layer to the input layer resulting from the total errors, the BP neural network is called Error Back Propagation network.

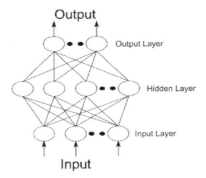

Fig. 1. BP Neural Network structure

The joint weights should be revised many times during the progress of the error propagating back in BP networks. The variation of joint weights every time is solved by the method of gradient descent. Because there is no objective output in hidden layer, the variation of joint weight in hidden layer is solved under the help of error back propagation in output layer. If there are many hidden layers, this method can reason out the rest to the first layer by analogy.

The variation of joint weights in output layer is as following:

$$\Delta w_{ik} = -\gamma \frac{\partial E}{\partial w_{ik}} = -\gamma \frac{\partial E}{\partial o_k} \cdot \frac{\partial o_k}{\partial w_{ik}} \tag{1}$$

$$\Delta b_{ki} = -\gamma \frac{\partial E}{\partial b_{ki}} = -\gamma \frac{\partial E}{\partial o_k} \cdot \frac{\partial o_k}{\partial b_{ki}} \tag{2}$$

The variation of joint weights in hidden layer is as following:

$$\Delta w'_{ik} = -\gamma \sum_{k=1}^{n} \frac{\partial E}{\partial w'_{ik}} = -\gamma \sum_{k=1}^{n} \frac{\partial E}{\partial o_k} \cdot \frac{\partial o_k}{\partial o'_i} \cdot \frac{\partial o'_i}{\partial w_{ik}} \tag{3}$$

Where $\Delta b'_{ki} = \gamma \delta_{ij}$, $\delta_{ij} = e_i f'_1$, $e_i = \sum_{k=1}^{n} \delta_{ik} w_{ik}$.

2.2   Clusters Indexing

As introduced before, STC method has some obvious advantages. Actually, these good qualities should all be ascribed to the smart way of forming base clusters by creating an inverted index of phrases for the documents. For the sake of convenience, we call this kind of way indexing base clusters hereafter. The precise meaning of indexing base clusters can be formulated as below: Let $F = \{F_1, F_2, ..., F_n\}$ be a document collection and $I = \{I_1, I_2, ..., I_n\}$ be a set of chosen indexes. Document F_i will be placed in the cluster indexed by I_i if and only if the number of times that I_i occurs in F_i exceeds a predetermined threshold T.

We take advantage of document characters in forming indexes. In natural language processing, a document is generally represented as a vector of words [4]. The weights of words are usually calculated using statistical techniques (such as "tf.idf"). Nevertheless, the document characters of words themselves should also be given enough attention. Take POS (Part-of-Speech) for example, intuitively, words with different POSes should have different contributions to characterizing a document [5]. Usually, nouns and verbs are the most indicative. Adjectives and adverbs are less valuable. Function words have little or no influence and should be excluded as stop words. In addition, NEs are more discriminative than normal words and should be assigned higher weight. In DCIC, we form the indexes by utilizing NEs coupled with important nouns and verbs.

2.3   The Pretreatment of BP Network Model

In the event of training BP model, the input vector P and objective vector O of topical concept should be determined firstly. And then we should choose the construction of neural network that needs to be designed, say, how many layers is network, how many neural nodes are in every layer, and the inspired function of hidden layer and output layer.

The training of model still needs the vector added weight, output, and error vector. The training is over when the sum of square errors is less than the objection of error. Or the errors of output very to adjust the joint weight back and repeat the training.

Topical concept depends on the context to judge the meaning of topical concept. So the input of model should be the topical concept and the contextual words round them. In order to vector the words in the context, the Mutual Information (MI) of topical concept and context should be calculated. So MI can show the opposite distance of topical concept and contextual words. MI can replace every contextual word. That is suitable to as the input model. The function of MI is as follow:

$$MI(w_1, w_2) = \log \frac{P(w_1, w_2)}{P(w_1) \cdot P(w_2)} \tag{4}$$

$P(w_1)$ and $P(w_2)$ are the probability of word w_1 and w_2 to appear in the corpus separately. While $P(w_1, w_2)$ is the probability of word w_1 and w_2 to appear together.

It should be based on context to determine the sense of topical concept. The model's input should be the vector of the topical concept and context words. It is well-known that the number of context words showing on the both sides of topical concept is not fixed in different sentences. But the number of vectors needed by BP network is fixed. In other words, the number of neural nodes of input model is fixed in the training. If the extracting method of feature vector is (-M, +N) in context, in other words there are M vectors on the left of topical concept and N vectors on the right, the extraction of feature vectors must span the limit of sentences. If the number of feature vectors is not enough, the topical concept on the left and right boundaries of whole corpus do not participate in the training.

According to the extracting method of feature vector (-M, +N), the vector of model input is as following:

$$V_{input} = \begin{cases} MI_{11} & MI_{12} & \cdots & MI_{1i} \\ MI'_{11} & MI'_{12} & \cdots & MI'_{1i} \\ MI_{21} & MI_{22} & \cdots & MI_{2i} \\ MI'_{21} & MI'_{22} & \cdots & MI'_{2i} \\ MI_{j1} & MI_{j2} & \cdots & MI_{ji} \\ MI'_{j1} & MI'_{j2} & \cdots & MI'_{ji} \end{cases} \tag{5}$$

Where, MI_{1i}, MI'_{1i} are the MI of context and the first meaning of topical concept MI_{2i}, MI'_{2i} are the MI of context and the second meaning of topical concept MI_{ji} MI'_{ji} are the MI of context and the third meaning of topical concept. MI1i, MI2i and MI3i are the feature words of topical concept on the left and MI of topical concept. MI'_{1i}, MI'_{2i} and MI'_{ji} are the feature words of topical concept on the right and MI of topical concept.

Every ambiguous word has three meanings, totally eighteen meanings for six ambiguous words. Every ambiguous word trains a model and every model has three outputs showed by three-bit integer of binary system, such as the three meanings of ambiguous wordW are showed as followed: $s_{i1} = 100$ $s_{i2} = 010$ $s_{i3} = 001$.

According to statistics, when $(-M, +N)$ are $(-8, +9)$ using the method of feature extraction, the cover percentage of effective information is more than 87%.

However, if the sentence is very short, collecting the contextual feature words on the basis of $(-8, +9)$ can include much useless information to the input model. Undoubtedly, that will increase more noisy effect and deduce the meaning-distinguish ability of verve network.

This article makes an on-the-spot investigation of experimental corpus, a fairly integrated meaning unit, which average length is between 9~10 words. So this article collects the contextual feature words on the basis of $(-5, +5)$ in the experiments, 10 feature words available that calculate MI with each meaning of ambiguous word separately to get 30 vectors. All punctuation marks should be filtered while the feature words are collected. The input layer of neural network model is regarded as 30 neural nodes. The triple-layer neural network adopts the inspired S function. From that, the number of neural nodes in hidden layer is defined as 12 on the basis of experimental contrast, and 3 neural nodes in output layer. Hence, the structure of model is 30 × 12 × 3, and the precision of differential training is defined as 0.3 based on the experimental contrast.

2.4 DCIC Algorithm

Step 1: Document is submitted to sequential preprocessing modules including word segmentation, POS tagging, and NE recognition. In this step, stop words are removed. Here, a stop list containing punctuations, common used words and some news specific words is maintained.

Step 2: In the stage of document representation, Vector Space Model (VSM) is employed. The vector terms here contain only NEs, nouns, and verbs. The tf.idf is used for weighing the vector terms.

Step 3: Forming indexes and creating base clusters: An index used in DCIC consists of two parts: an NE-part and a keyword-part:

Definition 1: Let F be a document, $A = \{a_1, a_2, \dots a_n\}$ be a set of NEs occurring at least twice in F, $B = \{b_1, b_2, \dots b_n\}$ be a set of keywords (nouns and verbs) in F whose tf.idf weights exceed a preset threshold T. $\forall a \in A$ and $\forall b \in B$, the two-tuple of (x, y) is defined as one of D's indexes.

If the size of A and B are x and y, then document F has x × y chances to be indexed by any of its indexes. This makes it possible that a single document can be indexed on different topics and put into several base clusters. With the well-designed indexes combining NEs and keywords, DCIC constructs base clusters by merging documents that share common indexes.

Step 4: Combining base clusters into clusters: The base clusters formed in the last section overlap a lot. Hence we combine base clusters to reduce duplication and form more complete clusters. Let c_i, c_j be two base clusters. If their distance is less than a preset threshold T_i, then they will be combined. In order to measure the distance between two base clusters, the centroids of them have to be calculated. The distance measure used in the combination algorithm is the cosine measure:

$$\cos(c_i, c_j) = \frac{c_i \cdot c_j}{\|c_i\| \cdot \|c_j\|} \tag{6}$$

3   Experiments

3.1   Data and Metrics

The method is evaluated using a collection of 2271 computer sciences documents collected from the web. From these documents, we have manually identified 300 topics, whose maximum size is 28 documents and minimum is 4. In this paper, the precision and recall are computed respectively in evaluation.

Given a particular topic T_i of size y_i and a particular cluster C_i of size y_i, suppose y_{ij} documents in the cluster C_i belong to T_i, then the precision of this topic and cluster is defined to be:

$$\text{precision}(T_i, C_i) = \frac{y_{ij}}{y_i} \tag{7}$$

The precision of T_i is the maximum precision value attained at any cluster in the cluster set C:

$$\text{precision}(T_i) = \max_{c_j \in c} \text{precision}(T_i, C_i) \tag{8}$$

The overall precision is computed by taking the weighted average of the individual precision:

$$\text{precision} = \sum_{i=1}^{N_T} \frac{y_i}{Y} \text{precision}(T_i) \tag{9}$$

Where N is the total number of documents and N_T is the number of topics. Similarly, the recall of the entire clustering results can be defined as

$$\text{recall} = \sum_{i=1}^{N_T} \frac{y_i}{Y} recall(T_i) \tag{10}$$

Where $recall(T_i) = max_{c_j \in c} precision(T_i, C_i)$ and $recall(T_i, C_i) = \frac{y_{ij}}{y_i}$.

3.2   Comparisons with Other Methods

We conduct two sets of experiments. In the first set, the DCIC is compared with K-means Clustering (KMC), STC and Agglomerative Hierarchical Clustering (AHC).

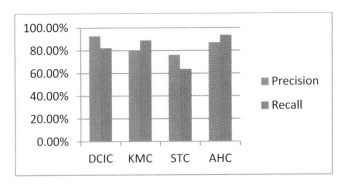

Fig. 2. Precision and Recall of four clustering methods

The stopping criterion for AHC and KMC is set to 300, which is the factual number of topics. First of all, the precision and recall of the above four methods are computed and compared in Figure 2. As expected, the DCIC method scores highest. We believe that this positive result is mainly due to DCIC's well-designed indexes which can identify topics more accurately.

3.3   Comparisons of Different Parts

In the second set of experiments, we evaluate the contributions of DCIC's different parts. Firstly, we try to find out whether the indexes involving both NEs and keywords work better than using only NEs or keywords.

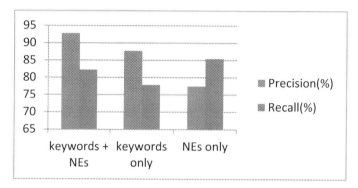

Fig. 3. Comparison of DCICs using different kinds of indexes for creating base clusters

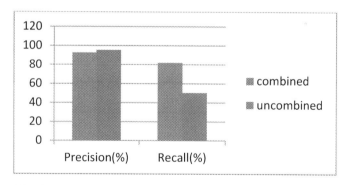

Fig. 4. Comparison of DCICs having base clusters combined or not

Figure 3-4 compares three runs of DCICs which use different kinds of indexes. It is obvious that the DCIC indexed by keywords plus NEs performs much better than that indexed by keywords in both precision and recall. This indicates that it is not enough to describe topics using keywords alone. We can also see that the run using NEs alone achieves the highest recall but the lowest precision. This is because, in the experimental data, more than one topic may be related to the same NE entity. Thus a single NE may index several topics. This results in many "large" base clusters which should be responsible for the high recall and low precision. We can conclude that combining base clusters can improve recall significantly.

4 Conclusions

In this paper, we propose a novel clustering method DCIC. This paper presents an input model of Neural Network that calculates the Mutual Information between contextual words and ambiguous word by using statistical method and taking the contextual words to certain number beside the topical concept according to (-M, +N).We conduct a number of evaluations that compare DCIC with some traditional clustering methods, including AHC, KMC and STC. Results show that the DCIC method can achieve a higher precision while maintaining an acceptable level of recall. Considering the "information overload" on the web, precision is more important than recall. Thus we can conclude that DCIC is effective.

References

1. Deerwester, S.T., Dumais, T.K., Landauer, G.W., Furnas, R.A.: Indexing by latent semantic analysis. Journal of the Society for Information Science **41**(6), 391–407 (2012)
2. Lee, D.L., Chuang, H., Seamons, K.: Document Ranking and the Vector-Space Model. IEEE Software **14**(2), 67–75 (2009)
3. Daniel, F. : An analysis of recent work on clustering algorithms. Technical Report, University of Washington (2004)
4. Zamir, O., Etzioni, O.: Web Document Clustering: A Feasibility Demonstration. In: Proceedings of the 21st International ACM SIGIR Conference on Research and Development in Information Retrieval, pp. 46–54 (2008)
5. Gusfield, D.: Algorithms on Strings, Trees and Sequences: Computer Science and Computational Biology. Cambridge University Press, Cambridge (2007)
6. Macskassy, S.A., Banerjee, A., Davison, B.D., Hirsh, H. : Human performance on clustering web pages: a preliminary study. In: Proc. of KDD, New York, NY, USA, pp. 264–268 (August 2008)
7. Maedche, S., Staab, A.: Ontology learning for the semantic web. IEEE Intelligent Systems **16**(2) (2011)
8. Miller, G.: WordNet: A lexical database for english. CACM **38**(11), 39–41 (2012)
9. Neumann, G., Backofen, R., Baur, J., Becker, M., Braun, C.: An information extraction core system for real world german text processing. In: Proceedings of the Conference on Applied Natural Language Processing, Washington, USA, pp. 208–205(2007)

The Research of Reducing Misregistration Based on Image Mosaicing

Yi Ding[(✉)] and Xian Fu

Hubei Normal University, Huangshi 435002, China
{teacher.dingyi,teacher.fu}@live.com

Abstract. Image mosaicing has been collecting widespread attention because it has become as an important tool for several different areas. Among other methods, homograph-based methods are the most accurate in the geometric sense. This is because these methods use planar projective transformation, which considers perspective effects as a geometric transformation model between images. We propose an automatic image mosaicing method which can construct a panoramic image from a collection of digital still images. These methods, however, have a problem of misregistration in the case of general scenes with arbitrary camera motion. Our method has been tested with several image sequences and comparative results are presented to illustrate its performance.

Keywords: Image mosaicing · Overlap extraction · Feature correspondence

1   Introduction

Image mosaicing has become an active area of research in the fields of photogrammetry, computer vision, image processing, and computer graphics. Application includes the construction of aerial and satellite photographs, photo editing and the creation of virtual environments. One conventional method is a cylindrical panorama that covers a horizontal view for creating virtual environments. However, this method limits the camera motion to horizontal rotation around the optical center, forcing the user to employ a tripod. Several other methods try to avoid this limitation, by using a planar projective transformation (homograph). These methods, however, fail in the case of general scene structures and general camera motions. Therefore, they have to restrict the geometric property of the scene to planar or the camera motion to rotation alone. We would like to allow users to shoot any scene with any camera motion, using only an ordinary hand-held still camera. To achieve this goal, we propose a feature-based method that can reduce misregistration. The method applies the trilinear constraint among a triplet of images to reduce the misregistration of mosaics. Our method consists of the following three steps. First, it makes feature correspondences between consecutive images by using optical flow estimation. Second, it computes a geometric

The work is supported by the S&T plan projects of Hubei Provincial Education Department of China (No.Q20122207).

Z. Zeng et al. (Eds.): ISNN 2014, LNCS 8866, pp. 629–636, 2014.
DOI: 10.1007/978-3-319-12436-0_70

transformation when the shooting scene can be regarded as planar. Third, when we cannot regard the scene as a planar surface, it divides the scene into several triangles, by using the popular constraints between images to obtain each transformation for each triangle. Then we transform images with the transformations.

2   Related Work

Given two images taken from the same viewpoint, or images of a planar scene taken from different viewpoints, the relationship between the images can be described by a linear projective transformation called a homograph. Once we obtain a homograph between two images, we can construct a panoramic image by transferring one image to the other with the homograph matrix.

Conventional techniques for obtaining lens distortion parameters can be categorized into two types. The first type uses special calibration patterns. Several techniques require patterns for providing 3D coordinates [4]. A recent, technique proposed in [6] is more flexible. However, it still requires a planar calibration pattern. The second type does not require any calibration pattern. Stein has proposed a method which uses only images of the scene [2]. This method, however, requires a computer driven rotating table to give a rotation angle. Stein has also proposed a method which does not need calibration patterns nor rotation tables [3]. This method, however, cannot be applied directly to image mosaicing because its purpose is for 3D reconstruction. Another problem of this method is that the computational cost is high because it is based on a non-linear minimization framework. Sawhney et al. have proposed a method for image mosaicing [5]. This method incorporates the lens distortion parameter into the homograph computation between images. We propose a feature-based method for image mosaicing that does not require any calibration pattern and is faster than previous methods.

3   Feature-Based Image Registration

We use small rectangular regions such as corners, which we call point features, to obtain a homography or a trifocal tensor. Here we show how to make correspondences by using the Lucas-Kanade gradient-based optical flow estimation.

The Lucas.Kanade method is one of the best regarding optical flow estimation, because it is fast to compute, easy to implement, and controllable because of the availability of tracking confidence [1]. The major drawback of the gradient methods, including the Lucas. Kanade method is that they cannot deal with large motion. Coarse-to-fine multi-resolution strategy has been applied to overcome this problem. A major problem, however, still remains. Low-textured regions have unreliable results. We solved this problem in [3], with a dilation-based filling technique after threshold unreliable estimates, at each pyramid level.

3.1   Overlap Extraction and Correspondence

We need to extract the overlapping region before applying the optical flow estimation, when the difference between images is large. This is because optical flow estimation may fail, even though our flow estimation incorporates a hierarchical multi-resolution technique. We find a rough displacement by maximizing the score of the normalized cross correlation, in the overlapping region, defined by the following equation:

$$C(d) = \frac{\Sigma_w(I_1(x)-\overline{I_1})(I_2(x+d)-\overline{I_2})}{(\sigma_1\sigma_2)^{\frac{1}{2}}} \quad (1)$$

Where C is the cross correlation coefficient of an overlapping region w (size: $M \times N$) between images I_1 and I_2 having displacement d. $\overline{I_i}$ and σ_i are the mean and the variance of the overlapping part, respectively, in each image. We make low-resolution images for faster computation, because the following optical flow estimation will obtain finer displacement. Fig.1. shows the original images and extracted overlapping region.

Fig. 1. Original images and Overlap extraction

Fig. 2. Selected features

Fig. 3. Corresponding features

After obtaining the overlapping region of the two images, we first select prominent features in the first image from their image derivatives and the Hessians [10]. This method finds small rectangular regions such as corners automatically. Next, we estimate sparse optical flow vectors for small rectangular patches such as 13×13 pixels by using the improved Lucas. To obtain the point correspondence, we interpolate the results of nearest four patch bilinear in sub pixel order. We discard false correspondences by measuring the cross correlation coefficients. Fig.2 and Fig.3 shows an example of the selected and corresponding features.

3.2   Planar Projective Transformation

When the scene is a planar surface or when the images are taken from the same point of view, images are related by a linear projective transformation called a homograph. Fig.4. illustrates the principal of the planar projective transformation. When we see a point M on a planar surface from two different viewpoints C_1 and C_2, we can transform the image coordinates $m_1 = (x_1; y_1)^{\frac{1}{t}}$ to $m_2 = (x_2; y_2)^{\frac{1}{t}}$ using the following 3×3 planar projective transformation matrix H [5]:

$$km_2 = Hm_1 \tag{2}$$

Where k is an arbitrary scale factor and image coordinates m_1 and m_2 are represented by homogeneous coordinates. This relationship can be rewritten using the following equations:

$$\begin{cases} x_2 = \frac{h_0 x_1 + h_1 y_1 + h_2}{h_6 x_1 + h_7 y_1 + 1} \\ y_2 = \frac{h_3 x_1 + h_4 y_1 + h_5}{h_6 x_1 + h_7 y_1 + 1} \end{cases} \tag{3}$$

We solve a homograph from feature correspondences. We can solve this homograph from four pairs of feature correspondences, because one pair of feature correspondence provides two equations for eight unknown parameters. If we have more than four pairs of feature correspondences, we can use a least-squares method for obtaining a homograph.

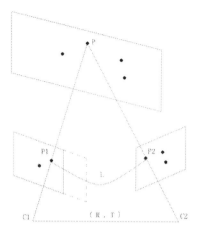

Fig. 4. Planar projective transformation

3.3   Homograph-Based Image Mosaicing

By using planar projective transformation, we can extend the image plane virtually. The dotted line in Fig.4 shows an extended region where it is invisible from viewpoint C_1, but visible from viewpoint C_2. Pixels in this area at viewpoint C_1 can be obtained by transferring pixels from C_2 with the homograph between the two images. Extending image plane corresponds to placing a wide-view lens whose focal length is short. We blend pixel intensities in the overlapping region. Averaging the intensities in that region causes a seam due to the brightness difference between the two images. Therefore, we blend intensities with the weights depending on the distance to each image. When we have three or more images, we cascade transformation iteratively. We select the middle image as a reference image by judging from the rough alignment by using normalized correlation described in the paper. We obtain the transformation for each image to the reference image, after obtaining those between consecutive images. We can obtain a transformation to the reference image with the following equation, even though the image does not have any overlapping portion with the reference image:

$$H_{0N} = H_{01}H_{02} \ldots H_{N-1N} \tag{4}$$

4   Nonplanar Scene

We propose a method that can construct a panorama from nonplanar, arbitrary depth scenes with unrestricted camera motion. If the assumption for homographs is not satisfied, that is, taking images of a nonplanar, arbitrary scene from different viewpoints, there will be misregistration caused by motion parallax. We use the Delaunay triangulation method. This is a well-known triangulation method which is based on Voronoi diagrams [5].

4.1   Piecewise Homograph Computation

We need at least four points to obtain a homograph since it has eight independent parameters. (It is a 3×3 matrix and is invariant to scaling.) We describe a novel technique to compute a homograph from three points using the epipolar constraint. The epipolar constraint between two images is described by the fundamental matrixF, with a point m on an image I and the corresponding point m' on the other image I' as follows:

$$m'^T Fm = 0 \qquad (5)$$

Recently a good method has been developed to obtain a fundamental matrix between uncelebrated cameras [10]. Using it, we first obtain the fundamental matrix between images from their corresponding points. By substituting Eq(2) into Eq(5), we have

$$m^T H^T Fm = 0 \qquad (6)$$

Since $H^T F$ means the outer product of vector m, it should be skew symmetric:

$$H^T F = \begin{pmatrix} 0 & a_1 & -a_2 \\ -a_1 & 0 & a_3 \\ a_2 & -a_3 & 0 \end{pmatrix} \qquad (7)$$

We obtain 6 equations from Eq(7), because diagonal elements and the additions of skew symmetric elements should be 0. We already have 6 equations from three sets of corresponding points with Eq(2). We can compute a homograph by least squares with three points and the fundamental matrix, because a total of 12 equations are available for eight unknown parameters.

4.2   Homograph Test

We can determine whether we should use a single homograph or divide the scene into several triangles by the following test. We measure the re-projection error D by using a homograph with the following equation:

$$D = \frac{1}{N} \Sigma_i (p'_i - p_i)^2 \qquad (8)$$

where p_i is a coordinate of a feature point, p'_i is a coordinate of the rejected feature point by the homograph, and N is the number of feature points. If D is larger than a predefined threshold, we divide the scene into triangles. Here we show how to make image mosaics with triangular patches. First, we obtain an average homograph H computed from all of the feature correspondences. Then we obtain each homograph H_T for each patch after triangulation. In the overlapping region, we use homographs H_T for triangles. Outside the overlapping region, we use the generic homograph H.

5 Experiments

This section introduces the results of applying our feature based techniques to image mosaicing. These images were taken with a hand-held digital still camera with no tripod.

Fig.5. shows an example. It demonstrates that our method does not restrict camera motion to horizontal rotation. Our method can make a panoramic image, even though the camera motion includes not only panning but also tilting or twisting.

We compared the processing time with one of the previous methods, namely, Szeliski's method. This method uses the following sum of squared intensity differences E between two images I and I' to obtain the homograph. It finds the optical flow vectors and the homograph by minimizing the cost E with the Levenberg-Marquardt nonlinear method.

$$E = \sum_i [I'(x'_i, y'_i) - I(x_i, y_i)]^2 \tag{9}$$

Fig. 5. Tilting or testing images

However, Szeliski's nonlinear minimization method has the following two problems, although it has been widely used. First, it is computationally expensive, requiring iterative warping. Second, it easily falls into a local minimum. This problem is common to the nonlinear minimization framework method. Then it is very important to provide a good initial estimation.

We measured the accuracy of our overlap extraction described in the paper. We used 150 pairs of images which were taken with a hand-held camera. The image size ranges from 640×4800 to 1920×1080. A size of 320×240 was used for overlap extraction. We counted the number of successful image pairs versus the correct answers by hand. We conducted an experiment not only with normalized correlation but also with the sum of squared difference (SSD). The image pairs that SSD failed but correlation succeeded, had large brightness differences. Using normalized correlation took 2 seconds to process, which was 0.5 second longer than that of SSD, with an image size of 1920×1080.

6   Conclusions

In this paper, we have illustrated a feature-based image mosaicing method. The greatest advantage of our method is that we can make a panoramic image of a nonplanar scene with unrestricted camera motion. First, we showed that our feature-based method is faster and more robust than previous featureless methods, because it is based on linear techniques. Second, for a scene that we cannot assume to be a single plane. Taken from different viewpoints, we described a method of dividing images into triangles with corresponding features. Our method provides the homograph for each triangle from three points using the epipolar constraint, although existing methods require four points to compute. The technique is fast and robust, because it is linear. In future work, we plan to develop a technique using line features which we can expect to have more accuracy. We are also interested in creating a virtual environment that can provide a novel view from any viewpoints in 3D space.

References

1. Baron, J., Fleet, D., Beauchemin, S.: Performance of optical flow techniques. Int. J. Comput. Vision **12**, 43–77 (1994)
2. Chen, S.E.: QuickTime VR-an image-based approach to virtual environment navigation. In: SIGGRAPH 1995, pp. 29–38 (1995)
3. Chiba, N., Kanade, T.: A tracker for broken and closely spaced lines. System Compute Japan **30**, 18–25 (2009)
4. Kumar, R., Anandan, P., Hanna, K.: Direct recovery of shape from multiple views: A parallax based approach. In: Proc. ICPR, pp. 685–688 (1999)
5. Faugeras, O.: Three-dimensional computer vision: A geometric viewpoint. MIT Press (2011)
6. Lucas, B., Kanade, T.: An iterative image registration technique with an application to stereo vision. In: 27th Int. Joint Conference on Artificial Intelligence (IJCAI-2009), pp. 674–679 (2009)
7. Szeliski, R.: Video mosaics for virtual environment. IEEE Computer Graphics and Applications, 22–30 (2012)
8. Stein, G.P.: Accurate internal camera calibration using rotation with analysis of sources of error. In: Proc. ICCV 2008, pp. 230–236 (2008)
9. Sawhney, H.S., Kumar, R.: True multi-image alignment and its application to mosaicing and lens distortion correction. IEEE PAM1 **21**(3), 235–243 (2007)
10. Zhang, Z.: Flexible camera calibration by viewing a plane from unknown orientations. In: Proc. ICCV 2006, pp. 666–673 (2006)

A Parallel Image Segmentation Method Based on SOM and GPU with Application to MRI Image Processing

Ailing De, Yuan Zhang, and Chengan Guo[✉]

School of Information and Communication Engineering, Dalian University of Technology,
Dalian 116023, Liaoning, China
{deailing,dianzizhishi}@mail.dlut.edu.cn, cguo@dlut.edu.cn

Abstract. This paper presents a parallel image segmentation method based on self-organizing map (SOM) neural network by extending the authors' former work from serial computation to parallel processing in order to accelerate the computation process. The parallel algorithm is composed of a group of parallel sub-algorithms for implementing the entire segmentation process, including parallel classification of the image into edge/non-edge pattern vectors, parallel training of an SOM network, and parallelly segmenting the image by using the trained SOM model with vector quantization approach. In the paper, the parallel algorithm is implemented on GPU with OpenCL program language and applied to segmenting the human brain MRI images. The experimental results obtained in the work showed that, compared with the original serial algorithm, the parallel algorithm can achieve a significant improvement on the computation efficiency with a speedup ratio of 64.72.

Keywords: Image segmentation · Parallel algorithm · SOM neural network · Vector quantization · Graphical processing unit (GPU)

1 Introduction

Image segmentation has been a very active research topic over several decades due to its essential role played in image analysis and computer vision. Many segmentation methods have been proposed in the existing literature (e.g., see [1]-[6]). The aim of image segmentation is to divide an image into some meaningful parts for further uses. There are two basic requirements for an image segmentation method: one is the segmentation accuracy that is desired to be as accurate as possible, and the other is the processing speed that is to be as fast as in need for applications. Usually the two requirements are contradictory each other. In order to meet the first requirement, both the gray value information and the spatial/structural information of the image must be effectively exploited by the segmentation algorithm that will result in heavy computational load. In addition, since in usual the data amount of an image is quite big, it is a challenging problem to process the data in real time with a complex algorithm. Recently, progresses have been made in developing parallel segmentation algorithms based on multi-core processors or graphic processing units (GPU), which can speed

© Springer International Publishing Switzerland 2014
Z. Zeng et al. (Eds.): ISNN 2014, LNCS 8866, pp. 637–646, 2014.
DOI: 10.1007/978-3-319-12436-0_71

up the segmentation process significantly (e.g., see [7]-[9]). With the rapid developing of multi-core and GPU technologies, this is a very promising direction for solving the bottleneck problem. However, there are still a lot of difficulties to be overcome in this field since different segmentation problems usually need different parallel schemes. A segmentation problem is always related to the images to be processed in an application while the images vary diversely in modality, shape, structure, etc. in different applications. In fact, there is no universal parallel scheme suitable for all the segmentation problems, which perhaps is the reason why so many segmentation algorithms have been proposed so far. Therefore, it still remains challenging and will have a long way to go in the field.

In [6], we proposed an SOM-based segmentation method in which the SOM network and vector quantization method are integrated together and applied to segmenting the human brain MRI images with excellent performance. As pointed in [6], however, the computational complexity of the method is very high and it is in need to speedup the computation procedure. In this paper we design parallel algorithms for the SOM-based segmentation approach in order to improve its computation efficiency significantly.

The sequel of the paper is organized as follows: Section 2 describes the proposed parallel algorithms, including the overall parallel scheme, the parallel algorithm for vector representation of images, the parallel classification of edge/non-edge pattern vectors, the parallel training of SOM network, the parallel quantization of the non-edge pattern vectors, and the parallel classification of the edge pattern pixels. Experimental results with applications to MRI image segmentation are given in Section 3. Section 4 draws conclusions and points out the possible direction of the paper.

2   Parallel Algorithms of SOM-Based Segmentation Method

In this section, after a brief summary on the original SOM-based segmentation approach proposed in [6], we give an overall parallel scheme for the approach at first, then present the corresponding parallel algorithms and concrete implementation steps of the scheme.

2.1   The Overall Parallel Scheme of the SOM-Based Segmentation Approach

The original SOM-based segmentation method proposed in [6] includes the following computational processes:

(1) Divide the image to be segmented into small sub-blocks of $n \times n$ pixels and represent each sub-block with a vector of $n \times n$ elements.
(2) Classify the sub-block vectors into two patterns, known as the edge pattern and non-edge pattern, by using the edge detection algorithm based on the wavelet modulus maximum edge detection [10,11].
(3) Train an SOM neural network by using the non-edge pattern vectors as inputs.
(4) Cluster (quantize) the non-edge pattern vectors into K_C classes by using the trained SOM network with vector quantization (VQ) method.

(5) Classify the pixels of the edge pattern sub-blocks into the clusters obtained in the VQ procedure.

In this paper, we design parallel algorithms to implement the above computation processes. The overall parallel scheme is illustrated in Fig. 1 that includes 5 parallel computation modules corresponding to the above 5 processes.

Fig. 1. Block diagram of the parallel scheme for the SOM-based segmentation method

2.2   Parallel Algorithm for Vector Representation of Images

In the SOM-based segmentation method [6], the first step is to divide the image to be segmented into small sub-blocks and represent them with vectors. In this section, we design a simple parallel algorithm to implement this operation.

Suppose that the image to be segmented is denoted by $[f(i,j)]_{M \times N}$ and we want to represent the image with N_V vectors, $X = \{ X(k); k = 1, ..., N_V \}$, in which each vector is constructed with the pixels of a sub-block $[f(i_k, j_k)]_{n \times n}$ of the image, where M and N are the height and width of the image respectively, n is the height or width of the sub-block with $n << \min(M, N)$, and the number of the vectors $N_V = \lceil MN / nn \rceil$.

The parallel algorithm for realizing the sub-block dividing and vector representation of the image $[f(i,j)]_{M \times N}$ is given as follows:

begin
    for $k = 1$ to N_V in parallel do
        1) find the location of the k-th sub-block $[f(i_k, j_k)]_{n \times n}$ in $[f(i,j)]_{M \times N}$;
        2) construct vector $X(k)$ with pixels of the sub-block by

$$X(k) = [f(i_k, j_k), ..., f(i_k, j_{k+n-1}), ..., f(i_{k+n-1}, j_k), ..., f(i_{k+n-1}, j_{k+n-1})]^T ;$$

    end for
end

Obviously, for the above parallel algorithm, the parallel degree is N_V and the theoretical speedup ratio is also N_V, compared with the original serial algorithm.

2.3   Parallel Classification of the Vectors into Edge or Non-edge Patterns

The second processing module of the SOM-based segmentation method is to classify the vectors (sub-blocks) into two patterns, the edge pattern and non-edge pattern, by using the wavelet modulus maximum edge detection method [10,11]. In [6], we presented a realization algorithm for the method using the partial derivatives of the two-dimensional Gaussian function as wavelet functions that involves the following 6 computation steps:

• Step (1): use the two wavelet functions as filters to filter the image by the following convolutions:

$$W_2^1 f(x, y) = f(x, y) \otimes \phi_2^1(x, y) , \quad W_2^2 f(x, y) = f(x, y) \otimes \phi_2^2(x, y) \tag{1}$$

where $f(x, y)$ denotes the image function, $\phi_2^1(x, y)$ and $\phi_2^2(x, y)$ are the wavelet kernel functions, $\phi_2^1(x, y) = \dfrac{-x}{16\pi\sigma^4} e^{-(x^2+y^2)/(8\sigma^2)}$ and $\phi_2^2(x, y) = \dfrac{-y}{16\pi\sigma^4} e^{-(x^2+y^2)/(8\sigma^2)}$.

• Step (2): compute the modulus and angle parameters, $M_2 f(x, y)$ and $A_2 f(x, y)$, based on the above convolution results:

$$M_2 f(x, y) = \sqrt{\left|W_2^1 f(x, y)\right|^2 + \left|W_2^2 f(x, y)\right|^2} , \quad A_2 f(x, y) = \arctan\left[\frac{M_2^2 f(x, y)}{M_2^1 f(x, y)}\right]. \tag{2}$$

• Step (3): compute the threshold parameter T_M through calculating the mean m_M and variance σ_M^2 of $M_2 f(x, y)$ by:

$$T_M = m_M + \alpha \sigma_M \tag{3}$$

where α ($\alpha \geq 0$) is an adjustment parameter to be determined in experiments.

• Step (4): find the maximum wavelet modulus value, denoted by $M_{\max}(x_k, y_k)$, for each sub-block among its pixels and take the corresponding wavelet angle as the maximum angle, denoted by $A_{\max}(x_k, y_k)$.

• Step (5): for each sub-block, compute the mean value $\bar{M}(x_k, y_k)$ of the 3 wavelet moduli of the sub-block along the vertical direction of $A_{\max}(x_k, y_k)$.

• Step (6): for each sub-block, classify it to the edge pattern if $\bar{M}(x_k, y_k) > T_M$, and to the non-edge pattern otherwise.

In order to effectively conduct the above computations, we design three parallel algorithms given in the sequel subsections for accomplishing the following 3 tasks:

1) Task 1: parallel implementation of the filtering (convolution) defined by equ.(1) and the computation of $M_2 f(x, y)$ and $A_2 f(x, y)$ defined by equ. (2);
2) Task 2: parallel computation of m_M, σ_M^2, and T_M defined by equ. (3);
3) Task 3: parallel implementation of steps (4)–(6) for the edge/non-edge pattern classification of the sub-blocks.

Parallel Algorithm for Implementing Task 1. The task of this parallel algorithm is to perform the convolution operations between image $[f(x, y)]_{M \times N}$ and the two wavelet functions, $\phi_2^1(x, y)$ and $\phi_2^2(x, y)$. Note that the convolution operations can also be conducted in frequency domain by using FFT. However, since the size of the wavelet kernel functions is usually much smaller than the image size, the computational complexity of the convolutions is less than that of FFT. Therefore, we perform the convolution operations directly instead of using FFT.

Let N_P be the number of the pixels of image $[f(x, y)]_{M \times N}$ ($N_P = M \cdot N$). The parallel algorithm is given below for computing the convolutions, the wavelet modulus $M_2 f(x, y)$, and the angle $A_2 f(x, y)$:

> begin
>> for $k = 1$ to N_P in parallel do
>>> 1) for the k-th pixel $f(x_k, y_k)$ of the image $[f(x, y)]_{M \times N}$, conduct the two convolution operations defined by equ.(1);
>>> 2) calculate the wavelet modulus $M_2 f(x_k, y_k)$ and angle $A_2 f(x_k, y_k)$ by using equ.(2) for $f(x_k, y_k)$;
>>
>> end for
>
> end

For the above parallel algorithm, both the parallel degree and the theoretical speedup ratio are N_P.

Parallel Algorithm for Implementing Task 2. The main computation load of this task is the calculation of the mean and variance, m_M and σ_M^2, of the wavelet moduli of the image.

Let $M_f(k)$ be the k-th wavelet modulus of the image. Then m_M and σ_M^2 can be expressed with the following formulas:

$$m_M = \frac{1}{N_P} \sum_{k=1}^{N_P} M_f(k), \tag{4}$$

$$\sigma_M^2 = \frac{1}{N_P} \sum_{k=1}^{N_P} (M_f(k) - m_M)^2 = \frac{1}{N_P} \sum_{k=1}^{N_P} M_f^2(k) - m_M^2. \tag{5}$$

Suppose that two arrays have been set with $A_1(k) = M_f(k)$ and $A_2(k) = M_f(k) * M_f(k)$, which can be easily implemented in parallel. Then the above two cumulative operations for calculating m_M and σ_M^2 can be realized by using the following parallel reduction algorithm and then the threshold T_M can be directly obtained by equ.(3):

> begin
>
> $N_R \leftarrow \lceil \log_2 N_P \rceil$; $K_m \leftarrow N_P / 2$;   // initializing N_R and K_m;
>
> **for** $m = 1$ to N_R **do**   // conducting N_R reduction steps in total;
>> for $k = 1$ to K_m in parallel do
>>> $A_1(k) \leftarrow A_1(k) + A_1(k+K_m)$; $A_2(k) \leftarrow A_2(k) + A_2(k+K_m)$;
>>
>> end for k
>>
>> $K_m \leftarrow K_m / 2$;   // reduce K_m to half in each step;
>
> end for m
>
> $m_M \leftarrow A_1(1)/N_P$; $\sigma_M \leftarrow$ sqrt($A_2(1)/N_P - m_M * m_M$);   // got m_M and σ_M;
>
> $T_M \leftarrow m_M + \alpha \sigma_M$;   // got the threshold T_M;
>
> end

The above parallel reduction algorithm includes $\lceil \log_2 N_P \rceil$ parallel reduction steps. One can see that the parallel degrees of different steps are from 1 to $N_P/2$. The average parallel degree is $(N_P-1)/\lceil \log_2 N_P \rceil \cong N_P/\lceil \log_2 N_P \rceil$. The theoretical speedup ratio of the parallel algorithm is also about $N_P/\lceil \log_2 N_P \rceil$.

Parallel Algorithm for Implementing Task 3. The task of this algorithm is to conduct all the computations in the steps (4)–(6) described in the beginning of Section 2.3 for classifying the vectors $\{X(k); k=1,\ldots,NV\}$ into the edge/non-edge patterns. The parallel algorithm for fulfilling this task is given below.

begin
    for $k = 1$ to N_V in parallel do
        1) for the *k-th* vector $X(k)$, find the maximum modulus $M_{max}(x_k, y_k)$ among the corresponding wavelet moduli of $X(k)$ and take the wavelet angle corresponding to $M_{max}(x_k, y_k)$ as the maximum angle $A_{max}(x_k, y_k)$ of $X(k)$;
        2) compute the mean value, $\bar{M}(x_k, y_k)$, by using 3 wavelet modulus values of $X(k)$ along the vertical direction of $A_{max}(x_k, y_k)$;
        3) classify $X(k)$ into the edge pattern if $\bar{M}(x_k, y_k) > T_M$, and into the non-edge pattern otherwise;
    end for
end

For the above parallel algorithm, it can be seen that both the parallel degree and the theoretical speedup ratio are N_V, the number of the vectors or the sub-blocks of the image being processed.

2.4   Parallel Training of SOM Network

After the vectors (sub-blocks) of an image have been classified into edge pattern/non-edge pattern, an SOM network will be trained for segmenting the non-edge pattern sub-blocks by using vector quantization (VQ) method. In this section, we design a parallel algorithm for training the SOM neural network.

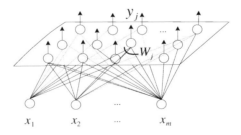

Fig. 2. Architecture of the SOM network being trained for VQ

Let $\{$ $X_{NE}(k)$; $k=1,\dots, N_{NE}$ $\}$ be the non-edge pattern vectors of the image being segmented, where N_{NE} is the number of non-edge pattern vectors. The architecture of the SOM network going to be trained is shown in Fig.2.

In Fig.2, the input vector $[x_1, x_2,\dots, x_m]^T$ is the non-edge pattern vector $X_{NE}(k)$ of $\{$ $X_{NE}(k)$; $k=1,\dots, N_{NE}$ $\}$, the weight vector between input $[x_1, x_2,\dots, x_m]^T$ and output neuron y_j is denoted by $W_j=[w_{j1}, w_{j2},\dots, w_{jm}]^T$, and the number of the output nodes is N_Y that is also the segmentation number for the image. Basing on the training algorithm described in [6], we design a parallel training algorithm for this SOM network and the overall parallel scheme is as follows:

(1) Parallel initialization of weight vectors $\{$ W_j; $j=1,\dots,N_Y\}$ using randomly selected vectors from training data $\{$ $X_{NE}(k)$; $k=1,\dots, N_{NE}$ $\}$;

(2) Parallel computation of the updating values for the weight vectors by using training data:

$$\Delta W_q(k) = X_{NE}(k) - W_q, \quad q = 1, \dots, N_Y, \quad k = 1, \dots, N_{NE},$$

where W_q is the weight with minimum distance between $X_{NE}(k)$ and the weights in $\{$ W_j; $j=1,\dots,N_Y\}$, i.e., $\left\| X(k) - W_q \right\| = \min_j \left\{ \left\| X(k) - W_j \right\| \right\}$;

(3) Parallel computing the cumulative updating values:

$$\Delta W_q = \sum_{k=1}^{N_{NE}} \Delta W_q(k), \quad q = 1, \dots, N_Y;$$

and parallel updating the weights:

$$W_q \leftarrow W_q + \alpha \Delta W_q, \quad q = 1, \dots, N_Y;$$

(4) Repeat steps (2)-(3) until the weights converged.

The detailed implementing operation algorithms for the above parallel computation modules are omitted here due to the limitation of the paper length, which are similar to those algorithms given in Section 2.2 and 2.3.

2.5   Parallel Quantization of Non-edge Pattern Vectors

Having the training process for the SOM network finished, the trained weight vectors $W_j=[w_{j1}, w_{j2},\dots, w_{jm}]^T$ ($j=1,\dots,N_Y$) are taken as the codebook of vector quantization and used for quantizing the non-edge pattern vectors of the image. We design the following parallel algorithm to implement the quantization process for speeding up the process.

    begin
      for $k = 1$ to N_{NE} in parallel do
          1) for the k-th vector $X_{NE}(k)$, compute the distances between $X_{NE}(k)$ and all the
              weights $\{$ W_j; $j=1,\dots, N_Y\}$: $d_j(k) = \left\| X_{NE}(k) - W_j \right\|$, $j=1,\dots, N_Y$;
          2) find the minimal distance of $d_j(k)$ and the index q by $q = \arg\min_j \{ d_j(k) \}$;
          3) quantize the vector $X_{NE}(k)$ with the weight W_q: $X_{NE}(k) \leftarrow W_q$;
      end for
    end

One can see that the quantization processing for the non-edge pattern vectors is also the segmentation of the non-edge pattern sub-blocks of the image. Through this quantization process, the non-edge pattern sub-blocks of the image can be clustered into N_Y clusters represented by the weights $\{ W_q; q=1,\ldots, N_Y\}$.

2.6   Parallel Classification of the Pixels of Edge Pattern Sub-blocks

After the non-edge pattern sub-blocks have been segmented by using the SOM-based VQ method given above, we process the pixels of the edge pattern vectors $\{X_E(k);$ $k=1,\ldots, N_E\}$ based on the VQ results by using the following parallel algorithm:

> begin
> > for $k = 1$ to N_{NE} in parallel do
> > > • compute the mean value of non-edge pattern vector $X_{NE}(k)$;
> > end for k
> > for $m = 1$ to N_E in parallel do
> > > • for the edge pattern vector $X_E(m)$, compute the differences between its pixels and the mean values of its neighboring non-edge pattern vectors;
> > > • classify each pixel of $X_E(m)$ into the class with the minimum absolute difference and replace it with the closest mean value;
> > end for m
> end

3   Applications to MRI Image Processing and Experimental Results

To verify the effectiveness of the proposed parallel segmentation method, we conduct the same experiments as conducted in [6] with applications to segmentation of human brain MRI images, and make comparisons between the experimental results obtained by the parallel algorithm and the original serial algorithm both on the segmentation performance and on the computation efficiency in this section. The parallel algorithm is implemented on a GPU platform using the parallel programming language OpenCL [12,13]. The GPU used in the experiment is the AMD HD 7950 which is composed of 28 computing units (CU) with 28×64 processing elements (PE) in total. The original serial algorithm is implemented on the PC of Intel E7500@2.93GHz and programmed with Visual Studio 2010.

Fig.3 shows the segmentation results of the human brain MRI image by using the proposed parallel segmentation algorithm. The segmentation processing is also conducted for the same MRI image by using the original serial algorithm given in [6] in the experiment. By comparing the segmentation results of the two algorithms, we have observed that they are giving almost the same segmentation performance except for a very few of negligible differences caused by different programming languages.

More experiments have been made on testing the computational efficiency for the parallel algorithm in the work. Table 1 gives the computation times of processing a

256×256 MRI image by using different algorithms in different computation modules. The computation times shown in Table 1 are measured by taking average over 1000 trials except for the training time of SOM network that is measured by taking average over 100 trials. From the experimental results given in Table 1, one can see that, for segmenting the image, the proposed parallel algorithm only takes 0.29686 seconds while the original serial algorithm takes 19.2133 seconds, in which the overall speedup ratio of the parallel algorithm is 64.72. This shows that the proposed parallel segmentation algorithm has achieved a significant improvement in computation efficiency. It should be noted that the speedup ratios obtained in the experiment for the parallel algorithms are far smaller than that of the theoretical analysis results given in Section 2. The main reason is that the parallel algorithm is implemented on the GPU and, for a single processing element (PE) of the GPU, its computational capacity is much less than that of a CPU. In addition, for a GPU, the parallel degree in data access is limited and only a few of PEs can read data from (or write data to) memories at the same time although all the PEs can do calculations parallelly.

Many other experiments have been done and similar results have been obtained in the work, which are not shown here due to the limitation of the paper length.

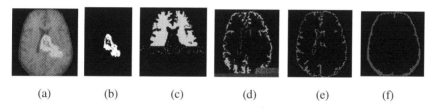

| (a) | (b) | (c) | (d) | (e) | (f) |

Fig. 3. Experiment result using the proposed method. (a) the original MRI human brain image with a brain tumor in the centre part; (b) the segmented tumor; (c) segmented white matter; (d) segmented gray matter; (e) segmented cerebrospinal fluid; (f) segmented skull.

Table 1. Computation times of different algorithms for processing a 256×256 MRI image

	Classification of edge/non-edge vectors	Training of SOM network	VQ + classification of edge pixels	Total processing time
Original serial algorithm	65.81 (ms)	19.1159 (s)	31.58 (ms)	19.2133 (s)
Proposed parallel algorithm	**0.577** (ms)	**0.296** (s)	**0.279** (ms)	**0.29686** (s)
Speedup ratio	**114.06**	**64.58**	**113.19**	**64.72**

4   Summary and Further Direction

In this paper, we develop a parallel segmentation algorithm for the SOM-based vector quantization method proposed in [6]. The parallel algorithm is implemented on GPU with OpenCL programming language and is successfully applied to segmenting the human brain MRI images. The experimental results show that this parallel algorithm can provide a significant improvement on the computation efficiency with an overall speedup ratio of 64.72 while the segmentation performance is kept unchanged, compared with the original serial algorithm.

It is noticed that, since the parallel capacity of GPUs in data accessing has become a bottleneck for further accelerating the GPU computation speed, more effort needs to be made to increase the parallel data throughputs of the process. This is a further research direction of the work.

References

1. Gao, X.L., Wang, Z.L., Liu, J.W.: Algorithm for Image Segmentation using Statistical Models based on Intensity Features. Acta Optica Sinica **31**(1), 1–6 (2011)
2. Zhao, J., Shao, F.Q., Zhang, X.D.: Vector-valued Images Segmentation based on Improved Variational GAC Model. Control and Decision **26**(6), 909–915 (2011)
3. Wu, Y., Xiao, P., Wang, C.M.: Segmentation Algorithm for SAR Images based on the Persistence and Clustering in the Contourlet Domain. Acta Optica Sinica **30**(7), 1977–1983 (2010)
4. Veksler, O.: Image Segmentation by Nested Cuts. In: Proc. of IEEE Conference on Computer Vision and Pattern Recognition, pp. 339–344. IEEE Press (2000)
5. Wang, S., Lu, H.H., Yang, F.: Superpixel Tracking. In: Proc. of IEEE International Conference on Computer Vision, pp. 1323–1330. IEEE Press (2011)
6. De, A., Guo, C.: A Vector Quantization Approach for Image Segmentation Based on SOM Neural Network. In: Guo, C., Hou, Z.-G., Zeng, Z. (eds.) ISNN 2013, Part I. LNCS, vol. 7951, pp. 612–619. Springer, Heidelberg (2013)
7. Fitzgerald, D.F., Wills, D.S., Wills, L.M.: Real-time, parallel segmentation of high-resolution images on multi-core platforms. Journal of Real-Time Image Processing (May 31, 2014)
8. Farias, R., Farias, R., Marroquim, R., Clua, E.: Parallel Image Segmentation Using Reduction-Sweeps on Multicore Processors and GPUs. In: 2013 XXVI Conference on Graphics, Patterns and Images, Arequipa, Peru, pp. 139–146 (August 2013)
9. Dessai, V.S., Arakeri, M.P., Ram Mohana Reddy, G.: A parallel segmentation of brain tumor from magnetic resonance images. In: 2012 Third International Conference on Computing, Communication and Networking Technologies, Coimbatore, India (July 2012)
10. Luo, Z.Z., Shen, H.X.: Hermite Interpolation-based Wavelet Transform Modulus Maxima Reconstruction Algorithm's Application to EMG De-noising. Journal of Electronics & Information Technology **31**(4), 857–860 (2009)
11. Liu, B., Huang, L.J.: Multi-scale Fusion of Well Logging Data Based on Wavelet Modulus Maximum. Journal of China Coal Society **35**(4), 645–649 (2010)
12. Gaster, B.R., Howes, L., et al.: Heterogeneous Computing with OpenCL. Elsevier(Inc.) (2012)
13. Munshi, A., Gaster, B.R., et al.: OpenCL Programming Guide. Pearson Education, Inc. (2012)

Author Index

Bian, Gui-Bin 200
Boulares, Mehrez 374

Cai, Qi 80
Cao, Jinde 27, 35, 69, 159
Chen, Guangyi 167, 423
Chen, Jiawei 43, 243
Chen, Jiejie 533
Chen, Lan 395
Chen, Liujun 43, 243
Chen, Yanjun 550
Chen, Yiran 150
Chen, Zijing 449
Cheng, Mingsong 602
Cheng, Quanxin 35
Cichocki, Andrzej 121, 459
Cui, Rui 338
Cui, Zhiwei 602

Da Cruz, Janir Nuno 524
Dai, Shuling 423
Dan, Yuanyuan 294
De, Ailing 637
Ding, Haishan 175
Ding, Yi 621, 629
Duan, Shukai 150

Fan, Jianchao 594
Fan, Yetian 19
Feng, Zhen-Qiu 200
Fu, Xian 621, 629

Gan, John Qiang 131
Gao, Jing 550
Gao, Jingying 252
Ge, Wentao 405
Ge, Zhanyuan 395
Greenhow, Keith A. 414
Gu, Guangyao 321
Gu, Shenshen 338
Guo, Chengan 637
Guo, Dongsheng 431

Guo, Yongxin 175
Guy, Tatiana V. 140

Han, Lu 439, 449
Han, Min 489
Han, Xinjie 395
Hao, Jian-Long 200
He, Hanlin 3
He, Ying 90
Hong, Haikun 499
Hou, Zeng-Guang 200
Hu, Cheng 611
Hu, Jianqiang 35
Hu, Jin 59
Huang, He 366
Huang, Tingwen 150, 366, 533
Huang, Wenhao 499

Jemni, Mohamed 374
Jia, Jie 602
Jiang, Haijun 611
Jiang, Jingqing 252
Jiang, Mei 3
Jiang, Ping 533
Jiang, Yunsheng 568
Jin, Long 286
Johnson, Colin Graeme 213
Johnson, Colin G. 414
Jumutc, Vilen 232

Kim, Chan Hong 479
Kim, Ji Eun 479
Kim, SooBum 479
Kim, Young-Soo 479
Krawczyk, Bartosz 358, 542
Kurzynski, Marek 469
Kárný, Miroslav 140

Le, Yaodong 321
Lee, Junghoon 584
Lee, Namgil 121
Lee, Sang Min 479

Lee, Seung-Phil   479
Li, Chen   321
Li, Dong-Juan   385
Li, Gaoyang   252
Li, Lin   516
Li, Ning   27
Li, Tieshan   312
Li, Wenye   262, 278
Li, Xiaojuan   150
Li, Xiaomeng   243
Li, Zhan   286
Li, Zhengxue   602
Lian, Cheng   270
Liang, Tian   12
Lin, Jie   303
Liu, Derong   51
Liu, Jinhai   110
Liu, Lei   110
Liu, Wei   516
Liu, Xiuchong   100
Liu, Yan   43, 243
Liu, Yan-Jun   385
Liu, Yingying   611
Liu, Zhe   558
Liu, Zhenwei   110
Lu, Hu   558
Lu, Yu   90
Lu, Yu-zhen   222
Luo, Ziyi   286

Ma, Jinwen   330, 568, 576
Ma, Yanhua   175
Macek, Karel   140
Mao, Jianqin   175
Miao, Baobin   312
Mo, Xia   366
Mou, Yi   439

Ouyang, Deqiang   611

Park, Gyung-Leen   584
Pei, Zhili   252
Peng, Jiao   338

Qiu, Binbin   431
Qu, Ming-hui   222
Quan, Longhu   100

Rizzi Raymundo, Caroline   213

Shi, Yongtao   508
Song, Guojie   499
Song, Xin   550
Song, Yuqing   558
Sun, Weiping   508, 516
Suykens, Johan A.K.   232

Tan, Hongzhou   286, 431
Tang, Huiming   270

Wan, Feng   524
Wang, Cuirong   550
Wang, Dejun   508, 516
Wang, Fei   12
Wang, Haixian   131
Wang, Jun   59, 184, 594
Wang, Lidan   150
Wang, Lihan   131
Wang, Siyuan   80, 395
Wang, Wenbiao   80, 395
Wang, Xinzhe   184
Wang, Yangling   69
Wang, Ze   524
Wang, Zhanshan   100, 110
Wang, Zhijie   192
Wei, Fajie   321
Wei, Hui   405
Wei, Qinglai   51
Wolczowski, Andrzej   469
Wong, Chi Man   524
Woźniak, Michał   358, 542
Wu, Chunguo   252
Wu, Jinghui   439, 449
Wu, Yuanyuan   159

Xie, Jinli   192
Xie, Kunqing   499
Xie, Wenfang   167, 423
Xie, Xiao-Liang   200
Xu, Xianyun   12

Yan, Lu   3
Yan, Yinhui   347
Yang, Dakun   19
Yang, Jian   303
Yang, Wenyu   19
Yang, Xiaopeng   439, 449
Yang, Xibei   294
Yang, Xiong   51

Yang, Yongqing   12
Yang, Zhi   431
Yao, Wei   270
Yu, Hualong   294
Yu, Shengsheng   508, 516
Yu, Shujian   439, 449
Yu, Ying   303
Yuan, Yulong   294

Zeng, Zhigang   270, 533
Zhang, Min   222
Zhang, Yu   459
Zhang, Yuan   637

Zhang, Yunong   286, 431
Zhang, Zhen   175
Zhao, Baojun   439, 449
Zhao, Dongzhi   594
Zhao, Jianyu   192
Zhao, Qibin   459
Zhao, Yixiao   439, 449
Zhao, Yue   330
Zheng, Danchen   489
Zheng, Mingguo   100
Zheng, Wenli   576
Zhou, Guoxu   459
Zhou, Mengzhe   27